The Apotheosis of Nullity

AMERICAN UNIVERSITY STUDIES

SERIES V
PHILOSOPHY

VOL. 226

This book is a volume in a Peter Lang monograph series.
Every volume is peer reviewed and meets
the highest quality standards for content and production.

PETER LANG
New York • Bern • Frankfurt • Berlin
Brussels • Vienna • Oxford • Warsaw

Bartosz Łubczonok

The Apotheosis of Nullity

A Transhistorical Genealogy of Human Subjectivity

PETER LANG
New York • Bern • Frankfurt • Berlin
Brussels • Vienna • Oxford • Warsaw

Library of Congress Cataloging-in-Publication Data

Names: Łubczonok, Bartosz, author.
Title: The apotheosis of nullity: a transhistorical genealogy
of human subjectivity / Bartosz Łubczonok.
Description: New York: Peter Lang Publishing, 2017.
Series: American University Studies V: Philosophy; vol. 226 | ISSN 0739-6392
Includes bibliographical references and index.
Identifiers: LCCN 2016040481 | ISBN 978-1-4331-3465-4 (hardcover: alk. paper)
ISBN 978-1-4539-1894-4 (ebook pdf) | ISBN 978-1-4331-3862-1 (epub)
ISBN 978-1-4331-3863-8 (mobi)
Subjects: LCSH: Subjectivity. | Nullity.
Classification: LCC BD222 .L83 2017 | DDC 126—dc23
LC record available at https://lccn.loc.gov/2016040481
DOI: 10.3726/978-1-4539-1894-4

Bibliographic information published by **Die Deutsche Nationalbibliothek**.
Die Deutsche Nationalbibliothek lists this publication in the "Deutsche
Nationalbibliografie"; detailed bibliographic data are available
on the Internet at http://dnb.d-nb.de/.

Contents

PART I

THE PRIMITIVE REGIME

PART III

THE PASSIONAL REGIME

PART IV

THE CAPITALIST REGIME

Acknowledgements

To enumerate the sheer number of people to whom I am indebted for their either direct or indirect assistance and support in allowing me to bring this manuscript to completion and into print would be a daunting enterprise. I would like to express my deep gratitude to Mieczysław Omyła, Professor Emeritus in the Department of Philosophy and Sociology at the University of Warsaw and a good friend, for his kind generosity in facilitating institutional support for my project. I likewise extend thanks to Professor Marcin Poręba, of the same institute, and Professor Adam Lipszyc, of the Polish Academy of Sciences, for their painstaking review and positive appraisal of this manuscript, as well as Professor Michał Herer for his affirmative review of my PhD thesis which *mutatis mutandis* comprised the first half of this final work.

This project would likewise not have been brought to fruition without the wonderful helpfulness and patient assistance of the editorial and production team at Peter Lang Academic Publishing, of whom Michelle Salyga, Jackie Pavlovic, Michael Doub, Meagan Simpson, Stephen Mazur, and Sophie Appel are but a few of the staff to whom I would like to express my most sincere gratitude.

Lastly but certainly not least, I extend thanks to my family and friends whose unflagging support and loving kindness in all things I cherish as the greatest of all divine gifts.

I have of late—but wherefore I know not—lost all my mirth, forgone all custom of exercises, and indeed it goes so heavily with my disposition that this goodly frame, the earth, seems to me a sterile promontory; this most excellent canopy, the air—look you, this brave o'erhanging firmament, this majestical roof fretted with golden fire—why, it appears no other thing to me than a foul and pestilent congregation of vapors. What a piece of work is a man! How noble in reason, how infinite in faculty! In form and moving how express and admirable! In action how like an angel, in apprehension how like a god! The beauty of the world. The paragon of animals. And yet, to me, what is this quintessence of dust? Man delights not me.

William Shakespeare, *Hamlet*, Act 2, Scene 2

Prologue

Eternity

1. The dark ground of the existence of God

For the Schelling of the *Weltalter* fragments, the anteriority of all anteriorities is the eternal and primordial ground of the existence of God. The ground hereto referred, an ultimate and groundless ground (and as such an *Ab/Ungrund*), is the Godhead—an eternal freedom, a pure will. This will's purity consists in its not being a will to something that it could apprehend as outside of itself, nor is it even a will that wills its own revelation. It neither covets what would otherwise be external objects destined for its apprehension, nor does it desire its own actuality. Rather, it is a will devoid of obsession or craving, a will that in fact—*does not will.*[1] More precisely, it does not will in any prosaic, mundane, or quotidian sense. Slavoj Žižek describes this groundless ground as "an abyss [*Ungrund*]—that is to say, 'absolute indifference' *qua* the abyss of pure Freedom which is not yet the predicate-property of some Subject but, rather, designates a pure impersonal Willing [*Wollen*], which wills nothing."[2] The absolutely primordial *Ab/Ur/Ungrund* of God's existence—that is, the Godhead—is thus a willing devoid of *both* subject *and* object. For Alberto Toscano, this purely vertiginous abyss is "the absolute, itself conceived as the unconditioned ground of indifference, the coincidence of producer and product, affirmation and affirmed, intuition and intuited, *natura naturans* and *natura naturata.*"[3] Toscano obviously has Spinoza in mind here. Jürgen Habermas glosses the Schellingian Godhead as the *Ein-Sof* of Lurianic Kabbala that the mystics considered "above God himself, pure joy, absolute bliss, grace, love and simplicity," a stratospheric freedom expressible only as "literally All and Nothing, the absolutely inaccessible."[4]

As Gershom Scholem notes, the very term *Ein-Sof*—as coined by the early kabbalists of Provence and Spain—is the unknowable aspect of the Divine and translates *inter alia* as "Infinite." It refers to the infinity of God or of his thought that "extends without end" (*le-ein sof* or *ad le-ein sof*). According to those in accord with Azriel of Gerona, *Ein-Sof*, rather than standing for "Him who has no end," points expressly to "that which has no end." This neutrality suggests that the attributes or personal epithets attached to the God of Scripture could not possibly appertain to the Godhead as such. Nor would *Ein-Sof* be subject to eulogies such as *Burukh Hu* or *Yitbarakh*.[5] Scholem goes on to delineate *Ein-Sof* as an absolute perfection entirely devoid of distinctions and differentiation, and even volition; avering moreover that *Ein-Sof* falls short of revealing itself in a way that would make knowledge of its nature at all possible, being itself inaccessible even to the most introspective thought (*hirhur ha-lev*) of contemplatives.[6] To the extent that—at least for the author of *Ma'arekhet ha-Elohut*—even the entirety of biblical Revelation and the Jewish Oral Law combined bears not a single reference to it, implying that *Ein-Sof* therefore cannot be an object of religious thought, and that only a vanishing handful of mystics have received the slightest hint of it. In the writings of *inter alia* the Gerona kabbalists, *Ein-Sof* is variously glossed as *mah she-ein ha-mahshavah masseget* ("that which thought cannot attain"), *ha-or ha-mit'allem* ("the concealed light"), *seter ha-ta'alumah* ("the concealment of secrecy"), *yitron* ("superfluity"—conjectured by Scholem to be a translation of the neoplatonic term *hyperousia*, which refers to the Beyond of Being itself), *ha-ahdut ha-shavah* ("indistinguishable unity"—in the sense of a unity devoid of opposition and differentiation), or as *ha-mahut* ("the essence").[7]

But let us forthwith return to Schelling. Here, from the outset, we must be careful: that the Godhead as pure will does not will in any prosaic sense does not in the least identify it with the Nietzschean will to nothingness—let alone with the nothingness of the will that the latter formally assumes in the case of the passive nihilism characterizing existential etiolation, leaden exhaustion and well-nigh sempiternal desuetude.[8] Instead, and quite in consonance with Habermas' characterization, this pure willing is at once nothing and everything. More precisely, on the one hand, it is nothing in the sense of its being entirely unbeknowst to itself—indeed, unknowing *tout court*—and also, in that it is bereft of Being (in the sense of not possessing the latter as a predicate); on the other hand, it is everything inasmuch as it is itself an unsullied happiness, a perfectly composed bliss that—by virtue of the foregoing (and in the absence of all anthropomorphism)—may be described as perfectly self-fulfilled. That the "emptiness" and seeming nullity of the Schellingian pure willing nowise arrays the latter with the yonder side of affirmation is expressed by Andrew Bowie as follows: "'the will which wills/wants nothing' can remain negative, in that it is absolutely within itself, because it does not entail any sense of possibility or development, but as such also does not entail any lack in itself, which is the highest affirmation."[9] Neither is this pure will (that is, the Godhead) a substance or a nature—though, as we shall later find, there does indeed *devolve* from God as eternal willing an eternal divine nature.

Schelling conjures the pure willing that is the Godhead as "the devouring ferocity of purity" in whose midst all that which bears Being as an attribute is instantly immolated.[10] It is "the most spiritual spirit, pure, inscrutable breath, the spirit of all spirit, so to speak."[11] This evocation chimes with some of Deleuze's more effusive thoughts on aspects of German expressionism, musings the tenure of which the less circumspect will readily confuse with a rupturous theological paean to the abysmal pure willing's ferocious incandescence: that the latter is an intensity "raised to such a power that it dazzles or annihilates our organic being, strikes terror into it," an intensity that "culminates in a fire, which burns us and which burns all of Nature," that "unleashes in our soul a *non-psychological life of the spirit,* which no longer belongs either to nature or to our organic individuality, which is the divine part in us, the spiritual relationship in which we are alone with God as light."[12]

Let us proceed with Schelling's further characterizations of the abyssal Godhead. The latter—one is reminded—is not *stricto sensu* eternal; for it is itself its *own* eternity, and thus—is eternity itself. It is not conscious, since it is consciousness itself. It cannot be said to have Being—that is, to possess Being as an *attribute.* Rather, it is Being itself. More precisely and moreover, it would notwithstanding be preferable to aver (as Schelling indeed does subsequently go on to do) that the Godhead neither has nor does not have Being, and is therefore "a sublimity beyond Being and Not-being"[13]—or, in Deleuze's expression, "the sublime, non-psychological life of the spirit."[14]

But there is more: for Schelling, the Godhead is outside of both the realm and reach of *necessity* and of what merely *is*. It is also, one may add, *outside of nature*: since nature (in the more prosaic and quotidian sense) is what of necessity belongs to that which possesses Being as attribute; whilst that which is outside of either having or not having Being is, in clear contradistinction, outside of all predication and, therefore, devoid of nature—or "the frightful non-organic life of things" alluded to by Deleuze.[15] The Godhead—devoid of necessity, Being, non-Being, and nature—remains thus ultimately, concludes Schelling, in its essence "the eternal freedom to be."[16]

This eternal subjectless and objectless pure willing, also goes by the name of the Absolute—and, more specifically, as Absolute Indifference. Christopher Groves alludes to Schelling's 1799 averment that, "Absolute Indifference considered in its natural aspect, as the in-itself of nature, that is, as the supra-temporal, non-conscious, ontological condition of our experience of natural objects, is somewhat like Spinoza's *natura naturans,* being an ontological unity which Schelling calls 'absolute productivity' or primordial life."[17]

Now, this gives us reason for pause—as the implications here are potentially vast. However, there is here also—at least ostensibly—a stumbling block, for we have only a few moments ago declared the Absolute (the Godhead) to be *devoid* of nature. How then may the Absolute be declared the in-itself of nature, let alone be possessed of a natural aspect? We shall only be in a position to resolve this aporia when our delineation of Schelling's theogony has progressed a little further. For the moment, let us rest content with the following anticipation: the Absolute, Godhead or eternal pure willing—owing to its being inhabited by an orginary and inherent dissonance—*precipitates* a (non-organic) nature within itself. It shall also emerge that, though the Absolute is anterior to the (non-organic) nature that it spawns within its own ambit, the strictly non-chronological character of the said anteriority will be seen to imply that the Absolute has *always already* generated the said nature (non-organic nature as such) from out of itself. Indeed, it will come to light that this non-organic nature has been begotten of the originary dissonance at the core of the Absolute *from all eternity.* But what is this originary dissonance that inhabits the Absolute?

It is the originary self-begetting of a second eternal will at the very heart of the Absolute. And it is to this that we presently turn.

2. The self-begetting of a second eternal will

The pure eternal willing that is the Godhead and eternity—as we have already asseverated—rests blissfully within its own plenitude. Yet, avers Schelling, "the more this composure is profoundly deep and intrinsically full of bliss, the sooner must a quiet longing produce itself in eternity, without eternity either helping or knowing. This is a longing to come to itself, to find and savour itself; it is an urge to become conscious *of which Eternity itself does not become conscious.*"[18]

This primordial internal dissonance within the Godhead finds poignant expression in the thought of the kabbalist Israel Sarug. Like Schelling, Sarug also conceives of the originary bliss of *Ein-Sof* as at once the source of its primordial and inexorably burgeoning endogenous unrest. As Scholem maintains apropos of Sarug, this paradoxically wrought dissonance within the Godhead is the case since primordially *Ein-Sof* pleasured in its own autarkic self-sufficiency, and this latter "pleasure" produced a sort of tremulous "shaking" (*ni'anu'a*) as the movement of *Ein-Sof* within itself.[19]

Returning to our Schellingian theogony, we may want to enquire as to the precise form which the primordial unrest, or the *ni'anu'a* of Sarug, within the Godhead as eternal willing assumes? Here Schelling avers that, in the very ambit of eternity's silent and completely unconscious seeking after itself, another will—a will other than the pure willing and entirely unbeknowst to it—begets itself. This latter will is self-sufficient and creates itself *wholly independently* of the first will. That is to say: the first will is eternity insofar as it—eternity—is a pure willing, whilst the second will is a will *in* eternity. It does verily stand to reason that the second will must be independent of the first: for inasmuch as eternity's self-seeking is entirely unconscious, such a purblind onanistic groping can only assume precisely the form of the said independent autogenesis of an entirely separate will that searches the eternity that the first will itself is. Put more succinctly: the unconscious seeking by the first will/willing is of itself identically its *being sought* by a second will that arises and seeks the first will independently of this first will/willing. That the second will's genesis is entirely autonomous and that it produces itself *out* of itself and *from* itself, means that this second will begets itself not *out* of eternity, but rather—and as indicated a little earlier—*in* eternity. From this, Schelling is able to conclude that the second will "is itself an *eternal will*" and indeed "the *eternal will* per se" whereas, on the other hand, "the will that wills nothing [the first will] was only the pure willing of eternity by itself."[20] That is to say: the first will is eternity insofar as it—eternity—is a pure willing, whilst the second will is a will subsisting or insisting within eternity's bosom. Yet, it is hereby nowise less eternal.

We have mentioned that this second will remains unbeknowst to the first will/willing. Notwithstanding, the second will—though it searches after eternity (the first will/willing) blindly and (at least initially) entirely unconsciously—nonetheless does have a certain modicum of awareness of the first will.[21] Schelling appears here to be—at least at first blush—contradicting himself. But let us follow him nonetheless. The said second will, he contends, "seeks eternity, driven not by knowledge but rather by divination, presentiment, and inexpressible longing."[22] The second will, moreover, "wills that the will that wills nothing [the first will/willing] become active and perceptible to itself *as such*"—whilst it itself (this second will) is "eternally only a will that wills and desires eternity."[23] On the other hand, the first will/willing—taken by itself (and as we already know)—is simply eternity itself.

3. The quandary of the second will and God's contraction of Being

The second will's quest (which is identically also that of the first will) shall, however, *fail.* Or, rather, it will only succeed on terms other than its own. This is so because the second will, though indeed the will *toward* eternity, *never finds or attains the latter.* It therefore "posits itself as the mere *willing* of eternity, and to that extent as negated. But in positing itself as negated, it is at the same time the self-negating will."[24] The second will negates itself in the sense of forthrightly positing itself as *not* eternity, as *not* Being, and thus as that which is *not* affirmation. It is, nevertheless, only able to do so by at the same time positing itself as *lack.* However, as essentially lacking (indeed as want itself), it cannot but desperately seek to satiate itself with Being—which latter it moreover inevitably fails to find either outside itself or inside itself. More precisely: on the one hand, it is unable to find Being *outside* itself in that, being essentialy a blind groping with scantly the merest presentiment of eternity (and thus of Being), it is—to begin with—unable to apprehend Being; on the other hand, it fails to find Being *inside* itself since a turning toward itself is at once and necessarily a turning away from eternity (and thus a turning away from exactly what is here being sought—Being). Thus, the second will—hungering and lusting as it does after Being—only succeeds in running itself into the direst quandary. Therefore, the only way out of this thankless dilemma is for the second will to actively posit Being (eternity, pure affirmation) as *completely outside of* (and moreover as *opposed* to) *itself.*[25]

Furthermore, through a curious reversal, the second will not only posits Being in the said manner, but in fact unconditionally (and in the manner of an autochthotonous *force*) *generates* Being—for lack of a better expression—"for the first time." That is to say, it is only with the second will's deadlock-resolving positing of Being as outside of and opposed to itself that—contends Schelling—Being is posited *as such.*[26] The second will posits Being by opposing it—at once positing itself as not-Being. Since Being is affirmation itself, the second will—in essentially positing itself as Being's opposite—is negation.

This positing by the second will is the movement through which negation injects itself in the midst of untrammeled affirmation. But this negation interposes itself solely to the ends and for the sake of affirmation. It is thus a recourse taken by affirmation to affirm itself in the first place—and certainly *not* the same as the inverting and perverting vicissitudes whereby the Nietzschean affirmative will to power finds itself irretrievably degraded to a negative will to power. In the latter Nietzschean scenario, negation obstinately interjects itself so as to thereupon serve as the ground of subsequent affirmations—thereby reducing these to mere pseudo-affirmations, runtishly obsequious redoublings and confirmations of a resounding and fundamental "No" whose tenure they merely serve to entrench in the mendacious guise of its antipode. In the Schellingian theogony, whose exposition we are here following, the *exact opposite* is the case: affirmation in the first will is *grundstimmung* whilst the desperately negating gesture of the second will is its handmaiden or adjunct. Indeed, put this way, the internal dialectic unearthed here between the first and the second eternal wills—is far closer (if not well-nigh identical) to Nietzschean double-affirmation: wherein pure becoming (*Werden*) as the first affirmation finds itself necessarily redoubled with a second affirmation that affirms the Being of becoming as becoming's return.[27] (The Nietzschean eternal return, which entails at its basis precisely this sort of double-affirmation, shall insinuate itself once more in our ongoing Schellingian discussion—and will in fact return again towards the very end of this book.)

But, for now, let us forge ahead and dwell more upon the second will itself. Since the second will's very action is to negate itself, the originary positing of Being herein effected—transpires by way of and as a *contraction*.[28] At the very start we described Being at its purest as the eternal Godhead—the most spiritual of spirits. The positing of Being by way of the second will's auto-contraction, insofar as this latter self-limitation is the manner in which the first will affirms itself, is therefore the movement whereby the Godhead (as precisely this first will/willing that wills nothing) *assumes Being*—and does so, likewise, by way of a contraction, i.e. by contracting *itself*. God contracts Being by way of an auto-contraction of himself. For indeed: "what is altogether first in God, in the living God, the eternal beginning of itself in itself, is that God restricts itself, denies itself, withdraws its essence from the outside and retreats into itself."[29] This contraction is a self-negation, a concealment of itself, a withdrawal, foreshortening and recession into the obscure. Anterior (though not in the chronological sense) to this, God—as the pure Godhead and thus indeed Being itself—had not yet found himself, had thus far not yet positively taken up the Being that he eternally is and was.

The Schellingian doctrine of the divine contraction derives from the kabbalistic thought of Isaac Luria. The latter's notion of the *ẓimẓum* ("contraction") in fact antedates Luria himself and is tracable to an earlier fragment from the circle of the *Sefer ha-Iyyun*. This contraction—conceived of as an anterior condition of the possibility of the creation of the world—is that by which God "gathers in and contracts (*meẓamẓem*) his breath" "so that the smaller might contain the larger; so He contracted His light into a hand's breath, according to His own measure, and the world was left in darkness, and in that darkness He cut boulders and hewed rocks."[30]

Luria likewise conceives of the *ẓimẓum* as the enabling condition for creation. (In Schelling's theogony, the creation of the world itself only transpires with God's upsurge as subject in the so-termed *Ent-Scheidung*. Nonetheless, the *ẓimẓum* still serves as a precondition for the *Ent-Scheidung* itself. Thus, it is not remiss to dwell upon Luria for a few moments.) This is so because *Ein-Sof* in its very essence leaves no space whatsoever for creation, for there is at this point no area which is not already God, as this would entail a limitation of his infinity.[31] From this Scholem adduces that the act of creation is solely possible by way of "the entry of God into Himself," that is, via an act of *ẓimẓum* through which God contracts himself and thereby enables the coming into existence of something which is not *Ein-Sof*. Some part of the Godhead must therefore withdraw and leave room for an ensuing creative process. Moreover, the *ẓimẓum* is in fact nowise the concentration of God's power *in* a place, but rather its withdrawal *from* a place.[32] The place from which God makes his retreat is actually a mere "point" in comparison with his infinity. From the human point of view, it however comprises all levels of existence—spiritual and corporeal alike. This place is primordial space, something which the Zohar terms *tehiru*.[33] For Luria, the actual coming to pass of the *ẓimẓum* is explained as follows. Prior to the *ẓimẓum* all the forces of God were contained and absolutely balanced within his infinitude sans separation. Even the forces of *Din* ("judgment") were stored in the Godhead without being as such distinguishable. The actual occurrence of *ẓimẓum* took place when *Ein-Sof* concentrated the roots of *Din* previously concealed within him in one place—the place at once vacated by the power of mercy. *Ẓimẓum* was for this reason an act of judgment and self-limitation on the part of *Ein-Sof*.[34]

But let us return to Schelling. The eternal self-negation by which God contracts Being is at once that by which eternal divine *nature* begets itself. God's auto-contraction of Being through the second will's self-negation *is the generation of eternal divine nature*. Put another way: we thus see here in retrospect that "this will [the second will], generated out of itself in the tranquillity of eternity, was *the eternal will to nature* [my emphasis]," and that eternal divine nature—as well as nature *tout court*—"is only built upon the ground of an eternal, self-negating will that returns into itself."[35] In regard of the generation of an eternal divine nature by way of God's self-withdrawal, Habermas avers: "God draws himself together, which means that he includes what he is as love within himself as nature."[36] Stéphane Mosès describes this auto-contraction of God (or Lurianic *ẓimẓum*) as "a return of God unto himself, a limitation imposed on his original expansion impetus, a tightening of his infinite essence that causes the appearance of a nature inside himself."[37] The latter nature that appears in God is eternal divine nature, or the Absolute or God *as* eternal divine nature.

Now, regarding this latter, we have already noted Schelling's assimilation of God's natural aspect to Spinoza's *natura naturans,* this latter being christened "absolute productivity," or primordial life. This would serve to suggest that the primordial dissonance at the very heart of the Godhead, or the Absolute, resolves itself as the begetting within God of something akin to Spinozan *natura naturans.* The latter—also termed Absolute Substance—is conceived by Spinoza as the in-itself of God, *natura* being God as nature in the evident and tangible sense. Comparing this configuration to the state of the Absolute/God anterior to the precipitation of divine nature within God by way of God's auto-contraction (*ẓimẓum*), it is evinced that only here does the distinction between *natura naturans* and *natura naturata* come into existence (bearing in mind Alberto Toscano's previously alluded to remark that the state of the Absolute anterior to this point was that of the perfect coincidence of these two).

However, in that the dissonance within the Absolute or God is truly originary, the precipitation into existence of the *natura naturans/natura naturata* distinction is likewise absolutely primordial. We may, thus, aver that the primordial dissonance within God/the Absolute *is* as the orginary appearance of the said distinction between *natura naturata* and *natura naturans,* these two appearing as (at least conceptually) separable not substantially but by way of a mere parallax shift. This is the case because all the relations of anteriority thus far espied are strictly non-chronological. Indeed, time as temporal succession has not yet emerged at this stage of the Schellingian theogony. For this, we must await the *Ent-Scheidung.*

But let us return to the eternal divine nature, or *natura naturans,* spawned here. According to Groves, Schelling has described the latter as "absolute productivity." This allows us to arrive at a crucial proposition: *the primordial dissonance within the Absolute/the Godhead has always already resolved itself as absolute productivity.*

Intermezzo 1: Eternal divine nature as Deleuzoguattarian desiring-production

(a) Organs-partial objects and the connective synthesis

It is this latter assertion which allows us to make our first sally into properly Deleuzoguattarian territory. But first, a swift propaedeutic. *Natura naturans* is Spinozan Absolute Substance (the "in-itself" of nature, God as immanent to nature), that which Groves earlier alluded to as "absolute productivity." However, *natura naturans* is inextricable from the infinite number of attributes through which it explicates itself—which attributes are its immediate powers/potencies/forces. This is where the link between Schelling, on the one hand, and Deleuze and Guattari, on the other, avails itself—inasmuch as the Spinozan attributes are wagered by these latter to be what they term "organs-partial objects."[38] And the foregoing "absolute productivity" is a characteristic of Deleuzoguatarrian "desiring-production." Bonta and Protevi define the organs-partial objects of Deleuze-Guattari as "mere points of intensity of matter-energy," each such organ also being termed "a 'desiring-machine,' that is, an emitter and breaker of flows" of production.[39] Each organ-partial object or desiring-machine is thus at once a "point" of intensity and an emitter/breaker of the flow of matter-energy.

However, though described as inter alia an intensive "point" (as we shall see, it is, in fact, a purely dispersive *multiplicity*), an organ-partial object is by no means hereby deemed to belong to Cartesian or Euclidean space. Indeed, at the level of Schellingian eternal divine nature and that of Deleuzoguattarian desiring-production alike, each here considered for itself, one may not yet speak of space in the sense attributed to *res extensa.* Space conceived in this latter manner has thus far not even emerged, and must await the Schellingian *Ent-Scheidung* whereby God emerges as subject. Notwithstanding, Schellingian eternal divine nature, Spinozan *natura naturans,* and the ambit of Deleuzian desiring-production, can all be said to "transpire" within space as *spatium,* as *intensio.* As Miguel de Beistegui avers, *spatium* differs *qualitatively* from extended space. More precisely, *spatium* is a *quantity* that is distinguished qualitatively from other spatial quantities. It is, however, not a quantity in the extensive sense—but rather a purely *intensive* one. In fact, it is the more quotidian extensive/numerical quantities such as length, height, width, that devolve from this intensive *spatium.* The latter itself is an originary "depth," the "absolute" in relation to which space-as-extension is merely relative.[40] This original depth is, as Deleuze expresses it, "space as a whole, but space as an intensive quantity: the pure *spatium.*"[41]

Regarding intensity itself, in the formulation of Constantin V. Boundas, "an" intensity is a difference *in itself.*[42] Intensity is a pure differential quantity, difference here being *internal* to intensity—and the very essence of intensity as such. The foregoing observation renders cogent Deleuze's own asseveration that, in fact, the "expression 'difference of intensity' is a tautology"—for indeed every "intensity is differential, by itself a difference. Every intensity is E-E′, where E itself refers to an e-e′, and e to ε-ε′ etc."[43] We thus have an infinitely ramified difference of intensity and may thus "call this state of infinitely doubled difference which resonates to infinity *disparity.*"[44]

Returning now to *spatium* itself, one may cursorily surmise this latter as a space anterior to and likewise subtending space-as-extension, possessed of *intensive ordinates* (as opposed to the extensive co-ordinates of Euclidean or Cartesian space). And it is in this latter sense that any given organ-partial object may be described as a "point" of *intensity*. In Deleuzoguattarian desiring-production, these organs-partial objects are related to one another by way of the so-termed "connective synthesis." Looking ahead a little to later stages of our present exposition, we may note that desiring-production consists of *three* interlaced and inherently inseparable syntheses—of which the said connective synthesis is but one. The other two are the *disjunctive* synthesis and the *conjunctive* synthesis.[45] These latter shall insinuate themselves into our discussion in due time. Presently, however, we move to dwell upon the connective synthesis. This latter synthesis is that of the production of *production*. Though, as we shall see, the three syntheses are merely aspects of desiring-production taken integrally, it is the connective synthesis that serves to foreground Deleuzoguattarian desiring-production (the equivalent of Schellingian eternal divine nature) *as* "absolute productivity" in the clearest manner. The connective synthesis entails the connection of desiring-machines (organs-partial objects) in an open paratactic linearly transverse formation. One machine emits a flow (*hyle*) of matter-energy into which another machine cuts, interrupting the flow and itself serving as a flow that is itself interrupted by a still further machine, and so on ad infinitum.[46] Hence, the continuity of the flows is paradoxically ensured by the breaks (*coupures*) in the latter. This is so because every machine is always already a machine of a machine. Deleuzoguattarian desiring-machines, or organs-partial objects, are thus systems of breaks-flows.[47] As intimated earlier, what is produced in the connective synthesis is production itself. That is, the connective synthesis is that of the production of production, in which producing and product are indistinguishable since the process of producing is endlessly grafted onto the purported product allegedly produced.[48]

Now, let us be reminded that, in that this all happens within *spatium* and not *extensio*, the organs-partial objects or desiring-machines caught up in the roiling ambit of this incessant desiring-production are "not partial (*partiels*) in the sense of extensive parts, but rather partial ('*partiaux*') like the intensities under which a unit of matter always fills space in varying degrees."[49] It is incidentally for this reason that Deleuze and Guattari characterize the connections between them as passive, or indirect.[50] Organs partial-objects refer not in the least to an organism that would otherwise function as a phantasmatic lost unity or coming totality.[51] In contradistinction, the extensive parts of integral wholes are merely pre-existing unities caught up within a still higher and more comprehensive unity. The emergence of such unities requires spatial circumscription. But within the ambit of Deleuzoguattarian desiring-production (Schellingian eternal divine nature) considered in and of itself, there is no extension. Thus, the organs-partial objects (desiring-machines) must constitute entirely open and dispersed multiplicities, that is, pure positive multiplicities entirely devoid of unification or totalization.[52]

The dichotomy *partiaux/partiels*, moreover, refers to a fundamental distinction made by Deleuze and Bergson between *continuous* (or non-metric) multiplicities and *discrete* (or metric) multiplicities. Bergsonian continuous multiplicities are effectively what Jonathan Roffe describes as intensive multiplicities, whilst discrete multilpicities are extensive. Crucially, discrete/extensive/metric multiplicities may be divided up indefinitely into parts (Bergson: "we may carry the division as far as we please; we change in no way, the nature of what is divided"); whereas continuous/intensive/non-metric multiplicities cannot be divided without at the same time changing in nature (Bergson: since they "are one with the successive moments of the act which divides" them).[53] This last-mentioned inherent change in the nature of a continuous (or intensive) multiplicity is an alteration in its intensive state—something we shall return to in due time.[54]

Manuel Delanda avers that continuous-intensive multiplicities fall within the ambit not of the analytical geometry of Descartes and Fermat but rather within that of the differential geometry of Friedrich Gauss and Bernhard Riemann—and moreover share certain of the traits of what the latter term a *manifold*: the most important of these common traits being that of "the absence of a supplementary (higher) dimension imposing an extrinsic coordinatization, and hence, *an extrinsically defined unity*."[55] Indeed, as Toscano observes, the internal differences that consistute intensive multilpicities are not measured externally or determined by a supplementary principle (whether objective or subjective), but instead entirely immanently.[56] More precisely, and in accordance with de Beistegui, their characterization as "non-metric" must not be seen as implying that continuous/intensive multiplicities are completely outside of measure; rather, each such multiplicity is "susceptible to measurement only by varying its metrical principle at each stage of the division."[57]

But let us return to the earlier point concerning the change in intensive state effected by the division of a continuous/intensive multiplicity. A change in the nature/intensive state of the latter triggers a *new individuation*.[58] And individuation as conceived here is a *becoming* (*devenir*); that is, intensive multiplicities are in essence marked by the imminence as well as the perpetual immanence of such becomings.[59] This stands to reason in that, insofar as intensive multiplicities are by definition bearers of internal difference and inasmuch as internal differences are marked by intensive individuality, continuous/intensive multiplicities are necessarily subject to constantly novel individuation—that is, becoming.[60]

But let us rephrase this conversely. Indeed, what is becoming? Cliff Stagoll's formulation is more than adequate: becoming (*devenir*) is "the continual production (or 'return') of difference immanent within the constitution of events, whether physical or otherwise."[61] This we may couple with Boundas' observation that intensities are "virtual yet real events."[62] Thus, an event is at once an intensity—or, more precisely, a pure intensive quantity. And, as already observed, the latter is internal difference—and, as shall presently be seen, what Deleuze terms a mobile *individuating factor*.

An individuating factor of this Deleuzian sort is a pre-individual singularity; and it is precisely the reprise of such pre-individual individuating factors that is entailed in becoming. Each such reprisal is an event. These pre-individual individuating factors, in the implacable and vertiginous movement that is called becoming, serve to engineer—but likewise break up—intensive multiplicities (organs-partial objects). Each such multiplicity is the individuation (or individual) wrought by the pre-individual individuating factors. But, since becoming is perpetually immanent to intensive/continuous multiplicities, the entailed constant reprise of mobile pre-individual individuating factors implies that these multiplicities are incessantly subject to categorically novel individuation. Thus, the very essence of continuous/intensive multiplicities is that they are perpetually becoming other such multiplicities, perpetually being re-individuated as these latter. This constant unfolding or becoming (*devenir*) of multiplicities Deleuze calls *differen*t*iation*—a term that shall in due time be distinguished from its near-doppelgänger: *differen*c*iation*. Otherwise stated, the process of differen*t*iation is, self-identically, that of individuation. The latter, as already contended, is effected by intensive quantities (pre-individual individuating factors). Deleuze: "The essential process of intensive quantities is individuation. Intensity is individuating, and intensive quantities are individuating factors."[63]

Now, the mercilessly unremitting character of the differen*t*iation of intensive multiplicities allows Delanda to contend/conclude that continuous/intensive multiplicities are fundamentally *meshed together into a continuum*, that they "blend into each other, forming a continuous immanent space very different from a reservoir of eternal archetypes."[64] This continuous and immanent space is precisely the purely intensive *spatium* mentioned earlier. The latter is the field of individuation as such, an extensionless depth in which—as Deleuze maintains—every "individuating factor is already difference and difference of difference," and in which individuating factors "endlessly communicate with one another across fields of individuation, becoming enveloped in one another ..."[65]

At this point we are able to, and indeed must before proceeding forth, draw a crucial distinction: that between the pre-individual individuating factors, on the one hand, and the continuous/intensive multiplicities, on the other. The former serve, in their infinite plasticity, to engineer (differen*t*iate) the latter as the multiplicities that they "are" and are becoming. Each such evanescent (and interiorly split) instant of engineering of "an" intensive multiplicity is an individuation, whilst the multiplicity itself thus wrought is an "individual." Obviously, in that Deleuzian becoming (*devenir*) implies the incessant return of pre-individual individuating factors, an intensive multiplicity is—in its very essence—constantly being individuated and re-individuated. Now, in that each individuation is a reprisal of pure intensive quantity (difference of difference, differen*t*iating difference), each such individuation may be viewed as a splitting or division of the multiplicity "subjected" to re-individuation. Continuous/intensive multiplicities, as we have observed, change in *kind* under the influence of such individuation/differen*t*iation/division. The perpetual immanence of becoming herein entailed, moreover, means that such multiplicities are *nothing beside* their constant re-individuation or differen*t*iation. In other words, over and above their constant change in kind, they *are* their own incessant transmogrification as regards kind. Thus, continuous/intensive multiplicities are conceivable as *ipseities* in constant breach of their very own ipseity, or always other ipseities, or never the same ipseities, or never their own ipseity, or multiple/disparate as regards ipseity. "Each" ipseity is a particular kind/nature. Intensive multiplicities are, in this sense, *qualitative* (this must however not be conflated with the qualities attributable to objects subsisting within extended space; that is, objects as arising and enduring within the volumetric space ushered in by the emergence of discrete/extensive multiplicities). Their being unremittingly re-individuated thus entails their incessant "re-qualification"; that is, the fact that they are always already of a different kind to their own singular kind.

On the other hand, with respect to the mobile pre-individual individuating factors, this very characterization (that of nature/quality/kind) has no purchase to begin with. Pre-individual individuating factors are purely intensive *quantities* (again, this must be clearly distinguished from quantity as associated with measure at the level of objects in extended space). They are infinitely ramified difference, differen*t*ing difference; and this latter differen*t*iation is precisely that which incessantly generates and re-generates, by individuation and re-individuation, intensive (qualitative) multiplicities as the singular individuals—perpetually becoming (*devenir*) other individuals—that these latter fundamentally are. However, the pre-individual individuating factors, though the infinitely plastic engineers of differences in kind, are themselves *of no kind whatsoever.* They are not even of a kind which at base perpetually changes in kind. They do not, therefore, attain the level of ipseity—even for the evanescence of a fleeting instant. Pre-singular individuating factors, as the pure differen*t*iating intensive quantities that they are, are therefore not ipseities but—in actual fact—*aseities.*

There it is. But let us now also include into consideration discrete/extensive multiplicities—for we are about to surmise. One recalls that the foregoing presuppose space as volumetric extension: that as multiplicities the latter are possessed of parts of integral wholes, that their nature remains unaltered by any amount of division since such division transpires within the fixed metric that governs such extensive multiplicities. Changes in discrete/extensive multiplicities effected by division and such like give rise merely to changes in *degree*—and not in kind. Extensive multiplicities are sempiternally of their own kind—or they are not at all. Contrarily, as pertains to the pre-individual individuating factors, these latter skirt the very arena of kind and of degree alike.

We thus have three distinct (though nowise unrelated) "tiers": (1) Continuous/intensive multiplicities, (2) discrete/extensive multiplicities, (3) pre-individual individuating factors. The first-mentioned are internal differences in kind that constantly change in kind. The second are multiplicities individuated from a vantage-point supplementary/external to them; herein implying that they are sempiternally of one and only of one kind, subject to perturbations whose affects can impinge upon these discrete/extensive multiplicities solely as regards degree. The third-mentioned subsist/insist "underneath" the level of the first two, repeatedly individuating intensive/continuous multiplicities as what the latter evanescently "are," are becoming (*devenir*), and will be.

Deleuze will break this supposed Gordian knot (if there is one). It is, after all, his. One proceeds as follows. On the one hand, there are continuous/intensive multiplicities: they are differences in kind. On the other hand, there are discrete/extensive multiplicities: they are differences in degree. Beneath/between these two, and more fundamental/originary than they, are "all the degrees of difference—beneath the two lies the entire nature of difference—in other words, the intensive. Differences of degree are only the lowest degree of difference, and differences in kind are the highest form of difference."[66] Anterior to (though not chronologically so), more primordial, more subterranean, more fundamental, and subtending the differences of kind and differences of degree, as Deleuze maintains, is "the *spatium*, the theatre of all metamorphoses or difference in itself which envelopes all its degrees in the production of each"—and whose ambit is "a world the very ground of which is difference, in which everything rests upon disparities, upon differences of differences which reverberate to infinity (the world of intensity)."[67]

We have thus investigated three levels. These, notwithstanding, are nowise mutually barricaded from one another. They are as follows. (a) Pre-individual individuating factors are at the most "subterranean" level: they are *aseities*, differen*t*iating (individuating) differences, outside of measure and of quality/kind. (b) Continuous/intensive multiplicities are purely dispersive, incessantly re-individuated, qualitative multiplicities whose very individuality and metric change at each re-individuation incessantly effected by the foregoing mobile pre-individual individuating factors. (c) Discrete/extensive multiplicities, devolving from continuous/intensive multiplicities (through the process of differen*c*iation), are self-identically and sempiternally of their own kind alone, change only in degree under the pressure of division, and are possessed of a unique and enduring metric which prosaically measures their volumetric properties.

Each of these three ontological "tiers" is, moreover, possessed of its own idiosyncratic temporality—or synthesis of time. These three distinct temporal syntheses, as shall become pellucid in good time, both presuppose and compromise/outflank one another. But we will not dwell upon this at this precise point in time.

(b) The body without organs (BwO) and the disjunctive synthesis

We have thus discussed the connective synthesis of production. What it produces is production itself. Indeed, all of desiring-production is the "absolute productivity" that Schelling attributed to the eternal divine nature that had precipitated itself into being with God's originary self-withdrawal or Lurianic *ẓimẓum* (attendant upon the likewise-originary dissonance that was the self-alienation of the supposedly tranquil eternal willing that was/is the Godhead).

We must, however, discuss the remaining aspects of this "absolute productivity"; that is, the other two (always mutually imbricated) syntheses that make up the cycle of desiring-production. The former we shall dwell upon presently, whilst the latter will inevitably insinuate itself into the Schellingian theogony before long.

Presently we proceed to discuss the disjunctive synthesis. In the course of the connective synthesis, suddenly—it is contended by Deleuze and Guattari—all production reaches a standstill for a vanishing fraction of a moment and spawns "an enormous undifferentiated object": this is the glorious and most hallowed body without organs, or BwO.[68] Immediately hereupon, however, the connective synthesis resumes.[69] Now, the BwO is not literally a body devoid of organs, but rather a body bereft of any form of organization (organ-ization). Also, like the organs-partial objects and unlike the organism, it is not extended spatially—does not "reside" within space as *extensio.* Rather, it "resides" in *spatium*; that is to say, in space conceived as a pure material density, the sheer intensity of unformed matter—the "space" of the aforementioned mobile pre-individual individuating factors and of the ceaselessly re-individuated qualitative-intensive multiplicities that these former incessantly engineer via differen*t*iation.

Matter fills up space to a given degree of intensity. Deleuze and Guattari asseverate that the BwO is matter at zero intensity, with all varying intensities of matter deriving from this null-intensity of the BwO. Indeed, they maintain that the BwO is comprised solely of the intensities that occupy, populate and circulate through it.[70] More precisely, the BwO is the very passing through itself of the intensities that it itself produces and distributes in *spatium*. The BwO is neither space, nor in space: it is matter occupying space to a particular degree, the degree corresponding to the intensities it produces. It is unstratified, unformed, intense matter, and in fact the very matrix of all intensity commencing from zero intensity onward. It is notwithstanding the case that there is naught negative concerning this "zero." That is to say, there are no negative or dialectically opposed intensities. They are all purely positive. For here matter and energy are exactly the same. The BwO is simply the production of the real itself as an intensive magnitude arising from zero intensity.[71]

Moreover, the BwO is not itself opposed to the organs-partial objects, but is rather produced as a part additional (supernumerary) to the said organs-partial objects. The BwO cannot be said to be the opposite of the organs. They are not its enemies. The enemy of the BwO is the organism. What the BwO is opposed to is not the organs, but the organization of the organs that one calls the organism.[72] That is to say, the BwO and its organs-partial objects are conjointly opposed to the organism, the latter being the "organic organization of the organs."[73]

In a particular modality of the body without organs known as the "full" or "catatonic" BwO, the interaction between the organs-partial objects of the connective synthesis and the BwO is such that the process of production itself is recorded on the BwO. More precisely, heterogenous chains of organs-partial objects enter into disjunctive syntheses on the surface of the BwO. Here, organs-partial objects attach themselves to the BwO as points of disjunction. They mark the surface of the BwO into a grid of co-ordinates.[74] The foregoing disjunctive synthesis is termed *inclusive,* in that all terms entering into the synthesis are simultaneously affirmed in their very difference from one another. Deleuze and Guattari maintain that this grid of disjunctive syntheses is one in which the disjoined terms are affirmed despite and even because of their mutual distantiation. This distance nowise entails an exclusion of one term from the other, or of the latter from the former. A massive paradox.[75]

This may all be so. It is, however, possible—and necessary—to characterize the BwO more pointedly, relating it to becoming (*devenir*), intensity, singularity, individuation, and absolute deterritorialization. In this connection, Peter Hallward maintains that the BwO is a virtual intensity that has nothing to do with space or place and all to do with pure potential and becoming.[76] For Keith Ansell-Pearson, the BwO is a "body" of energies and becomings that is constantly "permeated" by highly unformed and unstable matters—these latter describable as free intensities and nomadic singularities. The BwO is, moreover, the immanent field of desire *effecting* alike the production and distribution of these intensities. The becoming that is the BwO entails a play of individation through haecceity and singularity, an incessant production of intensities starting from degree zero.[77]

Moreover, the BwO is nowise pre-existent (such a notion would be the greatest error)—but rather constructed, and always in the process of being constructed. Responsible for the foregoing are processes of *absolute deterritorialization*. These latter may be described as movements of becoming that, vertiginously sweeping away (or transmogrifying) all individuals (continuous-intensive multiplicities), precipitate the reprise of pure intensities, that is, pre-individual individuating factors.[78] Related to this, movements of *relative deterritorialization* are those whereby individuals (continuous/qualitative multiplicities or organs-partial objects) are (re-)engineered. One does note, however, that there is nonetheless a perpetual immanence of absolute deterritorialization to relative deterritorialization. This makes perfect sense, in that the engineering of continuous multiplicities (organs-partial objects) by pre-individual individuating factors (via relative deterritorialization) is at once the reprise of these factors (via absolute deterritorialization). Otherwise put: absolute deterritorialization is the reprise of the pre-individual individuating factors; relative deterritorialization is the event of the individuation/re-individuation of continuous multiplicities (organs-partial objects). The former is in incessant immanence with regard to the latter.

There appear to be contradictions here. They do not need to, however, amount to such. The BwO is, on the one hand, said to be a becoming (*devenir*): which, as we know, entails the incessant reprise of pre-individual singularities (individuating factors) with which latter the BwO is "permeated." On the other hand, the BwO, is maintained to be a collection of purely distributed organs-partial objects (continuous/intensive multiplicities). From the just-discussed perpetual immanence of absolute to relative deterritorialization, however, it cannot but be concluded that the BwO is indeed the former *and* the latter.

Now, the approximate analogue in Schelling's *Weltalter* to the Deleuzoguattarian BwO has gradually come into increasingly clearer view. We have already regaled ourselves with Schelling's explicit assimilation of the natural aspect of the Absolute or God (i.e. eternal divine nature) to Spinozan *natura naturans*; that is, to the nature generated within God (or the Godhead) through his/its auto-contraction (Lurianic *ẓimẓum*). Schellingian eternal divine nature and, by implication, Spinozan *natura naturans*—were here both deemed the in-itself of nature (this latter term deployed presently in the prosaic and quotidian sense). Both are thus Spinoza's Absolute Substance. Now, for Deleuze and Guattari, it is precisely the BwO that is none other than the latter. For they contend that the BwO is indeed the fully Spinozan immanent substance (Absolute Substance or *natura naturans*), and that the organs-partial objects are the latter's attributes. These two terms do not, as said, exclude or oppose one another.[79] For the BwO is the pure potentiality of raw unformed matter, whilst the partial objects are its immediate powers or working parts.[80]

Spinoza's conception of substance, though possessed of the characteristics of immanence and *causa sui*, is—notwithstanding—fundamentally ossified by a certain intractable immutability, replete with intimations of stasis.[81] Therefore, whilst assimilating the BwO to Spinozan immanent substance, one must heed Claire Colebrook's qualification that substance in the usage of Deleuze and Guattari, far from being a noun, is indeed an *infinitive*; not some sort of ultimate being or supreme entity, but rather a power of creating.[82]

Let us take stock. We have thus far described the precipitation into existence of a nature (eternal divine nature) by way of God's auto-contraction (Lurianic *ẓimẓum*), this nature being closely assimilable to Spinozan *natura naturans*, affirmed by Schelling to be "absolute productivity," and wagered above to be akin to Deleuzoguattarian desiring-production (whose immanent principle is the BwO). The last of the three productive syntheses—that is, the conjunctive synthesis—comprising the latter remains to be discussed. However, we must presently return to our Schellingian theogony; that is, more specifically, to God's self-withdrawal.

4. The incessant rotary motion of eternal divine nature

The negation/contraction of God is merely the generative force by which eternal divine nature is initially posited. Yet, "this force never appears for itself alone, but only ever as the bearer of another essence [or force], fastening it down and holding it together. This other essence [or force] is expansive by nature, and it is thus volitilizing and spiritualizing."[83] Indeed, "without this overflowing and communicative essence [force], the attracting [and thus negating and contracting] force would be empty and genuinely ineffectual, unfulfilled and unbearable [*unleidlich*] to itself."[84] On the other hand, however, if "there were no negating force, then this other essence [force] would have nothing against which it could externalize itself and through which it could be put into effect."[85]

The immediate upshot of this is that a tension develops between the contracting-negating and the expanding-affirming forces (or powers/potencies), respectively. What now occurs, avers Schelling, is that the "negating power [or force] dislocates itself from itself in order to be, so to speak, its own complete being"—as does the expanding-affirming force—in such a way that a tension results in the two potencies (forces) arraying themselves in mutual opposition.[86] Eternal divine nature as the contracting force of negation stands in stark opposition to the same as the expansive force of affirmation—both claiming Being solely for itself to the reciprocal and wholesale exclusion of the other, both claiming to be indeed divine. Taken together, the resulting process of eternal divine nature is homologous to what Lurianic Kabbala describes as the dialectical "double beat of the alternately expanding movement of *Ein-Sof* and its desire to return to itself, *hitpashtut* ('egression') and *histalkut* ('regression'), as the kabbalists call it. Every movement of regression toward the source has something of a new *ẓimẓum* about it."[87]

One may conjure up a striking parallel here between the reciprocal antagonism between these forces of contraction-negation and expansion-affirmation of Schellingian eternal divine nature, on the one hand, and the tension that arises between what Deleuze and Guattari term the *paranoiac* and the *miraculating* machines at the unconscious level of desiring-production, on the other. These latter two are spawned as the result of two different modes of interaction between the Deleuzoguattarian body without organs (BwO) and its organs-partial objects. For Deleuze and Guattari, when the BwO experiences the organs-partial objects as a form of persecution wherein the partial objects seem to be attemping to break into the BwO, the latter repels the organs and thereby spawns a paranoiac machine. Here, the BwO can no longer tolerate the organ-machines. "In order to resist the organ-machines, the body without organs presents its smooth, slippery, opaque, taut surface as a barrier. In order to resist linked, connected, and interrupted flows, it sets up a counterflow of amorphous, undifferentiated fluid."[88] Clearly, the BwO's presenting of itself as a barrier to the organs-partial objects is its becoming taciturn in a movement of self-enclosure. In this sense, it is akin to the Schellingian contracting force (the first potency of eternal divine nature).

Now, on the other hand, in a different mode of interaction with the organs-partial objects, the BwO attracts these latter and appropriates them to itself, functioning as quasi-cause of the said organs—in which they appear to emanate from the BwO, seem to be miraculated by it.[89] Since the machine herein spawned—that is, the miraculating machine—is so by way of exactly the opposite process to that by which the paranoiac machine arises, the miraculating machine is hereby akin to the Schellingian second potency of divine nature: the force of expansion.

5. The emergence of the third potency as spirit amid the rotary motion

Presently, we return to Schelling's theogony itself. We ended up claiming that the contracting-negating and the expanding-affirming potencies stood in a relation of mutual antagonism, each alike arrogating to itself the dignity of alone being possessed of Being to the reciprocal exclusion of the other. Yet, eternal divine nature is in fact both of these—as it is also their "unity." That is to say, God—in his nature—is all of these foregoing three at once. And, in that eternal divine nature is that which in God is unavoidable, God is likewise separately all of the three *necessarily*.[90]

At this point, a confusion must not result: when it is said that God is alike all three divine potencies, this is *not* to say that the Godhead itself has been in wholesale fashion assimilated to these. For, most emphatically, the Godhead is *devoid* of nature. The three divine potencies are separately and conjointly God *only insofar as God has contracted Being*—that is, has acquired his eternal divine nature, the latter itself spawned by the generative force of the second eternal will (although, its genetic element is, more accurately, the dissonance within the first eternal willing that is at once this first willing's self-alienation as the quasi-duality of itself and another will: the second will). God *as Godhead*, meanwhile, is—for Schelling—wholly transcendent of the eternal divine nature that God contracts. (We will see soon enough whether this transcendence may still be allowed.) Moreover, whereas eternal divine nature is what God is of necessity, God considered as pure eternal Godhead is completely and utterly outside the realm of necessity, since it is quite contrariwise (and as priorly espied)—*the eternal freedom to be*. This will likewise be subject to subsequent scrutiny.

But let us now dwell specifically on the *third* divine potency. The latter is generated out of the mutual antagonism of the contracting force (the first potency) and the expansive force (the second potency) by way of an inner necessity, and moreover—as *spirit.* Despite its being thus characterized here by Schelling, the third potency is not to be conflated with the Godhead itself—for, though less corporeal and indeed a higher potency than the first and the second potencies, this third potency is nonetheless decidedly lower (more corporeal) than the absolutely pure spirit, the eternal and vertiginous "abyss of freedom" of the Godhead.[91]

Moreover, it is important to be mindful that the generation of the third potency out of the reciprocal opposition of the first potency and second potency—is precisely the manner in which the second eternal will *continues* to strive after eternity (which is, as we saw earlier, merely the way in which eternity unconsciously strives after itself), how the will in eternity (the second will) seeks to find the eternal willing that is the Godhead (the first will), or—most succinctly—how eternal divine nature strives in its relentlessly psychotic rotary motion to resolve the originary dissonance within the pure willing from whence it had always already arisen in the first place. And it is precisely due to the foregoing incessant and tireless quality of the second will's striving toward the Godhead, that the third potency calls forth again and again the mutual antagonism of the first two divine potencies—such that it (the third potency) may ever be generated anew from the latter.[92]

It is also by virtue of the third potency's fundamentally *voluptuous* nature that it finds itself repeatedly soliciting the renewal of this blind contraction-expansion antagonism. Indeed, for the third potency "the opposition serves as an eternal pleasure [*Lust*], since spirit [the third potency] only becomes sensible to itself in the opposition, and far from sublating this opposition, spirit seeks instead to constantly posit and confirm it."[93] The said antagonism is not only enjoyed by the third potency that arises from it, but is indeed *consumed* by the latter. That is to say: it is consumed by that which arises from it. Yet, though consumed by spirit, the antagonism between the first potency and the second potency reappears. In point of fact, the antagonism *eternally* "produces itself, in order always again to be consumed by the unity [the third potency] and the antithesis is eternally consumed by the unity in order always to revive itself anew."[94] Schelling is thus entirely within his wits to describe this incessant circulation that is eternal divine nature as "an unremitting wheel, a rotary movement that never comes to a standstill," as the "life that continually incinerates itself and again rejuvenates itself from ash."[95]

This blind rotary motion, moreover, *is bereft of beginning.* Certainly, in positing itself incessantly, it does begin—and repeatedly so. However, a beginning that begins again and again without extending itself into a duration that could thence in turn come to serve as "the ground of a steady progression," at least for

Schelling, does not qualify as an authentic beginning. Indeed, in eternity, such an authentic beginning would be entirely impossible—for eternity *as such* is bereft of duration.[96] In eternity, within the ambit of eternal divine nature, one instead finds "time, eternally commencing, eternally becoming, always devouring itself and always again giving birth to itself."[97] Thus, in eternity every emergent instant fails to hold onto itself—a failure that precludes such an instant from serving as one of a series of obdurate present moments whose sequential arraying would otherwise ground a temporal succession. Self-incinerating eternal divine nature partakes of precisely such a (subtemporal) temporality: the antagonism between the contracting force and the expansive force is instantly consumed by the spirit (the third potency) generated out of the latter, this consuming spirit itself vanishing out of existence in the very instant of its emergence from the antagonism that it so voluptuously devours. The third potency is, therefore, a pure evanescence constantly being born and reborn from the eternally returning opposition twixt the first potency and the second potency—a birth and rebirth that neither presupposes nor produces time in any chronological sense.

As such—that is, as a pure evanescence generated out of the antagonism that it consumes in the very instant of itself at once being extinguished from existence—the Schellingian third divine potency bears a remarkable resemblance to what Deleuze and Guattari call the *nomadic subject* at the level of desiring-production. This latter Deleuzoguattarian nomadic subject or nomadic self is, moreover, produced in a strongly homologous fashion to the mode of generation of the Schellingian third potency. The generation of the nomadic subject/self occurs as follows. The moments of generation of the paranoiac and miraculating machines give rise to an alternating movement whose oscillatory tension spawns the so-called celibate machine. This pulsating motion constantly being effected by the forces of attraction and repulsion between BwO and organs-partial objects gives rise to an open series of pure "intense *nervous states* that fill up the body without organs to varying degrees."[98] It is these intensive states, taken together, that precipitate the Deleuzoguattarian nomadic subject into emergence.

One must, however, immediately be alerted that the nomadic subject is nowise a fully-fledged autonomous agent. It is, in truth, far less than this. Indeed, it is at most an "excremental" subject that is born of each intensive state of the BwO and incessantly reborn of each ensuing state. In this procees, the nomadic subject consumes and consummates the intensive state from whence it arises (consumption-consummation). The crucial thing here, is that each intensive nervous state is primary in relation to the nomadic subject that lives it.[99] This vertiginous and abyssal ambit is one in which "at each moment, everything tends to be spread out into an instantaneous, indefinitely divisible *continuum,* which will not prolong itself into the next instant, but will pass away, only to be reborn in the following instant, in a flicker or shiver that constantly begins again."[100] The evanescent sub-ontology of the nomadic subject is thus somewhat analogous to what Sartre maintains of the utterly protean existence of Jean Genet, in the sense that its life is always already its own death and that it incessantly "dies its own life." It feels its own sub-sentient self to be both itself and that of another. Eternity is present in the "atom" of its sub-duration, in its own repeatedly fatal flitting instant. "Born of nothingness," its "being has the substantiality of nonbeing."[101] Here I have of course been paraphrasing Sartre on Genet.

Thus, and moreover, the nomadic subject—like the Schellingian third divine potency—is essentially evanescent and voluptuous. Indeed, in consuming and consummating each state through which it passes and of which it is incessantly born and reborn, the euphoric exclamation of the nomadic subject is "So *that's* what it was!"[102] It identifies with the emotional and delirious intensive states through which it passes (this is, of course, not in the least to be confused with an ego-identicification). Alphonso Lingis may be seen as glossing the fundamentally voluptuous nature of the Deleuzoguattarian nomadic subject in the following passage: "The excitation intensifies, is supported by its own substance, and explodes into itself. Forming the incandenscence of a reflection, an ipseity. This ipseity is not produced by a mirror-effect. It is produced by an intensity effect with itself, a pleasure tormented by itself, a torment complacent in itself, voluptuous, a lust luxuriating in itself."[103]

This evanescent quality of the nomadic subject appears also to belong to the Leibnizian monad. Here Leibniz avers that "all created or derived Monads are products and have their birth, so to speak, through continual fulgurations of the Divinity from moment to moment …"[104] This divinity (God), much as is the case with Spinoza's Absolute Substance or *natura naturans*—asseverates Leibniz—is the sole "primary unity or original simple substance" from whence the monad is generated.[105] Furthermore, much like the Deleuzoguattarian nomadic subject, the Leibnizian monad also surges up into existence *ex nihilo*: it "can only come into being or come to an end all at once; that is to say, it can come into being only by creation and come to an end only by annihilation …"[106] This reminds one of the observation made in quantum physics of subatomic particles suddenly coming into existence and similarly being instantaneously voided out of existence. Other aspects in which the Leibnizian monad and the nomadic subject are homologous are that they are both devoid of extension and form; and, moreover, "spiritual": Leibniz, in at least one instance, describes monads as inter alia "incorporeal automata."[107] Finally, in that the nomadic subject is perpetually being born, extinguished, and reborn of each intensive state of the BwO; it is never the same/self-same subject—and is therefore multiple. More precisely, it is at once all of its evanscent upsurges. That is to say, "the" nomadic subject is always already *all* nomadic subjects at once. Indeed, in his reading of Leibniz, Deleuze maintains that "the monad does not exist outside of other monads," that singularities "proper to each monad are extended as far as the singularities of all others and in all senses."[108] The nomadic subject and the Leibnizian monad are thus both fundamentally multiple in the most essential sense.

6. The third potency as traversing the entire ladder of future formations

In Schelling, the unremitting rotary motion of eternal divine nature, it is moreover averred, serves to incessantly generate "images," these latter ascending from the lowest to the highest potencies and from there on in unto the still concealed Godhead itself; that is, from the first through to the third potencies unto the Eternal itself, in an endless profusion.[109] These "images" are "the possibilities or spirits of things" that spirit (the third potency) runs through in its constantly being born and reborn, and which the Godhead itself beholds reflected in this spirit as "the visions of future spirits that are determined, along with the being of nature, for creation"—as indeed "everthing that someday should become actual in nature and then everything that someday should become actual in the spirit world."[110] In fact, the Godhead beholds herein "the entire ladder of future formations" that the eternally evanescing spirit (the third potency) itself traverses.[111] Nonetheless, all of this passes "before the eye of the eternal [the Godhead] only as a glimpse or a vision"—since eternal divine nature "has no actuality" and thus instead passes "again into becoming" with "nothing abiding, nothing stable," everything being therefore "in incessant formation."[112] Indeed, although Schelling describes the images generated by eternal divine nature as "archetypes," he deploys the latter term in a strictly non-Platonic sense: for these pseudo-archetypes are neither physical substances, nor vacuous genera, nor static or complete forms; but rather "eternally becoming and in incessant movement and generation."[113]

A comparison with aspects of kabbalistic thought invites itself at this point. The Schellingian quasi-archetypes—effectively pure becomings—tranversed by the envanescing third divine potency come to bear—upon closer inspection—a tantalizing proximity to the theory of the *Sefirot*, or divine attributes of God, as refined and transmogrified by the thought of *inter alia* Moses Cordovero. The *Sefirot* are said to subsume the archetype of every created thing. They are contained within the Godhead and impregnate each and every being outside the latter. In its original and more archaic forms, the theory of the divine *Sefirot* presents these latter as something akin to stases: static, atemporal forms. In this more archaic theory, there are precisely ten fundamental divine attributes, namely: *Keter Elyon* ("supreme crown"); *Hokhmah* ("wisdom"); *Binad* ("intelligence"); *Gedullah* ("greatness") or *Hesed* ("love"); *Gevurah* ("power") or *Din* ("Judgment," but also "rigor"); *Tiferet* ("beauty") or *Rahamim* ("compassion"); *Nezah* ("lasting endurance"); *Hod* ("majesty"); *Zaddik* ("righteous one") or *Yesod Olam* ("foundation of the world"); and *Malkhut* ("kingdom") or *Atarah* ("diadem").[114] Such a rigidly immobile conception of the *Sefirot* will, however, not do.

This earlier theory of the divine attributes, notwithstanding, undergoes considerable intrication with Moses Cordovero. The first complication here is that each *Sefirah* henceforth comes to be viewed as comprising all other *Sefirot*. The inherent ramification is that of an infinite reflection of all *Sefirot* within themselves—a *mise en abyme*. But Cordovero will exacerbate this initial complexification. For in addition to the foregoing notion of an unlimited embeddedness of all *Sefirot* within all other *Sefirot*, comes that of *behinot*. The *behinot* are the infinite number of aspects of each given *Sefirah*. Each *Sefirah* is infinitely internally divided into and by its *behinot*. Cordovero's insight on this point is, moreover, that every *Sefirah*'s division into a bottomless regress of endlessly ramified *behinot*, is at once each *Sefirah*'s internal connection to all other *Sefirot*. In Scholem's formulation, each and every *behinah* effects the rousing and manifestation of the next ensuing *behinah*—and so forth in infinitely ramified profusion.[115] In fact, each *Sefirah* falls into itself, precipitating an infinitude of internal reflections. In this process, the infinite regress of each *Sefirah* into itself at once causes the begetting and ushering into being from within itself of another *Sefirah*.[116] This infinite internal fracturing of the *Sefirot* and its immediate ramification as the genesis of supplementary *Sefirot* implies that any "given" *Sefirot* is *mutatis mutandis* a pure disparity—that is to say, a difference-in-itself. Differences in themselves, as already observed apropos of Deleuze, are simply pure becomings (*devenir*). Moreover, like Schelling's "possibilities or spirits of things," Cordovero regards the *Sefirot* to be the elemental structures of all beings whose process of emanation within the Deity the Zohar terms *istakluta le-fum sha'ata*, or the "fleeting vision of the eternal."[117]

A still more striking kabbalistic analogue to Schelling's mobile archetypes of future formations may be found in Israel Sarug's particular conception of the primordial Torah. Sarug seems to differ from Schelling in that he traces the emergence to these archetypes to a point genetically and logically prior to the *ẓimẓum*. Firstly, he avers that the primordial dissonance within *Ein-Sof* produces—or "engraves"—"primordial points" in the "power of *Din*," which points become the first forms to leave their marks in the fundamental essence of *Ein-Sof*.[118] Next, Sarug contends that the potency of these "engraved" points is "activated" by the "light" of *Ein-Sof*. This process produces and weaves the "primordial Torah" into the very substance of *Ein-Sof* itself. Scholem describes the primordial Torah as a sort of fundamental "linguistic" movement of *Ein-Sof* within its own self, and moreover as a *malbush* ("garment") completely inseparable from the divine substance—for it is woven within *Ein-Sof* in the same manner in which the clothing of a grasshopper is, as the Midrash has it, inextricably part of itself. For Sarug, the "length" of this divine garment comprises all the possible combinations of the letters of the Hebrew alphabet conjointly comprising the "archistructure" of divine thought, whereas its "breadth" is an elaboration of the Tetragrammaton according to the numerical value of the four possible permuted spellings of the names composed of the latter's letters—the foregoing being the "threads" and the "weave" originally placed in the very "hem" of the garment. The primordial Torah contains potentially everything that could possibly be revealed through the Torah that was subsequently to be given to humans on earth.[119] Scholem maintains it to be *mutatis mutandis* the kabbalistic equivalent of the platonic realm of pure forms.

Apart from the Schelling-Cordovero parallel here unearthed, one yet again glimpses another congruence between Deleuzoguattarian desiring-production and Schellingian eternal divine nature. As we have seen, the "images" generated by the incessantly renewed antagonism between the first potency and the second potency that the Schellingian third potency traverses are fluid "archetypes" in eternal becoming. Now, as has already been discussed in connection with Miguel de Beistegui apropos of Deleuzian onto-hetero-genesis, pure becoming (*devenir*) is the incessant reprise of pre-individual individuating factors; that is, the reprisal of differen*t*iating difference: the resurgence of pure *intensities* in and of themselves. The generation of images by the anatagonism between the two lower potencies is thus a production of such pure intensive states. The constantly evanescing third potency can therefore be said to perform a vertiginous—though purely intensive—journey through (likewise intensive) states generated by the opposition/oscillation between the forces of contraction-negation and expansion-affirmation. And it is precisely such a journey that is undertaken by the Deleuzoguattarian nomadic subject: for this latter passes through all the degrees of intensity in an endless migratory movement along the "surface" of the BwO. This journey in intensity is moreover, as Deleuze and Guattari go on to aver, one in which the nomadic subject "identifies" the states through which it passes with all the "names" of history: "*every name in history is I* . . ."[120] The foregoing vertiginous passage of the nomadic subject—the authors contend—is thus a veritable Nietzschean *eternal return*.[121]

In what sense may this be maintained? Firstly, it is paramount to urgently discard any notion of the eternal return as a return of the Same, of Being, or of the Identical. Such a conception—vacuous as it no doubt is—would leave us with nothing. Therefore, what returns in the eternal return is not Being, but indeed *becoming* (*Werden*). In fact, there is as such *no* Being—that is, no Being "in and of itself": for the only Being that may be cogently alluded to is Being inasmuch as it is affirmed of becoming. "Affirmed" in what sense? In the sense that Being is the *return* of becoming. Otherwise put: "Returning is being, but only the being of becoming."[122]

We have already observed that becoming (*devenir*) is the incessant production of events, each event entailing the upsurge of pre-individual individuating factors, that is, pure intensive quantities spawning categorically novel continuous-intensive-qualitative multiplicities. As the return of becoming, the eternal return would thus entail the return or reprisal of the said pre-individual individuating factors; that is, the return of differen*t*iating difference—or, more succinctly, of difference *tout court*. It therefore stands to reason that "eternal return does not cause the same and the similar to return, but is itself derived from a world of pure difference."[123] This implies that the eternal return entails "the assignation of difference as the origin, which then relates different to different in order to make it (or them) return as such"; and is hereby furthermore intimately imbricated with none other than the *will to power* in the precise sense that "the eternal return is indeed the consequence of a difference which is originary, pure, synthetic and in-itself (which Nietzsche called will to power)"—an infinitely taut recto-verso in which "difference is the in-itself, [and] the repetition in the eternal return is the for-itself of difference."[124] If will to power is the in-itself of difference, difference in its most pure and originary sense, the pure difference herein being alluded to can only be the purely intensive differen*t*iating difference of the pre-individual individuating factors. These latter, as repeatedly maintained, are pure intensive quantities. The reprisal of the said free intensive quantities (these latter comprising the vertiginous subterranean world of the will to power) inherent to the eternal return, implies that the latter entails not merely becoming (*Werden*)—but indeed an *intensive* becoming (*devinir*). (Indeed, all becomings—entailing as they do the recrudescence of pure intensive quantities—are inherently becomings-intense.) And it is in such an intensive becoming that the nomadic subject finds itself incessantly embroiled. As Deleuze maintains, this abyssal world of the nomadic subject is that of the eternal return and, at once and equivalently, that of the will to power: "the flashing world of metamorphoses, of communicating intensities, differences of differences, of *breaths,* insinuations and exhalations: a world of intensive intentionalities, a world of simulacra or 'mysteries.'"[125]

Belonging to and passing through this vertiginous world of pure intensive states, the nomadic subject is in many respects quite like the Schellingian third potency, which traverses "the entire ladder of future formations"—for the Deleuzoguattarian nomadic subject's intensive migratory movement is one in which "everything conmingles in these intense becomings [*devenir*], passages, and migrations—all this drift that ascends and descends the flows of time: countries, races, families, parental appelations, divine appelations, geographical and historical designations, and even miscellaneous news items."[126] And all of these things which conmingle on the BwO "designate regions on this body—that is, zones of intensities, fields of potentials. Phenomena of individualization and sexualization are produced within these fields. We pass from one field to another by crossing threshold: we never stop migrating, we become other individuals as well as other sexes, and departing becomes as easy as being born or dying."[127] This world of the eternal return and the will to power, of desiring-production whose immanent principle is the BwO—very much like Schellingian eternal divine nature—is unremitting in its evanescence, incessant in its sempiternal auto-incineration: for, as Deleuze maintains, repetition "in the eternal return never means continuation, perpetuation or prolongation, nor even the discontinuous return of something which would at least be able to be prolonged in a partial cycle (an identity, an I, a Self) …"[128] This is so because "eternal return, affirmed in all its power, allows no installation of a foundation-ground. On the contrary, it swallows up or destroys every ground," hereby making "us party to a universal *ungrounding.*"[129]

Ungrounding indeed. But let us pause briefly. For during the course of this elaboration a certain inconsistency has insinuated itself into our exegesis of the *Weltalter*. At almost the very outset, we assimilated the Schellingian pure willing (*Wollen*) to the Nietzschean will to power, contending that the very tranquility, non-striving, and subjectless and objectless character of the former—far from standing for complicity with the will to nothingness or a nothingness of the will—in fact, coincided with the highest affirmation. It has, however, emerged that will to power is *originary difference.* Therefore, it appears more correct *not* to directly identify will to power with the pure willing of the Godhead, but instead to assimilate the former to the originary dissonance (the independent emergence of the second will at the heart of the first which internally cleaves the said first will) that has always already disturbed the self-coincidence/tranquility of the latter willing. Moreover, the fact that eternal return is affirmed of the will to power—implies that this originary dissonance in the pure willing *is incessantly produced and re-produced*. And it is alike this latter perpetual reprisal (of pure intensive quantities, of differen*t*iating difference) that is at the basis of the unremitting pulsation of Schellingian eternal divine nature.

7. The *Ent-Scheidung:* the emergence of God as subject

As observed, eternal divine nature's incessantly reproduced opposition between contraction and expansion and its relentless pulsation—is precisely the form that the second will's seeking of the eternal Godhead assumes. This unremitting pulsation of eternal divine nature, when considered from *its* (eternal divine nature's) vantage point, is none other than eternal divine nature's mode of striving after Being: it is "not a true existence but only an eternal drive and zeal to be, without actual Being."[130] But, from within its own circumscribed ambit, eternal divine nature is *utterly incapable* of attaining Being, just as the second will cannot find eternity and just as—equivalently—eternity cannot find itself; instead, divine nature runs itself "to a standstill in desire, as an unremitting striving, an eternally insatiable obsession [*Sucht*] with Being."[131] Eternal divine nature is pure necessity and there is "nothing in that first nature except [its own] irresistible drive and insensate movement," a mad and maddeningly blind oscillation which—by itself—it can nowise thwart, resist, let alone transcend.[132]

Thus, the only manner in which it can possibly escape from this monotonous pulsation, to be brought outside its own insane ambit, is to be drawn out of itself by that which is in essence completely *external* to the blind rotary motion of eternal divine nature. That is to say: its "liberation and deliverance can only come through an Other that is outside of it and wholly independent of it and exalted above it."[133] Whilst still trapped within its own relentless rotary movement, eternal divine nature may be described as mere Being—but not as *having Being* (which latter it perpetually fails to attain in its incessant repetition of abortive beginnings). And it is only in and through such an "incontestably higher" Other that divine nature can come to eventually possess Being.[134] Now, this much elevated Other *is precisely the Godhead itself*. And quite certainly, "through its simple presence, without any movement (since it is still pure conation itself)," the Godhead already "rouses in that life [that of eternal divine nature] the yearning for freedom."[135] Indeed, this must be so: for eternal divine nature's "distress of pining" after Being is at once *inter alia*—as has already been mentioned—also eternity's searching for itself; that is, the pure willing's sisyphean ordeal at overcoming the originary dissonance by which it is always already inhabited (causing its, the pure willing's, self-alienation—a primordial self-estrangement at the very basis of the entire Schellingian theogony itself). Thus, contends Schelling, the pure presence of the Godhead induces in eternal divine nature the transmogrification of the latter's former headlong and wild obsession into a quiter yearning: that is, a yearning to ally itself with the Godhead "as if it were its own true or highest self, with the will that wills nothing [the first will], with eternal freedom."[136] Now, the *only* manner in which eternal divine nature is able to ally itself with a pure spirit (in this case, the purest and most spiritual of spirits), is through that which within itself is the most spiritual—that is to say, *through the third divine potency.*[137]

And this is what indeed occurs. Eternal divine nature's highest potency is drawn upward by the sheer mute presence of the Eternal (the Godhead) and "is elevated to [be] the immediate *subject* [my emphasis] of the pure Godhead."[138] Through this latter movement, eternal divine nature itself "becomes Being with respect to the highest [the Godhead]."[139] The obverse of this same movement is that *the Godhead*, until this crucial point a pure sublimity beyond the ambit of Being and thus outside of having or not having Being, *comes to have Being in eternal divine nature* by subjugating the latter to itself—herein rendering eternal divine nature a *substrate* in which it directly recognizes this same eternal divine nature as its *own.*[140] Moreover, that the third potency is elevated to the dignity of divine subject, means that God himself also emerges *as subject.*

There is more. We have here been speaking of God's coming to subjecthood from the vantage point of eternal divine nature's own liberation from its incessant oscillatory movement. From the purview of self-incarcerating eternal divine nature, the pure mute presence of the Godhead appears as liberator via something analogous to magnetic induction. (We must allow Schelling his foibles.) But let us now look at this liberation conversely: from the perspective of the Godhead itself. As seen, the Godhead is—at its most fundamental—the eternal freedom to be. The liberation of eternal divine nature from its unremitting cycle of contractions and expansions through which the Godhead also acquires Being, not in itself (not as such) but *as it (Being) comes to be reflected in the eternal divine nature liberated by it*, is accomplished through an absolutely (and vertiginously) free decision (*Ent-Scheidung*) on the part of the same Godhead. As Scholem contends apropos of Moses Cordovero's *Elimah*, the Godhead's decision to emerge from its own concealment into manifestation and creation—far from being "in any sense a process which is a necessary consequence of the essence of *Ein-Sof*"—must be maintained to be "a free decision which remains a constant and impenetrable mystery."[141] As shall moreover be seen, this decision is at once God's pronunciation of the Word.

The latter is also that by which the seemingly sempiternal repetition of abortive beginnings is finally terminated by a Beginning that truly begins. Indeed, as Žižek surmises apropos of Schelling, the "Beginning occurs when the Word is pronounced which 'represses', rejects into the eternal Past, this self-enclosed circuit of drives [the unremitting pulsation of eternal divine nature]."[142] The Slovene adds, moreover, that this "true Beginning is the passage from the 'closed' rotary motion to 'open' progress, from drive to desire—or, in Lacanian terms, from the Real to the Symbolic."[143] Furthermore, the *Ent-Scheidung* is that by which God for the first time discloses himself to himself, appears to himself.[144] The pronunciation of the Word by God is, thus, the primordial act by which God—as eternal pure willing, pure spirit—transforms himself from a mere "In-itself" into a "For-itself."[145] As we shall see more comprehensively later on, Žižek demonstrates that the mode of God's coming to subjecthood in the Schellingian *Ent-Scheidung* is strictly homologous to that of the Lacanian human subject as split subject ($).

Intermezzo 2: From the larval to the fully-fledged subject

What is of even greater interest to us here is, however, the Deleuzoguattarian genesis of human subjectivity (*molar* subjectivity). The latter is precisely a point of interest, in that we have already seen the striking homology between the Schellingian third divine potency and the Deleuzoguattarian nomadic subject—in both mode of generation and in nature. This homology, however, spawns further homologies.

(1) Toward the "larval" subject

How then do Deleuze and Guattari proceed from the nomadic subject to the fully-fledged human subject? Certainly not—that is, at least not in the primary sense—by way of any form of *Ent-Scheidung* imposed by that which purportedly transcends the purely immanent ambit of the said nomadic subject's endless precipitation. Let us begin by recalling that the nomadic subject, spawned by the oscillation between the paranoiac and miraculating machines, is generated as a pure evanescence born of the open series of intensive states produced in the BwO by the said oscillatory movement. An entire slew of ramified vicissitudes shall, however, insinuate themselves within this ambit.

In his creative exposition of the Deleuzoguattarian mode of genesis of human subjectivity, Brian Massumi contends that, in due course, a process of *etiolation* gradually besets the incessant engendering of the nomadic subject—whereby *fewer and fewer* intensive states are produced in the BwO. More precisely, certain of the BwO's intensive states are in fact amplified whilst the remaining ones are diminished almost to the point of extinction. Notwithstanding, this is the movement in and by which the overall degree of intensity of the BwO *falls*. In tandem, the evanescing nomadic subject hereby traverses increasingly *fewer* intensive states. Massumi avers that this etiolation reaches the point where the intensity of "all the states [traversed by the nomadic subject] but the one just left and the one about to come are muffled to the point that they are almost imperciptible."[146]

In that the intensive states produced in the BwO are consumed by the nomadic subject incessently born of them and that this latter subject is essentially voluptuous, this multiplicity of intensive states (of pure, communicating intensive quantities) taken together may at once be viewed as a *sensation*—and likewise termed a *perception*. Although, given that these latter are thus far not being referred back to a fully-constituted subject, it would be more accurate to use Deleuze's word "percept"—or at least "microperception."

We have already seen how—though with some hesitation—the nomadic subject may be assimilated to the Leibnizian monad. This shall help steer our ensuing discussion. Leibniz defines a monad as a simple substance—by which he means one devoid of extented parts.[147] We have already expressed our reservations about the use of the term "substance" in such a connection. The unextended character of the monad is notwithstanding the first point of convergence. Moreover, Leibniz proceeds to asseverate that "multiplicity in the unit [*unite*] or in the simple substance, is nothing but what is called *Perception*" and furthermore that "nothing but this (namely, perceptions and their changes) can be found in the simple substance."[148] Thus, the Leibnizian monad turns out to be—in a crucial sense—*multiple*. And this multiplicity in/of the monad concerns the latter's being a microperception/sensation. As seen, the nomadic subject is likewise multiple—its multiplicity consisting in it itself being at once all such upsurges from intensive states of nomadic subjects. Thus, instead of speaking of "the" nomadic subject/self, if would be far more accurate to refer to the "teeming plethora of nomadic subjects/selves." Crucially, this multiplicity of the nomadic subject is also its *being* a percept/sensation. This is the homology between the Deleuzoguattarian nomadic subject and the Leibnizian monad that we have sought here to establish.

But we may elaborate this still further. Leibniz goes on to contend that, as a perception, "each created Monad represents the whole universe," that it effectively "expresses the whole universe through the connexion of all matter in the *plenum*," and that thereby every "Monad is, in its own way, a mirror of the universe" in its entirety.[149] Apropos of Leibniz, Deleuze avers in this connection that every monad, though it indeed "expresses the entire world," does so only "obscurely and dimly because it is finite and the world is infinite"—that, though the microperceptions of monads are in truth "representatives of the world," they remain "minute, obscure, confused perceptions" that are in actual fact "lacking an object, that is, hallucinatory microperceptions."[150] Such characterizations therefore permit Leibniz to aver that (bare) monads are perpetually and literally "in a state of stupor."[151]

But, bare monads are not *all* stupor. Each monad, though it perceives—jointly and indissolubly with all other monads—the entire universe very indistinctly, is possessed of what is termed a "zone of clarity" or "clear zone" within whose compass its perceptions are distinct. Here, Leibniz contends that monads "all strive after [*vont à*] the infinite, the whole; but they are limited and differentiated through the degrees of their distinct perceptions"; and furthermore that, "although each created Monad represents the whole universe, it represents more distinctly the body which specially pertains to it, and of which it is the entelechy ..."[152] This nonetheless bears no salutory consequences for what are here being described as bare/naked monads and their state of effective *eternal damnation*, "their only single and clear perception being their hatred of God."[153]

The Leibnizian bare monad, conveying the entire universe through an inebriated haze, therefore resembles the Deleuzoguattarian nomadic subject traversing the abyssal and crepuscular world of the eternal return, and likewise the Schellingian third divine potency vertiginously running through the entire ladder of future formations (becomings) in its self-incinerating circular gauntlet. Since supposedly God does not survive the test of the Nietzshean eternal return and is thus included in it only as purportedly dead, whilst the unremitting wheel of Schellingian eternal divine nature is the carceral ambit within which God—desperate to supercede his own originary self-alienation—has ensnared himself; this thankless realm inhabited by the bare monad is truly that of the reprobate, which reprobate—for thinking God deceased—groans eternally under the wrath of the self-entrapped Deity he so offends with his mediocre and slavering hatred. But God shall survive this death: not only upon the ignominious gibbet at Golgotha—but (and prior to this and in a still more fundamental sense) in his very coming to subjecthood; that is, in his creation of the world.

But let us, right at this moment, return to Massumi. We have observed that the overall intensity of the BwO falls. The process of etiolation whereby a selected number of intensive states is amplified in tandem with the diminution of the remaining states—Massumi views as at once entailing a *contraction* whereby the "veritable infinity of impulses [intensive states/sensations], is contracted into a restricted set of higher order sensations."[154] He moreover adds that this restricted set is in turn itself contracted "into *a single retrospective sensation* [my emphasis]."[155] This latter higher-order sensation effectively constitutes a level of sensation/perception that superimposes itself onto the contracted sensations (or microperceptions).[156]

We noted earlier how the nomadic subject always consumed and consummated the intensive states of which it was continuously born. Homologously, the higher level retrospective sensation here effectuating itself is likewise a consumption-consummation—except at a level superimposed upon that of the teeming plethora of nomadic subjects. To this higher-level sensation/perception there must therefore accord a higher-level subject/self. Let us call it the "*larval*" subject or self.[157]

Now, as are the consumptions-consummations (sensations/perceptions) of the panoply of nomadic subjects recorded as intensive states on the BwO, so too is the higher level consumption-consummation (sensation/perception) of the larval subject likewise recorded. The recording of the consumption-consummation of the larval subject in conjunction with the fact that it contracts into itself a myriad of lower-level sensations/perceptions, implies—however—that a *feedback level* of sensation has been instituted. This also means that the circuit of states to be transversed by the larval subject may now be—to an increasing extent—*anticipated.* And an anticipation leads to a recognition, in the sense that the anticipation is the protention of which the recognition is the retention.[158]

This all being the case, the larval subject/self may be assimilated to a higher-order Leibnizian monad than the bare/naked monad. This is the so-called *soul* or *remembering monad.* Leibniz contends that souls exist at a higher level than do bare monads since they "act according to the laws of final causes through appetitions, ends, and means."[159] In terms of the larval subject, on the one hand, "appetitions" and "ends" would conjointly refer to the assuagement vouchsafed by the privileged organs selected into the said subject's anticipation-satisfaction circuit; whilst "means" would, on the other hand, consist in the bare fact of this assuagement having been protended. Moreover, Deleuze may be seen as in effect elaborating on the foregoing Leibnizian passage when he maintains that, in contradistinction to bare monads, remembering monads are possessed of "a zone of clear expression that is both more extensive and increasingly hermetic," whereas each of the "perceptions that comprise the zone is associated with others in the infinite process of reciprocal determination."[160] Here, when translating the nomenclature of the remembering monad or soul to that of the larval subject/self, "more extensive" would alike refer to, on the one hand, the fact of the privileged organs (on the anticipation-satisfaction circuit of the larval subject) more closely approaching the condition of actual objects than did the organs-partial objects associated with the nomadic subject (bare monad); whilst the implied move from the crepuscular dusk of a purely intensive *spatium* in the direction of the greater incandescence of *extensio*, on the other, refers to the larval subject's inhabiting an ambit of greater microperceptive clarity than that of the nomadic self. We may read "increasingly hermetic" to mean that, though the larval subject's zone of clear perception is broader than that of the nomadic subject, that which it perceives only a little less dimly is narrower than what the monadic subject perceives but obscurely. (Indeed, this stands entirely to reason, in that the very move from nomadic to larval subject had been effected through the superposition in the BwO of fewer intensive states than was formerly the case.) Finally, "reciprocal determination" may be seen as consonant with the instantiation of protention-retention circuits associated with larval selves. (Note: the coming into existence of such a protention-retention circuit, immediately presupposes the coextensive (or at least anterior) emergence of time as succession (of which more later). Thus, the scenario of the higher-level larval subject has already exceeded the ambit of Schellingian eternal divine nature and its rotary motion, as in the latter the chronological time of temporal succession has not yet emerged and must indeed await the *Ent-Scheidung* in order to come into existence. With the *Ent-Scheidung*, as has been established, God comes to subjecthood. Within the ambit of Deleuzoguattarian desiring-production, however, the larval subject certainly is not such a fully-fledged subject. More must happen in order for this to be the case.)

The emergence of each such anticipation-recognition circuit harbors a crucial corollary. This corollary is as follows: that a transition from the level of *resonance* to that of *redundancy* has taken place.[161] The resonance referred to here is that pertaining to the open series of purely intensive states traversed and at once spawning the evanescent nomadic subject. Since intensity is always already difference in intensity, talk of a purely intensive state is immediately that of a—likewise pure—intensive multiplicity. Thus, one speaks of a *resonance* of compossible intensive states—or one speaks of intensity naught at all. This is at the level, of course, of the nomadic subject whose temporality—that of *Aion* (of which more later)—is anterior to, or "beneath," the time of temporal succession (*Chronos*).

Now, the above move from resonance to redundancy entails the transition from the level of difference-in-itself of a purely intensive *spatium* (that of the incessant reprisal of the mobile pre-individual individuating factors entailed by the eternal return) toward one in which intensive states come to follow one another in a temporal succession, certain of which may be said to recur with an ever greater degree of probability. And it is precisely this ever more predictable *recurrence* of intensive states on the protention-retention circuit that Massumi refers to as redundancy.[162] Nonetheless, it must not be forgotten that beneath this level of redundancy of the larval subject, although submerged, there continues to subsist a vast miasmal plethora of nomadic subjects in resonance.

We must also recall that the multiple conjunctive syntheses out of which the nomadic subjects arose presupposed anterior connective syntheses of organs-partial objects. Anterior to the emergence of the anticipation-recognition circuit, the connections between the organs-partial objects were nonlocalized, passive, transversal and indirect. The foregoing connective synthesis was *inclusive.* (In general, avers Massumi, a synthesis is inclusive when it "multiplies." In particular, an inclusive *connective* synthesis concatenates the terms or repetitions that pass into it. Its logic is thus strictly additive.[163]) Organs-partial objects are themselves pure dispersions, continuous/qualitative multiplicities. The disjunctive syntheses (whereby part of the energy of the connective synthesis is siphoned off so as to record disjoined chains of organs-partial objects upon the zones of intensity of the BwO) were likewise inclusive—this latter meaning that any given disjunctive synthesis gathered together all heterogenous chains of connected organs-partial objects in such a way that the garnered chains all entered into synthesis sans exclusion. (Massumi asseverates that an inclusive disjunctive synthesis is one which spawns divergent series of individuals (organs-partial objects) that hereby coexist despite their heterogeneity. Its inherent logic is "this *and/or* that."[164]) To complete the picture of the three inclusive syntheses at the level of the seething plethora of nomadic subjects/selves, we must also not fail to omit the inclusive conjunctive synthesis. (Here Massumi's formulation is that an inclusive conjunctive synthesis assumes the *and* of the inclusive disjunctive syntheses' *and/or* so as to join individuals in potential mixtures wherein no given individual is a priori excluded. The logic of this synthesis is "*both* this *and* that.")[165]

Now, as already observed, the movement toward the emergence of the anticipation-recognition circuit of a resonance that has become redundancy entails a superposition of increasingly fewer intensive states. It is this movement away from the compossibility of intensive states that marks the process whereby the disjunctive synthesis becomes increasingly less inclusive and ever the more *exclusive.* (Broadly speaking, and in accordance with Massumi's able formulation, a synthesis is exclusive when it "subtracts." In particular, a *disjunctive* synthesis is exclusive when it engineers series of diverging individuals whose sole mode of coexistence remains abstract.[166] An increasingly exclusive disjunctive synthesis, in fact, insinuates itself into rendering the *connective* synthesis more exclusive as it serves to limit the number of ways in which individuals that may coexist and concretely connect with themselves.[167])

The corollary of the becoming-exclusive of the disjunctive synthesis is that it selects ever-fewer organs-partial objects, whilst—by implication—the conjunctive synthesis features a lesser and lesser profusion of teeming subjects (whilst, notwithstanding, beneath every single larval subject a plethora of nomadic subjects remain in endless upsurge and infinitely ramified proliferation). This means that the emergence of the larval subjects entails a partitioning out of the anterior myriad of nomadic subjects. The point is: the conjunctive synthesis becomes increasingly more *segregative*. The conjunctive synthesis generative of the larval subject is thus associated with a disjunctive synthesis featuring fewer organs-partial objects than in the case of the nomadic subject. That is to say, certain organs are privileged whilst the remainder, being "degraded," fall outside the ambit of the anticipation-recognition circuit associated with a given larval subject/self. This movement of privileging is the obverse of the process by which the organs-partial objects find themselves submerged in the wake of the emergence of ones which ever more closely approximate the status of organs in the more prosaic sense (that is, the condition of more closely resembling extensive parts of an integral whole).

(2) Toward the "fledgling" subject

The anticipation-recognition circuit reaches the point where, as Massumi avers, each larval subject is "associated with a[n] [intensive] threshold state featuring a privileged organ on the way to satisfaction through connection with another privileged organ. On the feedback level of recognition, there are always at least two organs in play, usually nominally belonging to distinct bodies: mouth and breast."[168]

Yet, the fall in the overall degree of intensity of the BwO shall continue. In the wake of this etiolation, the connections of organs-partial objects shall become increasingly more localized and direct; the disjunctive syntheses, all the more exclusive; the conjunctive syntheses, ever more segregative. At the level of the larval subjects, the number of corresponding anticipation-recognition circuits is still fairly vast. Each such circuit is at once what Massumi terms a *drive* (not to be confused with the Lacanian definition of the same word). Each drive is associated with a privileged partial object. Now, however, certain anticipation-recognition circuits come to predominate whilst the remaining ones find themselves relegated. Meanwhile, on each predonderant circuit, the privileged organs herewith associated themselves give away to one single metaphysicalized organ which now "seems to stand alone, the final cause of satisfaction, its end and origin in one, the preeminent image of a joyful future-past."[169]

Let us quickly pause to take stock of the present configuration. On the one hand, at the most dissipative level, one finds a vast plethora of purely evanescent nomadic subjects associated with organs-partial objects—or pure dispersive (continuous-qualitative) multiplicities. At this level, there are no privileged organs, no anticipation-satisfaction circuits, no drives. At the next (somewhat less dissipative) level, one may glimpse a somewhat less vast profusion of likewise fairly evanescent larval subjects—each associated with a few privileged organs within their anticipation-recognition circuits. At this level, though a small collection of organs have been exclusively selected into the circuit, no single organ among these latter is privileged so as to single-handedly substitute for all others. Nonetheless, any given anticipation-recognition circuit does *de facto* constitute a drive. However, in some of these circuits it happens that one single organ comes to be extolled—that is to say, metaphysicalized or sacralized—to the extent of standing in for all the organs within its anticipation-recognition circuit.

Now, it is this latter move that presages the emergence of a still higher (less dissipative) level. The precipitation into existence of this third level occurs as follows. Massumi avers that the organs-partial objects from the level of the teeming nomadic subjects/selves, in addition to the organs from the level of the larval subjects whose anticipation-recognition circuits thus far feature no sacralized organs, taken together, comprise what he terms a *nonlimitative BwO*. On the other hand, the sacralized organs-partial objects conjointly come to constitute a *limitative BwO*. A tension immediately develops between the nonlimitative BwO and the limitative BwO in which the "nonlimitative body without organs repels the sacred organs, and the limitative body without organs attracts them back, inducing rebel vibrations [intensive states] to recontract into a tame satisfaction," with the inevitable result that the sacralized/metaphysicalized organs "always manage to reimpose their supremacy—with the aid of reinforcements from even more powerful reactive forces."[170] In this latter process of reimposition/reinforcement, the intensive states produced within the nonlimitative BwO and those produced within the limitative BwO enter into resonance with the intensive states of the exterior milieu. Those of the limitative BwO inevitably resonate consonantly with those of the exterior milieu, whilst the more mutant and purely intensive states of the nonlimitative BwO clash with those of the said milieu and are in the process greatly attenuated. Massumi contends that the intensive states of the limitative BwO—that is, those that resonate consonantly with those of the surroundings—in fact "come back amplified into virtues (the genealogy of morals)," and that hereby these "sensations of the first feedback loop [anticipation-recognition circuit] are bumped up a level, contracting into sensations of pride or shame or guilt. Overflying the larval selves are new *fledgling* selves [my emphasis]."[171]

We have thus moved from the level of recognition to that of self-recognition; that is to say, to the level of *reflection.* Here, the anticipation and recognition of a sacred organ gives way to—or is transmogrified into—the anticipation and recognition of *oneself.* Objective anticipation becomes moral reflection. The moral level of reflection entails the introduction of a further redundancy: the anticipation of oneself necessarily doubles this self into a *moi* and a *je.* These two are held together by a proper name.[172]

We have thus far observed two orders of Leibnizian monad: naked/bare monads (first order) and remembering monads/souls (second order). The fledgling subject/self, on the other hand, closely resembles a third and still higher order of monad: the *rational soul* or *mind* (also termed the *reasonable* or *reflexive monad*). The latter are none other than human beings themselves: humans, avers Leibniz, "rise to *acts of reflexion,* which make us think of what is called *I,* and observe that this or that is within us"—differing as they do from souls or remembering monads furthermore in that "souls in general are living mirrors or images of the universe of created things, [whereas] minds are also images of the Deity or Author of nature Himself, capable of knowing the system of the universe, and to some extent of imitating it through architectonic ensamples [*échantillons*], each mind being like a small divinity in its own sphere."[173] The zone of clear perception of this third species of monad is still more extensive than that of souls/remembering monads, whilst its dim plethora of hazy microperceptions is yet further diminished than that of the latter. Our present Leibnizian intermezzo ends here.

(3) Toward the "fully-fledged" (molar) subject

Now, we progress to what Massumi maintains as a still higher order of subject—one who for the most part feels himself to be quite literally divine. Massumi proceeds as follows. The exterior milieu with whose intensive states the nonlimitative and limitative BwOs resonate/clash is the social field itself. More precisely, it is the *socius.* The socius is the societal correlate of the BwO. It is to social production what the BwO is to desiring-production. (We shall delineate the socius and social production more incisively in due time.) And, much as at the level of the BwO (over and above the more originary oscillation between the paranoiac and miraculating machines) there arises the tension between nonlimitative and limitative BwOs; likewise does the socius (as the BwO of the social field) comprise an analogous tension between its own nonlimitative and limitative BwOs. The nonlimitative BwO of the socius consists of the intensive states of the nomadic and larval subjects of all human bodies, whilst the socius' limitative BwO consists of the intensive states "proposed by a society for its individuals, the better to exploit their habit-forming potential."[174] The latter constitutes a "grid of abstract categories [that] systematizes images of suggested attractor states and maps the patterns of reproduction action and consumption they authorize."[175] The said grid is a "proliferating series of exclusive disjunctive syntheses adding up to a system of value judgment."[176] It is the application of this grid of value judgment that makes of the fledgling self a *fully-fledged* subject. Hereby, the somewhat tenuous correlation between the fledgling self that redoubles itself in the mode of reflection—now acquires the full obduracy of smug self-satisfaction.[177] We have entered the human, all too human arena of subzoology.

8. The immediate consequences of God's upsurge as subject

Thus, only now have we glimpsed the emergence into the light of day of the subject/subjectivity in the fullest sense. As shall become evident, it is approximately to this sense of subjecthood that the Schellingian God can be said to raise himself with the *Ent-Scheidung*. To be certain, the mode of coming to subjecthood of God and that of the Lacanian subject, on the one hand, and the mode of emergence of the Deleuzoguattarian fully-fledged subject (molar subject), on the other, appear quite different. However, in spite of their ostensibly dissimilar mode of generation; the Schellingian, Lacanian and Deleuzian subjects do (as shall be demonstrated soon enough) share strikingly similar features; whilst the divergence in the respective modes of genesis of the Schellingian-Lacanian and Deleuzoguattarian subjects shall likewise be seen to (at least for our overall purposes) come to naught, in the sense that the former mode of genesis may be explained from within the ambit of the latter.

But let us return to Schelling. With the *Ent-Scheidung,* God has acceded to subjecthood. We shall now explore the consequences of this colossal seismic jolt.

(i) The emergence of time from the deadlock of eternity

With the emergence of God as subject through the *Ent-Scheidung* (the decision whereby the Godhead takes up Being), eternity gives way to an entire "succession of eternities (eons) or times. But this succession of eternities is precisely what we, by and large, call time. Hence, eternity opens up into time in this decision."[178] In the incessant and abortive gesture of beginning enacted by the blind and haywire oscillationary movement of eternal divine nature, all that we had was merely an infinity of purely evanescent instants adding up to *no duration whatsoever*. In fact, there is no such "adding" to speak of to begin with. This is so because: since each beginning, in failing to establish itself as the inception of a duration, is *de facto* abortive and since in each such abortive beginning eternal divine nature incinerates itself completely, each such evanescing instant of abortive beginning is likewise *all* such evanescent instants taken together. We may thus view eternity (that of eternal divine nature's incessant rotary motion) as an uncountably infinite set of measure zero. It is uncountable, since the purely evanescent nature of each of its instants does not allow any such instant to be counted (it can never *stricto sensu* be said to be available for anything even remotely resembling a count). It is of measure zero, in that each instant—being evanescent—fails to extend itself into a duration. Moreover, since each instant is at once all instants, it is therefore infinitely internally fractured and thus pure dispersion itself.

An abortive beginning is, however, nowise tantamount to nothing at all. It is notwithstanding less than a truly accomplished beginning that may be punctually counted as the commencement of the enduring duration and temporal succession that it establishes. Such a truly accomplished beginning attains sufficient obduracy to be deemed present at its beginning and as the past present of the ensuing succession it spawns out of the duration into which it extends. The evanescent instant of an abortive beginning—since it is at once all other such instants—is thus all "past" abortive beginnings and likewise all "future" abortive beginnings. An abortive beginning, of course, cannot be deemed a past present—nor, for that matter, a future present. As past, it could only be a past that was never a present; whilst, as future, it would only be a future that itself never materializes. Thus, the evanescent instant of abortive beginning can only be—at once and alike—a past that was never present and a future that itself can never be present. Equivalently, it is a past that always already was and a future that is always arriving yet never arrives. We may thus speak here of a purely diachronous past and a purely diachronous future, with the evanescent instant—eternity itself—as the fulcrum of diachronic time; as opposed to the synchronous time of chronology, supercession and succession. In Delanda's apt formulation, this diachronous time—unlike "actual time, which is made exclusively out of presents (what is past and future relative to one time scale is still the living present of a cycle of greater duration)"—implies "a temporality which always *sidesteps the present*," and which "must be conceived as an *ordinal continuum unfolding into past and future,* a time where nothing ever occurs but where everything is endlessly becoming in both directions at once, always 'already happened' (in the past direction) and always 'about to happen' (in the future direction)."[179]

This latter diachronous temporality of self-incinerating Schellingian eternal divine nature may be readily assimilated to what Deleuze, apropos of the Stoics, terms the time of *Aion*. In the time of Aion, "only the past and the future inhere or subsist in time. Instead of a present which absorbs the past and future, a future and past divide the present at every instant and subdivide it ad infinitum into past and future, in both directions at once. Or rather, it is the instant without thickness and without extension, which subdivides each present into past and future, rather than vast and thick presents which comprehend both future and past in relation to one another."[180] This incessant, interminable and eternally unremitting subdivision of the present implies that, with Aion, the present is effectively elided and thereby perpetually deferred and likewise always already spent.

There is in Deleuze also another temporality distinct from that of Aion; that is, *Chronos*. Chronos is characterized by the hegemony of the present over the past and future, in the sense that "only the present exists in time. Past, present, and future are not three dimensions of time; only the present fills time, whereas past and future are two dimensions relative to the present in time. In other words, whatever is future or past in relation to a certain present (a certain extension or duration) belongs to a more vast present which has a greater extension or duration. There is always a more vast present which absorbs the past and the future. Thus the relativity of past and future with respect to the present entails a relativity of presents themselves, in relation to each other."[181] That is to say, with Chronos—far from the heterogeneity of past, present, and future being maintained—the present spreads its pincers in order to encompass any possible divergence from itself entailed by an overly independent past and future. The past and the future can thus only subsist as functions of the present; that is, respectively, as a past present and a future present. It is these properties which render Chronos identical to the time of succession—the form of quotidian temporality that emerges in the wake of the *Ent-Scheidung*.

With this transition to the time of temporal succession effected by the Godhead's acquisition of Being and coming to subjecthood having come to pass, the flickering evanescence of eternity gives way as the instant (that is eternity) loses its fleeting and flitting quality and holds onto itself in order thereby to serve as the *terminus a quo* of a duration upon which a temporal succession of punctual and punctiliously anchored present moments arrays itself in homogeneous series. This quotidian temporality to which the Deity freely raises itself, as de Beistegui avers albeit in a different context, is the "time of the world, the time of clocks, measurable and recognizable, the time of individuated phenomena," an extensive time which—pervasive as it may be—notwithstanding remains "merely the surface effect of a subterranean, or perhaps a truly earthly, reality," the mere superficies of an intensive time that, "beneath the extended time of the present, is the time of earth, the time of the shadow or the *oscuro* from which the *chiaro* of the phenomenon comes forward."[182] And this "obscurity of earth" that maintains its insistence beneath phenomenal synchronic time is precisely the crepuscular underworld of the eternal return, out of whose vertiginous depths the Godhead has—by way of the *Ent-Scheidung*—elevated itself to the actual time of temporal succession (Chronos). The emergence of the latter out of the disjointed temporality of Aion effects itself by way of the *contraction* of an immediate past and future into a living/lived present, whereby the said past and future—losing their heterogeneity—are subsumed under the present as mere functions of the latter.[183] This Deleuze designates as the *first synthesis of time*—that of "habit."[184] Accordingly, Ansell-Pearson relates that the first temporal synthesis "concerns a living present in which past and future do not designate separate instants but rather dimensions of a present that are involved in contraction."[185] The time of Aion, on the other hand, is the *third synthesis of time*: that of the eternal return and its universal ungrounding of time. The third synthesis is that which puts time "out of joint."

This picture is not, however, complete. For, though it contracts and assimilates the past and the future to itself, the subsumption effected by the first temporal synthesis does not imply that the present effectively ossifies time to the point of stasis. The present does itself pass. Indeed, the present is *ab initio* the passing present. For this to be the case, asseverates Deleuze, it is necessary for the present "to pass at the same time as it is present, at the moment that it is the present"; meaning (since the passing present is, more precisely, the present-becoming-past) that the present "has to be present and past, still present and already past, at once and at the same time. If it was not already past at the same time as present, the present would never pass on. The past does not follow the present that it is no longer, it coexists with the present it was."[186] But this contemporaneousness of past and present, in fact, has the effect of rending time asunder: time "splits in two dissymmetrical jets, one of which makes all the present pass on, while the other preserves all the past."[187] Crucially, however, this past that is preserved is nowise a past that was formerly—or at any point whatsoever—present. Rather, it is a purely immemorial and entirely virtual past; a past that is always already past. In this sense, it is akin to the mode assumed by the past in the third temporal synthesis. However, that this virtual immemorial past is *preserved* is what distinguishes it from the past as it figures in the purely diachronous and ungrounding time of Aion, that of the eternal return. For the latter categorically incinerates itself.

But in what sense is the virtual past said to be preserved? It is preserved as a "coexistence of circles which are more or less dilated or contracted, each one of which contains everything at the same time and the present of which is the extreme limit (the smallest circuit that contains all the past)."[188] Preserved in this manner, the virtual past is the very unconscious of the real itself. This is the *second synthesis of time*: that of "memory." As Keith Ansell-Pearson explains, this latter synthesis is presupposed by the first synthesis of time, in the sense that memory is the *sine qua non* of habit: memory is the "being" of the past, the enabling ground upon which the present passes. Habit itself entails, accordingly, the very passing of the present as such.[189]

But there is more. For the three syntheses of time may, respectively, be further characterized in yet another way. We have seen that the third synthesis of time (Aion or eternal return), as entailing pure becoming, is *the temporality proper to pre-individual individuating factors*.

Analogously, the second synthesis of time (the pure past of memory) is *the temporality proper to organs-partial objects (continuous/intensive multiplicities)*. Note that Deleuze also refers to the organs-partial objects as "virtual objects." In this connection, it is contended that virtual objects "are shreds of pure past," that—as regards its mode of temporality—the "virtual object is never past in relation to a new present, any more than it is past in relation to a present which it was. It is past as the contemporary of the present which it is, in a frozen present; as though lacking on the one hand the part which, on the other hand, it is at the same time; as though displaced while still in place."[190] And hence it may be surmised that: "Contemporaneous with itself as present, being itself its own past, pre-existing every present which passes in the real series, the virtual object belongs to the pure past. It is pure fragment and fragment of itself."[191]

Lastly, the first synthesis of time—being that of habit—accords with *the temporality of selves from the larval subject upwards*. In this latter connection, it has been observed that it was with the emergence of the larval subject or self at the level one higher than that of the nomadic subject that anticipation-satisfaction circuits first came into being. The existence of these protention-retention circuits implied that intensive states traversed by the larval subject (in contradistinction to the case of the more originary nomadic subject) could now be said to arrive with greater or lesser probability, whilst the corresponding connective syntheses came to be limited to a more narrowly circumscribed set of organs-partial objects. Clearly, the move toward habit was here at least incipient. Moreover, the contraction of myriad nomadic selves entailed in the emergence of each larval self may be seen as entailing the coextensive contraction of an immediate past and future in the burgeoning forth of the passing present of the first temporal synthesis. In the ensuing upward movement toward the fledging subject or self and thereupon unto the fully-fledged subject (molar subject); a progressive nesting of such passing present moments may be seen as transpiring, each more comprehensive than its predecessor, each encompassing its antecedent. The coextensive process of sacralization of organs on higher and higher-level protention-retention circuits, of course, entailed the further entrenchment and consolidation of habit. That, in the increasingly "teleological" movement toward satisfaction, the entailed connection of organs-partial objects became increasingly a matter of anticipation—clearly implicates memory as habit's pressuposition.

Let us return to Schelling and proceed to further consequences of the *Ent-Scheidung.*

(ii) The degradation of the eternal affirmative will into a negative will

We must be mindful that God's assumption of Being and coming to subjecthood, effects itself in accordance with his *highest* self; that is, in terms of God as the Godhead, as eternal freedom, and not from within the carceral ambit of eternal divine nature—which latter is governed solely by necessity. But, contends Schelling, as such, "it is precisely *as* this freedom that God is the eternal No of all external Being," that as quintessentially free he is the "the eternal negation of external Being."[192] Furthermore, God—as this negation of all Being that is external to him—becomes "the ground of the existence of the eternal Yes."[193] The pure will that is the Godhead and that previously willed nothing specific but was rather a restful eternal willing, with the *Ent-Scheidung*, becomes a positively negating will—a will that *positively willings nothing.* As such, it becomes the immediate basis of another will that produces itself in it: that is to say, a will that actively wills something.[194] However, precisely in that the latter's *grundstimmung* is a purely negative will and though ostensively affirmative on its own terms, as the mere pseudo-affirmation that it is, the latter cannot but merely runtishly and obsequiously redouble the deeper and more fundamental negation (divine freedom as the eternal No) from whence it initially arises and which latter irremissibly pervades and marks its general tenure throughout. This is precisely what in Nietzsche figures as the denigration in the immediate quality of the will to power (as differential and genetic element of force) from affirmative will to power to negative will to power—a most profound change for the *worse* that Nietzsche holds to stand at the very origin of history and from whence the concerted becoming-reactive of all things even remotely human forthwith prosecutes itself at a fairly impressive pace.[195]

(iii) The denuding of God as subject of actuality

Upon the ground of the will that positively wills nothing and its force of negation, avers Schelling, "the eternal Yes [is posited] only in accordance with possibility but not yet in accordance with actuality, that is, this force is God in accordance with possibility but not in accordance with actuality."[196] This means that, with the *Ent-Scheidung*, God voluntarily posits himself *only in the state of potentiality*. Thus, the Deity is not *stricto sensu* in the state of existence as may be ascribed to the intramundane. He fails to attain the ontic. Indeed, he scarely attains the ontological—or remains ontologically unaccomplished. The transition from God as the Godhead and eternal divine nature (prior to the decision) to God-the-subject (after the decision), is one from a God of pure Being who *happens* to also not to have Being—to a God of pure Being who in addition has *positively assumed* his non-possession of Being. This positive assumption of his not-having-Being is precisely the coming of God to subjecthood by way of his freely positing Himself, in the first place, as the Eternal No.[197]

But we may also explain all of this more incisively by recourse to aspects of the Lacanian conception of subjectivity. The *Ent-Scheidung* is indeed the pronunciation of the Word by God which breaks the deadlock of the unremitting movement of expansion and contraction that is eternal divine nature. Now, the Word is none other than "a *contraction in the guise of its very opposite, of an expansion*—that is to say, in pronouncing a word, the subject [in our case God] contracts his being *outside* himself; he 'coagulates' the core of his being in an external *sign*."[198] The transition effected here is that "from the Real to the Symbolic, from the realm of presymbolic antagonism (of contraction and expansion) to the symbolic order in which the network of signifiers is correlated to the field of meaning"; a transition which takes place "only by means of a paradoxical 'pure' signifier, a signifier without signified: in order for the field of meaning to emerge, that is, in order for the series of signifiers to signify *something* (to have a determinate meaning), *there must be a signifier (a 'something') that stands for 'nothing'*, a signifying element whose very presence stands for the absence of meaning (or, rather, for absence *tout court*)."[199] And it is precisely from the foregoing that Žižek concludes the pivotal point presently being made: "This 'nothing', of course, is *the subject itself*, the subject qua \$ (the Lacanian matheme, designating the subject with all content removed), the empty set, the void that emerges as the result of the contraction in the form of expansion: when I contract myself outside myself, I deprive myself of my substantial content."[200] God's coming to subjecthood is thus achieved by the latter only at the expense of his evacuating himself of all positive content, of all actuality. God's pronunciation of the Word resolves the tension of the pre-symbolic antagonism of potencies, but "at a price: the Word, the contraction of the self outside the self, involves an irretrievable externalization-alienation—with the emergence of the Word, we pass from *antagonism* to the Hegelian *contradiction* between \$ and S_1, between the subject and its inadequate symbolic representation."[201] God's acquisition of symbolic existence through his (the subject's) "contraction" of a signifier (the Word) exacts the merciless tithe of the subject (God) not being adequately represented/representable in the signifier. Thus, God's coming to subjecthood is his emergence as \$; that is, his subjecthood is itself nothing other than its incessant failure to find adequate representation in S_1 (i.e. the Hegelian contradiction between \$ and S_1 mentioned above is as such the *constitutive* split internal to \$ itself—that is, the split that constitutes and *is* \$). It is in *this* sense that God's upsurge as subject necessarily denudes God-as-subject of all content—that is, of all content adequate to God *as subject*.[202] The subject is thus a pure *form*.

From the foregoing, Žižek draws a startling—indeed quite shocking—conclusion: being in itself the residuum of its own auto-evacuation, "*the subject is* as such *the survivor of its own death*, a shell which remains after it is deprived of its substance"; and moreover that herein entailed is *not* "the rise of a new subject surviving its death/disintegration," but rather that "the subject as such is a 'second subject,' a formal survivor (the surviving form) of the loss of its substance, of the noumenal X called by Kant the 'I or he or it (the thing) that thinks.'"[203] The prototypical subject here is, of course, God himself; the aforementioned scenario being precisely the death that we earlier purported God to have survived. Here God primordially escapes from the macerating wheel of his own eternal divine nature, of the eternal return that he can nowise abide; coming to subjecthood by way of a death from which he rises—indeed most paradoxically—in order to emerge for the *first* time. But the auto-evisceration of all divine substance coextensive with God's upsurge as subject means that the Deity's first emergence (via primordial resurrection) is that of a empty hulk.

Is this voided form, the divine subject, not the same as the zero-level Neighbor—the *Muselmann*? If so, then God himself is the primordial *Muselmann*, in relation to whom the *Muselmann* of the *Sho'ah* plays the role of ectype to divine prototype; the Deity himself, the sole captive and sole survivor of the cosmic holocaust of his of own eternally begotten nature. *Dieu, mon prochain* …

The split nature of the Schellingian-Lacanian subject may also be accounted from the Deleuzoguattarian vantage-point. That is, the Deleuzoguattarian mode of generation of the molar subject (or fully-fledged subject) can explain the Schellingian-Lacanian subject's internal rivenness. First of all, Deleuze and Guattari do contend that the subject *is* indeed split. The contradiction between $ and S_1 is, however, *not primary*. That is, as Massumi argues: there are bifurcations *anterior* to this split—of which the latter is a mere ex post factum *recapitulation*. I doubt that Žižek denies this. He simply does not mention it in any foregrounded manner in his discussion of Schelling. We, however, will foreground it. The principal anterior bifurcation which the split between $ and S_1 recapitulates, is that between the non-limitative BwO and the limitative BwO—that is to say, the tension between the more dissipative level of teeming nomadic and larval selves, on the one hand, and the less dissipative level of those particular larval selves tethered to anticipation-recognition circuits featuring one sacralized organ each, on the other.[204] This tension already introduces a split at the level of the fledgling selves that it precipitates: that between the self as *moi* and as *je*—the redundancy implicit to the fledgling subject's anticipation and consequent recognition of itself. This level of the fledgling selves is, as we have seen, that of self-recognition—or better, moral reflection. It was also noted that a grid of value judgment, arising from the limitative BwO of the socius and superimposing itself onto the fledgling subject, was that which—effecting the move from moral reflection (a still precarious subjective configuration) to smug self-satisfaction—made of the said fledgling selves a fully-fledged subject (molar subject). The level of moral reflection at which the fledgling selves perambulated was precarious in the sense that these latter "were still plural and relatively localized, vacillating between staying on their own level and stepping down to larval status."[205] The precariousness inherent to this vacillation is rendered less so by way of the mechanism of over-coding imposed upon the potentially equivocating fledgling selves by the limitative BwO of the socius. The fully-fledged subject or "person"—always already minimally "bureaucratic" in nature—herein engineered is that of the selective aggregation of multiple fledgling selves.

The question is however: what, more precisely, is it that keeps the aggregated fledgling selves bound to their aggregate? We may answer this question by first noting that the initial precipitation into existence of the fledgling selves was by way of the sacralization of certain organs within the anticipation-recognition circuits associated with particular larval selves. It is these metaphysicalized organs that now form the *terminus a quo* from which the grid of value judgment fashions phantasmatic satisfactions (surplus value of enjoyment) for the fledgling selves, so as to keep them tightly bound within their aggregates. Each of these aggregates is, of course, a single fully-fledged subject, or categorical "person." We've already seen this apropos of Massumi.

This induced beholdenness of the fully-fledged subject (Deleuzoguattarian molar subject) to itself is strikingly homologous to the Lacanian suturing of the split subject to its phantasmatic identifications; that is, $◊a. To be sure, the latter suturing is rendered necessary only in so far as the subject ($) is itself split. At the most superficial level (that of the personality structure), this split appears as the gap between the self and its reflection as it transpires, or is evinced, in the imaginary register; that is, $ is the incessant failure of $◊a. However, "Lacanian splitting is a *retrospective projection* [my emphasis] of distinctions belonging to the personal level of the constituted subjected group [of aggregated fledgling selves making up the molar subject] onto the entire dissipative system of the body [the profusion of nomadic, larval and fledgling selves taken together]: the multitude of individuals [nomadic subjects and other more-or-less evanescent selves] that contract to produce the person is reduced to the one-two-(three) of self-other-(phallus), distinctions which can exist only on the second-order level of identity and identity loss."[206]

But—and this is the crucial point—identity and its repeated loss are merely "the end effects, not the foundation, of the process of individuation."[207] A deeper, antecedent, and far more fundamental cause of identity and its loss is the incessant oscillation between the nonlimitative and limitative BwOs of the body's dissipative system itself—whose unremitting tension is responsible for fledgling selves' propensity to vacillate between, on the one hand, vertiginous descent to the level of the seething plethora of larval selves down yonder and, on the other, their being the more inextricably bound within aggregates subsisting at the level of the fully-fledged molar subject. (Still deeper than this is, of course, the maddening oscillation between the Deleuzoguattarian paranoiac and miraculating machines at the most dissipated level of purely evanescent nomadic subjects—that is the same as the equally psychotic rotary motion of eternal divine nature born of the antagonism of the first potency and second potency of God.) It is this tension between the dissipative downward pull of the larval and nomadic selves below (and their propensity for recrudescence) and the coercive force of aggregation acting upon fledgling selves from above—that, from the purview of molar subjectivity, appears as an endlessly irresolvable dialectic: the contradiction between $ and S_1 that we espied a little earlier in Žižek's musings on the *Weltalter.*

The consequences of the latter shall indeed prove to be profoundly saddening. For the endless failure of the suturing $ to its perpetually inadequate phantasmatic projections ($◊a) and symbolic representations (S_1), describes a predicament much like "a revolving door: a single past that returns in the future (repetition-compulsion)," born of vicissitudes whereby the "jagged line of becoming buckles into a circle" and wherein desire, "the plane of consistency, is short-circuited by an infinite feedback loop of metaphysical redundancy" at whose center stands "a private sun that is and isn't there," a ubiquitous "ever-present absent object of overpowering attraction [that] inspires an impossible ritual quest for fulfillment."[208] This ritual quest is, ultimately, that of the subject's eternal quest for itself—a thankless, tragic and moreover futile endeavor whose *terminae a quo* and *terminae ad quem* are nothing but unremitting failure itself. From the *Ent-Scheidung* onwards, *this* shall be the fate of all subjects—whether human or divine.

(iv) The emergence of space, nature, the organism, and molar machines

(1) Extension and molar machines

The positively negating will that in the immediate wake of the *Ent-Scheidung* serves as the ground of all future affirmation, now assumes the guise of *inter alia* the delimiting force responsible for the positing of space as extension—in which visible, corporeal (for example, organic) forms, emerge. This delimiting force is that by which the entire cosmos is contracted "from the outside toward the inside and by which the cosmos first became spatial."[209]

In Deleuze there is a similar emergence of space as extension. However, unlike in Schelling, it does not emerge in so sudden and abrupt a way. Rather, the explication of *spatium* as *extensio* occurs at some point along the ascent from the larval subject to the fully-fledged subject (molar subject). This is the movement whereby intensity "is developed and explicated by means of the extension [*extensio*] which relates it to the extensity [*extensum*] in which it appears outside itself and hidden," and in which—despite being occulted by the extensity which veils it—intensity and its depth nonetheless remain implicated within the latter.[210] Here, the "explication of extensity rests upon the first synthesis, that of habit or the present; but the implication of depth [intensity] rests upon the second synthesis, that of Memory and the past."[211] We have in mind the first synthesis of time and the second synthesis of time in Deleuze. As de Beistegui elaborates it, the emergence of *extensio* occurs via "a concrete *physical* process in which an undifferentiated intensive space (that is, a space defined by continuous intensive properties) progressively differenciates [differenciation], eventually giving rise to extensive structures (discontinuous structures with definite metric properties)."[212] More pointedly, the non-metric Riemannian space in which, under the pressure of pre-individual individuating factors, continuous/intensive multiplicities unfold via a process of ceaseless differentiation—differenciates (*differenciation*) itself as the emergent metric Euclidean space of tangible intramundane spatial formations.

We have discussed the process in Massumi whereby the fully-fledged subject arose as the *terminus ad quem* of the ascent from the teeming plethora of nomadic selves, to the level of the larval selves, to that of the fledgling selves and beyond—each higher-level self emerging from the contraction of the panoply of selves on the level below it. At the level of the nomadic selves, one found an entire slew of organs-partial objects/desiring-machines—purely dispersive positive multiplicities. This level is also termed *molecular*. Now, on the ascent to the level of the fully-fledged subject, one finds the desiring-machines (organs-partial objects) caught up in aggregates that effectively constitute large-scale *molar* machines. Crucially, these latter machines presuppose the emergence of space as *extensio*. They exist on the macroscopic level and arise from the foregoing movement of statistical aggregation of desiring-machines according to the pressures of selection in tandem with the associated emergent forces of circumscription that bind the said desiring-machines together. The distinction between molecular and molar, as ontological domains, avers Massumi resides in the fact that whereas in "a molecular population (mass) there are only local connections between discrete particles"; in the case of a "molar population (superindividual or person) locally connected discrete particles have become correlated at a distance," molarity itself implying "the creation or prior existence of a well-defined boundary enabling the population of particles to be grasped as a whole."[213] As Deleuze and Guattari have it, the molar formations, which they consider forms of gregariousness (and thus at once of what Nietzsche terms herd instinct), are the result of the totalization and unification of molecular forces by way of a statistical accumulation abiding by the law of large numbers.[214] As statistical unities, the molar machines may be either organic, technical, or social. The socius, for instance, is a molar machine.[215] The desiring-machines and the molar machines are, notwithstanding, the same machines. That is to say, they do not differ in nature. For molecular desiring-machines are always already inherently the investment of molar machines, which latter are—at the same time—themselves configurations formed of desiring-machines.[216] They are, in fact, the exact same machines under determinate conditions, wherein the latter term refers to the statistically wrought forms into which the desiring-machines enter as stable forms of unification and structuring—proceeding as they do by way of large heavy aggregates.[217] So, whereas the two kinds of machines do not differ in nature, they do differ as regards *regime.*[218]

Along with this change in regime, the move from the molecular desiring-machines to the molar machines ushers in what we have already glimpsed in Massumi's earlier discussion of the passage from the nomadic subject to the fully-fledged subject: a change in the usage of the three syntheses. At the molecular level the connections were local, indirect and transverse; the disjunctions inclusive; and the conjunctions polyvocal. At the molar level the connections become direct, specific and global, the disjunctions exclusive, and the conjunctions biunivocal and segregative.[219] That is to say, the disjunctive syntheses now come to select certain terms whilst exiling others outside of those they aggregate, whereas the conjunctive syntheses cease to spawn residual multiple/fragmented nomadic subjects and instead engineer solidly circumscribed fully-fledged subjects possessed of an agency that they henceforth attribute to themselves alone. Here, that which lives comes to be structured by the statistical unities of its own "person," and each living organism manifests itself as a single obdurate subject.[220]

(2) The erection of the socius *and the move to social production*

In Schelling, in the very same motion that is God's emergence as subject, there occurs among the three divine potencies a *scission*: whilst the third potency rises vertiginously to the rank of immediate divine subject, the first two potencies suffer a comparative debasement in relation to this much extolled third. Having descended and fallen away from the latter, these lower potencies assume the role of substratum and prime matter for the said divine subject. In fact, the first potency is degraded even below the second.[221] Thus falling to the very bottom, the first potency becomes the ground of nature—that is, of the natural *world* that has now emerged into existence in the wake of the great scission. The second potency, on the other hand, comes to serve as the basis of the spiritual world that has come into existence with the natural world. The third potency becomes the "universal soul by which the cosmos is ensouled," herein henceforth functioning as "the eternal link between nature and the spiritual world as well as between the world and God," as the "immediate tool through which alone God is active in nature and the spiritual world."[222]

Now, we have already beheld the close homology between the mutual antagonism of the Schellingian first and second potencies at the level of eternal divine nature anterior to the *Ent-Scheidung* and the tension between the Deleuzoguattarian paranoiac and miraculating machines at the level of desiring-production. This latter oscillation generated the rapidly evanescing nomadic subject. Beyond this, the transition from the larval to the fledging subject was effected by the tension twixt the emergent limitative and nonlimitative BwOs that precipitated the latter. Massumi contends that a further oscillatory tension of this precise sort exists at the societal level: that is to say, a society is "a dissipative structure with its own determining tension between a limitative body without organs and a nonlimitative one."[223] The interaction between the latter two is the *socius* we espied a little earlier: the analogue at the level of social production of what the BwO is to desiring-production. The nonlimitative BwO of the socius is simply the collection of all the nonlimitative BwOs arising from the level of desiring-production, whilst its limitative BwO is the set of all limitative BwOs at the level of desiring-production.[224] Thus, the tension between all paranoiac and miraculating machines reappears as that which abides between the somewhat less vast panoply of nonlimitative BwOs and the limitative BwOs that oppose them, the latter opposition itself reemerging as the tension between the limitative BwO and the nonlimitative BwO of the socius.

In contradistinction to the massive transmogrification in the relation between the three Schellingian potencies of eternal divine nature that occurs in the immediate wake of the *Ent-Scheidung*, the relation between the limitative and the nonlimitative BwOs of the socius does not arise by way of a scission between the paranoiac and the miraculating machines, or that between the nonlimitative and limitative BwOs. Instead, all such antagonistic couples remain in reciprocally presupposing tension.

The socius may, however, be characterized from a different angle. As observed much earlier, the full BwO served as a surface for the recording of the flows of desiring-production. In this modality, the BwO attracted and attached the organs-partial objects to itself to the extent that they appeared as effluences of itself (it was their quasi-cause). Now, Deleuze and Guattari view the socius as the analogue of the full BwO at the level of social production.[225] The socius, or social machine, falls back upon (*il se rabat sur*) all of production, constituting in the process a surface over which all forces and agents of production find themselves distributed, in such a way that the entire productive process seems to emanate from the socius as from a quasi-cause, that is to say, in such a way that the said forces and agents seem to be "miraculated" (*miraculés*) by it (the socius).[226] The socius, moreover, inscribes and records the flows of desiring-production, in order that not a single flow exists that is not properly shackled, channelled, and regulated.[227] The socius, thus, harnesses the deterritorialized and decoded flows of desire so as to prevent them from escaping its ambit. The various different kinds of socius delineated by the authors—the full body of the earth, the full body of the despot and the full body of capital—accomplish this harnessing task in their own distinct ways.[228] A fair amount of the ensuing stages of this book shall be devoted to how this latter is effectuated.

Crucially, a very fundamental error must be avoided. The BwO is in no sense some kind of original primordial entity that subsequently projects itself into various kinds of socius.[229] This would be an utterly false statement. For the socius is in not a projection of the BwO, but instead the "ultimate residuum" of a completely deterritorialized socius. In this sense, the failure of the socius to successfully plug and harness the decoded and deterritorialized flows of desiring-production peregrinating along the BwO would effect the evisceration and dissolution of any socius whatsoever and the coextensive recrudescence of the BwO in the wake of this the socius' abject desuetude. It may therefore be said that the BwO, in fact, "haunts" each and every form of socius.[230] This is tantamount to averring that desiring-production situates itself at the limits of social production and becomes an *absolute* limit every time that the schizo-flows of the BwO overrun the socius. For the BwO is, "the deterritorialized socius, the wilderness where the decoded flows run free, the end of the world, the apocalypse."[231]

That the socius is nowise a projection of the BwO implies that social production is nothing other than desiring-production itself under specific determined conditions, and at the same time that the entire social field is immediately invested by molecular desire from the very outset. It is the "historically determined product of desire."[232] More precisely, it may be said that, on the one hand, social production—under determinate conditions—primarily derives itself from desiring-production; whereas, on the other hand, desiring-production is always already social in nature.[233] This is so, since it is impossible to separate the desiring-machines, as the "microphysics" of the unconscious, from the macroscopic social formations statistically constituted by them in accordance with the law of large numbers. For unconscious libidinal mircoinvestments of desire are always to be found lurking immediately beneath the surface of all conscious investments of formations of the economic, political, religious, or other, variety; meaning that the entire social field is not only immediately saturated by desire, but is at the same time statistically determined by the peregrinations of molecular desire.[234] Ultimately, desiring-production and social production are simply the inseparable recto and verso of one and exactly the same process of production.[235] This process of production is none other than the "production of the real" itself.[236] Twixt the two there is "nothing but the ongoing process of becoming [*devenir*] that is the becoming of reality."[237]

We have already noted that at the level of the molar machines, the three syntheses of desiring-production still persist, although their usage is altered from inclusive to exclusive. The socius is obviously one such molar formation. Now, the change in the usage of the connective syntheses in particular implies that at the level of the socius they cease to stitch the processes of production onto their product. Rather, there now appears a diremption between these two. In this connection, Toscano avers that at the molecular level desiring-machines (organs-partial objects or continuous-intensive multiplicities) "act so that the process of production is at one and the same time its product," whereas as regards the molar level it is always the case that "the product transcends its own production."[238] So, whereas at the molecular level of desiring-production, production was always and endemically that of production itself; with the shift to the molar level having come to pass, we move to the level of labor in the more prosaic sense of the term. That is to say, production comes to be a process that—as Marx would have it—extinguishes itself in its product. Marx: "Labour has become bound up in its object: labour has been objectified, the object has been worked on. What on the side of the worker appeared in the form of unrest [*Umruhe*] now appears, on the side of the product, in the form of being [*Sein*], as a fixed, immobile characteristic. The worker has spun, and the product is a spinning."[239] This, of course, also means that a purely impersonal, subjectless (in the sense of there being no molar subject) and objectless production gives way to a process of production enacted by the constituted human laborer. Moreover, whereas at the level of desiring-production there was an immediate consumption of the production produced in the connective synthesis and recorded as zones of intensity on the BwO by the disjunctive synthesis, in that at the level of the socius production and product come to be dirempt, a corresponding temporal hiatus interposes itself between products and their consumption by what are now molar subjects.

In our earlier discussion concerning the mode of emergence of the molar subject (fully-fledged subject), we spoke of the latter as inevitably fomenting a scenario closely homologous to what—in the Schellingian-Lacanian register—amounted to the endlessly reiterated failure of the split subject ($) to suture itself with its (perpetually inadequate) phantasmatic projections ($◊a) and symbolic representations (S_1) alike; a lugubrious and tragic scenario wherein the said subject stood for the sempiternal incarceration of desire within the strictures of an endless feedback loop at whose center resided a ubiquitously present yet perpetually absent object precipitative of an incessant and utterly hopeless quest for fulfilment, at once promethean and sisyphean. The very mode of emergence of the molar subject at once exacts the infinite burden of desire's being forever tethered to lack.

Now, this dire scenario is equivalently that which arises from the emergence of the molar machines. Here lack comes to be generated *ex post factum* from the interaction between the molecular machines (desiring-machines) and the molar machines. By way of the weight of large aggretates composed of organs-partial objects, the emergent molar machines are formed as statistical unities. And it is in being related to this incipient and abiding order that the molecular organs-partial objects come to appear as lack, whilst at the same time the whole itself is lacked by every partial object. It is this process and arrangement that fuses desire itself to lack, causing it to be a metonymy of a lack. Indeed, the statistical transmogrification of molecular multiplicity into a molar constellation is precisely that which organizes lack on an immensely large scale.[240] Lack (*manqué*) is created, planned and organized in and via social production, precipitating into existence a well-nigh interminable scenario in which desire comes to be the utterly self-degrading and masochistic fear of something perpetually being lacked.[241] This fusion of desire to lack shall have radical consequences for humanity.

To the fomentation of lack by molar machines such as the socius one must add a further important transmogrification ushered in by the latter. This is the bifurcation between form and function that is brought into effect by the movement from the molecular to the molar machines. In the case of desiring-machines use, function, production, and formation are one and the self-same process; for it is only the desiring-machines that produce the connections according to which they function, and in turn function through this improvisation and formation of these same connections.[242] They are formative machines whose functioning is completely indistiguishable from their formation; "chronogeneous" machines involved in their very own assembly (*montage*), machines which are solely and exacty "what they make of themselves."[243] This implies that desiring-machines, considered as forces, are strictly speaking *active* as regards their quality. As shall be seen in far greater detail in the following chapter apropos of Deleuze's reading of Nietzsche, active forces are inter alia characterized as autochthonous, plastic, self-affirming forces of metamorphosis.[244] Molar machines such as biological and social formations, in contradistinction to molecular machines (desiring-machines), are not formed in the same way in which they function. And, crucially for our present purposes, Deleuze and Guattari contend that only that which is not formed in the same way in which it functions can have a purpose and an intention.[245] The upshot here is that molar machines in general—and social machines in particular—being goal-driven as they are, are strictly speaking *utilitarian*. They, thus, imply the presence of *reactive* forces. They are indeed at least partly composed of these. As will likewise be seen in the ensuing chapters, reactive forces are distinguished from active forces in the sense that their prerogatives are those of adaptation, accomodation and utility.[246]

Now, a body—whether it be biological, social, political or anthropological—is a relation of forces.[247] Since such bodies are molar machines likewise by definition containing embedded molecular machines, they are necessarily comprised of a combination of active and reactive forces. This co-presence, co-habitation and co-implication of active and reactive forces is in no sense deleterious in and of itself. Nonetheless, it is a configuration fraught with danger and replete with the seeds of its own possible descent into dereliction. For the innate propensity of reactive force is to undermine and underhandedly pervert active force in such a way that the latter itself becomes reactive.[248] With this latter eventuality having come to pass, society and man shall forthwith and henceforth descend into and be swept up by an inexorable historical becoming-reactive of force and succumb to the resentment (*ressentiment*), bad conscience, guilt-slavery and other festering maladies that the hegemony of reactive force brings into being. As shall be seen, pre-State societies shall come replete with a mechanism enabling them to thwart the fall into this abyss. The terminal dereliction of this noble thwarting mechanism and the ensuing saga of history and its endless proliferation of State apparati and other forms of reactive gregariousness shall, however, inevitably hurl the species into the very depths of its festering moral sickness.

Notes

1. Schelling 2000, 24, 25.
2. Žižek 1996, 14.
3. Toscano 2004, 116.
4. Habermas 2004, 57.
5. Scholem 1978, 88, 89.
6. Ibid. 89.
7. Ibid.
8. Deleuze 2006, 148, 149.
9. Bowie 1993, 106.
10. Schelling 2000, 24, 25.
11. Ibid.
12. Deleuze 1986, 53, 54.
13. Schelling 2000, 26, 27.
14. Deleuze 1986, 54.
15. Ibid.
16. Schelling 2000, 26, 27.
17. Groves 1999, 30, 31.
18. Schelling 1997, 136.
19. Scholem 1978, 132.
20. Schelling 1997, 137.

21. Ibid.
22. Schelling 1997, 138.
23. Ibid.
24. Ibid. 138, 139.
25. Ibid.
26. Ibid.
27. Deleuze 2006, 23, 24.
28. Schelling 1997, 138, 139.
29. Schelling 2000, 16, 17.
30. Scholem 1978, 129.
31. Ibid.
32. Ibid.
33. Ibid. 130.
34. Ibid.
35. Schelling 1997, 140.
36. Habermas 2004, 58.
37. Mosès 1992, 41.
38. Deleuze and Guattari 2004, 358, 359.
39. Bonta and Protevi 2004, 120.
40. de Beistegui 2004, 313–315.
41. Deleuze 2001, 229, 230.
42. Parr 2005, 131, 132.
43. Deleuze 2001, 222.
44. Ibid.
45. Deleuze and Guattari 2004, 4.
46. Ibid. 5.
47. Ibid. 38, 39, 357.
48. Ibid. 5–8, 40.
49. Ibid. 339.
50. Ibid.
51. Ibid. 356.
52. Ibid.
53. Parr 2005, 176, 177; Bergson 1988, 206.
54. Ibid.
55. Delanda 2004, 10–12, 16.
56. Toscano 2006, 163, 165, 168, 169.
57. de Beistegui 2004, 249, 250.
58. Toscano 2006, 163.
59. Ibid. 170.
60. Ibid. 173.
61. Parr 2005, 21.

62. Ibid. 131.
63. Deleuze 2001, 246.
64. Delanda 2004, 16.
65. Deleuze 2001, 257.
66. Ibid. 239–241.
67. Ibid.
68. Deleuze and Guattari 2004, 8.
69. Ibid.
70. Deleuze and Guattari 1988, 153.
71. Ibid.
72. Ibid. 158.
73. Ibid.
74. Deleuze and Guattari 2004, 13.
75. Ibid. 84.
76. Hallward 2006, 98, 99.
77. Ansell-Pearson 1999, 153, 154.
78. Ibid. 172, 173.
79. Deleuze and Guattari 2004, 358, 359.
80. Ibid.
81. Spinoza 2002, 217–219, 224, 229.
82. Parr 2005, 273.
83. Schelling 1997, 139.
84. Ibid.
85. Ibid.
86. Schelling 2000, 9.
87. Scholem 1978, 131.
88. Deleuze and Guattari 2004, 10.
89. Ibid. 12, 13, 18.
90. Schelling 2000, 11.
91. Schelling 1997, 143.
92. Ibid. 143–145.
93. Ibid. 145.
94. Schelling 2000, 20, 21.
95. Ibid.
96. Ibid. 80.
97. Ibid. 20.
98. Deleuze and Guattari 2004, 20.
99. Ibid. 22.
100. Deleuze 1988, 86, 87.
101. Sartre 1983, 2, 19.
102. Deleuze and Guattari 2004, 19.

103. Lingis 1983, 30.
104. Leibniz 1898, 243, 244.
105. Ibid.
106. Ibid. 219.
107. Ibid. 217, 218, 229, 230.
108. Deleuze 2004, 86, 87.
109. Schelling 2000, 59, 60.
110. Schelling 1997, 154; 2000, 66, 67.
111. Ibid.; Schelling 2000, 59, 60.
112. Schelling 2000, 66, 67.
113. Ibid.
114. Scholem 1978, 105, 106.
115. Ibid. 114.
116. Ibid.
117. Ibid. 115.
118. Ibid. 132.
119. Ibid.
120. Deleuze and Guattari 2004, 22.
121. Ibid.
122. Deleuze 2001, 41, 42.
123. Ibid. 125.
124. Ibid.
125. Ibid. 242, 243.
126. Deleuze and Guattari 2004, 93.
127. Ibid. 94.
128. Deleuze 2001, 201, 202.
129. Ibid. 67.
130. Schelling 2000, 21, 22.
131. Ibid.
132. Ibid. 22, 23.
133. Ibid.
134. Ibid.
135. Ibid. 27, 28.
136. Ibid.
137. Ibid.
138. Ibid. 29, 30.
139. Ibid.
140. Ibid. 37, 38.
141. Scholem 1978, 91.
142. Žižek 1996, 13.
143. Ibid.

144. Ibid.
145. Ibid.
146. Massumi 1992, 70.
147. Leibniz 1898, 217, 218.
148. Ibid. 224, 228.
149. Ibid. 250, 253.
150. Deleuze 2004, 86, 87.
151. Leibniz 1898, 231.
152. Ibid. 250, 253.
153. Deleuze 2004, 92.
154. Massumi 1992, 72.
155. Ibid.
156. Ibid.
157. Ibid.
158. Ibid.
159. Leibniz 1898, 263.
160. Deleuze 2004, 92.
161. Massumi 1992, 72.
162. Ibid.
163. Ibid. 56, 57.
164. Ibid.
165. Ibid.
166. Ibid.
167. Ibid.
168. Ibid. 73.
169. Ibid. 74, 75.
170. Ibid.
171. Ibid. 75, 76.
172. Ibid.
173. Leibniz 1898, 233, 234, 265, 266.
174. Massumi 1992, 75, 76.
175. Ibid.
176. Ibid. 76.
177. Ibid.
178. Schelling 2000, 76, 77.
179. Delanda 2004, 126, 127.
180. Deleuze 1990, 164.
181. Ibid. 162, 163.
182. de Beistegui 2004, 327.
183. Delanda 2004, 109.
184. Deleuze 2001, 94.

185. Ansell-Pearson 1999, 100.
186. Deleuze 1989, 78, 79, 81, 82.
187. Ibid.
188. Ibid. 98, 99.
189. Ansell-Pearson 1999, 101, 102.
190. Deleuze 2001, 101, 102.
191. Ibid.
192. Schelling 2000, 77.
193. Ibid.
194. Schelling 1997, 168, 169.
195. Deleuze 2006, 138–141.
196. Schelling 2000, 86, 87.
197. Ibid.
198. Žižek 1996, 43.
199. Žižek 1997, 39.
200. Ibid.
201. Žižek 1996, 46, 47.
202. Ibid.
203. Žižek 2010, 307.
204. Massumi 1992, 83.
205. Ibid. 76.
206. Ibid. 83, 84.
207. Ibid.
208. Ibid.
209. Schelling 2000, 93, 94.
210. Deleuze 2001, 228.
211. Ibid. 230.
212. de Beistegui 2004, 262, 264, 265.
213. Massumi 1992, 54, 55.
214. Deleuze and Guattari 2004, 375.
215. Ibid. 315.
216. Ibid. 316, 317.
217. Ibid.
218. Ibid.
219. Ibid. 115, 316, 341.
220. Ibid. 316.
221. Schelling 2000, 29, 30.
222. Ibid. 37, 38.
223. Massumi 1992, 75.
224. Ibid. 84.
225. Deleuze and Guattari 2004, 34.

226. Ibid. 11.
227. Ibid. 35.
228. Ibid.
229. Ibid.
230. Ibit. 309–311.
231. Ibid. 192.
232. Ibid. 31.
233. Ibid. 35.
234. Ibid. 200.
235. Ibid. 373.
236. Ibid. 35.
237. Ibid. 38.
238. Toscano 2006, 182, 183.
239. Marx 1990, Vol. 1, 287.
240. Deleuze and Guattari 2004, 375.
241. Ibid. 29.
242. Ibid. 197.
243. Ibid. 315–317.
244. Deleuze 2006, 42, 55, 56.
245. Deleuze and Guattari 2004, 197, 317.
246. Deleuze 2006, 40, 41, 56.
247. Ibid. 40.
248. Ibid. 57, 64.

Part One

The Primitive Regime

1 A Cruel Mnemotechnics

1. The BwO of the earth

At the level of desiring-production—and more specifically as regards the disjunctive synthesis—we observed how the full BwO served as a surface upon which the connections of production were recorded and to which the organs-partial objects attached themselves. This modality of the BwO was that of the miraculating machine—a mode of operation of the BwO whereby it attracts the said partial objects to itself, to such an extent that they appeared to arise from it in the first place.

Though at this earlier point we spoke of the BwO in the most general terms, we now move to particularize. For Deleuze and Guattari also speak of the full BwO as the *earth*. As Alphonso Lingis has it, the "earth is in fact the original body without organs, closed in itself, full and warm, enormous vesticle suspended in the void, stuffed full of its warm flows, gurgling and belching from time to time, sufficiency itself."[1] The full BwO is the earth in the purely intensive sense that attaches to *spatium*. As such, the full body of the earth is an untrammeled genealogical force of pure *filiation*. It is the Numen.[2] This filiative Numen is the most fundamental energy of the intensive order.[3] More specifically, therefore, one must speak of the full BwO of the earth as a numenal energy of *intensive* filiation.

But in what sense may one speak here of filiation as such? Before answering this, a brief prodaedeutic is in order. We may recall that in our Prologue the inclusive disjunctive synthesis was described as *inter alia* the basis of the conjunctive synthesis of consumption-consummation wherein a portion of the Numen (disjunctive energy of recording) was siphoned off in the continuously reiterated upsurge of the nomadic subject. The said subject/self was, moreover, averred to be multiple in the sense that, inasmuch as the disjunctive syntheses facilitating its extrusion were inclusive, its upsurge was from a superimposition of multiple intensive states characteristic of the said syntheses' inclusive usage. Analogously, the same continues to be the case for the BwO of the intensive earth. For Deleuze and Guattari, this intensive filiation entails the limitless self-division of the same intensive "being" which, in an untrammeled profusion, shows itself to be completely fissiparous.[4] That which is multiple—and more particularly, a pure positive multiplicity—divides infinitely into itself. And does so in intensity.

But the sense in which one may here speak of filiation—and much more specifically, *intensive* filiation—shall only emerge shortly. Firstly, let us also be reminded of our earlier observation that the nomadic subjects incessantly rising up from endlessly reproduced conjunctive syntheses passed through all the intensive states produced in the BwO and that this veritable intensive becoming (*devenir*) was one in which the nomadic subject was likewise said to repeat entire intensive *genealogies* for itself. Now, the same is the case for the BwO of the earth. For the latter is traversed by a "non-excluded being" which moves through all the singularities and purely intensive states produced upon it by an incessantly self-reproducing and errant inclusive disjunctive synthesis. In this manner, the "cosmic egg" of the full BwO of the Earth is inscribed with the jagged lines of intensive genealogical filiation. Far more specifically, on the BwO as the full body of the earth, this profusion of nomadic subjects is that of primordial ancestors, where each such ancestor intensively reiterates an entire genealogy for his own self. This intensive filiation is, moreover, a veritable "litany of twins" simultaneously affirmed across all the "indivisible distances" which nowise separate them.[5]

Moreover, since the disjunctive syntheses are still inclusive and the conjunctions non-segregative (or polyvocal), the intense filiation on the full body of the earth nowise at this point comprises any distinction of persons, let alone one of sexes. Instead, it is merely one of pure pre-personal variations in intensity, which assumes the form of a "twinness" and "bi-sexuality" in varying degrees of intensity.[6]

Finally, this intensive filiative genealogy forms the object of a constantly vigilant *memory*; that is, it is the object of an intensive filiative memory, a memory moreover describable as "nocturnal" and "biocosmic."[7] This memory Deleuze and Guattari term the "great nocturnal memory of the intensive germinal filiation."[8]

2. The erection of the full body of the earth as *socius*

Whether that of the intensive earth or otherwise, the BwO abides—as observed earlier—at the molecular level of desiring-production. In the passage to the molar level of social production, however, the BwO must needs be submerged so as to give way to another full body: that of the molar social machine. The full body of the earth as BwO necessarily yields before the full body of the earth *as socius*. It is not, however, gone. This we shall see.

As likewise espied earlier, the movement from desiring-production to social production entails at once a change in the usage of *inter alia* the disjunctive synthesis from inclusive to exclusive. In a similar vein, as regards the BwO of the earth, Deleuze and Guattari refer to Marcel Griaule's observation about the Dogons, among whom there comes a moment at which—at the level of the eighth ancestor—the disjunctive syntheses, in a sudden discombobulation, switch their modality from inclusive to exclusive. In the same movement, the full BwO is deconstructed, the "twinness" of the intensive ancestors is eliminated, whereas the sexes come to be mutually dirempt through the rite of circumcision. In a coextensive counter-movement, however, the shattered BwO is recomposed. A new system of connections and conjugations enters into existence whereby human bodies find their mutual articulation through "a lateral corporeal inscription with articulatory stones of alliance, in short, a whole ark of alliance."[9]

What is at stake in the foregoing is the passing of the entire intensive system of the "naked" full BwO of the earth into extension as the "clothed" full body of the earth as socius. As intimated in what we have just mentioned, the foregoing transition entails several transformations.

(i) Corporeal and sexual disambiguation

One may note that the molecular level of the nomadic subject—wherein everything conmingled in a limitless intensive becoming—was at once that of sexual and corporeal ambiguity. It stands to reason that the passing of the full BwO of the intense earth into extension and of evanescent selfhood into fully-fledged molar subjectivity—that is, the usurpation of intensive states by "discernible persons"—at once entails the emergence of unambiguous corporeal and sexual individuation.

(ii) From intensive germinal filiation to extensive somatic lineage

The formerly intensive filiations themselves pass into extension as lineages comprised of distinct persons—a movement whereby the intense germinal order gives way to an extensive regime of somatic generations.[10] Succinctly put, the passage in question is one from intensive germinal filiation to phylogenetic lineage in the more prosaic sense.

(iii) The forging of alliance

In addition to extensive filiation, the passing into extension of the intense BwO of the earth introduces a further element: *alliance*. This latter entails the forging of lateral connections stretching across the phylogenetic lineages of extensive filiation. These lateral connections arise when the formerly polyvocal connections of production on the BwO of the earth themselves become extended in the form of a pairing of persons.[11] What emerges with these vicissitudes is, as Eugene Holland maintains, a primitive kinship system whose synchronic axis consists of marriage alliances and whose diachronic axis is comprised of extensive phylogenetic lineage filiations.[12] In this savage social organization, these marriage alliances extend throughout the entire social field, whereas the phylogenetic filiative lineages extend back a considerable number of generations. Through the pressure exerted by the densely woven patterns of alliance, the combination of kinship relations with myth serves to extend phylogenetic lineages as far back as the earth itself. At the same time, the network of lateral alliances becomes co-extensive with the entire full body of the earth as socius.[13]

(iv) From biocosmic to somatic and lexical memory

Coextensively and coterminously with the emergence of primitive social machine, and so as to accomplish its upsurge, there comes to pass an active repression of the gargantuan intensive memory of filiation.[14] This tremendous act of repression of the immense nocturnal memory of intensive germinal filiation serves to facilitate the creation of an extensive somatic memory, forged out of the vast network of henceforth extended filiative phylogenetic lineages and lateral marriage alliances that arise with them.[15] Crucially, it is specifically these alliances which now entail a particular kind of memory that is ineluctable for the functioning of the primitive regime: a memory of *words*.[16]

(v) Emergence of primitive forms of capital

Filiation and alliance now come to constitute two distinct, though closely related, forms of primitive capital. Here, Deleuze and Guattari maintain "filiative stocks" to be "fixed capital," and associate alliance with a perambulating capital entailing the coruscating movement of mobile blocks of debt.[17] As Ian Buchanan avers, fixed capital derives from the "right" to rule of a chief as arising from his descent from a long filiative lineage; whereas mobile capital arises from his need to maintain and further this standing by forging alliances outside his family through the inexorable adjunct of the circulation of gifts through which he induces others to be indebted to him and is enabled to convert perishable wealth into imperishable prestige.[18] This privilege, as filiative fixed capital, constitutes the potential energy of inscription upon the full body of the earth as the movement of gifts and counter-gifts; whereas the indebtedness of those with whom he is allied generates the kinetic energy and actual direction of this movement as determined by the respective trajectories of the gifts and counter-gifts traversing the same full body.[19]

3. Coding of flows, collective investment of organs, marking of bodies

As observed earlier, the whole of desiring-production is recorded on the full BwO of the earth whilst organs-partial objects are attached to its surface. The task of the full body of the earth as socius is analogous. What it records is, however, not desiring-production but *social* production. In the latter and in contradistinction to the case of the former, production ceases to graft itself onto its product. Instead, mutually discernible producers and products appear. Here, the entire process of production, that is to say, all the forces and means of labor of primitive man, as well as all the actual agents and products of labor are inscribed, recorded and distributed on the full body of the earth as socius.[20]

We have seen how the passing into extension of the full BwO of the earth as a socius inextricably entailed the repression of the great nocturnal biocosmic memory of intensive filiation of the former.

(i) The coding of flows

The manner in which this gargantuan act of repression is brought to bear is by way of the *coding* of the decoded and deterritorialized flows peregrinating along the surface of the full BwO of the earth. As Holland argues apropos of Deleuze and Guattari, through the codification, inscription and recording of the flows of desire, the socius aims to definitively ensure that not a single flow exists that is not adequately bottled up, regulated and channeled.[21] There is a very ostensible fear-forestalling aspect to this coding of the flows of desiring-production, in that it definitively aims to undercut and preclude the angst portended by the errancy of decoded flows.[22] These mutant and anarchic decoded flows are self-identically the great nocturnal "germinal implex" or "influx" of the full BwO of the earth; which, in the absence of being forcibly repressed through coding, would unleash absolute terror upon the primitive socius. Such incorrigibly runaway centrifugal fissiparity of decoded flows would spell doom for the socius. The mechanism of coordination of filiation and alliance would completely disintegrate and, instead of filiation passing into descent, descent would perpetually collapse back into the filiation from whence it had arisen in an endlessly aborted attempt at its own self-engendering. All that would remain in this catastrophic deluge would be an absolutely prodigal and destinerrant chaos of noncoded flows shifting and sliding upon the desert of the BwO violently extruding itself from beneath the torn integument and shattered crust of the full body of the earth.[23] This is so because the intensive germinal influx, being desire in its "free-labor" state or desire prior to "primitive accumulation," is intrinsically and of itself equivalent to the noncodable, noncoded, or decoded flows peregrinating along the surface of the body without organs.[24]

We have spoken of the "anguish" of decoded flows—and of these latter as the "terror" of the primitive socius. These mutant flows of desire rising from the bowels of the full BwO of the earth are experienced by the denizens of the primitive regime with fear and trembling as death extruding itself from the subterranean depths. Here, seemingly insignificant perturbations and vicissitudes carry the most inauspicious portents of catastrophe. The chthonian force of the female is a case in point. José Gil refers to Victor Turner's analyses of the Ndembu tribe in Central Africa as regards the social ramifications of women suffering from menstrual ailments such as menorrhagia (an excess of menstrual flow) which preclude her—at least temporarily—from bearing children. Such a hapless sufferer is viewed as having been afflicted or "seized" by the shade of a deceased relative or ancestor whom she had neglected to honor. Far from remaining at the isolated level of the ailing individual, the latter condition immediately sends seismic tremors that are felt on the social plane. More specifically, it instantly redounds upon almost the entire local matrilineal descent group of the patient in the form of pervasive feelings of anxiety and fear. In so doing, it opens up a breach in the social structure itself.[25] The very integrity of the Ndembu socius here comes under the imminent threat of mutant flows. For the menorrhagia of the sick woman serves to "unleash conflicts and incoherent processes that don't connect up with each other and that cannot be controlled by normal means. Because neither the patient nor the relative nor the social fabric that guarantees the course of events in normal life is up to the task of understanding or combating the pathological behaviour caused by the sickness, they run the risk of irrevocably rending the fabric itself."[26]

Pierre Clastres observes that among the Guayaki Indians, a daughter's first menstrual period, a woman's accidental or deliberate abortion, the birth of his wife's child, or a death in the tribe, are all experienced by a husband as attended by the release of death-wielding forces—forces directed at the husband himself.[27] Let us take the example of a birth. "By its very nature," Clastres avers, a birth is felt by the Guayaki community as "a provocation to social disorder and, beyond that, to cosmic disorder"—a disorder inextricably bound up with deathly premonitions for the father, in that "the order destroyed by a birth can only be re-established through a compensatory death."[28] A sort of cosmic parricide is here seen to well up in the collective Guayaki unconscious. In this occult drama, the bearers of death are jaguars (*baipu*), the newborn child's messengers, entrusted with killing the father as he hunts.[29] All mawkish Wordsworthian sentimentalism as regards nature would here be misplaced. The child, far from being the father of the man, is instead the death of the father. Or at least potentially. Espied conversely—that is, from the side of the father—the husband is *bayja*. That is to say, one who attracts living creatures. Owing to the recent birth of his child, he becomes an attracting center for the convergence of the inhabitants of the forest.[30] And it is jaguars who are the first amid wild animals to sense the presence in the woods of a man in a state of *bayja*. Hence the Indian saying *Ache bayja bu baipu tara ika*—"When a Guayaki is *bayja*, jaguars come in great numbers." The state of *bayja* is therefore one of being faced with the imminence of one's own death.[31] One Guayaki whom Clastres describes in his chronicle of the said tribe to have been *bayja* was Chachugi. His wife Pichugi had recently given birth. The anguish felt in the face of destinerrant flows of desire is here still further evinced from the fact that, around the time of his wife's delivery of her child, Chachugi would go as far as to remain at a strict distance from her and not allow his eyes to rest upon her. For, had he not been thus vigilant, he would have run the additional risk of being *pané*—that is, technically incompetent as regards hunting. In his particular case, Chachugi would run the risk of being both *bayja* and *pané*—a man not merely fated to return empty-handed from hunting, but indeed a dead man, a helpless man whose cadaver would be shredded by wild beasts.[32]

Another eventuality that threatens the very integrity of the Guayaki primitive socius with the terror and anxiety of decoded and deterritorialized flows, would be a man stubbornly bent upon overextending the frivolity of his youth. At the age of fifteen, with the reception of the *beta* that pierces his lip at his *imbi mubu* ceremonial, a Guayaki boy passes from childhood to youth—and thus into several years of relatively carefree existence known as the time of *pravoty*, that is, of "seducing women." During these years as a *betagi*, he is at liberty to sport freely with unmarried girls. Flocks of this variant of quarry are to be espied by the riverside. Yet a Guayaki desirous of overextending his existence as a *betagi* by continuing to seduce women well into adulthood and incessantly recoiling before the prospect of marriage, would swiftly become a source of disturbance and conflict in the tribe. Adultery and its corollaries are here the most tangible dangers.

His clinging to irresponsible bachelorhood at a time when it is encumbent upon him to take a wife would serve to inject serious disorder into the community. In obstinately refusing to become a responsible adult, the incorrigible *betagi* would embody and portend death disgorging itself from the entrails of the earth. He would be a blue jaguar threatening the social and cosmic order in moving to devour the sun of the moon.[33]

The integral perdurance of the primitive socius is likewise countenanced with a deadly threat at the hands of *death itself*. That is to say, the dead. Or, more precisely, those who have died recently. For the Guayaki, a *manove*—that is to say, dead person—is a source of utter dread. He inspires intense fear. The recently deceased are invisible and aggressive. They are wickedly perverse specters desirous of killing the living. Ceasing to be a person, a dead Guayaki acquires the mode of existence of a jaguar—the metaphor for all the mortal dangers lurking beneath the surface of the regime. It is the souls (*Ove)* of the dead that are turned into jaguars. *Ove ro baipu o*—the soul is transformed into a jaguar. The ghost of a deceased Guayaki has a jaguar's head.[34] In their invisible presence, the dead take hold of the living and tear away their *pakryra*—mental and bodily equanimity, and plunge them into the quivering mire of anguish. This anguish is the anguish of death itself. It is the manner in which a death effects a considerable seismic disturbance in the Guayaki primitive regime, sending tremors through the very entrails of the socius. The death of a member of the tribe causes unbearable anxiety and pulpitation of the heart. Having lost its own body through death, the soul of the deceased moves to infiltrate the bodies of the living. If such an occupation comes to pass, the consequence is always death.[35]

(ii) The collective investment of organs

Returning now to Deleuze and Guattari, the way in which the full body of the earth as socius effects the coding of the productive flows of desire is by way of the collective *investment* of the organs of human producers. Here, homologously to the manner in which organs-partial objects affixed to the BwO at the molecular level of desiring-production ensured the ideal continuity of the flows of desire by at once breaking and producing them, the productive organs of primitive man are likewise attached to and distributed upon the full body of the earth as socius.[36] As Buchanan maintains, through this collective investment of the organs, the flows of desiring production are made part of the very infrastructure of the primitive social machine, aggregating the BwOs of all producers into the integral whole of the full body of the earth as socius.[37] Paraphrasing Derrida concerning what he maintains as transpiring in Artaud's theater of cruelty, this collective investment of the organs is a rigorous and "totalitarian" codification which submits human bodies to an inexorable necessity. It sutures all the gaps within the social machine. Hereby, cruelty "pacifies itself" through a reduction of becoming within the "perfection" of a self-contained economy. The fully body is reassembled into its visible presence, at once terrifying and pacifying.[38]

(iii) The marking of bodies

Still further, the manner in which the collective investment of organs is effected is by way of the *marking* of bodies—directly in the flesh. For the primitive socius—that of the full body of the earth—is a socius fundamentally characterized by, as Philip Goodchild maintains, the inescapable necessity of a most visceral inscription in bare flesh.[39] The way in which this essentially recording and inscribing socius co-opts the productive forces and distributes agents of production is brutal. For it deploys terrifyingly bloodcurdling operations such as: tattooing, excision, incision, carving, scarification and outright mutilation.[40] These violent corporeal operations are typically brought to bear on the body in the initiation rituals of savage peoples. For Pierre Clastres, such initiation rites "often constitute a basic axis around which the whole social and religious life of the community is organized."[41] Clastres tells of George Catlin's having witnessed an exemplar of such a rite among the Mandan Indians over the course of four uninterrupted days. What horrified Catlin was that the ceremony laid hold of the body in no merely pusillanimous manner. In fact, the body subjected to the ritual effectively undergoes a treatment no less than *torture*. Holes are pierced in the body, skewers forcefully driven through its wounds, the flesh torn, the body subjected to hanging, its limbs amputated, and so forth. The resources of cruelty appeared well-nigh inexhaustible. In the case of the Mbaya-Guaycuru of the Paraguayan Chaco, young men being ritually admitted into the ranks of warriors have their penises and other body parts pierced through with the aid of a sharpened jaguar bone. Extreme pain—literally torture—is in fact the very essence of initiation rituals among primitives peoples. Indeed, as Clastres observes, those carrying out the initiation rituals ensure that the suffering undergone by initiates is pushed to the very highest and most exigent pitch. For instance, among the Guayaki, the use of a simple bamboo knife would be quite sufficient to slice into the skin. It would however be insufficiently painful. Thus, a stone with somewhat of an edge, but one which is coarse rather than sharp, is used so that the skin is torn rather than cut. Catlin notes this same proclivity for the infliction of intense suffering among the Mandan:[42]

> An inch or more of the flesh on each shoulder, or each breast was taken up between the thumb and finger by the man who held the knife in his right hand; and the knife, which had been ground sharp on both edges, and then hacked and notched with the blade of another, to make it produce as much pain as possible ...[43]

Yet Clastres' most impressive description of the pain endured in savage initiation rituals is that of the Guayaki *jaycha bowo*—the carving of tattoos, or scarification of the back. After about six or eight years of living as a *betagi*, at approximately the age of twenty-two or twenty-three, or *buta provi bu* ("when one has a little beard"); a young Guayaki man is old enough to pass beyond the dissolute frivolity of youth and into responsible adulthood. That is to say, he is ripe for being ensconced into alliance—in the form of marriage.[44] Clastres:

> When the time comes, an old man in the tribe, most often the one who pierced the young man's lip, says to him: "You are no longer a *betagi*. You are a full-grown man; for this reason, I want to carve the tattoos. I will be the one who carves your back." This is not a request but a statement: the thing must be done now, and the *betagi* must go through with it. The man goes off to find a stone. Sometimes he has to go far away because it can only be found in the beds of certain streams. One side of it must be very sharp, but not as sharp as a splinter of bamboo, which cuts too easily. To choose the right stone, therefore, requires good judgment. This is all the equipment needed for the new ceremony: one stone.
>
> When he has found what he wants, the *bowaregi* cuts down a young tree with a smooth trunk in the middle of an open space in such a way that when it falls it leans on the branches of another tree and remains in a tilted position. Everything is ready; the operation (and it is that—almost a surgical operation) can begin. They wait for the sun to reach its zenith. The young man stretches out along the tilted tree trunk and puts his arms around it; he is on his stomach, with his back exposed and his skin good and tight. The "carver" takes his stone and cuts from the top of the man's shoulder to the highest point on his buttocks. The incision is not a light and superficial one, as it would be as if a razor-sharp splinter of bamboo were used, for example. This is really a crevice in the back that cuts through all the layers of the skin. In order to make this kind of furrow without using the knifelike edge of the bamboo splinter, one must have a strong grip, because the skin resists. The *bowaregi* presses down with all his strength; he tears the skin. The entire surface of the back is plowed this way with straight parallel lines, from one shoulder to the other; there are at least ten incisions. The blood flows freely; the young man is covered with it, as are the arms of the man cutting him, the tree, and the ground all around him. The Atchei say that the pain is atrocious: it is much worse than the piercing of the lip, which is hardly felt. The *jaycha bowo* almost kills the person who undergoes it. But just as during the *imbi mubu*, the young man does not moan or cry out: he would sooner lose consciousness—though even then his teeth would still be clenched. This silence proves that he is brave and worthy of being considered a full-grown man.[45]

The sheer immensity of pain and mutilation, however, attains its aim. For the traces left on the body by the torturous ordeal of the initiation rite are indeed permanent. The primitive regime has imprinted its indelible mark on the body of the initiate. Inscribed as they are in the deepest layer of the skin, these scars cannot be effaced. In that they are a clearly visible and perpetual testimony to others and moreover a constant reminder to the permanently scarred of the fear and trembling he had undergone, the marks of the initiated inexorably render him a member of the group.[46] Indeed, after having worked the back—to ensure and preemptively exacerbate the visibility of the scars—the *bowaregi* takes charcoal produced by the *Kybai* (meaning "man") tree and, crushing it into a fine powder which he mixes with honey, rubs the resultant paste into the wounds. Apart from preventing the back from haemorraghing, the paste also becomes encrusted in the furrows so that, once healed, they will assume a striking bluish-black hue. Otherwise the operation is considered a failure. Moreover, during his supine recovery period—such that the blackening of his tattoos is not thwarted—the patient may not eat meat and abides on a diet of honey water and naught beside. From now on the newly initiated will be a *bretete*, or full-grown hunter, and a *kybai gatu*, or a good adult—and thus primed for entering into a marriage alliance.[47] This savage ritual of cruelty is described by Clastres as an "assertive pedagogy" passing as it does from the tribe to the initiate whose hyper-stoical silence under torture is self-identically his consent to being absorbed into the group as a fully-fledged member of the primitive community.[48]

In Deleuzoguattarian terms, this ritualistic cruelty is that through which primitive man's working organs, through the forcible repression of the massive biocosmic intensive filiative memory, are marked and "hewn" into the socius and hereby given over to the collective. Homologously to what transpires as regards the BwO at the level of desiring-production, the full body of the earth functions as quasi-cause, in accordance with which these organs are "miraculated" as purported effluences of the same surface.[49] In this, the primitive apparatus of cruelty transforms men and their organs into the operational parts, cogs and wheel of the social machine.[50] It is through this violent stitching of the organs of the initiates into the earth that the Guayaki compose their primitive regime and—in so doing—prevent the dissolute *betagi* from deluging the socius with the mutant flows of the blue jaguar (*baipu*). Through the scaring of the skin they at once scar the earth, imbuing the latter with the sutured integrity of a socius, in which cosmic and social subversion alike are forcibly repressed beneath the *bowaregi*'s coarse stone.[51]

4. Cruelty, alliance, debt, lexical memory

Thus, the repression of the full BwO's intensive germinal influx is effected by the coding of decoded and deterritorialized flows, which is in turn itself brought to bear by the cruel inscription of human organs as cogs and working parts of the socius which they comprise. The latter inscription (marking of bodies) is the originary and reiterated act from whence the filiative stocks, or fixed capital of tribal elders' lineally conferred prestige, arise and from whence the mobile alliance debts enforcing and maintaining this prestige through trans-lineal indebtedness draw their movement across the socius. For it is this directly physical repressing cruelty that divides and selects the desiring influx into a portion which will be arrested and coagulated into fixed capital of prestige and that which will be permitted to flow through as the pure ruinous expenditure of gifts bestowed by the prestigious giver in the relations of alliance.[52]

(i) Debt

Debt is both the effect of and direct means employed by the savage inscription procedure or visceral marking of bodies.[53] It serves to cement and articulate the extensive matrix of lateral marriage alliances with the vast panoply of extended filiative lineages through the prestigious bestower's debt-inducing generosity with respect to his allies by which his mythical lineal descent is in turn bolstered.[54] Conceived on the more individual level of particular human bodies considered for themselves, cruelly inculcated debt is self-identically the mode of forming and ensconcing man into the primitive debtor-creditor relation.

(ii) Lexical memory

The directly physical and cruel inscription of the flesh whereby man's working organs are given over to the collectivity and by which he is indebted is—as briefly intimated earlier—also that through which a memory of words is burned into him. This latter lexical memory—born as it is of the repression of man's deep nocturnal and intensive biofiliative memory—shall henceforth serve as the means whereby he will be enabled to discharge his debt to those with whom he is allied and to primitive society as a whole at some subsequent point in time, or in other words, that by which he shall be rendered a good debtor—or promising animal. The procedures employed, as Deleuze and Guattari fail not to observe, do ostensibly involve a considerable measure of atrocity, stupidity, caprice, and what would seem to us as outright sadism. Yet, with their task discharged, they successfully *breed* man by forming him in the debtor-creditor relation through the inculcation in him of a memory which at base strains toward the future.[55]

5. The primitive semiotic: a savage theater of cruelty

The savage inscription of the body and the futurally inclined memory of words which it instills in man may be regarded as two, out of a total of three, of the fundamental elements of what shall shortly emerge as the primitive regime's *semiotic*.

(i) The phonographic couple

Deleuze and Guattari link, on the one hand, the futural memory of words to the voice and, on the other hand, the cruel inscription of the body to what they term a certain corporeal "graphism." Lexical memory is instilled by alliance whilst corporeal inscription associates itself with extended filiation since, as already intimated, it is this latter that is of the order of *somatic* generations. Otherwise expressed, the semiotic of the savage social machine is comprised of two heterogeneous elements: voice and graphism. The former is the enunciation of sonorous and corporeally textured words and is the voice of alliance; whereas the latter, functioning within the ambit of extended filiation, serves to represent the human body.[56] As Holland relates, the visceral graphism and the intoning voice form two entirely independent subsystems in which neither serves as signifier for the other. Indeed, no bi-univocalization of the graphic sign has occurred here, as shall transpire in the despotic regime. Rather, the primitive sign is at once polyvocal and an authchonous "position" and production of desire, in the sense that it assigns organs a definite place within allied groups.[57] Now, in that desire, conceived as desiring-production, for Deleuze and Guatarri, is coextensive with and immanent to the movement of life itself, this polyvocal graphism of the primitive regime is likewise homogeneous with life. It is, thus, also Antonin Artaud's "theater of cruelty," which too is self-identically life itself. As anterior to signification in the sense of representation, this polyvocal graphism is at the same time the unrepresentability of life. As Derrida maintains, the theater of cruelty, whilst falling short of being representation, notwithstanding remains the nonrepresentable origin of representation.[58] The theater of cruelty entails a "space" of "original" representation which is at once the primordial manifestation of life. It is a space generated from within its own self and not from a vantage-point external or supplementary to it.[59] This is so because, in the theater of cruelty, the sign is in no sense distinct from force, and thus not separate from desire. In fact, the sign is not yet a sign in the manner of, as shall indeed verily come to be the case in the despotic regime, a biunivocal correlation of signifiers with their perpetually deferred signifieds. Instead, the primitive sign is here somewhere between being a mere thing and a signifier as lying outside the thing. It does not transcend force so as to be invested with the somber impassivity of a distinct and separate work. There is here no separation between force and meaning as text.[60] As a form of "writing," though not in the sense of linearized writing in a book, this polyvocal graphism is a writing of the body itself in a "hieroglyphic" system of signs, a physical language of signs. The words annunciated by the voice are here, as in Artaud's theater of cruelty, themselves still physical signs which fall short of proceeding toward concepts. They are instead incantatory, magical, and inextricable from their shape and sensuous emanations.[61] The word is laid bare in its flesh, sonority, intonation and intensity. Its enunciation is not yet coagulated into the articulations of discursive language and logic, and remains a unique and irreplaceable gesture refractory to conceptual generality. This spoken effusion is homologous to "glossopoiea," which latter is neither an imitative language nor does it fully entail nomination. Rather, speech here occupies a strange interstitial zone in which articulation is no longer a mere shout, but where words have only barely come to have been formed and are not yet discourse. There is, thus, within speech not yet a distinction between speech and gesture, concept and sound, the pneumatical and the grammatical, master and slave, or God and man.[62]

(ii) The beholding eye

Be they as they are, these two elements—"voice-audition" and "hand-graphics"—alone cannot possibly suffice in the constitution of the primitive semiotic. Something else must be introduced to bridge the abyss between them. More precisely, a third element must be introduced in order to apprehend, measure and appraise the conjunction between the intoning voice of alliance, on the one hand, and the polyvocal sign manually carved into the flesh of the human body being ensconced into the primitive debtor-creditor relation, on the other. Here, as Holland argues, graphics and voice are made to converge upon one another in a festive public ritual conducted under the gaze of the "collective" or "divine" eye of either a shaman, or a specific designated group, though ultimately the entire savage collective. Through the pain endured by the scarified body and the surplus value of code, or voluptuous *jouissance*, extracted by the beholding eye from the disinterested malice ventilated upon the suffering body, man is integrated into the community.[63] In the same manner as Derrida maintains of Artaud's theater of cruelty, the visibility of this suffering undergone by the body being ensconced into alliance is, in no sense one mounted by the intermediary of a master discourse, as shall be the case for the despotic regime. Rather, the "seeing" by the roving and appreciating eye is, to it, an autopresentation of pure visibility and pure sensibility.[64] This eye is in no sense a spectator abstracted from the totality of the sensory milieu who constructs the spectacle of suffering flesh as an object to be apprehended. Rather, the eye is dispersed amid the festival in which it luxuriates. Its attitude is nowise divorced from force and in no sense addicted to exaltation. The ethicometaphysical prohibitions constituting the distinction between representer and represented, stage and audience, author and actor, and between text and interpretation are absent in this savage theater of cruelty.[65]

(iii) The savage triangle

The triplicate comprised of the sensuously intoning voice, the blood-stained sign ingrained in naked flesh, and the eye which siphons off surplus-enjoyment or *jouissance* from the pain inflicted, Deleuze and Guattari view as constituting the three independent, though not mutually exclusive, sides of a "savage triangle."[66] In this savage theater of cruelty, the function of the eye is not merely the extraction of surplus enjoyment from the conjunction of the other two sides of the magic triangle. For it in addition "evaluates" the suffering inflicted by the graphism. The hiatus twixt these two other sides is straddled by the eye, which latter "sees" the word without actually reading it—since, as seen in connection with Derrida, we are here at a point anterior to the bifurcation between the historical dichotomies of text and meaning, interpretation and tradition, reader and author, audience and stage. This eye appraises the pain emanating from the graphism applied to the flesh: it "jumps." All together, the savage theater of cruelty comprises an order not of subordination, as shall emerge in the despotic regime through its mutilation, but one of *connotation*. In this primitive system, the graphism (cruel corporeal inscription) constitutes signs in conjunction with the bodies and things designated by words, whereas the function of the eye is to rove and oscillate incessantly between these two, so as measure the visibility of the former against the pain of the other (surplus value of code).[67]

(iv) Activity

Crucial to what will follow, is the essentially *active* nature of what transpires in these cruel rituals of ensconcement of primitive man into alliance. Here it is the case that strictly active force is involved. As shall be seen later, active forces are those which always act to the very limit of their power—to the point of self-transgression, to the extent of being in excess of themselves. Otherwise, a specific subcategory of reactive force is involved. That is to say, reactive forces that—though they, according to their intrinsic nature, interiorly limit themselves and thus always function strictly within their own self-imposed bounds—nonetheless distinguish themselves by at least reacting to anterior active force by way of forming a riposte. More succinctly put, these particular reactive forces, though they are by no means active, do nevertheless *act* their reactions. They are thus *re*-active. Now, within the ambit of the savage semiotic triangle (primitive semiotic) of the primitive regime's corporeal inscription (cruel mnemotechnics), all forces are active, acted upon, or reacted to: one has the action of the voice of alliance, the re-acting passion of the body of filiation, and finally the reaction of the appreciative eye which evaluates the declension of the foregoing two elements.[68] Everything within this ambit is in fact a matter of use and function.[69] That all forces in the primitive theater of cruelty are acted implies that, as in the Artaud's variant, the primitive socius harnesses the totality of existence and interposes no distinction between actor and action. There is no division between force and sign that would separate the actor from meaning and from his own life. For the savage theater of cruelty maintains man within the autarchic immediacy of his own body and speech.[70] At this stage, man's speech has not yet been spirited away from him, nor has his body been purloined at the behest of and laid out supine and abandoned before the stern inquisitive gaze of a reactive, obdurate, stolid and cold-hearted tribunal of gray, mildew-faced apparatchiks and their alterations of pensive silence and stentorian opprobrium.

Notes

1. Lingis 1983, 32.
2. Deleuze and Guattari 2004, 168, 169.
3. Ibid. 170, 171.
4. Ibid. 168, 169.
5. Ibid.
6. Ibid. 170, 171.
7. Ibid. 170.
8. Ibid. 175.
9. Ibid. 170.
10. Ibid. 171, 175.
11. Ibid. 169, 170.
12. Holland 1999, 71.

13. Ibid.
14. Buchanan 2008, 94.
15. Ibid. 95.
16. Ibid.
17. Deleuze and Guattari 2004, 161.
18. Buchanan 2008, 100.
19. Deleuze and Guattari 2004, 164.
20. Buchanan 2008, 95, 96.
21. Holland 1999, 71.
22. Buchanan 2008, 93, 94.
23. Deleuze and Guattari 2004, 178.
24. Buchanan 2008, 94.
25. Gil 1998, 61, 62, 71.
26. Ibid. 65, 66.
27. Clastres 2000, 38.
28. Ibid. 40.
29. Ibid.
30. Ibid. 33.
31. Ibid. 35.
32. Ibid. 36, 37.
33. Ibid. 178, 180, 181.
34. Ibid. 300, 301.
35. Ibid. 334.
36. Buchanan 2008, 95.
37. Ibid.
38. Derrida 2001, 242.
39. Goodchild 1996, 91.
40. Ibid.
41. Clastres 1994, 180.
42. Ibid. 181–183.
43. Cited in Ibid. 183.
44. Clastres 2000, 178.
45. Ibid. 178, 179.
46. Clastres 1994, 184.
47. Clastres 2000, 179, 180.
48. Clastres 1994, 184.
49. Buchanan 2008, 95.
50. Deleuze and Guattari 2004, 159.
51. Clastres 2000, 181.
52. Deleuze and Guattari 2004, 180.
53. Goodchild 1996, 91.

54. Ibid. 92.
55. Deleuze and Guattari 2004, 207, 208.
56. Ibid. 221, 222.
57. Holland 1999, 72.
58. Derrida 2001, 294.
59. Ibid. 300.
60. Ibid. 238.
61. Ibid. 240.
62. Ibid. 302, 303.
63. Holland 1999, 72.
64. Derrida 2001, 300.
65. Ibid. 308.
66. Deleuze and Guattari 2004, 206, 207.
67. Holland 1999, 72, 73.
68. Deleuze and Guattari 2004, 206, 207.
69. Ibid. 222, 223.
70. Derrida 2001, 238, 239.

2

Toward an Ethics of the Primitive Regime and Beyond

1. Debt, pain, and the absence of *ressentiment*

(i) The absence of ressentiment *and revanchism*

Though the primitive semiotic inflicts great pain upon even the innocent, this cruelty is nowise motivated by *ressentiment* or any spirit of revenge on the part of those who mete it out. Neither those whose hand inscribes its torturous graphism in the flesh of the patient being forcibly ensconced into alliance, nor those whose role it is to behold the spectacle of epidermal scarification (savage inscription), are in any sense possessed of such ignoble motivations. Nor does this ritual cruelty instill such sentiments in him who is made to suffer. In fact, as shall be rendered pellucid a little later, quite the reverse is the case. For, if anything, the cruel mnemotechnics actually serves to forestall the onset of *ressentiment* and its appurtenances in the man who suffers it. Indeed, as Nietzsche maintains apropos of the long and numerous centuries prior to the history of man as *homo historia*, it "did not enter the consciousness of those who judged and punished that they were dealing with a 'guilty party'. Instead, it was a question of someone who had caused harm, an irresponsible piece of fate. He himself, the recipient of punishment, which again descended like a piece of fate, felt no 'inner pain' beyond what he would feel if something unforeseen suddenly happened, a terrible natural disaster, a boulder falling on him, where resistance is futile."[1]

Catlin's observations of the cruel initiation rites of the Mandan Indians bears this out. Even beyond the pure absence of inwardly felt pain, evinced here is the remarkable hyper-stoicism and even joyful acceptance of the physical torture meted out:

> The unflinching fortitude with which every one of them bore this part of the torture surpasses credulity; each one as the knife was passed through the flesh sustained an unchangeable countenance; and several of them, seeing me making sketches, beckoned me to look at their faces, which I watched through all this horrid operation, without being able to detect anything but the pleasantest smiles as they looked me in the eye, while I could hear the knife rip through the flesh, and feel enough of it myself, to start involuntary and uncontrollable tears over my cheeks.[2]

Primitive man is, moreover, afflicted with these violent bodily marks not so much because he is a priori suspected of being a future bad debtor. It is rather, as Deleuze and Guattari aver, the converse which is closer to being the case. That is to say, it is the cruel corporeal inscription ritual of the primitive regime that is at fault instead of the bad debtor himself. In other words, he has defaulted on his debt to society because the grisly visceral scars of corporeal mutilation administered by the primitive theater of cruelty had insufficiently emblazoned themselves on his sensibility—almost to the extent that he had never even been submitted to the ministrations of this gory ritualistic procedure. It was merely the case that, in him, the gap separating the voice of alliance and the body of filiation had been widened to such an extent that it became necessary to reestablish the severed link by way of an increase in pain.[3] The cruel mnemotechnics of the savage machine is thus at no stretch an enactment of pre-emptive vengeance. Nor may it be considered in any sense an ex post factum revanchism directed against him who, having already been ensconced within alliance and its debt, has transgressed the bounds of alliance and defaulted on his debt obligations. The eye that beholds the increased pain of the transgressor is nowise a vengeful eye. Nowise are its voluptuous wallowings in the spectacle of cruelty born of the spirit of revenge. Rather, the malice of the wronged creditor's enjoying eye was that of a wan detachment. Nietzsche indeed hesitates not to marvel at "how naive and innocent" primitive man's "need for cruelty appears, and how fundamental is that 'disinterested malice' (or to use Spinoza's words, the *symphathia malevolens*) they assume is a *normal* human attribute—: making it something to which the conscience says a hearty '*yes*'!"[4]

In fact, as Nietzsche unequivocally contends, such vengeful motivations could scarely even begin to be in play, since primitive societies were entirely bereft of even the very notion of individual responsibility and individual guilt. Indeed, a very "*high* level of humanization had first to be achieved, so that the animal 'man' could begin to differentiate between those much more primitive nuances 'intentional', 'negligence', 'accidental', 'of sound mind' and their opposites, and take them into account when dealing out punishment." In fact, as he moves on to argue, "'the criminal deserves to be punished *because* he could have acted otherwise', is actually an extremely late and refined form of human judgment and inference; whoever thinks it dates back to the beginning is laying his coarse hands on the psychology of primitive man in the wrong way. Throughout most of human history, punishment has *not* been meted out because the miscreant was held responsible for his act, therefore it was *not* assumed that only the guilty party should be punished …"[5] During the vastly greater part of human history, "it did *not* enter the consciousness of those who judged and punished that they were dealing with a 'guilty party'. Instead, it was a question of someone who had caused harm, an irresponsible piece of fate."[6]

In this connection, Deleuze and Guattari maintain that the purpose of punishment was to extract from the pain administered and contemplated a "surplus value of code" that compensated for the severance in the link between the voice of alliance and the mark or savage inscription that had not adequately taken to the transgressor's body. Punishment served to mend the break in the phonographic connection wrought by his crime.[7] The latter procedure, as José Gil has it, may be viewed as homologous to the manner in which a medicine man—in order to show the observers of the public ritual of the magical cure that his patient's sickness has verily been assuaged—extracts from the patient's body a leaf, a piece of flesh, or a strip of bloody cloth. It is in exhibiting this said "insubstantial, though unique, sign, stripped of all symbolic charge" to the gaze of the collective eye that the curer "holds the material proof of his effectiveness"—hereby suturing the rent Deleuzoguattarian phonographic connection and the broken alliance entailed.[8]

(ii) The cruel Nietzschean equation

Though there is no inkling of individual responsibility that would render individual guilt a real consideration for prehistoric societies, one may notwithstanding speak of the existence of primitive justice. It functions according to the cruel Nietzschean equation of debt: "injury done = pain to be suffered." More explicitly, the meaning of the latter is that every injury has an *equivalent* and can in point of fact be paid back, even if solely through the pain of the culprit.[9] The equivalence inhering within the cruel Nietzschean equation, however, necessarily invokes a moment of excess: that which attaches to the beholding eye mentioned earlier. For this equivalence consists in the creditor receiving, not so much a compensation for injury in the form of any tangible items possessed of use or exchange value, but rather in the fact that "a sort of pleasure is given to the creditor as as repayment and compensation,—the pleasure of having the right to exercise power over the powerless without a thought, the pleasure '*de faire de mal pour le plaisir de le faire*', the enjoyment of violating …"[10] This voluptuous pleasure is precisely what is entailed in the surplus value of code extracted by the beholding eye in the primitive regime's theater of cruelty. Continuing in the same vein, Nietzsche asseverates that suffering did indeed serve as an equivalent for the injury incurred by an undischarged debt, in the sense that "to make someone suffer is pleasure in its highest form, and to the degree that the injured party received an extraordinary counter-pleasure in exchange for the injury and distress caused by the injury: to make someone suffer,—a true *feast* …"[11]

But the ethics of the primitive regime goes deeper than this. It is true that its mechanism—a cruel mnemotechnics—endows man with a futural lexical memory so as to render him a good debtor, a man who can keep promises. It is also the case that, for all its painful savagery, the theater of cruelty at the epicenter of primitive ethics is nobly bereft of *ressentiment* and the spirit of revenge. There is, however, more. For a crucial accomplishment of the primitive regime's savage theater of cruelty is likewise the *forestalling of the onset of ressentiment.*

However, before this issue may be addressed with full vigor, it is necessary for us to embark upon a propaedeutic: the body as conceived of by Deleuze's Nietzsche.

2. The Nietzschean conception of the body

(i) Force

As Ronald Bogue maintains apropos of Deleuze, a body is a relation of forces. That is to say, every relation of forces comprises a body—and it matters not whether the said body is chemical, biological, social or political.[12] Moreover, as Pierre Klossowski relates, the body is entirely the product of chance or fortuity. It is merely the "locus" where a plethora of what he terms "impulses"—and what Deleuze calls forces—confront one another so as to produce an "interval" which effectively constitutes human life. Consciousness, on the other hand, is merely an organ wrought by the fortuitous confrontation of impulses/forces and situated at the highest extremity of the body. Through forgetfulness, consciousness disassociates itself from those impulses/forces which constitute the body of which it is the extreme point, selects only those which preserve its cerebral activity, and—in a curious reversal—adopts the body as if it were its own product. Equally paradoxically, consciousness, as the most fragile of all organs, precisely because of its own fragility, henceforth comes to dominate the body from whose fundamental anterior somnolence it had emerged in the first place as the mere gossamer outer layer that it is.[13] Fragile and superficial as it is, consciousness remains only the capacity for receptiveness to external stimuli.[14] It is ultimately only a succession of highly evanescent and discontinuous states of intensity, whose apparent perdurance consists solely in the deceptive fixity it acquires from the ossified structures of language.[15]

(ii) Active and reactive

As regards the forces that comprise a body, they are *quantified* as either dominant or dominated. A force is quantitatively *dominant* in relation to another force if it commands this latter force. In such a relation, the latter subordinated force is quantified as *dominated.* It is said to "obey" the force which dominates it.[16] But apart from being quantified as dominant and dominated, forces are also *qualified* as either active or reactive. Here it is the *difference in quantity* between two related forces which confers upon each respective force its quality. A dominant force standing in relation to a force which it dominates is automatically qualified as *active.* Analogously, a dominated force related to a force that dominates it is immediately qualified as *reactive.*[17]

But the difference in nature between active and reactive forces may also be characterized in another manner. That is to say, active forces are distinguished by the fact that they always act to the limit of their power and even go as far as to exceed themselves. They are, in other words, intrinsically *self-transgressive.* Active force, moreover, asserts itself in the sense of affirming its difference and making the latter an object of enjoyment and affirmation. Reactive forces, in contradistinction, are inherently *self-limiting* and therefore do not act beyond the limits which they themselves set or which are set on their pseudo-activity by other forces or relations of force. Whereas active forces affirm their constitutive difference, in vast contradistinction, the very defining characteristic of reactive forces is to deny *ab initio* the difference which constitutes them from the very outset and to moreover invert the differential element from which they derive and render up a deformed image of it.[18] Finally—and as shall be observed in due course—the tendency of reactive force is likewise to encroach upon active force so as to render the latter itself *stricto sensu* reactive.

(iii) Will to power

Forces, in order to enter into relation and in order to possess their own particular and respective quantities and qualities, need a sort of inner motor or will for the latter to be the case—and indeed to be forces in the first place. This inner will is the Nietzschean will to power. The will to power is in fact at once immanent to and transcendent of force. It articulates the quantitative difference between forces (the quantification of forces into a relation in which some are dominant and others dominated) and serves at once as that from which the distinct qualities of force arise (active or reactive).[19] Unlike the monolithic Schopenhauerian Will (which Nietzsche considers too general), the will to power is a fundamentally *plastic* principle that is no broader than what it conditions, that itself changes together with the particular forces whose relations, quantities and qualities it determines. The foregoing does not however mean to conflate will to power and force. For the two are certainly distinguishable in the main sense that while force is what "can," will to power is what "wills" (*La force est ce qui peut, la volonté de puissance est ce qui veut*).[20]

(iv) Affirmative and negative

For qualities to devolve into force, the will to power—the genetic element which qualifies forces—must itself be possessed of certain corresponding qualities. Here, Deleuze maintains that whereas *active* and *reactive* are the original qualities of force; accordingly, *affirmative* and *negative* designate the primordial qualities of will to power. Firstly, it stands to reason that every action is possessed of affirmation and that there is negation in every reaction. On the other hand, however, action and reaction are means or instruments of the affirmation or negation of will to power, which latter extend beyond action and reaction since they are none other than the immediate qualities of becoming itself. By this is meant that affirmation is not action but the "power" of *becoming active*; whereas negation is not simply reaction but a *becoming reactive*. Affirmation transports us into the gloriously exalted world of Dionysus, the world of the Being of becoming. Negation, on the other hand, hurls us down into the festering depths from whence hideous reactive forces emerge.[21]

(v) Sense and value

The will to power is also that (or the "one") which *interprets* and *evaluates*. Interpretation pertains to the *sense* of a thing or phenomenon, whereas evaluation concerns itself with the *value* of the same. Furthermore, sense is related to force whilst value is associated with will to power. More precisely, one speaks here of the "signification" of a sense: the latter resides in the quality of the force expressed in a thing, and is concerned with whether this force is active or reactive—as well as its particular "nuance." On the other hand, the value of a value resides in the quality of will to power expressed in the thing thereunto corresponding, and is concerned with whether the said will to power is affirmative or negative—as well as its given nuance. Simplifying somewhat, a thing or a value is deemed *noble* or *high* if it is associated with active force and an affirmative will to power; whereas, on the other hand, a thing or value is *base* or *vile* when associated with reactive force and a negative will to power.[22] We shall return to the full implications of this in due time—as they are quite vast.

(vi) Sensation and sensibility

The will to power must somehow *manifest* itself. Deleuze argues that—since the relation twixt forces in each case is determined insofar as each force is *affected* by other, inferior or superior, forces—it follows that will to power is manifested as *a capacity for being affected*. More specifically, will to power is manifested as the determinate capacity of *force* for being affected. This capacity for being affected is nowise necessarily a passivity but an *affectivity*, a sensibility, and a sensation.[23] Now, since it brings forces into affective relation, the will to power—as regards affectivity—is rightly characterized by Nietzsche as *the absolutely originary and most fundamental affective form*. On this reading, the will to power is in no sense maintained to be a being, nor even a becoming (*Werden*), but in fact a *pathos*.[24] Deleuze adduces from this that will to power is manifested as the *sensibility of force*, that the differential element of force manifests itself as force's differential sensibility.[25]

Moreover, although he argues that force's capacity for being affected is always and necessarily brought to bear, Deleuze nonetheless contends that this capacity is not fulfilled unless the associated force enters into a history or process of *becoming* (*Werden*). This is because the differential and genetic element of force, the will to power, is either the power of *becoming* active or that of the *becoming* reactive of force. Thus, recalling that the will to power is the primitive sensible form, one is permitted to aver that all sensibility is as such solely *a becoming of forces*. Therefore, one may conclude that it is not merely that the will to power manifests itself as the sensibility of force. For it also appears as the *becoming-sensible* of force.[26]

Finally, since—as observed earlier—the relation of forces consitutes a body, the capacity of being affected of force is instantly the capacity for being affected of a body—and thereby furthermore the sensation and sensibility of the same body.

3. The becoming-reactive of force

Nietzsche avers that history—as the universal history of man—is the very becoming-reactive of force in which reactive forces come to prevail over active forces and, in the process, bodies composed predominantly of reactive forces establish a hideously perverse and much perverted sort of ascendancy over those composed primarily of active forces. It will, moreover, be seen that the emergence of man as *homo historia* is itself coextensive with the growing preeminence of reactive force.

(i) The perverse ascendancy of reactive force

Crucially, reactive forces do not triumph over active forces by sheer mass of numbers. Instead, reactive force comes from some point onward to claim a *perverse* kind of ascendency over active force, by *divorcing it from its own inherent potentiality*. Ceasing to act to its limit, active force itself becomes reactive. What likewise transpires in this process is that reactive forces *deny* the difference between themselves and active force, and moreover that active forces *are* by definition and *ab initio* qualitatively different from them.[27]

(ii) Negativity of the will to power

This becoming-reactive of active force, as Deleuze maintains of Nietzsche, is of course accompanied by and coextensive with the transmogrification in the quality of the will to power that attends it; that is to say, a hitherto affirmative will to power becomes negative—and hereby ensconces itself as the power of becoming-reactive of force. To be still more precise, by divorcing active force from its own inherent potentiality, reactive forces betray it to the will to nothingness and thus to a becoming-reactive more profound, more pervasive and excessive of these forces.[28] As observed, it is history itself which is the temporal ambit of the becoming-reactive of force.

Evidently, since the origin of universal history is concomitant with the emergence of man as its preeminent actor (*homo historia*), the becoming-reactive of force is in fact constitutive of man as such, of the very humanity of the said being. During the course of this history—claims Nietzsche—man passes, respectively, through the stages of being fundamentally determined in his essence by, firstly, *ressentiment*, then by bad conscience and, finally, by passive nihilism—the desire for passive exinction.

4. The forestalling of *ressentiment*

Notwithstanding, the cruel mnemotechnics of the primitive regime has the essential function of thwarting such an eventuality; that is to say, of warding off man's fall into the historical mire of becoming-reactive. While it is indeed the case that history is the arena of the becoming-reactive of force, savage societies are *not* strictly historical: their stage is occupied by *homo natura*, and not *homo historia*. Indeed, when referring somewhat earlier to man as fundamentally determined in his essence by a becoming-reactive, we were speaking of *homo historia*. Here the emergence of man as a historical being is identical with that of his incipient becoming-reactive. More precisely, historical man is in the first instance determined by *ressentiment*. Nietzsche and Deleuze, however, nowise conceive of *ressentiment* as a mere superaddition to some kind of basic nature which man already possesses. Nor have they in mind a merely psychological trait. Rather, *ressentiment* is thought of as *ab initio* constitutive of the very humanity of man.[29]

Yet, before we move to describe how primitive mnemotechnics thwarts the onset of the dereliction of *homo natura* as *homo historia*, a crucial divagation will be made to delineate the functioning of what Deleuze terms the "reactive unconscious" and its relation to consciousness, force, memory, action and reaction. The said propaedeutic will serve to elucidate in more detail the fall of prehistoric man into the interminable historical mire of *ressentiment*.

5. The reactive unconscious and the positive faculty of active forgetting

Freud's "topical hypothesis" maintains that one and the same system could not possibly be able to prosecute the double task of fresh receptivity by the human animal to present excitation from the exterior milieu, on the one hand, and the recording of the latter as memory traces, on the other. Thus, each side of this double task must necessarily be alloted to a separate system. More precisely, the perception-consciousness system (Pcpt.-Cs) is tasked with receiving perceptions while at the same time retaining none of them, whereas the permanent memory traces corresponding to these excitations are stored in the unconscious.[30] In *A Note upon the "Mystic Writing Pad,"* Freud conjures a device whose purpose is to illustrate the co-functioning of the perception-consciousness/unconscious double-system. The contraption consists of a wax tablet overlaid by a double-layered sheet consisting of an upper layer of transparent celluloid and a lower layer of thin translucent waxed paper. These two layers are completely detached from one another, except at the two ends at which they are bound. A pointed stylus, which does not deposit any ink, chalk or other material, causes impressions to appear on the celluloid upper layer owing to its grooving of the waxed paper into the wax tablet at the bottom of the entire device. If the celluloid and wax paper are both removed from the tablet, one finds that this surface is entirely voided and once more capable of receiving fresh impressions from the stylus, whereas permanent traces of the impressions made by the stylus are still retained by the wax tablet.[31]

Freud claims that this device *mutatis mutandis* describes the homologous co-functioning of the perception-consciousness/unconscious system, where the celluloid and waxed paper are the system Pcpt.-Cs and the wax tablet is the system Ucs. Moreover, the appearance and immediate disappearance of the writing, as visible through the celluloid, is analogous with the evanescent emergence and disappearance of consciousness in the process of perception. To draw this analogy still closer, Freud envisages that the mystic writing pad is operated by two hands: the first writing upon the celluloid outer layer whilst the second immediately hereupon raises the two-layered covering sheet over the wax tablet. In this way, consciousness instantly vanishes into non-existence as soon a perceived impression from the exterior milieu creates it *ex nihilo*, resulting in an endless series of purely evanescent upsurges of consciousness at the behest of cathectic innervations between perception and the unconscious.[32] We may reasonably conjecture here that it is precisely the second hand operating the mystic writing pad that ensures the autonomy of the two subsystems: celluloid-cum-wax-paper (perception-consciousness) and wax tablet (the unconscious). This second hand would be an active force of repression, at once ensuring the perennial freshness and receptivity of the perception-consciousness system to external stimuli and its non-encroachment by memory traces rising up from the depths of the unconscious—hereby vouchsafing the salubrious preclusion of any confusion between the former and the latter.

Deleuze attributes a similar system to Nietzsche, except that in the case of the latter the unconscious-consciousness apparatus is constituted by the reactive forces of the body. The Freudian unconscious, when transferred into Deleuze's Nietzschean ambit, is therefore termed the *reactive unconscious.* Consciousness, as is evident, is also reactive.[33] In the case of the reactive unconscious, reactive forces attach themselves to and invest the indelible memory traces and are thus akin to the furrows dug into the wax tablet by Freud's stylus. As for consciousness, reactive forces are employed so as to render reaction to present excitation or to direct images of objects of perception possible and actual.[34]

Much as in the case of the mutual autonomy of Freud's waxed-paper-cum-celluloid and the wax tablet, so as to preclude the incursion of unconscious memory traces into consciousness and the resultant confusion of mnemonic trace and immediate excitation, the independence of these two subsystems must be somehow maintained.[35] For this, something analogous to the second hand which constantly lifts Freud's celluloid and waxed paper layers immediately after the pressure of the stylus is brought to bear upon the celluloid must be figured into the Nietzscheanized variant of the mystic writing pad. To such an end, reactive forces co-opt *active* forces so as to enable the reactive forces comprising consciousness to act their reactions to stimuli from the exterior milieu in the form of a *riposte* and, at the same time, *repress* the mnemonic traces of the reactive unconscious.[36]

These specially-deployed active forces stand at the basis of what Deleuze's Nietzsche terms a positive faculty of *active forgetting.* For the said active and hyper-conscious faculty is one of the forgetting of memories.[37] It is this latter faculty which prevents mnemonic traces from becoming perceptible to consciousness. Crucially, however, as Nietzsche himself renders pellucid, this foregoing faculty of forgetting is not to be associated with some form of etiolation or mental dereliction. In this connection, he maintains that forgetting "is not a *vis inertiae*," but rather "an active ability to suppress," which ensures that past memory "no more enters our consciousness during digestion (one could call it spiritual ingestion) than does the thousand-fold process which takes place with our physical consumption of food, our so-called ingestion. To shut the doors and windows of consciousness for a while; not to be bothered by the noise and battle with which our underworld of serviceable organs work with and against each other; a little peace, a little *tabula rasa* of consciousness to make room for something new, above all, for the nobler functions and functionaries, for ruling, predicting, predetermining," where active forgetfulness is "like a doorkeeper or guardian of mental order, rest and etiquette," in the absence of which "there could be no happiness, cheerfulness, hope, pride, *immediacy*," within the psyche.[38]

However, for Nietzsche—as in the case of the great nocturnal biocosmic memory of intensive germinal filiation—in that the last-mentioned is effectively repressed, the memory of traces *must be supplanted by a different sort of memory.* This—precisely as is the case with the lexical memory instilled in man by the mnemotechnics of the theater of cruelty and its savage inscription—is the memory of *words* (lexical memory) which, unlike that of traces (which is directed backwards towards the past), is a *futural* memory associated—again as in the primitive regime's cruel mnemotechnics—with man's capacity to be trained so as to be enabled to keep promises (that is to say, to be a good debtor).[39] In that it is orientated to the future, the corollary of this cruelly imputed lexical memory is nowise a mere miredness in an affective past from which man is unable to liberate himself. Rather, it is "an active *desire* not to let go, a desire to keep on desiring what has been, on some occasion, desired, really it is the *will's memory*: so that a world of strange new things, circumstances and even acts of will may be placed quite safely in between the original 'I will', 'I shall do' and the actual discharge of the will, its *act*, without breaking this long chain of the will."[40]

6. Dereliction of the reactive unconscious-consciousness system

(i) Atrophy of the faculty of active forgetting

As noted, the positive faculty of active forgetting is what ensures the non-encroachment of the reactive unconscious upon consciousness. Deleuze, however, warns that this faculty, though an active force—in that it must work with the reactive forces of consciousness for the latter to act their reactions to present excitations from the exterior milieu—must borrow the energy of riposte-forming reactive forces to accomplish its task. Working in such proximity to reactive forces and for ends abstracted from activity, these active forces fulfilling the task of Freud's second hand are prone to myriad functional disturbances.[41]

Inevitably, these "subservient" active forces cease to be able to draw the requisite energy from reactive forces to renew consciousness and repress the entirely passive reactive forces of the unconscious. The faculty of active forgetting simply decays.[62] Suddenly, consciousness undergoes sclerosis. Freud's second hand is amputated. Celluloid and waxed paper sink into the wax tablet. Excitation from the exterior milieu becomes confused with indelible memory traces surging up from the unconscious, reaction to which is now perceptible to consciousness. Divested of the material conditions of their functioning, active forces supporting active forgetting—divorced by reactive forces from their own potentiality—themselves become reactive. Within the consciousness-unconsciousness system *all reaction ceases to be acted*, as a roiling antagonism solely between reactive forces ensues—the consequences of which shall bear upon the whole of activity in an evermore pervasive trans-historical becoming-reactive of force.[43] This irrepressible subversion of all activity by the incipient deluge of virulent reactive forces, means that any resurrection of the faculty of active forgetting is completely out of the question.

(ii) Ressentiment *1: topological and typological aspects*

This displacement of reactive forces bearing indelible mnemonic traces from the reactive unconscious into consciousness, in conjunction with the cessation of reactive forces' ability to act their reactions, is what Deleuze terms the *topological* aspect of *ressentiment*. That is, a reaction which at once becomes perceptible and ceases to be acted.[44] He in whom this topological displacement has been effected, the man of *ressentiment*, is henceforth only able to react to his own festering memory traces. His incorrigible confusion between mnemonic trace and external stimulus, implies that he is permanently unable to react to the latter.[45] This condition may be described as a sort of mnemonic indigestion in which consciousness is unable to extricate itself from memories arising from prior excitations. As Nietzsche has it, in him the entire apparatus of repression—that is to say, that of active forgetting—being irrevocably damaged, has failed. He cannot be done with any past experience. He is a dyspeptic.[46] Consciousness hereby becomes mired in the past. It is here that the historical product of *ressentiment* comes into view for the first time. This product—at once canine and dyspeptic—is *the man who cannot "have done" with anything*. Such a man—indeed the *terminus a quo* and first distinct manifestation of *homo historia* as such—is possessed of a prodigious memory. Perpetually mired in reminiscence, he is unable to extricate himself from his traumatic past. As Jean Améry formulates it, resentment "nails every one of us onto the cross of his ruined past," and thereby "blocks the exit to the genuine human dimension, the future."[47] Peter Sloterdijk avers that this preclusion of the future is the result of a cerebrally palsied kind of memory "through which the new introduces itself compulsively, and like an addiction, into older episodes of pain."[48]

Deleuze glosses the topological aspect of *ressentiment* as being its raw *content*. This "raw content" entails the manner in which reactive forces manage to escape the action of active force. Inextricable from this topologial aspect, there is, however, also the *typological* aspect of *ressentiment*. The latter refers to the manner in which the foregoing raw content of *ressentiment* assumes *form*. This "form" is, more precisely, the *type*—that is, typical character—that arises from *ressentiment*'s topological aspect.[49] To the displacement of reactive forces bearing searing mnemonic traces from the unconscious into consciousness, the typological aspect of *ressentiment* adds the actual internal diremption of active force by reactive forces which divorces active forces from their own inherent potentiality.[50]

(iii) Ressentiment *2: the spirit of revenge*

Owing to the miscegenation of direct impression and painful mnemonic trace, the dyspeptic man of *ressentiment* henceforth reacts to excitation from the external milieu solely in the form of accusing the corresponding object or other for his pain and the object's unattainability.[51] He experiences this object or other as a direct personal affront and holds it responsible for his incapacity to psychically invest anything but his own painful memory traces.[52] One, as Nietzsche has it, can never get rid of anything—cannot get over, transcend or jettison anything so as to clean the slate of one's sensibility. Both other people and things alike always obtrude much too closely. One is struck far too deeply by all experiences, and memory festers like a wound—like some kind of metastatizing cancer.[53] Everything contaminates the soul and is an immediate object of disgust. As Peter Poellner argues, the psychological predicament of the man of *ressentiment* is fundamentally one of extreme pain, discomfort and frustrated desire. Subject to so many impossibly cruel sufferings, he blames the object and other subjects for his own inability to liberate himself from his poisonous mnemonic traces, in that he sees them as the cause of his misery.[54] However, since he is unable to act his reactions, the revenge exacted by the man of *ressentiment* cannot be of a directly physical nature. It is thus necessarily confined to the purely mental level. Indeed, the revenge implicit to *ressentiment*, even when it is realized, remains entirely "spiritual," completely imaginary.[55] This revenge is enacted by the divorce of active forces from their own innate potentiality wrought through a *projection* by reactive forces of a supersensible world that stands in contradiction to the material world, through which reactive forces mendaciously represent themselves as purportedly *superior*.[56]

7. Toward a savage ethics of the primitive regime

The means through which primitive man is bred to act his reactive forces Nietzsche terms "culture" or the "morality of customs." Culture, thus conceived, is the means whereby man is rendered obedient. That is to say, man is reared as an animal capable of obeying laws by way of the already mentioned training that his reactive forces receive—to the ends that they may be acted. Deleuze also terms culture the *generic* species activity of man.

As Deleuze maintains apropos of Nietzsche, human species activity or the "morality of customs" is generic in the sense that it serves to instill in man the capability to obey laws *in general*. As opposed to the *actual* laws specific to the parochial ambit of races, peoples and classes which Nietzsche regards as arbitrary, limited, stupid and grotesque and as involving the worst *reactive* forces, generic species activity entails the deployment of *active* force.[57] Whereas that *which* is obeyed is historically contingent and therefore always provincial and reactive, the *fact* that law is obeyed is universal and active.[58] It is in this sense that the morality of customs may be regarded as the generic species activity of man.[59] And is so in the sense of instilling in the human animal the capacity and propensity for obeying laws in general. Describing culture—the "morality of mores"—as the process of breeding an animal (man) with the "right" (not in the politically correct sense) of making promises, Nietzsche goes on to maintain that the latter task takes hold of and presupposes as a propaedeutic "the more immediate task of first *making* man to a certain degree necessary, uniform, a peer amongst peers, orderly and consequently predictable. The immense amount of labour involved in what I have called the 'morality of custom', the actual labour of man on himself during the longest epoch of the human race, his whole *prehistoric* labour, is explained and justified on a grand scale, in spite of the hardness, tyranny, stupidity and idiocy it also contained, by this fact: with the help of the morality of custom and the social straitjacket, man was *made* truly predictable."[60]

This is the activity of man as *homo natura* prior to his dereliction as *homo historia*. Generic human species activity, or the morality of customs, *precedes* history as the universal history of man, replete with its linear writing, traditions, interpretations, and virtually interminable series of State apparatuses. It is a cruel mnemotechnics that serves to instill in man the capacity for obeying laws through the inculcation in *homo natura* of a faculty of active forgetting of mnemonic traces, coextensive with the gargantuan repression of the nocturnal biocosmic memory of intensive filiation and its replacement with a futural memory of words. It is the latter memory—burned as it is by the savage theater of cruelty directly into the flesh—that allows man to keep promises and thereby be primed for the obedience of law in general. This breeding of a creature capable of keeping promises is at once the rearing of a good debtor, as Deleuze and Guattari make clear in connection with the rituals of cruelty of the primitive machine. This brutal and grisly act of repression serves to avert the miring of consciousness in an affective past by thwarting the ascent of reactive forces attached to mnemonic traces from the unconscious into its perceptive ambit, whilst at once training the reactive forces of consciousness to act their reactions. In doing so, it forestalls the descent of the species into the roiling subterranean morass of *ressentiment.*

The morality of customs, or generic human species activity in its pre-historic element, is thus a means of *selection,* for it breeds a particular type of man: a man who acts his reactive forces, is equiped with a memory which represses the past and remembers the future, and a man who is therefore capable of keeping promises in the future by being responsible for the future assuagement of his alliance debt to society.[61] As observed, pain in the cruel Nietzschean equation "injury done = pain suffered" is inflicted not in the least part as vengeance for unpaid debt or upon the suspicion of a likely future defaulting on debt. Rather, it is an action performed to forge or to re-cement and consolidate the alliance between man and man, the debtor-creditor relationship, a compensatory measure wrought in direct consequence of the conviction that the bodily marks ensconcing man into alliance had hitherto not penetrated the flesh emphatically enough. The pain inflicted here is a means of retraining the reactive forces which, having failed to be adequately cajoled into acting their reactions, had led to a malfunctioning of the memory of words at the basis of man's inculcated capacity to keep his promise and in so doing discharge his debt. As Lawrence Hatab maintains, the morality of customs inflicts pain as a means of ensuring that man does not default on his debt-obligations since, in the absence of formal contracts between legal parties in primitive society, there does not as yet exist a notion of repayment in kind on the part of debtors. Thus, instead of restorative justice as characteristic of liberalism, the primitive regime's sole means of justice is physical punishment.[62]

Pain is, most fundamentally conceived, a *reaction.* As seen, reaction takes as its object either memory traces, in which case pain is internalized and foments *ressentiment*, or present excitation from the external milieu. Now, as Hatab maintains of the savage theater of cruelty of the primitive regime, the pain that it inflicts as means of punishment cannot be simply understood as a form of sadistic pleasure to be gained by the wronged creditor. For, from the standpoint of the latter, it is not pain as such that is the central preoccupation but rather the pleasure arising from an elevated experience of superiority over the transgressor. Moreover, pre-historic public displays of punishment, unlike the solemn and sequestered executions transpiring under the aegis of State apparatuses, are raucous and celebratory festivals. The punished debtor, on the other hand, in no sense interiorizes the pain he undergoes. In that it is not yet spiritualized and "deified," as shall be the case under certain forms of masochistic Christian practice, the meaning he ascribes to his pain is precisely that of the boisterous enjoyment it accords the wronged creditor and the savage society against which he has transgressed.[63]

As observed, in the triadic dialectic (primitive semiotic) of the savage theater of cruelty comprised of the voice intoning words whose futural memory facilitates the keeping of the promise entailed within responsibility for debt, the suffering body whose reactive forces were trained to be acted by the cruelty of savage inscription, and the evaluating eye through which the first two elements are brought into proximity and coherence, it is the last of these that seals the transaction by way of the extraction of enjoyment from the pain inflicted on the suffering body presently being violently ensconced into alliance—and does so through the extraction of a surplus value of code from the other two elements. This surplus value of code siphoned off by the beholding eye is an aneconomic excess, a pure expenditure, and entails thereby the transgressive nature of active force—force which goes to the limit of its power. The pleasure felt in the infliction of pain or its contemplation, as Deleuze renders pellucid, is indicative of an active standpoint and involves active forces, whose task is the training of reactive forces for the sake of their own pleasure.[64] This enjoyment of the party that inflicts or contemplates pain is precisely the sensibility of active force, the affective manifestation of an affirmative will to power.

Apart from thwarting as opposed to instilling *ressentiment* in man, Nietzsche likewise asserts that the pain inflicted by primitive justice in no sense produced any feeling of guilt in the culprit. For, if we consider "those centuries *before* the history of mankind, we can safely conclude that the evolution of the feeling of guilt was most strongly *impeded* through punishment,—at any rate, with regard to the victims on whom the primitive measures were carried out. Nor must we underestimate the degree to which the mere sight of the judicial executive procedures inhibits the criminal himself from experiencing his act, his mode of conduct, as reprehensible as *such* ..."[65] Guilt, as we shall see to an even greater extent in Chapters 6 and 7, is a development of *ressentiment*. That the latter is successfully averted by primitive punishment bears within itself the immediate corollary that guilt too cannot possibly arise from the depths of the primitive regime.

8. The sovereign

If the product of culture as generic species activity—or the morality of mores—is specifically not the man of *ressentiment*, then what is it? In the most immediate sense, we have already seen that it is the man who can keep promises and discharge his alliance debt. This capacity is effected through, as Brian Leiter argues, a conjunction of behavioral regularity and reliable memory, both of which are the result of a cruel mnemotechnic breeding that allows the primitive human animal to be answerable for its own future.[66] As observed, it is true that human species activity renders man responisible; that is to say, responsible for his reactive forces—forces that militate against his effective ensconcement in alliance and moreover against the coherence of extensive filiation with lateral alliance. The cruel mnemotechnics of man's prehistoric labor in the primitive regime is that which, through the creation of debt and the means of its assuagement via the memory of words, entrenches man into the debtor-creditor relationship of alliance and renders him responsible by training him as a good debtor accountable for his reactive forces—forces whose innate tendency is to undermine alliance. Species activity constitutes man as an animal indebted to an active tribunal for his reactive forces.

This breeding of a man capable of such responsibility is, however, itself only a *means* to a far grander end. Indeed, the *telos* or ultimate product of the prehistoric species activity of man is the *sovereign individual* (*souveräne Individuum*). Nietzsche maintains that, if one envisages the scenario of placing oneself at the very end of the tremendous process entailed in human species activity, where "society and its morality of custom finally reveal what they were simply *the means to*: we then find the *sovereign individual* as the ripest fruit on its tree, like only to itself, having freed itself from the morality of custom, an autonomous, supramoral individual (because 'autonomous' and 'ethical' are mutually exclusive), in short, we find a man with his own, independent enduring will, whose *prerogative it is to promise*—and in him a proud consciousness quivering in every muscle of *what* he has finally achieved and incorporated, an actual awareness of power and freedom, a feeling that man in general has reached completion."[67]

Above all, this latter "finished product" of generic human species activity is *not* the responsible man or the moral man. For, as Keith Ansell-Pearson contends, he is the man who has outrun the sphere of the morality of customs (*Sitte*) and morality in all of its forms (*Sittlichkeit*). He is no longer responsible before the law or even an active tribunal such as that of the primitive regime, since he is autonomous, supra-moral, supra-ethical (*ubersittlich*), master of a free will that is entirely his own, like only unto himself, and answerable only to himself.[68] As Nietzsche renders pellucid, in the sovereign "proud knowledge of the extraordinary privilege of *responsibility*, the consciousness of this rare freedom and power over himself and his destiny, has penetrated him to his lowest depths and become an instinct, his dominating instinct …"[69]

Deleuze and Guattari compare the Nietzschean sovereign (*souveräne Individuum*) as the ultimate product of generic human species activity, or morality of mores, to the "schizo" and even to Zarathustra himself. Like Zarathustra, these "men of desire" who are averred by the authors to have thus far not yet come into existence "know incredible sufferings, vertigos, and sicknesses. They have their specters. They must reinvent each gesture. But such a man produces himself as a free man, irresponsible, solitary, and joyous, finally able to say and do something simple in his own name, without asking permission; a desire lacking nothing, a flux that overcomes barriers and codes, a name that no longer designates any ego whatever. *He has simple ceased being afraid of becoming mad* [my emphasis]."[70]

The Nietzschean sovereign, thus, may be seen as homologous to that of Bataille. As Derrida maintains of the latter, the sovereign (*souverain*) is devoid of both identity and selfhood. Neither do the Hegelian epithets of for-itself, toward-itself or near-itself appertain to him. For the sovereign does not, in any prosaic sense, govern. He does not subjugate, nor is he subordinated to anything or anyone. Rather, the sovereign expends himself without reserve. He is bereft of interiority and festering memory, and does not assimilate meaning to himself. The sovereign is the ultimate subversion of Hegelian lordship, as he in no sense seeks to be recognized.[71] This is so because he dispenses with all mediation and is not conditioned by anything. Neither does he seek to propagate himself, instruct or endure, but instead expiates himself. Crucially, the sovereign, though prepared by the tremendous labor of generic human species activity, in no sense recognizes, let alone gives thanks for the latter. He does not even recognize himself or give thanks for his own labor. In this fundamental sense, the sovereign is profoundly *ungrateful*. This is so, because he takes no efforts subordinate to goals outside of himself. Rather, all his efforts are always already ends in themselves. The sovereign does not only exceed the ambit of man-as-subject, but likewise exceeds history itself: in the sense of transgressing both the history of meaning, the meaning of history, and thereby the entire project of knowledge which ties the latter two together. His prerogatives in this regard fall within the ambit of "unknowledge." All responsibility that had taken itself seriously within the compass of absolute knowledge, with the supra-historical advent of the sovereign, is betrayed. The sovereign betrays this epistemological responsibility by exceeding it and simulating it into boisterous Dionysian play.[72]

Now, as Leiter argues, since the post-historical product of the morality of mores, or sovereign individual (*souverane Individuum*), is the last fruit of this immense prehistory of man's breeding; the tree itself must wither and be left behind. This is so because the sovereign individual is "autonomous" and "supra-ethical" (*ubersittlich*) and therefore has no more need for the morality of custom (*Sitte*). Hence, the attainment of the sovereign as the absolutely definitive *telos* of generic species activity must coincide with the *cessation or abrogation of culture as the morality of customs*, in which both creditor and debtor liberate themselves from the process which had heretofore trained them.[73] Furthermore, since the morality of customs is the generic species activity of man as such and thus makes man, the disappearance of the latter in its post-historical product must coincide with the actual suppression of the human species. This may be adduced from Leiter's observation that the sovereignty of the Nietzschean sovereign individual (*souverane Individuum*) entails no conscious "ego" which contributes anything to his self-mastery. Instead, his sovereignty is merely the effect of certain drives or forces over whose hierarchy he presides. This dissolution of the ego clearly entails, as obverse, the destruction of man as a moral agent standing before the law. For the post-historical product of generic species activity transcends ethics since, as we shall see again in Chapter 16 in connection with Rosenzweigian "Character" extruding itself from beneath the human-all-to-human veils of third-person relations, it is beyond the moral world.[74] *Homo historia*, as a type primordially determined by the triumph of reactive forces in the form of the malady of *ressentiment* and its ramifications, must also pass out of existence to make room for the sovereign's advent. The goal of prehistoric human species activity is therefore necessarily beyond both *homo natura* and *homo historia*. This product can, therefore, be reached neither within mankind's prehistory nor during history. The *telos* of generic species activity is thus only attained at the very *end* of universal history. The sovereign individual is therefore the *post-historical* product of culture. We have here hints of the Overman. The latter shall, however, only be reached and investigated in the final chapter of this book.

9. History's hijacking of generic species activity

It is the detour of the becoming-reactive of force to which we must now turn in dwelling upon human species activity from the perspective of history itself. Even worse than being what Hegel termed "the slaughter-bench of reason," history is the blighted terrain upon which human species activity—or the morality of mores—is grotesquely perverted and terminally diverted from its goal. Indeed, in history, culture assumes a meaning very different from its own intrinsic essence. As Stephen Mulhall argues, generic species activity's interlinked aspects of cultivating man's will to remember his debt and his being rendered calculable, which in their prehistoric element were mere preconditions for the ultimate potential creation of the sovereign as post-historical product, within the ambit of history, become *ends in themselves*. Here, the strict enforcement of routinization and regularity serve, through the imposition of equivalences among men, to hound out each man's capacity and striving for singularity and authentic autonomy. This is attained through the grafting onto species activity of organizations, associations, classes, and communities of a strictly *reactive* character: *primus inter pares* the State apparatus. Through the imposition of these parasites, human collectivities are produced as *herds*.[75]

This superimposition of heavy molar aggregates onto species activity prevents it from reaching and vanishing in its goal. History co-opts culture's violence, passing it off as belonging to its own peoples, races and States. Nietzsche inveighs with massive polemical force against what he sees as transpiring within this gory and gloomy Statist arena of *homo historia*. In what he declares "the greatest piece of immorality that has yet existed on earth," the perverted instinct of the herd is unleashed to crush the strong and independent. Through narrow-mindedness, falseness and out-and-out cruelty, mediocrity prevails over the singular and exceptional (WP 274). In a generalized crass leveling of the human animal, individuals are forced to completely renounce their own independent ethical valuations and self-worth, while it is ensured that all others do the same. Value is accorded to the individual only from the perspective of the heavy molar whole and for the sake of this ponderous totality, as the hatred of all gregarious and docile herd animals is set loose upon those who even try to be different (WP 275). In a colossal perversion, morality gears itself solely to the utilitarian benefit of the herd, as all noble distinguishing features of rarer and higher men are deployed to the wicked goal of their diminution (WP 276). In a grotesque inversion, the mediocre comes to be valued as the highest and most valuable, whereas the exception is treated as something intrinsically harmful and to be slandered and riddled with guilt (WP 280). In a singular maligning of nature itself, a generalized hatred and inculpation is unleashed by the ugly and ill-constituted against all that is beautiful, proud and joyful, against the independent, and against free spirits and skeptics (WP 283). Armed with a foul instinct for mere preservation and utility, the herd trains its depreciative eye against both that which towers over it and against eccentrics, whom it indicts as "degenerate" (WP 285). In a unilateral usurpation of generic species activity, the cold and imperious State apparatus overpowers the individual, forcing him to act against his very nature to serve its ends (WP 718). Through an unparalleled degree of stupidity, the State maintains a commonality of ignoble rules and valuations to keep herds docile. With mind-numbing inquisitorial oversight, mistrust and intolerance, it deals with exceptional individuals as criminals, causing them to suffer their singularity with bad conscience and as a disease (WP 726).[76]

Countenanced with this, it is not solely what would appear here to be merely the crushing repressive force of Asiatic despotism against which Nietzsche hurls his disdain. For, privy to the knowledge that Deleuze and Guattari—consummate Nietzsheans that they are—regard all States within history as variants of the great Asiatic despotic State of yore, such configurations as extant socialism and the liberal bourgeois mode of governance likewise fall very directly under his interdict. As Ansell-Pearson renders clear, the same inversion of means and end characterizing history's perversion of generic human species activity and its inherent diversion of the latter from its goal is utterly preponderant in modern politics. Dominated by nationalist and militarist prerogatives, Nietzsche regards the modern polity as culturally degenerate, in that it fails to see politics as a mere means to what ought to be its true end: the self-overcoming of man as the production of veritably singular and self-deriving individuals. Liberalism does not escape this interdict. In fact, utterly dominated and perverted as it is by greedy capitalist imperatives and thus deprived of an authentic communal and ethical life, it likewise figures within the irresistible Nietzsche polemic as yet another instant of the perversion of generic human species activity by outright philistinism. Lacking a conception of culture, liberalism relies solely on an abstract and ahistorical conception of individual self-realization. Within its ambit, the expression and realization of authentic individuality is well-nigh impossible. Whereas modern liberalism does to a degree emancipate the "private person," being culturally cretinous, it singularly fails to render up the conditions of arrival of the true self-legislating sovereign individual.[77]

Despite their obvious differences, liberalism and socialism, for Nietzsche, remain mere attempts at the austere economic management of the herd in which generic human species activity is corrupted by a cold utilitarian calculus, through which culture is suborned at the behest of a society dominated by a bureaucracy of sullen gray cardinals to breed incorrigible mediocrity in the name of public utility. Societies constituted on the basis of social contract theory such as that of Hobbes, wherein the beastly rabble surceases its *bellum omnium contra omnes* by voluntarily submitting to an absolute despot; that of Locke, in which a more placid yet semi-lawless population acquires cohesion through limited representative government; or that of Rousseau, in which a quaint coterie of nullities sentimentally marked by "natural goodness" expresses a bovine sort of sovereignty through collective autonomy, all fall afoul before the Nietzschean hammer and stand indicted as mere reflections of a mendacious slave morality whose perverse aim is the seduction of the strong and their conversion to the decrepit morality of the inane. Liberalism cultivates a freedom conceived as merely that of the private self, which it detaches from concrete action. But as Nietzsche and indeed Hannah Arendt maintain, this sort of freedom is merely an effluence of slave morality. The freedom of bourgeois man is far too overdetermined by the preponderance of narrow-minded utilitarian self-interest, and ignores the vital expressivist, poetic, and aesthetic dimensions of the human animal. It is far too prosaic, hidebound, unromantic, and thus essentially ignoble; breeding as it does a very small, petty, at once sullen, literal-minded, self-righteous, obsequious, and boring hobbit.[78] Or as Nietzsche has it, "a smaller, almost ridiculous type, a herd animal, something eager to please, sickly, and mediocre has been bred, the European of today."[79] Moreover, unlike culture, these societies, seeing nothing superior to their own laws, *refuse* to pass out of existence. They fail to heed Zarathustra's advice: "Let yourself, therefore be overthrown."[80]

Zarathustra propounds a rather obscure—yet wonderful—image for species activity: the fire-dog. The earth has two sicknesses. Man is the obvious one but, oddly, so is the fire-dog itself—or, more precisely, the fire-dog as representing species activity as co-opted by reactive forces, that is, by history.[81] This fire-dog is possessed of a "salty eloquence" that evinces his proclivity to "drink copiously from the sea"; he is a "hound of the depth" that takes his "nourishment too much from the surface"; a "salty, mendacious, and superficial" scum-devil who knows how to "bellow and to darken with ashes"; first among "the best braggarts and great experts in the art of making mud seethe"; chief avatar of all "that is spongy, cavernous, compressed."[82] Yet, he is in truth merely "the earth's ventriloquist."[83] As shall be observed in the impending chapter—and indeed *ad nauseam*—the primary apparatus responsible for the grafting of reactive forces onto human species activity or morality of mores, is the State. And Zarathustra himself makes clear to the fire-dog that the latter dog shares all of the State's incorrigible mendacity:

> But be still, hypocritical hound! You surely know your kind best! Like you, the state is a hypocrital hound; like you, it likes to talk with smoke and bellowing—to make himself believe, like you, that he is talking out of the belly of reality. For he wants to be by all means the most important beast on earth, the state; and they believe him too.[84]

But Zarathustra also has another fire-dog:

> And that I may continue to be right, let me tell you about another fire hound. He really speaks out of the heart of the earth. He exhales gold and golden rain; thus his heart wants it. What are ashes and smoke and hot slime to him? Laughter flutters out of him like colourful clouds; nor is he well disposed toward your gurgling and spewing and intestinal rumblings. This gold, however, and this laughter he takes from the heart of the earth; for—know this—*the heart of the earth is of gold*.[85]

Whereas the first fire-dog represents species activity or the morality of mores as perverted by history in the form of societies governed by reactive forces and cretinous laws which have no wish to die and is inhabited by the gregarious and ethically palsied herd animal who blithely submits to them, the second fire-dog is species activity as taken in its post-historical element, its product—the sovereign (*souveräne Individuum*), emerging at the end of universal history in the self-destruction of primitive justice itself.[86] Gil provides an intimation of the laughter of the second fire-dog in his comments on the laughter of savages. This laughter he describes to be at once "profound" and "a belly laugh." It is moreover "endless," "without premeditation," and "as undefinable as a rhythm." Its function is to "guard against seriousness and the possible heaviness of signs. Like dance it is ready to irrupt and defuse the risk of the petrification of gestures listing with the weight of meaning."[87] This dour seriousness, as we shall see, shall arrive and entrench itself when the descent into the historical mire has come to pass.

10. The demise of the primitive regime

History shall, of course, be that of the first Nietzschean fire-dog—the ignoble one. As Emil Cioran would have it, it will be a veritable monstrosity called up by humankind against itself, an inescapable fatality,[88] a virtually interminable series of mind-degrading and idolatrous temples erected on the basis of utterly ignoble pretexts. It will be a lurid hybrid of the trite and the apocalyptic, the blighted and leprous hotbed upon whose terrain man shall at once exalt, pulverize and likewise contaminate his own soul.[89]

Evidently, the transition from generic human species activity in its original prehistorical form to the properly reactive guise it assumes within history itself, shall and must entail a fundamental change in regime. That is to say, the primitive regime—wherein the second and more noble fire-dog may still be heard intoning the song of the earth—with the commencement of the universal history of man as *homo historia*, shall succumb to dereliction. For the reversal in the hierarchical relation of active to reactive forces cannot but entail the passing into desuetude of the regime itself and its usurpation by the perverse hegemony of reactive force.

But how shall this demise come to pass? Such a demise—at least from the endogenous vantage point of primitive justice's savage theater of cruelty—seems difficult to conceive. For has it not already been observed apropos of Deleuze and Guattari that "as primitive justice, territorial representation has *foreseen* everything"? Certainly, that with regard to which the savage regime exercises all of its vigilance and the full brunt of its apparatus of repression is the ubiquitous and interminable threat of the entire primitive full BwO as socius being shredded to pieces by its being deluged from all sides by deterritorialized and decoded flows of desire. As observed, it is through cruel mnemotechnic justice that the primitive socius effectively codes flows that would otherwise rip it to shards. This latter ever-present threat of mutant flows—whose tendency to run amock is perpetually held in check by the gargantuan repression of primitive man's great nocturnal biofiliative memory—is felt as death rising from within the full BwO of the earth as socius. But the savage regime—through the conjunction of coding, inscription, debt, and the futural memory of words—effectively and repeatedly staves off the eventuality of this endogenous death. In fact, savage coding shores up all the fissiparous tendencies of mutant flows to such an extent that the primitive regime exercises, in Clastres' words, an "absolute and complete power over all the elements of which it is composed"; holding as it does "all the internal movements—conscious and unconscious—that maintain social life to the limits and direction prescribed by the society."[90] The savage socius may thus be described as "a society from which nothing escapes, which lets nothing get outside itself, for all the exits are blocked. It is a society, therefore, that ought to reproduce itself perpetually without anything affecting it throughout time."[91] The primitive socius plugs all the holes through which the destinerrancy of mutant flows portends an endogenous death.

It does *not*—and indeed *cannot*—however resist a death that comes from *outside* its own ambit. Deleuze and Guattari: "Coding pain and death, it has foreseen everything—except for the way its *own* death would come to it from without."[92] In this connection, Clastres avers that only "one structural, cataclysmic upheaval is capable of transforming primitive society, destroying it in the process: the mutation that causes to rise up within that society, or from outside it, the thing whose very absence defines primitive society, hierarchical authority, the power relation, the subjugation of men—in a word, the State."[93] This death, this destruction, this utter demise—and with it the wholesale dereliction of the entire mechanism of primitive justice and its consequent fall into the reactive mire of history and its well-nigh interminable succession of State apparatuses—shall come to the savage regime by way of military conquest. It shall arrive with untimely swiftness—and tremendous force of blows.

Notes

1. Nietzsche 2007, 55, GM II.14.
2. Cited in Clastres 1994, 181, 182.

3. Deleuze and Guattari 2004, 208.
4. Nietzsche 2007, 42, GM II.6.
5. Ibid. 39, 40, GM II.4.
6. Ibid. 518, GM II.14.
7. Deleuze and Guattari 2004, 208, 209.
8. Gil 1998, 104, 105.
9. Nietzsche 2007, 40, GM II.4.
10. Ibid. 40, GM II.5.
11. Ibid. 42, GM II.6.
12. Bogue 1989, 20.
13. Klossowski 1997, 26, 27.
14. Ibid. 40.
15. Ibid. 41.
16. Bogue 1989, 21.
17. Ibid.
18. Ibid.
19. Ibid. 21, 22.
20. Ibid. 22.
21. Ibid. 22, 23.
22. Ibid. 23, 24.
23. Ibid. 23.
24. Nietzsche 1968, 338, 339, WP 635.
25. Deleuze 2006, 62, 63.
26. Ibid. 63.
27. Bogue 1989, 25.
28. Ibid. 24.
29. Deleuze 2006, 64.
30. Freud 1925, 208, 209.
31. Ibid. 209–211.
32. Ibid. 211, 212.
33. Deleuze 2006, 112.
34. Ibid. 113.
35. Ibid.
36. Ibid. 113.
37. Ibid.
38. Nietzsche 2007, 35, GM II.1.
39. Ibid.
40. Ibid.
41. Deleuze 2006, 113.
42. Ibid. 114.
43. Ibid.

44. Ibid. 114, 115.
45. Ibid. 115.
46. Nietzsche 2007, 35, GM II.1.
47. Améry 1980, 68.
48. Sloterdijk 2010, 47.
49. Deleuze 2006, 124.
50. Ibid. 125.
51. Ibid. 115.
52. Ibid. 116.
53. Nietzsche 2000, 686, EH 1.6.
54. Poellner 2011, 123.
55. Ibid. 128.
56. Bogue 1989, 26.
57. Deleuze 2006, 133.
58. Ibid.
59. Ibid.
60. Nietzsche 2007, 36, GM II.2.
61. Deleuze 2006, 134.
62. Hatab 2008, 85, 86.
63. Ibid. 86, 87.
64. Deleuze 2006, 135.
65. Nietzsche 2007, 36, GM II.14.
66. Leiter 2011, 106.
67. Nietzsche 2007, 36, 37, GM II.2.
68. Ansell-Pearson 1994, 135.
69. Nietzsche 2007, 37, GM II.2.
70. Deleuze and Guattari 2004, 142.
71. Derrida 2001, 335.
72. Ibid. 340, 341.
73. Leiter 2011, 107.
74. Ibid. 110.
75. Mulhall 2011, 260, 261.
76. Nietzsche 1968, 156, WP 274; 157, WP 275; 175, WP 276; 158, WP 280; 160, 161, WP 283; 162, WP 285; 383, WP 718; 385, WP 726.
77. Ansell-Pearson 1994, 10, 11.
78. Ibid. 40–43.
79. Nietzsche 2000, 266, BGE 62.
80. Nietzsche 1976, 244, Z II "Of Great Events."
81. Deleuze 2006, 139.
82. Nietzsche 1976, 243, Z II, "Of Great Events."
83. Ibid.

84. Nietzsche 1976, 244, Z II, "Of Great Events."
85. Ibid.
86. Deleuze 2006, 139.
87. Gil 1998, 171.
88. Cioran 2013a, 34.
89. Cioran 2012, 3, 4.
90. Clastres 1994, 212.
91. Ibid.
93. Deleuze and Guattari 2004, 209.
93. Clastres 1994, 202, 203.

Part Two

The Despotic Regime

3

Emergence of the Despotic Machine

1. Bronze-eyed artists

Nietzsche alludes to a fateful and absolutely earth-shattering moment in the distant past of humanity, at which time a sudden and utterly unpredictable and unforeseen, bewildering and lighting-flash invasion transpires. Those who prosecute this invasion, possessed as they are of "that terrible artist's egoism that has a look of bronze," through their depredations, come to forcibly evict "a tremendous quantity of freedom" from the world under their "hammer blows and artist's violence."[1]

Who are these men, these artists who come with their mad—and maddening—hammer blows? They are "some pack of blond beasts of prey, a conqueror and master race which, organized for war and with the ability to organize, unhestitatingly lays its terrible claws upon a populace perhaps tremendously superior in number but still formless and nomad."[2] These swift invaders arrive at an unprecedented speed and subjugate the savages and their primitive machine upon whose now blighted terrain they come forthwith to erect a new and imperial formation—a despotic regime. These men are the founders of the State, "the coldest of cold monsters." The State, as Zarathustra sternly warns, is the "death of peoples." Whenever the State claims to be the embodiment or representative of the people, it is simply lying. It is "destroyers" who invent and preside over the State apparatus. The State is the panoply of "snares," the "sword," and the "hundred desires" which these annihilators hang over the populace. It "lies in all languages of good and evil." It perpetrates and perpetuates at once the greatest theft in history and the greatest mendacity in history. When it bites, it "bites with stolen teeth." It embodies the "will to death" and is "invented for the superfluous." These surplus men, are lured, devoured, chewed and re-chewed by it with rancid teeth which themselves have been purloined by this gargantuan apparatus of death. In relation to generic human species activity taken in its prehistorical element, it is hated by any authentic non-Statist people as "the evil eye and sin against custom and law." The State is an out-and-out idol and, one may add, *the origin of all idolatry*. The common man must worship it. Under it, "universal slow suicide is called—life."[3] This frigid monstrosity—that is to say, the State—aver Deleuze and Guattari, appears to rise up from the ground instantaneously and not in any gradual or progressive fashion. Its sudden emergence is a "stroke of genius, the birth of Athena."[4] This at once seismic and miraculous event may be seen as standing at the border between prehistory and history.

There are, of course, other such shattering genealogical breaks: for instance, ineluctably Statist enterprises such as democracy, socialism, capitalism, and even Christianity.[5] These breaks, notwithstanding, invariably presuppose an originary grand hiatus, in that "*there has never been but a single State,* the State-as-dog that 'speaks with flaming roars.'"[6] This principle break standing at the very entrance into the reactive quagmire of history—of which subsequent historical breaks are mere echoes—is the event of origination of what the authors term "the primordial *Urstaat*," which serves as the transhistorical model of all extant and possible State formations.[7] In relation to the "Mycenean manuscipt" that is the *Urstaat*, each particular historical State form remains a mere "palimpsest."[8]

As Ian Buchanan relates, the *Urstaat* itself, in fact, possesses only a virtual existence. It is a "vanishing mediator" which conditions all historical Statist formations, whether material or spiritual, whose existence notwithstanding remains abstract and only realizes itself as an abstraction. It is, therefore, akin to the *khora* in Plato's *Timaeus*: a groundless ground which facilitates emerging forms whilst becoming none of them. In this sense, it is the *Ab/Ur/Un-grund* of all social machines from the primitive machine all the way up to the modern capitalist machine, yet never concretely instantiated in any of these. This means that it has never *actually* existed and never will.[9] The *Urstaat*'s historical passage, therefore, traces a kind of zigzag from *ille tempore* into an impossible and perpetually deferred diachronic future that never arrives, in the process sidestepping all social machines that it erects in its wake.[10] Notwithstanding, the conclusion adduced here is, of course, that *all* Statist societies are despotic at least in the purely formal sense—if not, as is often enough the case, in the outrightly pejorative sense.

Presently, however, let us return to the act of subjugation of the primitive regime by the blond beast of prey and its immediate consequences.

2. Double incest

One of the bronze-eyed artists—whose rabid pack had collectively set its claws into the now moribund savage regime—shall become the *despot* or king of the new empire, establishing himself at the pinnacle of this incipient regime in the most extraordinary manner: by committing double incest![11] As Philip Goodchild avers apropos of Deleuze and Guattari, he begins by entering into a forbidden endogamous marriage with "the sister" of the tribe being subjugated by the inbound counter-signifying regime of bronze-eyed beasts.[12] He contracts this marriage in a liminal wilderness zone beyond the perimeter of both the sister's tribe and his own tribe, and does so by exploiting a privilege accorded the so-termed "ritual initiator" or "sacred deflowerer."[13] This first incestuous marriage allows him to *overcode* all the existing alliances of the subjugated primitive regime with the *new alliance* he has forged with his now-princess.[14] The bronze-eyed artist now proceeds to contract an incestuous marriage with the mother of the subjugated tribe as she exists *in* her tribe, in so doing "countersecting" the extended filiative lineages of the assailed primitive regime with a *direct filiation*—the mother becoming queen as he soars to the exalted apex-predator position of despot.[15]

The aim and effect of the double incest is that of *overcoding*. It is, as Deleuze and Guattari contend more specifically, to overcode each and every one of the existing flows, so as thereby to ensure that no primitive code and not a single underlying flow can escape the overcoding wrought by the despotic machine that now comes into existence with the incestuous marriages.[16]

Through the incestuous marriage to the sister, which represents the spatial divergence from the subjugated primitive regime, the bronze-eyed artist appropriates the entire patchwork of mobile alliance debts; whereas, by way of the incestuous marriage to the mother, which represents the temporal divergence from the vanquished primitive regime as the hiatus between the somatic generations of savage lineage, he accumulates all primitive filiative stocks.[17]

This massive overcoding does not, however, mean that the old alliances and filiations are in any sense jettisoned, abrogated, let alone destroyed. As Holland maintains, the savage non-power mobile blocks of alliance debt and filiative stocks of prestige still subsist to a limited degree and as remnants.[18] Herein, the old network of savage alliance debts is converted into the infinite debt of the new alliance between the subject peoples and the despot; whereas the extended filiations of prestige are subsumed under the direct filiation, in the form of allegiance owed by the masses to the same.[19] We shall expound precisely what is entailed in this "infinite debt" in due time.

3. The despot-deity filiation

The new alliance is that which links the supremely extolled hero-become-despot to not only specific former tribal clans and families but to the entire subject-people milling down below as a single gray, undifferentiated mass; whereas the direct filiation is that entered into by the despot with his characteristically monotheistic or at least henotheistic deity.[20] The overcoding effected by the doubly incestuous hero-become-despot "countersects" all the alliances and filiations, and prolongs them so as to cause them to converge into the direct filiation of despot and deity, and into the new alliance between the despot and what are now his much diminished and cowering subjects.[21]

It is specifically the direct filiation of the despot with the deity that shall occupy us presently. This most exalted apex-predator position of the despot and his filiation with the deity allows him forthrightly to roar: "On earth there is nothing greater than I: the ordering finger of God am I."[22] The despot-deity filiation takes the form of the despot's unique and privileged standing before the god, or gods, presiding over the despotic regime.

The degree of intimacy of this relation between despot and deity varies amid concrete historically existing regimes of the broadly despotic sort. The highest degree of such proximity is attained when the despot is viewed as the literal incarnation of the principle or sole divinity of the pantheon of gods, demigods, or such like. Such a characteristic almost invariably bears the corollary of the despot's being deified after his prosaic natural death, prior to this moment, or as from the outset standing well outside the need for deification by virtue of being intrinsically divine from the very start.

(i) The Hittites

In the State ideology of the Hittites, it is held that the king is loved by a great god. Indeed, he (the king) is the vicar of the gods on earth. Furthermore, after his death, the Hittite sovereign undergoes deification whilst a carved statue to his imputed and enduring divinity is placed in the temple; whereas subsequent sovereigns cement their own despot-deity filiation by bringing this effigy diverse offerings.[23]

(ii) Vedic India

In Vedic India (before the advent of Gautama Buddha), the horse was at once a royal and a divine being: it was a representative of royal power (*kṣatrā*) but also identified with Yama, Aditya (the sun), and Soma (the sovereign gods). Eliade therefore avers that the horse may be viewed as, in a sense, a substitute for the Vedic king. This implies that the horse, above all, can be interpreted as an index of the direct filiation between the Vedic sovereign and the divine realm.[24]

(iii) Zoroastrian Persia

In the first inscription engraved at Naqš-i-Rustam in new Persepolis at the behest of Darius, Ahuramazdā is extolled as the supreme divinity whom *inter alia* made Darius king of Persia; whereas the latter's (Darius') prime responsibility is careful maintenance of Ahuramazdā's creation.[25]

(iv) Shang China

With regard to the ancient Chinese civilization of the Shang Dynasty, Neolithic documents bespeak the preeminence of the supreme celestial god Ti (Lord)—or Shang Ti (the Lord on High), who controls the basic cosmic rhythms and natural phenomena. Ti grants the Shang despot military triumph and agricultural abundance. It is the king alone who is able to invoke Shang Ti in these paramount matters, and he is himself only able to do so through the ineluctable intermediary of his ancestors. Indeed, the despot-deity filiation of the Shang regime ultimately consists in the fact that the first ancestor of the king was the direct descendant of Ti.[26]

(v) Ancient Mesopotamia

In the religions of ancient Mesopotamia, from the time of the Sumerians until the final liquidation of Assyro-Babylonian civilization, royality was conceived to have been of divine and, moreover, celestial origin. Indeed, even before his birth, the king was divinely predestined to imperial sovereignty. A supernatural light shone from his cranium. He was considered to be of earthly descent (in the quotidian sense) but also as a "son of god." Indeed, Hammurabi perforce declared himself the progeny of the deity Sin, Lipit-Ishtar, and that of the god En-lil. Mircea Eliade avers that the herein implied dual descent of the Mesopotamian despot allowed him to be the intermediary between the gods and his subjects par excellence. The king, who had moreover lived in fellowship with the gods in the garden containing the Tree of Life and the Water of Life, was at once the envoy of the divinities and the shepherd of the people. In this vein, Lipit-Ishtar was, "the humble shepherd of Nippur" called by the gods Anu and En-lil to establish justice in Sumer and Akkad. Furthermore, the Mesopotamian sovereign—the texts aver—though it is conceded that he is as such not a god, is a geminate being—partaking of both human and divine existential modalities.[27]

(vi) Ngũgĩ's "Marching to Heaven"

A far more ridiculous version of the despot-deity filiation may be encountered in Ngũgĩ wa Thiong'o's picaresque and Rabelaisian political satire *Wizard of the Crow* about a fictitious blood-soaked African dictatorship known as the Free Republic of Aburĩria. Here the entire populace of the country unanimously decides to erect a massive architectural structure never attempted in all of history since the failed construction of the Tower of Babel. The regime would raise a tower sufficiently vertiginous for the Ruler to be able to visit God daily in Heaven in order to greet him in the morning and in the evening and enquire as to his happiness. Through his proximity to God, the Ruler would be constantly privy to divine advice and, through the latter, ensure that Aburĩria would rapidly grow to heights never since conceived by the entire human race. This project went by the name of "Heaven-scrape," or "Marching to Heaven."[28]

(vii) Kantorowicz's Norman Anonymous

The despot-deity filiation assumes a most intriguing form in the juridico-theological writings of an anonymous Norman writer dwelt upon by Ernst Kantorowicz in *The King's Two Bodies*. The degree of sophistication likewise increases. The Norman Anonymous recognizes in the king "a *twin person,* one descending from nature, the other from grace.… One through which, by the condition of nature, he conformed with other men: another through which, by the eminence" of his "deification and by the power of the sacrament" of consecration, "he excelled all others. Concerning one personality, he was, by nature, an individual man: concerning his other personality, he was, by grace, a *Christus,* that is, a God-man."[29] The Christian sovereign, according to the Anonymous, was a *christomimētēs,* the paradigmatic imitator and sole vicar of Christ, whose dual nature he likewise reflected. Indeed, the despot-deity filiation consists here in the near-total similarity between celestial prototype and terrestrial ectype, which two are perfectly isomorphic—apart from the crucial difference that Christ is King and *Christus* by nature and eternally, whereas his earth-bound deputy was king and *christus* only by the grace conferred upon him as an effluence of anointment by virtue of which the earthly king was apotheosized for the temporally finite span of his reign.[30]

However, having effected the latter dichotomy between Christ as "God by name" and the sovereign as "god by grace," the Anonymous proceeds to blur and therein partially close the cleft twixt the two he had himself engendered. For he contends that the difference between Christ and sovereign ends at the level of power (*potestas*), that the essence and substance of power are the same in both God and king, regardless of whether the power is possessed by nature or acquired as an effect of grace: "The power of the king is the power of God. This power, namely, is God's by nature, and the king's by grace. Hence, the king, too, is God and Christ, but by grace; and whatsoever he does, he does not simply as a man, but as one who has become God and Christ by grace."[31] The Christian sovereign is thus a *gemina persona*, a "twinned" being.

The liturgically and (supposedly) ontologically grounded dichotomous nature of the sovereign, we shall encounter again in fairly short order in the form of the Elizabethan legal and theological doctrine of the king's two bodies unearthed by Kantorowicz from Edmund Plowden's *Reports*.

4. Infinitization of debt and the ruinous consumption of the despotic caste

(i) A debt of the existence of the subject himself

We have already observed—albeit cursorily—that, by way of despotic overcoding, the vast existing horizontal patchwork of alliance debts of the primitive regime is converted into an infinite debt associated with the new alliance by which the said patchwork is subsumed. The despot, as Holland relates, seizes the vast patchwork of finite alliance debts and, by transmogrifying the general "syntax" of the primitive communities below him, imposes a debt on his subjects that is at once unilateral and infinite.[32]

This infinite debt, as Goodchild avers apropos of Deleuze and Guattari, is one that proves to be intrinsically *unpayable*. Paradoxically, this occurs through the despotic State's *abolition* of the existing finite debts of the vanquished primitive regime and the concurrent circulation of a new system of *credits*, through which the subject is effectively chained to an interminable and unnatural service to the State. Still worse, this infinitized debt becomes a *debt of existence* of each and every subject, which swiftly reaches the point at which the creditor has not yet lent while the debtor shall never quit repaying.[33]

We may also view this infinitization of what was formerly an assuageable debt from the vantage point of history's perversion of generic human species activity. As observed in connection with the primitive regime, savage man was held responsible for his reactive forces before an active tribunal (comprised of the scarifying hand, the intoning voice, and the contemplating eye). His ensconcing into alliance at once meant that he was responsible for the future discharge of a debt fomented by the same painful breeding. His debt to primitive society was assuaged according to the cruel Nietzsche equation "injury incurred = pain suffered."[34]

However, with the abstraction of species activity from its essence by the tepid bog of history and its becoming-reactive of force, the active tribunal of the primitive regime is swept away and replaced by reactive instances: *primus inter pares* the State. Henceforth, it is no longer a case of the effective discharge of debt in accordance with a cruel mnemotechnic justice, but rather a *deepening* of the debt. Pain no longer assuages and pays one's debt. Instead, suffering now serves to *shackle* the subject to his debt and—in so doing—renders him an *eternal* debtor. Suffering now merely pays the *interest* on the debt. It is the very form of responsibility that has changed categorically: the transition is that between responsibility-debt to responsibility-*guilt*.[35] Debt is henceforth the relation between a debtor who will never cease paying and a creditor who will never finish using up the interest on this debt. It soon becomes an infinite debt toward the despot's deity.[36]

(ii) The subject sucked dry by the despot

An upshot of this infinitization of the subject's debt under despotism is that henceforth all productive and desiring flows, all the sweat-drenched labors of the toiling populace, converge into a great torrent that constitutes the sovereign's consumption—a precipitous and seemlingly unhindered vertical movement in which the savage filiative stocks of prestige are accumulated so as to forge the despot's standing as well-nigh divine, whilst the finite blocks of alliance debts now form an infinite tribute owed to the latter.[37] The tribute perforce extracted from the subject peoples flows into the bottomless coffers of the despot and his caste.

As Ellen Meiksins Wood has it, this forcible extraction of surplus value (the surplus labor of subordinated man) is—at least in the more typical archaic historical despotisms—attained predominantly by way of draconian taxation: an extra-economic mode of surplus-absorption, effected primarily with the aid of superior authority in conjunction with the coercive force of the military.[38] Elias Canetti—as regards the *modus operandi* of the despot in relation to his subjects—pointedly asseverates that the despot's ultimate aim is to "incorporate" them into himself and to suck all of the substance out of them.[39]

Yet, this unidirectional flow of surplus toward the royal caste does not of itself implicate anything resembling capitalist primitive accumulation on the part of the despot. Rather, one encounters here the exuberant phenomenon of ostensible and ruinous consumption of the surplus extracted from the toiling subjects below. In this connection, Georges Bataille avers that the "sovereign [*souverain*], epitomizing the *subject,* is the one by whom and for whom the moment, the miraculous *moment,* is the ocean into which the streams of labour disappear. The sovereign spends festively for himself and for others alike that which the labour of all has accumulated."[40] Bataille's insistence here on sovereign (despotic) life as inherently bent on spending itself in the evanescent incandescence of the moment, stands in opposition to the servile life of the world of labor whose comparatively processual nature is that of subordinated man's necessary projection of himself into his future. This latter characteristic of the subjugated man of labor is either lacking or, more realistically, vastly more attenuated in the case of the despot and his caste.[41]

The foregoing immediately touches upon Bataille's conception of the subject. He conceives of the subject at its purest as precisely such an evanescence in and of the moment (that of the moment's evanescence)—as opposed to the object as proper to and of the order of the profane world of labor. That is, the object is of the prosaic realm of enduring self-identity of the thing, ensured by the prolongation into the temporal continuum of duration of that which would otherwise have been a mere event. It is for this reason that, as regards the unlimited and ruinous consumption of the royal caste, Bataille contends that "when the *present time* is operative, even if I speak of *objects,* of the consumed products on which it bears, those objects are *destroyed,* consumed, and the preference granted to the *moment* corresponds to contempt for the objective world."[42]

Though contempt for the servile and prosaic world of labor is the case here, one would not be remiss to contend that the said prosaic world is the mere obverse of the more affirmative and positively enjoined *raison d'être* of ruinous consumption: that profitless consumption aims at a return to the intimacy of the world anterior to that of utility, measure, labor, and the general poverty of things. Affirming that ruinous consumption is not merely the province of the subject but is indeed the subject *as such*, Bataille concludes wonderfully: "The world of *intimacy* is as antithetical to the real world as immoderation is to moderation, madness to reason, drunkenness to lucidity. There is moderation only in the object, reason only in the identity of the body with itself, lucidity only in the distinct knowledge of objects. The world of the subject is the night: that changeable, infinitely suspect night which, in the sleep of reason, *produces monsters.*"[43] Such is the milieu of the Bataillean sovereign (*souverain*).

The ruinous sacrifice of the surplus by the despotic caste also (and fairly often quite paradigmatically) assumes the form of *sacrifice* to the divine. Herein, it reveals itself as the foremost *raison d'être,* inextricable adjunct, and perhaps even *sine qua non*, of the despot-deity filiation (direct filiation) and beyond this the despot-people alliance (new alliance) as such. It must also not escape purview that this ruinous consumption of the subaltern's surplus, as propitious sacrifice to the deity, aims at the fecundity of the land and the fluid regeneration of the regime (often at once considered as the repetition of an entire cosmogony).

(iii) The Vedic aśvamedha

In Vedic India, for instance, ruinuos consumption of the laboring populace's surplus by the despotic caste assumed *inter alia* the form of the horse sacrifice, or *aśvamedha*. This ritual could only be performed by a king who had attained military victory. Though preliminary ceremonies occurred over the expanse of a year, the core ritual took place over three days: numerous domestic animals being slaughtered on the second day, whilst the horse itself suffered death by suffocation on the third day—after which the principle wife of the monarch lay down beside it, simulating carnal union. The dead animal (together with other victims) was hereupon cut up. Later in the same day, the attendant priests received their "four queens."[44]

We observe here, apart from the principle overt aim of the *aśvamedha* being the re-establishment of the obduracy and permanence of the regime as that of its constitutive social classes in their paradigmatic rigid apartness, that after being killed, the stallion incarnates the god *Prajāpati* who himself is ready to sacrifice himself.[45] This latter exemplifies our earlier point concerning the inherence of the despot-deity filiation in the phenomenon of the despotic caste's dissipation of surplus via religious sacrifice.

(iv) Monotheism on the horizon

There is, however, one particular upshot of the despot-deity filiation (direct filiation)—particularly as this latter appears in ramified conjunction with the despotic regime's infinitization of debt—that is of epochal consequence. That is: the infinitude of the debt owed to the despot, coupled with his direct filiation with the deity, inexorably implies that there is always a monotheism lingering on the horizon of despotism: the debt becomes a *debt of existence*, a debt of the existence of the subject himself.[46] In this connection, Nietzsche renders privy what he observes as the directly proportional and one-to-one relation between the movement within civilization toward universal empires and that straining toward universal deities, the indebtedness to which has burgeoned across human history in lockstep with the ever clearer and increasingly lucid and monumental conception of God and man's feeling for the divine—from which it may be adduced that despotism necessarily prepares the way for a monotheism of some kind. It is not long hereafter, as Nietzsche fails not to notice, that one reaches the point at which a God emerges who punishes the very existence of human beings in accordance with the injunctions of a vast and obscene array of perverse plenipotentiaries—plenipotentiaries who, so as to fully entrench the ironclad and implacable inexorability of the infinite debt of existence itself owed to the despot's hauntingly absent and dread deity, project an eternal and unremitting wheel of an afterlife in which the henceforth immortal soul of the cowering subject shall forever writhe under the full inquisitorial brunt of this monstrous punishment machine that has now come to encompass the entire universe within the ambit of its infernal machinery of torture. Good and evil, in a sempiternal fixity ensured by the pulverizing force of superegoic refrains veering well past the point of sadomasochism, become the ineluctable pylons crucifying the subject onto the cross of a permanent mutilation terrifyingly coincident with eternal damnation and slow universal suicide prior to actual decease. The entire world is now polluted with the promiscuous concept of inevitable punishment, under which life is defamed, slandered, execrated, voided of all breath, and converted into a foul and pestilential contraption for the sempiternal maceration and disembowelment of all spirit.[47] But more of this later.

5. The full body of the despot as *socius*

As part of the very same process of overcoding facilitated by the hero's double incest and coextensive with the establishment of the new alliance between the despot and his subjects and likewise of the direct filiation forged between the despot and his deity, there comes to pass the upsurge of a new socius, a new full body: that of the despot himself. Whereas the primitive social machine was entirely immanent; facilitated as it is by the wholly abstract actualization of the purely virtual Urstaat, that of the despot comprises a transcendent unity appropriating all productive forces of human labor to itself.[48]

The despotic regime, in fact, subjugates not one but an entire plethora of primitive regimes, and subsumes these "autochthonous" rural communities as its "cogs," "wheels" and "working parts," overcoding and "bricking over" their intrinsic coding with its own inscription so as to extract its requisite share of surplus value.[49]

The despotic State apparatus co-opts and integrates the primitive communities as working parts and productive organs, and falls back upon them by ushering in conditions for the prosecution of major Statist work projects that transcend the closed and parochial ambit and material capacities of these semi-autonomous territorialities.[50] In this sense, the despotic State may be considered a vertically transcendent unity that collates "relatively isolated subaggregates, functioning separatedly, to which it assigns a development in bricks and a labor of construction by fragments."[51]

Analogously to the manner in which—at the level of desiring-production—the BwO as miraculating machine attracts and attaches its organs-partial objects to itself, and likewise analogously to the way in which working human organs are "hewn" into the earth as primitive socius; in the despotic regime, all the organs of the subjugated subjects—including even rather intimate body parts—are attached to the despotic full body, "as though to a peacock's tail of a royal train"—an "intense multiplication of organs and their corresponding inscription on the new full body."[52]

As observed in earlier chapters, the organs-partial objects of desiring-production appeared to be miraculated by the full BwO; that is, seemed *ab initio* to arise from it. Likewise, as earlier espied, the working organs of the primitive machine appeared as the natural effluence of the full body of the earth as socius. Something fundamentally analogous is afoot in the case of the despotic regime. Here, though it is the case that the human working organs are the actual agents of production—whose labor sustains the regime at the agricultural and infrastructural levels and is moreover the sole source of surplus value—all of this appears for all practical purposes to be the natural effluence of the despot. "*He* is the sole quasi cause, the course and fountainhead and estuary of the apparent object movement."[53]

(i) The Vedic king

The latter mystification—and all the gargantuan ideological ramifications that follow as its corollaries—may be glimpsed in the elaborate ritual comprising the consecration of the Vedic king, the *rājasūja*. This multi-phase rite successively brings about the future monarch's regression to an embryonic state, followed by a year long gestation period, after which he is at once mystically reborn as cosmocrator and thus incarnates the entire cosmos. Subsequently, it is revealed that all the vicissitudes of State—and indeed of the cosmos—are adjuncts of the installed Vedic sovereign. Thus, when "the sovereign raises his arm, the gesture has a cosmogonic meaning: it symbolizes the erection of the *axis mundi*. When the king is anointed, he stands on the throne with his arms raised: he incarnates the cosmic axis fixed in the navel of the earth—that is, the throne, the 'centre of the world'—and touches heaven. The aspersion [anointing with water] is connected with the primordial waters that descend from heaven along the *axis mundi*—represented by the king—to fertilize the earth."[54]

(ii) The Egyptian pharaoh

The same is the case with the Egyptian pharaoh, who also functions as quasi-cause. The vizier Rekh-mi-Re long ago averred that the pharaoh is a god who endows all Egyptian subjects with life and movement. The solemn rites performed by the pharaoh vouchsafe the stability not only of the State but also of the cosmos itself. Furthermore, as Eliade has it, the "pharaoh's political activity repeats Re's exploit: he too 'repels' Apophis [the serpent who attempts to derail the sun's rising], in other words he sees to it that the world does not return to chaos."[55] "At each New Year the cosmogonoy was reiterated, and even more paradigmatically than Re's daily victory, for a vaster temporal cycle was involved."[56]

(iii) Plowden's Reports

The foregoing being as it may, Kantoriwicz finds a more intricate and thorough delineation of the full body of the despot (or at least something akin to it)—and what the latter could moreover possibility entail—in Edmund Plowden's *Reports* on Tudor legal theorists' characterizations of the king's two bodies. Obviously, the despot's actual human body is to be distinguished from the full body of the despot as the *socius* of the despotic regime. In the various formulations made by Elizabethan lawyers in Plowden's text, a similar distinction is made between the king's mortal body and the king's body conceived as the body politic of his kingdom.

To posthaste assimilate the king's body as body politic directly to the despotic full body as socius would, however, be crude and erroneous—especially in that, whereas the despot's full body is a thoroughgoing materiality (the social counterpart of the numinous BwO of desiring-production), the king's body as body politic is a legal and theological fiction. Notwithstanding, before we proceed to establish the precise relation between the legal fiction and the materiality, it is encumbent to see exactly what Plowden's text says about the king's two bodies. This is as follows. The king has, so to speak, within himself two bodies: a body natural and a body politic. Concerning the king's latter body: "his Body politic is a Body that cannot be seen or handled, consisting of Policy and Government, and constituted for the Direction of the People, and the Management of the public weal, and this Body is utterly void of Infancy, and old Age, and other natural Defects and Imbecilities, which the Body natural is subject to, and for this Cause, what the King does in his Body politic, cannot be invalidated or frustrated by any Disability in his natural Body."[57]

However, as Kantorowicz avers, the king's two bodies—though distinguished from one another in the said manner—form an indivisible unity wherein each is contained in the other. Indeed, the Tudor lawyers have it that the king "has not a Body natural distinct from and divided by itself from the Office and Dignity royal, but a Body natural and a Body politic together indivisible; and these two Bodies are incorporated in One Person, and make one Body and not divers, that is the Body corporate in the Body natural, *et e contra* the Body natural in the Body corporate."[58] Through its conjunction in indissoluble union with his body politic, the inherent moral and physical imperfections and fragilities of the king's body natural are, moreover, to a degree etiolated—or even voided. Again, the Elizabethan laywers: "His Body politic, which is annexed to his Body natural, takes away the Imbecility of his Body natural, and draws the Body natural, which is the lesser, and all the Effects thereof to itself, which is the greater, *quia magis dignum ad se minus dignum.*"[59] Powden's *Reports* now go on to state that "when the Body politic of [the] King of this Realm is conjoined to the Body natural, and one Body is made of them both, the Degree of the Body natural, and of the things possessed in that Capacity, is thereby altered, and the Effects thereof are changed by its Union with the other Body, and don't remain in their former Degree, but partake of the Effects of the Body politic ..."[60] This we must read in tandem with the later assertion which states that the members of the king's body as body politic are his subjects.[61]

The upshot here is that the natural bodies of the subjects, being members of the king's body politic, are (or at least *their* parts are) its operating parts, its working organs. On the other hand, insofar as anything that enters into conjunction with the king's body politic partakes of the latter's degree and effects (and one may add *a*ffects), to the extent that (assuming here the limit case) that which at first seems to merely partake of the body politic is in actuality from the first miraculated by the king's body politic. Thus, the king's members (his subjects) are in this fashion miraculated by the king's body politic. The king's body politic therefore functions as their quasi-cause, and herein they in fact *are* the said body politic's degrees and effects, immediate powers, working parts.

So, the king's body politic, as conceived of by Plodwen, is seen here to converge with the Deleuzoguattarian notion of the full body of the despot. More precisely, the convergence is between the king's body as body politic and the full despotic body *considered as quasi-cause of its working organs and their productive activity.* The king's body as body politic, as *corpus mysticum,* remains here nonetheless strictly a fiction, though crucially a *real* fiction. This is so, not least of all because the Deleuzoguattarian despotic full body *as quasi-cause* is itself likewise a real fiction. For, as has been maintained earlier, the productive labor of the organs partial-objects is the latters' own and not that of the despotic full body. In this sense the full despotic body as quasi-case and the king's body as body politic possess the structure of the *fetish*. We will return to Kantorowicz later to see how this fetish is undone.

6. The despotic State-formation

(i) A hierarchical State formation

What results from the lightening-flash ursurpation by the blond artists is a completely new social arrangement. This new social machine, the State, is a "functional pyramid" with the despot at its apex as its "immobile motor," the bureaucratic apparatus as its "lateral surface and its transmission gear" replacing the former network of inter-tribal alliances, and the collection of parochial village-dwelling peasants as the working parts or partial objects located at its base.[62]

Ellen Meiksins Wood considers the Chinese imperial State as the paradigmatic exemplar of the pattern of State-formation as existed in more preponderant civilizations of the pre-capitalist world: "a bureaucratic hierarchy descending from a monarch to administrative districts governed by royal functionaries and fiscal officials, who extracted surplus labor from subject villages of peasant producers for redistribution up the hierarchical chain."[63] Wood, however, avers that fairly kindred State structures existed in Bronze Age Greece, New Kingdom Egypt, and even much further a field in the cases of the empire of the Incas and in absolutist France.[64]

(ii) Transmogrification of the three syntheses, from geodesy to geometry

All of this obviously implies a new position and functioning of the three syntheses. The connective synthesis of the primitive machine, being that of lateral alliance cutting across extensive filiation, is now a connective synthesis of the old savage alliances with the new alliance through which the despot assumes ownership of all property; the disjunctive synthesis of primitive filiation transforms into a disjunctive synthesis that involves the extension of the old savage filiations of lineage and prestige into the direct filiation of despot and deity, through which the despot co-opts all power and stands over the masses; whilst the old conjunctive synthesis, formerly that of the extraction of a voluptuous surplus value of code by the eye contemplating the suffering entailed in the initiation of man into alliance and the debtor-creditor relation, now becomes that of the appropriation of surplus value created by the agrarian communities into the bottomless coffers of the despot and his ruling caste.[65] These alterations, moreover, usher in a fundamental change in the newfangled regime's relation to the earth. The primitive regime treated the earth as, first and foremost, matter for the inscription of phylogenetic extended lineages. This inscription would serve to constitute a sort of "geodesy." The movement toward the despotic regime is, on the other hand, at once one from geodesic space to astronomical and geometric space. Herewith, the earth ceases to be co-opted and deployed in conjunction with lineage. Instead, it is now an object of appropriation for the State apparatus, in which process lineages and land find themselves subsumed within the greater transcendent arc of a astronomical space and geometric extension which overcodes them.[66]

What precisely was to be done with the preexisting lineages upon which this despotic structure erected itself has been one of the main preoccupations of many a State apparatus in history. Chief among the potentially noisome tendencies of overly resilient lineages, in the form of kinship systems or extended family networks, has been their predilection to erode the foundations of despotic rule. State apparatuses have therefore tended to systematically fragment, break up, or co-opt these latter.

(iii) Denigration of lineage 1: Chinese Legalism, Mao Zedong

As Francis Fukuyama relates, early Chinese State structures, such as that of the Qin and the former and later Han, sought to accomplish this task through the deployment of an explicitly antifamilistic political ideology. Here, Chinese despots strove to forcefully undermine the power of entrenched families and kinship groups so as to create and consolidate the hegemony of a permanent class of trained bureaucrats and administrators. In accordance with Shang Yang's doctrine of Legalism, kinship networks were seen as a formidable impediment to Statist power. They were, thus, replaced en masse so as to tie individuals as directly as possible to the State. In a newfangled family registration system, Shang Yang divided the sprawling kin networks of traditional China into groups of five and ten households, which were moreover forced to mutually supervise and even spy and snitch on each other. A modified version of this system would return to the fray during the Ming Dynasty in the form of the bao-jia system. Indeed, all

Chinese State apparatuses operating in concord with the ideology of Legalism sought to inexorably tether citizens to the State by breaking asunder the bonds of familial moral obligation in all its forms. This strategy of eviscerating the anachronistic remnants and hold-overs of primitive lineage recrudesced in striking form in the social engineering advanced by the Chinese Communist Party after 1949. Mao Zedong, whose ideological sympathies lay unswervingly on the side of Legalism and anti-Confucianism, launched intensive and singularly brutal campaigns specifically aimed at fragmenting extended families, overcoding these sprawling networks with the new heavy moral aggregates of party, State and commune.[67]

(iv) Denigration of lineage 2: Catholicism, widowhood, spinsterhood, celibacy

Not so much for theological reasons, but rather so as to reinforce the reterritorializing sway of its Statist overcoding; since at least the time of Pope Gregory I, the Catholic church would seek to undermine the so-termed "strategies of heirship" formerly deployed by vast networks of kinship groups across Europe in order to maintain property within their keep across generations. To this end, the Church proceeded to undermine cross-cousin marriage (which kept property in the hands of close family members), the levirate (where the marriage of a deceased husband's wife to his brother consolidated his property with that of his siblings), adoption of children, concubinage, and divorce—hereby cutting off all available avenues through which families could pass property onto their descendants. Through what turned out to be an effective demise of unilineal descent, the purportedly selfless extension of property rights to women, and the eliciting of donations from Christians who died without heirs—such as childless widows and spinsters—the Church was able to subsume the landed property of a vast patchwork of lineages it had itself brought to well-nigh terminal obsolescence. Through these stratagems, the kinship structures of German, Norse, Magyar and Slavic tribes dissolved within a mere two to three generations of their conversion to Christianity. This all enabled the Church to accede to the position of a formidable Statist power, replete with a firmly entrenched bureaucratic hierarchy of priests, bishops, abbots, priors, and other offices.[68] Finally, with the imposition of celibacy on priests under Pope Gregory VII, in that this move removed all patrimonial awarding of benefices to kinsmen and descendants and extricated the priesthood from the micro-political machinations of localized clans, the Church was enabled to exert direct jurisdiction over territories in the manner of a despotic State apparatus.[69]

(v) Denigration of lineage 3: Plato, Ottomanism, and military slavery

In Plato's *Republic*, Socrates stridently advocates the necessity of the State's suppression of lineage and kinship relations. Here, it is specifically the class of military guardians of the ideal State, so as to ensure their unswerving loyalty to the latter and its material and spiritual security, that must be held at a remove from all familial bonds. The guardians are strictly forbidden to marry and are instead to procreate with multiple intermittent partners. Their offspring are then spirited off and reared in common by the State. The children of these sporadic carnal unions are told the noble lie that they are "children of the earth," and not products of biological parents.[70]

One would plausibly imagine that such a ridiculously impracticable notion would never accede to actuality at any point in history. Yet, something similar bestirred itself under the Islamic Abbasid caliphate and reached its apogee in the Ottoman Empire. In its assertion of centralized Statist hegemony, Islam had drawn military strength for its blistering geographical expansion from the underlying tribal networks upon which it had erected itself, yet was at the same time threatened by the fissiparous tendencies of the latter's innate nepotism and parochial fickleness. The Prophet's teachings would undermine tribal loyalty, overlaying kinship networks with the universal *umma* and direct loyalty to God. The early caliphs overcoded tribal affiliations and alliances with military units of tens and hundreds transcendent of tribal bounds. After its accession to power, the Abassid caliphate eviscerated the old Umayyad lineages which it had usurped.

The Ottoman strategy of denigrating lineage was particularly ingenious. At its height during the middle of the sixteenth century, the Ottoman Empire would annually snatch three-thousand of the most physically and mentally pristine Christian boys between the ages of twelve and twenty from its provinces in the Balkans. This was the *devshirme*, or levy of Christian boys. The top ten percent were intensively trained as senior administrators in the palaces of Istanbul and Edirne; others were recruited for the sultan's elite Janissary infantry corps; with the remainder filling high-ranking positions in the household cavalry, the sapahis of the Porte. These slave-soldiers could ascend right up the ranks to become generals, senior officials, provincial governors, and even grand vizier. However, so as to undercut the potentially anti-Statist machinations of the familial networks from which they had arisen and others that they could potentially forge, neither their offices nor their land were ever their own private property and thus could not be passed down to relatives. To further undermine lineage, they were forced into permanent celibacy and could be executed upon the slightest whim. Indeed, throughout their entire existence, these slave-soldiers and asexual bureaucrats were kept in strict isolation from the rest of society. This brilliant, yet morally deplorable, institution strongly enforced Muslim States' overcoding of the inherently fractious and potentially seditious tribal networks upon which their despotic regimes rested.[71]

7. The despotic semiotic

(i) The loss of independence between voice and graphism

A fundamental change occurs in the mode of inscription or semiotic as we pass from the primitive to the despotic regime. This upheaval concerns the transmogrification in the relation between voice and graphism at the basis of the transition from an essentially oral society to one characterized by the preponderance of writing. As seen, in the primitive regime, the polyvocal graphism inscribed by the scarifying hand upon the suffering body being ensconced into alliance was independent of the sensuously intoning voice. In the despotic regime, this independence largely disappears: the graphy loses its polyvocality, aligns itself upon the voice, and in so doing subordinates itself to the latter.[72] Paradoxically, however, through this subordination of itself to the voice, graphism henceforth comes to usurp and spirit the voice away, as the latter becomes denatured and hauntingly absent—as Derrida avers is the case for all writing in the Western tradition.[73]

(ii) From polyvocal graphism to linearized writing

This all implies that, through its alignment with the voice, the hitherto polyvocal graphism is converted into writing in the quotidian sense.[74] More precisely, as Deleuze and Guattari maintain, graphism's loss of autonomy from, alignment upon, and subordination to the voice, enables it to "extract" from the voice a "deterritorialized abstract flux" that it retains and makes "reverberate" in the linear code of writing.[75]

With this having come to pass, graphism, in entering into dependence with the voice and becoming writing, induces a "mute" voice from "on high"—or more precisely, a voice no longer pertaining to primitive alliance, but one of the new alliance between the despot and his subjects. This is a "fictitious" voice from a transcendent "beyond" that expresses itself in the linearized flows of writing, and does so in terms of the direct filiation between the incorrigibly absent yet ubiquitous despot and his still more hauntingly absent and ubiquitous deity—both of whom only speak to man through bureaucratic decrees and scripture.[76] We may here, by extension, infer that this denaturing and spiriting away of the "divine" voice likewise infects the subjects milling under the towering aegis of the despot and his deity. Paraphrasing Derrida apropos of Antonin Artaud's critique of the classical Western theater of alienation, what the semiotic transmogrification effected by the despotic regime ushers in at the behest of linearized writing is the *disincarnation* of the human voice, the purloining of the subject's breath (*souffle*), which also dispossesses him of his own body and mind, and indeed his entire existence—as his speech is no longer his breath, his body no longer his gestures, and his gestures no longer his life. The purloiner here, for Artaud, is quite simply God, or the "great Furtive One," or what for us is the despot's deity or Lacanian theo-political "big Other," who has stolen the subject's body as early as the moment of his birth whilst at the same time stealing this very birth from the now self-estranged self—to the extent that his birth has from the very outset smelt of dying. The despot's deity, as the "great invisible Other," effects this deprivation of the subject of the despotic regime of his speech and nature, and the bifurcation between the subject and the subject's own death, by furtively getting in his own silent enunciation prior to the subject being able to speak, hereby taking possession of the subject's own innateness. Relating to the maternal orifice from whence the subject had been born, the usurpation by the despot's deity of the subject's birth and divinity effecting his absolutely originary self-alienation is at once the induction of the subject into his own terminally ineluctable sense of being intrinsically obscene, sullied, of incorrigibly wallowing in his own filth. The prenatal theft by the despot's deity of the nascent subject's own ur-value and its consequent transmogrification into false dematerialized value, in that it recurs incessantly every time the subject cedes a part of himself in the form of excrement, makes of this spurious deity a scato-theological entity and its history identical to the history of defecation and this history of defecation a genealogy of all human surplus value stolen by the divinely excrementitious despotic State apparatus. Only appearing through the metaphorical and metonymic translations of linearized writing, this dematerialized Statist deity is, through the interpretive drift of his bureaucratic perverts, a pseudonymous, usurping counterfeit and counterfeiter of the existential coinage born of the discursively wrought prenatal diremption of the cowering subject.[77]

(iii) The biunivocalization of the sign

This absented deity is the effect of a "flattening" of the polyvocal graphism of the primitive regime onto the now purloined voice which causes a transcendent object to jump outside the signifying chain—a mute voice upon which the entire chain henceforth appears to depend, and in relation to which it comes to be linearized.[78] This detached object assumes the role of a signifier which overcodes the entire territorial signifying chain via its two-fold pervasion throughout this latter: firstly, it functions as the master signifier from whence the entire chain derives; secondly, it insinuates itself within the incipient chain in the guise of the biunivocal correlation of each signifier with its signified. That is to say: this proliferation of the master signifier throughout the territorial chain of the now-usurped and superceded primitive machine is that which transmogrifies the existing polyvocal connections of the transcursive motley of asignifying signs of the said primitive machine into a linearized signifying chain in which the former multiply-connective arrangement yields to a biunivocal alignment of signs.[79]

(iv) The endless deferral of the signified

The pervasion of the master signifier throughout the biunivocalized signifying chain, moreover, serves to evacuate all forms of content. The immediate upshot of the latter is that the signified is indefinitely deferred. Each signified thereby becomes merely another signifier. As Deleuze and Guattari have it—in that every sign now refers solely to another sign, and so forth ad infinitum—one may ultimately forego the very notion of the sign. For what remains is not fundamentally the sign's relation to the state of things it purportedly designates, or to an entity that it signifies, but only the purely formal and unidirectional relation between signs strung together into a virtually interminable signifying chain. The sign finds itself replaced by the limitlessness of "signifiance."[80]

The massive movement of relative deterritorialization which clears the ground for the erection of the despotic regime and the full paraphernalia of its State apparatus is at once that which deterritorializes the primitive polyvocal sign, in the process giving birth to the signifier. The signifier is nothing other than the redundancy of the sign with itself. At the same time, it may be conceived as the incessant movement of referral from sign to sign in a limitless network sans beginning or end that projects its "shadow" onto an "amorphous atmospheric continuum."[81]

It is precisely this amorphous atmospheric continuum that is all that henceforth remains of the signified under the pressure of the master signifier's overcoding of the entire now-linearized signifying chain. More succintly put: the newfangled amorphous atmospheric continuum has assumed the "role" of the signified. In these terms, the ceaseless deferral of the signified observed earlier is homologous to the fact that the amorphous continuum continually "glides beneath" the signifier, for which it now serves only as a "medium" or "wall" in which all forms of content find themselves dissolved. Deleuze and Guattari refer to this erosive process as the "atmospherization or mundanization of contents," a withering away and abstraction of content; the corollary of which is the coextensive emergence and perdurance of a world that "begins to signify before anyone knows *what* it signifies," that is to say, a world in which the signified is *given without being known*.[82]

(v) Faciality

As observed, the purely formal redundancy of the signifier with itself at once spells the erosion of all content. Despite this, the signifier must nonetheless have a certain *substance* of expression. This substance is the "face" of the despot. The face "emits" and "receives," "releases" and "recaptures" signifying signs. It is the "body" of "the center of signifiance" to which all deterritorialized signs affix themselves. It likewise marks the limit of their deterritorialization. In relation to the signifier, whose pure formality and nullity of content is the effect of deterritorialization, the face is that through which a compensatory reterritorialization is wrought. The face is the "Icon" proper to the despotic regime, and functions as the reterritorialization "internal" to the system. It is the face on which the signifier reterritorializes itself. Conversely put, the signifier is always "facialized," and "faciality" "reigns materially" over the entire constellation of signifiances and interpretations.[83] The authors aver furthermore that the despot's face is a mask that he makes for himself, a mask which—far from obscuring his face—is in fact identical with it. This mask-face, in fact, turns out to be the despot's whole body. The despot or his deity "brandishes the solar face that is his entire body, as the body of the signifier."[84]

8. Despotic paranoia

Through double incest, the abolition of debts, the circulation of credits, the rendering infinite of the debt, the vast and incessant accumulation of filiative stocks, and so forth, the well-nigh stratospheric despot lords over a populace milling far below. However, the artist-hero—though having so vertiginously risen to the despotic heights of an unprecedentedly elevated and domineering position—does nonetheless fall into a quagmire of his own. For, owing to his ubiquitous appropriation of all productive forces (to the point of their being henceforth parts or effluences of his own numinous body), the despot—via a curiously wrought reversal—suddenly finds himself confronted with the perpetual and well-nigh maddening threat that any single organ might flow outside his full body and thereby break away and escape. "*Suddenly the despot sees rising up before him, against him, the enemy that brings death—an eye with too steady a look, a mouth with too unfamiliar a smile; each organ is a possible protest* [my emphasis]."[85] This interminable vigilance that foists itself upon the now verily beleaguered despot—in spite of his overarching power—shall henceforth render his subjective position intrinsically *paranoid.* The invading bronze-eyed artist—hitherto oblivious to and unencumbered by the gnawing worm of overburgeoning conscience, agenbite of inwit, or suchlike progeny of the swamp[86]—through a most dramatic involution, now finds himself sempiternally assailed from all sides by the endless proliferation of working parts of the State machine: parts that are for all intents his "own" numinous flesh but which, precisely on this count, he can nowise allow to stray without loosing his very sanity. "*Better not a sole survivor than for a single organ to flow outside this apparatus or slip away from the body of the despot* [my emphasis]."[87] The despot thus becomes a killer, a mass murderer on a potentially genocidal scale.

The otherwise caustic Ambrose Bierce is therefore a little too optimistic in maintaining that a dictator is "The chief of a nation that prefers the pestilence of despotism to the plague of anarchy."[88] Regardless of this, the inherently paranoiac position of the despot—that is, despotic paranoia—may also be explained from the vantage point of the despotic semiotic itself. As already observed, the erosion of forms of content coextensive with the incessant deferral of the signified brings into being a world in which the signified is given without being known. It is felt that such a world signifies a great deal. But one does not know what precisely is being signified. This is the crux. The foregoing is strongly related to the despot's paranoia in the sense that the paranoiac shares this "impotence of the deterritorialized sign" which assails him from each and every direction in the "gliding atmosphere."[89] Born of the evanescing atmosphere, this feeling of impotence cannot but have its origin in the above mentioned conjunction of the signified's simultaneous givenness and unknowability. Moreover, this latter conjunction is itself the corollary of the dissolution of all forms of content attendant upon the signified's incessant deferral. This latter deferral—furthermore, and as afore espied—is in turn the immediate consequence of the master signifier's overcoding of the entire transcursive signifying chain of the now-vanquished primitive regime. As observed, the master signifier emanates from the voice of the despot silently resounding from on high. One may therefore conclude that the despot's constantly being besieged by the gliding atmosphere of the dissolved signified, since it is the immediate corollary of the master signifier's overcoding, in fact only serves to give him still better access to the "superpower" of the signifier. Deleuze and Guattari refer here to the "royal feeling of wrath" of the despot as "master" of the semiotic network which spreads through the atmospheric continuum. He suspects that he is constantly being attacked and made to suffer, but is utterly unabashed in his confidence that he knows exactly what his assailants are plotting—and is moreover convinced that he is at least one step ahead of them, that he has always known their subterfuges, that he has power even in his own utter impotence, and—finally—that he will certainly bring them under the full physical brunt of his power.[90]

Thus far, we have seen despotic paranoia to have arisen from two factors. On the one hand, it is a corollary of the despot's exclusive ownership of all property and his unilateral numinous assimilation of all working organs. From the vantage point of the despotic full body, the working organs of the entire peasant base are the despot's own whilst their labor appears as a mere intrinsic effluence of the said full body itself. On the other hand, despotic paranoia is the upshot of the master signifier's abstraction of all contents as conjoined with the ubiquitous givenness of the unknown signified. As observed, this second factor consists essentially in the incessant deferral of the signified along a signifying chain of endless referral. In that it is interminable, such a self-referential signifying chain may be described as circular.

In turns out, however, that there is far more than just one such circle. Indeed, there necessarily exists a clear distinction between circles since, though all signs refer to all others only to the extent that they are all deterritorialized, and are thus oriented toward the same center of signifiance and distributed across the regime's gliding amorphous continuum—they do notwithstanding have different *speeds* of deterritorialization. These different degrees of deterritorialization refer to different geographical places of origin of signs within the regime. Also, the signs have differential relations which maintain the distinction between circles or which constitute "thresholds" in the atmospheric continuum.[91] But there is more. For, in what is here an effective multiplicity of signifying chains arrayed in ever expanding concentric circles, one finds that each sign refers not solely to other signs within the same circle, but likewise to the signs in other circles. It is precisely here that one may espy the proper locus of paranoia's residence. For as Deleuze and Guattari argue, it is owing to the mutual "resonance" and taut interconnectedness of circles that something like a social disturbance, or even a very localized domestic incident immediately calls into question the entire system of the regime in its integrity, which regime's various levels are united by multiple correspondences.[92] This means that even the most trifling and local fluctuation within the circularly irradiating network of the despotic regime comes to reverberate throughout the entire despotic full body.

(i) Canetti and the poetics of paranoia

The latter is likewise why the despot must strive for omniscience. As Elias Canetti has it, the despot in his paranoiac omniscience "creates empty space all round him which he can survey," and moreover "observes and assesses every sign of approaching danger."[93] Indeed, he "does this on all sides, for he knows that he is dealing with many who may simulataneously advance against him," and "has eyes all round him"—implying that "not the slightest sound must escape his attention, for it might conceal a hostile intent."[94] The despot thus partakes of the salient characteristics of the first of the two types of persecution maniacs found in insane asylums, as delineated by Sartre. As the first of these two maniacs, the despot sees himself as a perpetual victim of conspiracy. He feels that the entire gamut of the world is secretly engaged in trying to annihilate him and the full body of the despotic State apparatus. Even passers-by are spies, provocateurs. All people are hard-bent on dishonoring and assassinating him.[95]

The immediate corollary of this is the paranoiac despot's "continual urge to unmask enemies. These the paranoiac sees everywhere, in the most peaceful and harmless disguises: he has a gift of seeing through appearances and knows exactly what is behind them."[96] In Mario Vargas Llosa's *The Feast of the Goat*, the gaze of Trujillo, tyrannical dictator of the Dominican Republic, is described as unendurable by anyone who did not lower his eyes to evade it. One was inexorably "intimidated and annihilated by the force radiating from those piercing eyes that seemed to read one's most secret thoughts and most hidden desires and appetites, and made people feel naked."[97] This gaze made him upon whom it fell feel as if it was dredging up his very conscience.[98]

The upshot of this the despot's constant and untiring surveillance and obsession with unmasking purported foes is that "the whole world shrinks. The wealth of appearance comes to mean nothing; all variety is suspect. All the leaves on all the trees are the same, and all dry as dust; every ray of light is extinguished in a night of suspicion."[99] This constant and obsessive vigilance is an inexorable adjunct and perpetual corollary of his inherently paranoiac position, which invariably prefers the death of a thousand men to the possibility of the flight of but a single organ from the despotic full body. In Solzhenitsyn's *The First Circle*, whenever the minister of State security Viktor Abakumov was to appear before Stalin, his ears would first turn ice-cold and then flush with burning heat. His terror would increase exponentially with the thought that the redness of his ears might arose the Autocrat's suspicion. For the slightest thing would arouse Stalin's suspicion—even if anyone in his presence put his hand into his inside pocket.[100] Indeed, when Stalin looked at Abakumov, as when countenanced with other zealous and fawning subordinates, his first thought was whether he could trust him. His second thought was whether the time was ripe to execute him. He was suspicious of Abakumov even though he could read him like a book. Suspiciousness was, in fact, Josif Djugashvili's dominant character trait. He was suspicious of his own Party members and his comrades in exile, the peasants, the workers, the intelligensia, and the soldiers and generals.[101]

As already observed, this absurd paranoia of the apex predator makes of him a killer. Canetti reasons that no ruler "can be permanently certain of the obedience of his subjects. As long as they allow themselves to be killed by him he can sleep in peace, but as soon as anyone evades his sentence he is endangered."[102] Indeed, he "can only calm his fears by making an example of someone. He will order an execution for its own sake, the victim's guilt being almost irrelevant. He *needs* executions from time to time and, the more his fears increase, the more he needs them."[103] In this connection, Solzhenitsyn avers of Stalin that his entire career was founded on the firm conviction that punishments such as demotion, persecution, incarceration in a lunatic asylum, life imprisonment, or deportation were simply inadequate. The only feasible punishment was death. All lesser penalties did not exist for him. And when Stalin's lower eye-lids would pucker, the official before him knew that his verdict was death.[104]

Canetti goes on to contend of the great paranoiac of the despotic regime that "from every execution for which he is responsible, some strength accrues to him. It is the strength of *survival* which he gains from it. His victims need not actually have challenged him, but might have, and his fear transforms them—perhaps only retrospectively—into enemies who have fought against him. He condemns them; they are struck down and he survives them. The right to pronounce sentence of death becomes in his hands a weapon like any other, only far more effective. Many barbarian and oriental rulers have set great store on this heaping up of victims round them, where they can actually see them all the time; but, even where custom has been against such accumulation, the thoughts of rulers have been busy with it."[105]

Such a consoling accumulation and stockpiling of the dead is readily discernible in Ngũgi wa Thiong'o's aforementioned novel. Here the second Ruler of the Free Republic of Aburĩria has a special chamber in the State House whose walls and ceiling are made of the skeletons of all the students, teachers, workers, and small farmers he had killed in various regions of the country, "the bodies of his victims falling down to his left and right like banana trunks."[106] From the walls and ceiling hung the skulls of his most hated enemies. Every morning the Ruler would bathe in the specially preserved blood of his demised enemies and, bestriding the chamber with staff and fly whisk, sedulously gaze upon each skull one by one whilst making "mocking gestures of triumphant contempt, at the dark holes and grinning teeth where once eyes and mouths had been."[107] Alternately guffawing at the skulls with derisive laughter and pointing at them menacingly with his staff whilst foaming at the mouth, he berated them for having tried to assail his sovereignty: "I plucked out your tongues and tore your lips to show you that a politician without a mouth is no politician at all."[108]

As Foucault maintains of the despotic regimes of the Classical age, any actual transgression of the Law by a subject, however minor, is viewed and felt as a direct and personal assault on the sovereign himself, since the Law represents the sovereign's own individual will. Through his infraction, the offending subject touches the very person of the despot. Such an infraction, in fact, represents an almost physical attack on the great paranoiac. In the face of this, a public execution is a ceremony which in fact serves to reconstitute the very sovereignty of the despotic full body gravely injured and threatened by a single subject's infraction. And this sovereignty restores its integrity only by manifesting itself through a public execution that is singularly spectacular. It must unequivocally render pellucid before all eyes that it is an absolutely invincible force. This brutal despotic show of force aims not so much to re-establish harmony or parity between the sovereign and the populace. Such an operation would be woefully insufficient. Rather, it aims to confirm and render palpable the gross dissymetry that should exist between the impotence of the criminal and all potentially recalcitrant subjects, on the one hand, and the purportedly absolute and irrevocable power of the sovereign State apparatus, on the other.[109] This over-reaction by the despotic State risides in its fundamentally paranoiac comportment, according to which every offence, however slight, is experienced and treated as a *crimen majestatis* and even the least criminal looms as a potential regicide.[110] Every form of disobedience constitutes an act of ostensible hostility to the despot, and is in principle even treated as a foreboding of civil war.[111]

One observes in Canetti another angle from which the despot necessarily becomes a mass murderer. He argues that the despot's position as one who issues commands, coupled with the fact that he receives orders from no one whomsoever, induces in the despot an *anxiety of command.* This is so because a command "which threatens death and then does not kill leaves the memory of the threat.

Some threats miss their target, but others find it and it is these which are never forgotten; anyone who has fled from a threat, or given in to it, will invariably revenge himself when the moment comes."[112] The despot, whose inherent role is to threaten, must therefore make as certain as possible that such an eventually does not and indeed cannot come to pass. He inevitably senses that "all those he has threatened with death, are still alive and still remember. He is always conscious of the danger he would be in if they all united against him," and this is what makes of him a potential or actual homicidal madman who would prefer that a thousand men be put to death at his behest rather than see a single organ flee the purview of the despotic full body.[113]

(ii) Semelin and the politics of paranoia

The imaginary constructs for social destructiveness which Jacques Semelin sees as the *terminus a quo* of massacres, purges, or genocidal operations, may be seen as pertaining to despotic paranoia in general. For they are born of paranoia. One of Semelin's points of departure is the despotic State's transformation of a general and vague atmosphere of anxiety into an intense and concrete fear felt in the face of an (often imaginary) enemy that is depicted as a threat to the integrity of the entire body politic. It is here that the despot or despotic caste, impelled by paranoia, moves to rhetorically identify this heretofore amorphous enemy with particular figures or groups. The latter are denounced as a cancer metastatizing within the very bowels of the regime. These enemy figures are forthwith demonized as at least frightening and quite frequently diabolical. Through this transmogrification of what was hitherto merely insidious anxiety into fear concentrated upon a specific hostile internal enemy, the despotic caste foments hatred against the aforesaid malign "Other." This paranoid transformation of free-floating anxiety into hatred of individuals or groups seen as tubercular microbes threatening the obduracy of the socius, is what elicts the desire to annihilate the purported object of fear. The destruction of this malignant "Other" inhabiting and polluting the regime is, moreover, that by which the said regime—at the behest of its despotic caste—aims to regenerate itself and reassert its own supposed omnipotence.[114]

So, the paranoid despotic caste calumniates specific individuals or groups as constituting a malignant threat to the integrity of the socius. But there is more. For this demonization of the said evil-minded "Other"-in-excess inhabiting the very bowels of the State apparatus may well assume the form of a veritable *beastialization* of the same. As observed, the aforementioned mere denouncement of foes fabricated by the perverse rhetoric of the despotic caste was the prelude to their potential—or real—annihilation at the hands of roiling hatred running amock. Evidently, this will be even more so when calumny turns into animalization. Indeed, as Semelin has it, such animalization of maligned individuals or groups can likewise proceed in conjunction with the upsurge of murderous violence against these latter. For a massacre ensues precisely all the more readily when its prospective victims have been dehumanized—these subhumans whose vertiginous subterranean Deleuzoguattarian becomings-animal have exceeded the degree of deterritorialization of the signs consistutive of the despotic regime itself. That which has ceased to be human—or never was human to begin with—may be killed with impunity. (We have here shades of Giorgio Agamben's *homo sacer*—of which much more later.) And the dehumanization of the human is effected precisely through its bestialization.[115] "There is no doubt," avers Semelin, "that dehumanization indeed takes place via this animalization of the Other by placing him outside the realm of human relations. The killing starts with the use of words disqualifying his humanity."[116]

The foregoing is even echoed in the etymology of the very word "massacre." For ever since the Medieval period, "massacre" meant putting an animal to death. In French, a hunted boar's head exhibited in the main room of a château is called a "*massacre*." Within the context of the calumniating rhetoric of despotic paranoia, however, it is more specific species of animal that are conjured up; that is to say, domestic pests. Thus we have Nazi paranoiacs referring to Jews as "rats" or "lice," and Hutu paranoiacs maligning Tutsis as cockroaches ("*Inyenzi*").[117]

Yet, the paranoid sense of uncleanliness inherent to the despotic caste's animalization of the other implies that the maligned beast shall forthwith be degraded further to still lower organisms. Hence Hitler's "Jewish germ," or "Jewish cancer."[118] The paranoiac Lenin likewise degraded his purported enemies to a level below that of the beast: "The Russian land must be rid of its harmful insects"; "Either the lice will defeat socialism, or socialism will defeat the lice!"; the contention that "parasites (and typhus) are the main scourges assailing socialism"; or the smug declaration that "Russian cities have been cleansed of that whiteguard vermin."[119] Since they are impure, these parasites have to be eliminated.

The despot Pol Pot was likewise assailed by the paranoid delusion that Cambodia's internal enemies were impure. Duch, the head of the notorious Tuol Sleng (S21) prison in Phnom Penh—where in excess of forty-thousand political prisoners were subjected to torture and death—likened the menace to the integrity of the Kampuchean State apparatus to the manner in which woodworms bore through wood. These were loathsome germs from the CIA, or from Vietnam, bent on infiltrating and eroding the body politic. Once infected, an individual would almost certainly infect others. Unless these microbes were flushed out by the most brutal and draconian force, the socius itself would suffer a precipitous demise and fall into sempiternal desuetude.[120]

Given the sheer direness of such an all too imminent possibility within the delusional imagination of the despot, the paranoid regime has no recourse but to incessantly multiply and diversify its real and completely imaginary enemies. A vast multitude of internal enemies of State follow in rapid profusion. The traitor has a well-nigh infinite number of forms and disguises and lies hidden everywhere—even within the very midst of the despotic caste itself. He is two-faced and plays a double game and may have betrayed the regime already. The despot's very definition of this innately treacherous enemy whose clandestine ubiquity verges on the absurd posthaste becomes ever more vague, and increasingly broad, constantly subsuming new groups—even ones that do not exist, never did exist, and never will or can come into existence. The upshot of this is that the despot will precipitously move to denounce an entire group for fear of missing an allegedly real traitor supposedly lurking within their midst.[121] Whence the infamous Khmer Rouge dictum: "It is better to arrest ten innocent people than to let one guilty one go free."[122] The result: massive indiscriminate violence.[123] Better not a sole survivor than for a single organ to flow outside or slip away from the body of the despot.

Semelin also refers to the more psychoanalytical approach of Franco Fornari. Here we may compare the paranoia of the despot or despotic caste with the paranoid process undergone by the subject swept up in the maelstrom of war. Substituting the despot for Fornari's belligerent subject, it may be said that the former "perceives the object as a threat to his own existence, a threat in and of itself illusory, but psychologically real."[124] The despot's paranoid violence here stems from the psychotic illusion that murdering those whom he perceives or misperceives as enemy others serves to preclude his own death. The death of the purported foe is at once the preservation of the despot's—or despotic caste's—life.

In this sense, the despot's homocidal or genocidal paranoia is related to the destructive fantasies of early childhood. More specifically, the relation here is to the paranoid-schizoid position of the infant—a position that the child attains prior to the advent of any real threats emanating from real enemies. Here Semelin avers apropos of psychologist René Spitz that an eight-month-old child will illusorily project malign intent onto an unknown person, or stranger. Without the latter having ever shown a properly hostile attitude, the infant will take him for an inveterate enemy. Thus, by again substituting the despot for societies in the thrall of war, we may paraphrase Fornari as follows: the despot is conditioned from a psychological and political point of view to situate himself at the level of anxiety of an eight-month-old child.[125] The latter may be seen as a defense mechanism of the despot and his caste that "aims concretely at the disintegration of the evil social object—sometimes by destroying it—to restore the primal position of omnipotence in early childhood."[126]

For a more extensive foray into the psychological—rather than the more overly political—determinants and aspects of despotic paranoia and its propensity to the destructiveness of human life, we presently turn to the analyses of Erich Fromm.

(iii) Fromm and the psychopathology of paranoia

Though he does not make use of the word itself, it is evident enough when Fromm dwells upon the matter of malignant aggression that his characterization of the latter falls well within the ambit of despotic paranoia. Let us take as a metapsychological starting point the Freudian "anal character." As is well known, the latter manifests itself in such character traits as stubbornness, excessive orderliness, parsimony, punctuality and cleanliness. Freud's is however a libido theory, whereas Fromm desires to approach the anal character from the perspective of *mode of relatedness*. Here, the hoarding character's mode of relatedness is one of reserve, control, rejection of the other, and the hoarding of goods. Yet, the hoarding character is orderly not merely with things, but also with his thoughts and feelings. Needless to say, this orderliness—being excessive—proves to be sterile and rigid. It swiftly reaches the point at which he is unable to bear having things out of place. Through this orderliness he controls space, whilst by way of his hypertrophied punctuality he controls time. By way of compulsive cleanliness, he aims as far as possible to undo the contact he had with a world which he regards as dirty and hostile. Here already evinced is the essentially paranoiac position of the anal-hoarding character.[127]

For, as Fromm renders pellucid, the "hoarding character experiences himself like a beleaguered fortress; he must prevent anything from going out and save what is inside the fortess. His stubbornness and obstinacy is a quasi-automatic defence against intrusion."[128] Like the paranoid despot, he is himself confronted with the omnipresent danger that a single working organ might flow outside his multiply assailed narcissistic citadel. He sees rising up before him, against him, an eye with too steady a look, a mouth with too unfamiliar a smile. For this reason, the anal-hoarder's relations with others must be ones of either remoteness or of dominance. Intimacy is a threat. He is suspicious. Owing to this his paranoia, the anal-hoarding character can only feel secure in his relations with the world if they are ones of possession and control.[129]

Though we need not look beyond the anal-hoarding character to find most of the salient psycho-political traits of the paranoid despot, they do nonetheless exist within him in a relatively *benign* form. On the other hand, a far less benign form of this malign character is that of the non-sexual sadist. In this sense, the sadistic character may be considered an intensification or development of the anal-hoarding character. As the hoarder transmogrifies into the sadist, his paranoia and despotism shall likewise be exacerbated. The more immediate difference between the two paranoid forms is, however, that whereas the anal-hoarder will merely maintain either a reserved distance to others or aim to control them to some degree—in the sadist this propensity to dominate others spirals out of all proportion. For the sadist desires nothing short of *absolute* dominance of other men—to the point of subjugating, reifying, and crippling them. As Fromm has it, "the core of sadism, common to all its manifestations, is *the passion to have absolute and unrestricted control over a living being,* whether an animal, a child, a man, or a woman. To force someone to endure pain or humiliation without being able to defend himself is one of the manifestations of absolute control, but it is by no means the only one. The person who has complete control over another living being makes this being into his thing, his property, while he becomes the other being's god."[130] In this sense, the sadist is quite like the paranoid despot who maintains a vast inchoate population of preemptively mutilated slaves in continued subjection (the new alliance) whilst at once communing directly with his god (direct filiation). "For the sadistic character everything living is to be controllable; living beings become things. Or, still more accurately, living beings are transformed into living, quivering, pulsating objects of control."[131]

The sadist's hypertrophied need for an extraordinary degree of control over circumstances and over the living (which has now become absolute), bears the corollary that even the slightest perturbation in the regime of hegemonic orderliness will rile the sadistic character to the core. His psycho-political position, therefore, shows itself to be *still more* paranoid than that of the anal-hoarder—to the extent that he lives in well-nigh permanent fear of life itself. "The sadistic character is afraid of everything that is not certain and predictable, that offers surprises which would force him to spontaneous and original reactions. For this reason, he is afraid of life. Life frightens him precisely because it is by its very nature unpredictable and uncertain."[132] The sadist is likewise "xenophobic and neophobic—one who is strange constitutes newness, and what is new arouses fear, suspicion, and dislike, because a spontaneous, alive, and non-routinized response would be required."[133] The sadist is therefore the paranoid despot who, countenanced with the merest scintilla of otherness, sees rising up before him, against him, the enemy that brings death—an eye with too steady a look, a mouth with too unfamiliar a smile.

We have thus observed that the non-sexual sadist is a more rabid version of the anal-hoarder. Beyond the sadist, however, there lies a still further stage in the development that we have been tracing—a stage which at once represents an even deeper descent into the mire of despotic paranoia. This is the *necrophile*. Given our present purposes, we have in mind specifically the non-sexual necrophile. Notwithstanding, the necrophilous character is better understood as not merely an intensification of the sadistic character, but as born of the exacerbation of certain defining traits of *both* the anal-hoarding character *and* the sadistic character.

Firstly, the relation between the anal-hoarding character and the necrophilous character resides in the former's preoccupation with feces—something that the anal infant scarely wishes to cede. His preoccupation is, therefore, with *dead* matter. As Fromm has it, this concern with feces is "the symbolic expression of the interest in all that is decayed or putrid, all that is not alive."[134] A similar obsession with feces in particular and dead matter in general is to be found in the necrophile. For Fromm has been privy to many a necrophilous person's dream in which the latter sees "parts of dismembered bodies floating or lying around, sometimes in blood, often in dirty water, together with faeces."[135]

This being said, the point of *difference* between the anal-hoarding child's preoccupation and that of the necrophile resides in the fact that, while both remain redolent of feces, that of the former remains at the level of mere dead matter whereas that of the latter comes to specifically involve *the dead themselves*. This of course brings us to the very definition of necrophilia which, etymologically speaking, refers not so much to the love of death but, more primordially, to the "love of the *dead*." The Greek *nekros* means "corpse." It does, however, at the same time refer to those who had attained corpsedom by unnatural means; that is, the murdered. This nuance is specifically borne out by the Latin *nex* or *necis*, meaning violent death or death by murder.[136] In this vein, the non-sexual necrophile's preoccupation with and love of the dead assumes the form of his "desire to handle, to be near to, and to gaze at corpses, and particularly the desire to dismember them."[137]

But this last-mentioned propensity to dismemberment points to the ultimate *raison d'être* of necrophilia. Here, Fromm concludes that necrophilia manifests its essence "in acts of the pure passion to destroy."[138] The said destructiveness is necrophilia's obsession and, moreover, its aim.[139] This brings us to Fromm's final characterological definition of necrophilia—as "*the passionate attraction to all that is dead, decayed, putrid, sickly; it is the passion to transform that which is alive into something unalive; to destroy for the sake of destruction; the exclusive interest in all that is purely mechanical. It is the passion to tear apart living structures.*"[140] Thus is it illustrated that the necrophilous character may be conceived as a vertiginous and most grotesque development of the anal character.

But we have maintained from the outset that the necrophilous character is likewise born of an intensification of the *sadistic* character. This, as observed, was presented as the second of the two points of ramification from the necrophile. How so? The sadist is fundamendally rankled to the core by the sheer contingency of life's seemingly inscrutable vicissitudes. In that perhaps a mere smidgen of the said vicissitudes escape the sadist's purview, his fear of the latter instantly finds itself magnified to a fear of life itself. The sadist has not, however, proceded on this account to precipitously assume a sempiternally disqualifying stance against the entirety of all living matter in and of itself. Also, though the sadist's relation to others is immeasurably perverse and premised on a will to their cruel subjection, his intention nonetheless remains their contingent preservation. He may, to be sure, resort to killing them. But this would be a mere consequence, final resort, or by-product, of his contumacious cruelty.[141]

This, however, ceases to be the case with the necrophile. For the latter incarnation of despotic paranoia wills the terminal extermination of all potential and prospective obstacles to the despot's purportedly deified narcissistic omnipotence. The necrophilous character ups the ante, he short-cuts the circuit of the sadist. What appeared in the sadist as a mere by-product, a corollary of the sadistic impulse, in the vast historical panoply of necrophiles manifests itself *ab initio* as their preordained aim. "Their aim is to transform all that is alive into dead matter; they want to destroy everything and everybody, often even themselves; their enemy is life itself."[142] The difference between the two is that the sadist "would demand *surrender*; only the necrophile demands *annihilation.*"[143] More precisely, the sadist is only moved to acts of slaughter after failing to induce surrender. The necrophile moves *directly* to mass murder. Notwithstanding, sadism necessarily remains present within necrophilia. The intermingling of these two aspects is present in the spectacular public violence with which the criminal was tortured and executed by the Classical despotic regimes dwelt upon a little earlier by Foucault. Here, to the ends of the restoration and ostensible reaffirmation of an easily threatened despotic State sovereignty, the subject's body is "effaced, reduced to dust and thrown to the winds, a body destroyed piece by piece by the infinite power of the sovereign" paranoiac in all his sado-necrophilous brutality.[144] Hence, the grisly torture and death of the eighteenth century would-be regicide Damiens whose flesh was torn from his breasts, arms, thighs and calves with red-hot pincers, and molten lead, boiling oil, burning resin, wax and sulphur poured into his wounds, before he was drawn and quartered by six horses, his surgically severed limbs and torso incinerated at the stake and reduced to ashes.[145]

Despotic sadism very often assumes the form of vastly disproportionate, and hence sadistic, punishments for infractions against State Law. In pharaonic Egypt, parricides had their flesh whipped from their body with bundles of reeds, were thrown into thorns and rolled back and forth until nearly flayed, or burned alive. A man convicted of having committed adultery with a slave was sentenced to as many as one-thousand lashes. Women convicted of adultery had their noses cut off.[146] In Babylon, Hammurabi's code decreed that adulteresses be thrown into a river to drown. Arsonists and robbers of houses were burned.[147]

The Mosaic code decreed that the penalty for eating meat that still had blood in it, blasphemy, pregnancy outside wedlock and the refusal to obey one's father, was execution by burning, crucifixion, stoning, and being buried alive in a tower of ashes. Prostitutes and daughters of priests found guilty of adultery were burned alive. Kind David had the citizenry of the city of Rabbah sawn in half or buried up to their necks in the earth and driven over with plows.[148]

When the Jews of the Seleucid Empire lived under the tyranical yoke of the despot Antiochus Epiphanes, anyone observing the Sabbath was ordered to be burned alive. One boy who refused to eat pork was tied to a wheel above a bed of livid coals where his joints were dislocated, his bones broken and his flesh torn with red-hot pincers. For the same offense, the eldest of seven brothers was broken on the wheel and his limbs hacked off, the second had his limbs severed and his still living trunk incinerated in a gigantic frying pan, the third skinned alive and disembowelled, the forth his tongue ripped out and roasted alive on a split, the fifth burned at the stake, the sixth and seventh boiled in a cauldron of water, and their mother burned alive.[149]

The Greek tyrant Draco declared that the punishment of all crimes, no matter how slight, ought to be death. In Athens, minor infractions such as public drunkenness led offenders to the pillory. The dictator Phalaris of the Greek city State Agrigentum had those convicted of capital crimes locked inside a bronze life-size figure of a bull which was heated red-hot with a blazing fire until the victim's death cries blared through the bull's mouth and nostrils. Lucian writes of a young woman being sewn inside the carcass of a freshly slaughtered donkey which shrank and began to rot, attracting worms and insects that gawned at her flesh and compounded her tortures of hunger, thirst and exposure. Aristophanes describes a condemned man locked in a pillory and smeared with milk and honey as an enticement to insects. In Sparta, men who became too fat were publically whipped. Those who remained unmarried for too long were publically branded with a hot iron. The Spartan tyrant Nabis condemned malefactors to the deathly embrace of an iron statue of his wife Apega, the inner face of whose hinged arms and chest were covered with an plethora of long, sharp spikes.[150]

In the Roman Empire, even fairly mild offenses were punishable by flogging with a whip made of plaited strips of parchment which flayed the skin off victims' backs, or with a multi-thonged *plumbatae* whip with small lead balls on the ends of its thongs, or the *ungulae* which had thorns or bits of sharp metal braided into its thongs. The latter two would slice to the bone, whereas bullwhips would simply kill victims. Other victims were strapped to the outer surface of massive drums and rolled downhill until they were crushed to death, slowly burned to death on the same contraption mounted on an axle over a fire, or ground to pieces when the drum was mounted with spikes. Parricides were enclosed in large canvas sacks with poisonous snakes and thrown into the sea. Under the tyranny of Nero, Christians were dressed in the skins of animals and set upon by packs of dogs, hung by their thumbs and roasted over a slow fire, skinned alive, crushed in wine presses, shoved into red-hot suits of chain mail, impaled on stakes, or their stomachs ripped open and their entrails devoured live by wild animals.[151]

With the Chinese Tang Code of the seventh century, which still found application even into the twentieth century, petty theft, public drunkenness or insulting of superiors was punished by flogging imposed immediately in the courtroom with a whip made of lengths of split bamboo which reduced the prisoner's entire back and rump to a mass of subcutaneous blood-blisters, and forced transgressors to wear a massive pillory-like collar—known as the *fcan*, *hao*, *cangue*, *tcha* or *ea*—mounted with iron spikes which prevented the victim from lying down or resting his head. Translators who mistranslated their employer's words were forced into a kneeling position whilst a guard stood on a thick bamboo rod placed behind their knees, causing excruciating pain. Women convicted of prostitution had slivers of wood placed between their fingers and a heavy chord drawn increasingly more tightly around the fingers, crushing the knuckles against the wood. Buddhist monks convicted of having sexual relations had a hot iron rod driven through the neck muscles, a length of chain threaded through the raw, charred hole, by which leash they were led naked through the street, begging for money. Other non-capital offenders were blinded with lime, which ate their eyes away in minutes.[152] For crimes such as treason, sedition or plotting rebellion, the offender was stretched on a rack-like device, kicked and stamped on by guards until his bones were broken, and then beaten to death with heavy clubs. Parricides were subjected to the *Ling Che*, or "Death by a Thousand Cuts," wherein the condemned was hauled in public, tied to a table or framework, and slowly carved up with a knife.[153]

For capital crimes, the Japanese employed a punishment almost identical to the *Ling Che* but likewise burned offenders to death in bundles of twigs. In seventeenth century Japan, those accused of being Christians were stripped naked and thrown from a cliff, tossed into an active volcano, torn limb from limb by oxen, or burned alive en masse.[154]

William the Conqueror advocated the maiming of law breakers, contending that their eyes be torn out and their testicles cut off.[155] In the thirteenth century, Henry III of England began to have recalcitrant subjects hung, drawn and quartered. Victims were pulled from the ground by a noose around the neck and left to dangle and lose consciousness, brought down and revived, castrated, their stomachs ripped open and their entrails pulled out and tossed onto a fire before their still living eyes, and then decapitated. Under Henry's son Edward I, prisoners refusing to plead to charges against them were chained face down to a dungeon floor, fed on near-starvation rations, and pressed beneath increasing amounts of weight until they either lodged pleas or died.[156] From the end of Richard II's reign, petty criminals were placed in stocks in almost every town and village across England and exposed to the contempt and derision of the public, or locked in iron collars fastened to the walls of local churches. Those who continued vagrant existences were stripped naked, tied to the back of a cart and hauled through the streets whilst being whipped until the blood dripped to the ground. Common shrews and scolding housewives were tied to chairs and led through town whilst taunted by passers by and pelted with objects. Small iron cages, with a metal protrusion which immobilized the tongue and caused severe lacerations of the cheeks and gums, were fastened over their heads. Women of loose virtue were strapped to stools and submerged in local ponds. Under the despotic sadism of Henry VIII, when a sailor and his common law wife was accused of stealing the king's gold, he ordered them wrapped in chains and suspended from the Thames embankment to be drowned by the rising tide. A guard suspected of complicity in the crime was hung by chains from the Tower of London to die of exposure. Vagabonds and hobos were burned through the right ear with a hot iron. For their second such offense, they were hung. Slightly worse crimes led to tongues being ripped out and hands being chopped off. When the Bishop of Rochester died of food poisoning, his cook was boiled alive at London's Smithfield Horse Market. Those caught failing to attend Henry's new Church regularly had their ears cut off. Anyone denying the king's right to decree the word of God could be hanged. Anyone adhering to either traditional

Roman Catholicism or Protestantism would be burned as a heretic. Anyone caught eating meat on Friday or denying the doctrine of transubstantiation was burned as a heretic. A merchant caught in possession of the writings of Luther was stoned to death. Those even heard uttering a word against the king were liable to be tortured and burned at the stake.[157] Under the Catholic reign of "Bloody Mary," the chief persecutor of English Protestants, Edmund Bonner, Bishop of London, burned the palms of one witness' hands in public court with a candle flame and pulled a barbed arrow backward between the tied thumbs of another. Heretics were pulled apart on the rack in the basement of the Tower of London almost every day. Bonner had a Protestant minister accused of possessing scandalous books locked in a pillory, whipped until his back looked like half-digested meat, his nose split with a knife, both his ears hacked off, and his cheeks branded with red-hot irons.[158] Under Queen Elizabeth I, the rack, the gibbet and the noose were in constant use. Cutpurses had one ear hacked off on their first conviction, the remaining ear sliced off in their second conviction, and were hung on their third offense. Prostitutes likewise lost one ear at a time and their nose on the third offence.[159] Under the sway of the Puritanism of Cromwell and others, on account of their religion alone, Quaker parliamentarians were sometimes pilloried, had both ears cut off, were physically branded as seditious libellers, and sentenced to life imprisonment. Some were incarcerated in underground sewers, others in a hole in the wall twelve feet above the ground. Under this regime, anyone convicted of being a theatrical actor could be publically flogged, whereas anyone caught having fun of any kind could be stocked, pilloried, or whipped. One non-Puritan, who insisted that an unmarried couple could have conjugal relations provided they went to the same church, was pilloried, whipped through the streets of London, again pilloried, had a hole bored through his tongue, was physically branded a blasphemer, carted off to Bristol where he was again flogged and placed in solitary confinement in a prison for an indefinite period of time.[160] Likewise during this time, being convicted of perjury, one Titus Oates was paraded around the courts of Westminster and the Royal Exchange with a placard describing his crime, flogged through the streets from Newgate Prison to Tyburn, and locked in a pillory in Westminster, Charing Cross and Tyburn, respectively, on three specified days of each month for the rest of his life. In the same year, a woman who was convicted of petty theft was hauled from court to Tyburn, scourged bloody, stripped in the cold, and hung.[161] Until the end of the eighteenth century in England, but likewise throughout much of Europe, villians were stuffed into human-shaped iron cages known as gibbets, and allowed to swing in the breeze until they died of exposure. Urban and suburban panoramas abounded in iron and wooden cages attached to the outsides of town halls and ducal palaces, halls of justice, cathedrals and city walls. The naked victims locked into these cages, frequently tortured and mutilated beforehand, were assailed by storm, cold, heatstroke and sunburn, and died of hunger and thirst. Their putrefying cadavers were usually left in place until the bones fell apart.[162]

In the Netherlands, for assassinating Prince William of Orange, Balthasar Gérard was hustled off to prison and mercilessly racked and tortured for eighteen successive days. He was hauled to a public scaffold, where his offending hand was thrust into a cauldron of boiling oil and, on the following day, publically amputated. For the next eleven days he was publically mutilated, his skin ripped piece by piece from his body with red-hot pliers. On the eighteenth day he was tied to a burning stake for a few minutes, cut down and his charred and blackened though still living body broken on the wheel for six hours, before he was strangled to death.[163]

The sadists of the Spanish Inquisition bound and gagged all suspects who resisted arrest by forcing a pear-shaped metal object into their mouths. With the turn of a small crank the device expanded, wrenching open the jaw and often breaking teeth and dislocating the jawbone.[164] At massive *autos da fé*, thousands of heretics were publically shamed before thousands of spectators and tied to burning pyres, on which they died over the next hour, shrieking and screaming as the fat oozed through their charred and cracked skin, their hands and feet reduced to blackened stumps while they still lived.[165]

In seventeenth century Massachusetts, a major Puritanical stronghold, idolatry, blasphemy and adultery were all punishable by death. Those who denied the existence of God had a hole bored through their tongue with a hot iron, whereas those caught smoking tobacco or wearing unseemly clothes could be fined, stocked or pilloried. A certain sea captain named Kimble, having returned from a long sea voyage, was clamped into stocks for kissing his wife on the doorstep of his home.[166]

The North Vietnamese communist regime nowise retreated from this sadistic logic. In the labor camps the supposed laziness of the subject was punished by his incarceration in a small metal box for at least two days and likely death after a week in the box.[167] All incarcerees were deprived of sleep, starved of strength, and caused to die slowly.[168] They were forbidded to urinate, on pain of being castrated. Those caught furtively consuming insects had their left thumb tied to their right toe and their right thumb tied to their left toe and, in this contorted posture, left to the mercy of millions of mosquitoes who would suck out all of their blood such that they would die with face sunken and mouth wide open.[169] Men were dragged along throught the dirt like animals, spat on in the face, kicked in the stomach, and executed.[170]

In the case of the Cambodian Khmer Rouge regime, despotic sadism reaches Dantesque proportions. Body parts were amputated, women disemboweled, fetuses aborted. Locked in a filthy jail cell with his feet locked in leg-irons with scores of other prisons on the suspicion of having hoarded a small quality of food, Haing Ngor was convinced that he and the other inmates had become *pret*, that is, spirits of the damned condemned to gory and everlasting tortures by the folkloric King of Death. This conviction became a grisly reality when he found himself tied to a cross with flames and smoke licking his legs from a fire burning just beneath it. The surroundings were a scene of infernal torment with countless rows of crucified men and women offering a collective cacophany of agonized moans amid a smoldering and incinerating landscape.[171]

9. The impossible lust of corporeal dismemberment: Sade

From the Cambodian killing fields, we move onto the despotic sadism depicted by Sade himself. *The 120 Days of Sodom* features a despotic regime comprising a fortress presided over by four libertines: the Duc de Blangis, the Bishop of X, the Président de Curval, and Durcet. The cowering subjects milling below their sadistic gaze were sixteen young men, ranging in age from twenty to thirty; sixteen boys between the ages of twelve and eighteen, who would assume the office of women; twenty young girls between the ages of seven and fifteen; and a rabble of the foulest creatures imaginable.[172] The despotic quartet's anal-hoarding paranoia is evinced in the fact that, so as to preclude the flight of organs from the reterritorializing sway of the fortress, all the gates and passages whereby the château might be penetrated were shut and cut off so as to absolutely preclude even the least egress to a deserter. The premises were so thoroughly barricaded that no trace whatsoever remained of where the exits had formerly been.[173] The fortress was, moreover, in the depths of an uninhabitable forest, high amongst naked mountains, whereas the paths that had brought the subjects to it had already been destroyed as they advanced along them.[174] The despotic quartet drew up a series of laws in which the sado-necrophilous aspect of the regime is already evident. According to the foregoing, it was inter alia strictly forbidden to relieve onself anywhere save in the chapel. The penalty for the contravention of this rule was death.[175] Any subject who refused to comply with a demand, even when incapacitated or when the request was impossible, would be punished severely. The least display of mirth or the least evidence of a lack of submission during debauches would be cruelly punished. Any man caught *flagrante delicto* with an unauthorized woman would be punished with the amputation of a limb. The slightest religious act would be punished with death. The punishment for fulfilling duties of bodily cleanliness would be severe. The penalty for attempted evasion during any sitting of the assembly would be instant death.[176]

Sadistic intent is immediately evident in opening remarks made by the Duc de Blangis: "Feeble, enfettered creatures destined solely for our pleasures, I trust you have not deluded yourselves into supposing that the equally absolute and ridiculous ascendancy given you in the outside world would be accorded you in this place; a thousand times more subjugated than would be slaves, you must expect naught but humiliation, and obedience is that one virtue whose use I recommend to you: it and no other befits your present state.... Useless to conceal it from you: your service will be arduous, it will be painful and rigorous, and the slightest delinquencies will be requited immediately with corporal and afflicting punishments; hence, I must recommend to you prompt exactness, submissiveness, and a total self-abnegation that you be enabled to heed naught but our desires; let them be your only laws, fly to do their bidding, anticipate them, cause them to be born. Not that you have much to gain by this conduct, but simply because, by not observing it, you will have a great deal to lose."[177] He continued that the killing of women was, within this regime, as insignificant as the crushing of a fly and that the most fortunate thing that could befall a women was to die young. The subjects would even be harshy punished if they failed to cotton onto the very slightest gestures, glances, or internal feelings through which the despotic quartet would indirectly suggest their desires.[178]

From the twenty-sixth day in the veritable fortress of the Châteaux of Silling, those failing to loudly utter blasphemies upon invitation would instantly be inscribed in the book of death. To accelerate the condemnation of the pious, in the manner of Soviet or Chinese communists, the cowering subjects of this despotic regime were encouraged to lodge complaints against those unwilling to explicitly berate God. These reporters of the infractions of others did not need to provide tangible proof of others' transgressions. The mere fact that they had hinted at such possibilities sufficed to call down the cudgel upon the accused. Whether the accusation was true or false, the accused would be condemned immediately. Those who squealed against their brethren, notwithstanding, nowise escaped punishment themselves.[179] As Sade maintains of the quartet holding sway over the despotic regime at Silly, "In a word, over this small population they established all the vexation, all the injustice one could imagine, certain in the belief that the more harshly their tyranny was exercised, the greater the sum of pleasures they would derive from their privileged situation."[180] One of the despotic libertines, Curval, declared that if he had government of the country he would borrow their law from the inhabitants of Formosa, where pregnant women under thirty were, together with their fruit, ground under a huge mortar.[181] He goes on to say: "What the devil difference can it make to Nature whether there are one, ten, twenty, five-hundred more or fewer human beings on earth?"[182]

The limitless and massively inventive torture of the hapless innocents is, of course, here seen to follow without abatement.

As shall be discussed in far greater detail in later chapters, the paranoid and specifically the sadistic aspect of the regime resides in the despotic *Law* which, owing to the conjunction of its paranoiac and manic depressive traits, is purely formal and inherently vacuous—its content only emerging through and in the punishment meted out upon the supine body of the transgressor. This emptiness of the Statist Law is at once that shared by the Kantian moral Law. As opposed to any positive law, which accords with the concrete filling of external reality, the Kantian pure and transcendental moral Law is a law of the Lacanian Real, entailing as it does an absolutely unconditional imperative entirely oblivious to the limitations imposed by extant situations. Its unconditional injunction is one which imposes a compulsive and impossible demand: *Du kannst, denn du sollst!* ("You can, because you must!"). Žižek argues in this connection that, from the Lacanian perspective, the ethical imperative of the Kantian pure moral Law conceals within itself an obscene superego injunction emanating from the voice of the Other—in our case, the hauntingly absent yet ubiquitously, oppressively and perversely proximate deity of the despot—which commands the bureaucratic pervert of, or the subject enslaved by, the regime to follow his duty purely for the sake of duty, in abstraction from whatever self-interest or pathological motivations he may have. This obscene injunction is a traumatic irruption of an unconditional demand for an impossible *jouissance* wholly excessive of the homeostasis of the Freudian pleasure principle and reality principle alike. It is Lacan's renowned "Kant avec Sade": that the obscenity of the Kantian moral Law resides not at all in the "pathological" empirical residues still befouling its purity, but rather in its purely formal vacuity itself—a vacuity which compels the bureaucratic pervert to obey the Law simply because it is the law and for no other reason, to be its unselfish instrument of execution. This superego obscenity, as inexorable obverse of the despotic Law, is the surplus-enjoyment (*plus-de-jouir*) produced by the renunciation attendant upon the unconditional obedience to the pure moral Law. It demands the sacrifice of a body to be tormented, not for an extraneous goal, but as an end in itself.[183]

Now, in Kant, the ability of the will to completely fulfill the unconditional injunction of the moral Law requires the postulate of the "immortality of the soul." The latter is the presupposition that the moral being possesses a necessarily infinitely enduring existence by which it is enabled, through continued and unending progress, to asymptotically approach the condition of complete fitness of its will to the moral Law or, otherwise put, "holiness."[184]

However, Alenka Zupančič, for one, questions the obvious flaws in this line of argumentation: on the one hand, being outside of space and time, an immortal soul would be able to attain holiness not progressively but indeed *immediately*; whereas, on the other hand, "continuous and unending progress" would be impossible for the immortal soul, as this would require it to be *inside* space and time. For this reason, Zupančič is right to replace Kant's postulate with that of the *immortality* of the *body*. Such a body would be indestructible and *sublime*, and would thus be able to submit itself to and endure the sadistic tortures meted out by the bureaucratic perverts of the despotic regime for all eternity.[185] In *The 120 Days of Sodom*, as readily observed, the panoply of torments of the body does purportedly approach infinity. A woman is hung head downward until she vomits. Another woman is given an enema of boiling water after being sodomized. A scullery maid receives two-hundred lashes at the hands of two men. A kitchen servant is whipped eight-hundred times, two-hundred by each of the libertine despots. A girl has her finger amputated the same night, the wound cauterized with fire. Another woman is plunged lengthways into a fire, whilst her nipples are burnt. Another has her anus pricked with a needle until it is covered with blood. A boy is flogged until he streams with blood. His nose is tweaked repeatedly, until it bleeds. A woman is given a powder that makes her cramp in agony, after which one of her fingers is cut off. After being dealt a hundred lashes by each of the four despots until she is bloodied, a woman's breasts and the palms of her hands are burnt, as drops of molten sealing wax are spilt upon her thighs, belly and navel. Her pubic hair is burnt after being doused with cognac. Another has a finger severed from each of her hands.[186]

Notwithstanding, the problem that confronts even Sade is that his torturers are only able to torment the bodies of their victims for a *finite period* of time, that is, until such time as they die. Moreover, apart from the obvious obstacle of death, the degree of intensity of suffering of the victims, though constantly increased to an ever more excruciating level of visceral affliction, encounters a limit in the fact that the human body has only a finite capacity for enduring *jouissance*. This limit is self-identically the Freudian pleasure principle, which imposes a plateau of homeostasis upon the limitlessness of *jouissance* unconditionally demanded by the sadistic superego underside of the despotic Law.[187] To transcend this limitation and thereby effect a complete apotheosis of the unconditional injunction of the despotic Law and the infinite *jouissance* of torture unbounded in both intensity and temporal duration that it demands, necessitates the fantasy of an immortal body. Only then can the cowering subject subjugated by the draconian machinery of torture of a henceforth divine and eternal despotic regime be condemned to ETERNAL DAMNATION.[188]

10. Infinite suffering: Dante and Milton

From Sade's necro-sadistic fortress and its hellish vicissitudes, we move onto Hell itself. In Milton's *Paradise Lost,* Satan and the rebel angels—having exceeded the coefficient of deterritorialization of the divine despotic sign—are hurled out of Heaven by Michael the Archangel like:

> … a herd
> Of goats or timorous flock together thronged
> … thunderstruck, pursued
> With terrors and with furies to the bounds
> And crystal wall of Heav'n, which op'ning wide,
> Rolled inward, and a spacious gap disclosed
> Into the wasteful deep; the monstrous sight
> Strook them with horror backward, but far worse
> Urged them behind; headlong themselves they threw
> Down from the verge of Heav'n; eternal wrath
> Burnt after them to the bottomless pit.
> Hell heard th' unsufferable noise, Hell saw
> Heav'n ruining from Heav'n and would have fled
> Affrighted; but strict Fate had cast too deep
> Her dark foundations, and too fast had bound.
> Nine days they fell; confounded Chaos roared,
> And felt tenfold confusion in their fall
> Through his wide anarchy, so huge a rout
> Encumbered him with ruin: Hell at last
> Yawning received them whole, and on them closed,
> Hell their fit habitation fraught with fire
> Unquenchable, the house of woe and pain. (*Paradise Lost,* VI.856–877)

There they:

> Chewed bitter ashes, which th' offended taste
> With spattering noise rejected: oft they assayed,
> Hunger and thirst constraining, drugged as oft,
> With hatefullest disrelish writhed their jaws
> With soot and cinders filled … (*Paradise Lost,* X.566–570)

Moreover, however, Raphael's narration to Adam and Eve protended the subsequent and consequent precipitation of humanity itself into the sado-necrophilous cauldron of Hell as a corollary of Satan's breaching of the coefficient of deterritorialization of the divine despotic sign. Milton:

> Thus measuring things in Heav'n by things on earth
> At thy request, and that thou may'st beware
> By what is past, to there I have revealed
> What might have else to human race been hid;
> The discord which befell, and war in Heav'n
> Among th' angelic Powers, and the deep fall
> Of those too high aspiring, who rebelled
> With Satan, he who envies not thy state,
> Who now is plotting how he may seduce
> Thee also from obedience, that with him
> Bereaved of happiness thou may'st partake
> His punishment, eternal misery;
> Which would be all his solace and revenge,
> As a despite done against the Most High,
> Thee once to gain companion of his woe.
> But listen not to his temptations, warn
> They weaker; let it profit thee to have heard
> By terrible example the reward
> Of disobedience; firm they might have stood,
> Yet fell; remember, and fear to transgress. (*Paradise Lost,* VI.893–912)

After his cardinal offense against God, Adam indeed laments aloud to himself the possibility of the foregoing infinitely extended divine necro-sadism:

> . . . Will he draw out,
> For anger's sake, finite to infinite
> In punished man, to satisfy high rigour
> Satisfied never? that were to extend
> His sentence beyond dust and Nature's law,
> By which all causes else according still
> To the reception of their matter act,
> Not to th' extent of their own sphere. But say
> That death be not one stroke, as I supposed,
> Bereaving sense, but endless misery
> From this day onward, which I feel begun
> Both in me, and without me, and so last
> To perpetuity; ay me, that fear
> Comes thund'ring back with dreadful revolution
> On my defenceless head; both Death and I
> Am found eternal, and incorporate both,
> Nor I on my part single; in me all
> Posterity stands cursed. (*Paradise Lost,* X.801–818)

With Adam and Eve's original transgression in the garden of Eden, and the consequent movements of deterritorialization in excess of the coefficient of detterritorialization of the despotic sign, the divine Damocles' sword of death would henceforth bare down upon all humanity in sado-necrophilous fashion. Michael renders this privy to Adam:

> … Death thou hast seen
> In his first shape on man; but many shapes
> Of death, and many are the ways that lead
> To his grim cave, all dismal; yet to sense
> More terrible at th' entrance than within.
> Some, as thou saw'st, by violent stroke shall die,
> By fire, flood, famine; by intemperance more
> In meats and drinks, which on the earth shall bring
> Diseases dire, of which a monstrous crew
> Before thee shall appear; that thou may'st know
> What misery th' inabstinence of Eve
> Shall bring on men. Immediately a place
> Before his eyes appeared, sad, noisome, dark;
> A lazar-house it seemed, wherein were laid
> Numbers of all diseased, all maladies
> Of ghostly spasm, or racking torture, qualms
> Of heart-sick agony, all feverous kinds,
> Convulsions, epilepsies, fierce catarrhs,
> Intestine stone and ulcer, colic pangs,
> Demonaic frenzy, moping melancholy
> And moon-struck madness, pining atrophy,
> Marasmus, and wide-wasting pestilence,
> Dropsies, and asthmas, and joint-racking rheums. (*Paradise Lost,* XI.466–488)

And with this entrance of death into the world effected by the actual excessive deterritorializations of Satan in the garden of Eden, which precipitated the fall of man, the sado-necrophilous cauldron of Hell immediately opens its jaws to infernally torment sinners for all eternity:

> … The aggregated soil
> Death with his mace petrific, cold and dry,
> As with a trident smote, and fixed as firm
> As Delos floating once; the rest his look
> Bound with Gorgonian rigor not to move,
> And with asphaltic slime; broad as the gate,
> Deep to the roots of Hell the gathered beach
> They fastened, and the mole immense wrought on

Over the foaming deep high arched, a bridge
Of length prodigious joining to the wall
Immovable of this now fenceless world
Forfeit to Death; from hence a passage broad,
Smooth, easy, inoffensive down to Hell. (*Paradise Lost,* X.293–305)

This anticipation of the eternal damnation of mankind in Hell, or at least in the roiling sadistic cauldron of Purgatory, due to his deterritorializations in excess of the divine despotic sign induced and following in the wake of Satan's original and abiding breach of the same, continues in the vast despotic regime envisaged by Dante in his *Purgatorio.* Firstly, those men who died excommunicated from the Holy Church, though repented of their sins prior to their decease, are detained in mourning on the first terrace of ante-Purgatory for thirty years for single every year of their contumacy.[189] Those who remained within the fold of the Church though postponed repentance until their last hour (*in articulo mortis*) either through indolence, through a sudden violent death which left them unshriven at decease, or through too overweening a preoccupation with worldly affairs, suffer a pining and lugubrious detention in the second terrace of ante-Purgatory for a length of time equal to the entire duration of their earthly lives.[190] On the first cornice of Purgatory, the proud and vainglorious are seen bowed down and bent double toward the ground with knees cramped to the chest as they carry heavy stones on their backs.[191] On the second cornice, the envious, clothed in the coarse hair-cloth of beggars imploring passersby for alms, have their eyelids stitched and sealed with iron wires, squeezing slow tears to bathe their cheeks as they weep.[192] On the third cornice, the wrathful are blanketed, blinded and suffocated by a foul and acrid cloud of thick smoke.[193] On the forth cornice, the slothful are harried into ceaseless frenetic activity.[194] On the fifth cornice, the covetous lie outstretched and motionless with their hands and feet fettered and their faces turned to the ground, weeping and sighing.[195] On the sixth cornice, the gluttonous are subjected to such acute hunger that the pallid skin on their faces is molded to the bone, the ravages of emaciation having withered their famished, shrivelled, scurvy-ridden and husk-like forms to the very hide, discolored their hue as by a dry leprosy, and reduced the sockets of their dark, hollow and weeping eyes to mere rings.[196] These shades, resembling things twice dead, chew the very air for hunger.[197] On the seventh and final cornice of Purgatory, the lustful are burned alive by great belching sheets of fire.[198]

Descending now below the terrestrial ambit of Purgatory into the subterranean bowels or the *Inferno* of the gargantuan divine State-like apparatus conjured up by the Florentine scribbler, one witnesses in the first and most superficial circle of Hell—that is, Limbo—the damned souls who had lived a life neither of blame nor of praise offering up a cacophany of sighs, groans, shrill cries and shrieks of lamentation aimlessly harried by a whirlwind and ceaselessly stung by hornets and wasps, their faces running with streaks of blood that mix with their tears and drip to their feet as hideous maggots collect in the pus.[199] In the second circle of Hell, one hears the shrieks, laments and anguished cries of the damned souls of the lustful as they are endlessly buffeted and driven in pitch black darkness in crowded flocks in various directions like cranes in flight by raging storms.[200] In the third circle of Hell, the damned souls of the gluttonous are eternally mired and drowned in filthy stinking muck and battered by cold and torrents of rain, thick hail and snow, beneath which they howl like dogs.[201] In the fourth circle of Hell, the damned souls of the prodigal and the miserly, incessantly screaming out vitriol, clash huge rolling weights against each other, are killed, and resurrect from their tombs to brawl again for the rest of eternity. In the fifth circle of Hell, the damned souls of the wrathful, naked and with faces scarred by rage, headbutt themselves and mangle and tear each other limb from limb with their teeth in a dark bog. In the same circle, the damned souls of the slothful lay submerged beneath the black muck of the swamp, their sighs causing bubbles at the surface as they gobble mud.[202] In the sixth circle of Hell, the damned souls of heretics lie buried and lamenting in their tombs, scorched by infernal flames and exuding a vile stench of odiferous fumes.[203] In the first ring of the seventh circle of Hell, often sunken to their eyelids, the shrieking and damned souls of those who did violence to others are boiled in a river of crimson blood, thousands of galloping centaurs firing arrows at those who deign to emerge too far above the surface.[204] In the second ring of the seventh circle of Hell, the damned souls of suicides are immured in black, twisted and entangled trees with poisonous thorns which, already echoing wails of unending grief, weep dark blood and hissing sap when broken whilst crying out for pity, incessantly feasted upon by repulsive harpies.[205] In the third ring of the seventh circle of Hell, the damned souls of blasphemers lie stretched out supine on an arid wasteland of burning sand rained upon by flakes of fire and irrigated by boiling blood, whilst on the same molten surface those of usurers crouch tightly haunched and the souls of sodomites wander incessantly on tormented feet.[206] In the first ditch of the eighth circle of Hell, the naked and damned souls of pimps and seducers walk in an endless circular path at the bottom in opposite directions whilst being lashed on the back by horned devils with enormous, stinging whips. In the second ditch of the eighth circle of Hell, the whimpering, snorting, grunting and damned souls of flatterers are mired in steaming excrement and slimy mold.[207] In the third ditch of the eighth circle of Hell, the damned souls of simonists are stuffed head first into the fissures of livid-colored rock like stakes in the ground, the soles of their exposed feet scorched by flames, with countless other damned simonists similarly stuck still deeper in the earth.[208] In the fourth ditch of the eighth circle of Hell, the damned souls of soothsayers walk in silent weeping, their faces twisted around as if in a palsy fit to look down upon their backs, the anguished tears of their eyes streaming down to their buttocks and all of the ground.[209] In the fifth ditch of the eighth circle of Hell, the damned souls of grafters squirm in boiling tar, huge chunks of their flesh torn out by a hundred devils' pitchforks whenever they chance to surface for air.[210] In the sixth ditch of the eighth circle of Hell, the damned souls of grief-stricken and tearful hypocrites walk in slow motion, wasted with fatigue by their ponderous lead-lined cloaks of everlasting weariness, hoods pulled down low to cover their eyes. The damned soul of the arch-hypocrite Caiaphas lies eternally naked, crucified and transfixed with three stakes to the ground.[211] In the seventh ditch of the eighth circle of Hell, the damned souls of thieves run terrified, naked and hopeless, their limbs and torsos bound and constricted by repulsive serpents, whilst others instantaneously burn to ash by the strike of a snake and are thereupon reconstituted to suffer more. In monstrous couplings, serpents coil around, partially devour and melt into the damned souls, the heads of the two fusing, the features of their faces flowing and blending into one face, the stomach and chest sprouting new limbs, the snakes' skins growing soft and hairy and the thieves' scaly, the beasts' hind feet turning into a man's member, the damned souls members growing reptilian legs.[212] In the eighth ditch of the eighth circle of Hell, the damned souls of deceivers are eternally incarcerated and burned within moving fires.[213] In the ninth ditch of the eight circle of Hell, the damned souls of the sowers of scandal and schism are repeatedly ripped open from the chin to the groin by devils, their throats slit, noses amputated, tongues hacked off, the gory stumps of their handless arms dripping blood, their guts, heart, and other vital organs spilling out.[214] In the tenth ditch of the eighth ring of Hell, amid piercing shrieks of lamentation and the livid stench of decaying flesh, the damned souls of falsifiers languish in strewn clumps, eternally scraping off the leprous scabs covering them from head to foot with their nails, others disfigured by bloating dropsy and steaming with infernal fever.[215] In the ninth circle, or the very pit of Hell, the damned souls of traitors are immured to various degrees in a gigantic frozen lake, some dripping congealing tears from faces bowed toward the ice, others facing upwards, their eyes frozen by their weeping. Often two heads are immured in a single hole, the one devouring the skull and brain of the other with the teeth and jaws of a rabid dog. Certain damned souls of traitors are completely buried in ice, like straws worked into glass. Some lie flat, some perpendicular, head up or head down, some bent head to foot like a bow. Finally, at the epicenter of Hell at the very center of the Earth, the greatest sinners in history—Judas, Brutus and Cassius—are constantly devoured by the three diabolical mouths of SATAN HIMSELF.[216]

Concerning Satan, he simply vastly exceeded the coefficient of deterritorialization of the divine despotic sign. He was, of course, also rather vile. The effects of this seditious movement, which by definition effected the foregoing excessive movement of deterritorialization, is described by Milton. Note how Milton describes the rebel angels' place of banishment as being an atopological non-place away from the center of the entire universe (away from the apex of this the absolute, all-encompassing despotic regime—as delineated by Dante) and, at the same time, as entering into a morass, which he assimilates with pure chaos in and of itself. This ex-centric chaos is, well, Hell. Milton:

> Satan in the serpent; who revolting from God, and drawing to his side many legions of angels, was by the command of God driven out of Heaven with all his crew into the great deep. Which action passed over, the poem hastes into the midst of things, presenting Satan with his angels now fallen into Hell, described here, not in the centre (for heaven and earth may be supposed as yet not made, certainly not yet accursed) but in a place of utter darkness, fitliest called Chaos: here Satan with his angels lying on the burning lake, thunder-struck and astonished, after a certain space recovers, as from confusion … (*Paradise Lost,* I.4–14)

Milton continues to castigate Satan:

> Th' infernal Serpent; …
> … what time his pride
> Had cast him out from Heav'n, with all his host
> Of rebel angels, by whose aid aspiring
> To set himself in glory about his peers,
> He trusted to have equalled the Most High,
> If he opposed; and with ambitious aim
> Against the throne and monarchy of God
> Raised impious war in Heav'n and battle proud
> With vain attempt. Him the Almightly Power
> Hurled headlong flaming from th' ethereal sky
> With hideous ruin and combustion down
> To Bottomless perdition, there to dwell
> In Adamantine chains and penal fire,
> Who durst defy th' Omnipotent to arms.
> Nine times the space that measures day and night

To mortal men, he with his horrid crew
Lay vanquished, rolling in the fiery gulf
Confounded though immortal: but his doom
Reserved him to more wrath; for now the thought
Both of lost happiness and lasting pain
Torments him; round he throws his baleful eyes
That witnessed huge affliction and dismay
Mixed with oburate pride and steadfast hate:
At once as far as angels' ken he views
The dismal situation waste and wild,
A dungeon horrible, on all sides round
As one great furnace flamed, yet from those flames
No light, but rather darkness visible
Severed only to discover sights of woe,
Regions of sorrow, doleful shades, where peace
And rest can never dwell, hope never comes
That comes to all; but torture without end
Still urges, and a fiery deluge, fed
With ever-burning sulphur unconsumed:
Such place Eternal Justice had prepared
For those rebellious, here their prison ordained
In utter darkness, and their portion set
As from the centre thrice portion set
As far removed from God and light of Heav'n
As from the centre thrice to th' utmost pole.
O how unlike the place from when they fell! (*Paradise Lost,* I.34–75)

Milton continues his diatribe concerning Satan's eternal damnation. Here, the bedraggled, woe-betided and rabidly fulminating Satan polemicizes with fury:

To adore the Conqueror? who now beholds
Cherub and Seraph rolling in the flood
His swift pursurers from Heav'n gates discern
Th' advantage, and descending tread us down
Thus drooping, or with linkèd thunderbolts
Transfix us to the bottom of this gulf. (*Paradise Lost,* I.323–329)

And the rebellion against the transcendent throne of the Deity was such that the recalcitrants had to be rendered politically null in relation to the grand cosmic Empire, to the point that their identities would be erased:

And Powers that erst in Heaven sat on thrones;
Though of their names in Heav'nly records now
Be no memorial, blotted out and razed
By their rebellion, from the Books of Life. (*Paradise Lost,* I.360–363)

Milton continues to harp on the eternal damnation of Satan and his rebel angels. Satan laments:

> To our destruction: if there be in Hell
> Fear to be worse destroyed: what can be worse
> Than to dwell here, driv'n out from bliss, condemned
> In this abhorrèd deep to utter woe;
> Where pain of unextinguishable fire
> Must exercise us without hope of end
> The vassals of his anger, when the scourge
> Inexorably, and the torturing hour
> Calls us to penance? … (*Paradise Lost,* II.84–92)

Milton moves on to ponder the imminent possibility that the mere existence of Hell itself is a direct cause of foreboding deterritorializations exceeding the coefficient of deterritorialization of the cosmic-divine sign. This threat of fully-fledged recalcitrance, sedition and out-right insurrection is expressed as follows:

> … What if all
> Her stores should spout her cataracts of fire,
> Impendent horrors, threatening hideous fall
> One day upon our heads; while we perhaps
> Designing or exhorting glorious war,
> Caught in a fiery tempest shall be hurled
> Each on his rock transfixed, the sport and prey
> Of racking whirlwinds, or for ever sunk
> Under yon boiling ocean, wrapped in chains;
> There to converse with everlasting groans,
> Unrespited, unpitied, unreprieved,
> Ages of hopeless end; this would be worse. (*Paradise Lost,* II.174–186)

Then Beelzebub holds forth on humankind itself, and the possibilities of inducing this species into deterritorializations excessive of the coeffient of deterritorialization of the divine despotic sign. He alludes here to man as the potential breacher of the strictures of divine despotism par excellence. The prospective and idealistic rebellion against God himself is likewise here glimpsed:

> And where their weakness, how attempted best,
> By force or subtlety: though Heav'n be shut,
> And Heav'n's high Arbitrator sit secure
> In his own strength, this place may lie exposed
> The utmost border of his kingdom, left
> To their defence who hold it: here perhaps
> Some advantageous act my be achieved

By sudden onset, either with Hell fire
To waste his whole Creation, or possess
All as our own, and drive as we were driven,
The puny habitants, or if not drive,
Seduce them to our party, that their God
May prove their foe, and with repenting hand
Abolish his own works. This would surpass
Common revenge, and interrupt his joy
In our confusion, and our joy upraise
In his disturbance, when his darling sons
Hurled headlong to partake with us, shall curse
Their frail original, and faded bliss,
Faded so soon. Advise if this be worth
Attempting, or to sit in darkness here
Hatching vain empires. Thus Beëlzebub
Pleaded his devilish counsel, first devised
By Satan, and in part proposed: for whence,
But from the author of all ill could spring
So deep a malice, to confound the race
Of mankind in one root, and earth will Hell
To mingle and involve, done all to spite
The great Creator? But their spite still serves
His glory to augment. The bold design
Pleased highly those infernal States, and joy
Sparkled in all their eyes; with full assent
They vote … (*Paradise Lost,* II.357–389)

A little later, Satan continues to speak of his incarceration in Hell:

O progeny of Heav'n, empyreal Thrones,
With reason hath deep silence and demur
Seized us, though undismayed: long is the way
And hard, that out of Hell leads up to light;
Our prison strong, this huge convéx of fire,
Outrageous to devour, immures us round
Ninefold, and gates of burning adamant
Barred over us prohibit all egress. (*Paradise Lost,* II.430–437)

Satan also effects massive deterritorializations transgressive of the divine despotic sign and regime through the fomenting amongst men of a Hobbesian state of war of all against all. Listen to what the subzoological Satan strives to achieve:

O Shame to men! Devil with devil damned
Firm concord holds, men only disagree

> Of creatures rational, though under hope
> Of Heav'nly grace: and God proclaiming peace,
> Yet live in hatred, enmity, and strife
> Among themselves, and levy cruel wars,
> Wasting the earth, each other to destroy:
> As if (which might induce us to accord)
> Mad had not Hellish foes enow besides,
> That day and night for his destruction wait. (*Paradise Lost,* II.496–505)

Satan continues the impetus toward an absolute deterritoriaziation against the divine despotic regime through the incessant seduction of man to be scurrilously torpid and sin without surcease:

> But he once passed, soon after when man fell,
> Strangle alteration! Sin and Death amain
> Following his track, such was the will of Heav'n,
> Paved after him a broad and beaten way
> Over the dark abyss, whose boiling gulf
> Tamely endured a bridge of wondrous length
> From Hell continued reaching th' utmost orb
> Of this frail world; by which the Spirits perverse
> With easy intercourse pass to and fro
> To tempt or punish mortals, except whom
> God and good angels guard by special grace. (*Paradise Lost,* II.1024–1033)

And, it is, above all, the Son of God who is weary that the divine despotic regime—through the diabolical agency of Satan (the "Adversary")—will, of necessity, cause most human beings to be damned to Hell:

> Or shall the Adversary thus obtain
> His end, and frustrate thine, shall he fulfil
> His malice, and thy goodness bring to naught,
> Or proud return through to his heavier doom,
> Yet with revenge accomplished and to Hell
> Draw after him the whole race of mankind,
> By him corrupted? … (*Paradise Lost,* III.156–162)

As a result of his Satan-induced perversion and imminently portended flight from the despotic regime of God himself, man must—of necessity—be killed, in the sense of being subjected to mortality itself. This is what scripture intimates. Milton's words:

> But yet all is not done; man disobeying,
> Disloyal breaks his fealty, and sins

> Against the high supremacy of Heav'n,
> Affecting Godhead, and so losing all,
> To expiate his treason hath naught left,
> But to destruction sacred and devote,
> He with his whole posterity must die ... (*Paradise Lost,* III.203–209)

As observed, the angst-ridden paranoid delusion of the despot as anal-hoarder was that a single working organ might perchance flow outside the despotic body, that it might break away or escape. That of the sadist consisted in his seeing rising up before him, against him, the enemy that brings death—an eye with too steady a look, a mouth with too unfamiliar a smile, each organ a possible protest. Finally, the paranoid despot as necrophile brings these foregoing moments to consummation: better not a sole survivor than for a single organ to flow outside this apparatus or slip away from the body of the despot. Immediate death to all. How crippled is this species. Unto himself, man shall henceforth be a beast.

Notes

1. Nietzsche 2000, 522, 523, GM II.17.
2. Ibid. 522, GM II.17.
3. Nietzsche 2003, 75–77, Z I "Of the New Idol."
4. Deleuze and Guattari 1988, 359.
5. Deleuze and Guattari 2004, 209.
6. Ibid.
7. Ibid. 237.
8. Ibid.
9. Buchanan 2008, 102, 103.
10. Ibid. 104.
11. Deleuze and Guattari 2004, 218.
12. Goodchild 1996, 94.
13. Ibid.
14. Ibid.
15. Ibid.
16. Deleuze and Guattari 2004, 218, 219.
17. Goodchild 1996, 94.
18. Holland 1999, 74.
19. Ibid. 75.
20. Ibid.
21. Deleuze and Guattari 2004, 214.
22. Nietzsche 1977, 161, Z I "On the New Idol."

23. Eliade 1978, Vol. 1, 141, 142.
24. Ibid. 219.
25. Ibid. 318.
26. Eliade 1985, Vol. 2, 7, 8.
27. Eliade 1978, Vol. 1, 76.
28. wa Thiong'o 2006, 16.
29. Kantorowicz 1957, 46.
30. Ibid. 47, 48.
31. Ibid.
32. Holland 1999, 74, 75.
33. Goodchild 1996, 94, 95.
34. Deleuze 2006, 141, 142.
35. Ibid.
36. Ibid. 142.
37. Holland 1999, 74, 75.
38. Wood 2003, 52.
39. Canetti 1978, 210.
40. Bataille 1991, Vols. 2 and 3, 240, 241.
41. Ibid.
42. Ibid.
43. Bataille 1988, Vol. 1, 57, 58.
44. Eliade 1978, Vol. 1, 218, 219.
45. Ibid.
46. Deleuze and Guattari 2004, 215.
47. Nietzsche 1968, 90, 91, WP 141.
48. Deleuze and Guattari 2004, 160.
49. Buchanan 2008, 105.
50. Deleuze and Guattari 2004, 215, 216.
51. Ibid. 216.
52. Ibid. 229.
53. Ibid. 212.
54. Eliade 1978, Vol. 1, 222.
55. Ibid. 91–93.
56. Ibid.
57. Kantorowicz 1957, 7.
58. Ibid. 9.
59. Ibid. 9, 10.
60. Ibid. 11, 12.
61. Ibid. 13, 14.
62. Buchanan 2008, 105.
63. Wood 2003, 27, 28.

64. Ibid.
65. Deleuze and Guattari 2004, 215, 216.
66. Deleuze and Guattari 1988, 388, 389.
67. Fukuyama 2012, 116–121.
68. Ibid. 237–239.
69. Ibid. 265–267.
70. Ibid. 199, 200.
71. Ibid. 189–192.
72. Holland 1999, 77.
73. Ibid.
74. Ibid.
75. Deleuze and Guattari 2004, 220.
76. Buchanan 2008, 106.
77. Derrida 2001, 225–229.
78. Deleuze and Guattari 2004, 224.
79. Deleuze and Guattari 1988, 112.
80. Ibid.
81. Ibid.
82. Ibid.
83. Ibid.
84. Ibid.
85. Deleuze and Guattari 2004, 229.
86. Nietzsche 2000, 522, 523, GM II.17.
87. Deleuze and Guattari 2004, 232.
88. Bierce 2011, 473.
89. Deleuze and Guattari 1988, 112.
90. Ibid.
91. ibid. 113, 114.
92. Ibid.
93. Canetti 1978, 231.
94. Ibid. 231, 232.
95. Sartre 1963, 20.
96. Canetti 1978, 378.
97. Llosa 2010, 37.
98. Ibid. 38.
99. Canetti 1978, 378.
100. Solzhenitsyn 1988, 104.
101. Ibid. 109.
102. Canetti 1978, 232.
103. Ibid.
104. Solzhenitsyn 1988, 115.

105. Canetti 1978, 233.
106. wa Thiong'o 2006, 10.
107. Ibid. 11.
108. Ibid.
109. Foucault 1991, 47–49.
110. Ibid. 53.
111. Ibid. 57.
112. Canetti 1978, 308.
113. Ibid. 308, 309.
114. Semelin 2007, 16, 17.
115. Ibid. 38.
116. Ibid.
117. Ibid. 38, 39.
118. Ibid.
119. Cited in Semelin 2007, 39.
120. Semelin 2007, 39.
121. Ibid. 40.
122. Cited in Semelin 2007, 40.
123. Semelin 2007, 41.
124. Cited in Semelin 2007, 19.
125. Semelin 2007, 19, 20.
126. Ibid. 20.
127. Fromm 1997, 390, 391.
128. Ibid. 391.
129. Ibid. 391, 392.
130. Ibid. 384, 385.
131. Ibid. 388.
132. Ibid. 388.
133. Ibid. 388, 389.
134. Ibid. 463.
135. Ibid. 438.
136. Ibid. 433.
137. Ibid.
138. ibid. 437.
139. Ibid.
140. Ibid. 441.
141. Ibid. 463.
142. Ibid.
143. Ibid. 540.
144. Foucault 1991, 50.
145. Ibid. 3–5.

146. Donnelly and Diehl 2011, 28.
147. Ibid. 30, 31.
148. Ibid. 32, 33.
149. Ibid. 35.
150. Ibid. 38, 39.
151. Ibid. 41–45.
152. Ibid. 145–147.
153. Ibid. 149, 150.
154. Ibid. 151, 152.
155. Ibid. 53.
156. Ibid. 58, 59.
157. Ibid. 64–71.
158. Ibid. 74–76.
159. ibid. 79.
160. Ibid. 100–102.
161. Ibid. 116.
162. Ibid. 98, 99.
163. Ibid. 79, 80.
164. Ibid. 84.
165. Ibid. 92.
166. Ibid. 118.
167. Ngoc and Huynh 1994, 54.
168. Ibid. 57.
169. Ibid. 71.
170. Ibid. 87.
171. Ngor and Warner 1988, 216–223, 240–244.
172. Sade 1990, 194–210.
173. Ibid. 240, 241.
174. Ibid. 250, 251.
175. Ibid. 242, 243.
176. Ibid. 248, 249.
177. Ibid. 250.
178. Ibid. 251.
179. Ibid. 518.
180. Ibid.
181. Ibid. 520.
182. Ibid. 534.
183. Žižek 2008c, 88, 89.
184. Zupančič 2011, 79.
185. Ibid. 80.
186. Sade 1990, 628–639.

187. Zupančič 2011, 80, 81.
188. Ibid. 81.
189. Dante 1971b, Canto III.136–140.
190. Ibid. Canto IV.103–105, 130–132, V.52–57, VII.91–136.
191. Ibid. Canto X.115–139, XI.67–78.
192. Ibid. Canto XIII.37–84.
193. Ibid. Canto XVI.1–36.
194. Ibid. Canto XVIII.97–108.
195. Ibid. Canto XIX.70–126.
196. Ibid. Canto XXIII.19–65.
197. Ibid. Canto XXIV.4–29.
198. Ibid. Canto XXV.112–124.
199. Dante 2003, Canto III.22–69.
200. Ibid. Canto V.27–49.
201. Ibid. Canto VI.7–19, 52–57.
202. Ibid. Canto VII.25–35, 55–59, 100–129.
203. Ibid. Canto IX.115–131, XI.4–12.
204. Ibid. Canto XII.47–48, 73–75, 101–105.
205. Ibid. Canto XIII.4–45, 100–102.
206. Ibid. Canto XIV.13–48, 77–90.
207. Ibid. Canto XVIII.24–36, 65–81, 103–126.
208. Ibid. Canto XIX.1–4, 13–30, 46–47, 73–81.
209. Ibid. Canto XX.4–24.
210. Ibid. Canto XXI.16–36, 46–57, XXII.16–30, 70–72.
211. Ibid. Canto XXIII.58–72, 91–102, 109–120.
212. Ibid. Canto XXIV.79–105, XXV.49–117.
213. Ibid. Canto XXVI.31–48.
214. Ibid. Canto XXVIII.21–42, 64–69, 101–105.
215. Ibid. Canto XXIX.40–89, XXX.52–54, 91–99.
216. Ibid. Canto XXXII.22–54, 124–135, XXXIII.76–78, 91–99, XXXIIV.10–15, 55–69.

4

Transhistorical Sadeo-Deleuzian Fugue, 1. Despotic Paranoia

221–210 B.C. (Qin Shi Huangdi)

The paranoid nature of the Qin State apparatus of the first Chinese emperor Qin Shi Huangdi may be gauged from its embrace of the Legalist school of statecraft—a Machiavellian and fascist system of dogma according to which the populace was to be held in check by brutality and terror.[1]

Fearing the flight of working organs from its reterritorializing sway portended in the still abiding strains of older philosophies of State refractory to its own, the paranoiac Qin State apparatus resolved to simply remove such ideas from the heads of future scholars. Thus, the government monopolized knowledge itself, burning all books not kowtowing to its hegemonic ideology.[2]

Qin forensics aimed not to seek justice, but to apportion blame and exact punishment. Even accidental deaths and suicides were treated as murder and scapegoats fabricated.[3]

This despotic paranoia rapidly assumed its sadistic aspect. State interrogators tortured not only suspects but also witnesses into confessions.[4] Male criminals had their eyebrows shaved, were tattooed, and often had their nose and feet hacked off, or were sent off to work in labor gangs in extremes of mountain cold or searing desert heat.[5] Pardoned criminals, tax-dodgers, fugitives from justice, convicts, the disenfranchised, and even State authorities deemed lax were rounded up like cattle and packed off to toil upon the southern frontier or on the Great Wall.[6]

The paranoiac Qin State apparatus was necrophilous from the start. Those who spoke in favor of foreign States were cut in two. The punishment for incest, even between half-siblings, was death. The punishment for a slave raping his mistress was death.[7] The punishment for false accusations of murder was death. The punishment for officials allowing murderers to escape was death.[8] The punishment for hoarding or discussion of banned literature was death. The punishment for officials allowing the foregoing to transpire was death.[9]

Seeing assassination attempts rising up from all quarters, the paranoiac and necrophilous emperor falsely indicted four-hundred and sixty scholars across the land for illegal dabbling in paranormal schemes and buried them alive.[10] When a recently fallen meteorite was found to bear a graffito portending the death of the emperor and the division of the Qin full despotic body into its former separate kingdoms, everyone in the region was tortured and executed.[11]

The emperor's fear of the dissolution of the despotic full body extended even beyond the grave. To avoid being posthumously tormented by grave robbers, he secretly installed a vast terracotta army in the series of underground chambers he was to occupy in the afterlife—the construction of which involved the piling up of the corpses of the tomb's builders and mass murder of the despot's concubines.[12] Better not a sole survivor than for a single organ to flow outside this apparatus or slip away from the body of the despot.

14–37 A.D. (Tiberius)

Seutonius relates a stanza of a satirical verse on the Emperor Tiberius describing how the tyrant had ceased to thirst for wine, as he had hitherto, and instead acquired a taste for a far more potable cup: one filled with the blood of murdered men.[13]

In the first third of the first century after the death of Christ, the dour sadistic paranoia of Tiberius swiftly exacerbated itself. Suspecting himself assailed form all sides upon the very slimmest of rumors and trifling portents, the turgid paranoiac despot was seldom bereft of a waking moment suffused with the dread foreboding of his numinous body politic dissolving and being rendered asunder right before his very eyes. Whence, and such that not a single working organ be permitted wilfully to flee the Tiberian full despotic body, his reign of terror commenced and death was to follow upon death.

After his two adoptive grandchildren through Germanicus, Nero and Drusus, recent appointees to the Senate, had had their names added to the New Year prayers for the safety of the Emperor; Tiberius openly despised the youths, incessantly arranged that false charges be brought against them, that witnesses were omnipresent when the pair vituperated against his schemes, and finally had the Senate declare them enemies of the State to be sentenced to death. Nero was forced to commit suicide when an executioner sent by the Senate came with the noose on which he would be hung and the hooks that would drag his lifeless corpse to the Tiber. His body and that of Drusus, starved to death in a Palace cellar, were then hacked into so many pieces that burial was rendered well-nigh impossible.[14]

Of Tiberius' newly formed council of foreign affairs consisting of in excees of twenty men, only two or three died natural deaths.[15]

Inquiring as to the dialect in which a certain Greek named Zeno was discoursing, Tiberius mistook the Greek's reply that it was Doric as a taunt referring to his prior exile to the island of Rhodes and swiftly banished him to the Aegean island of Cinari. Having discovered that Seleucis, a literature professor, had been finding out from imperial servants what books he was reading, Tiberius estranged him and then forced him to commit suicide.[16]

When a humorist hailed the corpse at a funeral procession, mockingly asking the latter to tell the ghost of Augustus that his dues to the commons had not yet been paid; Tiberius instantly ordered the man to be arrested and brought before him and, promising that he would give the humorist his due without delay, immediately ordered him to be executed.[17]

But Tiberius's despotic paranoia was soon to burgeon to previously inconceivable levels. After he had a man examined under torture and sentenced to death for allegedly decapitating an image of Augustus with the intention of substituting another head, the slightest and most ludicrously minute misdemeanors or eccentricies could incite the homocidal wrath of the State apparatus. People could be executed for mistreating a slave, changing their clothes before an image of Augustus, or for carrying a ring or coin bearing the latter's profile into a privy or a brothel.[18]

From this moment onwards, not a single day passed without an execution. Men were put to death with their children, some were accused by these same, whereas relatives where forbidden to go into mourning. The well-nigh ubiquitous myriad of informers were always believed in their accusations of the innocent. Every crime became a capital crime, even the slightest careless utterance. A poet who presented King Agamemnon in a bad light was swiftly put to death. A historian who descried Caesar's assassins, Brutus and Cassius, as "the last of the Romans" was put to death. Soon as many as twenty prisoners were executed every day, their corpses flung on the Stairs of Mourning and dragged to the Tiber with hooks. Virgins were raped by executioners and then strangled to death by the same.[19]

A deluge of lifeless human shapes precipitated itself into the sea from the top of a cliff in Capri, whilst the bodies hurtling down where bludgeoned with oars and boot-hooks by marines stationed below to ensure that they were dead.[20]

211–217 A.D. (Caracalla)

According to Cassius Dio, the notorius Roman emperor Caracalla "wished not only to know everything but to be the only one to know anything, and he desired not only to have all power but to be the only one to have power."[21] Caracalla's despotic paranoia reached such a feverish pitch that—sick alike in body and mind—he was assailed by distressing visions, often thinking himself pursued by men armed with swords.[22] His crippling paranoia was often aroused even by men without a recalcitrant bone in their bodies.

But it is precisely because these men asked nothing of him that they aroused his suspicions. Such men he censured and rebuked thus: "It is evident from the fact that you ask nothing of me that you do not have confidence in me; and if you do not have confidence, you are suspicious of me; and if you are suspicious, you fear me; and if you fear me, you hate me."[23] Maddened with anxiety by the inscrutability of men and stricken to the core by unfounded fears of an imminent internal coup against his rule, he arrested without charge the slaves, freedmen, and intimate friends of senators, and had them tortured to reveal his standing in the eyes of these and other senators.[24] He spent a gargantuan thirty-million sesterces on the purchase of massive quantities of various poisons from Upper Asia in order to secretly kill great numbers of men.[25]

For, as Herodian has it, Caracalla's incorrigible paranoia caused him to be perpetually suspicious that well-nigh everyone was plotting against him. To the end of unmasking this vast panoply of perceived conspirators, he even went as far as to immerse himself in the occult. He consulted oracles and summoned wise men, astrologers and sacrificial interpreters from every land. When he posthaste became suspicious of even these men, he ordered his most intimate friend Maternianus to seek out the best seers and, by consulting the souls of the dead, to divine the subterranean burgeonings of sedition within the entrails of the State apparatus.[26]

Upon rising to the position of sole emperor, and roiled by a singularly necrophilous paranoia, he immediately commenced a general slaughter of the recently deceased former co-emperor Geta's household, friends, and palace staff. All servants, even children, were murdered and their corpses mutilated, dragged around the streets, and conveyed outside the city where they were cremated in a great heap or simply discarded at random. Not a single person who had been even remotely associated with Geta survived. Senators of distinction were executed on the most trifling or even non-existent charges. Anyone connected with the principate by birth or descended from a patrician family in the senate was killed. Governors and procurators in the provinces were destroyed on the allegation of having been friends of Geta.[27]

He instantly put to death twenty-thousand imperial freedmen, men and women, who had formerly been with Geta wherever they chanced to be found in the palace.[28] Inexorably roiled by his rampant paranoia, he even accused those who had never had the slightest dealings with Geta of seditious intent. The number of these men was vast. Anyone who dared to even write or utter the name of Geta was immediatedly put to death, whilst those whose wills contained the name forthwith suffered wholesale confiscation of property.[29]

He buried the Vestal Virgins alive on the spurious grounds of a violation of chastity. When a crowd jeered at a charioteer whom he supported, he was so rankled by paranoia that he immediately ordered the cuprits to be killed, whereupon a multitude was slaughtered indiscriminately. Every night men of all classes were murdered.[30] Another charioteer, Euprepes, met his end for supporting the opposite faction to the one favored by the emperor.[31] In violation of all law, he put adulterers to death. Caecilius Aemilianus, the governor of Baetica, suffered death for consulting the oracle of Hercules at Gades.[32]

He murdered his friends and slayed all the men he wished to slay without distinction, whether guilty or innocent.[33] In his rabid necrophilous fury he killed many men in the open, whilst the others he exiled to climatically extreme outlying provinces where they perished in obscurity of searing heat or hyperborean cold, so that his hands were likewise soiled with the blood of these.[34]

Having been privy to reports that the Alexandrians were lampooning him in their usual light-hearted manner, Caracalla's necrophilous paranoia reached its apogee.[35] Soon people were being murdered indiscriminately in all parts of the city by night and by day, perishing wherever they were found according the dictates of chance. Foreigners were expelled, their property plundered, their shrines despoiled. To thwart what he bethought an omnipresent threat of rebellion against his despotic rule, Caracalla abolished all possibility of large communal gatherings such as spectacles and public messes, and ordered that Alexandria be divided by a massive cross-wall and permanently occupied by guards.[36]

Under the cover that he wished to enroll a phalanx in honour of Alexander, he issued an edict ordering all young Alexandrian men to assemble on an open piece of ground. With the young men assembled, Caracalla's army encircled them. Trapped like defenseless animals in a net, the young men were forthwith slaughtered *en masse* and cast into and buried in huge pits, often half-alive; so that a vast communal burial mound was swiftly raised upon this plethora of corpses, and the mouth of the Nile and the entire surrounding seashore was blotted red with vast streams of human blood.[37]

527–565 A.D. (Justinian I)

In the words of Procopius of Caesarea, the emperor Justinian was "the author of calamities for the Romans—calamities so serious and so manifold that in all the history of the world probably no one previously had ever heard their equal. For he used to proceed with the lightest of hearts to the unjust murder of men and the seizure of other men's money, and for him it was nothing that countless thousands of men should have been destroyed, though they had given him no grievance."[38] So livid with sadistic paranoia was the imperial despot that "not one living person of the entire Roman world had the fortune to escape" his wrath, "but, like any other affliction from Heaven falling upon the whole race, he left not a single soul wholly untouched."[39]

So singularily inexorable and rabidly necrophilous was his fury that "to state exactly the number of those who were destroyed by him would never be possible, I think, for anyone soever, or for God. For one might more quickly, I think, count all the grains of sand than the vast number of whom this Emperor destroyed."[40]

Procopius goes on to maintain that the wrath of Justinian had, in particular, purged the populace of Libya to the extent of reducing the land to a veritable Scythian wilderness.[41] For under Justinian mankind was subjected to the despotic paranoia of a "demon who had become incarnate" and who "by means of a hidden power" and "a diabolical nature" ensured "that the Deity, detesting his works, turned away from the Roman Empire and gave place to the abominable demons for the bringing of these things to pass in this fashion."[42]

Scarcely ten days on the throne, he had the director of the palace eunuch's, Amantius, slain merely for having spoken a hasty word against John, the chief priest of the city. Suspected of having been offended, he likewise executed one Vitalian.[43] Should anyone seek to gain forgivenness for purported offenders through the intercession of prayers and supplications, "'enraged and shewing his teeth,' he would seem to be ready to burst," and still more bent on the swift annihilation of the accused.[44] Likewise was it the case with the empress Theodora, whose constant suspicion failed even to be assuaged by the deaths of those that it caused—passing instead to their offspring, even up to the third generation.[45]

Stricken by paranoia and festering suspicions of seditious plots metastatizing within the bowels of the State apparatus, Justinian and his royal consort lent an eager ear to all manner of slanders and—obviating the need for thorough investigation—immediately moved to act on the words of calumniators and precipitously issued orders for the capture of towns, the burning and dismantling of cities, the death of thousands of innocent men, and the enslavement of entire populations.[46]

Such was the shared sadistic and necrophilous lust of the royal despots that reports of even the most trivial and irrelevant pseudo-offenses led the empress to posthaste fabricate a litany of exaggerated and utterly spurious accusations, whence the accused was fleeced of his property, and punished with banishment and death.[47]

Within the great paranoiac State apparatus of Justinian, a throng of spies was constantly deployed to report to the empress what was said and done in the market place and even in the homes of the citizenry. Those found to be of too loose a tongue would be bundled up, shackled and conveyed to the uttermost parts of the empire to suffer a prolonged and miserable existence, and at length die of privation or exposure.[48]

Among other imaginary tubercular microbes threatening the integrity of the emperor's numinous body politic were the so-called "heresies" of inter alia the Montani, Sabbatiani, and especially the Arians. Here Justinian, "with the one thought in mind that the earth should by many a device be filled with human blood," proceeded forthwith to confiscate and despoil *en masse* the properties and wealth of the votaries of diverse sects, repressing all resistance with measures so draconian that "the whole Roman Empire was filled with murder and with exiled men."[49] When the residents of Caesarea were likewise perforce compelled to orthodoxy but were instead espied by the despot to be inclining to the practices of the Manichaeans and the Polytheists, one-hundred-thousand perished in torrents of blood, leaving the land entirely purged of farmers.[50] The suspicious and paranoid eye of Justinian thereupon forthwith cast itself upon the "Greeks" who, though they had wholly come to espouse the religion of the empire at least in outward appearance, nonetheless had their properties plundered and, being overtaken and seized at their libations, sacrifices and other acts repugnant to the sight of the State apparatus, met their own destruction.[51]

The despotic full body of Justinian suffered the threat of a detachment of its working organs when the ordinarily opposed Blue and Green factions united in fomenting the Nika riots. Conspirators were releasing prisoners incarcerated on suspicion of sedition, killing government officials, setting buildings alight, and looting property.[52] They declared Hypatius, the nephew of the late emperor Anastatius, emperor against his will, crowning him with a golden necklace in the Forum of Constantine.[53] Jolted into a singularily sado-necrophilous paranoiac posture, the Roman State apparatus posthaste summoned the generals Belisarius and Mundus from their fronts abroad, together with the full military brunt of battle-hardened spearmen, guards and Erulian barbarians. The generals forthwith entered the hippodrome through the Bronze Gate and the Gate of Death, and descend upon the teeming partisans of Hypatius congregated there, slaughtering the recalcitrants to the number of thirty-thousand dead—whereupon Hypatius was executed and his lifeless body hurled into the sea.[54]

996–1021 A.D. (Al-Hakim bi-Amr Allah)

Though considered by some to be divine, the Fatimid caliph al-Hakim bi-Amr Allah ("He who commands in the Name of God") was a demented and despotic tyrant irrationally given to murdering his subjects upon a whim.[55]

One of the movements of deterritorialization threatening the full despotic body was associated with deviations from of the Shi'a State religion. This despotic paranoia swiftly progressed toward its sadistic and necrophilous aspects. A Syrian man who did not acknowledge 'Ali was arrested and, when upon release persisting in contumacy, was hung and crucified. Another man was hung for cursing A'isha and her husband. Groups of men who had prayed the forbidden Sunni fore-noon *duha* prayer were arrested, beaten and publically paraded on camels prior to confinement. Some who had prayed the prohibited sunni *tarawih* prayer during Ramadan were executed.[56]

In its anal-hoarding aspect, the despotic paranoia of al-Hakim also assumed the form of the outlawing of the selling and making of *fuqqa'* beer. Homes were searched for intoxicating beverages and a great deal of containers for it were smashed. It was forbidden to transport wine and another type of beer called *mizr*. Places which sold wine were raided. Groups found making or possessing *fuqqa'* were arrested, beaten and paraded. Those found selling wine were jailed. The selling and transportation of raisins, the prime ingredient in the alcoholic drink *zabib*, was forbidden. Gargantuan quantities of these raisins were thrown into the Nile or burned in public. The selling of honey, used to make mead-wine, was forbidden. Warehouses for honey were sealed, and thousands of honey jars were seized and sunk in the river. Grapes, whose sale was forbidden, were thrown into the Nile, ripped from vines in many regions and crushed under the hoofs of cattle. Groups connected to *fuqqa'* and *zabib* were arrested, jailed and ordered to be hung.[57]

The paranoia of al-Hakim's despotic regime, in its anal-hoarding aspect, was so extreme that it extended to the prohibition of foodstuffs with merely a symbolic association. *Mulukhiyya*, or Jew's mallow, was forbidden as it was eaten by Mu'awiya, the first Umayyad caliph and arch-enemy of the Shi'a. *Jirjir*, a variety of watercress, being associated with the Prophet's wife A'isha, an opponent of 'Ali, was forbidden. *Mutawakkiliya* was forbidden due to its connection with the Abbasid caliph al-Mutawakkil. It was forbidden to eat a small scaleless fish called *dalinis*. This culinary paranoia would, however, soon enter into its sadistic aspect. Fishermen were arrested for the contravention of the latter prohibition. Those caught with *mulukhiyya* and *dalinis* were beaten, paraded, jailed, and ordered to be hung.[58]

The paranoid comportment of the full despotic body of al-Hakim likewise effected a slew of prohibitions on the movements of subjects. All subjects were to remain at home following the evening prayer. It was forbidden to perform songs publically on the banks of the Nile, listen to them in houses or to drink in boats. Public houses were raided and many arrested. It was forbidden to erect tents among the tombs of saints. People were not to congregate in the desert or on the banks of the Nile, whereas women were forbidden to go to the baths, frequent places where they would be in proximity to men, or to sing or chant. Eventually, women were forbidden to leave their homes at any time whatsoever, being effectively sentenced to interminable house arrest, under pain of assault and incarceration. Once when overhearing the clamor of bathing women, the sadistic al-Hakim ordered the exits blocked and the fire heating the water stoked so high that all of the women inside died without having committed any crime at all.[59]

The forewarning eye of al-Hakim's despotic regime would constantly survey the baths to see if anyone entered without a loincloth, or *mi'zar*. Those caught were beaten and paraded. Displays of mourning and grief at funerals, as well as the public following of funeral processions were forbidden. Songs and public amusements were likewise prohibited. The game of chess was outlawed. Chess pieces were collected wherever they could be found and burned. Those caught playing chess were beaten and jailed. The necrophilous aspect of the regime's paranoia even extended to animals. More than thirty-thousand dogs were killed at the caliph's command and their bodies dumped in the desert and along the banks of the Nile.[60]

In order not to escape the purview of the great paranoiac, Jews were ordered to don the *zunnar* (a waistband) and Christians the *ghiyar* (a badge), both of which items had to be black—a symbol of the seditious Abbasids. Failure to do so resulted in those caught being beaten. Jews and Christians could not enter the baths unless they wore a bell and a cross, respectively. It was forbidden to sell slaves to Christians and Jews. Christian governmental clerks were often arrested and jailed, and sometimes executed. Crosses were seized from churches and burned at the doorways of mosques and at police stations. Monasteries and shrines were destroyed, and the patriarch of Alexandria was secretly put to death. Over thirty-thousand churches, convents, shrines, synagogues and non-Muslim cemeteries were demolished and looted by mobs, and the associated land along with tenements and other properties seized by the State. This included the church of the Holy Sepulchre.[61]

The despotic paranoia of al-Hakim's regime would likewise redound upon the necks of his kin, governmental officials and rebels. Having sent spies to a party attended by his cousin, who overheard an astrologer merely joking to the latter that he should be caliph, the despot had the cousin and all of his companions from the party executed and their bodies burnt. After his personal physician had accidentally drowned himself at a gathering attended by the chief executive and the chief justice, though they were innocent, al-Hakim had them killed in addition to the chief executive's sons, whose heads were sent to Cairo. When the sailor al-Allaqa revolted at the city of Tyre, he was hauled away in chains to the capital, crushed by a lead dunce cap, flayed alive and crucified with all his associates. When one Abu Rakwa declared himself the Umayyad messiah and revolted, he was captured, strapped to a camel whilst being beaten on the head by a monkey and pelted by bricks, hung and crucified. His head, together with over thirty-thousand of his accomplices were loaded onto one-hundred camels and thrown into the Euphrates.[62]

1209–1242 A.D. (Albigensian Crusade)

For Bernard de Clairvaux, though he regarded persuasion as a necessary first recourse, obstinate heretics were to be excommunicated and killed. Charles the Bald regarded heresy as worse than simple unbelief, in that it most insidiously distorted and corrupted orthodoxy itself. In that a Medieval socius was held intact by the uniformity of Catholic *Weltanschauung*, heresy was no less than a spreading poison eroding the body politic from within. St. Thomas Aquinas, comparing heretics to counterfeiters of the coinage, maintained that if the latter were executed for undermining the secular foundations of society, likewise should heretics be burned for undermining the spiritual foundations of society. Heresy was, thus, considered a threat to the Christian State itself. For this reason, canon lawyers in the twelfth century started to apply the Roman law of *lèse-majesté* to heresy—whilst theologians, alluding to the inherent rebelliousness of heresy, invited the swift intervention of the secular arm.[63]

For more than fifth years prior to the first decade of the thirteenth century, the quasi-manichaean heresy of the Cathars had been metastatizing in the lands of Languedoc. For failing to supress the spread of Catharism, Raimon VI, count of Toulouse, was himself under suspicion of heresy.[64]

Against this backdrop, on January 8, 1198 the full despotic body of the Roman Pontiff became that of Innocent III. Called "to reign over kings from the throne of glory" and being the first pope to use the title of "vicar of Christ," Innocent's despotic tendencies were present from the outset. He had a politician's autocratic bent. Princes were to be the agents of his vengeance.[65] Under Innocent the church was to be a veritable State.

On January 14, 1208 papal legate Peire de Castelnau was murdered in the lands of Raimon. Upon hearing of the legate's demise, the numinous full body of Innocent braced itself into a distinctly paranoiac posture. Convinced that the entirety of the landmass between the Rhône and the Garonne was poisoned by heresy, that Raimon was Castelnau's murderer, and thus seeing a seething plethora of working organs rising up against him—he vilified the count as a protector of heretics, thieves, and assassins. Lest a single organ would perchance slip away from the full body of the church, the pope immediately called upon all knights and barons to take up the cross and, like angels of death, "drive the heretics from amongst the virtuous" in a rampant crusade. Castelnau had been slain by "one of those mercenaries of Satan." His murder was the sign of heretical gangrene rising up within the very bowels of the transnational Catholic State apparatus. And so, Innocent beseeched his *crucesignati*, "in the name of the God of peace and love," and with "our promise of remission of sins," to posthaste "root out perfidious heresy" and to purify and disinfect the land. "Attack the followers of heresy more fearlessly than even the Saracens," he roared, "since heretics are more evil!"[66] These crusaders he enjoined to exterminate the serpent and cancer of heretical depravity—an injunction that was at length to be reiterated within the next twenty years by his successors Honorius III and Gregory IX.[67]

The sheer extent of Innocent's despotic paranoia is evinced by his delusional conviction that the sly foxes and venomous serpents of heresy, after centuries of concealment, were now poised on massacring Christians. These rabid beasts were teeming in Languedoc. All persons in the tempestuous zone between the Rhône and the Garonne were the microbes of a malignant plague countenancing the full body of Christendom with its own terminal dissolution. "Indeed," Innocent III thundered, "such pestilential Provinçales not only strive to devastate all we possess but they strive to annihilate us! Truly, not only are they sharpening their hands to annihilate our souls, they are raising their hands to annihilate our bodies!" These "perverters of souls and putrefiers of bodies" festering like a leprosy of the full papal body's very entrails, if not immediately obliterated, would spell the final and conclusive desuetude of all Christian existence. These perverse and putrescent heretical microbes would therefore have to be preemptively assailed with the full military brunt of the crusading war machine and be utterly destroyed.[68] The paranoid State apparatus appears as a machine of necrophilous extermination. Better not a sole survivor than for a single organ to flow outside this apparatus or slip away from the body of the despot.

Such forthwith was the necrophilous bent of the crusading machine that there ensued from the very outset of the holy purge an irrevocable obligation to mass murder. The runaway paranoia of Innocent III translated into an pervasive atmosphere of intense suspicion wherein any plea of ignorance on the part of the populace appeared as wilful complacency with heresy, if not a palpable symptom of infection. Even that which appeared orthodox on the surface was seen as a mere facade beneath which lurked the sempiternal menace of pullulating heresy. The *sine qua non* was thus indiscriminate slaughter.[69]

Forthwith did Simon de Montfort's *crucesignati* descend upon the Cathar vermin. At Béziers, persons—irrespective of age, station, or sex—were beaten to death or slain with knives and cudgels, as rivulets of gore spilled down the streets, whilst the remainder were incinerated together with the entire city.[70] The streets and alleys of Carcassonne were left strewn with the desiccated bodies of men, women, and children, the fetor of whose swollen corpses mingled with the stench of rotting animal carcasses.[71] At Bram, Monfort dug out one-hundred-and-ninety-nine eyes and hacked off one-hundred noses.[72] At Minerve, a plethora of heretics and another one-hundred people were burned in a massive flaming pyre and their corpses buried in mud.[73] Most of the inhabitants of the town of Lavaur were herded into a meadow and torched in a colossal fire whilst it's garrison was put to the sword.[74] The widowers, widows, and little girls of les Cassés were incinerated.[75] The inhabitants of las Toellas and Hautpoul were slaughered.[76] The meadow of Muret was turned into a field of naked bodies, thousands of whom were buried in cloying river mud.[77] The defenders of Casseneuil were slaughtered and its walls razed.[78] At Marmande, flesh, blood, brains, torsos, limbs and faces were carved and ripped; livers and guts torn out and strewn across marshes and meadows soaked red with blood.[79] Even when the periodically relaunched crusade against the Cathars officially ended in 1229, inquisitors scoured every last corner of Languedoc for heretics from 1234 onwards, burning the obstinate and even parading the exhumed remains of heretics convicted after their own deaths through the streets of cities.[80]

The final denouement of the Cathar tragedy came with the *Götterdammerung* on the vertiginous mountain fortress of Monségur atop an almost sheer rock rising five-hundred feet out of the ground, where on March 16, 1242, more than two-hundred were incinerated in a massive pyre of flaming wood.[81]

1547–1584 A.D. (Ivan the Terrible)

In March 1553, when Tsar Ivan IV, or Ivan the Terrible, was lying infirm from a fiery sickness, terrible suspicions germinated in his exhausted mind. Though always inclined to view his subjects in the worst possible light, during his illness he had grown still more suspicious.[82] He believed himself dying among traitors.[83] These days of sickness were fraught with a debilitating anxiety that would never quite leave him. The alleged treachery he had perceived all around him while he thought himself dying had reduced him, the Tsar of all Russia, to utter impotence. He had felt his throne usurped. He had even imagined his wife and son dead. Though his sickness lasted a mere week, it would leave an indelible mark on his soul. Henceforth he would trust no one, not even his closest intimates.[84] When in 1560 his wife Anastasia had died of illness, he was convinced that she had been poisoned and that her death was the work of the Boyars.[85] Later, in 1566, he would complain in desperation to Philipp Kolychev: "Do you not know, that my people desire only to devour me? There are people close to me who are preparing to destroy me!"[86] He was absolutely certain that plotters lurked everywhere. Danger was ubiquitous.[87] There were even sorcerers and witches all over Russia casting spells against him.[88] He saw rising up before him, against himself, the enemy that brings death—an eye with too steady a look, a mouth with too unfamiliar a smile, each organ a possible protest.

Hitherto, his incorrigible desire to rule imperiously as he saw fit, without the interference of anyone whomsoever, as a pure autocrat, clearly evinced a deep-seated anal-hoarding paranoia.[89] He had after all contended to Prince Andrey Kurbsky that he was absolute Tsar ruling by divine right, and that there thus was no limit that could be set to his earthly powers over his dominions.[90]

With the infernal ordeal of his sickness and the passing of his wife, this endemic despotic paranoia rapidly advanced toward its sadistic and necrophilous forms. Henceforth, violence would become his *modus vivendi*, and he would murder his companions. Seeing the dead around him was his solace. He attempted to destroy everyone—the Boyars, even his friends. Power became above all the power to kill. The Tsar as mass murderer entered the stage. Yet this man of boundless murderous power struck out against a plethora of imagined enemies since he was himself at the mercy of uncontrollable fears. He saw evil rising up everywhere and was relentlessly determined to destroy it with monstrous murderousness. If need be, he would destroy the whole of Russia to save his soul. And thus for nearly a quarter of a century, his regime was one of unmitigated violence and terror.[91]

He became a pure tyrant, wallowing in his own ferocity, given to breathtaking acts of the most obscene violence, killing real and perceived enemies wherever they chanced to be.[92] He luxuriated in torture, pools of blood, stiffening corpses, screamed boisterously when the blood of the tortured besmirched his face, as though his only joy was the infliction of pain and the degradation of defenseless innocents.[93] He assuaged his paranoia by the contemplation of men dying in agony. At the same time as the teeming plethora of imaginary enemies multiplied all around him, he continuously invented new and more excruciating forms of torture and ways of putting men to death.[94]

This necrophilous murderousness of the despot led Kurbsky to conclude that the Tsar was induced by none other than the Devil, and was himself the Anti-Christ. Yet it is here that the sheer magnitude of paranoia impelling the bloody necrophilia of the tyrant is evinced. For Ivan answers this opprobium by forcefully maintaining, on the contrary, that he is the most holy of all Russians and in fact the only man who stands between Russia and Satan. Not only is he the most godly and innocent, but to deny this innocence is to infallibly exhibit oneself as a traitor.[95] And there was no treacherous devil that he had slain who did not deserve to be killed a thousand times.[96] To the Boyars he once fumed: "I shall make it my business to destroy all of you root and branch, so that not even a memory of you remains!"[97] He destroyed blindly, not even caring who he was killing. Like a maniac.[98]

All around him the paranoiac Tsar created a wilderness, a no-man's-land, where no recalcitrants could dwell. Princes, Boyars, nobles, and commoners were executed simply because the despot felt safer when they were dead. He pined to annihilate his enemies to the very last man. In ever-widening circles he would reduce Russia to destruction.[99] Better not a sole survivor than for a single organ to flow outside this apparatus or slip away from the body of the despot.

In his despotic paranoia, Ivan believed that Russia was and had always been crawling with traitors desirous of dethroning him, that secret agents lurked in his court, and that his officers were continuously negotiating with foreign powers.[100] Continually threatened with the perceived threat of a flight of working organs from the Tsarist full body's reterritorializing sway, he formed a State security arm: the *Oprichniki*. The paranoiac State apparatus turned sadistic and necrophilous. A seven year reign of terror commenced during which hundreds of people were continuously being rounded up by men carrying witches' brooms and severed dogs' heads and hurled into the prison chambers of Alexandrova Sloboda where they were tortured and killed, often before Ivan's gloating eye. Scarcely a day passed without the murder of between twenty to forty people. They were strangled, drowned, burned, decapitated, had their tongues cut out, were hacked to pieces, as the stench of unburied corpses became unbearable.[101]

Not only the suspects themselves were killed, but also their entire families and all of their servants, whilst their houses were pillaged and burned to the ground, and their wives raped.[102] The reign of terror spread across the entire country, and not a single person, rich or poor, noble or peasant, was immune to it.[103]

When a petition against State violence was presented by a group of Boyars, a hundred and thirteen men were executed in Moscow's Red Square and buried in a mass grave, and their wives and daughters clubbed to death and hurled into the river.[104] With the mere separateness and individuality of the inhabitants of Novgorod already smacking of treason, the unearthing of a nonexistent plot headed by Archbishop Pimen aimed at bringing all Novgorod territories under the sway of King Sigismund Augustus of Poland jolted the despotic Tsarist full body into a still more rabidly sadistic and necrophilous paranoiac posture than heretofore.[105] In two days, nine-thousand Livonians and exiles from Polotsk living in Tver were summarily executed and pushed under the ice of the Volga River. More than five-hundred monks, abbots, abbesses and nuns in and around Novgorod were clubbed to death. Citizens of Novgorod were tortured into confessing imaginary crimes of sedition, made to implicate one another, dragged through the snow with ropes, and hurled into the Volkhov River were they were hacked and chopped with axes, thrust with pikes and hooks, until more than thirty-thousand—a third of the population—died. Another twenty-thousand perished in the ensuing famine, while the butchered corpses choking the river fomented a plague threatening to engulf the entire nation.[106]

1873–1874 A.D. (Robespierre)

Even before he left his native Arras in 1789, Robespierre had already been intensely preoccupied with the sempiternal threat of internal enemies to the State, having given one of his pamphlets the title *Les ennemis de la patrie démasqué* (*Enemies of the Country Unmasked*).[107] From the outset of the Revolution itself, he was convinced that every quarter of Paris was crawling with enemy spies, would-be murderers of patriots, and the agents of an aristocratic plot against the young Republican State apparatus. So acute was his paranoia that he even feared the presence of his name on an envelope would attract malicious attention.[108] He would greatly exaggerate purported plots and conspiracies that were or were not afoot. This paranoia would become a political obsession and posthaste intensify to the point of nausea.[109]

In his journal *Lettres à ses commettans*, addressed to his constituents, he contended with force that "the enemies of the republic are those dastardly egoists, those ambitious and corrupt men. You have hunted down kings, but have you hunted out the vices that their deadly domination has engendered among you? Taken together you are the most generous, the most moral of all peoples … but a people that nurtures within itself a multitude of adroit rogues and political charlatans, skilled at usurpation and the betrayal of trust."[110] In a later speech in the Convention, Robespierre was almost incoherent with paranoia: "Slander, arson, poison, atheism, corruption, starvation, and murder—they have been prodigal in every sort of crime: but there still remains assassination, assassination and again assassination."[111] The despotic revolutionary regime sees rising up before itself, against itself, the enemy that brings death—an eye with too steady a look, a mouth with too unfamiliar a smile, each organ a possible protest.

With the full despotic body of the Republic rankled by such suspicions and many others, Robespierre persuaded the Convention to revive the Revolutionary Tribunal, with its extraordinary powers to condemn people to death. Capital punishment, he demanded, ought to be applied to counter-revolutionary acts of any kind. There would, moreover, be no appeals against the Tribunal's judgments.[112] What rendered it still more draconian was that the jury's decisions were to be intuitive rather than reasonable, implying that the accused could be convicted not merely for their actions, but also for their perceived attitudes and dispositions.[113] This was Robespierre's infamous trial by "conscience," according to which "Whoever trembles is guilty."[114]

The Tribunal was placed in the hands of the Committee of Public Safety, which augmented the existing Committee of General Security. Notwithstanding,

Robespierre was more paranoid than ever, obsessed that the Republic would die, and convinced he was about to be assassinated.[115] That this paranoia of the full body of the Convention had already reached its necrophilous *telos* is evinced in Marat's rabid words at the Jacobins: "No! We are not going to die; we will give death to our enemies, we will erase them!"[116] The utmost necessity for such necrophilous measures is easily gauged from the sheer extremity of Marat's feverish paranoia, as exemplified in an almost demented effusion from his *The Friend of the People*: "Your emenies need only triumph for a moment, and blood will flow in torrents. They will murder you without compassion, they will rip open the bellies of your wives, and in order to choke within you the love of liberty, their bloody hands will explore the entrails of your children to find their hearts."[117] Despite such sentiments, Robespierre was still confident of being one step ahead of the potential sowers of sedition: "Must we despair of the safety of the Republic? No! Tyrants unmasked are nothing. The French people are only betrayed because they want to be; the French people are stronger than all their enemies. One republican who knows how to die can exterminate all the despots."[118] He would later reiterate the same sentiments: "Would to God that my physical strength were the equal of my moral fortitude! I might then, this very day, confound the traitors and call down national vengeance on every guilty head."[119] In a succinct formulation of paranoiac despotic vengeance, Saint-Just asserted the existence of "three sins against the republic: one is to be sorry for State prisoners; another is to be opposed to the rule of virtue; and the third is to be opposed to the Terror."[120] The punishment for all of these sins was death. This necessity of death to all purported vermin coruscating within the bowels of the State apparatus was rammed home by Robespierre in yet another paranoid and necrophilous rant at the Jacobins during which he asseverated that "my spirit is more disposed than ever to unmasking traitors and tearing off the masks with which they still dare to cover themselves … We swear by the daggers already reddened with the blood of the Revolution's martyrs, and recently sharpened for us too, to exterminate every single one of the criminals who want to rob us of happiness and liberty."[121] Better not a sole survivor than for a single organ to flow outside this apparatus or slip away from the body of the despot.

On September 17, 1793, the terrifying Law of Suspects was passed by the Convention. According to the law, anyone could be arrested and put to death who "either by their conduct, their contacts, their words or their writings, showed themselves to be supporters of tyranny, or federalism, or to be enemies of liberty."[122] Under this law everyone found not carrying a certificate of *civisme*—at once an identity card and stamp of civic virtue—could be arrested.[123] This draconian legislation was further exacerbated by the infamous Law of 22nd Prairial, which inter alia ushered in a new official category of criminals: the enemies of the people. These tubercular microbes included any who sought to reestablish the monarchy, betray the Republic, discredit the Convention, communicate with foreign enemies, shelter conspirators, interfere with food provisions, corrupt officials, speak ill of patriotism, spread false news, mislead the people, deprave the public conscience, insult morality, abuse public office, steal public property, or plot against the liberty, unity, and security of the State. The punishment for all these infractions was death. The formulation of the law and its vast panoply of crimes, made it possible to execute people for verbally insulting the nation, for producing sour wine, for hoarding, writing, or attempting to communicate with the English. This law menaced absolutely everyone, on the most spurious grounds, and without recourse to any form of legal defense.[124]

The paranoia of the regime was such that enemies of the State were even suspected to be lurking within the upper echelons of government. Before long the Girondon faction of the Convention, who according to Robespierre's fevered imagination were conspiring against the regime with General Dumouriez, property-owners, the nobility, and even British Prime Minister William Pitt, came under suspicion.[125] Brissot, whose federalist ideas menaced Robespierre's organicist notions of nationhood, and twenty other Girondins were arrested *and guillotined*, those still eluding the regime were declared outlaws and hunted down, while Pétion's body was found in a field half-eaten by wolves.[126] After proponents of violence Hébert and Carrier announced a state of insurrection that never materialized, the Hébert faction was swiftly denounced, charged with conspiracy, arrested, tried, and all but one sent to the *guillotine*.[127] Lenient Jacobins Camille Desmoulins, Danton, and Fabre—on the basis of Camille's calls for freedom of the press and the cessation of the Terror, Fabre's financial corruption, and Danton's former purportedly treacherous links with monarchists and Girdondins—were all implacated by the paranoid and necrophilous mind of Robespierre in a counter-revolutionary plot, arrested in the middle of the night, thrown into solitary confinement, charged with attempting to overthrow the government, given truncated and rigged trials, *and guillotined*.[128] When the so-called "federalist" revolt against the revolutionary government had begun to spread through the departments of the Midi; the paranoiac State apparatus retorted with a campaign of brutal necrophilous repression, including mass executions by grapeshot fired from cannons in Lyon and group drownings in the Vendée.[129]

Throughout the nine months of the Revolutionary Terror over sixteen-thousand people were formally condemned to *death* and *guillotined* whilst many more unofficial victims *died* in custody or were lynched like *dogs* without trial.[130]

1915–1916 A.D. (Armenian Genocide)

Already in 1906–1907, rising Young Turk leaders and ideologues Mehmed Nazim and Behaeddin Shakir referred to the Armenian populace dwelling within the waning Ottoman Empire as "tubercular microbes" that were contaminating the State. Ziya Gökalp, a leading theoretician of the pan-Turanian movement, averred that the Armenians effectively comprised "a foreign body in the national Turkish state." By the outbreak of the second world war, a Turkish nationalist slogan contended that the "Balkan dogs are trampling on Islam," whilst a pro-CUP (Committee of Union and Progress) newspaper ominously threatened that "it will be impossible to spare eastern Turkey," that is to say, the Anatolian home of most Armenians. Even prior to assuming the reigns of government, the Young Turk movement was already seeking to vastly diminish the far greater tolerance afforded to racial and religious minorities by both traditional and liberal versions of Ottomanism. Soon the CUP would ban national minority associations and clubs and impose a policy of language Turkification. On August 6, 1910, at a secret top-level CUP conclave in Saloniki, it was decided to entirely eschew the older imperial framework of formerly autonomous millets in favor of the coercive and "complete Ottomanization of all Turkish subjects." The suggested methods of implementing these draconian designs given by other party branches included "deportation," "uprooting," and even "extermination of all Christians hostile to young Turkey."[131]

In 1913, with the Young Turk seizure of the reigns of power, the full despotic body of the Ottoman Empire became that of the triumvirate of the three Pashas: Enver, Talât, and Jemal. Seeing rising up against itself, against themselves, what they perceived to be an enemy bringing death—an eye with too steady a look, a mouth with too unfamiliar a smile, the despotic regime of the Young Turk State braced itself into an increasingly paranoid posture. The escalating paranoid sadism of the Pashas was soon reflected in a panoply of vitriolic documents appearing in the Turkish press. According to a letter of November 1913, "You Armenians … you accursed ones have brought many perils on the head of our esteemed government [and] paved the way for foreign assault.… Turkish youth … shall not delay the execution of their assigned duties." Another letter that reached the Armenian press threatened: "The Turkish sword to date has cut down millions of *gâvurs* [infidels], nor has it lost its intention to cut down millions more hereafter." The Turks, the letter continued, would "clean up the Armenian *gâvurs* who have become tubercular microbes for us."[132]

Swiftly, the sadistic paranoia of the Young Turk State apparatus would transmogrify into a singularily necrophilous bent. A secret high-level CUP conference—presided over by Talât Pasha, Behaeddin Shakir, and Mehmed Nazim—was held in December or January 1915, at which a strictly confidential ten-point document was discussed. The document opened with the instruction to "close all Armenian Societies, and arrest all who worked against the Government at any time among them and send them into the provinces such as Baghdad or Mosul, and wipe them out either on the road or there." The document's third point called on officials to "provoke organized massacres as the Russians did at Baku." The fifth point ordered the application of "measures to exterminate all males under 50, priests and teachers," and to "leave girls and children to be Islamized." The eighth point enjoined: "Kill off in an appropriate manner all Armenians in the Army—this to be left to the military to do."[133] The rabidly genocidal intent of the Young Turk despotic regime is all the more clearly evinced from a September 15, 1915 cable addressed by Talât to his prefect in Aleppo, averring that "the Government ... has decided to destroy completely all the indicated persons living in Turkey ... Their existence must be terminated, however tragic the measures taken may be, and no regard must be paid to either age or sex, or to any scruples of conscience."[134] Better not a sole survivor than for a single organ to flow outside this apparatus or slip away from the body of the despot.

By February 25, 1915, all Armenians in the Ottoman army were disarmed and hurled into labor battalions, from which the vast majority of them, numbering in the tens of thousands, were soon shunted off to secluded areas and massacred. The regime then proceeded to release hordes of hardened criminals from jails across Turkey and recruit them into killing squads that the clandestine Special Organization molded into a twelve-thousand-strong army of godless murderers.[135] What followed was described by Henry Morgenthau, American ambassador to Constantinople, as the "systematic attempt to uproot peaceful Armenian populations and through arbitrary arrests, terrible tortures, wholesale expulsions and deportations from one end of the Empire to the other accompanied by frequent instances of rape, pillage, and murder, turning into massacre, to bring destruction and destitution to them."[136] It was a concerted attempt to destroy the entire Armenian race.

At the hill of Margada, starving, sick and naked men, women, and children tied together with ropes were hurled down into the river below to drown in their thousands. In Balajik on the Euphrates, soldiers crushed the skulls of children in front of their mothers.[137] At Mush, after being herded like cattle into stables and dragged to a bridge, men where lined up in rows from morning to nighfall, shot dead, and their lifeless corpses thrown into the river below. At Bitlis, in excess of nine-hundred women were drowned in the Tigris. The slaughter near the town of

Erzinjan was so great that the thousands of death bodies in the Euphrates formed a barrage that forced the river to change course for over a hundred meters.[138] Even old women, many of them carrying babies, were driven into the Habur river to drown. So choked with human remains was the Habur that it moved over one kilometer toward the east.[139] East of the village of Shedadi in the Syrian desert, in the middle of what is now an oilfield, in excess of five-thousand people were thrust into a subterranean cave over one kilometer deep and asphyxiated to death with smoke.[140] In a ravine outside Chunkoush, people were undressed, their hands tied, their throats slit and their heads shattered with axes, and their bodies hurled into pits.[141]

In the dead lands of Armenia around Lake Goeljuk corpses lay piled on rocks at the foot of cliffs, submerged in water and in sand. Corpses filled up huge ravines. Women's corpses lay on their backs, barbarously mutilated by bayonets. Survivors beheld all around them a surreal landscape strewn with corpses with whitened bones, worm-infested eyes and mildewed cheeks; unburied and forlorn anatomies with naked eyes and grotesquely distorted expressions scattered over a vast expanse; an endless morass of putrid flesh exuding the loathsome stench of a gargantuan slaughter house.[142]

1928–1953 A.D. (Stalin)

Stalin fervently believed in the absolute ubiquity of conspiracies in the Soviet Union. From the very outset, he was given to suspecting that plots abounded even when direct evidence was entirely lacking. When the Soviet regime was beset by difficulties, the cause had invariably to be an agency of deliberate maleficence working out its diabolical schemes behind the scenes. Traitors therefore had to be omnipresent, even in the leadership of the People's Commissariats in Moscow.[143] As Robert Service has it, he "perceived enemies everywhere; his whole cognitive tendency was to assume that any slight problem in his personal or political life was the result of malevolent human agency."[144]

Stalin's distrust extended even to his subordinates and allies. No one was more suspicious.[145] His proclivity was to regard anyone who failed to show him respect as an enemy of the people.[146] He always watched people as if they were his mortal foes. While believing steadfastly in communism, he by no means trusted communists.[147] And those persons whom he felt he could not trust he was implicitly convinced were working assidiously and conspiratorially against the State.[148] Suspicion even fell upon those skeptical of the quasi-kabbalistic Kremlin conviction that a plethora of fiends were employing diabolical methods for the subversion of the State. These skeptics immediately became the heretics and infidels of the regime.[149] In fact, Stalin's almost feverish suspiciousness implied that any given perceived enemy would immediately multiply without end in his delusional imagination, to the extent that he "was seldom content to allow for the possibility that a particular" recalcitrant "might have been acting alone. His preference was to link his 'enemies' with a conspiracy spread out across the world and connected with the intelligence services of hostile foreign powers."[150] So paranoid was the despot that he followed the progress of political trials lasting months or even years, and could't sleep until confessions of guilt were secured.[151] He saw rising up before itself, against himself, the enemy that brings death—an eye with too steady a look, a mouth with too unfamiliar a smile, each organ a possible protest.

This utterly paranoid suspiciousness of Stalin was consonant with the pervading sentiment of Alexander Qazbegi's story *The Patricide*, according to which betrayal by intimates is guaranteed from the very outset. For this reason, the prerogative was all the more so that of revenge—in particular, preemptive revenge. He who is betrayed and takes revenge is the *abrek*—such as the main character Koba. *Abreks* pursue their quarry relentlessly and to the death—and it is with Koba that Stalin identified himself. Since his youth, Stalin saw himself as a second Koba, called himself Koba, and never wanted to be called by any other name.[152] And Koba writ large to the level of the body politic is the terrorist State. This State terrorism was installed as a permanent feature of Stalin's mind from a very early stage. His predilection for it was at once psychological and ideological. Violence was to be swiftly adopted by the Stalinist Soviet State for its primary ends.[153] The more imperiled he felt by the gangreous cells of treachery proliferating within the entrails of the State, the more lethal would be his deployment of State terror.[154]

He condoned the State brutality of Asiatic despots, but deemed even Ivan the Terrible insufficiently ruthless and systematic in the repression and destruction of purported recalcitrants. He humiliated, traumatized, and killed his subordinates.[155] No price in lives was too high for the maintenance and enchancement of the full body of the Stalinist Koba-State. Heads had inevitably to be broken.[156] For Stalin, Voroshilov, and other top echelon apparatchiks, the inveterate betrayers of Koba, these omnipresent tubercular microbes, were "swine," "horrible little invididuals, traitors, finished people," "poisonous and miserable scum" "to be annihilated."[157] Annihilated they would have to be. Stalin: "And we will annihilate every such enemy, even if he were to be an Old Bolshevik! We will annihilate his entire clan, his family! We will mercilessly annihilate everyone who by his actions and thoughts (yes, thoughts too) assails the unity of the socialist state. For the total annihilation of all enemies, both themselves and their clan!"[158] The paranoiac Stalinist State ruled as if conducting a war against its own people.[159] Better not a sole survivor than for a single organ to flow outside this apparatus or slip away from the body of the despot.

The result was a State whose sadism verged on and frequently coincided with necrophilia on a gargantuan scale. Stalin, as Martin Amis avers, "loved grinding people together, pestling them together, leaving them without air and space, without recourse; he had always loved hemming and cooping them, penning them, pinning them."[160] He, of course, also loved killing them. The untrammeled use of the death penalty was something for which Stalin had an almost physical, visceral need.[161]

In the dekulakization drive, the richest kulaks were arrested and shot or imprisoned, and their families exiled.[162] The entire families of hundreds of thousands of supposed kulaks were deported to remote locations where they often died of starvation or were shot.[163] The State visited anarchy, plunder, mania and sadism on the countryside.[164] Torture was used to force peasants to reveal grain reserves. Petty theft or trifling damage to State property was punishable by ten years hard labor or death. Orphaned children were shot en masse.[165]

Through the Terror-Famine, raging punitive food-requisitioning induced mass starvation, as thousands of children lay dying on straw beds in huge hangars whilst others were moved out in goods trains and abandoned to die many miles outside their towns.[166] Pilferers of food were sentenced to ten years in the Soviet Gulag or shot. Those caught using the very word "famine" were likewise shot. "In essence," as Amis avers, "people were being killed, quickly, for the capital crime of saying that they were being killed slowly."[167]

During the Great Terror, the paranoid move to obliterate all possible opposition to totalitarian rule continued until even the temporary prisons, schools and churches, were all full, and the courts were sitting twenty-four hours a day, and until not a family in the country remained unaffected—and all people were enemies of the State.[168] So as to terrorize the entire populace, denunciation was extolled as the sacred duty of every Soviet citizen—with the result that, for fear of being denounced by another or for fear of being denounced for not denouncing enough, denunciation and consequent arrest reached epidemic proportions and even Old Bolsheviks were confessing to a plethora of phantasmagoric crimes in show trials and being shot.[169] Indeed, those refusing to plead guilty were dead certain to die, either dying under torture or by being shot. Those admitting everything, on the other hand, were almost certainly shot.[170]

Families of captured wartime officers were arrested and often shot. Trapped or fleeing army units were tried and shot. Soldiers taken prisoner were immediately declared traitors to the motherland and in many cases shot. Soldiers who had fought their way out of encirclement were sent to the gulag or shot. Hundreds of thousands of soldiers were shot for cowardice.[171] Eighty percent of close to two-million returning prisoners of war were hurled into the Soviet Gulag or—shot.[172]

Shortly prior to his death, against the backdrop of increasingly overt State anti-Semitism whose paranoia led up to the so-called Doctors' Plot wherein a group of Jewish doctors were falsely accused of medical murder, Stalin was rabidly bent on launching a nationwide pogrom against Jews beginning with the public hanging of the doctors in Red Square and culminating in the deportation of swathes of Jewish people to Birobidzhan on the Chinese border and other parts of Siberia.[173]

Responding to the despot's spasms of raging paranoia, innocents accused of political crimes were arrested, tortured into false confessions, confined in dark closets where bedbugs drank their blood, beaten with truncheons, mallets and sandbags for days on end, starved and forced to lie half-naked in pits, and subjected to chronic insomnia and death in astronomically overcrowded cells.[174] Interrogators urinated into detainee's mouths. Prisoners had the soles of their feet and their spines beaten with leather straps to the point of internal haemorrhaging and until they jerked about like patients in the last stages of typhoid fever. Wives were killed and had their eyes cut out.[175] Husbands had their eyes gouged out and their eardrums perforated before their wives. Mothers were tortured to death before their children, who were sent to the Soviet Gulag.[176] Thousands of women covered in boils, blisters and hideous red spots, and suffering from skin disease, stomach ailments and dysentry, were crammed into open cages in the holds of slave ships as if in a gigantic poultry farm.[177]

Political prisoners were hauled off to hard labor camps where they were forced to work in temperatures below fifty degrees Fahrenheit; confined to isolation chambers where they starved, shivered, screamed and suffered the ravages of dysentry; worked to death in mines; crammed upright into sheds for several days and then shot; trussed and stacked like logs in vans and executed.[178]

Stacks of corpses lay outside the cemeteries of every major city. Hundreds of thousands of bodies lay buried in necropolises strewn across the country, whilst thousands of skeleton-like cadavers with twisted fingers and frost-bitten toes disgorged themselves from the mass grave at Kolyma.[179]

1933–1945 A.D. (Hitler)

In the pages of *Mein Kampf* the paranoia felt by Hitler in the face of the figure of the Jew is extreme, manifesting itself as a morbid fear of uncleanliness, filth, and disease threatening the very integrity of the German State apparatus. As this paranoia burgeoned into a conspiracy theory, the Jew was soon seen to be associated with every of the many evils that he perceived: the liberal press, cultural life, prostitution, and Bolshevism.[180] Jewry was likewise linked in Hitler's mind with international finance capital and its treacherous machinations. Being strongly associated with both of these immense and immensely wicked well-nigh global forces, the Jew was for him a ubiquitous and all-pervasive source of deep-seated fear. For he loomed within the delusional and paranoid rationality of Hitler in the guise of a "Jewish world conspiracy" that appeared almost insuperable—even for the National Socialist movement.[181] He saw rising up before him, against himself, the enemy that brings death—an eye with too steady a look, a mouth with too unfamiliar a smile, each organ a possible protest.

Yet this anal-hoarding aspect of paranoia, in which what appears below the reach of surveillance and overt control is felt as a disconcertingly miasmal drift refractory to the reterritorializations of the ego and State alike, had—in the case of Hitler, and almost as if bypassing the quasi-intermediary stage of nonsexual sadism—already transmogrified into its necrophilous form. This well-nigh immediate transmogrification may be seen as inherent to his bacterial imagery of the Jew: as a germ, the Jew ought to be treated in the same manner as germs, that is, suffer extermination. Evidently, this bacterial metaphor for world Jewry betrayed ever more the paranoia felt by Hitler in its face. For it suggested the insidious agitations of the quasi-molecular and of becomings-molecular excessive and subversive of the coefficient of deterritorialization of the signs constitutive of the despotic regime.

Already in August of 1920, Hitler had railed in favor of combating this "racial tuberculosis" through the extirpation of its "causal agent, the Jew."[182] Four years later, again in *Mein Kampf*, he averred that, "aside from all the positive struggle for the soul of our people, their international poisoners"—the Jews—must be "exterminated."[183] He even deployed the necessity of extermination retrospectively, contending that if twelve to fifteen-thousand "Hebrew corrupters of the people" had been put under poison gas at the beginning of the First World War, then "the sacrifice of millions at the front would not have been in vain."[184] The poisoners of the German body politic should be and should have been themselves poisoned.

In June 1922, Hitler proclaimed that the inexorable mission of the German State was to destroy Bolshevism, and together with it "our mortal enemy: the Jew."[185] By October of the same year, he was already proclaiming that the threat of the Judeo-Bolshevik bacillus was so acute that a war of extermination loomed large. It would be a struggle in which the vanquished would be the annihilated.

The threat was no less that the "Bolshevization of Germany" and thereby "the complete annihilation of the entire Christian-western culture."[186] The steadfast aim of National Socialism had thus to be nothing short of the "Annihilation and extermination of the Marxist *Weltanschauung.*"[187] But since Marxism and the Jew were by this stage already synonymous in Hitler's mind, this at once meant the complete eradication of all international Jewry. In July 1924, he asseverated that this Judeo-Bolshevist menance was still worse than being an imminent threat to the very integrity and continued existence of the German body politic, as it was in fact a deep existential concern for all peoples: "for Juda is the world plague."[188]

Much later, in the same month as the *Reichskristallnacht* pogroms against the Jews, the SS organ *Das Schwarze Korps*—portraying Jewry as comprising an unseemly, parasitic and criminalized subterranean infestation germinating within the very bowels of the Nazi State apparatus—contended on November 24, 1938, that the incumbent regime is "faced with the hard necessity of eradicating the Jewish underworld just as we are accustomed in our ordered state to eradicate criminals: with fire and sword! The result would be the actual and final end of Jewry in Germany, its complete annihilation."[189]

With the movement toward a world conflagration drawing ever more vertiginously and suffocatingly close, Hitler portended in a blistering speech to the Reichstag on 30th January 1939 the imminence of a bloody war as the effect of the poisonous Jewish bacillus. In the rabid tones of necrophilous paranoia, the Führer contended: "I believe that this once hollow laughter of Jewry in Germany has meanwhile already stuck in the throat. I want today to be a prophet again: if international finance Jewry inside and outside Europe should succeed in plunging the nations once more into a world war, the result will be not the bolshevization of the earth and thereby the victory of Jewry, but the annihilation of the Jewish race in Europe!"[190]

With the war raging and Jews already being put to the sword in their numbers, this paranoia of a despotic State seemingly assailed from all sides by Semitic microbes was clearly echoed in Goebbels' paranoid and aghast October 1940 intimation to Hitler of the Jews as "so terrible and brutal" "that your blood clots in your veins," and that one can only "shrink back at the sight of such brutishness. This Jewry must be annihilated."[191] He continued: "It's indescribable. Those are no longer human beings. They are animals. So it's not a humanitarian but a surgical task. Otherwise Europe will perish through the Jewish disease."[192] He would describe them as "the lice of civilized mankind. They had to be somehow eradicated, otherwise they would always again play their torturing and burdensome role. The only way to cope with them is to treat them with the necessary brutality. If you spare them, you'll later be their victim."[193]

On 30th January 1942, this necrophilous paranoia of the Nazi State apparatus was again rendered pellucid by another vitriolic speech by Hilter at the Sportpalast. "We are clear that the war can only end either with the extermination of the aryan peoples or the disappearance of Jewry from Europe."[194] He continued, rabidly averring that "this war will not come to an end as the Jews imagine, with the extermination of the European-aryan peoples, but that the result of this war will be be annihilation of Jewry. For the first time the old Jewish law will now be applied: an eye for an eye, a tooth for a tooth … And the hour will come when the most evil world-enemy of all time will have played out its role, at least for a thousand years."[195]

Yet, it was only with the Wannsee Conference of 20th January, at which the Final Solution was decided upon, that the relative preponderance of sadism and necrophilia within Nazi despotic paranoia shifted decidedly and inexorably in the direction of the latter. For here the hitherto preponderant proclivity toward a portended deportation programme aimed at the annihilation of the Jews through forced labor and starvation in the occupied territories of a defeated Soviet Union swiftly gave way to the rabid desire to systematically destroy the Jews in death camps in the territory of the General Government well before the end of the war.[196] These deaths were thus to be considerably hastened and effected sans the intermediary of a sadistic labor regime. Death henceforth loomed far more clearly as the direct objective of the paranoiac Nazi regime.

With the industrialized mass murder in death camps now well under way, this now vastly fast-tracked necrophilia of the Nazi State apparatus was clearly reflected in Reichsführer Heinrich Himmler's October 4, 1943, speech in Posen: "We had the moral right, we had the duty to our people to destroy this people which wanted to destroy us … We do not want in the end, because we have exterminated a bacillus, to become ill through the bacillus and die."[197] As regards the Nazi killing of women and children, Himmler contended in typically paranoid fashion two days later in the same venue that the failure to bring death also to these would be tantamount to allowing "the avengers in the shape of the children to grow up for our sons and grandchildren. The difficult decision had to be taken to have this people disappear from the earth."[198]

As Hitler himself intimated in a May 26, 1944, speech in the Platterhof, had the decision to annihilate this bacillus not been taken, "the German people would be eradicated. Bolshevism would slaughter millions and millions and millions of our intellectuals. Anyone not dying through a shot in the neck would be deported. The children of the upper classes would be taken away and eliminated. This entire bestiality has been organized by the Jews."[199] "The entire state" would be "undermined and corroded, Jews everywhere, even in the highest places Jews and more Jews, and the entire state covered, I have to say, by a seamless web of agents and spies" and other vermin through whose metastatic pervasion the Nazi full body would perish from necrosis.[200]

Yet as the Nazi despotic regime found itself ever the more besieged on all sides by the well-nigh infinitely pervasive international Jewish bacillus sustaining a rapidly encroaching Allied war machine, Hitler's increasingly paranoid suspicions henceforth came to fall more and more upon the *German* people themselves. With the imminent prospect of the dissolution of the Nazi body politic in ever more nauseating proximity, he felt increasingly that this dire scenario was due to the pusillanimity of his own kinsfolk. The German subjects themselves had been irrevocably infected by the Semitic tubercular microbes.[201] The collapse of the Reich could ultimately "only be caused through the weakness of the people," he averred.[202] "But if the German people turned out to be weak," he continued, "they would deserve nothing else than to be extinguished by a stronger people; then one could have no symphathy for them."[203]

His suspicions fell upon his military leaders. They were all disloyal, dishonest, opposed to National Socialism, reactionaries. Any German subjects who exhibited the slightest hint of defeatism were immediately seen as internal enemies, as traitors.[204] Following the foiled July 20, 1944, von Stauffenberg plot against the Nazi State, Hilter's suspicions extended from Jewish Marxists to the officers of the Wehrmacht who were henceforth seen as deliberately undermining the war effort, of stabbing the nation in the back.[205] In early 1945, he contended: "I've been betrayed. Since July 20 everything has come out that I didn't think possible. Precisely those were against me who have profited most from National Socialism. I spoilt them all and decorated them. That's the thanks."[206]

Thus, ultimately it was not merely the entirety of international Jewry that had to be completely and utterly annihilated. Having proven themselves to be infected by the Jewish bacillus, the German people themselves would likewise have to suffer wholesale destruction. Rather than capitulate to the Allies, Hitler would rather that every living German on earth perish. Ultimately, the destruction of an entire cross-section of humanity was thought preferable to the dissolution of Hitler's Nazi body politic. "We'll not capitulate. Never. We can go down. But we'll take a world with us."[207] Better not a sole survivor than for a single organ to flow outside this apparatus or slip away from the body of the despot.

1949–1976 A.D. (Mao Zedong)

In March 1949, Mao Zedong adjured the Central Committee of the Seventh Congress of the Chinese Communist Party with the following words: "After our armed enemies have been crushed, there will still be our unarmed enemies, who will try to fight us to the death. We must never underestimate their strength."[208] In another pronouncement, the Great Helmsman asserted that "spies are as numerous as the hairs on a head."[209] This intense despotic paranoia is echoed in Gang of Four member Zhang Chunqiao's July 1968 contention: "If a few people are wrongly accused, it does not really matter; but it would be disastrous to allow any of the guilty to escape."[210]

The despotic character of the Maoist regime may already be gathered from Mao's tireless extolment of the first Chinese emperor Qin Shi. The latter was explicitly taken as a model of governance by Mao.[211] For he adopted the approach of the Legalist school which prevailed under the Qin State, and which advocated that the State must affirm its omnipotence through unabashed terror.[212] Mao himself affirmed his own omnipotence to such an extent that he was often known during his life as the Red Emperor.[213] Yet the sheer gargantuan scale on which and systematic intensity with which the Maoist State apparatus deployed terrorist tactics against its own subjects reduced the otherwise rapacious bloodlust of Qin Shi Huangdi to comparative pusillanimity. The necrophilous aspect of this despotic regime's paranoid posture is evinced by Mao's November 1950 personal sanction of mass murder: "We surely must kill all those reactionary elements who deserve to be killed."[214]

The anal-hoarding aspect of Maoist despotic paranoia is evinced inter alia in the manner in which Party discipline was enforced. The latter was effected through the *cheng feng* sessions during which Party members were criticized for their behavior or inclinations. Upon being criticized, the sole right of the accused was to repent his "errors." Any protestations of innocence or insufficient repentance was met with renewed attacks. The ultimate aim of the *cheng feng* technique was to ensure that even the personal and intimate thoughts of all were known to all others and were thus immediately available for public scrutiny. The method was extensively deployed in purges of the Party ranks that were launched in July 1943 in Ya'an province. There ensued a massive spate of criticisms and public self-criticisms, selective arrests leading to confessions that themselves led to more arrests. This tirade of paranoid machinations rapidly turned sadistic and thereupon necrophilous as the slew of arrests led to public humiliations, beatings, torture and death. Only ten percent of all those arrested had in fact been guilty.[215]

With the Communists having seized power in 1949, "mass campaigns" such as the "Recitification" of 1943 came to be launched almost every year by the paranoiac Maoist State apparatus—except that these now spread across the entire country, whereas certain social strata were subjected to massacres on a hitherto unprecedented and genocidal scale. People were crammed by the hundreds of thousands into the new *laogai* archipelago of labor camps.[216] Landowners were declared traitors before village assemblies, humiliated, beaten, sentenced to death, and immediately executed. Poor peasants who had shown solidarity with the rich were beaten. People were tortured to death with red-hot irons in attempts to force them to reveal the whereabouts of purported treasure. Families of executed people were tortured and the tombs of their ancestors destroyed. Rich peasants were incarcerated in labor camps, put under perpetual surveillance, or murdered.[217]

Soon violent purges spreach to the cities where entire groups such as intellectuals, the bourgeoisie, small bosses, non-Communist militants, and overly independent Communist cadres were seen as threatening the integrity of the Chinese Communist full body. There was a general crackdown on criminality, prostitution, gambling, opium dealing. Millions of bandits were liquidated and incarcerated. In the towns, families were placed under close surveillance by neighborhood committees who were themselves in turn watched by district committees. All nocturnal visits and strangers were recorded by residents' committees. Everyone became a police informer whilst police numbers rapidly swelled. Prisoners were overworked, given starvation-level rations, and subjected to the imminent threat of physical brutality and beaten with rifle-butts. Seditious camp inmates were massacred and often buried alive.[218]

In July 1950, the campaign to eliminate "counterrevolutionary elements" was launched; followed in 1951 by a campaign targeting waste, corruption, and bureaucracy; an anti-bourgeois campaign against bribery, fraud, tax evasion, lying, and revealing State secrets; and a campaign to "reform thought," targeting Westernized intellectuals. Suspected members of the last venomous group were subjected to regular periods of "reeducation" under the auspices of the local labor collective. So intense was the anal-hoarding paranoia of the Maoist State apparatus that, in order that not a single working organ would escape its reterritorializing sway, the very definition of what constituted a counterrevolutionary was rendered so vague and broad that any past or current position that diverged even slightly from the Party line sufficed to bring condemnation. Penal legislation punished even past counterrevolutionaries through retroactive legislation. Even if the accused had not transgressed a given law, he could nonetheless be condemned for committing an analogous crime. Most found guilty were imprisoned for about twenty years. "Capitalist tigers" were forced to open their accounts to public scrutiny and control and intimidated into denouncing their colleagues. Intellectuals were forced to confess their errors at their workplace and show that they had abandoned "liberalism," "Westernism," proclivity to doubts about the regime, and the predilection for independent thoughts. Their only possible escape was suicide.[219]

In 1955 it was "hidden counterrevolutionaries," or *sufan*, whose elimination was sought. The Intelligentsia in particular was targeted, inclusive of any former Party member and sympathizers who had shown a modicum of independence. In the Hundred Flowers Campaign of May and June 1957, Mao tricked hundreds of thousands of cadres into revealing "evil thoughts." These, including at least ten percent of all Chinese intellectuals, technicians, and engineers, were branded "right-wingers" and incarcerated for twenty years in prisons or camps in remote regions, whilst millions of cadres and students were deported to the countryside.[220]

As mass campaign followed mass campaign, the purported enemies of the paranoiac Maoist State apparatus swelled rapidly as the number of "black" categories grew and formerly neutral categories shifted inexorably to "black." The anal-hoarding and sadistic paranoia of this despotic regime is further evinced from the fact that labels stuck no matter how repentant individuals were. Even after official rehabilitation, a right-winger would remain a target of mass campaigns and would never be allowed to return to the city. There were always enemies to hunt down and murder. The number of enemies of State would also be increased by an expansion of potentially incriminating traits and by the search for people who had supposedly fallen back into their old ways.[221]

Mao was perpetually convinced that peasants were undermining the State economically by concealing private stocks of grain. In autumn 1959, gripped by necrophilous paranoia, the Maoist despotic regime launched a military offensive against the peasants. In Henan province, thousands of peasants were imprisoned and died behind bars, thousands of detainees were systematically tortured, whilst children were killed and even boiled and used as fertilizer. In Anhui province, with the intention of keeping the red flag flying even if almost all of the population died, peasants were tortured with red-hot irons and buried alive, whereas villagers fleeing into towns were mowed down with machine guns.[222]

So that not a single organ should slip away from the anal-hoarding and sadistic Maoist State apparatus, in excess of ten-thousand detainees per year disappeared into the *laogai* concentration camps, where some twenty-million died during their incarceration. After being incarcerated in the *laogai* for about twenty years, or in the *laojiao* for somewhat less, inmates were often kept in the *jiuye* camps in semi-penal conditions until death. Thus, under Mao, any sentence was effectively a life sentence. To further cement the reterritorialization of working organs onto the full body of the despot, Catholic techniques such as meditation, confession, and repentance were reinvented as Marxist-Leninist practices so as to destroy individual personality, and to dissolve the individual into a mass submissive to the State. The potential recalcitrance of prisoners was further undercut by systematic underfeeding and permanent indoctrination, whereas prisoners who inadvertently spoke of their prison experiences after eventual release were reimprisoned.[223] The plethora of denunciation boxes in the streets of most cities made it almost impossible to hide anything of significance about one's past, whilst the simple fact of arrest was taken as unquestionable proof of guilt.[224] Life-long incarceration could easily be the result of a mere momentary error.[225]

With Mao having suffered a loss of support from cadres and intellectuals in the wake of his purges of 1957 and that of the rural masses in the famine of 1959–1961 and having been forced to hand over power for two years in 1962 to the President of the Republic, Liu Shaoqi, the Maoist full despotic body braced itself into a paranoiac posture from whence the purges of the Chinese Cultural Revoluion ensued. In a gargantuan mass mobilization of those too young and too urbanized to have experienced the horrors of the Great Leap Forward, millions of sadistic youths and students recruited as Red Guards conducted rabid purges of intellectuals and others across the land. Liu Shaoqi was thrown into prison and died under torture, communist cadres were purged en masse. The Maoist despotic regime turned sadistic, necrophilous, and even cannibalistic. Beginning on June 1, 1966, professors and teachers were denounced as "monsters and devils," "evil geniuses," "bovine ghosts," "reptilian spirits," "poisonous snakes," "carnivores," "lizards," and "rats that run through the streets." "Class enemies" had notices stuck on their backs, were dressed up in ridiculous clothes, forced into grotesque and painful positions, had their faces smeared with ink, were forced to bark like dogs on all fours, made to eat grass, tied to plows, whipped and forced to plow fields, beaten or boiled to death, and sometimes even *eaten.* Those with long hair were forcibly shaved, tight trousers were torn up, high-heels were ripped off, whereas "black" families refusing to reveal supposed treasures were often beaten to death.[226]

So acute was the paranoia of the Maoist despotic regime that the purging Red Guards themselves came to be purged. In the second half of 1968, the Red Guards were disbanded, millions of young people deported into the countryside, three-million cadres were incarcerated, campuses were annihilated with heavy artillery and napalm, rebels were executed en masse, and blood and oxygen supplies of rebel patients in hospitals siphoned. The number of public executions increased, sometimes occurring before massive crowds. The Maoist regime entered into a dark and intensely paranoid era of nonexistent plots, nonexistent subversive groups, and absurd conspiracy theories, in which arrests, torture, and forced confessions, and death by execution occurred on a massive scale across the entire country.[227] Better not a sole survivor than for a single organ to flow outside this apparatus or slip away from the body of the despot.

1975–1979 A.D. (Khmer Rouge)

The Democratic Republic of Kampuchea was possessed of as paranoiac a State apparatus as any in history. CPK party leaders were firmly convinced that traitors were lurking absolutely everywhere.[228] Pol Pot and his henchmen invariably ascribed the economic and military disasters that were increasingly besetting the regime to incessant acts of treachery or sabotage by purportedly extant exploiting classes and their allies.[229] In July 1978 the Party monthly, *Tung padevat* (Revolutionary flags), contended: "There are enemies everywhere within our ranks, in the centre, at headquarters, in the zones, and out in the villages."[230]

Moreover, according to Pin Yathay, despite the marked scarcity of seditious movements, "what the Khmer Rouge feared was the anger that might surface in their new people if they eased up on repression. Because they were haunted by the possibility of revolt, they decided to reproach us for our impassivity and to make us pay for it. We were afraid of persecution; they were afraid of insurrection. They were also afraid of the ideological and political maneuvring of their comrades-in-arms."[231] The regime saw rising up before itself, against itself, the enemy that brings death—an eye with too steady a look, a mouth with too unfamiliar a smile, each organ a possible protest.

It is this misplaced anxiety and suspicion of Angkar that the State was constantly being assailed on all sides by recalcitrant and potentially insurrectionary elements that lay behind the regime's almost immediate movement toward the sadistic and strictly necrophilous aspect of its essential paranoia. This latter is borne out by slogans such as: "One can always make a mistake and arrest the wrong person, but one should never let the wrong person go"; or "And we would rather kill ten friends than allow one enemy to live."[232] Yet, the rampant necrophilous propensities of the Kampuchean despotic regime are most poignantly—though inadvertently—evinced in the opening lines of its national anthem:

> Bright red blood that covers towns and plains
> Of Kampuchea, our motherland,
> Sublime blood of workers and peasants,
> Sublime blood of revolutionary men and women fighters!

The irony here is that this superabundant blood that was liberally spilled was not that of revolutionaries, but that of the Cambodian people by a subrevolutionary despotic regime. Better not a sole survivor than for a single organ to flow outside this apparatus or slip away from the body of the despot.

The preponderant tubercular microbes threatening the integrity of the Kampuchean despotic full body were the so-called *prasheashon thmei*, or "New People," who were seen as the "lackeys of capitalist imperialists." So as to preclude potentially seditious links between these latter and the *praesheashon shah*, or country people, the New People were forced to live in separate areas of villages, and were prohibited from speaking to, let alone intermarrying with the former group. Educated New People such as civil servants and intellectuals were slowly purged until they entirely disappeared; whilst, in order to weed out those of the remainder with individualist tendencies, the uneducated were fleeced of their possessions and deported to areas were conditions were more dire.[233]

The burgeoning nature of the regime's paranoia meant that purges soon reached the upper echelons of the political hierarchy. Anyone associated with the Vietnamese, the earlier royalist government of Norodom Sihanouk, with the Chinese, or with the West, were tortured in barbaric fashion and executed without trial and without clear charges being brought. Any quality whatsoever that might threaten the preeminence of Pol Pot led to fierce repression. All that was required for an arrest was three denunciations as a "CIA agent." Confessions were extorted by any means possible. Imaginary conspiracies abounded and more purported subversive networks were constantly being unearthed. All the people living in the eastern region were labelled "Vietnamese in Khmer bodies" and condemned to death. From May to December 1978 hundreds of thousands of these were massacred, whereas after July any survivors were deported and progressively exterminated. Just before the fall of the regime, women, children, and old people were massacred together with the young men, whilst the country people were killed together with the New People.[234]

Almost all monks across the country were systematically eliminated.[235] After being forbidden to leave the country after mid-1976, according to a April 1, 1977, directive from the Center, all Vietnamese were to be arrested—as were their friends and anyone else who spoke Vietnamese. In Kratie province, those with a Vietnamese ancestor were themselves liable to arrest. Almost half of all

Cambodian Catholics were eliminated. In February 1977 all of Pol Pot's guards of the Jarai ethnic group were arrested and eliminated. In 1976 all cadres with Cham origins were fired, whereas in the northwestern zone, people were sometimes even killed for speaking Cham. In May 1975 Muslim dignitaries were executed for having gone to pray or for having campaigned for the right to religious wedding ceremonies. After mid-1978 a number of Cham communities were systematically exterminated, including women and children, as the regime attempted to completely extirpate Islam. In 1977 the wives of men who had already been executed often much earlier were themselves killed. Whole families were slaughtered, and sometimes entire villages.[236]

Perpetually threatened with the flight of working organs from its reterritorializing sway, the regime banned all public displays of affection, arguments, insults, complaints, and tears. To undercut people's capacity for individual thought, association with others, and their propensity to run away, they were systematically starved; whereas the individual searching for food often resulted in execution. Sick people were considered malingerers and saboteurs, whereas the infirm were hospitalized only so as to be eliminated. Deserters were sometimes forced to eat their own ears before being executed. With family ties considered a source of recalcitrance to the regime, husbands and wives were often separated for weeks or longer, children were kept from their extended families, whilst adolescents sometimes went six months without seeing their parents. On the other hand, so that the sadistic regime maintained its monopoly on violence, men could be executed for striking their wives or denounced by their children for hitting them. All forms of cultural expression other than revolutionary songs and poems were banned. People were killed for losing cattle and tortured to death for having struck a cow. At compulsory self-criticism meetings each person's past was scrutinized, whereas the slightest doubt about the veracity of one's admissions resulted in arrest and torture. Nothing escaped the omniscient vigilance of the authorities. Children and young adolescents were employed as spies to hide under houses to listen to private conversations and hunt out forbidden stores of food, or to track the political views of their parents and relatives. The slightest infraction of regulations was considered an act of opposition and a "counterrevolutionary crime." Such was the despotic paranoia of the regime that even an involuntary slip amounted to treachery. Breaking a glass, losing control of an ox, or plowing a crooked furrow sufficed for one to be brought before a ramshackle court. It was even forbidden to speak about the dead, who were inevitably traitors or saboteurs. Even the word became taboo. Anything that was not explicitly permitted was de facto forbidden. One could be imprisoned or executed for hoarding a banana.[237]

The vast prison system combined the sadistic and necrophilous proclivities of the paranoiac regime. People were incarcerated for theft, extramarital relations, complaints about inadequte food, jokes about the hymn to the revolution, purported prior bourgeois employment, breaking a spoon in a canteen, speaking English, visiting loved ones, or supposed laziness. Prisoners were constantly kept in chains around the ankles and elbows tied to an iron bar fixed to the floor. Almost all died within three months, either from conditions, from interrogation under torture, or else by execution with a knife or iron bar after having confessed all and named all accomplices. Imprisoned juvenile delinquents were kicked by guards whilst hanging by their feet from the roof and subjected to near-drowning before at length dying. Disgraced cadres and fallen leaders were hurled into Tuol Sleng (S21) prison—a black hole out of which there was almost no chance of emerging alive.[238]

The necrophilous paranoia of the Khmer Rouge regime was so acute that, as Jean-Louis Margolin has it, murder was in fact a means of government. In fact, under Pol Pot, violent death by government was more frequent than death from disease or senility. Even imprisonment was a mere stay of execution during which prisoners were tortured into confessing their plots and naming their accomplices before being killed. In the last year of the regime, concerted military purges massacred teams of disgraced cadres, entire villages of "suspects," and even entire populations. Almost every crime merited the death sentence. People caught pilfering food were shot in the head or beaten to death on the spot with pickaxe handles and left to rot. Whole families were killed for having shared out a cow. The punishment for overly frequent visits to one's family was death. The punishment for arguing was death. The punishment for extramarital relations was death. The punishment for the consumption of alcohol was death. Trance ceremonies were punishable by death. The punishment for defeatist remarks about the regime or complaining that one was hungry was possible death. The punishment for insubordination was immediate death. For failing to complete their tasks, the handicapped, the maimed, and the mentally ill were often killed. Madmen who argued incoherently were usually shot. Men and women were killed and buried in mass graves to make fertilizer. Those condemned to death were mostly beaten to death with iron bars, pickaxe handles, or agricultural implements, whilst some were hanged, asphyxiated with plastic bags, or had their throats slit. Disgraced cadres were buried up to the chest in ditches filled with firebrands or had their heads doused with gasoline and set alight.[239]

1979–2003 A.D. (Saddam Hussein)

The paranoid despot Saddam Hussein was always irrevocably convinced that Baghdad was awash with plots and conspiracies. No person or institution was to be trusted. Not even his own family. He was dead certain that an entire plethora of assassins who bayed for his blood perpetually lurked beneath the surface.[240] He claimed that "I can tell simply by looking in someone's eyes whether they are loyal or a traitor."[241] He saw rising up before him, against himself, the enemy that brings death—an eye with too steady a look, a mouth with too unfamiliar a smile, each organ a possible protest.

It is this perpetual fear of the subversion of the Iraqi despotic full body that lay beneath the sadistic aspect of Saddam's paranoia. With the regime invariably threatened in its very integrity by the ubiquity of gangreous cells within the very bowels of the State apparatus, Saddam determined from the very start that his presidency would be run on the basis of absolute terror. No dissent was too trivial or slight in his eyes. Even the merest suggestion of opposition to the great tyrant's formidable will was to be crushed ruthlessly and with the utmost brutality.[242]

That this sadistic tenor of Saddam's despotic paranoia was already pregnant with a rapidly burgeoning necrophilous aspect is evinced inter alia by his unequivocal conviction that: "It is far better to kill an innocent man rather than to allow a guilty man to survive"; or in his approving citation of Stalin's maxim that "If there is a person then there is a problem, if there is no person then there is no problem."[243] Though Iraq under Saddam was by no means a communist State, his Baathist regime employed the same methods as those used by the Soviets to strike fear into the citizenry and to forcibly and sadistically mold the cowed and terrified massess into the party's organizational machinery. The Baathist despotic State apparatus likewise turned necrophilous in its incessant political purges of real and perceived recalcitrants.[244]

The Baathist full body already perceived itself threatened with a flight of working organs from its reterritorializing sway immediately upon Saddam's accession to the presidency in July 1979. At an extraordinary conference of senior Baath Party members in Baghdad presided over by Saddam the details of "a painful and atrocious plot" involving members of the Revolutionary Command Council and aimed at overthrowing the president so as to enable a Syrian-Iraqi union were revealed. A total of sixty-six people were denounced, including a number of senior Baathists very loyal to Saddam. Fifty-five were convicted, of whom twenty-two were blind-folded, forced to kneel with their wrists tied behind their backs in the courtyard of the very same building in which they had allegedly plotted, and shot in the head.[245] With the sadistic and necrophilous paranoia of the Baathist despotic regime rapidly burgeoning in intensity, the purge immediately spread across the entire country as hundreds of party workers and military officers were fired, tortured, imprisoned, and even executed.[246]

Seeing himself assailed on all sides, Saddam immediately replaced the old Mukhabarat with the newly created Amn al-Khass—at once the State's security arm and his own personal security force.[247] In 1984 the Republican Guard, drawn from imposing Sunni peasant stock and completely dependent on Saddam for their existence and expected to defend him to the death, became his own army and replaced the Baathist one.[248] The sadistic paranoia of the regime manifested itself moreover in the vast panoply of torture techniques employed against the citizenry. People were beaten, subjected to bastinado, had their hair pulled, their limbs twisted until they broke, were electrocuted, placed in solitary confinement for extended periods, put in cells until their limbs froze, had their body parts set on fire, their limbs amputated, were anally raped with foreign objects, and had their relatives raped before their own eyes. Suspects were held without charge, women were tortured in front of their families, husbands and children were tortured in front of their wives and mothers, and children were put in sacks with starving cats.[249]

Yet this hitherto sadistic despotic paranoia rapidly moved toward its necrophilous aspect. By the mid-1980s a total of twenty-four offenses carried the death penalty. In that the definition of these offenses was deliberately vague and since almost all information in Iraq on government, economy, and society was considered a State secret, any unauthorized disclosure of information could be interpreted as treason—the penalty for which was death. The penalty for the propagation of Zionist or Masonic principles was death. The penalty for rape, arson, armed robbery, and sodomy was death. The penalty for having contracted the AIDS virus was death. There was no possibility of appeal on sentences passed by the Revolutionary Court in Baghdad. When perceived political recalcitrants could not be brought to trial, they were poisoned. Hundreds of Iraqi writers and intellectuals were imprisoned and tortured to death. The penalty for membership in any opposition party was death. The penalty for any criticism of Saddam was death.[250] In the summer of 1982, upon articulating constructive criticism as to how the war against Iran ought to be prosecuted, more than three-hundred high-ranking officers were executed together with party officials sharing their views. War deserters were hurled into unsanitary and pitch black cells in Abu Ghraib prison, tortured and executed, whilst Dawa and Communist Party members and other oppositionists were kept in basement dungeons.[251]

When in 1983 the Kurds were pursuing their total independence and had spurned Saddam's offers for limited autonomy; the Iraqi despotic regime, threatened with a flight of working organs from its reterritorializing sway, braced itself into a necrophilous paranoiac posture. Eight-thousand Kurdish prisoners were executed. By the end of 1988, more than half of the towns and villages in Kurdistan were razed and their populations deported to the main towns and concentration camps in the southwestern Iraqi desert. In May 1987 twenty Kurdish villages were gassed. In March 1988 five-thousand men, women, children, and babies in the village of Halabja were gassed with hydrogen cyanide.[252] A few months later about sixty-five Kurdish villages likewise came under chemical attack as Saddam attempted to impose a "final solution" in the face of Kurdish desires for independence. In excess of five-thousand people died.[253]

In the wake of various suspected coup attempts between 1988 and 1990, more military officers died in a single year of helicopter crashes contrived by the regime than had died in eight years of war with Iran. With the Iraqi despotic full body threatened in its integrity by the bonds of comradeship formed during the war years, a brutal purge of the military was launched in which not even officers of the same provincial background or those directly related by clan or marriage to Saddam's ruling circle were spared the paranoid despot's sadism and bloodlust. Those alleged to be disloyal were tortured and executed.[254]

Following Operation Desert Storm, the Baathist despotic full body was again jolted into a sadistic and necrophilous paranoiac posture by the spread of revolts from the Shi'a heartlands of sourthern Iraq to nearby Sunni cities and to Kurdistan in the north, and the mass flight of working organs this portended. The Republican Guard was unleashed against the rebels. In the holy cities of Najaf and Karbala thousands of clerics were arrested and hundreds were summarily executed. Any turbaned or bearded man in the street risked being arrested and shot. People were tied to tanks and used as human shields. Women and children were indiscriminately shot. Karbala looked as if it had been hit by an earthquake. With the main cities of Kurdistan pounded by government forces, Kurds fled to the mountains en masse and starved to death in their refuge at the rate of a thousand people per day.[255]

When in the summer of 1992 two mechanized Republican Guard brigades were linked to a quite possibly nonexistent plot against Saddam, six officers were immediately executed whilst another four-hundred were arrested.[256] In 1995, with the integrity of the Iraqi despotic full body threatened by an alleged coup by the London-based Iraqi National Accord headed by a Baathist defector, in excess of eight-hundred suspects from various security services were hurled into prison, the majority of whom were tortured and executed.[257] Better not a sole survivor than for a single organ to flow outside this apparatus or slip away from the body of the despot.

1949–? A.D. (Kim Dynasty)

"Was Kim Il-sung an East Asian Saddam Hussein?" This is the rhetorical question that Bradley Martin asks himself.[258] We shall see that North Korea under the Kim dynasty was and is, in all likelihood, even worse than Iraq under the aegis of the Butcher of Baghdad.

According to the North Korean despotic regime's former chief ideologue, Hwang Jang-yop, the Great Leader Kim Il-sung—in the manner of a Byzantine Emperor—increasingly came to regard the government as his own personal property: "In a situation where all means of production actually belong to the Great Leader, the economy itself naturally serves the interest of the Great Leader before all else. The national economy is nothing more than the household economy of the Great Leader. North Korea's economy exists first and foremost to serve the Great Leader."[259] Here, we see in a palpaply tangible manner the full body of the despot as the unengendered miraculating surface of which the regime's working organs appear in the first instance as natural effluences. This, as discussed much earlier, is the veritable wellspring of the despot's paranoia. One of the ways in which this paranoia manifested itself was through strict adherance to the dynastic principle of succession. To guard against posthumous deprecation and also to obviate challenges to his power from nonfamily underlings impatient to take over, Kim Il-sung was determined to name his own flesh and blood as his successor.[260]

The principle means of the reterritorialization of working organs onto the full North Korean despotic body was mind control. As Martin maintains: "Serving the very conservative goal of protecting and perpetuating the existing regime, North Korean mind control soon surpassed in thoroughness all other twentieth-century totalitarian political movements."[261] This mind control, and at least the anal-hoarding aspect of North Korean despotic paranoia, may be seen as embodied in the enforcement of *juche*, or "self-determination," ideology—the cornerstone of which is compete loyalty to the supreme leader, or *Suryŏng*, Kim Il-sung. This ideology is primarily deployed as a form of both nationalism and control over the population. *Juche* differs fundamentally from Marxism-Leninism in that, far from standing for the eventual whithering away of the State, its primary goal is the latter's maximal reenforcement as a transcendent full body. Here, *juche* is not a mere nameplate, whereas the regime effects its reterritorialization by a combination of brute terror and ruthless suppression. Whilst the latter is obviously the case; in that it pervades all aspects of life so profoundly, *juche* ideology is already a mechanism of reterritorialization of potentially runaway deterritorializations in and of itself. It forms the backbone of the State's control. In fact, without this ideology, the State could not survive.[262]

The effectiveness of *juche* as a means of reterritorializing control for the regime resides inter alia in its co-optation of already deeply entrenched Korean notions of Confucianism, with its emphasis on respect and hierarchy, as the basis of order and social harmony. More specifically, Kim Il-sung redefined collectivism in a Confucian rather than Marxist context. Here, the masses would serve the State leader just as children would show filial piety to their parents. The rhetoric of *juche* propaganda played heavily on this filial piety. The masses are portrayed as yearning for the "bosom" of the leader's wisdom just as a child feels comfort in the bosom of his mother. Kim Jong-il wrote about how the State was the "loving mother" of all the people. Military song lyrics refer to how the "country I call my mother I now know is the General's bosom." So as to further consolidate the conflation of political loyalty with filial piety, portraits of Kim Il-sung deliberately accentuated the softer, more effeminate features of Kim's face and depicted workers with youthful, rosy-cheeked loving expressions. This served to conflate disagreement with or disrespect for the State with disrespect of one's parents. One could therefore not even contemplate such heinous eventualities.[263]

The reterritorializing force of *juche* is, moreover, cemented and recemented by its being seared into the minds of every subject through daily repetitive indoctrination sessions. Here, it was repeatedly inculcated that the North Korean State was a biological organism, in which the Great Leader (*Suryŏng*) Kim Il-sung was the brain, the party the nerves, and the people the arms, muscle, and bone. One of the messages of obedience herein entailed are that without the brain the rest does not function; hence, there must be complete loyalty to the despot. Secondly, independent thought is unneccessary, as this was already handled by the brain. Therefore, the only critical thinking that is allowed is self-criticism based on guilt for failing to serve the despotic will. The latter assumes the form of self-criticism sessions, taking place in the Korean Workers' Party but likewise in the various social institutions which function as mechanisms of mass control. The most tangible outward manifestation of *juche* control is the Kim Il-sung badge, which every subject is duty-bound to wear every day, and which is a constant reminder of where one's allegiances lie. These reterritorializations are further consolidated by the plethora of portraits of the Great Leader and the Dear Leader (Kim Jong-il) in every office, school and workplace, and the tens of thousands of public monuments, statues, murals, which cumulatively leave no psychic space to think outside of *juche* ideology.[264] In school, about a third of the curriculum is devoted to the personality cult of Kim. Children are taught that Kim gave them their clothes, toys, and books, and that they must love Kim more than they love their parents. They are taught that they can live without their parents but that they cannot live without love for and undying loyalty to Kim Il-sung. Furthermore, individualism in North Korea is taught by the State to be a normatively bad trait, since the rights and duties of the citizen are based on the collective. The regime maintains the tightest grip on its internal workings, rendering it impossible to find an individual around whom to organize a liberatory cause.[265]

The merely anal-hoarding facets of *juche* paranoia flow seemlessly into its sadistic aspect through the Ch'ŏllima movement and mentality, according to which the worker is cajoled and brainwashed into working for the Great Leader beyond his physical limitations. This forcing of workers to labor for longer hours is likewise an instrument of control.[266]

A reconsolidation of the reterritorializing force of *juche* ideology commenced about three years prior to the death of Kim Jong-il. There is presently a return to the Cold War era heyday of *juche*, in what Victor Cha terms "*neojuche* revivalism," except with an even harder line and more dogmatic posture than before. This new, more conservative, strain of *juche* inter alia rejects and discards what is now seen as the deviancy and spiritual impurity of certain reform policies with which the government flirted from the mid-1990s to the mid-2000s, and has inter alia resuscitated the paranoid sadism of the Ch'ŏllima movement and its tyrannical overexertion of the human body.[267]

In addition to, but not apart from *juche*, another mechanism of reterritorializing human working organs onto the full body of the North Korean despotic regime is the ration system whereby food and other goods are supplied to the citizenry. This form of political control is reinforced by the State's repeated unilateral redenomination of the national currency. The latter, together with the limitation of exchange by citizens to only a fixed amount, effectively wipes out whatever savings and extra cash North Koreans can earn from entrepreneurial activities—thereby making them once again dependent on State handouts.[268]

Whatever changes that may exist in economic laws are designed solely to strengthen the government's central control. All legal statutes that had at some prior point allowed for a glimmer of innovativeness and entrepreneurship at the local level have been completely erased. Moreover, all residents of Pyongyang are legally bound to carry their government-issued residency cards at all times.[269] In addition to this, the average North Korean is not permitted to travel within the country. Even leaving one's own village requires a special travel pass.[270]

Another reterritorializing mechanism employed by the State is the almost airtight control of information as filtering in from the external world. Here, the government works exceedingly hard to deprive the people of any information about the living standards outside of the country. There is, moreover, no freedom of assembly in North Korea, whereas the only organized and armed social institutions with tools of enforcement belong to the State. The police, internal security forces, and military ruthlessly extinguish any sign of social disorder.[271] In the face of the demonstrations in Egypt in 2011, the regime braced itself into a paranoid posture in banning all news of the public revolt that toppled the Mubarak government whilst at the same time prohibiting all forms of public and private gatherings, including coalescence in restaurants and open-air markets. The uncomfortable parallels to their own situation that Kim Jong-un and his despotic caste would doubtless have seen in the Arab Spring, and the herein portended flight of working organs from the North Korean full despotic body, was certainly the trigger for these paranoid measures.[272]

Propaganda alone would, of course, be nowise sufficient to enable Kim senior and junior to maintain unprecedented control over their subjects. This maintenance likewise necessitates a massive State police apparatus. The latter is modelled on the Soviet one, with similarities to that of the Imperial Japanese. It is, however, even more pervasive and thorough. Agents from the Ministry of Public Safety and the Ministry of State Security are stationed at every level of the administrative agencies, right down to the lowest level. Even the smallest act that is out of the ordinary is a cause for arrest. People attracting the attention of the authorities become sacrificial lambs. Agents at the post office intercept and inspect letters and packages. Bugging and wiretapping are used to keep tabs on even high-ranking officials. People know that they must not voice their innermost thoughts even at home. Agents moreover supervise and monitor all university departments and administrative units. They even persuade students to watch one another, to the extent that one out of every five students is a secret agent of the State Security branch. In fact, entire communities, all neighbors, are made to survey one another.[273] The upshot of the foregoing is that sons often rat on their fathers to the authorities, whereas wives report on their husbands.[274]

Students who return after studying abroad are subjected to two months of mandatory re-indoctrination. From here on in, they are forcibly accompanied by a member of the security apparatus, and are moreover forced to sign legal contracts pledging that they will not utter a word about what they had seen overseas.[275] Those who have experience overseas, enough intelligence to think about the regime, or are part of the old elite, are placed under an even stricter form of surveillance. They are made to talk as much as possible so as to eventually slip up and utter something that could be used as grounds for indictment and arrest. Some persons falling within this category have been under concerted surveillance for decades on end.[276]

Teams of young morale-builders even come to schools and inspect the children. They hold a meeting each day, take one child at a time and make him or her confess publically to purported misdeeds. These teams are in reality snoops and spies, who spy on ordinary citizens, factory managers and government officials.[277] At Central Committee meetings, Kim Jong-il enforced compulsory mutual criticism of members, at which those who refrain from criticizing others are denounced, whereas those who loudly and harshly criticize others are praised for their loyalty to the Dear Leader. The more party members criticize themselves and fight among themselves, the greater does the head of State's authority become. A measure of the utter anal-hoarding despotic paranoia of these mutual denunciation sessions is that even the smallest defect is blown completely out of proportion into a serious incident. Kim Jong-il, for one, would secretly watch these sessions on closed circuit television.[278]

Another manifestation of the anal-hoarding sadism of the North Korean regime, at least under the aegis of Kim Jong-il, was that—so as to forestall deterritorializations piloting a flight of working organs from the despotic purview—the latter categorically forbade any relationship that did not expressly revolve around himself. For this reason, family orientation and regionalism are condemned by the regime as hotbeds of sectarianism, whereas all forms of socializing inclusive of class reunions are anathema. Even the forming of bonds based on teacher-student or senior-junior relationships are forbidden. It is demanded that subjects maintain close relationships with those close to the head of State and keep those not close to the Dear Leader at arm's length. Allied to this despotic paranoia is the paranoia of what was Kim Jong-il's extreme secretiveness. Here, it is contended that maintaining secrets is the essence of the life of the party, whereas everyone is forbidden to reveal anything more than what is reported in the papers. Inter alia, for fear that they would leak party secrets while at work, he forbade the wives of party officials higher than vice-director from holding any job whatsoever.[279]

These anal-hoarding and sadistic aspects of North Korean despotic paranoia have, notwithstanding, long since been counterpointed by the sado-necrophilous trait of the regime's paranoia. In order to consolidate his power upon assuming its reigns, Kim Il-sung had already purged all potential opposition leaders by

1949, either through public executions or by hurling them into newly created labor camps that were the precursors to the present-day regime's gulags.[280] As the Korean War raged, Kim attributed its continuing failure to the disloyalty of certain members of his own party. These purged elements included both loyalists and factionalists. They were exiled, condemned to hard labor, or killed. Those to be purged were in fact promoted and later removed with the suddenness of a lighning bolt out of the blue. The sectarian elements that he was purging, Kim—in a typically necrophilous vein—compared to rats that required complete extermination. The paranoid comportment of the State is here evident: "If we don't catch the rat, it will give birth to young. They will gnaw holes through the walls and finally destroy the whole house." Clearly, in accordance with this logic, the rats portended the terminal dissolution of the North Korean despotic full body.[281]

Likewise almost from the outset, in order to forestall potentially runaway deterritorializations that political and religious influences could foment, Kim Il-sung took draconian measures. Over two-thousand Buddhist temples and Christian churches were burned. With missionaries calumniated as Western spies who branded Korean children with hot irons and sold their blood, over a hundred-thousand Christians were incarcerated.[282] To this day, the government bans organized worship as grounds for treason against the State.[283] Even praying in secret can be enough to get one thrown into a camp. Almost all religious believers in North Korea who are caught are thrown into a gulag.[284] Through the destruction of others, the great paranoiac made himself the sole Creator of everything material and spiritual in the North Korean State—hence ensuring his auto-deification. Nothing existed before him.[285]

The sado-necrophilous aspect of North Korean despotic paranoia has, in fact, exacerbated itself more recently. The government's public distribution system provides far less than the minimum UN requirement for nutrition, inducing widespread malnourishment that will be evident for generations.[286] Bradley Martin avers that it is entirely possible that the paranoid Pyongyang regime artificially devised the great famine of the mid-90s as a deliberate genocidal policy ensuring that all or most of the people classified as disloyal would die from hunger, whereas the survivors would specifically include those whose loyalty was considered essential to the regime's survival—so that the regime would in fact emerge from the famine stronger than before.[287] In particular, during the famine, there was a conscious policy to starve certain groups, such as political prisoners and members of banished families more systematically than had been the case in pre-famine times.[288] People started to eat tree bark. Ravaged and delirious from hunger, individuals would resort to infanticide and cannibalism. Lots of persons were to be seen dead on the streets. Dead bodies lay beside train tracks.[289] The starving would freeze to death. In general, those who were and are not critical to the survival of the regime, such as coal miners, factory and cooperative workers, are subject to starvation.[290] A lot of people die on trains, their bodies disposed by railroad workers and buried in mass graves by delinquents.[291]

Since 2012, the number of public executions has more than tripled. The number of prisoners in the *kwalliso*, or prison camps, is on the rise. Shoot-to-kill orders are in effect for refugees attempting to cross the Yalu or Tumen Rivers into China. The use of Chinese-made mobile phones—that is, ones that can dial outside of the country—is punishable by death.[292] For this reason, most "elite" citizens of Pyongyang are forced to line up in long queues for public telephone booths to make their one phone call of the day.[293] During the famine of the mid-90s, public executions became an integral part of North Korean life, with local villagers, including children, and even elementary-school teachers with their whole classes, regularly being forced to witness them. People were executed for theft, slaughtering livestock for food, and attempted defection. Those executed were either hung, shot, garroted by being strangled with a chain, wire or rope, and even burned alive with relatives forced to light the pyre.[294] At one point the regime would even dig up the remains of the dead and subject them to ritual execution by firing squad—a modern resuscitation of the feudal custion of exhuming and decapitating the corpse of a posthumously disgraced official.[295]

So paranoid is the comportment of the North Korean full despotic body that subjects can be thrown into a gulag for six months of hard labor merely for humming a South Korean pop song. If the portraits of Kim Il-sung and Kim Jong-il in one's home or office are not properly dusted or are hung off-center, one can be sent to jail. Even if one's conduct is flawless and exemplary, one evening the secret police can knock on one's door, strip one of one's entire family and worldly possessions, and throw one into a gulag (*kwalliso*) because it was discovered that someone in one's family from a previous generation was a collaborator with Japanese colonial authorities.[296] In this vein, Kim Il-sung warned potential factionalists that their seditious actions would destroy three generations of their family. He and his son made good on this threat. Even failure to obtain permission before boarding a train or being absent from work were offenses that can land a citizen in a detention centre.[297] One person who criticized the underwear in a North Korean shop was hurled into a political prison.[298] Pronouncing the names of the Great Leader or the Dear Leader without using the necessary honorific prefix can likewise result in swift incarceration in a gulag. Also, if any family member commits suicide, all other family members including the spouse, are immediately branded traitors and may be hurled into a camp.[299]

Anyone seen as a potential political threat to the Kim family is thrown into the camps, regardless of whether they are from the party, military, or bureaucracy. Students who have studied abroad or diplomats and who are suspected thereby to be "polluted" by outside ideas are also potential gulag incarcerees. The prison camp system is a fundamental pillar of the North Korean despotic regime's repression and political control strategy, and an indispensable aspect of the Kims' totalitarian rule.[300] Attempting to defect is likewise a major crime punishable by anything from a minimum of seven years in a gulag to death. Once forcibly repatriated, these offenders are hurled into labor/interrogation camps along the Sino-Korean border. They are sandwiched into small overcrowded cells and interrogated whilst being beaten with chains, belts, and sticks over the course of a number of weeks, subsisting on kernels of corn and salty broth, and forced to clean latrines with their bare hands. As a measure of the despotic paranoia of the regime, Pyongyang began to hold the forcibly repatriated separately from the general prison population, for dissemination of what they may have seen while in China. Of the hundreds of thousands that have attempted to escape from the country, almost all have been tortured to death.[301]

Hundreds of thousands of nameless and faceless men, women, and children waste away in the camps. Prior to being hurled into the camps, individuals are subject to arbitrary arrest and imprisonment, without trial or any semblence of judicial process. Upon being brought there, they face torture until they confess to whatever crime they are accused of, often regardless of whether or not they had even committed it. To preempt and stave off deterritorializations piloting an exodus of working organs from the despotic full body, the camps are surrounded by four-meter-high walls, topped with barbed and razor wire, or electrical fencing, and surrounded by minefields. Intermittent guard towers are staffed by thousands of heavily armed guards accompanied by viciously trained dogs and equiped with machine guns and grenades, and ordered to shoot and kill anyone who appears to be making an escape attempt. The camps are divided into a number of zones. Those incarcerated in "total control zones," as opposed to "revolutionizing zones," are imprisoned for life and will never be freed until they are on the very verge of death, at which time they will be released to go off and die elsewhere.[302] Many camp inmates die of dehydration and hunger within the first six months. They die from eating wallpaper glue. They die from sucking the excremental water on toilet mops. Driven to delirium by hunger, some swallow pebbles or pieces of dirt and die immediately.[303]

Conditions in the gulags (*kwalliso*) are subhuman. Inmates are reduced to a condition lower than slaves. Men and young female prisoners are subjected to hard physical labor in incredibly dangerous conditions, resulting in large numbers of work-related deaths and a shockingly high count of inmates who end up as amputees, cripples, hunchbacks, and other generally deformed hominids. Prisoners are forced to work ceaselessly and made to fulfil extremely strict quotas, which are enforced brutally. For working too slowly or failing to make a quota they are hurled into solitary confinement, physically abused or tortured. They are made to witness public executions, in which mostly prisoners who have tried to escape are either hung, or shot, or stoned, or dragged behind a moving car until they are dead. The hideously emaciated, disfigured, and discolored inhabitants of the camps toil in permanent situations of deliberately contrived semi-starvation, being forced to eat grass, weeds, the bark of trees, bugs, beetles, snakes and rats.[304] Starving prisoners even make incursions into pigsties to eat the pig feed.[305] They are crammed into cells with eighty or ninety other prisoners, sleeping on the hard floor without blankets whilst aching from ravages of frostbite.[306]

As former prisoner Kang Chul-hwan avers of life in the gulags (*kwalliso*): "You're nothing, an animal, a savage. The guards have the right to kill you without penalty because you're just an animal. If you disobey them or talk back, the guards hit you."[307] For talking without permission, prisoners are often bludgeoned with an iron pole. They are tied to a stake and their jaws are smashed.[308] Prisoners are forced to engage in daily "self-criticism" sessions in which they must confess to the wrongs, whether real or contrived, that they have committed against the State during the past day—for which they are punished with even more draconian work quotas and longer hours. They are beaten every day with sticks. They are shot dead at random by security guards. Those who attempt to escape have their hands tied behind their backs and are hung on the wall for up to a week. Female prisoners are beaten and have their genitalia and breasts prodded with brooms during interrogation. Pregnant women are forced to have abortions, or their children are suffocated to death upon birth with vinyl or wet towels before their mothers' eyes and then discarded into a plastic box with other dead infants to be buried.[309] Alongside the camps, in the mountain areas, there are hundreds of thousands of unmarked graves.[310]

Notes

1. Clements 2007, 102.
2. Ibid. 149, 150.
3. Ibid. 109.
4. Ibid. 105, 106.

5. Ibid. 114, 115.
6. Ibid. 116, 145, 146.
7. Ibid. 106.
8. Ibid. 110, 111.
9. Ibid. 150.
10. Ibid. 152, 153.
11. Ibid. 155.
12. Ibid. 191, 192.
13. Suetonius 1958, III.59.
14. Ibid. 136, III.54.
15. Ibid. 136, III.55.
16. Ibid. 137, III.56.
17. Ibid. 137, III.57.
18. Ibid. 138, III.58.
19. Ibid. 139, 140, III.61.
20. Ibid. 141, III.62.
21. Dio 1927, Vol. 9, 303, LXXVIII.
22. Ibid. 317, LXXVIII.
23. Ibid. 321, LXXVIII.
24. Ibid. 343, LXXIX.
25. Ibid. 351, LXXIX.
26. Herodian 1969, Vol. 1, 443, IV.12.3, 4.
27. Ibid. 403–405, IV.6.1–4.
28. Dio 1927, Vol. 9, 285, LXXVIII.
29. Ibid. 307, 308, LXXVIII.
30. Herodian 1969, Vol. 1, 403–407, IV.6.4–7.
31. Dio 1927, Vol. 9, 279, LXXVIII.
32. Ibid. 321, 322, LXXVIII.
33. Ibid. 291, LXXVIII.
34. Ibid. 305, LXXVIII.
35. Herodian 1969, Vol. 1, 423, IV.9.1, 2.
36. Dio 1927, Vol. 9, 335–337, LXXVIII.
37. Herodian 1969, Vol. 1, 425–427, IV.9.4–10.
38. Procopius 1960, 75, VI.vi.19–21.
39. Ibid., VI.vi.23, 24.
40. Ibid. 213, VI. xviii.2–4.
41. Ibid. 213, 214, VI.xviii.5–21.
42. Ibid. 223–225, VI.xviii.36–38.
43. Ibid. 77, VI.vi.26–28.
44. Ibid. 157, VI.xiii.3.
45. Ibid. VI.xv.4, 5.

46. Ibid. 101, VI.viii.28–30; 157, VI.xiii.2.
47. Ibid. 181, VI.xv.19–23.
48. Ibid. 193–195, VI.xvi.14–17.
49. Ibid. 135–137, VI.xi.13–23.
50. Ibid. 137–139, VI.xi.24–29.
51. Ibid. 139–141, VI.xxiv.31–33.
52. Procopius 1954, 219, I.xxiv.1, 2, 7–10.
53. Ibid. 227–229, I.xxiv.22–24.
54. Ibid. 233–239, I.xxiv.40–58.
55. Walker 2009, 185.
56. Ibid. 192, 194.
57. Ibid. 197–199.
58. Ibid. 200, 201.
59. Ibid. 202–204.
60. Ibid. 204, 205.
61. Ibid. 205–211.
62. Ibid. 160–173.
63. Sumption 1978, 40, 41.
64. Ibid. 16.
65. Ibid. 66, 67.
66. Pegg 2008, 5–7.
67. Ibid. 60.
68. Ibid. 61.
69. Ibid. 78.
70. Ibid. 76, 77.
71. Ibid. 86.
72. Ibid. 100.
73. Ibid. 104, 105.
74. Ibid. 110.
75. Ibid. 113.
76. Ibid. 120.
77. Ibid. 132, 133.
78. Ibid. 138.
79. Ibid. 164.
80. Sumption 1978, 230, 231.
81. Ibid. 240.
82. Payne and Romanoff 1975, 143.
83. Ibid. 145.
84. Ibid. 147.
85. Ibid. 198.
86. Ibid. 239.

87. Ibid. 242, 243.
88. Ibid. 344.
89. Ibid. 218.
90. Ibid. 210, 213.
91. Ibid. 198, 199.
92. Ibid. 209.
93. Ibid. 237.
94. Ibid. 249.
95. Ibid. 210, 211.
96. Ibid. 213.
97. Ibid. 300.
98. Ibid. 249.
99. Ibid. 264, 265.
100. Ibid. 225, 226.
101. Ibid. 236, 237.
102. Ibid. 247, 252.
103. Ibid. 261.
104. Ibid. 300–304.
105. Ibid. 267, 268.
106. Ibid. 272–283.
107. Scurr 2006, 276.
108. Ibid. 116.
109. Ibid. 125.
110. Cited in Ibid. 209.
111. Cited in Ibid. 294.
112. Scurr 2006, 234, 235.
113. Ibid. 261.
114. Cited in Ibid.
115. Scurr 2006, 236, 237.
116. Ibid. 238.
117. Marat 2003, 84.
118. Cited in Scurr 2006, 238.
119. Cited in Ibid. 276.
120. Cited in Ibid.
121. Cited in Ibid. 294, 295.
122. Cited in Ibid. 258.
123. Scurr 2006, 258.
124. Ibid. 297.
125. Ibid. 239.
126. Ibid. 261, 262.
127. Ibid. 277, 278.

128. Ibid. 281, 282, 284, 286.
129. Ibid. 246, 256.
130. Ibid. 258.
131. Kiernan 2007, 401–405.
132. Ibid. 406.
133. Ibid. 408.
134. Cited in Fisk 2006, 390.
135. Kiernan 2007, 408.
136. Cited in Fisk 2006, 395.
137. Fisk 2006, 390–391.
138. Ibid. 394, 395.
139. Ibid. 396, 397.
140. Ibid. 398.
141. Ibid. 432.
142. Ibid. 400.
143. Service 2004, 170, 171.
144. Ibid. 343.
145. Ibid. 285.
146. Ibid. 173.
147. Ibid. 337.
148. Ibid. 354.
149. Ibid. 309.
150. Ibid. 343.
151. Amis 2003, 17.
152. Service 2004, 27, 28.
153. Ibid. 158.
154. Ibid. 322.
155. Ibid. 332–334.
156. Ibid. 338.
157. Ibid. 339.
158. Cited in Service 2004, 340.
159. Amis 2003, 30.
160. Ibid. 235.
161. Ibid. 176.
162. Ibid. 123.
163. Ibid. 63.
164. Ibid. 126.
165. Ibid. 129.
166. Ibid. 137, 138, 140, 141.
167. Ibid. 240.
168. Ibid. 176–178.

169. Ibid. 142, 143, 172.
170. Ibid. 238, 239.
171. Ibid. 203.
172. Ibid. 211.
173. Ibid. 220–222.
174. Ibid. 60–62.
175. Ibid. 16, 17.
176. Ibid. 101, 102.
177. Ibid. 66, 67.
178. Ibid. 69, 70, 76, 78, 79.
179. Ibid. 57, 58.
180. Kershaw 2009, 38.
181. Ibid. 150.
182. Cited in Kershaw 2009, 149.
183. Cited in Ibid.
184. Cited in Ibid. 150.
185. Cited in Ibid.
186. Cited in Ibid.
187. Cited in Ibid.
188. Cited in Ibid. 151.
189. Cited in Ibid. 468, 469.
190. Cited in Ibid. 469.
191. Cited in Ibid. 525, 526.
192. Cited in Ibid.
193. Cited in Ibid. 671.
194. Cited in Ibid. 697.
195. Cited in Ibid. 697, 698.
196. Kershaw 2009, 696.
197. Cited in Kershaw 2009, 776.
198. Cited in Ibid. 776, 777.
199. Cited in Ibid. 802.
200. Cited in Ibid.
201. Kershaw 2009, 743.
202. Cited in Kershaw 2009, 743.
203. Cited in Ibid. 743, 744.
204. Kershaw 2009, 747.
205. Ibid. 850.
206. Cited in Kershaw 2009, 887.
207. Cited in Ibid.
208. Cited in Margolin 1999b, 463.
209. Cited in Ibid. 475.

210. Cited in Ibid. 536.
211. Margolin 1999b, 468.
212. Ibid. 467.
213. Ibid. 465, 466.
214. Cited in Margolin 1999b, 480.
215. Margolin 1999b, 474, 475.
216. Ibid. 476.
217. Ibid. 479.
218. Ibid. 481, 482.
219. Ibid. 482, 483.
220. Ibid. 484, 485.
221. Ibid. 486.
222. Ibid. 492.
223. Ibid. 497–503.
224. Ibid. 509, 510.
225. Ibid. 512.
226. Ibid. 516–523.
227. Ibid. 534, 536.
228. Margolin 1999a, 633.
229. Ibid. 623.
230. Cited in Margolin 1999a, 586.
231. Cited in Ibid. 633.
232. Cited in Ibid. 633, 608.
233. Margolin 1999a, 585, 586.
234. Ibid. 584–587.
235. Ibid. 591.
236. Ibid. 591–597.
237. Ibid. 598–607.
238. Ibid. 611–615.
239. Ibid. 608–611.
240. Coughlin 2003, 201.
241. Cited in Coughlin 2003, 165.
242. Coughlin 2003, 164.
243. Cited in Coughlin 2003, 48, 166.
244. Coughlin 2003, 75.
245. Ibid. 156–161.
246. Ibid. 163.
247. Ibid. 167.
248. Ibid. 202.
249. Ibid. 167, 168.
250. Ibid. 168–171.

251. Ibid. 198.
252. Ibid. 224, 225.
253. Ibid. 242.
254. Ibid. 239.
255. Ibid. 276, 280, 281.
256. Ibid. 287.
257. Ibid. 303, 304.
258. Martin 2006, 482.
259. Cited in Martin 2006, 193.
260. Martin 2006, 192.
261. Ibid. 236.
262. Cha 2013, 37–39.
263. Ibid. 40, 41.
264. Ibid. 42–44.
265. Ibid. 165, 166.
266. Ibid. 44, 45.
267. Ibid. 58–61.
268. Ibid. 93.
269. Ibid. 107.
270. Ibid. 181.
271. Ibid. 208.
272. Ibid. 106.
273. Martin 2006, 262–264.
274. Ibid. 547.
275. Ibid. 433.
276. Ibid. 294–296.
277. Ibid. 273, 274.
278. Ibid. 283.
279. Ibid. 286, 287.
280. Ibid. 71.
281. Ibid. 94–96.
282. Cha 2013, 73.
283. Ibid. 164.
284. Ibid. 599.
285. Ibid. 73.
286. Ibid. 9.
287. Ibid. 558.
288. Ibid. 566.
289. Ibid. 618, 619.
290. Ibid. 561.
291. Ibid. 624.

292. Cha 2013, 9.
293. Ibid. 164.
294. Ibid. 197.
295. Martin 2006, 575.
296. Ibid. 8.
297. Ibid. 290.
298. Ibid. 296.
299. Ibid. 606, 607.
300. Cha 2013, 172.
301. Ibid. 181–183.
302. Ibid. 170–175.
303. Martin 2006, 612.
304. Ibid. 172–174.
305. Martin 2006, 598.
306. Cha 2013, 172.
307. Cited in Martin 2006, 299.
308. Martin 2006, 598.
309. Cha 2013, 174–177.
310. Martin 2006, 299.

5

Trial by Ordeal and *Homo Sacer*

1. The fusing of desire to the Law and the Law's essential vacuity

The circulation of credits and the new alliance's co-optation of the old mobile debts—which renders the debt an infinite and existential one owed to the despot and his deity—and the extension of savage filiation into direct filiation, in concomitance with the linearization of the signifying chain wherein the despot functions as signifier of the transcendent voice of the deity; fundamentally changes the nature of desire.

(i) Desire as the desire of the Other

The territorial polyvocal sign engraved by the graphy of the primitive regime on the suffering body being ensconced into alliance is always already a self-positing and self-validating, multiply-connected "position" of desire. Nowise is it a sign subordinated to another sign that would render desiring a desire of another desire.[1] In the despotic regime, this is no longer the case. The fact that now the sign has become a sign of a sign in a biunivocalized signifying chain in which the despot functions as the master signifier animating the entire chain, implies that desire—far from being (as it formerly was) its own self-sufficing validation and positioning of circulating objects of desire—becomes the desire of the desire of the despot or his deity. Indeed, with this having come to pass, desire is now too cowardly to desire of its own accord. Linearized writing no longer designates organs of production or reproduction whilst distributing their positions on the surface of the socius, and instead merely reacts to the written Law promulgated by the absent yet ubiquitous despot and his bureaucratic perverts.[2] It is therefore within the despotic regime that the early—basically Hegelian—Lacanian conception of desire assumes preeminence: desire as the desire of the Other and of the Other's desire. In the despotic regime, the Other par excellence is, of course, the despot himself.

(ii) The fusion of desire to Law

Attendant upon all of this is the fact that desire becomes fused to Law—though it shall soon be evinced that the precise meaning of this ought not be seized at with excessive haste. In the appropriation and distribution of all the territorial machines, productive agents, products and organs of the primitive machine, the new despotic full body integrates and circumscribes the latter under the higher and transcendent unity of the State formation. This circumscription acts as a transcendent Law which presides over parts, in the process partitioning them off from one another, organizing them as "bricks," measuring their respective distances and categorically forbidding their mutual communication. This regimental distribution of subaggregates as noncommunicating "blocks" or "bricks" under the auspices of the State is what Deleuze and Guattari term the "paranoid-schizoid trait" of the Law.[3] The paranoid-schizoid trait of the transcendent Law is precisely consonant with the paranoiac position of the despot; that is to say, the great paranoiac's endlessly furtive vigilance that not one of the "organs" or "blocks" may communicate with another or jump outside the position accorded to them by the distribution imposed by the higher unity of the State, since this would entail a protest and flight of the said organ or organs from the full body of the despot.

(iii) The Law's essential vacuity

There is, however, a second aspect to the Law. It is brought into effect by the fact that, in the despotic biunivocalized signifying chain, no signifier arrives at a terminal and conclusive signified. This second aspect is the so-called "manical-depressive trait," in accordance with which the Law reveals nothing and is entirely bereft of any knowable object. The upshot herein entailed is that the Law's verdict has no prior existence to the penalty it metes out, and that the statement of the Law's content has no prior existence to the verdict it pronounces.[4] Devoid of content as it is, the Law is bereft of any interiority. It is *purely formal*, that is, all that can be said about it is that obeying it is merely *necessary*. In this connection, Kafka's "Before the Law" is most germane. Here, the man from the country is unable to pass through the door of the Law, is unable to "gain admittance" to the Law, precisely insofar as this door is already open and in that the Law prescribes nothing. For Žižek this purports that the ultimate secret of the Law is that it *does not exist*—which is analogous to asserting that that there is no big Other (in the Lacanian sense). But, contends Žižek, this "non-existence [of the Law as such], of course, does not simply reduce the Law to an empty chimera; rather, it makes it into an impossible Real, a void which nonetheless functions, exerts influence, causes effects, curves the symbolic space."[5] As Giorgio Agamben avers, it is precisely this vacuity and purely formal character of the Law that allows the Law to impose and affirm itself with the greatest force (and, one may add, in the first place and—*at all*): "The man from the country is delivered over to the potentiality of law because law demands nothing of him and commands nothing other than its own openness."[6]

But there is more. For one observes that this formal vacuity of the Law corresponds to State terror, insofar as "life under a law that is in force without signifying resembles life in the state of exception [wherein the law, which otherwise is replete with positive content, is suspended], in which the most innocent gesture or the smallest forgetfulness can have the most extreme consequences."[7] For Žižek, the Law conceived in this sense is as a "meta-Law, the Law of the state of emergency in which positive legal order is suspended, the 'pure' Law, the *form* of ordering/prohibition 'as such', the enuncuation of an Injunction deprived of any content."[8] We shall dwell more intently upon this and diverse corollaries of the formal emptiness of the Law fairly shortly. For now, let it suffice to be mentioned that this aspect of the Law is precisely what Deleuze and Guattari above termed its manical-depressive trait.[9]

This manical-depressive trait of the Law is also that of the subjects in the despotic regime, just as the paranoid-schiziod trait of the Law is at once the paranoiac position of the despot. As Solzhenitsyn avers vis-à-vis the psychological predicament of the denizens, or zeks, of the Gulag Archipelago (a despotic regime in its own right):

> … as soon as a zek is left in a state of inaction, by himself, buried in his thoughts, his neck ceases to support the weight of his head, and his shoulders and his back immediately express an inveterate stoop, as if he were born with it …[10]

This stoop and depressed demeanor will persist when he approaches his master, whom he will not look in the eye but rather onto the ground—though if he is forced to look at one, such a one "will be surprised by his blank, stupid stare," and how in discourse with others the zek "will be laconic, speaking without expression—either monotonously and dully or else with servility—if it is necessary to ask you for something."[11] The paranoiac despot (or sovereign) and his depressive subjects.

2. The eye that forewarns

Obviously, the modifications in the mode of inscription from the primitive to the despotic regime bear considerably upon what in the primitive regime was termed the magical triangle of voice, graphics and eye. To be sure, far from being completely obliterated by despotic overcoding, this triangle subsists as a "base" and as a "brick," inasmuch as the primitive territorial machine still continues to operate under the aegis of the despotic machine, albeit in a much distorted and mangled form. With the latter, this triangle becomes the "base" of what is now a pyramidal structure, in which the vocal, the graphic, and the visual all converge in the direction of the higher unity of the supernal despot and his hauntingly absent yet ubiquitous voice-purloining deity.[12]

Nonetheless, what in the primitive machine was a plane of connotation—voice and graphics connoted along with each other upon the suffering body whose pain was appraised by the beholding "collective" or "divine" eye extracting surplus *jouissance* from this boisterous proto-sadomasochistic spectacle—has in the despotic regime become a plane of subordination: signifieds subordinated to signifiers, the entire biunivocalized chain subordinated to the master signifier, the voice to writing, subjects to the despot, and desire to Law and to the desire of the despot or Other. Moreover, in this movement from the primitive plane of connotation to the despotic plane of subordination, the very function of the individual elements of the magic triangle changes diametrically. Heretofore having intoned the words of corporeal and sonorous alliance, the voice now only serves to dictate and pronounce decrees; what was formerly a polyvocal graphism animating bodies, has become prosaic writing set in manuscripts and other Statist documents; whereas the eye has ceased to appraise and luxuriate in its own enveloping *jouissance*, and henceforth merely reads written letters.[13] The mouth has ceased to speak, and instead "drinks the letter." Having, under savagery, been inscribed like the earth with the visceral marks of cruel corporeal graphism, the body now lies in supine prostration before the dictatorial engravings of the despot. The plight of the now cowering subject under the almost immortal might of the despot and his dread spectral deity may be adduced from what Derrida avers apropos of Antonin Artaud's critique of Western representation, which is invariably Statist in character. Here, it is not only that the body is subjected to supine prostration at the behest of the transcendent deity, or "great Furtive One," but this latter moreover finds itself charged with responsibility for stealing the subject's very body and mind, and of effecting the terminal diremption of the two. This theft depreciates and soils the subject in terms of the separation between his body and his own excrement, which latter on this account comes to disgust him and render the history of the furtive God that dominates him and steals his surplus value a history of defecation. In this interminable theft of both the subject's body and the incessant exploitation of its labor, the subject, now reduced to a mere metaphorical name for God, is linguistically dispossessed of his very being by this void, purely consumptive, sybaritic and bone-idle locus of the despotic master signifier, in an excremental movement of purloining of the subject's works. Dead as they are, the stolen works of the subject nowise allow him to stand upright to his full stature.[14] Instead, he cowers in supplication before the imperial Law.

The voice no longer intones the sonorous and breathy words of savage alliance, but pronounces the subject *guilty* prior to the commision of any crime. The eye has now ceased extracting a surplus value of code from the joyous spectacle of suffering at public enactments of primitive justice. Indeed, it no longer evaluates. Instead, its function is henceforth to "forewarn" and hold watch—meaning that its role is now that of a sentinel, ensuring that no surplus value whatsoever can escape the overcoding of the despotic State apparatus.[15] It forewarns of the danger of but a single organ detaching itself from the full body of the despot, and moreover suffers. It is the eye of the guard peering repeatedly through the peephole onto prisoners undergoing interrogation, the eye scrutinizing from the heights of the observation tower that the perimeter wall is not breached by dissident inmates.

The political function of this forewarning eye, and indeed of other organs of sense, is to be seen in the structure of the Aburĩrian State apparatus unfolded in Ngũgi wa Thing'o's novel mentioned earlier. Here the Ruler's minister of foreign affairs, Machokali, had travelled to London in order to have his eyes enlarged to the size of electric light bulbs, so as to make them "ferociously sharp" and capable of spotting the enemies of the Ruler in the most remote and secluded hiding places. The minister of State in the Ruler's office, Silver

Sikiokuu, on the other hand, had travelled as far as Paris to have his ears enlarged to be even larger than those of a rabbit and "primed to detect danger at any time and from any direction," so that he could more easily eavesdrop on the most private conversations of husbands and wives, children and parents, students and teachers, priests and parishoners, and psychiatrists and their patients. He was in charge of spying on the citizenry and presided over the secret police. Taken together, Machokali and Sikiokuu were the eyes and ears of the Ruler and his paranoid despotic regime.[16]

3. Preemptive vengeance of the despotic caste

It is despotic paranoia, the despot's propensity to rabid mass murder, that is at the helm of the transmogrification of the role and nature of punishment in the passage between the primitive regime and the despotic regime. Punishment completely ceases to be a festive event, with the beholding eye extracting a pleasurable surplus value of code in the magic triangle of alliances and filiations. It is now a form of *vengeance*. It is the vengeance of the voice, the hand, and the eye conjoined. It is the vengeance proper to the new alliance, the publically displayed vengeance of the despot. It is the despot, and his caste, who now perpetually threaten to strike with multiple avenging daggers. The State, thus, as Zarathustra sternly warns, becomes the "death of peoples," invented as it is by the "destroyers" who preside over it and assail a populace of small-eyed, scampering and terrified quarry reared for slaughter. These annihilators, the despotic caste, now hang a vast panoply of "snares," the "sword," and a "hundred desires" over the teeming populace. The State proceeds immediately to bite "with stolen teeth." It comes forthwith to embody the "will to death." Invented for the superfluous, it shall henceforth lure, devour, chew, and re-chew these surplus men with rancid teeth purloined from elsewhere by its gargantuan apparatus of death. For this nefarious and necrophilous purpose, it has devised "a cunning device of Hell," a "horse of death jingling with the trappings of divine honours."[17]

Elias Canetti's thoughts on the preemptive vengeance of the despotic caste are not to be passed over in silence. The very nature and structure of sovereignty *ab initio* demands it, for anyone "who wants to rule men first tries to humiliate them," so as to undercut from the start "their capacity for resistance, until they are as powerless before him as animals."[18] Deleuze and Guattari go even further than this. For here it is not only the case that the despotic State resorts to a form of pre-emptive humiliation of its citizenry. To render the masses powerless, despotism shall have recourse to an a priori *mutilation* of the masses. Indeed, the State apparatus tries to ensure that the mutilation, and even death, of its subject peoples, is its first and founding act. It well-nigh needs people to almost be "born" in this manner. They must be pre-accomplished cripples and zombies. In this connection, the authors aver that the zombie-myth, far from being one of war, is in fact one proper to labor, to work. To be sure, mutilation is in a vast number of cases a consequence of war. Yet it remains a necessary condition and presupposition of the State apparatus' designs to render its subjugated populace a malleable laboring mass.[19] "The State apparatus needs predisabled people, pre-existing amputees, the stillborn, the congenitally infirm, the one-eyed and the one-armed."[20]

In the case of the Soviet Gulag, preemptive despotic vengeance is vented upon the subjugated political prisoners by setting loutish criminals (who, unlike the politicals, are the class *allies* of the Soviet functionaries) against the politicals. Caged in a Stolypin wagon, the archetypal prisoner transport, the political prisoner is harassed by thieves (*urki*) malingering just outside the bars. Solzhenitsyn is here tormented by one of these hideous ghouls who:

> … jabs two protruding fingers, spreading into the "V" of the slingshot, right into your eyes—not even pausing to threaten you but starting to punch them out then and there. And this gesture of theirs, which says, "I'll gouge out your eyes, crowbait!" covers their entire philosophy and faith! If they are capable of crushing your eyeballs like worms, what is there on you or belonging to you that they'll spare?[21]

He continues:

> An emissary of the ugly snout descends, most often a vicious boy whose impudence and rudeness are thrice despicable, and this little demon unties your bag and rifles your pockets—not tentatively, but treating them like his very own.[22]

Indeed, such preemptive haranguing by the despots through the agency of their thuggish minions, entails the subject being treated as:

> … a rubber dummy around which superfluous things are wrapped which can easily be taken off. Nor can you explain anything in words, nor deny, nor prohibit, nor plead with that evil little skunk or those foul snouts up above.[23]

Such hideous little tykes would make a point of kicking a runaway's head when the latter would be down and vulnerable, and would not shy from booting a piece of bread out of the mouth of a grey-haired old man in handcuffs.[24]

As Venko Markovski relates it, in the despotic regime of the Yugoslavian Gulag on the Adriatic island of Goli Otok, new convicts were forced to run a guantlet of two rows of men whilst being prodded in the ribs, kicked, punched, pelted with buckets of filth, stones, saliva, whipped and chased like a herd of wild animals by convicts, commanders, policemen, and agents.[25]

It was, however, long before these travails, immiseration, sadistic vindictiveness, and slow deaths doled out in regular installments in humanity's vast and sprawling archipelago of labor camps, death camps and foul, morally deplorable "black sites," that a God had emerged to preside over the death-wielding machinery of the State, a dread deity who punishes the very existence of human beings in accordance with the injunctions of an obscene array of perverse plenipotentiaries and mildew-faced apparatchiks—plenipotentiaries and apparatchiks who, so as to fully entrench the ironclad and implacable inexorability of the infinite debt of existence itself owed to the despot's hauntingly absent and dread deity, had, to the ends of an unparalleled sadism, projected an eternal and unremitting wheel of an afterlife in which the suddenly immortal soul of the cowering subject would forever writhe under the full inquisitorial brunt of the monstrous punishment machine that so impressively came to encompass the entire universe within the ambit of its infernal machinery of sordid spitefulness. Good and evil, in a sempiternal fixity ensured by the pulverizing force of superegoic refrains veering well past the point of sadomasochism, become the ineluctable pylons crucifying the subject onto the cross of a permanent mutilation terrifyingly coincident with eternal damnation even prior to actual decease. The entire world was polluted with the promiscuous concept of inevitable punishment, under which life in all its variegated aspects is defamed, slandered, execrated, voided of all breath, and converted into a foul and pestilential contraption for the sempiternal maceration and disembowelment of all spirit.

4. Trial by ordeal

But let us return to the functioning of the Law in the despotic formation. Here, it may moreover be said that, since the transcendent imperial Law is devoid of content (vacuity of the Law), the verdict can have no possible existence prior to the penalty imposed by the State. This implies that the glorious savage festivals of primitive justice—in which man is marked, ensconced into the debtor-creditor relation, and equiped with a futural memory of words so as to enable him to keep promises—gives way to trial by ordeal. Deleuze and Guattari here make the familiar, though still highly pertinent, reference to the torture machine in Kafka's "In the Penal Colony," with which the penalty writes both the verdict and the rule that has been violated. Then they turn lugubrious: "In vain did the body liberate itself from its characteristic graphism [polyvocal graphism] in the system of connotation, for it now becomes the stone and the paper, the tablet and the currency on which the new writing is able to mark its figures, its phoneticism, and its alphabet. Overcoding is the essence of the law, and the origin of the new sufferings of the body."[26]

The gruelling and slow ordeal of the interrogation of political prisoners ripe for the Soviet Gulag partakes of this logic. Solzhenitsyn describes the purgatorial process of Soviet interrogation antecedant upon sentencing (more precisely, they have already been sentenced) to the ensuing horrid years of hard labor, whose aim is the eliciting of a confession to charges that are fabricated in advance, as a grinding of souls in:

> … the great Nighttime Institution [interrogation sessions almost invariably took place at night], when our souls are pulverized and our flesh hangs down in tatters like a beggar's rags … [under] … those pale night executioners who torture us.[27]

This trial by ordeal, from the outset, takes the form of the immediate bundling of the newly arrived prisoner into a stand-up punishment cell so narrow that, when he loses the strength to stand, must sag and support himself with bent knees propped up against a wall. To break their resistance, prisoners were kept sandwiched in this way for more than a day without pause. Following upon this, the deliberately infrequent and mis-timed lavatorial access, day after day, would "torment you the whole day long, oppress you, rob you of the inclination to talk, read, think, and even of any desire to eat the meagre food."[28] Indeed, having:

> … passed through the meat grinder of political interrogation, the human being was physically crushed in body: he had been starved, he hadn't slept, he had frozen in punishment cells, he had lain there a beaten man. But it wasn't only his body. His soul was crushed too.[29]

This logic of trial by ordeal may be seen in Foucault's analysis of the functioning of the despotic Law during the Classical age in his *Discipline and Punish.* Here it is through the torture of the body of the suspected transgressor that the truth of the crime that possibly had been committed was produced.[30] In this procedure, a certain degree of punishment already corresponded to a certain degree of guilt. In the unusual economy of judicial torture, the ritual which produced the truth of the crime moved in lockstep with the ritual which inflicted punishment upon the subject in what were the recto and verso of a single, unitary process. The point of application, convergence and coincidence of these two aspects of the judicial procedure was the body of the suspect. This private judicial torture seemlessly extended itself into the public torture and execution of the subject whose purpose was to produce the truth of his crime in the public sphere.[31]

The production of legal criminal truth by the body of the accused quite likely originated in Classical Greece. The rack and a version of the wheel, in which the suspect was tied to a cart wheel and spun around until he confessed, were employed. Both Demosthenes and Aristotle were unequivocally in favor of trial by ordeal as a means of generating the truth of crime, the latter averring that torture provided a sort of evidence to which one may attribute absolute credibility.[32]

Trial by ordeal continued under the Roman Empire. The suspect would be repeatedly hoisted into the air and thereupon dropped violently to the ground into a pile of sharp rocks that ripped his flesh to shreds and shattered his bones, or be torn limb from limb by pulleys arrayed in the form of an improvised rack.[33] As Darius Rejali relates, for the Roman jurist Ulpian, torture was used by judges "to unearth crimes."[34] "By quaestio [torture], we are to understand the torment and suffering of the body in order to elicit the truth."[35] Both Greeks and Romans, in fact, did not consider certain judicial testimony as true unless it was coerced under torture. The torture of slaves, for instance, was not only customary, but indeed compulsory. In a similar manner, in the early seventeenth century, Italian jurist Sebastian Guazzini defined torture as "the distress of the body devised for extracting truth."[36] It was "invented by the Civil Law, as a mode of discovering truth, for the sake of the public welfare, to the end that crimes might not remain unpunished. It is called a species of evidence substituted to supply the lack of witnesses."[37]

Due to the influence of the new inquisitional system that had come into being in Europe in the late middle ages, judges and prosecutors began to prize confessions. To this end, the Roman practice of torture, as a method of generating criminal evidence, was revived. Medieval judicial systems, in fact, had an almost irrational fetish for confessions. Confession was the queen of proofs.[38] Under the despotism of the Spanish Inquisition, the trial of suspected heretics started with an ordeal which was not immediately physical. The accused was free to confess his guilt prior to being tortured. The problem for him was, however, that he often had not the slightest idea what he had been accused of, and therefore found it well-nigh impossible to either confess or deny his guilt. The foregoing fact, notwithstanding, had the crucial corollary that the inquisitorial procedure laid a particularly singular and pronounced burden on the body of the suspected heretic for producing the truth of his crime. This production would commence with the heretic being hauled into an underground room before an official inquisitor, an inspector, a secretary, and a torture master. After the heretic was threatened with torture, thereupon shown the instruments of torture, and finally blindfolded; the production of legal truth by the heretic's body would commence: he would be repeatedly tortured until he confessed his guilt, or stubbornly protest his innocence until the end and die. Tortures employed were manifold. Heretics were pinned to the ground by several men and mercilessly flogged with chords dipped in melted pitch until their backs were stripped of flesh and reduced to one huge ulcer.[39] The Jewish physician, Isaac Orobio, after being arrested and hurled into a dungeon for three years, had his thumbs tied together with a chord until blood spurted from under the nails, was slammed against a wall, and suspended by the arms, hands, legs and feet by ropes from the ceiling. The Englishman William Lithgow had water poured into his stomach via a funnel thrust down his throat until it nearly burst and was then hoisted upsided down by ropes tied to his big toes until he vomited. Heretics were likewise subjected to the *garrucha*. So as to produce the truth of heresy, the suspect was hauled with his hands tied behind his back into the air by a rope run through his wrists and then repeatedly allowed to plummet, resulting in the excruciating dislocation of the shoulder joints. Heretics were likewise flayed alive inside giant drums festooned with chunks of razor-sharp iron. They were strapped to a chair with their feet locked in stocks as a red-hot brazier gradually burned their feet. After confessing under torture, the heretic was bound over for trial. If he exhibited recalcitrance during the latter, he would be resubmitted to torture.[40]

In England, under the reign of William the Conqueror's second son Henry I, the body of the suspected criminal was again allotted the role of producing the truth of his purported crime—trial by ordeal being reintroduced by the said monarch. There was inter alia trial by water, in which the suspect was hurled into a pool and deemed guilty if he failed to sink; trial by fire, according to which the suspect was forced to grab a hot iron bar whilst walking nine paces and deemed guilty if blisters appeared after three days; and trial by a combination of fire and water, in which a suspect plunged his hands into a cauldron of boiling water and was once more declared guilty if blisters appeared after three days. Priests accused of civil crimes were ordered to eat bread coarsened with feathers and deemed guilty if they choked.[41] Under Henry VIII, the Star Chamber, a court answerable neither to the laws of Parliament nor to the dictates of Common Law, employed a vast array of means and devices to extract confessions of guilt. The thumb screws would crush a suspect's thumb knuckle; the boot would crush the ankle bone; the scavenger's daughter would crush the body until blood squirted from the nose, mouth, ears and finger tips; and the rack would dislocate the joints of stretched limbs. Suspects were incarcerated in cells in the Tower of London so small that they were forced to constantly remain in a crouched position, or in cells filled with voracious rodents and insects which ate the flesh from the arms and legs of prisoners while they slept.[42] Under the reign of Queen Elizabeth I's successor King James VI, a certain John Cunningham and forty accomplices were arrested and charged with attempting to bewitch the king to death. In order to produce the truth of Cunningham's crime, royal torturers ripped off his fingernails and drove needles into the bleeding flesh of his finger tips, and crushed his lower legs with the boot until blood and bone marrow oozed from the edges of the iron shoe. This all, however, failed to elicit a confession and Cunningham was strangled and burnt at the stake. The trial by ordeal of a purported Scottish witch the following year did induce a confession. The witch's fingers were locked in a clamp-like device called the pilliwinks while her head was lashed with ropes and jerked violently back and forth. The legal truth of sorcery was ultimately produced by the detection of a witch's mark on her throat, whereupon the witch confessed and was executed.[43]

The use of trial by ordeal to manufacture the truth of witchcraft was accelerated in the late fifteenth century when, at the behest of Pope Innocent VIII, German Dominican monks Heinrich Kraemer and Johann Sprenger compiled a witch hunters' guide called *Malleus Maleficarum, Maleficas, & earum haeresim, ut phramea potentissima conterens,* that is, "The Hammer of Witches which destroyeth Witches and their heresy like a most powerful spear." The *Malleus* was widely circulated among both ecclesiastical and lay authorities. In Lausanne, a woman charged with poisoning and attempted kidnapping was subjected to torture, and within two days confessed not only to the original charges, but also to attending witches sabbaths, fornicating with the Devil, creating storms and murdering children. She was condemned on the strength of her forced admissions. The body—via trial by ordeal—served as the site of production of the truth of witchcraft. It was searched for the *stigmata sagarum*, or "Devil's marks". These could be a mere mole, wart, birthmark or third nipple. The witch was stripped naked, shaved, and searched for these evidences of a pact with the Devil. Failing this, her Devil's mark was assumed to be invisible and thus either insensitive to pain or unable to bleed when pierced. To find this invisible Devil's mark, the witch was repeatedly jabbed with a sharp, heavy needle, until a spot without a nerve ending was found or until shock caused the nervous system to shut down, making the witch insensitive to continued punctures. In this manner, the pricking ordeal induced the witch's body to yield up the truth of her collusion with Satan.[44] Whilst not using the precise procedures of the *Malleus*, Matthew Hopkins, the greatest witch hunter in all history, would notwithstanding succeed on numerous occasions in teasing out admissions of the said diabolical collusion in mid-seventeenth century England. The lonely, cantankerous, one-legged crone Elizabeth Clarke was arrested, stripped naked, deprived of food, drink and sleep, and forced to hobble up and down her cell on her crutch until she collapsed. Her confession of witchcraft was submitted to court. Everywhere he went, Hopkins arrested and tortured witches into confessions that were readily accepted by the authorities. Subjected by Hopkins to trial by ordeal, the septuagenarian Reverend John Lowes confessed to giving birth to and suckling four little demons as well as conjuring up a storm that caused a ship to sink. These admssions were accepted by court and Lowes was hanged.[45]

The Chinese, faithful to the dictates of the Tang Code, would likewise utilize the torments of the body to usher in the truth of the subject's crime. In one such procedure, the prisoner's thumbs and big toes were bound together behind the back, forcing the entire weight of their kneeling body to fall on the toes and knees. Hereupon, a coil of sharp-edged chain was placed under his knees, inflicting excruciating pain and lacerating the knees, sometimes cutting so deep as to sever the tendons. Even in the absence of this form of torture, a few days or weeks in a Chinese jail almost invariably elicited the confession of even the most obdurate prisoner. During the night, the incarceree was chained, manacled and stocked to his bunk in such a manner that any movement was impossible.[46]

In Japan, judicial torture assumed the form of inter alia flogging with the use of a split bamboo whip that sliced deep into the flesh of the victim. The flogging would continue until the victim volunteered to confess his crime, or would continue until the one-hundred-and-fiftieth lash—at which point the suspection would almost certainly be dead. Japanese trial by ordeal also assumed the form of the *Yet Gomon* torture, wherein the prisoner's wrists were bound behind their back and they were hoisted into the air and dropped violently so that their shoulders were dislocated.[47]

This visceral and archaic mode of production of legal truth reappears in twentieth-century communist despotic regimes. As part of interrogation, the political prisoner under Chinese communism would be subjected to forced standing aggravated by having the ankles bound in gauze strips, which forced them to contract against the swelling—thus causing pain more quickly. Also, as part of this trial by ordeal, the prisoner was forced to run between interrogations in leg chains that cut and bruised. He had his arms manacled behind his back, causing cuts on the arms which became infected, whilst the arms swelled to the point that any contact became agonizing. During questioning, interrogators put chopsticks and pencils between the prisoner's sensitive fingers. Sleep was made impossible for him. When lying on his side his sore shoulders were aggravated. When lying on his back his swollen hands were traumatized. He would be subjected to sleep deprivation by his own cell mates by being alternately pinched, poked and slapped. If he returned to his cell wearing manacles, he would be beaten, tortured and starved by them.[48]

Bao Ruo-Wang (Jean Pasqualini) intimates in his memoir of incarceration under the Chinese communist regime how the latter's interrogation techniques aimed to manufacture the truth of one's everyday life as de facto a criminal one, as "rotten and sinful" in relation to the pristine State. The sheer fact of the inmate's being arrested already conferred an initial portion of culpability upon the subject.[49] In a manner that was probably less gruesome though certainly more thoroughly comprehensive and scientific than in certain other communist regimes, the subject would be painstakingly forced to offer up gargantuan, drawn out, and meticulously detailed confessions of their life-long criminality against the State apparatus. Interrogation sessions eliciting confessions continued for five days a week.[50] Trial by ordeal inter alia assumed the form of a near-death starvation of the subject to the point that he was ready to confess nearly anything. This left Ruo-Wang's stomach completely caved in, his joints bruised, the skin on his posterior like the dugs of a old hag, and his vision blurred. His skin rubbed off like a dusty film, his hair fell out, as the ravaging hunger of the famished inmate rendered him almost insane. This was the perfect interrogation climate. He was interrogated for fifteen months.[51]

Another technique deployed within the *laogai* for producing the truth of the subject's crime was "struggling." This was a sort of intellectual gang-beating whereby the inmate was assailed by a "fierce and pitiless, crescendo of screams" demanding that he confess his infractions. This "ear-splitting din" went on for hours at a time, the subject being surrounded by jeering and hating faces, their fists swiping menacingly close to his face, after which he was led away in chains to solitary confinement and tormented by a guard who would constantly awaken him by jerking him up by the hair. After a few days of this, the inmate was already admitting to crimes he had never commited.[52] By the early eighties, the prisoner of Chinese communism would begin to be tortured with stun guns and electric cattle prods. From about the same time, the Taiwanese suspect is subjected to mecidal electrotorture in the form of electrical acupunture and more recently to electro-convulsive therapy.[53]

Moving on to the Soviet Union during Stalin's great purges, torturers utilized scarring torment. Interrogators beat and sweated suspects until the latter confessed. The joint panoply of techniques known as the "the Conveyor" did not, however, leave scars. The confessions elicited were thereupon used as evidence of guilt for show trials. Apart from the far greater stealthiness of the torture administered to the body, one sees here a recrudescence of the mode of eighteenth century judicial torture dwelt upon by Foucault. Though the political prisoner's physical torture does not extend into the public realm, as was the case with the ostensible display of royal power in the Classical age, the Stalinist show trial does serve to manifestly reinforce the physical and ideological power of the State through the public haranguing of the defendant.[54] Though political prisoners' confessions of their ideological infractions against the State as coerced by secret torture were initially private, this was by no means the end of their ordeal. For Soviet trial by ordeal induced defendants to stick to their confessions before the cameras and press. In order to prepare them for public confession before a tribunal, suspects were subjected to nocturnal interrogation and sleep deprivation. Interrogators would bully, question, insult, humiliate and threaten suspects hour after hour, day after day. To further frustrate the prisoner, each interrogator would start the case from the beginning and go over it repeatedly. The prisoner would answer questions under blinding floodlights, whereas in his cell a bright light would fall on his face. In what was termed the *vystoika*, he stood for hours while being questioned, causing the ankles and feet to swell to twice their size within twenty-four hours, movement to become agony, large blisters to develop, the heart rate to increase, the kidneys eventually to shut down. In his cell he was subjected to the *vysadka*, that is, to sitting without moving. When sleeping, he had to have his hands on his blanket and his face toward the light. His shins were kicked, his face was slapped, he was punched in the kidneys and put into chokeholds. He was placed in a cell which was either very hot, freezing cold, or dark and wet. In some cases, he would be put in a cell with several thousand men in suffocating heat and asphyxiating stink. Sometimes he would be put into a straitjacket, or had his head wrapped in a wet cloth which painfully compressed his head as it dried. He was subjected to freezing baths of cold water and fierce dehydration with meals of salted herring. Moving from private torture to the public show trial, Soviet interrogators used the so-termed "Yezhov Method" to hold prisoners to their confessions. According to this method, the interrogator would assess the prisoner's character, raise his fears, and manipulate his weaknesses. He would establish a relationship with the prisoner, sometimes even weeping along with the prisoner when he confessed. Interrogators would feign friendship with the prisoner or threaten aggressively. They rewarded cooperation and punished recalcitrance. They used the confessions of less important prisoners in order to force more important ones to confess, and then used these against the holdouts. This all ensured that the prisoner had a stake in confessing, and therefore stuck to his confession in public. In a manner which doubtless dwarfed the Inquisition, at the peak of the Stalinist purges the Soviet State employed over one-hundred-thousand interrogators to torture perhaps even millions of suspects in the mass scale ritualistic production of confessions.[55]

Other communist countries likewise employed trial by ordeal to render the body of the political prisoner the site of production of criminal truth. The Romanian political prisoner was beaten with rulers, flogged with wires and wet ropes, and had his teeth broken or removed. He was likewise subjected to the *falaka*, or the beating of the soles of the feet with a heavy cable. He was placed in manacles which cut into the wrists and tightened with movement, or held him in contorted positions for hours on end. He would be forced to wear a gas mask, which made breathing difficult. He was also forced to run around a small cell at a minimum speed for hours.[56]

Eugenia Semyonovna Ginzburg describes in her Soviet concentration camp memoir how she was subjected to uninterrupted interrogation, during which she was often deprived of sleep and food for an entire week at a stretch. This was the so-termed "conveyor belt," whose purpose was to wear out the nerves and weaken the body so as to force out the truth of the crime that the prisoner had purportedly committed.[57] This procedure, the author maintans, she felt as "a conspiracy of all the demons in hell" that would potentially turn her from a woman of thirty into a hundred year-old crone.[58] During her trial by ordeal she was confined in an underground punishment cell for days on end. Her arms were twisted behind her back, she was subjected to freezing cold which caused second-degree frostbite. It was impossible to sleep in the isolation cell due to the unbearable cold and the incessant scuttling of rats past her face.[59]

This same extortion of the truth of the subject's crime is in evidence in Eugen Loebl's memoir of his incarceration in a Czechoslavakian communist prison. Here, so as to extract the truth of his transgressions against the regime against which he had purportedly infracted, he was awakened forty to fifty times during the night and blindfolded to the point of disorientation. These ordeals were imposed to the end that he would confess his being a spy and agent of imperialists. When he would refuse to confess, he would be forced to stand stockstill before a wall for hours on end, and forced at the interrogator's desk to repeatedly confess that he was a spy, saboteur, and traitor hundreds of times.[60] Czechoslovakian interrogators pulled out nails with special pliers, and repeatedly submerged prisoners in a bathtub of ice water, nearly drowning them each time, prior to being kicked and subjected to still more violent torment.[61]

A somewhat more gruesome scenario in which the torturous ordeal of the subject's body serves to produce the truth of crime is to be found in Reinaldo Arenas' memoir of Castro's paranoiac despotism. Incarcerated in a prison cell, Arenas overhears the bloodcurdling screams of the man in the adjacent cell. This man was incessantly yelling out that he was going to die. He was on the point of suffocating from the steam that was being pumped into his tightly shut cell. Upon the point of almost dying of a heart attack, the neighboring prisoner was removed from his steam-suffused cell into the interrogation chamber where the truth of his crime was to be created. This torture of a semi-starved man continued for over a month. Arenas remains uncertain whether the incessantly tormented man confessed or simply died.[62]

5. The eternal *ressentiment* of the subjects

The henceforth passive role assumed by each of the elements of the magical triangle as disfigured by despotic overcoding—the voice which merely "drinks the letter," the eye which only reads and suffers, and the body which supplicates itself before the tablets of the law—implies that, apart from the preemptive vengeance of its functionaries, nothing in despotic "justice" is animated by active force in the Nietzschean sense. The body does not act its reactions but merely abandons itself to in cowering passivity before the despot.

This inability of the subjected body to act its reactions, as we saw in connection with the dereliction of the primitive regime's generic human species activity, implies that its reaction to present excitation becomes something which is above all felt—and far too deeply felt. In this regard, Deleuze and Guattari maintain the despotic regime to be one of "latency" and of "terror." In this system, that which has ceased to be active, en-acted or actively reacted to, becomes the very cannon-fodder and raw material for the suspendous growth of *ressentiment*. For it is an instinct for freedom that has, by brute force and otherwise, been rendered latent. It has been thrust back, incarcerated, and repressed to the extent that it is now able to discharge and vent itself only on its own self—in a protracted masochistic orgy of universal slow subterranean suicide. The bowels of the State apparatus are henceforth occupied by former men who are, for Nietzsche, now but "worm-eaten physiological casualties," "a whole, vibrating realm of subterranean revenge, inexhaustible and insatiable in its eruptions against the happy," whose sole aim is "*shoving* their own misery, in fact all misery, *on to the conscience* of the happy," so that the latter "start to be ashamed of their happiness and perhaps say to one another: 'It's a disgrace to be happy! *There is too much misery!*'"[63]

When this insidious, dank, fetid, and miasmal subterranean realm of roiling hatred inevitably, of its very own nature, vehemently turns against itself in a self-flagellating orgy of sickening, moralized bad conscience, there emerges into the garish and bleary-eyed light of palsied day a subzoological human variant unabashedly shackled to his own obscene "will to self-tormenting," to "the repressed cruelty" of the human animal "scared back into himself, the creature imprisoned in the 'state' so as to be tamed, who invented the bad conscience in order to hurt himself," "so as to drive his self-torture to its most gruesome pitch of severity and rigour." For this man, his incorrigible, implacable and nauseatingly horrifying guilt before a now pulverizing and supernal deity of immense vindictiveness and lividly necrophilous sadism shall henceforth and without surcease serve as "an instrument of torture to him. He apprehends in 'God' the ultimate antithesis of his own ineluctable animal instincts; he reinterprets these animal instincts themselves as a form of guilt before God (as hostility, rebellion, insurrection against the 'Lord,' the 'father,' the primal ancestor and origin of the world); he stretches himself upon the contradiction 'God' and 'Devil,'" he transmogrifies his utter denial of his own nature and of his entire self into an affirmation of "the holiness of God, as God the Judge, as God the Hangman, as the beyond, as eternity, as torment without end, as hell, as the immeasurability of punishment and guilt."[64] The endlessly downtrotten, hectored and pilloried subject will move to mentally vent his bile against his despotic oppressors, whom he may well come to regard as worse than ravenous and livid-brained beasts.

We observe political prisoner Solzhenitsyn in the Stolypin compartment, describing how:

> … suddenly you lift your eyes to the square recess in the middle bunk, to that one and only heaven above, you and up there you see three or four—oh, no, not faces! They aren't monkey muzzles either, because monkey's muzzles are much, much decenter and more thoughtful! No, and they aren't simply hideous countenances, since there must be something human even in them. You see cruel, loathesome snouts up there, wearing expressions of greed and mockery. Each of them looks at you like a spider gloating over a fly. Their web is the grating which imprisons you—and you have been had! They squinch their lips, as if they intend to bite you from one side.[65]

Such is the despotic story of vengeance and counter-vengeance: the unconsciously active bronze-eyed artist who founds the despotic regime with his hammer-blows becomes a sempiternally stricken paranoiac (despotic paranoia); whilst the savage whom he subjugates, formerly capable of acting his reactions, emerges as the first in a vast procession of lamentable anthropological products of the becoming-reactive of force—the man of *ressentiment.*

We shall dwell at far greater length on this subjugated man of *ressentiment* and his further pilgrim's regress to the still more fetid and inspissated mire of Nietzschean bad conscience somewhat later.

6. The Damocles' sword of death and *homo sacer*

Presently, more can be said about this subjugated man, especially as regards the change in the position and function of death as effected by despotic overcoding.

(i) The savage dread of dying

In the primitive regime, culture's training of reactive forces to act their reactions ensured that the threat of death instantly passed into action. Death was thus diverted from the possibility of assuming any kind of unconscious or latent state. Indeed, Freud avers in *Totem and Taboo* that, within the sensibility of savages, death as a simple termination of life and naught beside was simply nonexistent. This is evidenced by the meticulous rituals observed by savages who had killed an adversary. Here, the deceased was properly honored and mourned by his murderers so as to thwart the latter's ire—the ire of one who, far from being considered entirely extinct, was seen as merely having migrated elsewhere. He could therefore quite readily return to disturb the equanimity of the living in the form of a malignant ghost. This was so in that the otherwise biologically deceased remained lodged in collective tribal memory, within the spectral domain of the undead.[66] This indicates that the primitive savage is in no way strongly and preponderantly existentially defined in relation to death—if indeed death is conceived of here, in its primary determination, as the pure annihilation of the individual subject; that is to say, an annihilation feared by the individual subject as the termination of his ipseity and future possibility.

Bataille, on the other hand seemingly, avers that primitives have a *hypertrophied* dread of death. How can this be so? Have we not just stated the reverse? No. For the savage dread of death to which Bataille refers does not attach itself to death as simple annihilation, but rather to death as the decaying of the organism coextensive with the process of dying—in the midst of which there is a seething of the "universal ferment of life" in all its noisome and pullulating repulsiveness.[67] The sense of dread occasioned by death in primitive men consists, above all, in dying's fundamental ambiguity and inherent equivocation between the moments of organic decay and that of the renascence of proto-life at its heart. Death is "that putrefaction, that stench ... which is at once the source and the repulsive condition of *life*."[68] The disgust occasioned by the phenomenon of putrefaction is the case insofar as, through an existential perversion of logic, savages assimilate this feeling to "the cruel rancour and hatred visited upon them by death, which the mourning rites [for the deceased] are meant to appease."[69]

For Bataille, it is only when this noxious fetidness of decaying flesh gives way to the more palatable aridity of whitened bones, that the movement from death (as the interstitial and dread ambiguity of the dying and undead) to death (as cessation and demise) takes place: that is, the transition (though one which is never entirely devoid of a lingering equivocation) effected from death as the repulsive sign of life's corruption and ferment to death in its "solemn grandeur" wherein the formerly obtaining "excess of decay's active virulence" is (largely) thrust into the background.[70]

(ii) The despotic fear of death as terminal annihilation

It is upon the ground of this solemn grandeur of a death attained or attainable that, through a semiotic transformation, the despotic machine supervenes so as to massively foreground and exacerbate death as *primus inter pares* the perpetual threat of imminent annihilation that forthwith hangs over the terrified subject's cranium. In contradistinction to the dread of death as noxious decay, the anguish occasioned by death here comes first and foremost to be felt as a hypertrophied fear of "the bitter annihilation of being—of all that I am, which expects to be once more, the very meaning of which, rather than to be, is to expect to be (as if we never received being authentically, but only the anticipation of being, which will be and is not, as if we were not the presence that we are, but the future that we will be and are not)."[71] Here the despotic regime is already and *ab initio* redolent of the general atmosphere of terror characteristic of the *univers concentrationnaire*, in accordance with which Jean Améry avers that "the camp inmate did not live next door to, but in the same room with death. Death was omnipresent."[72]

This fear of death, not as the dread of dying and putrefaction but as a terminal annihilation eliminating the future possibilities of Heideggerian *Dasein*, is what the despotic regime foregrounds in the minds of its subjects. For in the despotic regime there transpires a "detachment and elevation" of the death instinct, which is now no longer coded in the dialectic of actions and reactions of the savage theater of cruelty. In the primitive regime, fatalism was still en-acted. Now, death has become the "sombre agent of overcoding," the detached object par excellence that henceforth dangles over each cowering subject. This means that death is now a latency inscribed into the very entrails of the despotic State apparatus.[73] Death thus, ceasing to be re-acted to, passes into a latent state and thereby establishes itself as an *instinct*, in the process also welding itself to desire. And since desire is, in turn, wed to the transcendent and vacuous Law, death itself is thereby imbricated with the Law—in the form of the persistent Damocles' sword hanging over each subject and constantly threatening to fall should but one organ abscond from the distribution of organs on the full body of the despot. The imperative of non-violation of this distribution is precisely the paranoiac Law of the State, that is, what Deleuze and Guattari have already termed its paranoid-schizoid trait.

The subject grovelling in nauseating terror beneath this thanatic regime would thus, in Améry's phrase, be one always "promised to death, already in the midst of life," the days he had heretofore perchance survived being "a period of false grace that could be revoked at any second."[74] He could only be "a dead man on leave, someone to be murdered, who only by chance was where he properly belonged," and whose "sole duty was to disappear from the face of the earth."[75] In this self-same vein, Sara Nomberg-Przytyk avers of her Nazi incarceration that even "tomorrow they would be dragging me through the block, a nameless dishonoured corpse, unmourned by anyone."[76] "I was thinking about my own death, feeling its cold breath washing over me."[77] In identical vein, Olga Lengyel, likewise a survivor of the *univers concentrationnaire*, alludes to the omnipresence and shocking proximity of death during her own incarceration by the Schutzstaffel (SS). One nod signaled the immediate end of the inmate. Even being late for roll call could easily result in being shot. For "the idea of death seeped into our blood. We would die anyway, whatever happened. We would be gassed, we would be burned, we would be hanged, or we would be shot."[78]

This notion of the subject as inherently promised to death in the very midst of his life, as living a life of false grace, and as a dead man on leave, is graphically illustrated in Ambrose Bierce's short story "An Occurrence at Owl Creek Bridge." Here a man stands upon a railroad bridge with his wrists bound together behind his back by a chord, a rope closely encircling his neck. He is a Confederate secessionist captured for sedition by Unionist forces during the American Civil War. He is about to be executed. As he stands on precarious footing, rapidly approaching the inevitable moment when the tilting of the plank beneath his feet will send him to his death, an equally rapid distension of internal time-consciousness transpires. Begazing a piece of driftwood caught up in the current of the stream below, he is increasingly perplexed by the apparent slowness with which the driftwood is moving. At the same time, striking through his burgeoning thoughts of his nearest loved ones, is "a sharp, distinct, metallic percussion like the stroke of a blacksmith's hammer upon the anvil" that seemed as "slow as the tolling of a death-knell," every new stroke of which he awaited with apprehension. The intervals of silence between each successive stroke grew progressively longer, as the delays became maddening and the strokes ever sharper—to the point of pulverising his ear "like the thrust of a knife." This was merely the ticking of his watch. As the trapdoor opened and sent him down through an opening in the bridge he lost consciousness as one already dead, yet was roused what seemed like ages later. It seemed to him as though the rope had broken, had fallen away, and that he had floated up to the surface of the stream he had plunged into and regained full possession of his senses. His would-be executioners were shooting at him profusely from the bridge but constantly missing as the current swept him further and further downstream onto a bank where he lay concealed. He rushes up the embankment, plunging into the forest and dashing through it until nightfall, urged on by thoughts of his family. Falling sleep whilst traipsing along a deserted road, he dreams of approaching the veranda of his house after having traveled the entire night, springing forth with extended arms to hug his wife. As he is about to fervidly embrace her, he feels "a stunning blow upon the back of the neck; a blinding white light blazes all about him with a sound like the shock of a cannon—then all is darkness and silence!" He was "dead; his body, with a broken neck, swung gently from side to side beneath the timbers of the Owl Creek bridge."[79]

His flight had been a pure illusion. Close to a full day had been interposed and hallucinated into being between the few seconds between the opening of the bridge's trapdoor and the noose's severing of his spinal cord. Here one may venture a thought experiment. What of dilating time still further to the length of an entire life time, or at least a sizable segment thereof? If the noose was the omnipresent Damocles' sword hanging over the protagonist's oneiric one-day escapade and existence, then the death-wielding Damocles' sword threatening the subject of many a despotic regime would preside in readiness to annihilate the subject throughout his entire life. This would be the extension of Owl Creek bridge to Stalin—whom we shall meet again shortly.

A similar thought experiment is to be found in Dostoyevsky's *The Idiot*. Here, during a soirée, Prince Myshkin tells the tale of a man that he knew and who had strangely escaped being executed. When this man was about to be executed and felt convinced that he had only five minutes before he would die, in a gargantuan distension of internal time consciousness, he suddenly experienced those five short minutes as a well-nigh infiite length of time during which he could live several lives.[80]

Now, returning to theory, the sempiternal despotic Damocles' sword of death may—Žižek would aver—be conceived of as the corollary of the obscene underside of the Law, its superegoic supplement: an unconditional exercise of Power assuming the form of the anterior and ubiquitous death-threat (which we shall soon see is in fact inherent to sovereignty itself) emanating from the sovereign (or from the very structure of sovereignty as such).[81]

Within the ambit of the despotic regime of the concentration camp, Solzhenitsyn knew that to "defend yourself in that savage world was impossible. To go on strike was suicide."[82] Twice every day, prisoner Solzhenitsyn's transport was marched along by guards, "tied together in silent and deadly brotherhood," for any one of the guards was at full liberty to shoot and kill any of the prisoners. Indeed, any wrong movement, any false step, they were duty-bound to interrupt with a bullet.[83] Better not a single survivor than for a single organ to flow outside this apparatus or slip away from the body of the despot.

As Solzhenitsyn likewise observes, this proximity of the subject to the Damocles' sword of death was also the situation countenanced by officials working under Stalin. If "you made a mistake with him, it was like mishandling a detonator—it was the last mistake in your life."[84] Stalin always ignored excuses and, when "a malignant gleam from his yellow, tigerish eyes and a slight puckering of his lower eyelids" was in evidence, he had already passed sentence of death. The subject would be arrested on the same night and by the following morning he would be shot. The only prospect for those officials who had ascended somewhat too high up the Soviet bureaucratic ladder was death. It was likewise fatal to work at full capacity for, thus being a threat to Stalin's absolute superiority, one would be shot.[85]

In the grand cosmic despotic regime of the Deity, owing to Satan having vastly exceeded the coefficient of deterritorialization of the divine sign in the garden of Eden, the Damocles' sword of death would henceforth hang over Adam:

> Till thou return unto the ground, for thou
> Out of the ground wast taken; know they birth,
> For dust thou art, and shalt to dust return. (*Paradise Lost,* X.206–208)

Immediately after the prelapsarian phase of man had ground to a halt, Satan would draw the scent:

> Of carnage, prey innumerable, and taste
> The savour of death from all things that live … (*Paradise Lost,* X.268–269)

Henceforth, there was a redolence:

> Of mortal change on earth. As when a flock
> Of ravenous fowl, though many a league remote,
> Against the day of battle, to a field,
> Where armies lie encamped, come flying, lured
> With scent of living carcasses designed
> For death, the following day, in bloody fight. (*Paradise Lost,* X.273–278)

(iii) The sovereign and homo sacer

What is evident is that the subject grovelling in terror under the yoke of the despotic regime is existentially determined by death in a different way than was his primitive counterpart. The despotic subject is, at least in this sense, a veritable Heideggerian *Dasein*—its existential circulation being defined by its projection onto its own dying, as being-toward-death (*Sein-zum-Tod*).[86]

To elucidate this point in more depth, it is propitious here to read the Deleuzoguattarian paranoid-schizoid trait (its propensity and predilection to strike with deathly violence upon the slightest glimmer of real or perceived subversion or infraction) of the Law together with its manical-depressive trait (its infinite openness and formal vacuity). The conjunction of these two facets, their intersection and convergence, forms the very essence of sovereignty as such. This we shall presently see through a foray into Agamben's interpretation of the very notion of sovereignty—that is, of the political sovereignty of the despot.

For Agamben, the sovereignty of the sovereign (*sovrano*)—somewhat counterintuitively and against the grain of what would seem ostensibly to be the case—stems from and consists not primordially in the sovereign's assidious maintenance of the Law but, in fact, in his *suspension* of the order of positive Law in the inauguration of what is termed the state of exception. The suspension of the entire juridical order that is the prerogative of the sovereign corresponds to the exclusion from the sphere of both divine and profane Law of that which through this exclusion (the sovereign ban) is herein consistuted as *homo sacer* (of which more later). This "exclusion" itself soon shows itself to be paradoxical, in that Agamben forthwith maintains it to be an *inclusive* exclusion: that is to say, it is a capturing under the purview of the sovereign of that which sovereignty banishes from the positive legal order in the very self-positing and constitution of this sovereignty that lies at its origin. So on Agamben's reading, it is certainly *not* the case that the sovereignty of the sovereign may in the first instance be defined as a normative vertically transcendent sphere presiding over legal subjects below. Instead, sovereignty is in essence a lofty instance exempt from the juridical order and holding sway over that which it itself deems extra-legal—that which it refuses admittance to the Law. The former naïve and erroneous characterization of sovereignty is only the case in a derivative sense and by way of corollary in relation to this latter more originary sense of the term.[87]

But what precisely is it that is *ab intio* captured under the domain of the sovereign (*sovrano*) through its very banishment by the same sovereignty? Agamben answers that it is that life which may not be sacrificed (the Roman *neque fas est eum immolari*: "it is not licit to sacrifice him") but that may be killed with *impunity*. It is a "form" of life that is therefore excluded from both the *ius divinum* and from the *ius humanum*. This life is neither that of Aristotle's *bios* (the qualified life of an individual or group) nor exactly *zoē* (which the same defines as immediate life in its "natural sweetness"). Rather, it is *bare* life, life conceived in its raw subsistence, life distinguished by nothing other than its *naked* existence. This all permits Agamben to surmise that the "*sovereign sphere is the sphere in which it is permitted to kill without committing homicide and without celebrating a sacrifice, and sacred life—that is, life that may be killed but not sacrificed—it is the life that has been captured in this sphere*" and therefore that that which "is captured in the sovereign ban is a human victim who may be killed but not sacrificed: *homo sacer.*"[88]

Since *homo sacer* may be killed without his death constituting homicide, he is, as Primo Levi maintains, "a man whose life or death can be lightly decided with no sense of human affinity, in the most fortune of cases, on the basis of pure judgment of utility."[89] In this connection, and dwelling upon the annihilation of the inmate of Auschwitz, Améry laments that whereas the soldier dies the hero's or victim's death, the latter dies the death of "an animal intended for slaughter."[90] "I am therefore dead," writes Genet of his incarceration. "I am a dead man who sees his skeleton in a mirror, or a dream character who knows that he lives only in the darkest region of a being whose face he will not know when the dreamer is awake."[91] In that it is merely bare, unqualified life, *homo sacer*—for Levi—is "nothing more than an involucre, like the slough of certain insects which one finds on the banks of swamps, held by a thread to the stones and shaken by the wind."[92] In like vein, Venko Markovski maintains of the inmates of the Yugoslavian Gulag on the island of Goli Otok that they are "not real human beings" but indeed "shadows" and "ghosts," each of which is "a loose page torn from a shattered life." These shadows are "reduced to things, to numbers" and "treated as mere quantities." They are likewise already dead in the midst of life: "As if speaking some mute, fateful language, everything says: we were once alive, but now we are dead."[93] That *homo sacer* is unadorned naked existence implies that he is bereft of the accoutrements that would otherwise bestow human dignity upon him, and thus teeters on the brink of the inhuman. That his death is not a homicide, his very life is almost already death. This allows Agamben to aver that he is the extreme threshold between life and death and between the human and the inhuman.[94] His death is therefore not socio-politically death but rather the production of a corpse without death.[95] It is thus impossible to distinguish between his death and his mere decease, between his dying and his merely being liquidated.[96] The death of this entity was thus, as Améry has it, simply a vanishing happenstance registered "with the set phrase 'substraction due to death.'"[97]

Now, the sovereign realm is that of, Žižek avers, Benjaminian divine violence.[98] Divine violence (not to be confused with the *ius divinum* of Roman law) is violence that is nonsacrificial and expiatory. It is thus precisely the form of violence that visits itself upon that bare unqualified life which is *homo sacer*, so as to annihilate the latter sans sacrifice and with impunity. Thus, importantly for our purposes, "human life is included in the political order in being exposed to an unconditional capacity to be killed."[99] Thus, for Agamben the sovereignty of the sovereign (*sovrano*), of the despot, consists *ab initio* in his capacity (and, one may well add, readiness) to kill those subject to him; whilst—by corollary—the life captured and thereby subjugated under this sovereign despotic sway is in its primary political and existential determination defined by its proximity to potential death, by the permanent foreboding of imminent death portended by the sempiternal Damocles' sword hanging over it. Indeed, as "the two extreme limits of the" political "order, the sovereign and *homo sacer* present two symmetrical figures that have the same structure and are correlative: the sovereign is the one with respect to whom all men are potentially *homines sacri,* and *homo sacer* is the one with respect to whom all men act as sovereigns."[100]

In this connection, one may reasonably venture an isomorphism between the couple composed of the sovereign (*sovrano*) and *homo sacer*, on the one hand, and the analogous pairing of torturer and tortured dwelt upon by Améry, on the other. Firstly, Améry avers that torture transmogrifies the tortured into pure "flesh." "Frail in the face of violence, yelling out in pain, awaiting no help, capable of no resistance, the tortured person is *only a body*, and nothing else beside that [my emphasis]."[101] The reduction of man to a mere subsentient corporeal entity, to pure *physis*, is of course his reduction to bare, unqualified life; that is, his reduction to *homo sacer*. Secondly, the diminution of man to a mere unadorned body sans accoutrements is at once his reduction to pure pain. And pain, the author maintains, is the pure premonition of death, according to the equation "Pain = Body = Death."[102] Thus, throught pain, the tortured is not merely *homo sacer* as bare semi-human life. He is likewise *homo sacer* as naked life to be annihilated in the manner of anonymous, subsentient, though potentially recalcitrant refuse. In this terrifying dialectic, the brutal sovereign reduces the subjugated and subaltern subject to flesh, herein already bringing him to the very precipice of death—from which point the tortured may imminently be "driven beyond the border of death into Nothingness."[103] In this scenario, which is that of sovereignty in its essence, the sovereign is utterly in control of *homo sacer*'s "scream of pain and death" and remains "master over flesh and spirit, life and death."[104] For is not the sovereign's a "dominion over spirit and flesh, orgy of unchecked self-expansion," the power of "one who can reduce a person so entirely to a body and a whimpering prey of death," and is not this sovereign hereby "a god or, at least, a demigod?"[105] Sovereignty reveals itself as "the power to inflict suffering and to destroy."[106] The power of the sovereign as torturer "under which the tortured moans, is nothing other than the triumph" of the former "over the one who is plunged from the world into agony and death."[107] The sovereign renders privy to the tortured that "those things one may, according to inclination, call his soul, or his mind, or consciousness, or his identity, are destroyed when there is that cracking and splintering in the shoulder joints."[108] For Améry, the sovereign makes him learn that "a living person can be transformed so thoroughly into flesh and by that, while still alive, be partly made into a prey of death."[109]

(iv) Homo sacer *and becomings-animal*

Rodolphe Jhering identifies *homo sacer* with the *friedlos* ("man without peace") of ancient Germanic law. As Agamben maintains, like *homo sacer*, the *friedlos* is defined in terms of his exclusion from the community, as being the bandit whose liminal status is moreover intimated by his assimilation to the *wargus*, the *garulphus*, the *loup garou*, and the *wulfesheud*—terms meaning wolf-man or werewolf. Further: "The life of the bandit, like that of the sacred man, is not a piece of animal nature without any relation to law and the city. It is, rather, a threshold of indistinction and of passage between animal and man, *physis* and *nomos*, exclusion and inclusion: the life of the bandit is the life of the *loup garou*, the werewolf, who is precisely *neither man nor beast,* and who dwells paradoxically within both while belonging to neither."[110]

Deleuze detects a homologous threshold of indistinction between man and animal in the work of Francis Bacon. He notes that Bacon's paintings sometimes replace a human head with that of an animal, whereas sometimes a dog is treated as the "shadow" of its master whilst the man's "shadow" assumes an "indeterminate" animal existence. What Bacon therefore brings into effect is a "*zone of indiscernibility or undecidibility*" between man and animal, in which man becomes animal and the animal becomes the "spirit" of man. Now, according to Deleuze, this zone of indistinction between man and beast is "meat." By this he does not mean that meat is simply dead flesh. Rather, this flesh is living and suffering flesh, a flesh of "convulsive pain and vulnerability." The reality of the becoming-man of the beast and of the becoming-beast of the man resides in the fact that "the man who suffers is a beast" whereas "the beast who suffers is a man," and moreover that "every man who suffers is a piece of meat."[111] Is not this Baconesque meat and its becomings-animal—whether it be a becoming-dog or a becoming-wolf—precisely the bare, raw, unqualified life that Agamben christens *homo sacer*?

These double coextensive becomings-beast of man and becomings-man of beast entail for Deleuze and Guattari certain utterly unnatural couplings twixt the human and the animal. Incidentally, there is here a rather sizeable point of contention between Deleuze, on the one hand, and Alain Badiou, on the other. That is, apart from the fact that Badiou does not believe that Deleuzian pure becoming (*devenir*) is tantamount to true change in any sense whatsoever, since such becoming remains wholly within the ambit of the closed economy of mere modifications presided over by the transcendental regime of the given world in which they occur (whereas true change always entails the deregulation of this very transcendental logic according to which beings appear in the said world); he at once likewise holds that pure becoming does not produce or even facilitate the emergence of any "truths." As we shall see in much greater detail in later chapters, the latter are, for Badiou, exceptional to worlds—though they can only emerge *within* worlds and from thence transmigrate to virtually all other worlds. A world without "truth," for Badiou consists purely of an organization of bodies and languages. Badiou terms such a world *atonic.* Now, since for Badiou pure Deleuzian becoming does not guarantee the emergence of "truth" from an "event" arising from an "evental site" (both necessary antecendents of the possible ensuing production of "truths"), such a world is one composed merely of bodies and languages. And the prototypical exemplar of such a "truthless" world is the one depicted in Pierre Guyotat's unbelievably violent, visceral, expectorating, vociferating, and bestially obscene hyper-pornographic novel *Eden, Eden, Eden.* By way of Deleuzoguattarian unnatural couplings, Guyotat's flat and merciless über-Sadean tableau launches a vast cascade of intersecting, criss-crossing, vertiginous, maddening, and singularly obscene lines of becoming-animal, in a searing narrative comprised of one single sentence stretching forth for nearly two-hundred uninterrupted pages. Through a merciless barrage of unnatural cross-species couplings, Guyotat launches an absolutely relentless fusillade of monstrous becomings-other: becomings-dog, becomings-bitch, becomings-antelope, becomings-cow, becomings-monkey, becomings-tarantula, becomings-goat, becomings-sheep, and even becomings-mosquito, or becomings-excrement.[112]

Returning to Agamben, the latter's identification of *homo sacer* with the wolf-man suggests here another (though equivalent) delineation of the essence of sovereignty: that of power over life and death (the Damocles' sword) in response to the subterranean becomings-animal which threaten the sovereign (*sovrano*) with a flight of organs from the despotic full body. These becomings-animal—amongst other vertiginous becomings—entail movements of deterritorialization rumbling within the very bowels of the despotic regime.

In this connection, it is however crucial to recall that the very erection of the despotic regime was in the first place effected by pervasive movements of relative deterritorialization. In relation to this, the sovereign ban placed upon *homo sacer*, especially as exemplified here in the figure of the wolf-man, indicates that there have been and are movements of deterritorialization in excesss of the coefficient of deterritorialization constitutive of the signs of the present signifying regime, intimating the presence of monstrous becomings-wolf.

This relation between *homo sacer* and the becoming-animal of the human may also be read into certain observations made by Deleuze and Guattari. As observed, the sovereign and *homo sacer* form the recto and the verso of sovereignty as such. In a homologous vein, Deleuze and Guattari maintain that the face or body of the despot or his deity is possessed of a kind of counterbody: the body of the tortured, or of the excluded.[113] The authors do, however, make a distinction between the former and the latter. Firstly, it is averred that the one who is tortured is at base he who loses his or her face, and in the same movement enters into a vertiginous becoming-animal, becoming-other, becoming-molecular, to the point of unlimited fissiparity.[114] This definition of the tortured would seem to, at first glance, accord with *homo sacer* as Jhering's already discussed *loup garou*—that is, as a being who is neither man nor wolf and engaged in becomings-wolf.

However, the authors immediately hereupon proceed to add that the one who is tortured and defaced is merely the first step in what shall emerge here as the scapegoating procedure. The second and ultimate step is that of exclusion. More precisely, these are the two facets of the becoming-animal of the scapegoat: firstly, a strictly expiatory animal is sacrificed; then a second one is driven off into the desert.[115]

It is here that one must be careful. For the sheer fact that this first animal is described as "expiatory" and is said to be "sacrified," seems to definitely disqualify attempts to identify it with *homo sacer.* As observed, Agamben's *homo sacer* is that bare life which cannot be an object of sacrifice—let alone serve as a means of expiation. It merely may be killed outright—and with impunity. It would therefore appear that the second animal, the one that is driven off, would seem a better candidate for *homo sacer* than the first. As observed, *homo sacer*—especially as glimpsed in the figure of the werewolf or the bandit—is one who is under the sovereign ban and thus excluded from the normal life of the State. In a similar vein to this, as concerns the second animal (or scapegoat), Deleuze and Guattari aver that in the despotic regime, the scapegoat stands for an intolerably burgeoning "entropy" within the semiotic of the despotic regime. It is blamed for everything "bad" in any specific period, that is to say, with everything that resists the coherent operation of the regime's signifying signs: at once, for everything which eludes the referral of sign to sign throughout the biunivocalized signifying chain of the various semiotic circles, for everything that prevents the recharging of the signifier upon the central face of the despot (faciality), and thus for everything synonymous with destinerrant spillage beyond the outermost circle. At base, the scapegoat incarnates a line of flight that the signifying regime cannot possibly tolerate. Such a line is by definition one of absolute deterritorialization, since it necessarily exceeds the degree of deterritorialization of the despotic signifying sign, however high this degree may be. The despotic regime must therefore deploy fairly brutal means to block a line of this sort, or redefine it in a completely negative manner. This line of flight, which directs itself at a tangent to the circles of signifiance and the center of the signifier, is thus placed under a veritable curse. The regime forges the oppositional pairing comprised of the goat's anus, on the one side, and the face of the despot or god, on the other. This goat's anus represents anything that threatens to put the system to flight. It will be marked with a "negative sign." It will be either killed or itself be put to flight.[116]

Here it may clearly be seen that the scapegoat's being cursed and its line of flight demeaned with a negative sign is homologous to *homo sacer*'s standing under the sovereign ban of Agamben's sovereign (*sovrano*). In this sense, both the scapegoat and *homo sacer* are excluded from the domain of the common citizenry. As likewise observed, the exclusion of *homo sacer* was, paradoxically, an *inclusive* one—in that it was that around which the sovereignty of the sovereign was able to constitute itself in the first place. In an analogous manner, the (likewise constitutive) integrity of the Deleuzoguattarian despotic regime rests upon the exclusion and indictment of movements of deterritorialization whose excessiveness would otherwise overtax the regime with an entropy precluding the recharging of the signifier on the solar face (faciality) of the despot. In that the scapegoat's deterritorialization becomes absolute, this its line of flight would likewise constitute a threat to the integrity of the State apparatus by way of the recrudescence of the BwO or plane of consistency of desire at the edge of the despotic socius. It was after all, through its overcoding of the mutant and destinerrant flows of desire of the full BwO of the earth that this latter arose for the first time. The Deleuzoguattarian scapegoat is thus a figure highly akin to *homo sacer.*

In Ngũgĩ wa Thiong'o's great novel dwelt on earlier, one may likewise espy the correlation between the scapegoat, the increasing entropy within the despotic regime, despotic paranoia, and the reduction of subjects to *homines sacri.* During a certain turbulent and bloody period within the history of the Free Republic of Aburĩria, so as do attenuate this excessive entropy within the regime, some elders decided to treat the evil in the manner in which epidemics were treated. That is, flies, representing the epidemic, were inserted into the belly of a he-goat through its anus, and the goat driven out beyond the borders of the regime. In a somewhat modified version of this procedure, led by a medicine man, the elders inserted some of the Aburĩrian despot's hair—secretly obtained from the Ruler's barber—mixed with grass, salt, and magic potions into a he-goat's mouth. With needle and thread, and starting with the anus, the medicine man began sewing up the seven orifices of the goat. The he-goat, however, escaped before the medicine man could sow up its mouth. The scapegoating procedure was therefore botched. Instead of being driven beyond the perimeters of the Aburĩrian despotic regime laden with all the evil responsible for the unacceptably high level of entropy besetting the regime's integrity, the he-goat roamed about the territory of the regime crying grief. In typically paranoid and despotic fashion, the Ruler felt the goat's anguished cry to be a curse. Indeed, a strange illness had come over the despot, and certain people theorized that it was the result of the despot having been cursed by the goat, bearer of all the entropy besetting the Aburĩrian despotic full body and threatening it with a flight of working organs from its reterritorializing sway. The Ruler imagined the cry of the goat to be a clarion-call for a coup. Forthwith the goat and all involved in botching its terminal flight from the Republic became *homines sacri.* Raining down

Damocles' swords on mortals below, the Ruler swiftly ensured that the goat, the barber, the medicine man, the elders, and even a number of soldiers, were given over to the crocodiles of the Red River. Much in the same way as the spectacular public torture and execution of subjects in the Classical Age aimed to effect the reconstitution of the despot's sovereignty in the face of potentially runaway deterritorializations portended by even the least infraction against the Law, so too did the Ruler signal the Aburĩrian full body's restoration heretofore threatened by the unexpelled goat by including a picture of the Red River where all culprits had met their demise on newly circulated notes of the Aburĩrian currency. Even this did not, however, quell the Ruler's despotic paranoia. For, observing that the goat already eaten by crocodiles had a beard, in a peculiar form of reverse-Talibanism, he decreed the so-termed Law of the Beard, according to which all goats and humans had to have their beards shaved off.[117] Unlike in Afghanistan, facial hair in the Free Republic of Aburĩria was a sinister portent of impending insurrection.

Let us summarize. We have delineated and characterized the existence at opposite ends of the despotic State apparatus of two diametrically heterogenous figures: (1) at the apex stands the sovereign (Agamben's *sovrano*), or despot, whose intrinsic position is paranoiac and whose primordial power and inclination is that over the life of his subjects (each of whom are therefore at all times potential *homines sacri*); (2) down below is the subject subjugated by way of forced laboring activity and to the ever present forebodings of a potentially imminent death, whose inherent subject position is depressive and moreover poisoned by the constantly roiling bile of *ressentiment*.

In relation to this second point, Bataille refers to what may here be termed the milieu of the resentful and death-anguished subject as the world of *practice*, in which the said subject dwells in the domain of things upon which he labors. This is a world of subordination that is "submerged in death like a city in a tidal wave."[118] The fearful and anguished representation of death, death as source of anguish, belongs intrinsically and ineluctably to the world of subordination and, reciprocally, subordination itself is necessarily grounded in the spurious (though most traumatizingly felt) need to flee death.

Let us now view this latter servile world from the perspective of the sovereign. For the Bataillean sovereign (*souverain*), this servile world of practice "smells bad; if it does not smell of death, it smells of anguish; its crowds sweat from the anguish provoked by shadows; death exists in it in a contained state, but fills it up."[119] In contradistinction to this, the world of the sovereign is fundamentally premised on the occlusion—if not downright rejection—of the fear of death and the abject corollary of what would have otherwise been the latter's successful inculcation:

servile existence. The obverse of this disdainful foreclosure of death-begotten anguish is the sovereign's violation of the prohibition forbidding the act of killing. The very sovereignty of the sovereign therefore entails his readiness to produce his subjects as *homines sacri* and let drop the Damocles' sword of death.[120]

We now commence the presentation of a deeper—or at least, different—historical genealogy of the emergence of this subterranean man of servility and his vicissitudes.

In the following chapter, Hegel shall enter.

Notes

1. Holland 1999, 77.
2. Ibid.
3. Deleuze and Guattari 2004, 231.
4. Ibid.
5. Žižek 2006, 39.
6. Agamben 1998, 49, 50.
7. Ibid. 52.
8. Žižek 2003, 105.
9. Deleuze and Guattari 2004, 231.
10. Solzhenitsyn 2007, Vol. 2, 507, 508.
11. Ibid.
12. Buchanan 2008, 106; Holland 1999, 77.
13. Ibid.
14. Derrida 2001, 226–230.
15. Deleuze and Guattari 2004, 230.
16. wa Thiong'o 2006, 13, 14.
17. Deleuze and Guattari 2004, 231; Nietzsche 2003, 75–77, Z I "Of the New Idol."
18. Canetti 1978, 210.
19. Deleuze and Guattari 1988, 425.
20. Ibid. 426.
21. Solzhenitsyn 2007, Vol. 1, 501, 502.
22. Ibid.
23. Ibid.
24. Solzhenitsyn 2007, Vol. 3, 221.
25. Markovski 1984, 35, 36.
26. Deleuze and Guattari 2004, 231.
27. Solzhenitsyn 2007, Vol. 1, 144.
28. Ibid. 204, 205.
29. Ibid. 504.

30. Foucault 1991, 35.
31. Ibid. 41–43.
32. Donnelly and Diehl 2011, 37.
33. Ibid. 42.
34. Cited in Rejali 2007, 36.
35. Cited in Ibid.
36. Cited in Ibid.
37. Cited in Ibid.
38. Rejali 2007, 50, 51.
39. Donnelly and Diehl 2011, 85, 86.
40. Ibid. 88–91.
41. Ibid. 54.
42. Ibid. 73.
43. Ibid. 95, 96.
44. Ibid. 102–105.
45. Ibid. 109–112.
46. Ibid. 143.
47. Ibid. 151, 152.
48. Ibid. 84.
49. Ruo-Wang and Chelminski 1975, 39, 40.
50. Ibid. 43.
51. Ibid. 46.
52. Ibid. 58–61.
53. Rejali 2007, 210.
54. Ibid.
55. Ibid. 79–82.
56. Ibid. 83, 84.
57. Ginzburg 1995, 83.
58. Ibid. 93.
59. Ibid. 217–220.
60. Loebl 1976, 82–88.
61. Rejali 2007, 98.
62. Arenas 2001, 203, 204.
63. Nietzsche 2007, 91, GM III.14.
64. Nietzsche 2000, 528, GM II.22.
65. Solzhenitsyn 2007, Vol. 1, 501.
66. Freud 1968, 36–41, 51–59.
67. Bataille 1991, Vol. 2 and 3, 80–82.
68. Ibid.
69. Ibid.
70. Ibid.

71. Ibid.
72. Améry 1980, 15.
73. Deleuze and Guattari 2004, 232.
74. Améry 1980, 85.
75. Ibid. 86.
76. Nomberg-Przytyk 1985, 22.
77. Ibid. 23.
78. Lengyel 1947, 157.
79. Bierce 2011, 10–19.
80. Dostoyevsky 2004, 70, 71.
81. Žižek 2006, 336, 337.
82. Solzhenitsyn 2007, Vol. 2, 391.
83. Solzhenitsyn 2007, Vol. 3, 219.
84. Solzhenitsyn 1988, 104.
85. Ibid. 105, 110.
86. Heidegger 2001, 279–311.
87. Agamben 1998, 81–83.
88. Ibid.
89. Levi 2003, 33.
90. Améry 1980, 16.
91. Genet 1971, 31.
92. Levi 2003, 48.
93. Markovski 1984, 30, 31.
94. Agamben 1999, 47.
95. Ibid. 71, 72.
96. Ibid. 76.
97. Améry 1980, 17.
98. Žižek 2008d, 198.
99. Agamben 1998, 85.
100. Ibid. 84.
101. Améry 1980, 33.
102. Ibid. 34.
103. Ibid. 35.
104. Ibid.
105. Ibid. 36.
106. Ibid. 39.
107. Ibid. 40.
108. Ibid.
109. Ibid.
110. Agamben 1998, 104, 105.
111. Deleuze 2003, 20–22.

112. Guyotat 2003, *passim.*
113. Deleuze and Guattari 1988, 115.
114. Ibid. 116.
115. Ibid.
116. Ibid.
117. wa Thiong'o 2006, 3–5.
118. Bataille 1991, Vols. 2 and 3, 220–222.
119. Ibid.
120. Ibid.

6

Despotic Dialectics of Subjective Self-Immural, 1. Hegel

The servile man living within the bowels of the despotic regime may be arrived at as a product of the Hegelian master-slave dialectic. The latter dialectic, of course, goes far beyond this, providing as it does rich detail as to the further internal psychological and ideological development of this subterranean man and what his ultimate fate may be and often is.

The Hegelian dialectic of master and slave is entirely germane here, since the slavish man of *ressentiment* generated under the aegis of the despotic regime is *homo historia*, whereas the said dialectic deals precisely with the mode of upsurge of man as a historical being. Crucially, the emergence of *homo historia* is, for Hegel, coextensive with the advent of self-consciousness.

The human animal, which prior to its subsumption under the aegis of the State apparatus lingers somewhere between the zoological and the anthropomorphic strata, becomes a desire underpinning an emergent self-consciousness only when it is directed at another desire. Kojève terms such a desire "anthropogenetic," since only a desire that has attained self-consciousness may be regarded a veritable *anthropos*, that is, as truly human. To be anthropogenetic, a desire must—unlike in the case of zoological desire—not aim at the mere incorporation of nature. Rather, it must aim at the subjugation of another desire.[1]

It does so in order to suborn the other desire into recognizing its own immeasurable superiority over it. Only through such an unreserved recognition (*Anerkennung*) can desire elevate itself from mere immediate nature and pre-reflexive consciousness, toward anthropogenesis and the upsurge of itself as self-consciousness.[2] For, as Hegel has it, self-consciousness (*Selbst-bewusstsein*) exists "in itself" and "for itself," only insofar as it does so for another self-consciousness; that is, it exists solely by being recognized.[3] As Hyppolite maintains: "human desire occurs only when it contemplates another desire, or, to put it in a better way, only when it bears on another desire and becomes the desire to be recognized and hence itself to recognize."[4] In the same vein, Gadamer argues that "if self-consciousness is to become true self-consciousness, then it must stand on its own and find another self-consciousness that is willing to be 'for it.' Thus the doubling of self-consciousnesses is a necessary consequence: self-consciousness is only possible as double."[5]

1. The life and death struggle

The problem, of course, is that the other desire that anthropogenetic desire wishes to subjugate likewise wishes to subjugate *it*. That is to say, both desires, beholding each other, aim at the recognition by the other of their superiority over the other. A quandary is therefore reached the resolution of which can, after an endless stockade of smashed skulls amid putrid flesh feasted upon by carrion birds is piled up by the millions, only assume a headlong struggle to the death.[6]

Peter Kalkavage terms this the primordial war of independence in which two selves, or radically singular ipseities, act like colliding bodies or forces that repel one another.[7] Jürgen Habermas expresses it thus: "The struggle for recognition they conduct as a life-and-death struggle. The abstract self-assertion of parties contemptuous of each other is resolved by the combatants risking their lives …"[8] Gadamer has it as follows: "The dialectic of recognition is experienced in a process, that is, in the life-and-death conflict, and in the determination of self-consciousness to prove its truth, its being recognized, even at the risk of its life."[9]

Be this as it may, Kalkavage contends that this primordial struggle is strictly speaking *combat* and not murder. For it is an *affaire d'honneur*, in which the aim is not the death of the adversary's body but the cancellation of his claim to supremacy. In the case of murder, the adversary would be no more to the slayer than an apple that has been devoured. The slayer himself would immediately sink back to the level of mere nature.[10] He would be anthropologically null.

2. The vanquishment of the proto-slave

On the other extreme, we cannot assume that this battle may end in parity, since the attainment of self-consciousness necessarily entails the ascendancy of one desire over another. Hence, as Kojève has it, in the struggle between two anthropogenetic desires, there must be a victor and a vanquished wherein *both* remain alive so that the vanquished comes to be subjugated and therein harried into recognizing the victorious party's absolutely crushing superiority.[11]

A unilateral cessation of hostilities, as Jessica Benjamin contends, can nowise come to pass, since "self-consciousness wants to be absolute. It wants to be recognized by the other in order to place itself in the world and make itself the whole world. The *I* wants to prove itself at the expense of the other; it wants to think itself the only one; it abjures dependency."[12] Shakespeare's Prince Hal captures this situation adriotly when he intimates to Hotspur: "Two stars keep not their motion in one sphere" (*Henry IV*, I.v.4). Selfhood in this its naked belligerence is the Hobbesian condition of uncompromising egotism and untrammeled self-assertiveness. A stalemate would simply be downright intolerable for a budding self-consciousness (*Selbst-bewusstsein*). As Žižek maintains apropos of Catherine Malabou, self-consciousness, as a "spiritual" individual, in that it is in no sense a mere member of the human genus, has an "infinite right" to universality. That is to say, when one nascent self-consciousness encounters another, there is an irreducible stopping point inherent to self-consciousness that ineluctably resists the two being reduced to mere members of the human species. Two universalities engage one another where there is only space for one.[13]

As it transpires, the belligerent party whose mode of desiring is less intensely anthropogenetic, more tightly tethered to its biological substrate, and who thus values continued bare subsistence within mere natural life more highly than the attainment of recognition of his superiority—owing to a preponderating cowardly dread of his own terminal biological annihilation—will submit.

A more anthropogenetically bent desire, on the other hand—in that it transcends this merely natural level more emphatically—is imbued with a greater resolve to go beyond the level of pure biological survival. It is therefore less fearful of biological death and hence more willing to gamble its own life to the end of recognition by another. This markedly more fearless adversary will therefore not submit. Indeed, he will subjugate.[14] This calls to mind George Armstrong Kelly's Hegelian observation that man—here exemplified in the form the proto-master—is the only creature who, under certain non-natural pressures, is willing to stake his own life. This, Kelly continues, is man's first creative act.[15]

Hyppolite summarizes the foregoing as follows: "One of the self-consciousnesses rises above animal life; able to confront death and not fearing the loss of its vital substance, poses abstract being-for-itself as its essence and seems thereby to escape the enslavement to life. This is the noble consciousness, that of the master, and it is recognized in fact. The other self-consciousness prefers life to self-consciousness: it chooses slavery."[16] Kojève's formulation of the same is: "The vanquished has subordinated his human desire for recognition to the biological desire to preserve his life: this is what determines and reveals—to him and to the victor—his inferiority. The victor has risked his life for a nonvital end: and this is what determines and reveals—to him and to the vanquished—his superiority over biological life and, consequently, over the vanquished."[17] The victor is "the man who went all the way in a fight for prestige, who risked his life in order to be recognized in his absolute superiority by another man. That is, to his real, biological life he preferred something ideal, spiritual, nonbiological: the fact of being *anerkannt*, of being recognized in and by a consciousness," and as shall be seen presently, "of bearing the name of 'master,' of being called 'master.'"[18]

Kalkavage refers to the greater "spiritedness," or *thymos*, of the proto-master—alluding to Socrates' description of the latter as the middle part of the soul which mediates between the "lower" appetites and the "higher" faculty of reason. If *thymos* had a natural place in the body, he avers, that place would be the chest. It is the Achillean part of the soul, the part which delights in victory and honor. It is that which longs to demonstrate its absolute superiority and to attain the recognition (*Anerkennung*) of this superiority by others. With Hegel, *thymos* ceases to be a mere part of the soul and is extolled to the status of a fundamental anthropogenetic drive, to a force that impels the self toward freedom. By contrast, he in whom this fundamental drive is lacking, or at least attenuated, is—in the phrase of C.S. Lewis—a "man without a chest."[19]

One may even aver that the more anthropologically-bent desiring proto-self-consciousness above *mutatis mutandis* approaches the condition of the Bataillean sovereign (*souverain*). For the latter—if we envisage here for conceptual purposes the limit case of a somewhat idealized and hypertrophied sovereignty—the very *representation* of death is impossible, in that to live absolutely sovereignly implies a life that dwells purely in the present, with the futural dimension elided. Here, it is precisely the foreclosure of the futural-anticipatory dimension for the hypertrophied sovereign man that inoculates the latter—if not from the brute fact of death itself—then certainly from the anguish of death, the anguish felt in the face of death's representation as portended by futural projection. For the hypertrophied sovereign, well-nigh spending himself in the exuberance of the moment, lacks such anxious protention. And therefore: "He cannot live in an anguish likely to enslave him, to determine the flight from death that is the beginning of servitude. He cannot die fleeing. He cannot let the threat of death deliver him over to the horror of a desperate yet impossible flight."[20]

3. Enslavement of the vanquished and the master's life of sovereign plenty

The victor *enslaves* the vanquished and forces him to *labor* on the thing that he has appropriated. He becomes the master of the vanquished, whereas the vanquished becomes his slave.[21] The master realizes and gives his superiority material permanence through the intermediary of the slave's labor. In that he does not have to work or to strive against nature to maintain himself within biological existence, his existence is one of effortless enjoyment (*Genuss*) and pleasure.[22] His material subsistence and corporeal sustenance is guaranteed by the slave. As Hegel expresses it, the master aligns himself solely with the "dependence of the thing," enjoying it sans qualification or reserve. Its "independent" aspect he leaves to the bondsman, who works upon it.[23]

In this sense, one may aver that the Hegelian master lives a sovereign existence—like that of Bataille's sovereign (*souverain*)—and that, on this account, he is this sovereign. As observed much earlier, the Bataillean sovereign is well-nigh sempiternally engaged in the ostensible and ruinous consumption of the surplus extracted from the toiling masses he has subjugated. The sovereign lives in the evanescence of the moment, and this moment is the ocean into which all the streams of productive labor vanish. In marked contrast, the servile life of the world of labor is comparitively processual in nature and thus entails the subordinated worker's necessary projection of himself into his future. Moreover, in that Bataille conceives of the subject at its purest as precisely such an evanescence in and of the moment, the sovereign epitomizes the subject. In vast contradistinction, the object is of the order of the world of labor, of the prosaic domain of enduring self-identity characteristic of the thing. Far from spending itself in the vanishing moment, the latter objective realm of things is perdurant. For this reason, contends Bataille, when the produced things of the world of labor enter into the realm of the sovereign—and thus when duration is infinitely shrunk into the evanescence of the instant—these objects are forthwith destroyed in the sovereign's ruinous consumption. The vast gulf separating the world of the sovereign master and the prosaic slave is that between intimacy and lucidity, immoderation and moderation, madness and reason, drunkenness and lucidity.

The Hegelian master, being this sybaritic and virtually limitless sovereign consumer, may also be compared to the despot of the Deleuzoguattarian despotic regime. As likewise observed much earlier in our text, under despotism the sum total of the productive flows of the subordinated laboring population milling below converge into a great river that constitutes the consumption of the despot and his caste. In that he is tantamount to the idle and domineering sovereign (*souverain*) described by Bataille, the despot is the Hegelian master par excellence.

4. The paradoxical advantage of the slave and the master's near atavism

It would seem that the master, in climbing to the despotic apex of sovereign existence, may be described as wholly and incontrovertibly satisfied (*befriedigt*). Yet, it is not a life of idle and sybaritic luxury for which he had staked his life in his antecedent belligerent struggle with the proto-slave, but instead the attainment of mastery through recognition (*Anerkennung*) as a self-consciousness (*Selbst-bewusstsein*) by another self-consciousness, that is, by another figure of the master.[24] As it is, however, the master is only recognized by a slave; that is, only by a dependent consciousness and not by an independent (masterful) one. His purported mastery is therefore completely vacuous.[25] For the truth of his self-consciousness, or certainty of self, is in fact the dependent, or unessential, consciousness of the slave.[26] On the other side of the equation, by dint of fear (*Furcht*), refusal to look death in the eye and consequent abdication from the goal of recognition in favor of mere biological preservation, the slave remains frustratingly dependent on nature and—in that his lord holds dominion over this latter—likewise dependent on the master.[27]

However, through the biological fear of death (*Angst*) and shockingly tangible foreboding of himself as nothingness which had initially caused him to surrender, the slave acquires a paradoxical ascendancy over the master.[28] For, having survived this rapidly impending death by submitting to the master and having in the last moment plucked himself from the yawning abyss into which he was fated to vanish, he henceforth comes to understanding his entire existence as an *aufgehoben* (overcome) nothingness, that is, as nothingness *maintained* in Being, as pure self-existent absolute negativity, as a nothingness that nihilates in Being.[29] For he has verily experienced this reality in its very core, since his consciousnesses, as Hegel contends, "was not in peril and fear for this element or that, nor for this or that moment in time, it was afraid for its entire being; it felt the fear of death, the sovereign master. It has been in that experience melted to its inmost soul, has trembled throughout its every fibre, and all that was fixed and steadfast has quaked within it. This complete perturbation of its entire substance, this absolute dissolution [*Auflösung*] of all its stability into fluent continuity, is, however, the simple, ultimate nature of self-consciousness [*Selbst-bewusstsein*], absolute negativity, pure self-referrent existence, which consequently is involved in this type of consciousness."[30]

Death's foreboding has thus jolted the slave's desire into anthropogenetic mode. As Žižek maintains of the foregoing passage, through the intimation of the infinite power of negativity in his experience of the fear of death during his confrontation with the proto-master, the slave recognizes this overwhelming force of annihilation as at once the absolute negativity residing with the core of his self. That is, what he feared in fearing death at the hands of the proto-master was the negative power residing within himself. It is this with which he now identifies in recognizing himself as a nothingness that nihilates in Being.[31] The master, on the other hand, in that the angstful representation of death as a premonition of his own nothingness does not occur to him, fails to grasp the human nothingness that lingers at the foundation of his natural Being and thus cannot undergo anthropogenesis.

Now, in that the nothingness nihilating in Being as which the slave apprehends himself is more precisely such by way of the negation of Being or of the natural given, and in that this transpires in relation to an idea that is still nothingness, the slave comes to understand himself as *labor*.[32] Through his labor, the slave not only transforms nature in accordance with an idea, but likewise transforms *his own* nature, in relation to the *ideal* of a master. As Žižek maintains, the "truth" of the slave's labor is at once the auto-purification of the laborer himself. It is at the same time the transformation of nature and the disciplinary self-formation (*Bildung*) of the slave.[33] For this reason, Hegel is right to aver that, by way of his service and work, the slave does move in the direction of erasing his dependence upon mere natural existence.[34] As Habermas has it, "the relationship of the one-sided recognition [*Anerkennung*] of the master by the servant (*Knecht*) is overturned by the servant's power of disposition over nature, just as one-sidedly acquired by labor."[35] It is the slave's desire that is truly anthropogenetic and not the master's. "Only if man," as Georg Lukács maintains, "places labor between his desire and its fulfilment, only if he breaks with the instinctual immediacy of natural man will he *become* fully human."[36]

Shlomo Avineri may be seen as providing ballast for this point when he avers that "labor is the transformation of the appetites from their initial annihilative to a constructive attitude toward the objective world; whereas primitive man, like the animals, consumes nature and destroys the object, labor holds up to man an object to be desired not through negation but through creation."[37] This, as Kalkavage maintains, entails at once the loss by man of his former cognitive piety toward objects. And it is solely in the slave that this piety is destroyed, for in work the objective world is experienced as having an essence only insofar as it is recreated by a laboring self.[38]

On this count, the master—bereft as he is of laboring praxis—remains a primitive, or is at least prone to atavistic rot. As regards this dire fate that here forbodes for the stagnant master and the eventual possibility of ascendancy of the slave, George Armstrong Kelly maintains that "the future belongs to the once-terrorized producer, progressively liberated by the spiritualized quality of his own labor, not to the seemingly omnipotent consumer, who treats both the servant and his product as mere dead things."[39] As Judith Shklar underlines it, it is precisely through this conjunction of the mortal fear to which the slave is exposed with unremitting productive toil that he is able to gradually attain self-discipline, self-mastery, and to raise himself to a consciousness of autonomy. For, "Work without fear is just vexing, whilse pure fear merely paralyzes."[40] Through labor, the slave comes to dominate the very same biological nature whose prior dominance of him in the belligerent struggle with the proto-master caused him to be a slave in the first, in so doing diminishing his dependence on the natural conditions of his existence and becoming conscious of his autonomy (*Selbständigkeit*) and freedom (*Freiheit*).[41]

Of course, this latter notion of freedom entertained by the slave is thus far nowise embodied in any concrete external reality. It is a merely mental freedom—and one which arises solely under conditions of real and abiding slavery. Immediately behind the slave's present free assent to slavery, the constant possibility and threat of coercion, the ubiquitous threat of imminent death at the hands of the seemingly omnipresent despotic Damocles' sword remains.[42]

Meanwhile, though unburdened and uncoerced, the master's concrete freedom is merely one supported by slaves. Sybaritic, idle, and thus bereft of a utopian ideal of liberty, his freedom is inert and vacuous.[43] He is swallowed by his own desires, and becomes dialectically irrelevant. He will never attain self-consciousness (*Selbst-bewusstsein*).

For the slave, on the other hand, the attainment of real Hegelian freedom is possible, as his abstract ideal of liberty derives from the notion of a master, whom he recognizes as a self-consciousness.[44] However, though he is able to bring himself into closer propinquity to his abstract ideal of liberty through labor as *Bildung*, the same pusillanimous fear of the bitter and terminal annihilation of his present and future being as had caused his original submission to the death-wielding master still keeps him cowering under the Damocles' sword of the master's despotic regime. For long, he will not revolt.[45]

5. The self-justificatory ideologies of the slave

Hemmed in on all sides, the slave will therefore have to somehow reconcile the ideal of freedom pulsating within him with the stubborn reality of his slavery. He will need to justify the painful bifurcation between these two in order to render his existence bearable. In order to do this, he will have to *delude* himself. He will inescapably live in a state of permanent denial by surrendering to a series of self-induced ideologies.[46]

(1) Stoicism: Solzhenitsyn and Beckett

The first of these ideoligies is *stoicism*. The stoic slave considers himself to be free simply because he finds himself possessed of the abstract idea of freedom. He ceases to see freedom as something which of its very nature exceeds the ambit of pure thought, and thereby convinces himself that this merely mental freedom of his is in fact consonant with or identical to actual freedom. What one has here is thus a conflation of abstract and concrete freedom, in which the real external conditions of existence are considered entirely irrelevant.[47] "The essence of this consciousness is to be free," as Hegel has it, "throughout all the dependence that attaches to its individual existence, and to maintain that stolid lifeless unconcern which persistently withdraws from the movement of existence, from effective activity as well as from passive endurance, into the simple essentiality of thought."[48]

The stoic's freedom, avers Hyppolite, is a mental self-mastery through which he regards himself as always free, irrespective of the hazards or circumstances of fortune.[49] He is a world unto himself, and as such he regards himself entirely self-sufficient and therefore absolute. As a purely intellectual being he considers himself infinitely worthy and in need of no other, let alone a god, to render his self-certainty real. Otherwise put, his integrity, concentrated in the limitless power of thought, is his god. Given his purely mental freedom, the stoic naturally cultivates freedom from the ravages of passion—or *apatheia*. His wish is to counteract the enslavement to passion and imagination, for to him the human life he sees all around himself appears as a condition of intense and widespread anxiety and suffering. To this end, he strives for *ataraxia*, or unperturbed indifference, by holding himself at a safe remove from life. He protects himself from the world's disordering influence of desire and fear, striving to maintain a pure inwardness that is absolute. He makes himself into a veritable fortress of self-contained invulnerability, an impregnable self-sameness, and does so merely through disembodied thought. He is at one with Aristotle's averment that thinking is the highest form of happiness since, out of all activities, it displays the highest degree of self-sufficiency—or *autarkeia*. In fact, he is at one with Parmenides' contention that thinking and Being are *the same*.[50]

The longer the slave is held in subjection by an obdurate and irremissible reality and the longer he adheres to his own stoic indifference to this reality, the more the very sensibility that accords with actual freedom is dulled in him. His is a quietism which serves to compound the stoic slave's stubborn adherence to his irenic miredness in his own self. Through this weightless quietism, he comes at length to acquire a strangely submissive equanimity hitherto unbeknowst even to himself.

Returning to Solzhenitsyn's zek—the stoic slave of the Soviet Gulag archipelago, the author contends:

> No matter how darkly circumstances may be stacked against him, he knits the brows of his rough and weathered face and says: "*They cannot drop me any deeper than a mine.*" Or they comfort each other: "*It could be worse.*" And in reality, this conviction "It could be worse!" clearly supports and encourages them in the most profound suffering of famine, cold, and spiritual depression.
>
> The zek is always *expecting it to be worse.* That is how he lives, constantly awaiting the blows of fate and stings of the evil spirit. And, on the other hand, he perceives every temporary relaxation as an oversight, a mistake. In this constant expectation of misfortune the austere soul of the zek matures, stoically hardened to its own fate....
>
> The zek's deviations from his state of equilibrium are very minor—either on the dark side or the bright side, either on the side of despair or on the side of happiness. This was felicitously expressed by Taras Shevchenko (who spent a little time on the islands in prehistoric times): "I now have almost neither grief nor gladness. On the other hand, there is a moral calm like the cold bloodedness of a fish. Is it possible that constant suffering can rework a human being in this way?"
>
> Indeed! Yes. Indeed in can. A *stable, indifferent state of mind* is the zek's indispensable defence so he can survive long years of grim island life. If he does not attain this smouldering, lustreless state of mind in his first year in the Archipelago, ordinarily he dies.

> Once he has attained it, he continues to live. In a word: If you don't kick the bucket—you'll become an adept.
>
> All the zek's feelings are dulled, and his nerves coarsened.[51]

The corollary of this sempiternally equanimous state of mind—born as it is out of the zek's hardening to his thankless circumstances—is the fatalism of the zek. Stoicism and fatalism march in pairs—for stoicism is ultimately fatalistic. Solzhenitsyn:

> The most prevalent world outlook among them is *fatalism*. This is their profound, universal trait. It is to be explained by their dependent situation, their total lack of knowledge of what will happen to them even in the most immediate future, and their actual inability to influence events. Fatalism is even necessary to the zek because it confirms him in his spiritual stability. The son of Gulag believes that the most tranquil course is to put his faith in fate. The future is a cat in a sack, a pig in a poke, and not understanding it clearly, and having no idea what's going to happen to you in the possible variations of life lying ahead, you don't need to strive too insistently for something or reject something too obstinately—be it a transfer to another barracks, or bridage, or camp. Maybe it will be for the better, maybe for the worse, but you are free of self-reproaches; let it be worse for you, but it wasn't done by your own hands. And this is how you preserve that precious feeling of dauntlessness, how you save yourself from fussing and ingratiating.
>
> The fatalism widely prevalent among them extends not only to their personal fate but also to *the overall course* of things. The last thing in their mind is that the overall course of events could be changed. They imagine that the Archipelago has existed *for all eternity* and that earlier it was even worse.[52]

Thus it may be seen how the acquired passivity, or learned helplessness, of the stoic or zek—held together as it is by a false equanimity born of the unbridled acceptance of circumstances—can only leave the slave mired in his subordinate station.

Indeed, as Charles Taylor avers apropos of Hegel, stoic freedom remains radically incomplete. It is a mere pseudo-freedom fit only for slaves. More precisely, it is a form of political impotence which reflects itself in the pure abstractness of disembodied thought. It is a strategy of retreat founded on the entirely misguided intuition that consciousness is a mere thinking essence.[53] In Hegel's formulation, stoicism maintains that consciousness is at base a mental reality and that whatever is essential for it, whether good or true, is so only when consciousness treats it from the vantage-point of a merely thinking being.[54]

To be sure, stoic man, though not in reality recognized by another self-consciousness (*Selbst-bewusstsein*), does notwithstanding acquire a modicum of certainty of self. The latter is, however, only attained at the cost of his retreating into himself. But a self-certainty that fails to externalize itself in concrete social and political forms remains entirely vacuous. Still worse, the man of stoic consciousness in fact turns out to be a potential plaything of fortune and of unbridled forces of political savagery. His certainty of self is a unilateral withdrawal into an inner citadel that ultimately can nowise be defended.[55]

Above all, stoicism—in remaining stubbornly at the level of thought untethered from external reality—moves purely within the ambit of vapid universal categories (such as the "true" and the "good," "wisdom" and "virtue") for which it is wholly unable to derive a determinate content. It cannot find a content embodied in external reality for its ethical thought. As observed, the strategy of stoicism in the face of this deficiency is to consider such content *inessential* to freedom. If thought by itself is freedom, the subject—conceived of as a thinking essence, as a mere *ego cogito*—is itself likewise free. The stoic is at base free as a thinking subject in abstraction from the concrete external circumstances and vicissitudes of life.[56] He is one in whom, as Hegel has it, freedom is entirely indifferent towards natural existence—this latter having been effectively jettisoned.[57]

But the stoic deludes himself in thinking that he is free. For, far from being merely freedom of thought, freedom—as Hegel conceives it—is being oneself, *bei sich* ("at home") in one's other, or object. In this sense, the stoic subject is certainly *not* free. For, in that the determinate reality of the world is something entirely foreign to the stoic subject and underivable from its thought, stoic consciousness cannot be *bei sich* in relation to this said reality.[58]

In fact, the stoic slave himself already admits that he is unrelated to his other. For, as Hegel argues, he considers himself free precisely because he feels himself solely in touch with himself: his essential reality is the undivided unity of his own self-existence, and his dealing with notions a process confined purely to himself.[59] But, as Charles Taylor has it, this wholesale withdrawal in the direction of pure inner self-identiy can nowise accomplish freedom for an embodied subject whose real freedom must necessarily be externally expressed in a *modus vivendi*. Stoicism, for this reason, contradicts itself, for it is a mere putative realization of freedom which is in actual fact its negation.[60] Freedom of thought, as Hegel avers of the quagmire of stoicism, assumes only pure thought to be its truth, thus lacking the concrete and tangible filling of life. It is, on this account, the mere notion of freedom and not living freedom as such. For it is, from the very outset, only thinking in general that is its essence.[61] The stoic slave has therefore emancipated himself, but only in the sense attributed to emancipation by Ambrose Bierce's dictionary: "A bondman's change from the tyranny of another to the despotism of himself."[62]

In stoicism, man behaves not as a nothingness nihilating through the negation of Being, but as a pure disembodied Cartesian *ego cogito* which dissipates externally directed self-liberatory praxis in favor of endless cogitation and discourse.[63]

This stoical self-entrapment of man within the mere confines of discourse not only becomes wearisome, but also collapses him back into mere natural Being, animality, thing-hood, and the inertia of static self-identity.[64] The stoic's purely contemplative emigration to the vacuity of his inner self, as Shklar avers, dooms him to a condition of gloomy spiritual emptiness and utter immobility.[65] As William James would have it, the stoic's resigned acceptance of the universe is simply too drab and discolored, too devoid of passion and exultation. Even the stoic himself eventually comes to suffocate in this his own self-induced arctic climate.[66]

This cul-de-sac which mires the Hegelian stoic slave may be considered as still more dire than even James suggests. As already intimated, the stoic is effectively an incessantly chattering solipsistic *ego cogito*. Beyond the burgeoning and lingering sense of drabness which stoicism instills in the self, Badiou contends that the subject immured in the enclosure of the cogito and its endless enunciation is thereby submitted to nothing short of torture. Badiou refers to Beckett's work (especially *The Unnamable*), in which a "motionless voice" is assigned residence in a "mutilated and captive" body serving solely as the "fixed localization of the voice."[67]

The "I" is, in fact, "doubly enclosed": firstly in the "fixity of the body" and secondly in the "persistence of a voice without either echo or response," through which it "endlessly strives to find a way to its identification." Through this striving, via an endless stream of statements, the subject aims to produce "the pure and silent point" from whence its enunciations arise. This "silent point" is the "subjective condition," or the "support," of all speech. It is through this latter that the solipsistic ego attempts to identify itself. This, avers Badiou, is the quest of the hero of Beckett's *The Unnamable.*[68]

Beckett, however, demonstrates that this quest is intolerable and fraught with anguish, moral exhaustion, and inevitably doomed to failure. If thinking wants to seize itself as the thinking of thought—a predicament into which the stoic inevitably drives himself—then a "reign of terror" ensues in the self. The stoic's imperative to inwardly withdraw to its own point of enunciation of its statements with interminably repetitive insistence is that by which it closes all exits for itself. The sheer impossibility of this necessity that it foists upon its voice is for it an intolerable torment. Thus, throughout *The Unnamable* the speaker's countenance streams with tears.[69]

Though wearied, anguished, morally exhausted, tormented and depressed by the mere flattery of elevated feeling and platitudinous edification, in that he remains captive to his anguished representation of death, the stoic cannot outwardly negate the material conditions of slavery underpinning his incarceration in the disembodied *cogito*, since he has introjected the despotic Damocles' sword as his overriding instinct.[70] Since he is unable to corporeally negate the debased and servile reality undergirding stoical self-enmiring, his only recourse is to negate external reality mentally in order to rationalize his stubbornly abiding slavery.[71]

(2) Skeptical nihilism: Pessoa

He moves forthwith to utterly execrate his existing social conditions—and indeed the entire world—as intrinsically debased and repugnant. The value, and in fact the very reality, of all that is external is completely denied. Through this act of mental negation, the stoic slave turns into the skeptic-nihilist slave. He adjudges all concretely embodied reality to be inherently contradictory in and of itself.[72]

The difference between the strategies of the stoic slave and the skeptical-nihilist slave resides in the fact that whereas stoicism considered the determinate content of reality to be inessential or of no import, skeptical nihilism—as observed—moves to assail this content and to call it into question.[73] For this reason, Charles Taylor regards skepticism as a "polemical consequence of stoicism."[74] Shklar delineates this difference as one between stoicism's merely passive, complacent, and dogmatic indifference of thought to external reality—and skepticism's far more active and aggressive denial of all sense experience.[75] The stoic slave remains indifferent to external reality. The skeptical nihilist derides it. He despises the world. As Hegel maintains, with skepticism, "Thought becomes thinking which wholly annihilates the being of the world with its manifold determinateness"; for it is a "negative attitude towards otherness, to desire and labour," a "polemical attitude towards the manifold substantiality of things" which "turns against them" entirely.[76]

As Kalkavage maintains, the actual aim of the skeptic is the same as that of the stoic: *ataraxia*—or unperturbedness. Like the stoic, he aims to liberate and pacify himself—or to liberate himself by pacifying himself. The difference is that, in that the skeptic denies all sense experience, he is fundamentally at war with dogmatism and with all forms of certainty. That is to say, his only certainty is that nothing is certain. Unlike the stoic, he actually practices a logic of refutation, through which he sees himself as achieving what he takes to be the highest form of human freedom: *epochē* or "suspension of judgment," by which the ultimate goal of mental tranquility is attained.[77] He is the imp of contrariety.

However, in order to thwart the onset of his own wholesale existential collapse, the skeptical nihilist must nonetheless spare a certain small portion of reality from the wrath of his contemptuous eye. This miniscule remainder will, of course, be his own circumscribed cogitating and polemicizing ego. In order not to collapse the floor that supports him, the skeptical nihilist must retain his *own* self-identical ego as the true domain and locus of an authenticity untainted by the contradictions that he considers endemic to external reality. The skeptic-nihilist's own self—and naught beside or outside it—shall now serve as his *terra firma.* The cost of this maneuvre will, however, be that he forthwith casts himself into the mire of an increasingly constraining and hermetic *solipsism.*[78] The skeptical nihilist's experience, Hyppolite avers, is that of a purportedly absolute mental liberty that dissolves every position with the exception of that of his own ego.[79]

He inevitably becomes Fernando Pessoa: quite possibility *primus inter pares* among solipsistic nihilists. Reality for Pessoa—being inherently contradictory—is devoid of meaning, bears no relation to the self, and abounds in maddening transitoriness and utter inchoateness.[80] It consists merely of:

> … circles on the surface when chance drops a stone, echoes of unknown voices—a collective mishmash of life.[81]

The world as such is:

> … but an unaesthetic nightmare, like the result, in dreams, of a mental indigestion.[82]

Other people are all:

> … of that metaphysical insignificance known as flesh and blood.… dead things bleeding as if they were alive …[83]

Given this paltriness of humanity and of reality itself, Pessoa is moved to concur with Horace that the ethically upright man will naturally remain completely unperturbed, even in the event of the whole world disintegrating all around him; and that for noble men "life should be a dream that spurns confrontations."[84] Pessoa experiences life as that in whose turgid rhythm humans reptate:

> … like worms when a rock is lifted—under the huge abstract boulder of the meaningless blue sky.[85]

Cognisant that this is all self-evidently the case, he is moved to aver that "Only Tedium, which is a withdrawal, and Art, which is a disdain," can possibly serve as sources of edification.[86] Incurably fatigued by and disdainful of the world, the skeptical nihilist is desirous to:

> … vanish in the fog as a foreigner to all life, as a human island detached from the dream of the sea, as a uselessly existing ship that floats on the surface of everything.[87]

He wishes quite definitively to abscond from society as a whole, contending that the:

> ... best and most regal course is to *abdicate* [my emphasis]. The supreme empire belongs to the emperor who abdicates from all normal life ...[88]

Owing to his consummate misanthropic bent and proclivity to wholesale abdication from the company of men, the solipsistic nihilist is made especially livid by the notion of social responsibility, and by the tacit assumption that it could possibility pertain to him at all. Its very tenets, he ridicules as "make-believe dishes filled with make-believe tidbits for the miniature table of a doll."[89] On account of this noisome pettifoggery, acting in concourse with others is:

> ... a metaphysically *morbid* [my emphasis] impulse.... The divine fact of existing shouldn't be surrendered to the satanic fact of coexisting.[90]

Incontrovertibly satanic, coexistence is death. This is extreme solipsism. But Pessoa will go still further. For he now proceeds to have quasi-Cartesian doubts (whether methodical or spontaneous or both) about his very *own* existence, contending that it is quite possible that his very identity is a self-wrought construct made of material extraneous to his own nature. He thus feels himself completely external to himself, to the point of at once no longer being able to recognize himself in his own self-forged unreality and, at the same time, not knowing what the purportedly true self behind it could possibly be.[91]

It is precisely here that one glimpses the source of the contradictoriness attributed to the world by the skeptical nihilist: that is to say, the contradiction that is inherent and internal to the solipsistic nihilist himself.[22] This contradiction internal to the solipsist is what he *imputes* to *the world.* In fact, so acute is the contradictoriness of the solipsistic nihilist's fractured self that he experiences the world as the perpetual and imminent threat of the said contradiction's further exacerbation—to the point of his own possible *self-dissolution.* In this vein, Pessoa relates that he consciously avoids living his life precisely in order to protect his artictifiality forged identity. In the absence of such caution and unilateral withdrawal, he feels he would certainly be *annihilated.* This pusillanimous comportment is what is entailed in his "aesthetics of abdication" and "aesthetics of artificiality."[93] When he does indeed commune with others, he feels himself as passing into a living death as:

> ... a pained and pale shade, which the first breeze will knock to the ground and the first physical contact dissolve into dust.[94]

Otherness for the skeptical nihilist at this level of solipsism is *death*. It is for him the sheer dread of alterity. This is why he finds it necessary to fashion an artificial self as the panacea and conduit through which exteriority is filtered, or even blocked, and as a prosthesis to artificially render salubrious the vicissitudes of the contradiction that he himself is.[95]

But this new ideology of the slave itself entails an inherent contradiction, since the very impetus animating his quest for a realization of his freedom as self-consciousness (*Selbst-bewusstsein*)—despite its presently glimpsed perverse solipsistic vicissitudes—at length proves itself to be merely a *function* of his anthropological desire to be recognized by *another self-consciousness*.

For Charles Taylor, this is so because, despite his retreat into an ever more insular inner citadel; as an embodied subject, the skeptical nihilist cannot but continue living in external reality. Though he may well declare the world's non-Being, it unceasingly and inexorably recrudesces. The upshot is that the solipsistic nihilist is prey to an incessant oscillation between the stubborn obduracy of his own hermetically closed self-identity and an equally acute sense of his beholdenness to and dependence on the shifting sands of a constantly changing external reality. As soon as he calls this highly mutable reality into question in order to experience himself as immutable and self-identical, the resulting inner vacuity to which he hereby condemns himself forces him to concede that he is after all embodied in the changeable and self-external.[96] This form of consciousness, as Hegel has it, is "the aimless fickleness and instability of going to and fro, hither and thither, from one extreme of self-same self-consciousness, to the other contingent, confused and confusing consciousness."[97] Yet, it does not allow itself to, or cannot, integrate these two thoughts within itself. It finds its freedom, at one time, in the form of an "elevation above all the whirling complexity and all the contingency of mere existence, and again, at another time, likewise confesses to falling back upon what is unessential, and to being taken up with that. It lets the unessential content in its thought vanish; but in that very act it is the consciousness of something unessential. It announces absolute disappearance but the announcement *is,* and this consciousness is the evanescence expressly announced. It announces the nullity of seeing, hearing, and so on, yet *itself* sees and hears."[98]

Notwithstanding, the skeptic-nihilist slave does at length come to perceive that the contradictoriness he heretofore ascribed to external reality does in actuality reside in himself. That is to say, he is forced to concede the fact of his painful inner division (*Entzweiung*) into an ideal immutable and self-identical being, on the one hand, and a self submerged by a world of interminable flux, on ther other.[99] Emil Cioran may be seen as describing this predicament when he complains that his attempts at disentangling himself from the merely apparent only envince his being ensnared in it. This causes him to feel that he is trapped in some sort of intermediate zone between appearances and that which serves to invalidate them. Yet, he feels himself compelled, almost as if by his very nature, to allow this equivocation to abide, since any decisive salvific break from it would certainly spell his destruction.[100]

However, much as in the case of the stoic slave, the fundamental contradiction internal to the skeptical-nihilist between his utopian libertarian ideal and the obdurate reality of his slavery—due to his terror-riddled fear of biological annihilation at the behest of the sempiternally threatening Damocles' sword of the despotic regime under which he toils—remains insurmountable.[101]

(3) The unhappy consciousness

As observed, skeptical consciousness wilfully vacillates between the assertion of its own fortuitousness and contingency, on the one hand, and that of its immutability and necessity, on the other. Though oscillating wildly between these two extremes, skeptical consciousness is for long unable to bring them together into a single unified stance that would serve to hold them together. Despite dialectically generating itself as a changeable self and an unchangeable self, it is not conscious of this contradiction as its own. Eventually, however, it will become conscious of this fact. The said predicament of skepticism, Hegel contends, is fated to vanish since it is in reality but one sole consciousness that is possessed of these two modes. Its new modality shall be one which is "*aware* of being the double consciousness of itself as self-liberating, unalterable, self-identical, and as utterly self-confounding, self-perverting; and this new attitude is the consciousness of this contradiction within itself."[102] Nonetheless, "the reconciliation [*Versöhnung*] of the two elements is not yet present. Hence the *Unhappy consciousness* [*unglückliches Bewusstsein*], the Alienated Soul which is the consciousness of the self as a divided nature, a doubled and merely contradictory being."[103]

The unhappy consciousness, or anguished soul, is man's painful consciousness of himself as an internal rivenness between his own immutability and his own contingency. The man of unhappy consciousness therefore strives to bridge this divide so as to effect a *reconciliation* between the immutable (universal) and mutable (singular) within himself. His strategy will be to contrive this reconciliation through the subsumption of the singular under the universal. The anguished soul will take the "simple unulterable" as essential, and the "manifold and changeable" as unessential.[104] It will take itself to be the unessential singular that is here to be subsumed. However, though it is to itself a changeable consciousness, by participation in the universal, it aims to free itself from its own unessentiality. That is to say, the anguished soul, in that it is the consciousness of the contradiction between itself as mutable and as immutable, takes upon itself the "aspect of changeable consciousness and is to itself the unessential; but as consciousness of unchangeableness, of the ultimate essence," is seeks doggedly to, at the same time, attempt to free itself from the unessential, that is, "to liberate itself from itself."[105] In other words, it attempts to free itself from its own self as unessential singular by striving to elevate itself and dissolve this its own singularity in the changeless universal. The anguished soul's will to assert iself and hold its own, as William James puts it, has been displaced by its willingness to close its mouth and "be as nothing in the floods and waterspouts of God."[106]

Though Hegel also associates it with a subjectivity which is not necessarily strictly monotheistic, the unhappy consciousness characteristically experiences the singular as man and the universal as God. It is, however, in fact itself the implicit unity of the human and the divine. Nonetheless, despite the fact that it is this union, the unhappy consciousness does not *experience* this unity as the reconciliation (*Versöhnung*) of the human and the divine. This is so because, as observed, it experiences itself as *solely* the unessential singular, in relation to which the essential universal appears to it as a transcendent divine Other. It views its self-otherness as a changeableness that renders it radically other than the unchangeable supersensible and thus unattainable Other. For this reason, and since it is desirous of liberating itself from itself by auto-dissolution in the universal, the unhappy consciousness becomes an *infinite yearning* for the Other. It is a yearning in which the anguished soul pines for God as the changeless, the pure, and completely self-identical. In relation to this pure self-identical divine self, it experiences its self-otherness and changeability as a beholdenness and enslavement to its own impure and base desires.[107] The unhappy consciousness—as William James would characterize it—is thus a feeling of unwholeness, of gross moral imperfection, of sinfulness, that irrefragably finds itself haunted by an incessant yearning after a well-nigh unattainable peace of unity with the divine.[108]

The anguished soul's relation to the transcendent divine Other is one of extreme estrangement. It is, paradoxically, likewise one of close intimacy. Here one may allude to the great personal propinquity which the Christian feels himself as bearing to the divine. For God is not merely Being, Mind, Truth, or the Unconditioned, but indeed *my* God. He is also like me: unique, universal, self-aware, intelligent, and individualized. He is, moreover, the embodiment of certain qualities which man longs to possess in himself: purity, constancy, and incorruptible wholeness. Yet, since this longed-for wholeness exists only within an unattainable Beyond, man's own comparative baseness is all the more starkly felt and all the more incorrigible. In this sense, the entwinement of estrangement and intimacy within the unhappy consciousness renders the anguished soul's estrangement from the divine all the more painful. For, to boundless longing, his beastly impurity adds humiliation and misery. The unhappy consciousness exemplified by certain—but in not sense all—forms of Christian experience is thus one of tormented subjectivity.[109]

What one thus observes concerning the unhappy consciousness in its Christian instantiation is the *reemergence of the master*. Having hitherto fallen into desuetude in the face of an inner retreat from all otherness effected by stoicism and skeptical consciousness, he now finds himself reborn as the transcendent divine Lord of mutable, inessential man—which latter is now still more deeply wretched than the earthly master's former slave.[110]

All of the foregoing is not, however, to say that Hegel is wholly against Christianity as such. He merely ascribes the unhappy consciousness to what may be termed "nascent" or "primitive" Christianity—a stage of Christian religiosity wherein the relation to God is merely abstract and entertained toward an essential universal. This consciousness is unhappy since it does not experience God as present in the religious community. The relation of this form of debased Christianity is therefore not one with God as he is truly, but merely with the divinity as it appears for the unhappy consciousness.[111]

Crucially, the overriding desire of the anguished soul is the attainment of union with the divine Other. As observed, however, it longs for this union specifically in order to liberate itself from its own changeable self, that is to say, to annul its own contingent singularity—which it sees as unessential consciousness. Thus, in order to be united with the divine Other, it must be at war with itself. It is here that one espies a strange transmogrification of the initial life and death struggle between the proto-master and the proto-slave: *what was previously a struggle between two separate antagonists, is now a battle transpiring within a single self.* Hegel maintains that, by way of a perverse involution, this battle is a strange struggle against an enemy (*Feind*), "victory over whom really means being worsted, where to have attained one result is really to lose it in the opposite."[112] In that the anguished soul experiences its own singular changeableness as at once its own incorrigibly vile baseness, this enemy assumes the form of the soul's own vices. Ultimately, however, the enemy is the singular existence of the anguished soul itself, its existence as *this* particular individual, as quiddity. Its consciousness of itself is "mere pain and sorrow over this existence and activity; for therein consciousness finds only consciousness of its opposite as its essence—and of its own nothingness."[113] The anguished soul rises up to God, only to confront its own sinful ipseity. It ascends to the divine as an unworthy nullity.

Something else is, however, afoot. For in ascending, consciousness suddenly experiences precisely this emergence of particularity in the unchangeable itself, and—vice versa—of the unchangeable in particularity.[114] This is the characteristically Christian moment of the unhappy consciousness, in which the unchangeable or God assumes the form of a singular man: Christ. Of course, this mutual imbrication of the singular and the immutable is what defines the unhappy consciousness to begin with. However, in that the anguished soul takes itself to be only the singular and experiences the immutable as radically external, it experiences the appearance of singularity within the unchangeable as positively miraculous. It fails to realize that this is an appearance generated by the unhappy consciousness itself. It does not realize that it has itself given rise to its unchangeable counterpart, and that the singular ipseity that it glimpses in the immutable is simply itself. As Hegel avers, that "the unchangeable, which assumes essentially the form of particularity, is *its own self*, the self which is particularity of consciousness—this is *not* established *for it*."[115] In that the singular it sees in the universal is the divine individual Christ, it likewise pictures Christ as essentially separate from God himself.[116]

The great problem of the unhappy consciousness is that, though it does posit singularity in the immutable, it fails to grasp the inner necessity of this divine manifestation. It merely takes the appearance of the divine individual, Christ, as a contingent fact. It fails to realize that the Incarnation is in fact a wholly *necessary* moment in the establishment of God's self-identity. Instead, it simply represents or pictures the Incarnation as "a contingent *happening*"—mere gratuity. It is oblivious to the fact that God's humanity is necessary to and inextricable from his divinity. The upshot of this unawareness is that the diremption between the singular individual and the universal persists, and that the individuality of Christ only exacerbates the misery of separation of the anguished soul from God, who remains an unattainable Beyond.[117]

We have seen that initially the anguished soul strove for union with the divine through the dissolution of its own singularity. However, once it sees singularity appearing within the unchangeable itself, it amends its strategy and henceforth strives to unite itself with the divine individual Christ, who is the universal incarnate and the divine Mediator twixt the sinful human individual and God as the pure universal Judge of mankind. Christ as divine Mediator becomes the ultimate object of desire, and is the concrete embodiment of what sinful consciousness attempts to become. The anguished soul longs for union with Christ so as to be united with the universal. Otherwise put, man struggles for union with the transcendent being as incarnate or "shaped" (*gestaltete*). In striving to accomplish this, it will pass through three consecutive phases, each one of which represents a different way in which it tries to forge this union.[118]

(a) Motions of the heart: monotheistic self-immolation in the Godhead

"*The heart as universe*—this is the deepest meaning of sainthood. *Everything happens in the heart:* that's mysticism and saintliness," avers Emil Cioran.[119] The anguished soul first strives for union with the divine through pure feeling. Though it is feeling, Hegel actually regards the latter as a certain kind of thinking. More precisely, it is an intense intellectual straining of the mind for contact with the divine. Thinking finds itself conmingled with fervid prayer and personal desire. Anselm, for one, rouses his longing for a mystical ascent to God, who dwells "in light inaccessible."[120] The anguished soul approaches God through "the inward movement of pure emotion (*Gemüth*)."[121] That is to say, he does so not through action but through feeling. Or rather, his feeling is his act. This "boundless pure inner feeling," as Hegel terms it, is a mystical, quasi-romantic relation between the anguished soul and the divine.[122] He describes this mystical union with the absolute as "the discordant jingling of bells, or a cloud of warm incense, a kind of thinking in terms of music," as it were.[123]

In dwelling upon mystical states, or the ideal state toward which devotional mysticism strains, William James delineates their basic characteristics as *inter alia* ineffability, noetic quality, and passivity. They are ineffable in that they are, as discussed in connection with Hegel, more like states of feeling than like states of intellect. They are nonetheless not mere feelings. For they are also possessed of a noetic quality. That is to say, they are likewise states of knowledge. They are states of insight into depths of truth outside the reach of the discursive intellect. Yet, though they remain discursively inarticulate, they are nonetheless illuminations and revelations and thus not outside of the intellect *tout court.* Finally, mystical states are passive in that, though they may be preceded by preliminary voluntary acts such as the focusing of the attention; when they have set in, consciousness feels as though its will has fallen into abeyance and that it has been annexed by a superior power.[124]

The attainment of this elevated state, which mystics take to be the absorption of the anguished soul's singular ipseity in the universal Godhead, however, requires an assidious pursuit of devotional practice. To at length subsume his changeable ipseity under the universal, the Sufi, for example, must purge his heart of all that is not God, follow a contemplative life of fervent prayer, and wholly immerse his heart in meditation upon the divine.[125] Here, dwelling upon Islamic mysticism in particular, Annemarie Schimmel contends that the Sufi mystic must liberate himself from the material attachments of the world. Such a practitioner "polishes the mirror of his heart." As in the case of the Christian mystic, he must follow the *via purgativa* in order to reach the *via illuminativa* in pursuit of the final goal of *unio mystica* in a *visio beatifica* in which the soul is "surrounded by the primordial light of God" and experiences a "lifting of the veil" that covers the essential unity underlying the ontological identity of God and man.[126]

This is a path of repentance and renunciation in which the subject must subdue the *nafs*—that is to say, his lower self, base instincts, or flesh. The *nafs* is the cause of sin and base qualities of the soul. It is the worst enemy of all and the Sufis christen the battle that must be waged against it with the familiar phrase: the Greater Jihad. Oddly, Qur'anic expression refers to the *nafs* as "soul." More precisely, however, it is *an-nafs al-ammāra bi's-sū'*—or "the soul commanding to evil" (Sūra 12:53). This is what the Sufi mystic must pacify and render *mutma'inna* (Sūra 89:27), or "at peace." Acting contrary to the wishes of the *nafs*, he must purge it of its evil attributes.[127] As Schimmel maintains it to be, the intrinsically impure *nafs* is inter alia to be purged through fasting and abstention from sleep. In this connection, she describes many a Sufi ascetic as *qā'im al-lail wa sā'im ad-dahr*—or "spending their nights upright in prayer and mainting a perpetual fast by day." The Sufi's purgative itinerary must be that of *qillat at-ta'am, qillat al-manām wa qillat al-kalām.* This is to say, "little food, little sleep, little talk," and likewise "loneline-less, keeping away from men." Mystics such as Shaqīq al-Balkhī and Rūmī speak of an "alchemy of hunger," wherein the soul receives the divine breath on an empty stomach. The body is here considered merely "a morsel for the tomb" and thus ought to be kept bony and lean. Some Sufis have exeggerated this fasting to the point of an "idolatry of the empty stomach." Some have even starved themselves to death.[128] The mystic must likewise spend extended periods of intense loneliness. Schimmel defines this as *qabd*, a compression of the soul "to make one's home in a needle's eye." *Qabd* is the "dark night of the soul" in which the "I" entirely disappears.[129] In this connection, when Mansur al-Hallāj maintains that "There is nothing good in love without death," he implies death as dying to one's qualities, or even corporeal death. In the Sufi tradition of "die before ye die" this would entail the slaying of the lower qualities of the soul, whereas for al-Hallāj it assumes a more extreme form of the pursuit of one's own actual physical death to destroy the "I" standing between the soul and the Deity.[130] For Abū-saīd al-Kharrāz, it is even during prayer (*salāt*) that one must reduce onself to "less than a mote of dust."[131] For Rūmī, during the ritual pronunciation of the *takbīrs*, or glorifications of God, the soul must offer its obdurate ipseity as a sacrifice upon the altar of the divine, in so doing "slaughtering the fleshly soul which ought to be killed ..."[132]

Within a Christian context John Cassian maintains that, such that union with the divine be attained, there must first of all be a wholesale purging of vice. To this end, the anguished soul must ensure that the "tottering and dead rubbish of the passions has been dug out." The "harmful shoot of fleshy lust and of avarice should be uprooted."[133] In a more involved manner, Saint John of the Cross maintains that the soul must pass through two kinds of night. The first of these nights consists of the purgation of the subject's sensual nature, of all his desires for the external things of the world, of the pleasures of the flesh, and of the satisfactions of the will. These desires must be mortified and put to sleep. The second night is worse. It is more spiritual, obscure, dark, and terrible. Yet the anguished soul must still pass through it so as to attain union with the Deity.[134] This is very similar to what Hugh of St. Victor terms the "fifth mode of divine knowing." In this mode, the subject must disregard the world and entirely reject himself so as to be "naked" and pure. The soul must pass through the three consecutive phases of what Job describes as suspension, John as silence, and Solomon as sleep. The author associates the first with purity, the second with charity, and the third with beatitude. In the first phase, the subject must terminate all that arises from the senses so as to inwardly return to himself as he was at the time when he was created. In the second phase, the soul must silence his mouth, his mind and his reason. When these two initial stages are accomplished, the soul is forthwith snatched away by a "sweet heavenly sleep" wherein it "dissolves and rests in the embrace of the Highest Light." Herein its reason, memory and will are put to sleep, and the soul slumbers, "dead to self and world."[135] Marguerite Porete speaks in perhaps even sterner terms as regards the self-abnegation the soul must undergo in order to attain absorption into the divine. The creature must abandon itself, destroy its own will. One must "crush oneself, hacking and hewing away at oneself to widen the place in which Love will want to be" and dwell. For Porete, this self-abnegation of the soul must reach the point where the creature realizes that it in fact does not even properly-speaking exist, and only exists as pure unadulterated evil. In comparison to this utter ontological and moral nullity of the anguished soul, it is only God who truly has Being.[136]

Generally speaking, as William James surmises, Christian mystics practice "orison," or the mystical elevation of the soul toward God. Here, much as in Sufism, the mind firstly aims to detach itself from its outer sensations by a graduated series of efforts to imagine holy scenes, until such time as it is entirely suffused by a semi-hallucinatory image of Christ. In the very highest raptures, all imagery falls away. The ultimate aim is to attain what Saint John of the Cross terms the "union of love," which is reached through "dark contemplation."[137] In the union of love, in which the Deity compenetrates the soul:

> The soul then feels as if placed in a vast and profound solitude, to which no created thing has access, in an immense and boundless desert, a desert the more delicious the more solitary it is. There, in this abyss of wisdom, the soul grows by what it drinks in from the well-springs of the comprehension of love.[138]

"In the orison of union," as Saint Teresa of Avila contends, the soul's changeable and singular ipseity finds itself entirely absorbed by and dissolved in the changeless universality of the Godhead, for:

> ... the soul is fully awake as regards God, but wholly asleep as regards things of this world and in respect of herself. During the short time the union lasts, she is as it were deprived of every feeling, and even if she would, she could not think of any single thing. Thus she needs to employ no artifice in order to arrest the use of her understanding: it remains so stricken with inactivity that she neither knows what she loves, nor in what manner she loves, nor what she wills. In short, she is utterly dead to the things of the world and lives solely in God....
>
> Thus does God, when he raises a soul to union with himself, suspend the natural action of all her faculties. She neither sees, hears, nor understands, so long as she is united with God.[139]

Elsewhere Teresa of Avila describes the experience of orison as one in which the Deity is felt as "an enormous and sovereignly limpid diamond," wherein all the actions of the anguished soul are contained in such a way that their essential sinfulness appears more starkly than ever before.[140] This absorption of the anguished soul in the Godhead calls to mind Meister Eckhart's description of the latter as a still desert, "where never was seen difference, neither Father, Son, nor Holy Ghost, where there is no one at home, yet where the spark of the soul is more at peace than in itself."[141] In a similar vein, Hadewijch of Antwerp felt that she "remained in a state of oneness with my Beloved so that I melted into him and ceased to be myself."[142] Bernard of Clairvaux describes this dissolution of the subject's singular ipseity in the universal Godhead as analogous to the manner in which "a drop of water seems to disappear completely in a big quantity of wine, even assuming the wine's taste and colour; just as red, molten iron becomes so much like fire it seems to lose its primary state; just as the air on a sunny day for the saints that all human feelings melt in a mysterious way and flow into the will of God."[143] This, in turn, is similar to what John of Ruusbroec maintains to be the "unity without difference" between the anguished soul and the Deity. Here the absorption of the subject's ipseity sinks into the Godhead in the manner of "a modeless abyss of fathomless beatitude." All the inclinations and distinctions of the creature pass away. All such enlightened spirits "melt away" and are "annihilated" in God's superessential essence. They "fall away from themselves and are lost in a bottomless unknowning."[144] For Teresa of Avila this "spiritual marriage" is "like rain falling from heaven into a river or stream, becoming one and the same liquid, so that the river and rain water cannot be divided," or "a streamlet flowing into the ocean, which cannot afterward be disunited from it," or "a room into which a bright light enters through two windows—though divided when one enters, the light becomes one and the same."[145] John of the Cross' analogue for the absorption of the soul into God is entailed in his observation that "when the light of a star or a candle is united to that of the sun, the light is not that of the star, nor of the candle, but of the sun itself, which absorbs all other light in its own." Orison entails a unification of the anguished soul with the Deity so acute and thoroughgoing that "the two natures are so united" that "without undergoing any essential change, each seems to be God."[146] In this union of orison, as Frances de Sales contends, "the soul is so strongly united to God that she cannot detach herself from him, she lives in God rather than in herself; as a crucified body belongs to the cross to which it is nailed and ivy to the wall whereon it is fastened."[147]

This absorption of the anguished soul in the divine would fall under what Annemarie Schimmel terms a "Mysticism of Infinity." Most strongly elaborated in Plotinus and in the *Upanishads*, particularly as in Shankara's *adviata* philosophy, the latter is present in Ibn 'Arabi's Sufism. Here the hyperousiological Numen is the absolute divine existence that is akin to an unbounded ocean in which the individual self vanishes like a drop of water.[148] Rābi'a al-'Adawiyya speaks of a loving union in which the subject has "ceased to exist and passed out of self. I am one with Him and altogether His."[149] For Ja'far as-Sādiq, the sixth Imam of Shi'a Islam, the soul is engulfed by "a divine fire that devours man completely."[150] Abū Yazīd Bistāmī, or Bāyezīd, describes it as the experience of *fanā*, or annihilation, in which he craved for the complete extinction of all traces of the self. He claims to have attained this state through extreme austerity, incessant mortification, and an emptying of himself. At length he had reached "the world of absolute unity" in which "lover, beloved, and love are one," and where "he himself is the wine, the wine-drinker, and the cupbearer."[151] For Abdū'l-Qāsim al-Junayd, the soul is likewise annihilated in God and, through a transformation in which his attributes are replaced by those of the Deity, no longer separate from Him.[152] For Jalāluddīn Rūmī, the soul is only admitted into the abode of his beloved God after he has become destroyed, for "there is no room for two I's in this narrow house."[153] This form of annihilating love is the same as the that of Mansur al-Hallāj: "Love is that you remain standing in front of your beloved, when you are deprived of your qualities and when the qualification comes from His qualification."[154] Hallāj likewise associates it with the "way of the moth," when the moth experiences *haqq al-yaqīn* in being burned and consumed by the flame.[155] Shāh Walīullāh of Dehli likens the soul thus dissolved to a piece of ice melting in a kettle of flames that becomes warm, boils, and is evaporated as steam.[156]

These are all ways in which *fanā* is described. Schimmel elaborates this matter somewhat by averring that it partakes of several stages. Firstly, man is annihilated and takes on God's attributes: *hadīth takhallaqū bi-akhlāq Allāh*, that is, "qualify yourself with the qualities of God" by the subsumption through incessant mental struggle of one's own base qualities with the elevated ones of the Deity. The next stage is annihilation in vision, wherein the soul finds itself surrounded by the primordial light of God. Finally, in a third stage, there comes an "annihilation from one's vision of annihilation," in which the soul is immersed in the *wujūd*, or existence of God. In this final state, the absolute sin of merely existing is in some sense erased. Here man recovers the state he possessed prior to the moment when he became endowed with individual existence by God. Admittely, here the annihilation of the soul in God is not complete but is simply lost in the "fathomless ocean of bewilderment" defined by Kalābādhi as the absence of the subject from his own attributes, to the extent that "he appears really mad and to have lost his reason."[157]

The best delineation of *fāna* Schimmel regards as being that of Toshihiko Izutsu, who describes it as "the total nullification of the ego-consciousness, when there remains only the absolute Unity of Reality in its purity as an absolute Awareness prior to his bifurcation into subject and object"—a state Sufis term *jam'*, or "unification, collectedness." In this state, the soul returns to the time when God was the sole existence, and nothing else existed. It is averred that though *fanā* is experienced by the soul, the soul is nothwithstanding not the subject of this experience. The subject is instead unified metaphysical reality itself. Beyond this state is *baqā*, that is, "persistence" or "subsistence" in God. In *fanā* and *baqā* the subject is nullified in the sense of being overcome by a "Black Light" of "bewilderment" in which all material things become "invisible." Everything is blacked out until the soul perceives this blackness to be the very light of the Absolute in and of itself, in which existence itself appears as pure nothingness. Whereas al-Hallāj uses the allegory of the moth and the flame to describe this state of the soul's unitive extinction (*anāl'l-Haqq*) in the Godhead, Rūmī compares the mystic in this state to a fragment of iron in a furnace wherein it regards itself as fire. Mystics from 'Attār to Mīr Dard have allegorized the anguished soul's experience of *fanā* and *baqā* to that of a dog or a donkey falling into the divine salt mine wherein the animal sheds its base qualities and is transformed into and preserved in the said element.[158]

Returning to specifically Christian mysticism, this matter of "orison" complicates itself in the case of a spiritual writer like Madame Guyon. Here one espies degrees of union with God. She distinguishes between the "union of powers" and "essential union." In the former the soul is united to God in a somewhat more superficial manner than in the latter. The union of powers can be the union of the soul's understanding with the Word. It may likewise be the union of the memory wherein is effected the absorption of the soul into God in which the soul comes to profoundly forget itself. This is the union with the Father. Thirdly, the union of powers may also assume the form of a union of the will bereft of vision or knowledge of anything distinct. This union is that of the soul with the Holy Spirit. Of these three forms of union with the distinct divine persons, Madame Guyon regards the last as the most intimate and consummate, closest to essential union and a prelude to it. The union of powers, nonetheless, remains a union of "betrothal." Essential union, on the other hand, is a "spiritual marriage." In the latter, the essence of the soul is united with the essence of God and their respective substances communicated to one another. The soul is united with God as a spouse, no longer through the persons of the Trinity, but in a completely immediate sense in which the soul is entirely possessed by God in his own unity. In essential union the soul flows into God "like a river into its original source" and is completely submerged and lost in Him. This is a "perfect fusion" in which sight and knowledge, and likewise division and distinction, cease to exist. Here "God is she and she is God," for the soul is entirely "absorbed into God and lost in him without power to distinguish or find herself again." This essential union of the pining soul with God is so extreme that it causes an "admixture of the soul with God so great and so intimate that she can distinguish and see herself no longer"—a fusion that "divinizes, so to speak, the actions of the creature arrived at this lofty and sublime position," at whose supernal summit the creature's soul is "melted" and "absorbed" into the Deity. In this essential union with God, the soul is in fact "annihilated" and freed from its own ipseity in a fusion strongly akin to that of the absorption of a single drop of water in an ocean. The problem, as we shall soon see, is however that, though the drop of water "has become ocean," it nonetheless "*forever remains a little drop* [my emphasis]."[159] This drop has thus failed to stymie the recrudescence of its own stubbornly obdurate ipseity. The self is thus still in a pickle, as shall be discussed a little later.

Richard of St. Victor's delineation of union with the Deity is broadly the same as that of Madame Guyon. What in the latter appeared under the name of "essential union" is homologous to what Richard of St. Victor terms the "third degree" of "violent charity." Here the soul "thirsts into God" and "through mental ecstasy" desires to pass entirely into the Deity so as to be completely oblivious to itself. In this form of orison, the mind is "so raptured into that abyss of divine light" that the intellect is rendered oblivious to all external things and even to itself. The soul is completely absorbed into the divine substance. In this state all carnal desires are utterly subdued and cast into a deep sleep. Surrounded "on all sides by the fire of divine love," the soul is "intimately penetrated and inflamed to such a degree that it totally puts itself of and puts on a kind of divine desire." It is liquefied as molten iron in a smelter. It is, in a sense, "put to death in God."[160] Saint John of the Cross maintains that in this state the "the divine fire" transforms the soul in such a manner that it is not merely "entirely burned up in this vehement fire" but indeed deified.[161] Marguerite Porete ups the ante here when she maintains that in orison the soul ceases to even see itself in any sense whatsoever. It in fact ceases to see that anyone else exists, and feels that only God exists. For in this infinitely intimate absorption of the soul into God, the soul is made privy to the fact that *only* God exists.[162]

Though it eschews the divine individual Christ as mediator, a similar striving for mystical union of the soul with God may be detected in various strains of kabbalistic thought. The task is to guide the soul back to its native home in the Godhead. For each *Sefirah* (divine attribute) the soul must attune itself to a specific ethical attribute such as to integrate itself into the mystical life of the *Sefirot* (divine attributes). This latter is the subject of Moses Cordovero's *Tomer Devorah*. The ultimate *telos* of the soul's mystical path, kabbalists unanimously agree, is that of *devekut* or mystical cleaving to God. According to Isaac of Acre, there exists an entire ladder of ascending ranks of *devekut*, such as *hishtavvut* ("equanimity," or the indifference of the soul to praise or blame) and *hitbodedut* ("equanimity," being alone with God). For the early kabbalists of Provence, *devekut* is both the process whereby the soul cleaves to his Creator and an ultimate goal of the mystical way. According to Isaac the Blind, the overriding task of the mystics (*ha-maskilim*) and those who contemplate His Name is expressed in the Deuteronomic injunction "And ye shall cleave unto Him" (Deut. 13:51).[163] Here, by way of prayer and the reciting of blessings, the soul strives to harmonize its thought with its faith as though it cleaved to the divine, "to conjoin God in His letters, and to link [*likhlol*] the ten *Sefirot* in Him as a flame is joined to a coal, articulating his epithets aloud and conjoining Him mentally in His true structure."[164] Commenting on

Deuteronomy 11:22, Nahmanides in his *Sha'ar ha-Gemul* defines *devekut* as the state of mind in which "You constantly remember God and His love, nor do you remove your thought from Him" "to the point that when" the soul "speaks with someone else, his heart is not with them at all but is still before God. And indeed it may be true of those who attain this rank, that their soul is granted immortal life [*zerurah bi-zeror ha-hayyim*] even in their lifetime, for they are themselves a dwelling place for the *Shekhinah* [divine presence]."[165] In Azriel of Gerona's *Sha'ar ha-Kavvanah*, the soul's attainment of *devekut* entails the effacement of the human will in the divine will or the bringing into conformity of the two wills. Inherent to this intimate union is a sense of beatitude. It is attained through the ascent of the soul to a state of ecstatic rapture through concentrated thought, meditation and fervid prayer.[166]

The actual psychological striving and pining of the anguished soul for union with and dissolution in the changeless universal is finely and dramatically exemplified in Carlos Fuentes' novel *Terra Nostra.* In a poignant scene, El Señor (King Philip II of Spain), having dragged his body toward the altar with his arms spread in a cross and touching a great painting of Christ from Orvieto with his bloodless fingertips, implores the divine. He implores Jesus for orison, for a true union with his Savior, a spiritual union purged and purified of all ruminations of the base part of his soul. Through this union, El Señor aims to become indifferent to whatever he possessed or did not possess in his terrestrial life. He beseeches Jesus to be suffused by, and inebriated and annihilated in the immense sea of softness and sweetness of the divinity's immediate touch, so as to be transported beyond himself and borne entirely in God. He pleads with his Lord to manifest himself so as to effect the passive union of his soul with God, wherein he may partake of the great mysteries narrated in the dark Songs of Solomon. His soul pines like a bride to enter the Mystic Cellars of its Blessed Husband to be infamed and intoxicated by the free-flowing wine of divine love. It sighs for the Deity like a virtuous wife, for the innermost Kingdom of Heaven, for the flowery bridal chamber of God as the Divine Husband and his Paradise of celestial delights. To this end, he begs Jesus to enter into binding matrimony with him right now even during his own mortal life, so that both the Deity and his creature may instantly and forever be consummated in the eternal felicity of Glory.[167] In fact, going beyond even this, he hereupon beseeches God to hasten his death to fast-track this orison:

> Grant me your divine presence and Your divine touch and the sovereign encircling arms of the Divine Husband; I can live no more apart from you; grant me a brief life to hasten my nuptials with you; my inflamed anticipation can bear no more; grant me eternal glory

> where I will have no need to wait longer, wait, for nothing, where I need not despair of the resolutions effected by the tyrant Time; oh, my Jesus, when will it be![168]

What further impels the anguished soul, or unhappy consciousness, to seek union with the divine through thinking is that, for it, the unchangeable is itself pure thought. It views that the very universality and perfection of God resides in the fact that he does not imagine or feel but merely thinks. That the divine is pure disembodied thought is his very perfection. God is intellectual light, and the anguished soul exiled from God dwells in darkness.[169] Moreover, as Hegel maintains, the anguished soul considers itself to be at least implicitly a being of pure intellection, "that its nature is a pure emotion of this kind, a pure thought," except that it is weighed down by the wordly accoutrement of corporeality and its own base desires.[170] These it longs to shed in order to be united with God in an act of pure comtemplation. It pines for a purity of heart that would enable it to see God in pure intellectual light. To this end, through devotion (*Andacht*), it struggles to purge itself of all corporeal hindrances to its intellection.[171] As seen, El Señor longs for the union of the spirit purged and purified of all sentiments of the base portion of the soul. This allows Hegel to aver that the anguished soul's immersion in a ritualized mysticism of pure feeling is ultimately that by which it "gives itself up to thought, devotes itself to thinking (*geht an das Denken hin*)."[172]

However, since the attainment of an absolutely unsullied heart is literally impossible, this thinking of the anguished soul "is the movement of an infinite Yearning."[173] It is merely an infinite yearning, since pure inner feeling inevitably fails to reach and thus attain unity with God incarnated as the divine individual. In the person of Christ, the Absolute remains "the unattainable 'beyond' which, in being seized, escapes or rather has already escaped."[174] Feeling fails to reach the Absolute, since it is ultimately only self-feeling. Unbeknowst to itself, in devotional feeling, the anguished soul or unhappy consciousness is in contact, not with the universal, but with the very singular and corporeal selfhood which it strives to transcend. It is thus a retreat by the anguished soul into its own quiddity and a turn away from the divine that it pines to reach. Instead of effecting a unification of the singular and the universal, of man and God, devotional feeling has in fact exacerbated their mutual estrangement.[175] St. Anselm complains that, in striving to ascend to God's light, he merely fell back into his own bleak darkness. The imploring devotions of El Señor before the altar only enlarge the gulf between himself and the divine. For, having heretofore faced forward, the figure of Christ in the painting from Orvieto begins slowly to turn its face away from El Señor.[176]

(b) Work and desire: deification of the world through labor

God has eluded the anguished soul. Notwithstanding, with the return of the feeling heart back into itself, a paradoxical advantage accrues to him. For he now comes into intimate contact with himself. He feels his own actual existence. The "withdrawal of the emotional life into itself," as Hegel maintains, "is to be taken in such a way that this life of feeling, in its own regard, *has* actuality *qua* particular existence. It is pure emotion which, for us or *per se* has found *itself* and satiated itself, for although it is, no doubt, aware in feeling that the ultimate reality is cut off from it, yet in itself this feeling is *self-feeling*; it has felt the object of its own pure feeling, and this object is its own self. It thus comes forward here as self-feeling, or as something actual on its own account."[177] In this sense, the anguished soul resembles the slave initially vanquished by the master who, repressed back into himself, discovers his own ipseity as a nothingness maintained in Being. In like fashion, being rejected by the divine, the feeling heart rediscovers its own being-for-self.[178] With this its return into itself, the anguished soul assumes its second attitude to shoring up the hiatus between itself and the divine: that of "desire and labor."[179]

With the withdrawal into itself, the anguished soul enters into a new relation with the actual, that is, with the material world of things. It appears as though the individual overcomes his exile from God by his transformation of the given, by being a laborer in the vineyard of creation. The anguished soul's labor now serves as a constant reminder of his place in God's world. It is, for him, a religious act. Yet, unlike the inner motions of the heart, it is an active production. Like the vanquished slave, he negates the given and thereby gains intimacy with his own existence as labor.[180]

In the case of inner devotion, God's singularity appeared to the anguished soul or unhappy consciousness as the divine individual Christ. Now it ceases to be Christ as such, and is rather the world itself as God's "offspring." The world, in that it is here stamped by the unchangeable, partakes of the sacredness of the divine. In this way, the anguished soul communes with the divine by working on and enjoying the things which God has created.[181]

But it is precisely for this last reason that the desire and labor of the anguished soul, unlike that of the initial slave, fail to attain their goal. For whereas the slave's active negation and transformation of the natural given remains his own active production, that of the unhappy consciousness is not. This is so because, in that he experiences the world as bearing the stamp of the unchangeable, it is ultimately not his world, but God's. Therefore, the anguished soul's enjoyment and transformation of the natural given in a sanctified world occurs only under the condition that the individual's faculties, talents, and transformative powers have been granted to him by God. This means that, unlike the slave, he does not really experience himself as an absolute negativity which nihilites in Being.[182] Everything that he is and does is purely the result of divine grace or, as Hegel has it, of "an external gift [*eine fremde Gabe*], which the unchangeable here hands over for the consciousness to make use of."[183] The anguished soul is therefore not the real author of his transformation of the world. He is, to be sure, the one who desires and labors. But the actual outcome of this effort is produced only inasmuch as God's singularity has appeared as a sanctified world. It is therefore not the anguished soul's own negativity that effects the transformation of the given.[184]

In that they increasingly appear to do nothing of themselves, the anguished soul is moved to renounce the worth and efficacy of its desire and work. As Hegel maintains, it "denies itself the satisfaction of being conscious of its independence, and refers the essential substance of its action to the 'beyond' and not to itself."[185] That is to say, it "gives thanks" for that which it regards as not really its own in an attempt to shed its own changeable individuality.[186] In so doing, as Hegel puts it, the anguished soul "repels itself from itself."[187]

Yet, on the other hand, the anguished soul in fact "*has been* desire, work, and enjoyment; *qua* consciousness it has willed, has acted, has enjoyed."[188] That is to say, he is quite aware that God has not literally done everything, and that even the act of thanksgiving is one that is his own, however much it is bent on crediting the divine as the ultimate or sole benificent source. But, in so doing, the anguished soul mires himself in his own terrible quandary. For in the very commission of a good deed or act of self-deprecation, he suddenly feels pride at having done so. He suddenly feels rising up against him the enemy within, that is, the recrudescence of his own obdurate singularity that he wishes to dissolve in order to effect subsumption by the universal. In negating himself, the anguished soul merely asserts himself, since it is he who negates.[189] "Consciousness," Hegel contends, "feels itself therein as this particular individual, and does not let itself be deceived by the semblance of its renunciation; for the real truth of that procedure is that it has *not* given itself up."[190] The anguished soul thus fails to shed his own changeable ipseity that separates him from union with the divine. God and the anguished soul remain in unreconciled opposition.

(c) Self-surrender: night of anguish and monotheistic masochism

For the anguished soul, this experience is absolutely harrowing. It has failed to negate the enemy (*der Feind*) that stands between itself and the divine Other with whom it craves union. This enemy is, of course, itself. This shattered experience intimates to it that there is absolutely no way in which it is able to divest itself of its own particular obdurate ipseity. This its own singularity clings to it even more stubbornly than plaque does to the coronary artery.[191] As Hegel maintains, "herein the enemy is discovered in its special and most peculiar form [*in seiner eigensten Gestalt*]."[192]

To be sure, under different circumstances, this incessant return of consciousness's own unfailingly recrudescing ipseity would produce happiness. For it would confirm beyond any doubt the individual's own singular ipseity, and show that he is true actuality—that he, as self-consciousness, is indeed himself. For the anguished soul, however, this fulfillment is felt as the most profound and tragic failure, since the unhappy consciousness posits its essential selfhood as abiding not in its own singular ipseity but in a transcendent, immutable Beyond. The more actual and singular the anguished soul senses itself to be, the more hideous does he appear to himself, and the more harrowingly distant from God he feels. His awareness of his own wretched individuality is for him an exquisite misery. It is a self-disgust that finds itself immediately and immeasurably compounded by the incessant thought of the pure and sacred Other, of God, from whom the anguished soul feels himself estranged by his own radically particular corporeal vileness. The wretched man of unhappy consciousness experiences his stubborn quiddity as thoroughtly impure. It is repulsed by what it takes to be its own excrementality.[193] For this reason, the anguished soul "discovers itself as this concrete particular in the functions of animal life," which, "since it is in them that the enemy is seen in his proper and peculiar shape, are rather an object of strenuous concern and occupation, and become precisely the most important consideration."[194]

The anguished soul is the consciousness of its own exquisite impurity, defilement, and thoroughgoing wretchedness in complete alienation from an infinitely pure divine Other before whom he feels utterly unworthy. It is "a personality brooding over itself, as unfortunate as it is pitiably destitute."[195] The unhappiness of the anguished soul assumes, as William James avers, the form of a well-nigh implacable "moral remorse and compunction," a deep sensation on the part of man of himself as "inwardly vile and wrong, and of standing in false relations to the author of one's being and appointer of one's spiritual fate."[196] This state is one of melancholy and the conviction of sin, and inevitably precipitates the anguished soul into the seething mire of self-loathing and despair wherein melancholy appears in the form of terminal self-condemnation.[197] He becomes extremely sensitive to his own inconsistencies and discords. Any inner confusion and miscegenation becomes intolerable. Anything that is unspiritual pollutes the pure water of the soul and is therefore repugnant.[198] Any violation of the inner smoothness and cleanness of his soul is torture.[199] In fact, such can be the overwhelming dread occasioned by the anguished soul's horrid sense of his own fundamental internal defilement that, for him, "the whole universe" is suddenly felt to be "coagulating about the sufferer into a material of overwhelming horror, surrounding him without opening or end."[200] So precipitously and irrefragably has the anguished soul falled into the mire of revulsion at its own self that it is irremissibly exposed to "the grisly blood-freezing heart-palsying sensation" of its own proximity to an evil that suffocates it with its own promiscuous and irrevocable viral presence.[201]

In the Sufi version of this irremissible sensation of the subject's moral and psychological revulsion at its own self, one may return to the soul's self-purifying battle against the *nafs*. Schimmel refers to accounts in which, having failed to consummately tame the *nafs*, the subject can suddenly come to see the latter outside its own body. Mystics have seen it as a black dog, or coming out of their throats in the form of a fox or of a mouse. The *nafs* has likewise been begazed in the form of a restive horse or mule that has to be starved and subjected to incessant mortification in order not to run rampant. It has also emerged in the form of a snake. The *nafs* may also be a pig. In Attār's poetry those who fail to mortify their own piglike nature can be turned into pigs.[202]

Owing to the anguished soul's feeling of his own nullity and vileness—he now proceeds to chastise his own desire through fasting, ascetic self-denial, and self-laceration.[203] His life will, owing to a rabid and malicious self-debasement, be one of what William James terms "grubbing in rat-holes instead of living in the light," an incorrigibly obsessive "manufacture of fears, and preoccupation with every unwholesome kind of misery" that chimerically assails the sick soul.[204] The ascetic self-denial of the anguished soul may become so acute as to overcome the ordinary inhibitions of the flesh and turn into outright self-immolation.[205] It is a "passion of self-contempt wreaking itself on the poor flesh," a "divine irrationality of devotion" that makes "a sacrificial gift of all it has."[206] Only through ascetic mortifications and torments is the anguished soul able to effect a modicum of assuagement of his sense of his own vileness.[207]

To this end, Saint John of the Cross enjoins that one's soul must always turn to that which is most distasteful, to that which disgusts, and from consolation to desolation; that one must aspire to the lowest and the most contemptible; that one must seek the very worst in all things, so as better to enter into complete destitution, poverty of spirit, and the absolute renunciation of everything in the entire world; that one must wholly despise oneself and wish to be despised by others; and that one must openly commit detraction against oneself and positively desire that others speak ill of oneself. All of this must be practiced to the end that, in order to attain and be all things, we must all willingly submit ourselves to being absolutely nothing, that is, to deliberately debase ourselves.[208]

The resignation, self-surrender and self-crucifixion of the anguished soul is what Lejeune describes as a:

> … sacrifice which man offers to God, and of which he is himself both the priest and the victim. By poverty he immolates his exterior possessions; by chastity he immolates his body; by obedience he completes the sacrifice, and gives to God all that he yet holds as his own.… The sacrifice is then complete and unreserved, a genuine holocaust, for the entire victim is now consumed for the honour of God.[209]

When Fuentes' El Señor sees the Christ-figure in the painting from Orvieto averting his countenance, he screams, stretches out his hand, grabs a penitential whip, and begins to lash his own back, his hand, and face.

Through these acts of self-surrender and self-abnegation, the Hegelian anguished soul "deprives itself, completely and in truth, of the consciousness of inner and outer freedom, or reality in the sense of its own existence for itself."[210] The man of unhappy consciousness, in this final act of absolute surrender, now gives up his entire selfhood and thereby degrades himself to the condition of a mere *thing*, to animality, to excrement. Self-consciousness self-destructs. It has plunged into abysmal abjection.[211] This auto-reification of the anguished soul's own self-consciousness is clearly evinced in Saint Ignatius of Loyola:

> In the hands of my Superior, I must be a soft wax, a thing, from which he is to require whatever pleases him … I must consider myself as a corpse which has neither intelligence nor will; be like a mass of matter which without resistance lets itself be placed wherever it may please any one; like a stick in the hand of an old man, who uses it according to his needs and places it where it suits him.[212]

Still beyond Loyola, one espies the fourteenth century mystic Suso who wore a hair shirt and an iron chain, causing himself to bleed. He made an undergarment studded with a hundred-and-fifty brass sharpened nails which drove into his flesh. On cold winter nights he exposed himself to the gnawing of vile insects, drove sharp tacks into his breast, and tore himself until his flesh festered. Day and night he bore a cross with thirty protruding iron needless and nails, which bloodied and seared his back. He struck this cross, driving the nails deeper into his flesh, until they stuck into his back. His feet were full of sores, his legs dropsical, his knees seared and bloody, his loins covered in scars, his body wasted, his mouth parched, his hands tremulous from weakness.[213]

The self-inflicted torments which James of Vitry, confessor of Mary of Oignies, attributes to this his penitent partake of the same gruesome and grim Suso-esque self-flagellating logic. He recalls of the latter that, upon being forced to eat meat and imbide a smidgen of watered wine in the wake of a grave illness, she punished herself out of sheer disgust at having felt a modicum of pleasure from the latter and likewise as recompense for former delights of any kind that this quaffing immediately called to mind. Riled to the core by the roiling and implacable guilt these supposed sins effected, she forthwith tormented her flesh, proceeding as far as to surgically amputate her nonesuch and bury it beneath the earth. To continue the self-torment, she for long forced herself to eat bread so hard, stale and black that even dogs could scarcely countenance. This caused her mouth to be torn up inside and to exude blood in copious profusions.[214]

Returning to Fuentes, already bleeding from his own penitential self-scourging, El Señor suddenly recalls a harrowing prior futural vision—on a day when he woke in a bedchamber soiled by canine slobber, tortured stools, animal placenta, the stains of urine and blood, semen and grease—of a tiny dot that, with shivers, he beheld transforming into a whitish, hairless, and amorphous being swimming in dark liquid, and then into a newly born wolf cub whom, with a growing tremor of horror, he observed being hunted on his own lands by his own descendants. This hunted wolf is *himself*.[215] His destiny, he concludes, "is to be an animal in Hell."[216]

A perhaps still more acute level of distress and dereliction is that attained by Angela of Foligno. She experienced the stubborn recrudescence of her indissoluble ipseity in the form of being incessantly tormented by demons, to such a degree that she felt herself to be like "a person hung by the neck, with his hands bound behind his back and blindfolded, who is hung by a rope on the gallows and yet lives, who remains without any aid, any kind of support or help."[217] Of this torment, she spoke thus:

> I beheld the demons hold my soul up in suspension in such a way that, just as a hanged person has no support, so too there seems to be no support left for my soul.[218]

An excruciating sense of her own moral impurity assailed her, to the point of it being an angry and self-lacerating masochism:

> All its virtues are overthrown while the soul sees, knows, and looks on. Then, when the soul sees that all the virtues have been overthrown and departed, and that she can do nothing about it, there is such great pain—such desperate pain as well as anger of the soul—that I am scarcely able at any time to weep over this desperate sorrow and anger. There are times that I cry without being able to stop. There are times when such great anger ensues

> that I am scarcely able to stop from totally tearing myself apart. There are also times when I can't hold myself back from striking myself in a horrible way, and sometimes my head and limbs are swollen. When the soul begins to see all virtues fall down and fall away, then there is fear and lament, and I speak to God, calling out again and again to God without letup, saying to him, "My Son, my Son, do not leave me, my Son!"[219]

Her utter dereliction was not merely spiritual and moral, but indeed likewise physical. For she experienced the vileness and digustingly particular quiddity of her own body in the guise of demons striking and punishing her in every single part of her body. The inexorable recrudescence of her own ipseity likewise assumed the form of the revival of all of her former vices. This she experienced as draconian punishment. Even gargantuan vices that had heretofore never even existed extruded themselves into full bore existence out of pure nullity. These too assaulted not only her soul but also her body, causing an extreme sense of persecution.[220] What may likewise be observed is that she, in pointedly masochistic vein, swiftly comes to *welcome* her own torments, in actual fact wishing that they be *compounded*. She laments:

> I see that I have been handed over to demons who revive and bring back to life vices long dead and add those that never were. And then, remembering that God was afflicted, despised, and poor, I wish that all my evils and afflictions were doubled.[221]

In particular, those vices by far the most difficult to purge—that is, those contrary to the virtue of chastity—came to rise up from the depths of her unconscious. Angela of Foligno was privy to diabolical temptation and its exquisitely lascivious vissicitudes:[222]

> While I am in the most horrible darkness of the demons it seems that any kind of hope of the good is lacking—it is a terrible darkness. The vices that I know were dead in my soul are brought back to life, and the demons rouse them up in the soul from the outside—and they even raise to life vices that were never there. They did the same in the body (where I suffer less) in three places, that is, in the sexual places.[223]

This sense of feeling disgustingly sullied to the core by her own corporeality again induces her hypertrophied superego to chastise her still further. She indeed ups the ante in this regard, her ever burgeoning masochism exacerbating itself into a rapidly and rabidly impending desire for not merely death—but, in fact, ETERNAL DAMNATION. She complains, moans and laments thus:

> There was so great a fire there that I used to use material flame to extinguish the other fire ... For the time during which I am in that darkness, I believe that I would rather be

> burned than to suffer those pains; indeed, I would then call out and shout for any kind of death that God would give me. Then I tell God that if he has to send me to *hell* [my emphasis], not to delay it, but to do it at once. And I say: "Because you have abandoned me, finish it and drown me."[224]

Did a sexual thought abide in the consciousness of Angela of Foligno for slightly too long, to the extent that it became a *mortal* sin, causing one to lose the grace of God and, in the event of dying in a sudden and unpredictable accident, render one radically susceptible to *torment in Hell for the rest of eternity*? Or is it rather the case that the anguished soul's mere fear and presupposition that he may perhaps have assented to and luxuriated in the sexual thought be a ruse whereby the Devil would cause the soul to surrender himself to the abyss of eternal damnation through an anxiety-riddled capitulation to pure unadulterated despair? The flipside of this coin is of course that the anguished soul may well think himself into the conviction that he had not really given into the sexual thought, whereas in reality he had. In a strong sense, the anguished soul cannot really tell which of the two possibilities he has actually fallen into. For this reason, he may well go completely *mad* and *curse the very day of his birth*.

Notes

1. Kojève 1996, 50.
2. Ibid.
3. Hegel 1966, 231.
4. Hyppolite 1996, 76.
5. Gadamer 1996, 156.
6. Kojève 1996, 50, 51.
7. Kalkavage 2007, 115, 116.
8. Habermas 1996, 129.
9. Gadamer 1996, 159.
10. Kalkavage 2007, 116, 118.
11. Kojève 1996, 51.
12. Benjamin 1996, 210.
13. Žižek 2013, 544, 545.
14. Kojève 1996, 51.
15. Kelly 1996, 270.
16. Hyppolite 1996, 80.
17. Kojève 1996, 52.
18. Ibid. 54, 55.
19. Kalkavage 2007, 117.

20. Bataille 1991, Vol. 2 and 3, 219, 220.
21. Kojève 1996, 51, 52, 55.
22. Ibid.
23. Hegel 1966, 236.
24. Kovèje 1996, 55.
25. Ibid.
26. Hegel 1966, 236, 237.
27. Kojève 1996, 56.
28. Ibid.
29. Ibid.
30. Hegel 1966, 237.
31. Žižek 2013, 198.
32. Kojève 1996, 57.
33. Žižek 2013, 202, 203.
34. Hegel 1966, 238.
35. Habermas 1996, 141.
36. Lukács 1996, 108.
37. Avineri 1996, 191.
38. Kalkavage 2007, 123.
39. Kelly 1996, 256, 257.
40. Shklar 1996, 292.
41. Kojève 1996, 58.
42. Ibid.
43. Ibid. 58, 59.
44. Ibid. 59.
45. Ibid. 61.
46. Ibid.
47. Ibid. 61.
48. Hegel 1966, 244.
49. Hyppolite 1996, 67.
50. Kalkavage 2007, 128–131.
51. Solzhenitsyn 2007, Vol. 2, 521, 522.
52. Ibid.
53. Taylor 1975, 158.
54. Hegel 1966, 244.
55. Taylor 1975, 206.
56. Ibid. 158, 159.
57. Hegel 1966, 245.
58. Taylor 1975, 159.
59. Hegel 1966, 243.
60. Taylor 1975, 159.

61. Hegel 1966, 245.
62. Bierce 2011, 481.
63. Kojève 1996, 61, 62.
64. Ibid. 62; Hegel 1966, 246.
65. Shklar 1996, 293.
66. James 1987, 44, 45.
67. Badiou 2008, 259, 260.
68. Ibid. 260.
69. Ibid.
70. Kojève 1996, 62.
71. Ibid.
72. Ibid.
73. Taylor 1975, 159.
74. Ibid.
75. Shklar 1996, 294.
76. Hegel 1966, 246, 247.
77. Kalkavage 2007, 133.
78. Kojève 1996, 62.
79. Hyppolite 1996, 67.
80. Pessoa 2015, 15.
81. Ibid.
82. Ibid. 37.
83. Ibid. 271.
84. Ibid. 147.
85. Ibid. 109.
86. Ibid. 62.
87. Ibid. 84.
88. Ibid. 100.
89. Ibid. 146.
90. Ibid. 186.
91. Ibid. 106.
92. Kojève 1996, 62.
93. Pessoa 2015, 106.
94. Ibid. 215.
95. Ibid. 107.
96. Taylor 1975, 159.
97. Hegel 1966, 249.
98. Ibid.
99. Taylor 1975, 159.
100. Cioran 2013b, 6, 7.
101. Kojève 1996, 62, 63.

102. Hegel 1966, 250.
103. Ibid. 251.
104. Ibid. 251, 252.
105. Ibid. 252.
106. James 1987, 49, 50.
107. Kalkavage 2007, 136–138.
108. James 1987, 187.
109. Kalkavage 2007, 138, 139.
110. Ibid. 141.
111. Ibid. 139.
112. Hegel 1966, 252.
113. Ibid.
114. Ibid. 253.
115. Ibid. 257.
116. Kalkavage 2007, 143.
117. Ibid. 144.
118. Ibid. 144, 145.
119. Cioran 1998, 6.
120. Kalkavage 2007, 145, 146.
121. Hegel 1966, 257.
122. Ibid.
123. Ibid.
124. James 1987, 343, 344.
125. Ibid. 364, 365.
126. Schimmel 1975, 4.
127. Schimmel 1975, 112.
128. Ibid. 114–116.
129. Ibid. 129.
130. Ibid. 134, 135.
131. Cited in Schimmel 1975, 150.
132. Cited in Ibid.
133. John Cassian 2006, 90, 91.
134. John of the Cross 2006a, 73, 74.
135. Hugh of St. Victor 2006, 337–339.
136. Marguerite Porete 2006, 174, 175.
137. James 1987, 366, 367.
138. Cited in James 1987, 368.
139. Cited in James 1987, 368, 369.
140. James 1987, 371.
141. Cited in James 1987, 376.
142. Hadewijch of Antwerp 2006, 104.

143. Bernard of Clairvaux 2006, 436.
144. John of Ruusbroec 2006, 448.
145. Teresa of Avila 2006, 457.
146. John of the Cross 2006c, 463.
147. Frances de Sales 2006, 470.
148. Schimmel 1975, 5.
149. Cited in Schimmel 1975, 39.
150. Cited in Ibid. 41.
151. Cited in Ibid. 49.
152. Cited in Ibid. 58.
153. Cited in Ibid. 60.
154. Cited in Ibid. 71, 72.
155. Schimmel 1975, 142.
156. Ibid. 134.
157. Cited in Ibid. 142–145.
158. Cited in Ibid.
159. Madame Guyon 2006, 42–45.
160. Richard of St. Victor 2006, 159–161.
161. John of the Cross 2006b, 215.
162. Marguerite Porete 2006, 178.
163. Scholem 1978, 174, 175.
164. Cited in Ibid. 175.
165. Cited in Ibid.
166. Scholem 1978, 175, 176.
167. Fuentes 2003, 153–155.
168. Ibid. 154, 155.
169. Kalkavage 2007, 146.
170. Hegel 1966, 257.
171. Kalkavage 2007, 146.
172. Hegel 1966, 257.
173. Ibid.
174. Ibid. 258.
175. Kalkavage 2007, 147.
176. Fuentes 2003, 155.
177. Hegel 1966, 259.
178. Kalkavage 2007, 147, 148.
179. Hegel 1966, 159.
180. Kalkavage 2007, 148.
181. Ibid.
182. Ibid. 148, 149.
183. Hegel 1966, 260.

184. Kalkavage 2007, 148, 149.
185. Hegel 1966, 261.
186. Kalkavage 2007, 149.
187. Hegel 1966, 261.
188. Ibid. 262.
189. Kalkavage 2007, 149, 150.
190. Hegel 1966, 262.
191. Kalkavage 2007, 150.
192. Hegel 1966, 263.
193. Kalkavage 2007, 150.
194. Hegel 1966, 263.
195. Ibid. 264.
196. James 1987, 157.
197. Ibid.
198. ibid. 266.
199. ibid. 272.
200. Ibid. 151.
201. Ibid.
202. Schimmel 1975, 112, 113.
203. Kalkavage 2007, 151.
204. James 1987, 152.
205. Ibid. 251.
206. Ibid. 278, 279.
207. Ibid. 272.
208. Cited in James 1987, 279, 280.
209. Cited in James 1987, 285, 286.
210. Hegel 1966, 265, 266.
211. Ibid. 266; Kalkavage 2007, 151, 152.
212. Cited in James 1987, 287, 288.
213. James 1987, 281–283.
214. James of Vitry 2006, 63.
215. Fuentes 2003, 148–155.
216. Ibid. 156.
217. Angela of Foligno 2006, 375.
218. Ibid. 376.
219. Ibid.
220. Ibid.
221. Ibid. 376, 377.
222. Ibid. 377.
223. Ibid.
224. Ibid.

7

Despotic Dialectics of Subjective Self-Immural, 2. Nietzsche

Soon enough, we shall move on to investigate certain rather poignant, incisive, and unexpectedly efficacious exit-strategies from the quagmire into which the very structures of the despotic regime incarcerate the human subject.

Notwithstanding, prior to this, we must be privy to the following observation. The Hegelian discussion in our previous chapter almost entirely omitted any mention of one crucial psycho-political defining feature of the subject subjugated under the draconian Statist auspices of the despotic regime. That is to say, *ressentiment.* For this reason, we need to rerun the entire passage of formation of the distinction between masters and slaves and its vicissitudes—specifically in terms of Nietzsche.

As with Hegel, for Nietzsche, the universal history of man begins with an initial act of subjugation, in which the hierarchy between masters and slaves is established. There is, however, a very fundamental difference between these two dialects or genealogies. For, in Nietzsche's view, the said originary violent subordination is *in no sense* born of a desire for recognition (*Anerkennung*) of one consciousness or desire by another. At the moment when the warlike bronze-eyed artists overrun the totemic savages of the primitive regime, neither of the two races define themselves in terms of recognition or self-consciousness (*Selbstbewusstsein*)—whether it be desired or already attained.

Death is likewise not *primus inter pares* a fear of biological extinction in the face of a superior adversary. As observed much earlier, death for the savage is rather the anguish of decoded and deterritorialized flows of desire threatening to submerge the primitive socius. This means that death for primitive man is already overridingly social and no longer crudely zoological. This dread of decoded flows felt by the savage is occasioned by certain pivotal events, such as birth, death, the prolonged philandering of the *betagi*, infertility, and so forth. As regards the warriors with a look of bronze, consciousness is too emphatically submerged by the ubiquity of unconscious action for death to acquire existential virulence. They are fearless and rabid.

Let us now transport ourselves to the hypothetical point just anterior to the moment of impact of the warrior race with the sedentary primitive savages. We observe that up until this point, everything within both regimes is acted. The warrior race are, as Nietzsche asserts, the most unconscious artists that ever lived,[1] and thus a fundamentally active type; whilst the savage system of primitive justice ensures that all the reactive forces of savage bodies are themselves likewise acted.

The tremendous impactual force dealt to the savage regime by the rapidly incoming warriors will, however, wreck and discombobulate the primitive system of justice. Suddenly, reactions that were acted in the savage become felt as memory traces ascending into consciousness. The formerly noble savage becomes reactive. He becomes the man of *ressentiment* who stands at the origin of history and occupies its ambit. In contrast—or at least for the time being—the invading warrior shall remain an active type.

As observed, Hegel conceived of the relation between proto-master and proto-slave—and then that between master and slave—as a dialectic of recognition (*Anerkennung*) played out twixt two adversaries aspiring to self-consciousness (*Selbst-bewusstsein*). For Nietzsche, on the other hand, the relation between master and slave is strictly *non-dialectical.* Moreover, though Nietzsche's and Hegel's slave are virtually identical, Nietzsche's master is not Hegel's master. For Nietzsche, the conquering artist-warrior shall stand as master vis-à-vis the vanquished savage which latter, due to the atrophying of the faculty of active forgetting formerly instilled by a now-defunct primitive justice, becomes the slavish man of *ressentiment.* Instead of two budding self-consciousnesses defining themselves in relation to one another, as in the Hegelian master-slave dialectic, it shall here be *only the slave* who comes to define himself in terms of the master—and not vice versa. Thus, the Hegelian master-slave dialectic is replaced in Nietzsche by a situation which runs as follows.

The victorious artist-warrior who subjugates the primitive regime is the Nietzschean master. The body of this artist is constituted predominantly of active force. If there are reactive forces in the composition of his body, they are necessarily acted. The inner will immanent to the artist's body is a fundamentally affirmative will to power. If there is any trace of a burgeoning negative will to power pervading his body, this negation by the artist shall subsist only at the service of affirmation. Reaction, in being acted, is always externalized in the form of outward aggression. The masterful artist therefore experiences life and himself as a foreground plenitude of active or acted force. This foreground plenitude he feels as untrammeled joy. For, as Nietzsche has it, in "the foreground stands the feeling of plenitude, of power which seeks to overflow, the happiness of high tension, the consciousness of a wealth which would and bestow."[2] The artist-master, Nietzsche furthermore maintains, is also powerful in the directly physical sense. He is a warrior, a hunter, a dancer, and generally one who engages in all vigorous and freely ribald and vivacious activities.[3]

1. The mode of evaluation of the artist-master

All of this has a fundamental bearing on the manner in which the artist-warrior evaluates himself and his environs. Filled with direct joy—or with what Nietzsche describes as "some fundamental certainty which a noble soul possesses in regard to itself, something which may not be sought or found and perhaps may not be lost either," the artist-warrior cannot but evaluate himself as "good."[4] Thus, the masters refer to themselves as *inter alia* "the truthful." Etymologically speaking, "truthful" is rooted in the Greek word *esthlos,* which signifies he who *is,* he who possesses reality directly within himself, he who is always already "actual." This term very quickly entertains a close propinquity with the word "noble."[5]

Moreover, being an artist of active force, the master is by definition a *creator*. All of the objects of his creation he sees as emanations of himself and even as part of himself and his boundless creative force and joy. He therefore glorifies himself in these objects and experiences their beauty as immediate and incontrovertible manifestations of his own nobility. Above all, as Nietzsche asseverates, the "noble type of man *feels* himself to be the determiner of values, he does not need to be approved of, he judges 'what harms me is harmful in itself', he knows himself to be that which in general first accords honour to things, he *creates values*. Everything he knows to be part of himself, he honours: such a morality is self-glorification."[6]

The artist-master, however, cannot possibly accord dignity to all objects and persons. What strikes him as impotent, ugly, derelict, decrepit and repugnant, will offend his fundamentally joy-suffused and joy-defined sensibility. The reprehensible, weak, sterile, slavish and prosaic man will therefore necessarily evoke his disdain. Far from it being the case that such a man will offend the artist due to the former's non-recognition of the latter, the artist will instead disparage the base man only in comparison to his *own* grandeur. He describes this lowly man with a veritable barrage of invective and vitriolic terms: *deilos* (cowardly, worthless, vile), *deilaios* (paltry), *ponēros* (oppressed by toils, good for nothing, worthless, knavish, base, cowardly), *mochtēros* (suffering hardship, knavish). The last two of these epithets designate the man of slavish consciousness as "the common man as slave worker and beast of burden."[7] In contradistinction to the truthfulness the master attributes to himself, the slave is seen by the master as a common liar—a kind of small town crook. The artist-master likewise evaluates the slave as *kakos.* This latter word is closely related to *deilos* (the plebeian in contradistinction to the *agathos*), through which designation the master effectively denounces the slavish man as a outright coward.[8]

As observed, the artist does not accord nobility to himself from the vantage point of some supposed recognition he would seek to gain from the decrepit man. As Stephen Mulhall argues, the master's evaluation of himself as "good" is absolutely primordial. It precedes slave morality and is a seigneurial privilege which serves as the originating evaluative medium of language as such. Just as God's first words immediately succeed his creation *ex nihilo*, so does the master's positive self-evaluation and self-appraisal arise from what he deems the intrinsic goodness of his actions.[9] As Nietzsche has it, the masters "did not need first of all to construct their happiness artificially by looking at their enemies," or by "*lying themselves into it*," as is the case with resentful men. For, "as complete men bursting with strength and therefore *necessarily* active, they knew they must not separate happiness from action," that "being active is by necessity counted as part of happiness."[10]

That is to say, no comparison with another consciousness interjects itself into this first and fundamental evaluation the master has of himself. The master's evaluation of himself as "good" is there from the very outset and stands alone. It is self-sufficient. It is made at a time before the reactive and slavish man has entered the master's purview. Indeed, the slave's presence or absence is entirely superfluous to the master's immediate positive self-appraisal. Thus, the master's move toward a comparison between himself and the base man is subsequent and ancillary to his *ab initio* self-glorification. As Lawrence Hatab argues, the master's deeming the slave "bad" arises out of feelings of contempt and embarrassment, but in no sense condemnation or denial. It is a secondary and consequent judgment used by the master to confirm his preexisting sense of noble rank, pathos of distance, and antecedent experience of self-worth.[11] In this consequential secondary evaluation, the noble artist-master divorces himself from those in which the opposite quality abides. He despises them.[12]

The master's full syllogism is thus: "I am good, therefore you are bad." Its premises are entirely positive—born as they are of enjoyment and foreground abundance. The conclusion is a negation—but only as arising from positive premises. Yet, it is a mere corollary to the premises and therefore inessential to them. It is only a post-production, a mere after-thought, a contrasting shade.[13] The conclusion is a mere supplement whose sole purpose is to augment the basic tenor of the already affirmative premises and to redouble their essential enjoyment. The master's conclusion only serves to confirm and compound his initial valuation of himself as noble. He searches for his opposite only to affirm himself the more graciously and victoriously. His purportedly negative concept "low," "common," "bad" is only a subsequently-invented and crepuscular, contrasting image in relation to his primordially positive basic concept. This latter is utterly replete and suffused with life and ribald passion. He has no doubt that he is noble, good, beautiful, and happy.[14]

In order to move toward a negation, the master requires two positive premises: I am positively good, the slavish man is positively base. It is only then that, in conjoining these latter two premises, the master is moved to negation. And, since in the master everything is externalized and acted, this negation—subsisting as it does within the greater pervasive ambit of an affirmative will to power—cannot but take upon the form of outwardly directed *aggression*. The master therefore subjugates the slave.

2. The mode of evaluation of the slave

Now, let us look at this whole situation from the perspective of the slave. The slave, the man of *ressentiment*—owing to the fall into desuetude of the primitive regime's cruel mnemotechnics—is no longer able to act his reactions. He confuses memory traces with present excitation. Every excitation received from the exterior milieu is irrevocably co-opted and polluted by a painful memory trace. Pain, which for the savage in the primitive regime possessed a fundamentally external character—that of the surplus joy experienced by him who inflicts or contemplates it—now acquires interiority: he is poisoned with rising bile. The base man is rankled to the core.

Since a wholesale indictment of himself—that is, an indictment as to the decrepitude that he has fallen into—would entail a complete existential collapse for the slave; he has no option other than to seek the cause of his irrevocable miredness *outside* himself. His sole recourse is to seek the origin of his consistutive impotence in the external milieu. In so doing, he must establish the point of departure of his own self-validation in another self—a self that is fundamentally unlike himself. He finds it in the bronze-eyed artist-master. In that the man of *ressentiment* is unable to act his reactions, he must—in order not to plunge himself further into his own pain-riddled quagmire—attribute an intrinsic repugnancy to the very defining characteristic of the artist: foreground action itself. Because of the confusion of excitation and memory trace, the slave fails to attribute his dereliction to himself (although he verily should), but instead blames that which is active. He blames the master's activity for his own decrepitude: the *spirit of revenge*.

But, of course, he cannot properly speaking strike out against this master in any tangible sense, for he is unable to act his reactions. Nonetheless, since he incorrigibly sees the bronze-eyed artist as the purported cause of his own corporeal dereliction, he still desires revenge against him. His revenge can, notwithstanding, only transpire at a purely inward, or spiritual, level. In that it is thus of necessity ideological, the reactive man's revenge against the bronze-eyed artist will not be immediately palpable but will instead assume an insidious form: that is, it will take the form of a *perversion of the value-positing eye*. This slave revolt on the terrain of morality, as Nietzsche argues, transpires when *ressentiment* itself "turns creative and gives birth to values: the *ressentiment* of those beings who, denied the proper response of action, compensate for it only with imaginary revenge."[15] As Sloterdijk has it, the resentful man's vengeful rage, in that it is prevented from expressing itself directly in the form of physical aggression, is "forced to take a detour through sublimation, internalization, transference, and distortion."[16]

The latter distortion will at base entail a transmogrification in the mode of evaluation. The slave begins by evaluating the artist-master as "evil." This is a mental negation of the master's active mastery, and is immediately followed by a second negation: in that he knows that he is not like the master, whom he considers to be evil, he thinks that he is—almost by definition—good. The slave resents the master, he absolutely hates him—and, in relation to this hatred of the master and indeed of all life that is active, he evaluates his truncated and reactive form of existence as "good." In this, "he has conceived of the 'evil enemy', 'the evil one' as a basic idea to which he now thinks up a copy and counterpart, the 'good one'—himself!"[17] From this much distorted perspective, the slave comes to view himself as "superior" to the artist-master.

In this sense, the slave is like one of the zeks from the horrid world of the Soviet Gulag. Solzhenitsyn perspicaciously observes that in the camps the:

> ... zeks inwardly consider themselves to be the superiors of their chiefs—both in literacy and in their mastery of work skills, and in their general level of comprehension of the circumstances of life ... but in their self-satisfaction: the zeks fail to see that, on the other hand, the administration of the islands possesses a *permanent* superiority over

> the natives by virtue of its *world outlook*.... As a generalization, the entire scale of values is topsy-turvy among the zeks, but this need not surprise us if we recall that it is always that way among savages.[18]

The zek's very mode of evaluation is none other than that of the slave.

But let us look more carefully at the slave's—or the zek's—syllogism and the way he uses it to justify his own inaction. As Aaron Ridley argues, in contradistinction to the master's syllogism of evaluation, which remains anchored in an originary, founding, positive and self-regarding affirmation of himself ("I am good") and where the negative ("You are bad") arises by way of contrast with and as a natural effluence of the former, the syllogism of the slave commences with an originary rejection and negation of the master ("You are evil") and only through a second negation ("I am not like you") manages to establish a positive self-appraisal ("I am good"). That is to say, the slave, so as to ultimately affirm himself, posits a purported "outside" or "non-self" in order to, through the rejection, reaction and negation of the latter, form himself as a positively self-appraised, salubrious, solid and contained moral self for the first time. Thus, in what may be termed a crude and lowly sub-Hegelianism, the slave requires two negations to reach a semblance of affirmation. As is evident, this affirmation is a mere counterfeit.[19]

The difference between the master's syllogism and that of the slave is likewise the hiatus that separates "ethical determination," on the one hand, and "moral judgment," on the other. The master evaluates in the former sense, the slave in the latter. The passage from ethics to morality, conceived in this way, is one in which "good" and "evil" —in a curious reversal—switch places. The "good" of ethical determination becomes the "evil" of moral judgment—and vice versa. As Nietzsche maintains, the foregoing definition of "bad" of noble origin and that of "evil" born out of the roiling cauldron of sheer frustrated hatred, are two entirely different things. The former is a subsequently wrought creation, a mere corollary and contrasting shade. The latter is absolutely originary. It stands at the very beginning. It is *the* distinctive deed. In both forms of evaluation, the epithets "bad" and "evil" have a completely different meaning. Yes, they do oppose the concept "good." But their conceptions of "good" are vasts apart, and do not have the same referent. For he *who* is "evil" in the morality of *ressentiment* (slave moraity), is exactly the "good man" of that other type of morality (master morality).[20]

Apart from the inversion of values, it is the very manner in which values are secreted in the two distinct modes of evaluation that is vasts apart. Lawrence Hatab refers in this connection to Nietzsche's parable of the bird of prey (the master) and the bleating lamb (the slave), where the latter deems the former "evil" for his necrophilous rapacity, that is, for acting in a way entirely consonant with his fundamentally active nature and, on the other hand, appraises himself as "good" for refraining from such activity.[21]

The reasoning of the bleating lamb is, however, deeply flawed. For it rests on the spurious metaphysical assumption that the bird of prey *could* in some way refrain from manifesting his force through abstention from activity, whereas the bleating lamb *could* give free reign from within himself to the *same* active rapacity as that of the bird of prey; that is to say, that the *same* potential or latent aggression is *voluntarily*, and *reflexively* held back in the bleating lamb and wickedly ventilated in the bird of prey.[22] In accordance with this syllogism, the lamb, or the slave, reviles the bird of prey as "evil" and appraises himself as "good."

But the syllogism of the lamb is no syllogism at all. It is in fact a *paralogism*. For the question remains: in that the capacity for action does not belong to the essence of reactive force (the lamb, the slave), how can the latter be said to voluntarily abstain from an activity of which it is inherently incapable in the first place? Nietzsche exposes the bankruptcy of the foregoing delusional rationality: for it effectively demands of strength that strength not express itself as what it is. It also requires that strength must not be a desire to overcome, may not be a desire to defeat enemies, may not desire to establish mastery, and may not desire to triumph. But this is, of course, absurd. For it is exactly as ridiculous as demanding of weakness that it should somehow manifest itself as strength—a strength that it by definition nowise possesses.[23]

As Hatab maintains, what lies at the basis of the peculiar syllogism of the bleating lamb is the illegitimate projection and superimposition onto an energetic field of forces of the linguistic distinction between nouns and verbs, wherein a continuous, undifferentiated and unbroken course of action suffers an internal diremption into the binary opposition between cause and effect.[24] As Keith Ansell-Pearson has it, this dichotomy between cause and effect is redoubled as that which comes to exist between "subject" and "object," wherein purely grammatical categories are deployed to grasp the relation between the human animal and its *umwelt*. In the same movement, that which serves at the level of linguistics as the ground of predications is *substantialized* as an obdurate and extant "ego" possessed of an agency entirely its own, at liberty to manifest its force through action or refrain from doing so through self-willed quiescence.[25] This is the spuriously insinuated terrain upon which the masterly bird of prey may be blamed for acting in the only way he can. For, as Raymond Geuss contends, it is precisely with the imputation of the fictitious construct of an "immoral soul" lying behind the rapacious actions that are his own, that the bird of prey is presumed as possessing the free agency at the behest of which his actions may be deemed voluntary, and thus voluntarily wicked and "evil." At the same time, likewise presumed of having the same free agency as the bird of prey, the bleating lamb is now deemed praiseworthy, virtuous and "good" for refraining from actions that he is *in reality inherently incapable of performing*.[26] As Nietzsche has it, "the majority of the dying, the weak and the oppressed of every kind could construe weakness itself as freedom, and their particular mode of existence as an *accomplishment*."[27]

3. The development of *ressentiment* and "Judaic" consciousness

Much earlier in our text, it was observed how *ressentiment* is possessed of two aspects: topological and typological. The topological aspect pertains to the manner in which *ressentiment* is constituted as "raw content." This involves the displacement of unconscious mnemonic traces into consciousness and the escape of reactive forces from the action of active forces through which these reactive forces were formerly enabled to act their reactions. The typological aspect refers to the way in which *ressentiment* acquires "form." It is here that the invasion of consciousness by painful memory traces in conjunction with reactive forces' non-action of their actions and the atrophy of the faculty of active forgetting congeals to form a typical character embodying the spirit of revenge and engaged in the enterprise of perpetual accusation. Whereas the first aspect of *ressentiment* entails a displacement between reactive forces, it is only with the second aspect that reactive forces separate active forces from what they can do. They do so by way of the projection of the fiction of a super-sensible world and, as Deleuze and Nietzsche have it, of "a god in contradiction to life."[28] As we shall observe through Bloch, Buber, and Walzer, the God of Exodus is not, strictly speaking, a deity of this precise kind.

Deleuze however avers that, in order to accomplish the movement from the topological aspect to the typological aspect of *ressentiment*, an "artist" of *ressentiment* is required. This artist is the "Judaic priest," who insinuates the paralogism of the bleating lamb in the consciousness of the slave, gives *ressentiment* form and elaborates its content, projects the fiction of a supersensible world in contradiction to material life of earth, conducts the disparagement of active force, and directs the profoundly disturbing and dangerous spiritual revenge of reactive force on active force.[29] It is he who "ventured, with awe-inspiring consistency, to bring about a reversal and held it in the teeth of the most unfathomable hatred" that only the poor, wretched and subjugated are good, and that "'you the noble and powerful, you are eternally wicked, cruel, lustful, insatiate, godless, you will also be eternally wretched, cursed and damned!'"[30]

The Judaic priests themselves are, however, not derived from the stock of the slaves. As R. Lanier Anderson argues, the priests in no sense suffer the painful psychological dyspepsia of slavish consciousness. Rather, they create the lamb-like mode of valuation *for* the slaves and insinuate it *into* slavish consciousness *without* themselves abiding by it as a self-valuation scheme. For their aim, though necessarily co-opting the *ressentiment* of the slaves, is entirely distinct from that of the latter. This aim is to avenge, by the most extraordinary means ever deployed in the entire history of human civilization, the antecedent political and military defeat they had suffered at the hands of the masterly warrior aristocracy and do so *by poisoning their consciences* such that they may be *miserable* and *indict their very existence as a foul pestilence worthy only of ETERNAL DAMNATION.*[31]

The Judaic priests are actually also of the masterly stock, except of a somewhat different caste. In fact, in them the internalization of their affirmative self-appraisal is *still* more pronounced than in that of the masterly warrior aristocracy. Indeed, owing to a truly profound pathos of distance, it is *so extreme* that it detaches itself entirely from all class domination, to the extent of comprising a sense of innate superiority that remains absolutely unparalleled. Whilst according with the basic "good/bad" valuation pattern of the warrior aristocracy, theirs had assumed and continued to assume the form "pure/impure," in which the warrior masters had *always* secretly been held in contempt as *impure*. For this reason, the Judaic priestly caste had *always* regarded themselves as incontrovertibly *superior*—more spiritual, more intellectual, more noble—that the warrior caste. In this, Nietzsche actually *agrees* with them and maintains that they were *right*. Regarding their political and military defeat—at the hands of what they considered brutish, crass, inarticulate and singularly stupid morons—as *profoundly* unjust, and indeed *reprehensible*, they would have to exact an *eternal* revenge against these idiots *by causing these same idiots to malign their own birth*.[32]

4. From *ressentiment* to bad conscience

As observed, when *ressentiment* has set in, active force is falsified by way of a fiction, deprived of the material conditions of its proper functioning and thus *divorced from its own inherent and inescapable potentiality*. But when the foregoing has come to pass, active force will be *turned back inside*, will *turn against itself*. In so doing, active force itself becomes reactive.[33] Nietzsche describes this as a process in which all undischarged instincts turn upon themselves inwardly. This is the "*internalization* of man," in which "all those instincts of the wild, free, roving man were turned backwards, *against man himself*."[34] This effective introjection of active force is the origin of *bad conscience*. It is an instinct for freedom that has forcibly been "forced back, repressed, incarcerated within itself and finally able to discharge and unleash itself only against itself."[35]

One must, however, in no sense conceive of bad conscience as ushering in a new type of man separate and entirely distinct in character from the man of *ressentiment*. For bad conscience is merely the consummate efflorescence of *ressentiment*, a more developed and ramified stage in the prosecution of its at once mendacious and incorrigibly insidious project. This project of *ressentiment* entails the accusation of active force and all on earth that is noble, high-minded, robust and beautiful. Here, as observed, active force is accused as blameworthy with the aid of the projection of the fiction of a supersensory world at the behest of reactive forces co-opted from the slavish man by the inestimably superior Judaic priest to the ends of first poisoning the mind of the warrior master on earth and then damning the body and soul of the latter in a place of eternal torture. The aim, however, of *ressentiment* is to lead the accused—that is to say, the comparatively crude and oafish warrior master but then, largely by historical mishap, potentially all men—to take over the prosecution and utterly demean, mock, ridicule and disparage *himself*. It is in this sense that *ressentiment*'s reactive *projection* extends itself into active *introjection*. The latter is a consequence and continuation of the former by more insidious means—to the extent that bad conscience hides the eternal *ressentiment* that it still harbors and always will harbor. For bad conscience is indeed the most sublime form of triumph attainable by *ressentiment*, originating as it does from the righteous indignation and virulent contempt of the Judaic priest and implanted into the base mind and hapless physique of the slave.[36] All of these men are "worm-eaten physiological casualties," "a whole, vibrating realm of subterranean revenge, inexhaustible and insatiable in its eruptions against the happy," whose sole aim is "in *shoving* their own misery, in fact all misery, *on to the conscience* of the happy," so that the latter "start to be ashamed of their happiness and perhaps say to one another: 'It's a disgrace to be happy! *There is too much misery!*'"[37] The infection must be spread.

5. The hypermultiplication of pain

But this is not all. For the introjection of active force that lies at the origin of bad conscience has a most dire upshot. In turning back upon itself, active force *produces pain*. This situation, as Bernard Reginster avers, arises from the fact that, under his condition of forced powerlessness, the frustration of the instincts of the slave has by now become *systematic*. Yet, they still do not cease to make their demands on the man of *ressentiment* and thus, failing to discharge themselves outwardly, now turn inward and against themselves. The mode in which the slave strives to liberate himself from these instincts is, paradoxically, by extracting enjoyment from his sheer overcoming of the suffering which they cause him. This commitment to the Sisyphean ordeal that he strives to perpetuate against himself, of course, necessitates that he preserve these self-lacerating instincts. Therefore, in a singularly voluntary masochistic internal orgy, he effectively incessantly overcomes pain through its exponential multiplication and hyper-impregnation.[38] Nietzsche refers to the exact same phenomenon when alluding to the "labour of a soul voluntarily split within itself, which makes itself suffer out of the pleasure of making suffer," [39]—a peculiar scenario in which pleasure, in a singularly sadomasochistic fashion, is experienced and actually "*looked for* and found in failure, decay, pain, misfortune, ugliness, voluntary deprivation, destruction of self-hood, self-flagellation and self-sacrifice."[40] The man who multiplied his pain with such prodigious industry become the rabid animal who "ripped himself apart, persecuted himself, gnawed at himself," "battered himself raw on the bars of his cage," made himself into "a torture-chamber," and invented "the worst and most insidious illness," "one from which mankind has not yet recovered; man's sickness of *man*, of *himself*."[41] Sloterdijk describes this voluntary hyper-multiplication of pain as one in which, through its very exaggeration, man's pain is made more bearable. It is an attempt on the part of the man of bad conscience to transcend his depressed suffering. He exacerbates his suffering "to the size of a mountain in order to be able to stand on its peak full of bitter triumph."[42]

Now, precisely as in the case of *ressentiment*, Deleuze contends that bad conscience likewise has a topological and a typological facet. Its topological facet, or raw material, is the hyper-multiplication of pain through the introjection of active force. Its typological facet, or elaborated content, is the new meaning which pain acquires through this interiorization of active force: moralized *guilt*, in which pain is at once the consequence and means of assuagement of *sin*.[43] Unlike the meaning accorded to pain in the primitive regime—that of rendering a compensatory excessive *jouissance* and heightened sense of levity to the collective eye luxuriously venting its spleen—the meaning of pain is now its spiritualization and the limitless auto-production of *still more pain*. That is to say, the man of bad conscience cures himself of his pain by infecting his wound.[44] This is precisely the phenomenon wherein Saint John of the Cross avers that "the divine burn of love heals the wound that love has caused, and by each application renders it greater. The healing that love brings is to wound again what was wounded before," to the extent that "he who is most wounded is the most healthy, and he who is all wound is all health." "O happy and most blessed wound! For you are inflicted only for the joy and comfort of the soul!"[45] One need not elaborate.

6. The development of bad conscience and "Christian" consciousness

As observed, the passage from the topological aspect of *ressentiment* to its typological aspect required the intervention of an artist of *ressentiment*: the "Judaic" priest. A similar sort of artist is necessary to enact the transition from the first to the second facet of bad conscience; that is to say, from the interalization of force (topological facet) to the internalization of pain (typological facet). This artist will be the "Christian" priest. He imposes form on the raw material of bad conscience by presiding over the internalization of pain. As observed, pain here is firstly assuaged by its exponential aggravation and then made the consequence of a fault of him who suffers it. The "Christian" priest effects this transmogrification *by changing the direction* of *ressentiment.* Formerly, the "Judaic" priest presided over the prosecution of active force, leading the man of *ressentiment* to blame all aspects of the active life for his own pain and suffering. However, when active force eventually turns against itself and itself becomes reactive, the accused must be found elsewhere. The reactive man who has made the transition from *ressentiment* to bad conscience must henceforth seek the cause of his suffering in *himself.*[46]

It is here that man, "suffering from himself," "like an animal imprisoned in a cage," seeks the cause of his suffering from "the ascetic priest" and finds it "within *himself*, in *guilt*, in a piece of the past" and "as a *condition of punishment*."[47] With the sorcerer's task discharged, the man of bad conscience is henceforth "a hen around which a line has been drawn. He cannot get out of this drawn circle: the sick man has been made into 'the sinner'."[48] Here we encounter the second of the two persecution maniacs that Sartre avers to be found in mental asylums. This second variant cannot even take refuge in himself, for within that inner citadel he finds his enemy already installed. His persecutors are no longer, or at least far less so, outside of his own psyche. Rather, they dwell within the innermost recesses of his own consciousness. "It is he himself who will be both the court and the accused, the policeman and the thief. It is he who will commit the offense and who will deliver sentence and apply it. If he tries to withdraw into himself in order to escape the censure of those about him, he will find an even more severe censure, his own. He will be a zealous self-tormentor and will henceforth experience his states of mind, moods, thoughts, even his perceptions, in the form of conflict."[49] "He fears himself as one fears a fire, a flood, an avalanche."[50] "He observes himself, spies on himself, foils himself, as if he were an odd instrument that one must learn to use. He struggles against an angel within him, an angel of Evil."[51] "Another animal, a monster, is digging tunnels, is going to get at him and devour him. This other animal is himself."[52]

We see here limping before us the "will to self-tormenting," the "repressed cruelty" of the human animal "scared back into himself, the creature imprisoned in the 'state' so as to be tamed, who invented the bad conscience in order to hurt himself," "so as to drive his self-torture to its most gruesome pitch of severity and rigour." His guilt before the Deity is "an instrument of torture to him. He apprehends in 'God' the ultimate antithesis of his own ineluctable animal instincts; he reinterprets these animal instincts themselves as a form of guilt before God (as hostility, rebellion, insurrection against the 'Lord,' the 'father,' the primal ancestor and origin of the world); he stretches himself upon the contradiction 'God' and 'Devil,'" he transmogrifies his utter denial of his own nature and of his entire self into an affirmation of "the holiness of God, as God the Judge, as God the Hangman, as the beyond, as eternity, as torment without end, as hell, as the immeasurability of punishment and guilt."[53] As Régis Debray would have it, what one sees here is the inscription into the psyche of "a materialized memory trace, a searing and tormenting refrain, ideal for the overly scrupulous" that "incises the superego in the ego" in a manner that "veers into sadomasochism."[54]

William James alludes to the case of John Bunyan as a paradigmatic exemplar of this merciless self-accusation, this will to self-torture. Bunyan was sensitive of conscience to a diseased degree—to the extent of being the victim of motor and sensory automatisms. Half-hallucinatory voices of a damnatory nature would beset his mind and buffet his soul often without surcease. He was a figure of melancholy self-contempt and despair:[55]

> If now I should have burned at the stake, I could not believe that Christ had love for me. … as to this act of sinning, I never was more tender than now; I durst not take a pin or stick, though but so big as a straw, for my conscience now was sore, and would smart at every touch.… I found myself as on a miry bog that shook if I did but stir; and was as there left both by God and Christ, and the spirit, and all good things.
>
> But my original and inward pollution, that was my plague and my affliction. By reason of that, I was more loathsome in my own eyes than was a toad; and I thought I was so in God's eyes too. Sin and corruption, I said, would as naturally bubble out of my heart as water would bubble out of a fountain. I could have changed heart with anybody. I thought none but the Devil himself could equal me for inward wickedness and pollution of mind. Sure, thought I, I am foresaken of God.…
>
> And now I was sorry that God had made me a man. The beasts, birds, fishes, etc., I blessed their condition, for they had not a sinful nature; they were not to go to hell-fire after death.…
>
> I was both a burthen and a terror to myself; nor did I ever so know, as now, what it was to be weary of my life, and yet afraid to die. How gladly would I have been anything but myself! Anything but a man! and in any condition but my own.[56]

It is this sort of relentless self-immolative and incessantly self-tormenting auto-flagellation of the soul that Nietzsche has in mind. What it calls to mind is the infinitization of man's existential debt and the consequent incapacity of any amount of pain to discharge the latter. This, of course, is one of the salient characteristics of the despotic regime already mentioned. Since pain now merely discharges the interest on man's debt to divinity, this pain must be ratcheted up sans surcease: "In this psychical cruelty there resides a madness of the will which is absolutely unexampled: the *will* of man to find himself guilty and reprehensible to a degree that can never be atoned for; his *will* to think himself punished without any possibility of the punishment becoming equal to the guilt; his *will* to infect and poison the fundamental ground of things with the problem of punishment and guilt so as to cut off once and for all his own exit from this labyrinth of 'fixed ideas'; his *will* to erect an ideal—that of the 'holy God'—and in the face of it to feel the palpable certainty of his own absolute unworthiness. Oh this insane, pathetic beast—man! What ideas he has, what unnaturalness, what paroxysms of nonsense, what *bestiality of thought* erupts as soon as he is prevented just as little from being a *beast in deed!*"[57]

And so does El Señor flagellate himself with venomous force, imagining himself reborn *as an animal in Hell.*

Notes

1. Nietzsche 2000, 522, 523, GM II.17.
2. Nietzsche 2014, 242, 243, BGE IX.260.
3. Nietzsche 2007, 17, GM I.7.
4. Nietzsche 2014, 272, BGE IX.287.
5. Nietzsche 2007, 14, GM I.5.
6. Nietzsche 2014, 242, BGE IX.260.
7. Nietzsche 2007, 20, GM I.10.
8. Ibid. 14, GM I.5.
9. Mulhall 2011, 250.
10. Nietzsche 2007, 21, GM I.10.
11. Hatab 2008, 43.
12. Nietzsche 2014, 242, BGE IX.260.
13. Nietzsche 2007, 22, GM I.11.
14. Ibid. 20, GM I.10.
15. Ibid.
16. Sloterdijk 2010, 86.
17. Nietzsche 2007, 22, GM.I.10.

18. Solzhenitsyn 2007, Vol. 2, 513.
19. Ridley 2011, 315.
20. Nietzsche 2007, 22, GM I.11.
21. Hatab 2008, 62.
22. Ibid.
23. Nietzsche 2007, 26, GM I.13.
24. Hatab 2008, 62.
25. Ansell-Pearson 1994, 17.
26. Geuss 2011, 20.
27. Nietzsche 2007, 27, GM I.13.
28. Deleuze 2006, 124, 125.
29. Ibid. 125, 126.
30. Nietzsche 2007, 17, GM I.7
31. Anderson 2011, 40, 41.
32. Ibid. 44.
33. Deleuze 2006, 127, 128.
34. Nietzsche 2007, 57, GM II.16.
35. Ibid. 59, GM II.17.
36. Deleuze 2006, 128.
37. Nietzsche 2007, 91, GM III.14.
38. Reginster 2011, 63.
39. Nietzsche 2007, 59, 60, GM II.18.
40. Ibid. 86, GM III.11.
41. Ibid. 57, GM II.16.
42. Sloterdijk 2010, 48.
43. Deleuze 2006, 129.
44. Ibid. 129, 130.
45. John of the Cross 2006b, 216.
46. Deleuze 2006, 131.
47. Nietzsche 2007, 104, GM III.20.
48. Ibid.
49. Sartre 1963, 21.
50. Ibid. 39.
51. Ibid.
52. Ibid.
53. Nietzsche 2000, 528, GM II.22.
54. Debray 2004, 76, 77.
55. James 1987, 147.
56. Cited in James 1987, 147, 148.
57. Nietzsche 2000, 529, GM II.22.

8

Exit Strategies from the Structures of Despotism, 1. Against Hegel

The Mosaic Revolution and St. Paul

Though Hell may perhaps exist, let us temporarily remove ourselves from this cauldron of roiling self-hatred. Returning to the broad movements of Hegel's line of reasoning, we shall forthwith proceed to argue against the Hegelian contention that Judaism and Christianity, through the purported erection of an absolute and transcendent master, necessarily condemn the subject to self-slavery. To be sure, they may well do so when conjoined with other political factors, despite themselves, inadvertently, or when they find themselves stripped of their own essence—as in the case of the unhappy consciousness. In that an entire slew of vicissitudes may beset a given religion and cause it to undergo self-alienation or distortion, it will be impossible given the present scope to speak so expansively as to treat these religions across their multivariegated internal mutations and hybridizations. Thus, we shall restrict ourselves solely to the two momentous opening salvos of Judaism and Christianity which specifically mark them as strikingly novel emancipatory movements. More precisely, we will dwell upon the revolution in politics and ethics alike effected by Moses and by St. Paul, respectively. It will be maintained that both represent exit strategies from the subject's beholdenness to the claustrophobic confines of the Hegelian master-slave dialectic and do so, not by simply ushering in the ultimate, most extreme and final version of slavery, but *by bringing into existence a relation of a fundamentally different order than that between masters and slaves.*

1. The Mosaic Revolution

We first move to scrutinize pivotal aspects of the biblical narrative of Exodus. As observed regarding the Hegelian master-slave dialectic, the transition to the unhappy consciousness was made from the vantage point of solipsistic nihilism. This was the circular ambit of Dante's Belacqua who, trapped in his own incorrigible self-perpetuating onanistic stasis, narcissistically mires himself backward. Backward into his own eternal stasis. The sempiternal stasis of his own perpetually mired self. In any case, even though the move toward the unhappy consciousness entailed the projection of a transcendent or other-worldly dimension for the first time, what this latter maintained of solipsistic nihilism was the wholesale disparagement of the intramundane world, of the world of the master.

In the Exodus narrative the world of the Hebrew slave's Egyptian taskmaster is likewise judged and condemned. However, as Michael Walzer renders pellucid, this condemnation is not one of the *entire* world. It does not bring to bear the full weight of its polemical brunt upon the entire gamut of external material reality. The unswerving judgment here is not one uttered against a global regime under the diabolical yoke of Satan. It is merely the world of Pharaoh that falls under the biblical interdict. And Pharoah is not Satan, whilst the world is not all Egypt. Indeed, that there is a material world outside Egypt means that a life devoid of Pharaonic oppression may verily be contemplated. This is what gives substance to God's promise of the possibility of such an existence. And it is from this promise or idea of a possible life free of oppression that the condemnation of Egypt stems and from which it acquires moral force. That is to say, the condemnation of Egypt is made only against the backdrop of a conviction as to the real possibility of the future assuagement and extirpation of the Egyptian mode of repression—more specifically, if not via the subversion of the Pharaonic regime itself, then certainly by exodus from it.[1]

To effect such an exodus, Yahweh calls upon Moses as liberator: "I have seen the misery of my people in Egypt. I have heard their outcry against their slave-masters. I have taken heed of their sufferings, and have come down to rescue them from the power of Egypt … I have seen the brutality of the Egyptians towards them. Come now; I will send you to Pharaoh and you shall bring my people Israel out of Egypt" (Exod. 3.7–10). And Moses begins this long and hard struggle for the liberation of the Hebrews.

That the possibility of actual material deliverance and redemption is unequivocally broached may also be evinced from the etymology of these words. For the Hebrew word for "redemption" derives from a legal term that means "to buy back." What is here to be bought back is the freedom of a slave. In a similar vein, the Hebrew noun for "deliverance" is derived from the verb meaning "to go out." One must go out of Egypt.[2] These observations of Walzer are crucial in that, in the absence of this sense of the possibility of real emancipation, the oppression would otherwise be experienced as an inescapable condition and therefore as somehow intrinsic to the world itself.[3] In such a case, the Hebrew slave laboring under his Pharaonic taskmaster would be in the position of the skeptical nihilist. Indeed, the Hebrew slave who had lost hope in this-worldly emancipation and had conflated Pharaoh and Satan may well have assumed a position akin to the Hegelian unhappy consciousness.

This, however, would certainly not have been the case for the vanguard that was to lead the massive contingent of Hebrew slaves along a line of flight out of their bondage under the Egyptian despotic regime. When inveighing against Judaism, Hegel had attributed the invention of the unhappy consciousness to Abraham. We will not argue for or against this particular point. What is, notwithstanding, certain is that the unhappy consciousness could not possibly have been a *Mosaic* invention.

Concerning the actual emancipation of the Hebrew slaves, two seminal points must be made.

Firstly, as Walzer argues, this emancipation cannot come about from within the ambit of the slaves themselves. That is to say, it will not arise endogeneously from or be effected by the man of slavish consciousness himself. In and of himself, the latter is literally hopeless. The liberation requires an *external intervention* by one who is himself not a slave and who is initially separate from the man of slavish consciousness. This man is Moses himself who, far from arising from slavery, is of the princely caste.[4] Martin Buber provides ballast for this point, maintaining that, in order that he could at length arise as the liberator of his nation, Moses has from infancy to be introduced into the very citadel of the Egyptian aliens, into the royal court by which the Hebrews are being enslaved. He must grow up there and thus, in a fundamental sense, be alien to the Hebrews themselves. For the kind of liberation which is here being bespoken can nowise be effected by anyone who has been brought up as a slave. Moses has not passed through the degrading and embarrassing forms of existence entailed within the slave status of the other Hebrews. This, of course, is not to say that he could matter-of-factly have been simply a non-Hebrew. For he remains ideologically connected with the slaves. He is of the slavish stock but, being brought up an Egyptian, not a man of slavish consciousness. He is a man of an enslaved nation but not one enslaved with them. He is equiped with all the crucial Egyptian wisdoms and powers necessary for his ultimate liberatory task. His immersion in the epicenter of this alien culture was the necessary antecedant of his being the future agent of the withdrawal of his people from the Egyptian despotic regime. Until such time, however, he has to remain hidden from his own people within the very maw of the monkey-faced Egyptian despotic regime. Moses will, in fact, retain his separateness from his own people for a long time. He will not become in all ways one with his brethren until the encounter with the Midianites. Until such time, he still wears Egyptian gear. For this reason, his passage involves a well-nigh shockingly direct transition from the loftiness of Egyptian court circles to the quasi-nomadic existence of former slaves.[5] He starts off as separate from the Hebrew slave and will later separate himself again.

In addition to Buber's and Walzer's observations concerning the separateness of Moses from those whom he would forthwith come to liberate, one may likewise bring into play Charles Taylor's conception of what he calls the "buffered" self. Although Talyor chiefly speaks of the latter within the context of the Protestant disenchantment vis-à-vis Catholicism, this may easily enough be extended to the Mosaic disenchantment in relation to the pagan Egyptian despotic regime. Generally, Taylor maintains that the "buffered" self is one defined by its awareness of the possibility of disengagement from a social world populated by subjects submerged within a consciousness of magical thinking. This magical realist milieu is that of the so-termed "porous" self.[6] The Hebrew slaves, for Moses, were immersed in and tethered to their slavery through the power of such "occult" forces. In comparison, the Mosaic revolution, though still necessitating an overwhelmingly forceful and liberating God, would have been seen as radically and shockingly secular in relation to the phantasmagoric polytheistic superstitiousness of the Pharaonic regime. But let us return to Walzer's musings on the Exodus narrative itself.

Secondly, the emancipation will in fact not be of the man of slavish consciousness himself. Rather, the deliverance will assume the form of the bringing into being of a future generation that will not know slavery since they will not be born into it. As in the case of the first point, the second point stems from the conviction that the present generation of Hebrews is so thoroughly immersed in slavish consciousness as to be literally incorrigible. For the alienation of the children of Israel was so extreme that, when the possibility of freedom was first broached to them by Moses, "they did not listen to him; they had become impatient because of their cruel slavery" (Exod. 6.9). Even after they had left Egypt and were threatened by the Pharaonic armies, they complained to Moses: "Were there no graves in Egypt, that you should have brought us here to die in the wilderness?

See what you have done to us by bringing us out of Egypt! Is not this just what we meant when we said in Egypt, 'Leave us alone; let us be slaves to the Egyptians'? We would rather be slaves to the Egyptians than die here in the wilderness" (Exod. 14.11–12). This miredness of the Hebrews in their own slavishness is still further illustrated by the fact that, even when they have escaped the territory of the Egyptian despotic regime itself, have seen the destruction of Pharaoh's army at the sea, and have trekked well into the desert, putting many miles between themselves and their former masters, the Israelites are still "sore afraid." For here they go as far as telling Moses that they actually prefer the security of slavery to the vertiginous uncertainties of the liberation process: "If only we had died at the Lord's hand in Egypt, where we sat round the fleshpots and had plenty of bread to eat" (Exod. 16.3). This pusillanimous and obstinate persistence in their slavish consciousness goes so far as to enrage God himself, who is only thwarted from utterly annihilating them by Moses' intercession.[7]

As Walzer contends, the forty years of wandering upon which God settles with Moses before the Hebrews are able to enter the promised land nonetheless has the purpose of indirectly exterminating most of the present generation. For the specific term is chosen so that all those Israelites who were twenty or more years old at the time of the flight from the Egyptian despotic regime will die natural deaths in the wilderness.[8] As Lincoln Steffens has it, the overriding political lesson of Exodus is that: "The grown-ups must die." He interprets the text to mean that "the Lord God killed off the whole of the Egyptian generation of the Jews."[9] This is not to say that there are not discrepancies. In Numbers 14, for one, it is averred that only some of the older generation die and that the greater number of these slaves ultimately survive the wandering period and come to raise a new generation of freeborn children. It is to this generation that Moses confers the laws and rituals of the new religion.[10] Whichever version one assumes, what remains incontrovertible is the fact that the Mosaic emancipatory project's ultimate beneficiaries are to be the offspring of the present generation, and not the present generation itself.

For this reason, likewise, it may be seen that the carceral bounds of the Hegelian master-slave dialectic can only be rendered asunder in a strictly non-dialectical manner. As observed, two radical discontinuities are required. The first is spatial: the external intervention of a non-slave, Moses. The second is temporal: the hiatus separating a generation defined by immersion in the consciousness of slavery, on the one hand, and their freeborn progeny, on the other. These latter represent a sudden upsurge of a new and now thoroughly political collective subject. They are a pure beginning completely outside of the contours and vicissitudes of the

Hegelian master-slave dialectic. This external intervention, bearing as it does the force of a limitless exposure to a pure outside, is experienced by the slaves as a miracle. For Buber, a miracle is something which intervenes in the life of an individual or group with the pure force of fate. For him to whom it occurs, a miracle renders asunder the security of his entire nexus of knowledge, exploding as it does the fixity of his fields of experience. A miracle is properly-speaking a singular *event* and as such is not included in a general chain of cause and effect.[11]

The temporal hiatus between the two generations also brings with itself another vital corollary: the hypertrophied and paralysing fear of death, which lay at the origin of the Hegelian slave's subjugation by his despotic Pharaonic master and which kept him subjugated, has been left behind. It has been cast into the past. As Walzer notes, this is evinced by the fact that the Israelites at the Jordan and the Israelites at the parting of the sea represent two fundamentally different subjectivities. For by the time they have reached the Jordan the old pusillanimity and slavishness has died with the older generation of Egyptian slaves. Suddenly, the Israelites have the courage to fight.[12] This enables Moses, in one of his last speeches, to maintain: "When thou goest out to battle against thine enemies, and seest horses, and chariots, and a people more than thou, be not afraid of them: for the Lord thy God is with thee, which brought thee up out of the land of Egypt" (Deut. 20.1). And they will not be fearful. For they have ceased to be "as grasshoppers" in their own eyes. They are now a mutually committed political society, purged of pusillanimity by a four decade long ordeal of desert wandering, and subsequently bound together by a covenant before the Deity. Though the portrayals of these now transmogrified Israelites—this completely new collective Jewish subject—in the biblical texts are frequently harsh, the new Israelites are nowise consistently described as servile. They are described as "stiff-necked" and stubborn. They are no slavish rabble. When marching away from Mount Sinai, they march not back into their former servility, but courageously toward the promised land.[13] The Israelites have accomplished what Charles Taylor maintains to be a "reversal of the field of fear." By facing down what they had heretofore feared the most, they have now been filled with courage and energy. Admittedly they have exchanged their fear of monkey-faced Egypt for the fear of God. However, whereas the former fear was debasing, the latter is exalting. They have unreservedly thrown themselves into the "naked" power of God, and it is through this oddly secularizing power that they have disentangled themselves from the sway of the Egyptian magic that had held them captive. Their liberation is a movement of disenchantment in which the God's might conquers an enchanted pagan world. He annihilates enchantment, voiding the world of it.[14]

Yet, is not the fact that the Israelites have entered into a covenant with God indicative of their merely having dispensed with their Pharaonic earthly master only to replace him with a transcendent spirit who shall henceforth erect himself as the great celestial and hyper-Pharaonic paranoiac holding dominion over man and nature from the supernal apex of a new quasi-spiritual despotic regime? Does this not entail an act of acquiescence to self-subjugation from whence even the new Israelite condemns himself to live in perpetuity as a slave-in-himself? Do we not have here a recommencement, or a version, of the Hegelian unhappy consciousness? Our overriding contention is that it is above all this that is precisely not the case at all. In entering into covenant with the Deity, the new collective and courgeous Israelite subject does not merely transfer the older Israelite slavish obedience from Pharaoh to God. For, as already intimated afore, the relation into which this new Israelite now enters is of a radically different order. Though the two are named by the same Hebrew word, the service of God into which the post-Sinai Israelite subject enters is very much unlike the Pharanoic slavery under which the older generation groaned. For whereas slavery—the Hegelian master-slave dialectic—is begun and sustained by coercion or the constant threat of force, service is begun and sustained by *covenant*. Covenant is the political invention of the Book of Exodus. It is unprecedented.

A further objection however arises: may it not be said that a covenant is likewise *imposed* upon a people? Walzer maintains that this is also not true. As observed, at the temporal and genealogical origin of despotism and of the Hegelian master-slave dialectic, there lies an act of violent subjugation from whence the distinction between masters and slaves arises and is maintained. The covenant entered into between the new Israelite subject and God, in contradistinction, waits until *after* this new generation of Israelites have *already* savored freedom. It waits until they have marched as far as Mount Sinai. For Spinoza, at this intermediate stage between an Egyptian servitude that has been left behind forty years since and their entrance into the promised land, the Israelites are presently in a state of *natural* liberty. This natural liberty is the immediate consequence of an already accomplished deliverance. It is, however, likewise practically unendurable for any significant length of time. For this reason, the Israelites *voluntarily* enter into a covenant with God so as to escape the inevitable cul-de-sac and sheer nullity of bare natural freedom. That the covenant is entered into freely by the new collective Israelite subject is evinced by the observation that it is not the elders but the people themselves who accept it of their own accord.[15] "And all the people answered together, and said, All that the Lord hath spoken we will do" (Exod. 19.8). The covenant thus remains a solemn commitment entered into by free men. Whereas in Egypt the Israelites constituted a people mainly by virtue of their shared experience of oppression, at Sinai a new Israelite nation comes into being through a covenant entered into by willing members. Their former identity was something conferred upon them by happenstance and an imputed consciousness of servility. Only with the covenant do they actively constitute themselves as a people in the strong sense; that is, as capable of sustaining their own moral and political history, and capable at once of obedience and stiff-necked resistance.[16] In Deleuzoguattarian terms, the difference may be seen as one between a subjected group and a group-subject.

At Sinai, the entire people commits itself sans the intermediary of representatives or proxies. Each individual assents in his own voice. This individualist character of the covenant and its explicit reliance on popular consent is insisted upon by rabbinical commentaries on Exodus 19 and 24. As Walzer has it, the covenant reflects the general will of the Israelites—a collective will, in the manner lauded by Rousseau, comprised of a plethora of independent and noncommunicating individual wills.[17] With regard to the line "And all the people answered together" of Exodus 19:8, one rabbinical commentary avers: "They did not give this answer with hypocrisy, nor did they get it one from the other, but all of them made up their minds alike and said: 'All that the Lord hath spoken we will do.'"[18] Moreover, it is also insisted that the consent of the Israelites was not arbitrary. For they had fully understood the requirements of the covenant prior to binding themselves. According to another rabbi referred to by Walzer, "Moses read aloud to the people all of the Torah, that they might know exactly what they were taking upon themselves."[19] Here, in vast contradistinction to God's forcible deliverance of a generation of slaves from the Egyptian despotic regime in which all decisions were made for them by the Mosaic vanguard, the covenant entered into by the newfangled Israelites at Sinai is indelibly marked by a distinctive voluntarism. Formerly God alone decided, now it is the people who decide. This means that they have not merely moved from servitude and deep immersion in servile consciousness toward natural liberty. For the covenent implies that they have in fact proceeded beyond this point and have actually acquired at least a semblance of free will.[20]

Admittedly, their entrance into covenant implies that they have surrendered a degree of natural liberty in being bound to the Mosaic law and the obligations which the latter entails. They have, however, bound themselves of their own volition—and are thus *freely bound.* Moreover, adherence to the Mosaic law cannot be compared to subjection to the infinitely more draconian and despotic law of Pharaoh. For this law is much easier to break, whereas God's taskmasters are far more lenient than Pharaoh's. In fact, the leaders of the Israelites are themselves more likely to disobey the commandments and sometimes even encourage or require disobedience rather than compliance with the ways of Moses. On the positive side, however, the freely bound nature of the Israelites covenanted at Sinai implies that, unlike the incorrigibly volitionless slaves grovelling under the Egyptian yoke of yore, the new generation are henceforth wholly amenable to the gradual process of their own deliverance through collective self-directed labor. They will do this since they aim to fulfill the covenant *not* for the sake of the convenant itself, of duty for duty's sake, but so that God's *promises* may become for them a reality. In other words, after the events of Sinai they resume their march onward to the promised land such that they may attain consummate liberty.[21]

But let us pause. For another possible objection rises up: though admittedly entered into willingly by free men, does not the covenant notwithstanding remain an act of self-subjugation to a transcendent spirit through which even the new generation of Israelites at Sinai fall prey to the Hegelian unhappy consciousness? This objection, in fact, appears once more not to be sustainable. For, as shall be evinced presently, God—as he appears to the Israelites at this moment of their history—does not seem to overly resemble an absolute transcendent spirit. In fact, if anything, his overt "vertical" transcendence with regard to the people is *less* than that of the despot of the despotic regime and his deity. Certainly, he is not as implacable and imperious. Notwithstanding, it would be a mawkish and hokey error to insipidly—and smugly—aver that he is in no sense violent. Yet, this violence is of an entirely different order than despotic, or Statist, violence. It is not the violence of mastery. Quite the opposite: it is a kind of violence through which mastery lays in tatters. This is, of course, not to say that God cannot be co-opted—in distorted form or in grotesque effigy—by subsequent instances of despotic regimes and their despots to the ends of Statist violence and still more heinous impositions of slavery.

As regards this refractoriness of God to both Statist violence and the Hegelian master-slave dialectic, Walzer alludes to the panoply of traditional tales and folk elaborations custered around the Exodus narrative which portray the covenant as a sort of bargain and God as a kind of salesman hawking his commandments across the globe and eventually coming to the Israelites after being rejected everywhere else. Moreover, in certain versions of the events at Mount Sinai, the Israelites have to be persuaded to accept the covenant—and by much more than mere promises. One particular early midrash compares the God of Exodus to a king entering into a province and asking the permission of the people if he may also be their king. After initially being rebuffed and asked by the people if he had done anything to benefit them so as to merit such elevation, he builds a city wall for them, provides a water supply, and fights their battles, and is only then accepted as their king. Likewise, continues the midrashic writer, does God bring the Israelites from out of the Egyptian house of bondage, divide the sea for them, send down manna for them, and fight the battle of Amalek on their behalf, and is only then accepted by them as their king.[22] In this sense, the God of Exodus and related texts is quite unlike the former Nietzschean artist with a look of bronze who swiftly and violently subjugates a comparatively formless and naively apolitical populace over which he henceforth proceeds to exert the full brunt of State terror from the absolute and absolutely paranoid position of despot.

One may find in Jack Miles' book on God another sense in which the God of Exodus differs from the despot's deity. When at the burning bush Moses asks God for his name, God replies with *'ehyeh 'ašer 'ehyeh*. This is the transliteration from the Hebrew that the Jewish Publication Society *Tanakh* gives. The Revised Standard Version bible translates this phrase as "I am who I am." More precisely, *'ehyeh* ("I am") *'ašer* ("who" or "that which") *'ehyeh* ("I am"). However, *'ehyeh* can mean both "I am" or "I shall be." It is ordinarily context that determines which translation to opt for. But God in the Exodus narrative provides no context. Thus, instead of saying that his name is "I am who I am," God could well be revealing his name to be "I am what I shall be." Miles goes on to complicate the matter. The original Hebrew text, in fact, only contains the consonants of God's name: *hyh 'šr 'hyh*. Now, by changing only one letter in the last of the three words, one obtains the archaic Hebrew root *hwh*. It is from the latter that the word *yahweh* is derived. Thus, changing *hyh 'šr 'hyh* to *'hyh 'šr 'hwh*, and thereupon interposing slightly different vowels, one ultimately arrives at the phrase *'ehyeh 'ašer 'ahweh*. What one notes is that the last word *'ahweh* is almost identical to *yahweh*. Now, Miles wagers that the latter word means "become" or "cause to become" or "make happen" or simply "act." Hence, the whole phrase *'ehyeh 'ašer 'ahweh* means "I am what I do." This imples that God is defined primarily by what he does and that he is a mystery that at base unfolds through his actions.[23] It therefore seems that the God of Exodus is a pure *infinitive*, on Mile's reading.

Ernst Bloch is entirely in concord with this ontological characterization of God as he appears in the book of Exodus. For he avers that God's very Being is defined in terms of the *futurum* and as open possibility. He argues that whereas the inscription on the temple of Apollo at Delphi translates as "Thou Art," Yahweh appears before Moses at the burning bush not as an overburdening presence, nor as a dread haunting absence, as a *Deus absconditus*—but as a *becoming*. Yahweh is intrinsically open within Himself to the not-yet.[24] His very mode of Being is the *futurum*.[25] Due to the revelation by the fiery vision on Moriah of itself as a pure becoming, Moses' memory is not henceforth linked with or inundated by a deity who is a tyrannical father-ego and whose throne is swaddled in thick darkness. His God is a redeemer defined in the future tense, a spatial and temporal wandering. This God is an unfixed futurity that is inherently and *ab initio* ahead of itself. Moses' Yahweh is likewise not a violent nature God raining thunderbolts upon mortals below from the astro-mythical constellations. He is rather a symbol and a flag of liberation from servitude, a source of leadership through time with a distant goal. He is born of the sublimation of an earlier God of pure subjugation into the liberating God with whom the Israelites shall covenant. Yahweh ceases to be an omnipotent autocrat making contradictory demands upon subjected serfs below.[26] He is not Aeschylus' Zeus who "is there as the punisher of all who bear themselves too loud and noisily, and his judgment is harsh."[27] He is, in vast contradistinction, an incipient and utopian God of human *revolt*.[28] The Exodus narrative brings to bear the gradual retreat of the notion of God as demiurge and as hypostatic supernal autocrat before God as the Spirit of the Goal.[29] Coming to the fore is a God with no fixed abode, "a God," as Bloch contends, "who is himself not yet what he is: who *is* only in the future of his promise-to-be—if he should keep his word—and in no other way."[30]

That God, as he appears in the Exodus narrative, differs fundamentally from the despot's deity is also rendered evident by Buber's observation that the God of Israel was originally a so-called "god of the way," though dissimilar in character to other gods of the way. A god of the way is one who accompanies and protects wandering nomads and caravans through the wilderness. In Mesopotamia this function was that of the moon, the god "who opens the way." During the time that the Hebrews were subjugated in Egypt this guiding function of their ancient clan God had fallen into desuetude, yet was revived by Moses in Midian when he bethought the possibility of bringing his people out of the Egyptian despotic regime and into liberty. It is here that Yahweh meets with Moses and expresses his intention to resume his former guiding function.[31] The God of Exodus is outside of and refractory to State apparati. As Régis Debray has it, "The Eternal irrupts by way of a wasteland for caravans, 'antisocial' fringes, outskirts where institutions collapse and taxes are not levied. In which plunderers, cattle thieves and reprobates roam."[32] He is a desert God, where the desert serves as "a refrain for dissidence, for monotheistic *desertion*. To be Godsick is to be stricken with the desert."[33] Vasts apart from the inherent xenophobia of the great paranoiac of the despotic regime, God is the paradigmatic celestial xeno*phile*, the "call from elsewhere," who demands his people to "outstrip the narcissism of the sedentary" and to hit the road with "soles of wind."[34] Like God, Moses himself is "the man of movement," of Deleuzoguattarian deterritorialization. Resisting settlement "like a temptation," his prerogative is to "tear himself loose" and engage in "a mystique of wandering."[35]

Thus, we espy another manner in which it may be seen that God diverges in nature from the type of sedentary über-despot that by nature presides as a dictatorial transcendent spirit over a downtrodden populace milling far below. Fundamentally-speaking, he is not proper to the despotic regime, for he is associated with migration, nomadism, lines of flight traversing a parched monochromatic landscape. He, in fact, belongs to an entirely other Deleuzoguattarian regime: the passional, or post-signifying, regime. And within this latter regime, God is nowise transcendent but, quite contrarily, radically immanent.

But there is more. For, as Walzer renders clear, it is also the type of society that the God of Exodus promises to bring into existence—for the new generation of Israelites and with their labor—that differs fundamentally from the regime presided over by the despot. For the promised land is to be the exact opposite of the Egyptian despotic regime. It is to be bereft of domestic bondage, slave labor, the fear of being enslaved by others, oppression, and exploitation. It is to be a land were the rights of the poor are defended and oppressors are punished; a land free of tyrants; a land where justice and freedom prevail. For Moses and the Levites, the overriding goal of the Exodus is the founding of "a kingdom of priests and a holy nation." That this nation is holy consists *primus inter pares* in its rejection of Egyptian bondage. Thus, the oppression of strangers, the denial of Sabbath rest to servants, and the withholding of wages from workers would have to be expirtated in such a nation. This nation would be a kingdom of priests also in the sense that it is a kingdom without a king and hence bereft of Pharaonic despotism and the tyranny of taskmasters. Due to this absence of gross inequalities between tyrants and subjects, and of taskmasters and slaves, the new society would be strictly egalitarian—and thus all Israelites would be priests and the nation itself holy.[36]

The recalcitrance to tyranny, exploitation, and slavery of the Exodus narrative continues and runs through the prophets Amos to Isaiah: "Wash ye, make you clean, put away the evil of your doings from before my eyes, cease to do evil, learn to do well, seek judgment, relieve the oppressed, judge the fatherless, plead for the widow" (Isa. 1.17). Exploiters and expropriators are an abomination to Yahweh: "and he looked for judgment, but behold oppression; for righteousness, but behold, a cry. Woe unto them that join house to house, that lay field to field, till there be no place, that they may be placed alone in the midst of the earth" (Isa. 5.7). Serfdom must be categorically extirpated: "And I will punish the world for their evil, and the wicked for their iniquity; and I will cause the arrogance of the proud to cease, and will lay low the haughtiness of the terrible. I will make a man more precious than fine gold; even a man than the golden wedge of Ophir" (Isa. 13.11–12). The oppressed will be liberated and tyrants smashed: "But thus saith the Lord, 'Even the captives of the mighty shall be taken away, and the prey of the terrible shall be delivered: for I will contend with him that contendeth with thee, and I will save thy children. And I will feed them that oppress thee with their own flesh, and they shall be drunken with their own blood, as with sweet wine: and all flesh shall know that I the Lord am thy Saviour and they Redeemer, the mighty One of Jacob'" (Isa. 49.25–26).

2. St. Paul

As observed, the liberation of the Israelite slaves from Egyptian bondage entailed two necessary radical discontinuities. Firstly, there was the spatial discontinuity of the Mosaic vanguard who, though arising from the ranks of the masters, breaks with them—and may thus be seen as having moved outside the confines of the hitherto prevalent master-slave hierarchy. This spatial discontinuity will in fact be repeated when Moses separates himself from the Israelites at Mount Sinai at the time of the Revelation. The external position of the Mosaic vanguard and that of Moses at Sinai gives the initial forcible liberation of the Israelite slaves, and likewise the dramatic events just prior to the covenant, the quality of an exposure to a pure outside, to the violence of divine grace—which is of the order of the event (the event of the initial Deliverance and that of the Revelation). A truth-event, as Alain Badiou defines it, is of the order of what occurs. It is radically singular, and absolutely indifferent to the state of the situation upon which it impinges, and likewise indifferent to any generality, extant Law, and—above all—refractory to the State.[37]

Secondly, we had the temporal discontinuity of the gap between the old slavish generation and the new freeborn generation effected by the forty years of nomadic destinerrancy—whereby the servile generation of Israelites largely died in the upsurge of a new collective Israelite subject heterogenous to the distinction between masters and slaves. The actual upsurge of the new collective subject, of course, only occurs with the covenant contracted in the wake of the aforementioned second spatial discontinuity (the event of Revelation). This is so since, as

Badiou has it, the subject of a truth-event only arises with a declaration testifying to a conviction relative to the said event.[38] The covenant entailed precisely this sort of subjective declaration, in the form of the assent of the Israelites at Mount Sinai. Viewed more succinctly, the two spatial discontinuities and the single temporal discontinuity effectively give us the following couplet: a truth-event and a pure subjective upsurge.

This last formula also defines St. Paul's problematic. Here the founding truth-event is the Resurrection. The Resurrection of Christ is a veritable event in that—being totally incalculable—it is irreducible to the state of the situation (of which much more later in our text), to the State, and to any Statist Law. This *ur*-event, however, only occurs for a potential subject arising in its wake if the self anterior to this new subject is inundated by an event that mimics it. In other words, the Resurrection-event only occurs for a potential subject if it happens to the latter as a pure encounter such as that of St. Paul on the road to Damascus. That is to say, this *ur*-event occurs for the self as the event of conversion. The latter marks the upsurgence, or re-surgence (resurrection), of the Christian subject. This emergence of the Christian subject is the Resurrection of Christ as the subject's resurrection within the self exposed to the Resurrection-event. The said subjective upsurge occurs only with the incipient subject's declaration of the Resurrection-event. In a manner homologous to the covenant at Sinai whereby the collective Israelite subject effected its own upsurgence or resurgence by declaring the Revelation-event (and by implication the still earlier Deliverance-event), the nascent Christian subject effects his own emergence or resurrection by declaring Christ risen from the dead.[39]

As observed, it is through the covenant that a collective Israelite subject outside the dichotomy of mastery and slavery emerges. For the Israelites at Sinai are not the same as the previous generation of slavish Israelites who left Egypt, whereas the newfangled Mosaic law and the utopian vision of the promised land are both alike shot through with an egalitarianism fundamentally refractory to Pharaonic despotism, tyranny, and oppression. Though it is true that when the Israelites at length become sedentary and erect for themselves a regime whose despotism shall in certain ways resemble that of Pharaoh, at the time of this initial liberated surge, Mosaic law and practice retains a relatively non-Statist character.[40]

By comparison, St. Paul's activities in the wake of the Damascus-event are still more markedly anti-Statist. For after it he does not go to see the authorities, the institutional apostles who knew Christ. Rather, he leaves his subjective upsurge outside of every official seal. He turns away from all authority, apart from the Voice that had personally summoned him, and leaves for Arabia for three years to repeatedly declare the Resurrection-event. His second period of wandering through Cilicia, Syria, Turkey, Macedonia, and Greece lasts a staggering fourteen years. This ex-centric, migratory, and nomadic character of St. Paul's activity likewise evinces his refratoriness to the trappings of despotism—which latter bases itself upon the topological centrality of an overarching State apparatus, however large the coefficient of deterritorialization of the despotic sign.[41] In a later chapter, we shall encounter an analogous quasi-anarchic—and likewise extralegal—nomadism in the seemingly interminable peregrinations of Kafka's protagonists. In any case, in this sense, Pauline nomadism—like the vertiginous peregrinations of the Exodus—entails movements of deterritorialization that have become absolute. This absolute deterritorialization, in the precise Deleuzoguattarian sense, is endemic to the upsurge of a subject of a truth-event, for such a subject constitutes a necessary and inescapable distance from the State.[42] Cutting diagonally across the State, St. Paul is inexorably forced to conceive of himself as the leader of a subversive party, faction, or vanguard.[43]

Due to the irremissibly Statist character of the Law, the post-evental subjectivity of St. Paul is likewise necessarily at odds with the latter. In the wake of the Damascus-event he is lucidly aware that the Mosaic Law (which had itself already ossified and acquired a distinctly Statist character) and the Roman Law—and indeed all Statist Law as a whole, ceases to be tenable. Not only is it untenable; it has become a figure of death for the ascendant truth issuing from the anterior Resurrection-event. In relation to this, and in vast contradistinction, the evental declaration is the truth-event's principle of life.[44]

The truth-event is *stricto sensu* illegal, and produces no (literal) Law. Unlike the Law—with its Statist and thus necessarily despotic character, the truth-event is refractory to any form of mastery.[45] This is so because the State is the ambit proper to the inauguration and unfolding of the Hegelian master-slave dialectic, and moreover the arena and machinery within which even the anti-State God of Exodus accrues an extraneous imperiousness and vertical transcendence—paving the way for his unwilling installation as the despot's deity and thus as the masterful and immobile spirit underpinning the Hegelian unhappy consciousness.

There is also a particular *discourse* organised under the auspices of the State. It is the discourse of the *Father*. This discourse serves to bind communities in a form of obedience to the Empire, to the despot's deity, and to the Law. It is a discourse of mastery. In contradistinction, the discouse ushered in by the Pauline declaration of the Resurrection-event is a discourse of the *Son*. It signifies God's sending of his Son as a radical external intervention in history, a rupture through which any discourse laying claim to the form of mastery finds itself undermined.[46]

It is especially the movement toward a preponderance of weakness over strength that effects the dissolution of mastery, in accordance with the Pauline formula contending that "the foolishness of God is wiser than men, and the weakness of God is stronger than men" (1 Cor. 1.25). Mastery, together with every one of its figures, is effectively expirtated.

For Christ is not a master in any Hegelian sense. He is certainly "lord" (*kurios*), and St. Paul his "servant" (*doulos*). Yet, what the subject serves is the truth procedure attendant upon the Christ-event, and not any figure of the master. Slavery under a master is precisely that from which the faithful service of Christ releases one. For the relation between lord and servant is absolutely refratory to that between owner and slave, and even that between master and disciple. As Badiou avers: "It is not a relation of personal, or legal, dependence. It is a community of destiny in that moment in which we have to become a 'new creature.'"[47]

Under the pressure of the Christ-event, the subject—formerly a slave under the discourse of the Father—finds himself *relieved* of the Law and becomes a *son.* The truth-event gives rise to a universal becoming-son wherein the dichotomy of masters and slaves gives way to filial equality.[48]

The subject who declares a truth-event does not enter into the logic of mastery, for he does not attest to any lack that would demand fulfilment by the figure of a master. He remains refractory to this logic, since the subject of an event—as opposed to a disciple-subject—is one for whom nothing is lacking. He lacks nothing since he is a pure emergence and thus nothing but *beginning.* His position is that of the son and is only born as such out of the event. In declaring an event, he becomes the son of the event. God himself likewise retreats from the logic of mastery by assuming the form of a son and hereby renders the son-subject possible. By consenting to the figure of the Son, God precipitates the universal proliferation of sons. "So through God you are no longer a slave but a son, and if a son then an heir" (Gal. 4.7). As the resurrected Son, God himself exposes the figure of mastery as a fraud and filiates all of humanity. He deposes the master and founds upon the latter's demise the equality of sons. These sons become *theousunergoi*, God's coworkers. The demise of the master gives way to an equality of workers engaged in an egalitarian endeavor which elaborates the effluences of the truth-event out of which they are born.[49]

Notes

1. Walzer 1985, 21–25.
2. Ibid. 24, 25.

3. Ibid. 21, 22.
4. ibid. 66.
5. Buber 1958, 35–38.
6. Taylor 2007, 42.
7. Walzer 1985, 66, 67.
8. Ibid.
9. Cited in Walzer 1985, 67.
10. Walzer 1985, 67, 68.
11. Buber 1958, 75–77.
12. Walzer 1985, 68.
13. Ibid. 68–70.
14. Taylor 2007, 73, 74.
15. Walzer 1985, 73–75.
16. Ibid. 76.
17. Ibid. 79, 80.
18. Cited in Walzer 1985, 80.
19. Cited in Ibid.
20. Walzer 1985, 80, 81.
21. Ibid. 97, 98.
22. Ibid. 76, 77.
23. Miles 1995, 98, 99.
24. Bloch 1972, 56.
25. Ibid. 81.
26. Ibid. 92–94.
27. Cited in Bloch 1972, 94.
28. Bloch 1972, 94.
29. Ibid. 95.
30. Ibid.
31. Buber 1958, 125, 126.
32. Debray 2004, 38, 39.
33. Ibid.
34. Ibid. 84.
35. Ibid. 92.
36. Walzer 1985, 101, 105, 107, 108, 109.
37. Badiou 2003, 14, 15.
38. Ibid. 14.
39. Ibid. 17, 18.
40. Ibid. 18, 19.
41. Ibid.
42. Ibid. 15.
43. Ibid. 21.

44. Ibid. 27.
45. Ibid. 42.
46. Ibid. 43.
47. Ibid. 63.
48. Ibid. 48, 49.
49. Ibid. 59, 60.

9

Exit Strategies from the Structures of Despotism, 2. Against Nietzsche

Job and St. Paul

1. The defiant Job: no Hegelian slave, no man of bad conscience

Despite Nietzsche's rather formidable diatribe, one may find a distinct counter-example within the Judeo-Christian scriptural ambit at once entirely refractory to the maladies of both *ressentiment* and bad conscience which Nietzsche imputes to the man who suffers.

The narrative is a well known one. Its interpretations, however, vary quite significantly. At the wily and saurian instigation of Satan, God permits Job to be cast into catastrophic destitution. His livestock are stolen, his servants slain, and his children die (Job 1.11–19). Yet, though knowing that this his calamitous suffering is ultimately the doing of God, Job's heart is in no sense rankled by *ressentiment*. Albeit given to mourning, as evinced when he rends his mantle and shaves his head, he otherwise maintains a stoical attitude to his suffering and does not accuse God for it (Job 1.20, 22). Falling to the ground and worshipping, he says:

> Naked came I out of my mother's womb,
> and naked shall I return thither:
> the Lord gave, and the Lord hath taken away;
> blessed be the name of the Lord. (Job 1.21)

Hereupon, however, and again at the instigation of Satan, God permits the Prince of Darkness to afflict Job with sore boils over his entire body, causing him to ceaselessly scrape himself with a potsherd whilst sitting down among ashes. Seeing all of this, his wife—rankled to the core by both God and Job alike—orders him to curse God and die. Job's comportment and retort is once more free of *ressentiment* of God and his machinations, and once more strictly stoical: "Thou speakest as one of the foolish women speaketh. What? Shall we receive good at the hand of God, and shall we not receive evil?" (Job 2.4–10).

Yet, though we have here described his equanimous comportment in the face of unbearable suffering as stoical, Job is strictly speaking not a stoic at all in the sense ascribed to the term by Hegel. In this connection, Antonio Negri, though he describes Job as a sort of "protostoic," contends that he is ultimately not a stoic at all. For, as we shall see repeatedly in what follows, there is not a smidgen of the Hegelian stoic slave's resignation in Job. Moreover, Job lacks the most fundamental aspect of stoicism; that is, the separation of ethics from ontology, the conviction that ethics transpires in spite of the vicissitudes of Being. Otherwise put, unlike the Hegelian stoic slave, Job does not consider the realization of his freedom to be indifferent to the content of external reality. This means that, for Job, ethics is inextricable from Being. In fact, for Job ethics *is* Being. That is to say, in Hegelian terms, for Job ethics must be concretely embodied in the real itself.[1]

Now, returning to the narrative of the book of Job, soon enough, upon hearing of Job's misfortune, his three friends, Eliphaz the Temanite, Bilbad the Shuhite, and Zophar the Naamathite, come to mourn with him, weeping, rending their mantles and sprinkling dust on their heads. They sit down with him for seven days and nights in sullen silence (Job 3.11–13).

Hereupon, the preachings and unbraidings of Job by his three interlocutors swiftly ensue. Job, however, retorts with polemical defiance.

Firstly, beholding the dire and calamitous dereliction into which Job has been plunged and having lent ear to his lamentations, the more salient part of the interlocutor Eliphaz's rationalizations of Job's plight are as follows:

> Remember, I pray thee, who ever perished, being innocent?
> or where the righeout cut off?
> Even as I have seen, they that plough iniquity,
> and sow wickedness, reap the same.
> By the blast of God they perish,
> and by the breath of his nostrils are they consumed.
> The roaring of the lion, and the voice of the fierce lion,
> and the teeth of the young lions are broken.

> The old lion perisheth for lack of prey,
> and the stout lion's whelps are scattered abroad. (Job 4.7–11)
> ...
> I have seen the foolish taking root:
> but suddenly I cursed his habitation.
> His children are far from safety,
> and they are crushed in the gate,
> neither is there any to deliver them.
> Whose harvest the hungry eateth up,
> and taketh it even out of the thorns,
> and the robber swalloweth up their substance. (Job 5.3–5)
> ...
> He disappointeth the devices of the crafty,
> so that their hands cannot perform their enterprise.
> He taketh the wise in their own craftiness:
> and the counsel of the froward is carried headlong.
> They meet with darkness in the daytime,
> and grope in the noonday as in the night. (Job 5.12–14)

Eliphaz, though resorting to generalities, is indirectly upbraiding Job, conjecturing that his numerous sufferings are the result of prior sins that he has committed. Hence the words "iniquity," "wickedness," "craftiness," and "froward." In Nietzschean terms, Eliphaz exhibits himself in his discourse as a man whose *ressentiment* has burgeoned and extended itself into bad conscience. For the man of bad conscience, moving beyond the resentful man's mere accusation of the other, now comes to employ the strategy of insidiously inducing the other to accuse himself. This is the psychical and ethical disease that Eliphaz is striving to impart unto Job.

The purveyor of bad conscience does so by attempting to inculcate the idea in the other than his pain is purely the consequence of his sinfulness. He strives to induce in his adversary the conviction that his suffering is a divine punishment for a guilt-riddled fragment of his past.

As Negri may be seen as contending about the above polemic, Eliphaz—as Job's other interlocutors shall likewise proceed to do, cognizes ethics in terms of the dogma of retributive justice. That is to say, it is conceived—in a kind of Calvinist manner—as justification through work.[2] But Job will have nothing of this theory of individual retribution. He argues that it does not hold water from any point of view, in that there is in truth no ethical measure. In fact, he will at a later point move to vociferate that his interlocutors' incessant dialectical efforts to demonstrate that justice functions in accordance with a "just measure" are not merely false, but also deceptive. Job's rebuttal of the erroneous thesis of retributive justice is as follows:

> That the wicked is reserved to the day of destruction?
> they shall be brought forth to the day of wrath.
> Who shall declare his way to his face?
> and who shall repay him what he hath done?
> Yet shall he be brought to the grave,
> and shall remain in the tomb.
> The clods of the valley shall be sweet unto him,
> and every man shall draw after him,
> as there are innumerable before him.
> How then comfort ye me in vain,
> seeing in your answers there remaineth falsehood? (Job 21.30–34)

Therefore, since the wicked are not necessarily punished whereas the innocent are often subjected to intolerable suffering, Negri concludes that value, labor, and justice cannot be apportioned according to a common measure. In the foregoing diatribe, Job has just revealed this commonality to actually be entirely meaningless. Still further, he in fact protests that the retributive model of justice amounts to pure and unadulterated charlantry.[3]

This is *inter alia* the reason why Job will have nothing of Eliphaz's accusations. He proceeds without delay to berate Eliphaz for his cruel words. He refuses to give into his skulduggery, and will not be cajoled into the latter's attempted imputation of bad conscience to his person: "you dig a pit for your friend" (Job 6.27) Indeed, Job moves to vehemently protest his innocence, fulminating that his present sufferings have nothing to do with any prior sins that he may have committed: "my righteousness is in it./Is there iniquity in my tongue? cannot my taste discern perverse things?" (Job 6.28–29). After all, did not God say to Satan: "Hast thou considered my servant Job, that there is none like him in the earth, a perfect and upright man, one that feareth God, and escheweth evil"? (Job 1.8). Job to Eliphaz:

> My brethren have dealt deceitfully as a brook,
> and as a stream of brooks they pass away;
> which are blackish by reason of the ice,
> and wherein the snow is hid:
> what time they wax warm, they vanish:
> when it is hot, they are consumed out of their place. (Job 6.15–17)
> …
> Teach me, and I will hold my tongue:
> and cause me to understand wherein I have erred.
> How forcible are right words!
> but what doth your arguing reprove?
> Do ye imagine to reprove words,

> and the speeches of one that is desperate, which are as wind?
> Yea, ye overwhelm the fatherless
> and you dig a pit for your friend.
> Now therefore be content, look upon me;
> for it is evident unto you if I lie.
> Return, I pray you, let it not be iniquity;
> yea, return again, my righteousness is in it.
> Is there iniquity in my tongue?
> cannot my taste discern perverse things? (Job 6.24–30)

Hereupon, Bildad the Shuhite—another man of *ressentiment*—interjects and, much in the same way as Eliphaz, strives to pollute Job's soul with the sickness of bad conscience. Again, Job's present suffering is traced back to his purported prior sinfulness—and likewise even that of his offspring:

> Doth God pervert judgment?
> or doth the Almight pervert justice?
> If thy children have sinned against him,
> and he have cast them away for their transgression;
> If thou wouldst seek unto God betimes,
> and make thy supplication to the Almighty;
> If thou wert pure and upright;
> surely now he would awake for thee,
> and make thee habitation of thy righteousness prosperous. (Job 8.3–6)

Again, Job protests against these attempted imputations of bad conscience. Over and above this, he renders privy to Bildad that he will address his protestations of innocence to none other than God himself:

> My soul is weary of my life;
> I will leave my complaint to myself;
> I will speak in the bitterness of my soul.
> I will say unto God, Do not condemn me;
> show me wherefore thou contendest with me.
> Is it good unto thee that thou should oppress,
> that thou shouldst despise the work of thy hands,
> and shine upon the counsel of the wicked?
> Hast thou eyes of flesh?
> or seest thou as man seeth?
> Are thy days as the days of man?
> are thy years as man's days,
> That thou enquirest after my iniquity,
> and searchest after my sin?

> Thou knowest that I am not wicked;
> and there is none that can deliver out of thy hand. (Job 10.1–7)

Here, Job not only announces his intention to confront God so as to express his bewilderment as to why the Almighty condemns him as guilty of prior sins as the antecedent cause of his present suffering. He, moreover, declares to Bildad his resolve to ask God whether he is not perhaps judging him as men judge—and specifically men of *ressentiment*, who ascribe sinfulness where it is absent.

Yet, Job's interlocutors will not cease to harangue and calumniate his character. Zophar the Naamathite is the next to ascribe fictitious guilt to Job, debunking Job's estimation of himself as pure and righteous:

> For thou hast said, My doctrine is pure,
> and I am clean in thy eyes.
> But O that God would speak,
> and open his lips against thee;
> And that he would show thee the secrets of wisdom,
> that they are double to that which is!
> Know therefore that God exacteth of thee less than thy iniquity deserveth.
> (Job 11.4–6)

Eventually, Job grows sick and tired of his three interlocutors' supposed counsel. He will have nothing of it, and moves to unequivocally denounce their self-righteous sense of haughty superiority and lack of moral probity—clouded and perverted as it is by their festering *ressentiment.* He accuses them of being liars and hypocrites, and declares his unswerving conviction that God will justify him:

> Lo, my eye hath seen all this,
> mine ear hath heard and understood it.
> What ye know, the same do I know also:
> I am not inferior unto you.
> Surely I would speak to the Almighty,
> and I desire to reason with God.
> But ye are forgers of lies,
> ye are all physicians of no value.
> O that you would altogether hold your peace!
> and it should be your wisdom. (Job 13.1–5)

On his appearance before the court of God, he contends:

> Though he slay me, yet will I trust in him:
> but I will maintain mine own ways before him.

> He also shall be my salvation:
> for a hypocrite shall not come before him.
> Hear diligently my speech,
> and my declaration with your ears.
> Behold now, I have ordered my cause;
> I know that I shall be justified. (Job 13.15–18)

Yet, Zophar—in a slightly later diatribe—holds these declarations by Job of his certain future salvation and justification in contempt. He, in fact, moves to vociferate that Job's sufferings for his purported prior and present sinfulness can only increase. Again, though he speaks of the dire plight of the sinner in generic terms, it is Job that he clearly has in mind:

> Knowest thou not this of old,
> since man was placed upon the earth,
> that the triumphing of the wicked is short,
> and the joy of the hypocrite but for a moment?
> Though his excellence mount up to the heavens,
> and his head reach unto the clouds;
> yet he shall perish for ever like his own dung.... (Job 20.4–7)
> ...
> His bones are full of the sin of his youth,
> which shall lie down with him in the dust.
> Though wickedness be sweet in his mouth,
> though he hide it under his tongue;
> Though he spare it, and forsake it not;
> but keep it still within his mouth:
> Yet his meat in his bowels is turned,
> it is gall of asps within him. (Job 20.11–14)
> ...
> He shall suck the poison of asps:
> the viper's tongue shall slay him. (Job 20.16)
> ...
> He shall flee from the iron weapon,
> and the bow of steel shall strike him through.
> It is drawn, and cometh out of the body;
> yea, the glittering sword cometh out of his gall:
> terrors are upon him. (Job 20.24–25)
> ...
> The heaven shall reveal his iniquity;
> and the earth shall rise up against him. (Job 20.27)
> ...
> This is the portion of a wicked man from God,
> and the heritage appointed unto him by God. (Job 20.29)

After the initial man of *ressentiment*, the polemically indefatigable Eliphaz, makes his final desperate attempt to infect Job with the festering malady of bad conscience, Job still more forcefully expresses his resolve to bring his case to God—before whom he is utterly convinced that he will be vindicated:

> I would order my cause before him,
> and fill my mouth with arguments.
> I would know the words which he would answer me,
> and understand what he would say unto me.
> Will he plead against me with his great power?
> No; but he would put strength in me.
> There the righeous might dispute with him;
> so should I be delivered for ever from my judge. (Job 23.4–7)
> …
> But he knoweth the way that I take:
> when he hath tried me, I shall come forth as gold. (Job 23.10)

Indeed, in an earlier retort to Bildad's dystopian prognosis of his inevitable demise in punishment of his purportedly having sinned, Job paints a utopian picture of what he believes will ultimately be his redemption from his present desuetude (Job 19.25–27).

Negri views the above declaration of hope as a fervid conviction on the part of Job that justice will be founded in a radically new form, that through the living God an alternative order of values will be founded—one utterly refractory to the measure of this world of injustice and pain. There shall transpire a transformation in and of the world. Out of Job's state of extreme prostration and his cry of pain and faith, exploited labor shall yield to the resurrection of the flesh.[4] This movement shall at once be that from the justification of pain—in the form of bad conscience's and retributive justice's sourcing of suffering in prior or present sinfulness, to the practical *transfiguration of pain*. Herein entailed is the associated realization by Negri that there is no pre-existing value, but merely the possibility of creating value. In the same trajectory, there emerges the realization that there is no judge as such, but rather merely the possibility of practicing justice.[5]

Thus, Job goes beyond the view of pain as the poisonous fruit of sin to its transmogrification. In this connection, Negri asseverates that this "beyond" is what appears in the book of Job as the "Messiah." The Messiah is a kind of divine conciliator who, before God, will take upon himself the burden of the defense of Job's being unjustly punished for non-existent past iniquities. Job:

> Also now, behold my witness is in heaven,
> and my record is on high.

> My friends scorn me:
> but my eye poureth tears unto God.
> O that one might plead for a man with God,
> as a man pleadeth with his neighbour! (Job 16.19–21)

This Messiah, Job is convinced, will redeem him and moreover allow him to finally directly behold God face to face:

> For I know that my redeemer liveth,
> and that he shall stand at the latter day upon the earth:
> And though after my skin worms destroy this body,
> yet in my flesh shall I see God:
> whom I shall see for myself,
> and mine eyes shall behold, and not another;
> though my reins be consumed within me. (Job 19.25–27)

In relation to the above, Negri maintains that through the Messiah man is able to come to experience his relationship with God outside all determination, that is, outside of all teleology and outside of all preordained purpose. In this way, the power of man is able to escape from the encroachment of the Power of the Deity.[6]

In particular, it is man's labor that, via the intervention of the Messiah, can escape the law of measure—a measure imposed upon human labor solely and exclusively by the overarching Power of the Deity. This measure foisted upon labor by divine Power Negri avers to be that of the exploitation of labor. However, through the mediation of the Messiah, labor can become value without measure, that is, autonomous human power. Hereby, labor is enabled to struggle effectively against exploitation—in which labor is humiliating, promethean and destructive *corvée*—and embark on the path toward becoming an expansion of free activities engaged in by men who create the world directly as equals.[7]

The Messiah, moreover, gives a divine sense to human labor, raising it up to spirit. The decisive passage in this regard is: "For I know that my redeemer"—that is, the Messiah—"liveth,/and that he shall stand at the latter day upon the earth:/and though after my skin worms destroy this body,/*yet in my flesh I shall see God* … [my emphasis]" (Job 19.25–26). This is what Negri maintains to be a resurrection of the flesh through the Messiah, in which the body tormented by a labor subject to divine Power's measure is modified and ennobled by a labor whose value henceforth lies outside of measure—a labor that has now become immeasurable human power.[8]

We know that most of the book of Job is about the latter's suffering and pain. Yet, as Negri goes on to declare, far from leading inevitably to a state of terminal demise, Job and man's suffering, through the Messiah, ceases to be the price of absolution from existent or non-existent sin, but serves instead as the path leading to salvation. Pain is that through which man struggles against expropriation at the hands of Power and refuses exploitative labor. It is, moreover, the foundation and of a democratic political society consisting of men fraternally bound in filial equality. The transfiguration of pain through the figure of the Messiah is therefore that by which the hierarchy of masters and slaves is subjected to a progressive dissolution. In vast contradistinction, the bifurcation of society into masters and slaves, as already seen somewhat earlier apropos of Hegel, is inaugurated and sustained not by pain but by *fear*.[9]

That is to say, due to his fear of being physically annihilated by the master, the slave submits to and comes to be well-nigh interminably subjected to a life of forced labor for his sybaritic and dialectically inert overlord. Now, if Negri's contention that Job is *stricto sensu* not a stoic slave holds—that is to say, that he does not separate ethics and freedom from ontology (meaning: from the content of external reality), then he cannot possibly move onto Hegelian skeptical consciousness and the subsequent three phases of the unhappy consciousness. This is so, because skeptical consciousness can only be arrived as a dialectical outcome of stoicism—whereas the unhappy consciousness is, in turn and similarly, a dialectical result of skeptical nihilism. Job is in pain, but—as evinced by his indefatigable audacity—he fears no figure of the master. He, therefore, does not even pass through *any* of the vicissitudes of the Hegelian master-slave dialectic.

Now, as observed in an earlier discourse, Job has expressed his unswerving resolve to bring his case before God. So, he does.

Notwithstanding, when Job eventually does come to stand before God and protests his innocence, God discourses not on ethics but holds forth at great length on his immense physical might (with which he is able to subdue gigantic monsters such as Behemouth and Leviathan), his gargantuan creative capacity (through which he had created the entire universe and all the creatures in it), and his unlimited omniscience and timelessness. In the wake of this, God inquires of Job whether he is within his wits to contend with, instruct, and reprove the Almighty (Job 38.1–40.2).

In this connection, Jack Miles avers that what God is doing here throughout his thunderous sermon uttered from out of the whirlwind is subsuming his justice to his Power. Through his thundering at him from out of the whirlwind—whose surging and crashing power Miles compares to the bludgeoning percussiveness of Stravinsky's *Rite of Spring*—God wishes to impress upon Job that might makes right.[10] As Miles goes on to maintain, what Job deals with in the whirlwind episode is a God who appears to have *erased his more recent history*. In the book of Exodus through to the Psalms, he was a God of liberation, revolt against despotic oppression, compassion for the suffering, and so forth. Now, he seems to have recurred to being the immoral Deity of the Flood.[11]

In a very similar vein, Ernst Bloch avers that the whirlwind theophany experienced by Job is so utterly foreign to the Bible that it is as if it were that of a completely different God: one that has nothing in common with Yahweh even in the latter's most draconian comportment. He is reminescent, rather, of Isis or of Baal. Here, it appears as though a regression has been made into the abyss of pre-prophetic and pre-Canaanite demonism.[12] Indeed, the Yahweh of the finale speaks to Job in the manner of a nature-demon.[13] Bloch, in fact, goes as far as to maintain that the God who appears to Job is an oppressive celestial Pharaoh.[14] With him, Jewish history appears to have gone backwards from Canaan to Egypt.

In the face of this overbearing display of infinite greatness on the part of God, Job's reply appears singularly self-debasing and defeatist:

> Behold, I am vile; what shall I answer thee?
> I will lay my hand upon my mouth.
> Once have I spoken; but I will not answer:
> yea, twice; but I will proceed no further. (Job 40.4–5)

Here, it seems evident that Job has finally contracted the disease of bad conscience upon the shocking realization of the overwhelming superior power of God as compared to his comparative impotence. He castigates his own character and proceeds to repent the purported iniquities that, through the *quid pro quo* retributive conception of justice tirelessly intimated to him by his human interlocutors, had precipitated him into his intolerable suffering in the first place. So it would seem.

But this is, in fact, not the case at all. For, as Miles explains, this response of Job to the Lord in fact bristles with ironic double entendres and a mock-deferential stance in relation to the very divine words which are aimed at intimidating him. Far from what would otherwise appear to be the case, Job concedes *nothing*. "What shall I answer thee?" is an evasion. "Once have I spoken;/Twice, but I will proceed no further" are utterances that defy God's demand that Job acknowledge his gargantuan Power.[15]

In fact, even in his being at length upbraided and subjected to the fierceness of the Lord, Job remains intransigent—stubbornly and indefatigably protesting his innocence, much as he had done in his earlier speeches to his human interlocutors. God has said and done nothing whatsoever to change Job's view of himself as free of sin. For Job has not deviated an inch from doggedly maintaining his present suffering to be the result of a wholesale perversion of justice and naught besides.[16] So, even the Almighty has failed to transmogrify Job into a man of bad conscience.

Yet, Job's polemical duel with the Deity does not cease here. For it moves to a final denouement in which Job actually comes to "see" God with his own eyes:

> ... I have heard of thee by the hearing of the ear:
> but now mine eye seeth thee.
> Wherefore I abhor myself,
> and repent in dust and ashes. (Job 42.5–6)

This "seeing" of God's utter superiority appears to, at least on a superficial reading, induce Job to contract the malady of self-hatred characteristic of Nietzschean bad conscience and, equivalently, the masochistic self-debasement endemic to the third and last phase of the Hegelian unhappy consciousness.

But we may be wrong yet again. For, another interpretation of this passage, and especially of the line "but now my eye seeth thee," is possible. This other reading is that of Negri. He maintains that Job's seeing of God tears God from his absolute transcendence. Moreover, that God finds it obsessively necessary to justify himself in a virtually interminable diatribe would seem to suggest that God is not God. In Nietzschean terms, God would appear to be dead.[17]

I would prefer to aver that God's protesting too much in his stentorian tirade from out of the whirlwind evinces, rather, that he is closer to the pole of immanence than to that of transcendence. This would imply that the purportedly astronomical and unbridgable gap separating God from man—as we have already observed regarding the closing of the hiatus between the Father and human beings through God's incarnation as the God-Man Christ and the consequent Resurrection alluded to by Badiou's St. Paul—is here shrunk to the point of a significant visual and aural propinquity.

Indeed, for Negri, all of this implies that Job, since he is now able to see and speak directly to God, comes to partake in divinity. In this immersion of Job in the divine substance, human power—instead of being subjugated to the transcendent Power of God in the form of the latter's imposition of measure on value and the entailed exploitation of labor—finds itself ontologically linked to the creative Power of God. This ontological symbiosis of human and divine potency vastly enhances man's capacity to create society through a now immeasurable labor freed from exploitation. This is the redemption effected by the pain-riddled, yet fearless, struggles of Job.[18] Job's "seeing" and consequent immanentization of the On-High is an explosion of intense light that renders life *powerful*. It activates action's potential energy. Job's "seeing" is an act which develops human power beyond the pain of existence.[19] It returns human existence and labor to the side of activity.

2. St. Paul

The becoming-reactive of force; the ubiquity of an all-pervading negative will to power; the accusation by reactive force of active force, of all activity, and the affirmative life itself; active force co-opted to the sole end of a self-impregation of pain, which suffering is made the immediate consequence of sin and sole means of its assuagement; and the movement of the center of gravity of the affirmative life, of life itself, to the Beyond of life, such that life may subsist only under the cipher of the negative and of death ... There would seem to be no way out of this terminal quagmire. All exits appear to be inexorably blocked. So it would seem.

As is known, Nietzsche attributes the transition to this stage of nihilism (negative nihilism) to St. Paul—that venomous priest who turned "Judaic" *ressentiment* inward so that the bad conscience could be born, who invented sin so that man (in an exquisitely masochistic orgy of self-accusation) would torment himself, and who used the belief in immortality so that life would henceforth suffocate and be mortified under the sempiternal death sentence passed on it by the Beyond.

According to this allegation, St. Paul emerged onto the scene as "the chandala hatred against Rome, against 'the world', become flesh, become genius, the Jew, the *eternal* wandering Jew par excellence."[20] In him was embodied "the genius in hatred, in the vision of hatred, in the inexorable logic of hatred."[21] He is purported to have "comprehended that he *needed* the belief in immortality to deprive 'the world' of value, that the concept of 'hell' would become master even over Rome—that with the 'beyond' one *kills life*."[22] For when "one places life's centre of gravity not in life but in the 'beyond'—*in nothingness*—one deprives life of its centre of gravity altogether. The great lie of personal immorality destroys all reason, everything natural in the instincts—whatever in the instincts is beneficient and life-promoting or guarantees a future now arouses mistrust."[23] And thus St. Paul is alleged to have ensured that "out of the *ressentiment* of the masses" Christianity "forged its chief weapon against *us*, against all that is noble, gay, high-minded on earth, against our happiness on earth. 'Immortality' conceded to every Peter and Paul has so far been the greatest, the most malignant, attempt to assassinate *noble* humanity."[24]

Certainly, at least in the view of Ernst Bloch, such an attempted assassination of humanity from and by an absolutely transcendent Beyond would seem to be the intention of Karl Barth. The maligning of man by Barth stems from the exaggerated transcendence that he attributes to God.[25] Turning God into an idol of the mythical On-high, Barth simply defines the former as "transcendence pure and simple."[26] He demarcates the sharpest, most abrupt, and most impenetrable boundary between man and this unreal divinity. He raises the heteronomousness of the Deity to such an ultimate degree that the correlation between the present world and the one to follow is next to null. Indeed, for Barth, the only possible point at which the paths of God and man intersect is the grim and shattering Moment. This is the only purported manner in which God, even in Christ, may enter the world and history.[27] Even here, the evanescent entry of God into history assumes the form of a mere point of intersection: "In Christ, the plane of worldly reality is cut vertically through by the plane of divine reality."[28]

Moreover, this encounter in the fleeting moment is invariably violent: God's entry into the world and human history occurs only as a tremendous blow from above, as a "bomb crater." Barth even refuses to draw the clear distinction made by Luther between *Deus absconditus* (God insofar as he is unknowable to us) and *Deus revelatus* (God insofar as he renders himself knowable to man through his word). Instead, *Deus revelatus* tends to be subsumed as much as possible by *Deus absconditus* so as to arrive at *Deus totaliter absconditus.* Within the same strategy of pushing the brute transcendence of God to the absolute maximum, Barth likewise blurrs Luther's distinction between the "fear of the law" and the "love of the gospel." Here again, the latter is largely subsumed under the sway of the former so as to render the fear of the law all the greater.[29] In fact, the unknowable, unattainable, and legalistic aspects of Barth's God imply that the universe he hereby erects is positively Kafkaesque.[30]

It is this massive qualitative gulf that the Barthian system places between God and man that is at the source of Barth's consequent assassination of the intramundane life of man from the Beyond. For as Bloch asseverates, Barth reduces man's activity to the most minute possible proportions in relation to the activity of God—which alone is of import.[31] He stretches this disparity to the point of stating that "The divine utters an unchanging No into the world."[32] The qualitative difference between man and God becomes infinite, whilst man himself is hereby reduced to nullity: "The true God dispenses with all objectivity, and is the origin of the crisis of all objectivity; he is the judge of the world's nothingness."[33] In consonance with these averments, Barth moves to utter the anti-Promethean words: "The reality of religion is man's disgust at his own self."[34]

Somewhat unexpectedly then, he notwithstanding maintains Christ to be a Yes to the world. Yet this apparent concession must not mislead us. For the purported Yes of Barth's Christ is merely the Yes of an eschatology which *terminates* the world. As Bloch contends, this eschatology bears no substantial relation to the world. It is not conceived as, for instance, the utopian *telos* of a Hegelian world-historical process of discontinuous emancipation, or even as the final home-coming of St. Augustine's *civitas Dei* after its prolonged wandering through history. Barth's hyper-transcendentalist assault is a war on history, the world, and on anthropocentric actualism. By way of an extreme myth of lordship, in which God's other-worldliness and sovereignty is redolent of the despot's deity, he minimizes to the point of extinction every achievement of the human spirit, as well as all facets of man's experience and thought. Barth's is a divinity whose countenance is entirely turned away from man. His divine order is one which incessantly and unswervingly utters an unchanging No into the world. And does so *ad nauseam*. His excessively hypostatized and lofty deity prevents itself from entering into the purview of human thought and human history. Its aloofness is incorrigible. It is a deity devoid of process, pregnancy, fecundity, and novelty. For this reason the Barthian system precludes and forbids any emergence of a new aeon, of an Eschaton within history. Even the shattering Moment wherein the divine makes its evanescent appearance in the world is strictly a-historical, and not an emancipatory surge that may be anticipated and pursued in the manner charted in the book of Exodus. Succinctly put, Barth's historically null, static, and alien transcendence is a vantage point from whence he assassinates life.[35]

That Karl Barth is a most willing and astute assassin of noble humanity nowise implies that theology in general, or Christian theology in particular, is a pure business of assassination. There are assassins within theology's ranks, but one would be entirely deluded in making the descent down to the slumming cant of contending that theology is—in and of itself—an unequivocal and dogged project of assassination. One who, in particular, is no assassin of noble humanity is Gustavo Gutiérrez. The disparity between Barth and Gutiérrez is so vast that one would be forgiven in believing that they are partisans of entirely different *Weltanschauungen*. But the more fundamental point to be made in this connection is that of the dichotomy between an imperious transcendence and the commitment to immanence. If Barth is the imperious and dread avatar of the former, then Gutiérrez seems to believe that what is done within the intramundane world is of paramount importance.

Beholding this world, the slum-dwelling Peruvian priest maintains, it is far from the case that the world beyond is unequivocally that of the true life. It is rather the transformation and fulfilment of the present world and of life. The absolute value of salvation does not devalue this world since salvation is already latent within and a product being wrought within the world. For salvation is not an a-historical or an extra-historical matter, but rather an intra-historical reality. It is the communion of human beings with the Deity and among themselves that orients, transforms, and guides history. For Gutiérrez, it is not the case that one may speak of sacred history in oppositional juxtaposition to profane history. Rather, there is merely one human destiny, that of Christ as an immanent Lord of history. Soteriological history is thus wholly coextensive with and immanent to human history itself.[36] It is not a lofty instance that intervenes imperiously from a draconian transcendent vantage point.

From the Gutiérrezian point of view, one may conclude that Barth stands guilty of an excessive spiritualization which stems from a wholly disincarnate notion of the spiritual, that itself arises from a too strongly wrought dichotomy between the spiritual and the material. This disincarnate spiritual (scornfully superior as it is to earthly reality and likewise the product of Western dualistic thought), avers Gutiérrez, is simply foreign to the biblical mentality. If this manner of conceiving the spiritual is dropped, then the opposition of a temporal promise to a spiritual promise shall likewise be overcome in the sense of being replaced by partial fulfilments through liberatory historical events. Christ himself does not spiritualize the eschatological promises in the Barthian manner. Rather, he gives them meaning in the quotidian present. Gutiérrez is quite aware that an overwrought and distant spiritual would merely serve to devalue temporal and earthly realities—as it would view them as mere contingent obstacles to be eliminated. For this reason, eschatology must be integrally welded with the temporal, earthly, and the historical. It is an intra-historical reality and as such nowise lessens the value of present life.[37] It indeed extols it.

This proclivity toward an assertion of the immanence of the spiritual itself on the part of Gutiérrez is far from surprising, in that it stems from what he espies as a progressive immanentization within the biblical narrative of God himself. The active presence of God in the midst of the people, his communing with them, is underlined.[38] In the first covenant, Yahweh says that he will "dwell in the midst of the Israelites" (Exod. 29.45). Through the lips of Ezekiel he maintains that his "sanctuary is in the midst of them forever" (Ezek. 37.28). From the relative distance of Mount Sinai, Yahweh's presence becomes closer when it is linked with the tent and the Ark of the covenant that accompanies the Israelites through the desert. These underscore his sharing of the historical vicissitudes of the people. This tendency reaches the point of absolute immanence with the Incarnation.[39]

Furthermore, as St. Paul maintains, each Christian is a temple of the Spiritus Sancti: "Surely you know that you are God's temple, where the Spirit of God dwells" (1 Cor. 3.17). "Do you not know that your body is a shrine of the indwelling Holy Spirit, and the Spirit is God's gift to you?" (1 Cor. 6.19). Furthermore, it is every human being that is the temple of the Spirit sent by the Father and the Son. And these men dwell in very concrete historical situations. As Gutiérrez asseverates, this process of immanentization of the Deity nowise implies any form of spiritualization. The indwelling of God in every man is no more spiritual than the presence of God on Mount Sinai or in the Temple. For, in fact, this movement—though it involves an increasing universalization—is one whereby God is rendered more *material*. It is a movement *away* from what was once the vertical transcendence of a domineering spirit. More precisely, the Incarnation and the linked procession and immanentization of the Spirit implies a concrete universal—in the Hegelian sense.[40]

But let us return to Nietzsche. Our question remains: Is Nietzsche judicious in his indictment of *St. Paul* as effecting a similar assassination of life to the one undertaken by Karl Barth and not undertaken by Gustavo Gutiérrez?

(i) Against ressentiment *and bad conscience*

It would appear that he is not. Firstly, what is St. Paul's position with regard to suffering as a means of the assuagement of sin? According to Badiou, suffering plays no such role in St. Paul's apologetic. For him, suffering possesses no intrinsically redemptive function. Suffering is merely seen as an inevitable worldly vicissitude. St. Paul does admittedly relate it to the distribution of consolation: "Our hope for you is unshaken, for we know that as you share in our sufferings, you will also share in our consolation" (2 Cor. 1.7). Yet, it is not that the foretold consolations and future glory arise at length out of the squalid workshop of suffering by way of a process of ramification or transmogrification. Instead, consolation and glory, on the one hand, and suffering, on the other, are incommensurable: "For these slight momentary afflications are preparing for us an eternal weight of glory beyond all comparison" (2 Cor. 4.17). His own sufferings St. Paul regards with a strictly militant distance and stoic ambivalence. Even Christ's suffering recedes into the netherground in relation to the preeminent truth-event of the Resurrection.[41] In both this particular case and as regards suffering in general, Badiou proposes the following formula: "in Paul, there is certainly the Cross, but no path of the Cross. There is Calvary, but no ascent to Calvary. Energetic and urgent, Paul's preaching includes no masochistic propaganda extolling the virtues of suffering, no pathos of the crown of thorns, flagellation, oozing blood, or the gall-soaked sponge."[42] No Nietzschean bad conscience, no Hegelian unhappy consciousness. The abject self-flagellation of El Señor is something he would certainly revile.

Now, if suffering fails to redeem sin, then—and this is our second point—what is the position of sin itself? At present we shall only answer this question very provisionally and *per via negativa*. For Badiou, St. Paul does not dwell upon sin from the vantage point of a pedantic, or venomous, moralist. That is, he does not regard it as *primus inter pares* a fault, or as an evil act.[43] He does, notwithstanding, regard sin to be a profoundly deep and burdensome ethical problem—although in quite a different sense. Of this, more later.

Thirdly, if sin has ceased to be an out and out evil act, the entire insidious mechanism of self-accusation which Nietzsche alleges St. Paul to have erected upon this notion—immediately collapses with its support. And, with it, likewise the necessity of the punishment of sinners. Badiou's contention here is that Hell, as a roasting spit were sinners are repeatedly incinerated, is of no interest to St. Paul. Neither as a place of damnation for the individual who torments himself with his own sinfulness, nor as a means of the eternal punishment of enemies. In the former sense, St. Paul falls short of being a purveyor of bad conscience. In the latter sense, as Badiou has it, St. Paul's discourse is free of vengeance and *ressentiment*.[44] Indeed, his agenda is entirely at loggerheads with that of the man of *ressentiment*. For what he rails against with poignant vehemence is precisely this latter's rancorous tendency to perpetual accusation: "You, why do you pass judgment on your brother? Or you, why do you despise your brother? … Then let us no more pass judgment on one another" (Rom. 14.10–13).

As we have seen earlier vis-à-vis Deleuze's Nietzsche, with *ressentiment* the accusation of active force and the depreciation of the active and affirmative life by the reactive tribunal was effected *inter alia* through the projection of a supersensible world. This is the sense in which Nietzsche purports St. Paul (who is alleged to sit at this tribunal) to have needed recourse to the notion of immortality, to the Beyond, and thus to have shifted the center of gravity of life to the yonder side of the grave—from whence the active, affirmative and noble life could be hatefully condemned, mortified, and mired in its own death by a negative will. Notwithstanding, to inquire as to the presence or absence of a doctrine of personal immortality in the teaching of St. Paul shall not be our present purpose.

(ii) Toward affirmation

Our contention will however be that, whether it is present or otherwise absent in his writings, St. Paul makes no use of it to conduct a prosecution of the active life. What is, notwithstanding, certain is that St. Paul *does* indeed effect a shift in the center of gravity of existence. But this shift, as Badiou argues, occurs neither in accordance with death, nor in accordance with hatred. That is to say, it is not a movement whereby the center of gravity of life is shifted beyond the grave in accordance with the principle of *ressentiment.* It is, in fact, a movement in exactly the *opposite* direction. For St. Paul inaugurates a movement *away* from *ressentiment* and bad conscience, so that the rights of affirmation may be restored. It is a movement beyond, though not into the transcendent Beyond of death from whence death could reign over life. Rather, it is a movement of life beyond the confines placed upon it by *ressentiment,* bad conscience, the negative and death—effected according to the principle of *overexistence* (an overexistence endemic to the Resurrection-event) inherent to the affirmative life itself.[45]

In this sense, the shift in the center of gravity of life ushered in by St. Paul is in fact quite similar to that desired by Nietzsche. For both wish for life to pass beyond its present nihilist decadence, its deep-seated resentment and guilt slavery. To break the shackles of these latter, Nietzsche conjures Zarathustra as the sovereign subject who will finally be able to utter something in his own name, whereas Badiou's St. Paul—as already observed—alludes to the subject of the truth-event, the son-subject who arises out of his own solemn act of declaration of the Resurrection as a pure beginning outside the confines of the master-slave dialectic. So, it is not that in himself declaring the Resurrection of Christ, as Nietzsche erroneously contends, St. Paul is effecting the subsumption of life by death and by the negative. Rather, the Resurrection is that on the basis of which life takes vengeance on death in the here and now, and that by which the center of gravity of life returns to life—so that the affirmative life, or life in accordance with the spirit, vanquishes and supercedes what was hitherto a negative life of the flesh mortified by death.[46]

St. Paul's revolutionary program, far from being an assassination of the affirmative life, is instead bent on the disarming and maiming of death: "O death, where is thy victory?" (1 Cor. 15.55). And with death effectively neutralized and brought to heel, the "no" of nihilism gives way to life's "yes," such that negativity is vanquished and the rights of affirmation are restored: "For the Son of God, Jesus Christ, whom we preached among you, Silvanus and Timothy and I, was not Yes and No; but in him it is always Yes" (2 Cor. 1.19). Succinctly put, St. Paul's itinerary is not a cult of death but a resolute surge toward the foundation of universal affirmation. His aim is to bring about, through the truth-event of the Resurrection, the untrammeled affirmation of life against the hitherto prevalent reign of death and the negative.[47] He unflinchingly pines for the moment wherein "death is engulfed in victory" (1 Cor. 15.54).

Here, far from being merely denied or subsumed dialectically, death must simply be *overwhelmed.* It must be abolished. Though the Resurrection-event—which eradicates all negativity—requires death as its condition, it cannot be reduced to this latter and remains a strictly affirmative operation. For the Resurrection is not arrived at by the simple negation of death. It is, rather, an extraction—or a *subtraction.* Christ is extracted from the dead (*ek nekrōn*).[48] As Badiou has it, "Resurrection suddenly comes forth *out from* the power of death, not through its negation."[49] And it is through this extraction of Christ from the mortal site that death ceases forthwith to hold sway over life: "But if we have died with Christ, we believe that we shall also live with him. For we know that Christ being raised from the dead will never die again; death no longer has dominion over him" (Rom. 6.8–9). The Resurrection-event reduces death to powerlessness.

(iii) Against the Law

But let us backtrack somewhat. What is it that—in the absence of the Resurrection, the upsurge of the son-subject, and the universal becoming-son—ensures the dominion of death over life? As Nietzsche has it, it is the other-worldly transcendence of a supersenory realm projected by reactive forces that ensures the hegemony of death over life. For St. Paul, on the other hand, it is the transcendent instance of the Law—that "minister of death carved in letters on stone"—through which life lies in deathly paralysis (2 Cor. 3.7).

The Law—in its transcendent and Statist character—is, however, maintained by another transcendence: that of God the Father. Under the Law we are still incarcerated by the discourse of the Father. Yet, with the sending of the Son, both of these transcendences are overcome. For the death of Christ names the effective renunciation of God's radical transcendence and, by setting up an immanentization of the spirit, erects the site for the occurrence of the Resurrection-event whereby the Law is superceded and the abyss separating us from God overcome.[50]

The son-subject who declares the truth-event of the Resurrection ceases to tarry under the Law, under which he had hitherto lain frozen in subjective death, and is himself resurrected in the pure beginning that he is. In so doing, he is relieved of the Law's deathly grip and henceforth lives not under it but under grace: *our gar este hupo nomon all' hupo kharin,* "for you are not under law, but under grace" (Rom. 6.14).

The truth-event is itself not history but *grace,* whereas the son-subject is constitued by evental grace and henceforth lives under it through his fidelity to the event out of which he is born. Under the Law the mortified ante-subject (or self) followed the way of the flesh (*sarx)*, whilst under evental grace the son-subject proceeds along the way of the spirit (*pneuma*).[51] Whereas the real of the path of the flesh is death, that of the spirit is life. The way of the flesh proceeds by way of a coordination of law and works, whilst the way of the spirit is that of the coordination of grace and faith.[52]

Flesh's adequation of law and works grants us another perspective from which it may be seen that the Law is on the side of death. The Law presides over a *unter*-subject that it deadens, since it functions inter alia through the strict prescription of works. Subjectification by evental grace, which is not of the order of prescription but that of its own pure gratuitousness and of the pure conviction or faith of the incipient son-subject who declares it, is thus *blocked* by the Law through objectification. It is blocked by the essentially Statist nature of the Law. The Law is fundamentally Statist in character, since it is that which "enumerates, names, and controls the parts of a situation."[53]

This broad definition of the Law accords with what Agamben avers to be St. Paul's take on the same. When St. Paul speaks of the Law, he deploys the direct etymological meaning of the Greek term *nomos.* The latter term derives from *nemō*, which means "to divide, to attribute parts." Agamben alludes more specifically to the passage on vocation in 1 Corinthians 7.17, in which St. Paul refers to the various conditions into which men find themselves divided: St. Paul writes *hōs emerisen ho kyrios,* or "in the part which the Lord has attributed to him."[54]

What Badiou refers to as the enumerative, nominative, and controlling aspect of the Law with regard to the parts which the State subsumes from the situation over which it presides is strongly homologous to what Deleuze and Guattari term the paranoid-schizoid trait of the Law. According to this latter, the despotic full body integrates and circumscribes productive agents, products, and working organs under the overarching unity of its State apparatus. As observed, this circumscription by the despotic State acts as a transcendent Law which governs parts which it partitions off and organizes as bricks whose distance it measures and whose mutual communication it forbids. Notwithstanding, whether expressed in Badiousian or in Deleuzoguattarian terms, the fundamental operation of the Law remains that of the enumeration, predication, control, and measurement of the parts of the integral whole which it circumscribes and governs—in which each part renders and is rendered its due. The Law governs a predicative multiplicity of extensive parts comprising an integral whole, each part of which it grants its due.[55]

It is for this reason that the Law thwarts the truth-event and is therefore an obstacle to grace, the way of the spirit, and thus life. For, unlike the Law, the truth-event and the living son-subject that it spawns is nondenumerable, outside of predication, and uncontrollable. Grace occurs without being couched in any predicate, and happens to us without assignable reason. It is radically heterogenous to Law, for it comes without being due.[56]

It also governs a fundamentally different sort of multiplicity. Where the Law presides over a discrete and extensive multiplicity that is measurable and counted as one by the operation of overcoding, evental grace implies a continuous and intensive multiplicity that is outside of measure and in excess of both itself and of the fixed distributions of the Law. Not comprising an integral whole, the parts of the latter kind of multiplicity are inherently refractory to the topology of place, thus implying what Badiou terms "a nomadism of gratuitousness."[57] It may therefore be argued that the movement from the Law and death toward grace and life is likewise one from the molar order to the molecular order. For this reason, it necessarily entails a recrudescence of the full BwO of the earth submerged and incarcerated by the full body of the despot as socius, and likewise a movement toward the preponderance of absolute deterritorialization over relative deterritorialization. But more of these deterritorializations later.

(iv) Against sin

We have now reached the point where we may address the Pauline notion of sin. As already intimated, St. Paul does not regard sin to be an evil act. At least not in and of itself. What is it then? An initial clue may be gleaned from a certain feature of the despotic regime. As observed much earlier, Deleuze and Guattari aver that the State apparatus effects a fusion of desire with Law. For St. Paul, this latter fusion animates desire in such a way as to render it sinful:

> Nay, I had not known sin, but by the law: for I had not known lust, except the law has said, "Thou shalt not covet." But sin, taking occasion by the commandment, wrought in me all manner of concupiscence. For without the law sin was dead. For I was alive without the law once, but when the commandment came, sin revived, and I died. And the commandment, which was ordained to life, I found to be unto death. For sin, taking occasion by the commandment, deceived me, and by it slew me. (Rom. 7.7–11)

Provisionally then, we may define sin as that which comes into being through the Law and which by way of the Law causes the ante-subject to languish on the side of death. Calling to mind that it is the Law that mortifies the ante-subject, a more succinct definition of sin would however be: sin is the manner in which the Law deadens the ante-subject. But what, more precisely, is this manner of deadening? St. Paul:

> For we know that the law is spiritual: but I am carnal, sold under sin. For that which I do I allow not: for what I would, that do I not, but what I hate, that do I. If then I do that which I would not, I consent unto the law that it is good. Now then, it is no more I that do it, but sin that dwelleth in me. For I know that in me (that is, in my flesh) dwelleth no good thing. For to will is present with me: but how to perform that which is good I find not. For the good that I would I do not: but the evil which I would not, that I do. Now if I do that I would not, it is no more I that do it, but sin that dwelleth in me.
>
> I find then a law, that, when I would do good, evil is present with me. For I delight in the law of God after the inward man: but I see another law in my members, warring against the law of my mind, and bringing me into captivity to the law of sin which is in my members. O wretched man that I am: who shall deliver me from the body of this death? (Rom. 7.14–23)

It is clear from the above that the deadening of the ante-subject is effected by way a forced disjunction between willing and thought, on the one hand, and agency and action, on the other:[58] "For that which I do I allow not: for what I would, that do I not, but what I hate, that do I." This disjunction itself is effected by the indwelling of sin within the ante-subject: "Now then, it is no more I that do it, but sin that dwelleth in me." As an incubator of sin, the ante-subject is dead: "For sin, taking occasion by the commandment, deceived me, and by it slew me." The life of sin is the death of the ante-subject. It is the Law that gives life to sin, though the place of residence of sin is the body of the ante-subject: "For I was alive without the law once, but when the commandment came, sin revived, and I died." Sin occupies this ante-subject as a parasite which causes it to be dead. Or as Badiou has it, "*with the law, the path of death, which was itself dead, becomes alive once more.* The law gives life to death, and the subject as life according to the spirit falls onto the side of death. The law distributes life on the side of the path of death, and death on the side of the path of life."[59]

Badiou refers to the subject deadened by the Law as the "self." The self is the death of life, whilst the life of death is sin. The ante-subject is dead in the sense that thought can no longer prescribe action: "if I do that I would not, it is no more I that do it, but sin that dwelleth in me." The ante-subject therefore does not will what he desires and desires not what he wills. Under the Law, seduction by commandment ensures the bifuraction of thought and action.[60] In this sense, sin *corrupts the will* by rendering it impotent and at the same time hijacks the ante-subject's desire by severing it from this same will. A will corrupted by sin is one which has suffered severance from its desire. In this diremption of will and desire, the will becomes powerless whilst desire becomes incorrigibly sinful.

Now, in that desire has—under the deathly paralysis of sin—moved outside of (and thus become autonomous from) the will, we may say that the desire of the ante-subject has become—from the perspective of the defunct will—quasi-*automatic*. This latter dirempt subject, "wherein the division lies between the dead self and the involuntary automatism of living desire, is, for thought, a figure of powerlessness. Basically, sin is not so much a fault as *living thought's inability to prescribe action* [my emphasis]. Under the effect of the law, thought disintegrates into powerlessness and endless cogitation, because the subject (the dead self) is disconnected from a limitless power: that of desire's living automation."[61] More succinctly, the self mortified by the Law "presents himself as a disconnected correlation between an automatism of doing and a powerlessness of thought."[62]

This now finally brings us to a consummate definition of sin. In Badiou's formulation, "*Sin is the life of desire as autonomy, as automatism.*"[63] More precisely, the "law's prohibition is that through which the desire of the object can realize itself 'involuntarily', unconsciously—which is to say, as life of sin. As a result of which the subject, de-centered from his desire, crosses over to the side of death."[64] With desire having ceded from the dead ante-subject's will and embarked upon a runaway automatism, sin—as this incorrigibly vagabond life of desire—henceforth becomes a veritable *doppelgänger* of the deadened ante-subject or self. To the latter's Dr Jeckyll it plays a most loathsome Mr Hyde.[65]

Still more graphically illustrative of the parlous vicissitudes of a dead subject ruined by the parasitic indwelling of living sin is James Hogg's novel *Confessions of a Justified Sinner*. Here the dead subject is Robert Wringhim Colwan, whilst sin is an enigmatic being who gives his name as Gil-Martin.[66]

The question notwithstanding remains: How then is sin to be *slain* so that the dead subject can return to life? As already intimated earlier, this can only come to pass by having done with the Law.[67] Michel Henry inveighs against the Law in a highly analogous manner to Badiou and St. Paul. The problem with the Law, for Henry, is its essential *unreality*. This unreality stems from the Law's fundamentally transcendent and exterior position to life. For this reason, life itself perceives the Law as beyond it. In that it is divorced from life, the Law is likewise sundered from that in life by which it may be fulfilled: action. This is precisely what renders the Law powerless. Owing to the conjunction of powerlessness and unreality in the Law, the subject placed underneath it is put into an untenable situation. On the one hand, the Law prescribes and proscribes in clearly perceptible and indubitable injunctions. On the other hand, however, the commandments that it enunciates are unable of themselves to produce the actions that suit them. For this reason the self who is subjected to the Law's prescriptions and prohibitions is himself unable to observe the Law. He lacks the force necessary to accomplish it. For this force resides neither in the self deadened by the Law nor in the Law itself. The Law projects the path which action should follow without granting the subject the power he would require in order to follow this path. The mire into which the Law casts the deadened subject is therefore one in which this same subject sees quite clearly what ought and ought not to be done and yet at once finds himself bereft of the ability to execute salutary actions or to abstain from insalubrious ones. And it is precisely through this very same lucidity that the Law confers upon the subject and by way of this self-same commandment with which it equips him that the Law mires the subject. A dramatic and desperate situation. The Law defines the infraction and the crime that the subject must circumvent to filfull it, yet without granting him the power to do so. The Law is therefore a *cursed* Law—and in turn inevitably and inexorably curses the subject, since it fails to furnish the latter with the power to follow it.[68] Succintly put, the Law multiples crime: "The law was added so that trespass might increase" (Rom. 5.20). The Law proliferates sin.

As observed, the Law—being transcendent—is powerless in granting the subject to which it addresses itself the force necessary to accomplish it, since it is not located in the domain proper to action: life. And, being foreign to life, the Law is likewise inherently foreign to action. As Michel Henry argues, though it presents itself as an ideal archetype for all righteous action, the Law is nonetheless incapable of producing such an action. The Law does nonetheless communicate something to the action that it motivates: its own unreality. This unreality is at once the same action's *inefficacy*. The powerlessness and unreality of the Law redounds upon the action it prescribes to the point of distorting it.[69] The Law renders action sinful.

(v) Toward a transliteral law of the spirit

As intimated, the powerlessness of the subject to act in any manner other than sinfully can only be assuaged by way of his being relieved of the Law. For Badiou, the Christ-event is the abolition of the Law, that empire of death: "Christ redeemed us from the curse of the law" (Gal. 3.13). The subject's resurrection from its lingering death under the auspices of the dead self can only transpire by way of its declaration of and fidelity to the Christ-event (a lawless irruption) out of which he, as an incipient son-subject, shall be born. In so doing, the subject embarks upon the path of salvation (*sōtēria*)—which latter is the unseparation of thought from agency and the instalment of a formerly impotent thought and defunct will as the power of action. The ruin of the hitherto prevailing disjunction between thought and action is rendered possible solely on the basis of the pure excess that is evental grace. It is only through the pure and brutal act of grace that the subject is raised from its lingering death under the Law and by which sin is slain. Evental grace, which is in excess of both the Law and the order of thought, is the *sine qua non* by which thought is raised up from its former impotence and, in a convalescence of the will, attains the capacity to prescribe action.[70]

An important objection needs at this point to be answered. In that the Christ-event entails the abolition of the Law, may it not be said that the son-subject is thereby entirely lawless? Certainly not. For, as Michel Henry has it, the fact that the Law has been eschewed as the directing principle of action is nowise synomynous with the abandonment of action itself to unmitigated contingency, evanescence, or the momentary whim of capricious subjectivism.[71] Indeed, when St. Paul is on the very brink of proceeding to depose the Law, he nonetheless proceeds to—in a certain sense—maintain it: "Wherefore the law is holy, and the commandment holy, and just, and good" (Rom. 7.12). He contends furthermore that the law is spiritual: "For we know that the law is spiritual: but I am carnal, sold under sin" (Rom. 7.14). But this law that is here described as holy, good, and spiritual—as Badiou maintains—is not the same sense of law. For the Law which the Christ-event abolishes may be described as "a legalizing subjectivation, which is the power of death"; whereas the law which St. Paul extols in the foregoing passage is "a law raised up by faith, which belongs to the spirit and to life."[72] This latter, Badiou terms "*a transliteral law, a law of the spirit.*"[73]

As Agamben has it, the transition that St. Paul makes here is one from *nomos tōn ergōn* (the Law of works) to *nomos pisteōs* (the law of faith). The former is the Law of the State, whereas the latter is Badiou's transliteral law of the spirit. Though the son-subject is nowise anarchic, the movement from the first to the second sense of law does entail a passing flirtation with lawlessness. For instance, in 1 Corinthians 9.21, St. Paul contends that he has made himself *hōs anomos*, or "as without law." That this flirtation is verily evanescent is evinced by the fact that he immediately proceeds to rectify his assertion, averring that he is not *anomos theou*, or "outside God's law," but rather *ennomos christou,* or "in the law of the Messiah." This latter law of the Messiah Agamben takes to be the law of faith, that is, the transliteral law of the spirit. Badiou is perhaps being somewhat crude in using the word "abolished" in relation to the Law that the spirit supercedes. St. Paul himself does not use this word in the same connection. Instead, he uses the verb *katargeō*. The latter is a compound of *argeō,* which in turn is derived from the adjective *argos,* meaning "inoperative, not-at-work (*a-ergos*), inactive." The compound *katargeō* therefore means "I make inoperative, I deactivate, I suspend the efficacy." St. Jerome renders the infinitive of the same word as *evacuari*, or "to empty out." This, above all, is not the same as "to abolish," or "to annihilate," or "to destroy," or "to make perish." The infinitive of *katargeō* means "to deactivate"—as St. Paul writes in Romans 7.6: "But now we are de-activated [*katergēthēmen,* 'made inoperative'] from the law." The Law is, thus, not so much abolished as *suspended.* What Agamben terms the messianic, or what Badiou terms the Resurrection-event, is not the destruction of the Law. It is its *deactivation.*[74]

But it is not merely its deactivation. It is also its *fulfilment.* For on the one hand St. Paul writes in 1 Corinthians 15.24 that the Messiah "will render all rule, authority, and power inoperative [*katargēsē*]," whereas in Romans 10.4 he avers that the Messiah is "the *telos* of the law." Yet, here it is deactivation which is in fact the condition of fulfilment. For it is to the extent that Christ renders *nomos* (the Law) inoperative that he is its end and fulfilment, in the sense that the Law is fulfilled solely by first being rendered inoperative. Indeed, in 2 Corinthians 3.12–13 it is written that Christ is *telos tou katargoumenou*, or "fulfilment out of that which has been de-activated," that is, *at once* deactivation and fulfilment. This means to say that that which is deactivated (the Law) is nowise annulled, but in fact conserved and preserved for its fulfilment. The latter seeming self-contradiction, Agamben avers, may be understood according to John Chrysostom's asseveration that what St. Paul refers to in *katargēsis* is nowise the destruction of Being (*aphanisis tes ousias*), but rather the progression to a different and superior state. Finally, the Pauline confluence of suspension and fulfilment as regards the Law is given a further gloss by Luther when the latter ex-monk translates the verb *katergein* as *Aufheben*: a term which harbors the double meaning of abolishing and of conserving (*aufbewahren* and *aufhören lassen*). A Hegelian word indeed.[75] The Resurrection-event is an *Aufhebung* of the Law in that its abolition of the latter is at once the Law's preservation *as* the transliteral law of the spirit.

(vi) Love

Now, the movement from the abolition of a legalizing Statist Law toward this new transliteral law of the spirit is at once that from faith to love. For Badiou, faith is the son-subject's declaration of the Resurrection-event whereby he is relieved of the Law of the State. Love, on the other hand, has faith as its anterior condition but goes further: "love names a nonliteral law" that "gives to the faithful subject his consistency, and effectuates the post-evental truth in the world."[76] Love, or *agapē*, is the universalisation to all subjects of the address of the truth-event declared by the son-subject. This now-universal address is consistent in accordance with the transliteral law of the spirit. With faith, the Law is deactivated and the subject returns to life. With love, on the other hand, *life itself* henceforth becomes the sole and universal law. In this sense, love does indeed restore law—but only as the articulation of life for all subjects.[77]

Without love as its necessary supplement, faith is far from sufficient for the salvation of the subject. It merely declares publically that the configuration of the Statist Law and its mortifying mode of subjectivation is only one of many and may thus be superceded. It declares that the subject *is* verily able to escape the Law-induced impotence of its thought to prescribe action. Yet, though it prescibes this newfangled possibility, faith is not and does not of itself effectuate this possibility for all subjects. Faith is the declaration of the possible power of thought to prescribe action, yet it is not this power as such. It is the declaration of the ruin of the diremption between thinking and doing, but it is not yet the *bringing into effect* of this ruin itself. It is a break with the Law, but not yet the fulfilment of this break in the transliteral law of the spirit. For this fulfilment of Law's abolition and its transmogrified instalment as the handmaiden of life itself is only brought to bear by love.[78]

The commandment to "love your neighbour as yourself" is not a commandment that could possibly arise from a Statist Law. It could not pertain to a subject mortified by the Law. For it requires faith as an anterior condition, "because *prior to the Resurrection, the subject, having been given up to death, has no good reason to love himself*."[79] Unlike the Law's seduction by commandment, this commandment of love entails no prohibition (a negation), and is therefore a pure affirmation. For it is based on the pure conviction (an affirmation) of the Resurrection-event and enjoins the love of others on the basis of self-love, which latter is in turn itself the pure self-affirmation of the newly upsurgent son-subject of this event. This commandment of faith, and of the new transliteral law of the spirit, is an injunction to the deployment of the power of self-love in the direction of others in the mode of universal address.[80] Faith commands love and is alone fulfilled through it. Without love, faith is merely hollow subjectivism, the mere bellowings of a thin and reedy subject devoid of consistency.[81] St. Paul:

> Though I speak in the tongues of men and of angels, and have not love, I am become as sounding brass, or a tinkling cymbal. And though I have the gift of prophecy, and understand all prophetic powers, and understand all mysteries and all knowledge: and though I have all faith, so that I could remove mountains, but have not love, I am nothing. And though I bestow all my goods to feed the poor, and though I give my body to be burnt, and have not love, it profiteth me nothing. (1 Cor. 13.1–3)

Salvation, as the effectuation or resoldering of the hitherto ruined unity of the subject's thought and action in the world, is attainable only through love. Faith—that is to say, subjectivation through faith—is not salvation (*sōtēria*) as such, but rather merely justification (*dikaiōma*). One is justified by faith, but saved only by love. The former and the latter are linked as the *terminus a quo* and *terminus ad quem* of liberation. Subjectivation by faith opens the space for liberation, yet only love and its intrinsically implied universality of address actually effectuates liberation itself. The evental grace unto which faith opens and through which the son-subject emerges merely points toward the possibility of salvation (*sōtēria*). Salvation itself is, however, only attained by a collective labor of love.[82]

We have thus seen that the transliteral law of the spirit is that of love. But why, more precisely, is this the case? Before answering this question, we need to probe the nature of this transliteral law more incisively. As Michel Henry avers, in order for law and action not to be dissevered, the law "*that commands acting cannot be of another order than acting itself, which belongs to Life and only deploys its essence within it.*"[83] Since the site of action resides in life itself, it may only be brought into effect from within life. As observed, the transcendent Law failed as a principle of action for the subject, in that it was essentially foreign to life. For this reason the principle of acting must be radically homogeneous with life itself. The new law, or the transliteral law of the spirit, if it is to serve as a principle for undistorted action, must for this reason be the authority and commandment of life and naught beside. This is the displacement of the principle of action effected by the Christ-event: the placement of the commandment *within* life and as the commandment *of* life itself.[84]

The commandment of the Badiousian transliteral law of the spirit is therefore, quite simply, life. Life itself constitutes the commandment, as the latter now ceases to be in any sense external to it. This commandment is henceforth radically immanent to and inside of life, one with life and with the movement of life. Thus, in the passage from the Statist Law to the transliteral law of the spirit, the very relation between law and life has effectively been reversed. It is no longer the Law that determines life, but rather life that determines law. Moreover, unlike the commandments of the old Law, the commandment of the transliteral law of the spirit is not bereft of power. For here the relation of commandment to the one who is commanded is now merely the relation of life to the living. More precisely, it is the relation of life to the son-subject.[85]

As observed, the sole commandment of the new law is the nonprohibitive injunction to neighborly love on the basis of self-love. But why should the commandment of the transliteral law of the spirit be this precise one? Because it is the commandment of life—and life *is* love. By life Michel Henry means *absolute* life, that is, God. God is life and likewise love. Life is love because it experiences itself without reserve, as pure affirmation. In that life is in essence love, it of necessity commands love. Life commands love by generating its livings as Sons, or as incipient Badiousian son-subjects—that is, as "*those who, feeling themselves in infinite Life's experience of self and its eternal love, love themselves with an infinite and eternal love, loving themselves inasmuch as they are Sons and feeling themselves to be such*—in the same way as they love others, inasmuch as they are themselves Sons and feel themselves to be such."[86] The commandment of the transliteral law of the spirit is that of self- and filial love, in that it itself arises out of and remains tributary to love itself, to the love that life bears itself.[87]

It may thus been seen that the new law of the spirit arises out of and orients the son-subject in the direction of affirmation and activity, in that its commandment is immersed in an absolute life which experiences itself with an infinite enjoyment and love of self. The commandment of love is immanent to life's process of self-generation and likewise to its generation within itself of the son-subject. Since the commandment of the law of the spirit is identical with the process of generation of the son-subject and the creation of life in him, there can be no hiatus between the commandment of love and the action of this son-subject, since his action is at once the acting of life itself within him.[88] The son-subject is therefore all foreground action and, proceeding as he does from the pure affirmation that is absolute life, limitlessly shot through by an affirmative will to power.

Notes

1. Negri 2009, 25.
2. Ibid. 31.
3. Ibid. 36.
4. Ibid. 46, 47.
5. Ibid. 68.
6. Ibid. 70.
7. Ibid. 71.
8. Ibid. 72.
9. Ibid. 89, 90.
10. Miles 1995, 314.
11. Ibid. 316.
12. Bloch 1972, 112, 113.
13. Ibid. 117.
14. Ibid. 122.
15. Miles 1995, 317.

16. Ibid. 324.
17. Negri 2009, 96.
18. Ibid. 96, 97.
19. Negri 2009, 99.
20. Nietzsche 1976, 649, A 58.
21. Ibid. 617, A 42.
22. Ibid. 650, A 58.
23. Ibid. 618, A 43.
24. Ibid. 619, A 43.
25. Bloch 1972, 42.
26. Cited in Ibid.
27. Bloch 1972, 45.
28. Cited in Ibid.
29. Bloch 1972, 45.
30. Ibid. 47.
31. Ibid. 44.
32. Cited in Ibid.
33. Cited in Ibid. 44.
34. Cited in Ibid. 46.
35. Bloch 1972, 46, 49.
36. Gutiérrez 1988, 85, 86.
37. Ibid. 96.
38. Ibid. 106.
39. Ibid. 107.
40. Ibid. 109.
41. Badiou 2003b, 66, 67.
42. Ibid. 67, 68.
43. Ibid. 83, 84.
44. Ibid. 96.
45. Ibid. 61.
46. Ibid. 61, 62.
47. Ibid. 72.
48. Ibid. 72, 73.
49. Ibid. 73.
50. Ibid. 69.
51. Ibid. 63.
52. Ibid. 75.
53. Ibid. 76.
54. Agamben 2005, 47.
55. Badiou 2003b, 78.
56. Ibid. 76, 77.

57. Ibid. 78.
58. Ibid. 83.
59. Ibid. 82.
60. Ibid. 83.
61. Ibid.
62. Ibid. 84.
63. Ibid. 79.
64. Ibid. 80.
65. Stevenson 1992, *passim.*
66. Hogg 1992, *passim.*
67. Badiou 2003b, 81.
68. Henry 2003, 177.
69. Ibid. 178, 179.
70. Badiou 2003b, 84, 85.
71. Henry 2003, 181.
72. Badiou 2003b, 87.
73. Ibid.
74. Agamben 2005, 95–98.
75. Ibid. 98, 99.
76. Badiou 2003b, 87.
77. Ibid. 87, 88.
78. Ibid. 88, 89.
79. Ibid. 89.
80. Ibid. 89, 90.
81. Ibid. 91.
82. Ibid. 91, 92.
83. Henry 2003, 181.
84. Ibid. 182, 183.
85. Ibid.
86. Ibid. 186.
87. Ibid.
88. Ibid. 187, 188.

10

Exit Strategies from the Structures of Despotism, 3. Kafka as St. Paul

We have seen how in St. Paul, by way of the Resurrection-event, the Law is suspended and preserved only as ancillary to life itself. Here, insofar as life was contended to be love, the transliteral law of the spirit commanded this love. Under the pressure of the truth-event, the transcendent instance of the Law gave way to the immanence of grace, under which the subject was relieved of the Law and sin alike.

A homologous operation is seen to transpire in Deleuze and Guattari's reading of Kafka. Here, the transcendent Law likewise gives way to immanence. As shall be seen, this latter immanence is that of the peregrinations of molecular desire subtending the molar order of legality itself.

There remains, notwithstanding, a slight divergence between the terrain scrutinized by the Pauline and the Kafkan projects, respectively. This concerns the different emphasis they place as regards the Law. When St. Paul speaks of the unobservability of the Law, he is referring to the Mosaic Law—replete with its full panoply of precisely specified *misvot*. The kind of Law which faces Kafka's characters is likewise unobservable. The point of difference is, however, that the unobservability of the Law in Kafka stems above all from its essential *unknowability.* Exactly as in the case of the Law of the despotic regime itself, this Law is an empty form devoid of content expressible only through the sentence pronounced upon the subject—a sentence which, in turn, itself is only knowable through the punishment of the subject. No one knows the Law in "The Penal Colony"—wherein the torture machine inscribes the condemned man's sentence on his body. This is also the case in "The Great Wall of China." So, whereas the Mosaic Law and Kafka's inscrutible Law are both shot through with a fundamentally Statist character and both incessantly multiply crime, Kafka's Law is far more abstract. In fact, it is so abstract as to have no tangible existence. Nonetheless, it remains in force—as an irremissible practical necessity.[1]

1. The transcendent Law

We have seen how in the despotic regime, the despot occupies a transcendent position of power in relation to his subjects, as does the Law. This fuses desire to the Law and likewise to death. This is also the case in all readings of Kafka—such that of Max Brod—which see in the latter a fundamental negative theology in which the articulation of the desire and the distribution of all subjects is determined by a central *absence*, a sort of *Deus absconditus*, and in which the subject is viewed as a priori guilty.[2] Here, as in the case of the despotic regime, it is precisely the transcendence of the Law that immediately appears to instill this a-prioriness of guilt. As observed earlier, according to the paranoid-schizoid trait of the despotic Law, all subjects are assumed guilty from the very outset—whether verily guilty or innocent. Upon such a reading, just as much as the mute voice of the despot and his deity communicates the Law only through the written decrees of the despot's bureaucratic perverts, so too Klamm in *The Castle* is forever unattainable for K. In *The Trial* the highest court is never reached. This much is intimated to K by the painter Titorelli, who avers that the highest court is wholly inaccessible to anyone at all,[3] and likewise by the priest in the cathedral, who maintains that K's case will never proceed beyond even a lower court and that his guilt is entirely a proven fact.[4]

Žižek terms this sort of interpretation of, in particular *The Trial*, the "obscurantist-religious" approach, according to which K is effectively guilty precisely due to his incessant protestantion of innocence. Here, in his perpetual questioning of authority, K is seen to arrogantly rely on naïve enlightenment-style rational argumentation in a way which blinds him to the mysterious machinations of bureaucratic and Statist power. It is from this his "subjectivist" perspective, which the obscurantist-religious approach views as distorted, that the tribunal before which he stands appears as an inscrutible and obscence agency bombarding him with irrational and unseemly accusations.[5]

This negative theological conception of the Law—as we shall see in the argument that Deleuze and Guattari unfold, will gradually be seen to be marginalized and reduced to a radically secondary and derivative position.

Our argument shall immediately commence with the overtly architectural terms in which much of the authors' interpretation of Kafka's novels is couched. Firstly, however, it is beneficial to recall that the trascendent unity and formal vacuity of the Law emerged directly from the molar unity of the socius, or despotic full body, upon which all the working organs of the subjects found themselves distributed. The transcendent unity of the Law was seen to be the direct effect of this statistical unity of the full body of the despot, whilst its legal and pragmatic force lay precisely in the inviolability of the distribution of organs circumscribed under the same transcendent despotic full body.

As observed earlier, this conception of the Law accords—at least from this angle of analysis—with that of St. Paul, who refers to the Law with the term *nomos*, which we saw apropos of Agamben as deriving from *nemo*, meaning "to divide, to attribute parts." That is to say, the Law is that which enumerates, names, and controls the parts over which it presides.

As seen, for Deleuze and Guattari, the very emptiness of this despotic Law resides precisely in the fact that it functions as an instrument serving, in the first instance, merely to maintain the distribution of the organs on the full body—and naught beside. To this is added the observing eye, whose function is to forewarn of the imminent danger of but a single organ taking flight from the full body and thereby violating the distrubution imposed by the despotic socius' disjunctive synthesis of recording as deployed in its exclusive usage. The function of the hand was the preemptive vengeance exacted upon the bodies of the subjects. Here, owing to the paranoiac position of the despot, or despotic paranoia, every organ immediately represents an imminent danger to the precarious unity of the despotic full body. It is only in the preemptive violence of despotic vengeance that the Law acquires a content, the verdict upon the vagrant organ—and the subject to which the latter is attributed—acquiring existence only in and through the act of punishment. As in the recently mentioned mainstream readings of Kafka, the main upshot of this depotic paranoia is that all subjects, from the standpoint of the Law, must from the very outset be presumed guilty.

2. The transcendent Law and the astronomical state of architecture

Deleuze and Guattari go so far as to link the transcendent Law with a particular state of architecture. We recall that the old primitive territories are maintained under despotism as bricks or blocks on the new socius, under this new full body.

Here, every time that "power presents itself as a transcendental authority, as a paranoid law of the despot, it imposes" what may be described as "a discontinuous repartition of blocks, with spaces between each one. In fact, the transcendental law can only regulate pieces that revolve around it at a distance from it and from each other. It is an astronomical construction."[6]

More specifically, in *The Castle* one finds an architectural arrangement comprising a discontinuous outer circle of buildings surrounding an extremely tall central tower whose apex verges out of sight, seemingly to infinity. This astronomical architectural state Deleuze and Guattari define by the following set of parameters: high-angle/low-angle, stairs, craning up/craning down, discontinuity of block-arches, distant/close.[7] "High-angle/low-angle" evidently enough pertains to the line of sight traced by the eye from the vantage point of the outer circle vertically from the base of the tower to its apex; whereas "craning up/craning down" cannot but refer to the vertical movements required to ascend or descend the tower. "Discontinuity of block-arches" is associated with the paranoid-schizoid trait of the despotic Law, according to which this latter by definition presides over a myriad of parts whose mutual contact and communication it regulates or even forbids.

"Distant/close" presents a more interesting distinction. There is a coexistence of the distant and the close in the astronomical architectural model. On the one hand, the blocks on the outer circle that form arches are close to one another in that they form couples. One the other hand, they are distant from one another since there are gaps between them that will never be bridged or filled. There is, moreover, a second co-implication of the "distant" and the "close" in this model. The seemingly infinitely tall tower, and likewise the transcendent Law, is infinitely distant from each block; whereas it at the same time incessantly sends messangers to each block and is therefore almost suffocatingly close. This second imbrication of the distant and the close is entriely consonant with the position of the great paranoiac who, though at an infinite remove from the lowly and bedraggled subject of the despotic regime, never ceases to harangue the latter through his pestering and over-present bureacratic perverts.[8]

Finally, to all of these spatial aspects of the astronomical architectural model, there also corresponds a *temporal* aspect: homologously to the manner in which the blocks on the outer circle are separated by unfillable gaps, the transcendent Law enforces a discontinuous succession of time periods.[9] These latter correspond to the sequence of phases during which the subject is alternately arrested, acquitted and re-arrested. As shall be seen, this alternating sequence functions according to what we shall presently see as the formula for superficial acquittal.

3. The astronomical state of architecture and the three possibilities in the Law

The characteristics of the transcendental and astronomical state of architecture link with certain possibilities within the functioning of the Law. In *The Trial* the painter Titorelli reveals to K three possible eventualities within the machinations of the Statist Law: definite acquittal, ostensible or superficial acquittal, and indefinite postponement. In the first of these, case documents are completely discarded and permanently vanish from the record. These include any documentation of not only the charges, but also the proceedings and even the acquittal itself. Such a disappearing act can, however, only be effected by the highest court. Since, as observed, this court is inaccessible, definite or actual acquittal is an outright impossibility.[10] No one, let alone K, has the least influence on it. Titorelli likewise confesses that he has never seen any instance of it.[11] Apart from this it would entail the cessation of desire itself. One may also argue the impossibility of definite acquittal from the observation that, owing to the ubiquitous threat of vagrant organs to the unity of the despotic full body, even if the subject is temporarily exculpated, the paranoia of the despot and of the transcendent Law must continue to assume that all subjects are intrinsically and incorrigibly guilty.

Titorelli's second possibility—that is to say, ostensible acquittal—is, however, possible. Here, the lower court judges are able to temporarily disengage the accused from his charge. The charge does, notwithstanding, hang over the subject and may at any subsuquent moment be reactivated by a higher court. For the file with the charges, certificate of innocence, and the acquittal is not destroyed as in definite acquittal, but continues to circulate between the higher and the lower courts: it "oscillates between the two regions with greater or lesser swings and longer or shorter pauses." For this reason, the accused may be arrested again, and then once more acquitted, and then arrested and acquitted a third time, and a fourth time, ad infinitum. Indeed, renewed arrest is already presaged within the judge's very concept of ostensible acquittal.[12] Ostensive or superficial acquittal, as Deleuze and Guattari aver, pertains to the temporal mode of functioning of the transcendent Law observed moments earlier, according to which the subject is submitted to an alternating succession of periods of repression and of reprieve.[13] The third of Titorelli's three possiblities—that is to say, indefinite postponement—shall be attended to later.

4. The subterranean state of architecture

There is, however, something different afoot. *The Castle* and *The Trial* evince that, burgeoning within the bowels of this first astronomical architectural configuration of the despotic State apparatus, is a fundamentally different sort of architectural arrangement. Here, Deleuze and Guattari argue that, rather that being distributed around a circle as a number of discontinuous arches—as in the case of the astronomical state of architecture—the "blocks" now align themselves as "segments" on the "continuous" and "unlimited" straight line of a hallway or corridor.[14]

Furthermore, each block-segment possesses a door or opening onto the hallway and likewise a back door; whereas the doors of the respective blocks which open onto the hallway are faraway from one another, the blocks' back doors are in fact contiguous—making the blocks themselves contiguous. This sort of configuration is encountered repeatedly in *The Trial*. For instance, when opening the door of a miniscule room close to his office at the bank, K suddenly finds himself in the judicial site where two inspectors are being whipped for complaining about the examining magistrate.[15] Though Titorelli's abode is almost at the diametrically oppose side of the town to the offices of the court, K espies that there is a door at the back of the artist's room which leads into exactly the same judicial site.[16]

To this second state of architecture, proliferating itself within the innards of the first state, Deleuze and Guattari attribute the characteristics of "frontal view/corridor view," "low ceiling," "wide-angle/depth-of-field," "unlimitedness of the immanent hallway," and "faraway/contiguous."[17] "Frontal view" obviously pertains to the view from the hallway onto a series of doors entering onto their corridors, whilst "corridor view" is that experienced whilst passing through the narrow passageways of the corridors—each of these being possessed of a "low ceiling." The "unlimitedness of the immanent hallway" is quite clearly the seeming boundlessness of the latter's thoroughfare.

The opposition "faraway/contiguous" deserves more thorough consideration than the foregoing pairs. As seen, in the second state of architecture the offices were "faraway" from each other from the vantage point of the hallway, but "contiguous" from the perspective of the back doors which rendered them virtually next to one another. Yet the terms of the couple "faraway/contiguous" are not merely opposed to one another. For, in fact, "faraway" is equally opposed to both "close" and "distant"; whilst "contiguous" is opposed to "distant" just as much as it is to "close." This may be adduced from the fact that, being separated from one another by the formidable length of the hallway, the offices are far away from one another and therefore not very close; whereas, on the other hand, the contiguity imposed upon them by their back doors implies that they are not particularly distant. It remains tempting to raise the objection that it is somewhat obtuse to oppose the synonyms "faraway" to "distant" and "contiguous" to "close" to one another, as if they were antonyms. The apparently paralogistic and counter-intuitive nature of this line of reasoning is, notwithstanding, assuaged by noting that the pairings "close/distant" and "faraway/contiguous" belong to two different dimensions. The former belongs to the dimension of height, whereas the latter belongs to that of length.[18]

In addition to the above, Deleuze and Guattari also distinguish the two states of architecture, respectively, in terms of the following sets of opposed terms: "infinite-limited-discontinuous-close and distant" (for the astronomical state of architecture) and "unlimited-continuous-finite-faraway and contiguous" (for the subterranean state of architecture).[19] In the former, "infinite" pertains to the infinite transcendence of the apex of the vertiginous tower and thereby to that of the despot and his transcendent Law, "limited" concerns itself with the limited length of the outer circle of arches, whilst "discontinuous" describes the separation of the arches or blocks on the circle. "Close and distant" has already been described. In the latter state of architecture, "unlimited" pertains to the longitudinally unbounded length of the hallway, "continuous" relates to the untrammeled depth of the corridors, whilst "finite" is an inherent property of the immanent nature of the second and earthly architectural arrangement. The distinction between "close and distant" and "faraway and contiguous" has already been delineated.

5. The subterranean state of architecture in relation to the transcendent Law

The question to pursue now is how this indwelling of a fundamentally heterogenous and subterranean architectural formation burgeoning beneath and within the more hierarchical formations of the astronomical architectural model, bears upon the transcendence of the paranoid despotic Law and its machinations. Returning to our earlier discursus on St. Paul, we recall that the Epistle writer's critique of the Law was that, under it, the subject languished in a sin-addled mortification wherein thought and action were inexorably dirempt. For this reason, the Law was unobservable. For Michel Henry, the Law remained unobservable in that, being transcendent, it was foreign to life and thus outside the domain proper to action. For Kafka, on the other hand, the Law is unobservable due to its unknowability. For St. Paul, the exit strategy from this quagmire was to deactivite the Law and move toward the immanence of a transliteral law of the spirit whose locus was life.

The Kafkan critique of the Law, though not opposed to the foregoing, differs somewhat. Kafka's strategy is to demonstrate that the Law is, in fact, *not transcendent to begin with.* This is so because the unrecognizability of the Law nowise stems from its being occulted by its purported transcendence. Rather, the Law is and remains unrecognizable since it is denuded of interiority (the formal vacuity of the Law). In the parable of the law in *The Trial*, the door-keeper to it in fact allows the man from the country to peer through the door onto the Law—who, notwithstanding, sees nothing. This door to the Law cannot be opened to the man in the country, in that it is always open—and, in fact, only closed to him at the very end of his life.[20] When the priest in the cathedral intimates to K that the door-keeper himself does not know the interiority of the Law, we should adduce here that this is so because this "interiority" is *as such* unutterable, in that it does not exist.[21] The Law is not utterable by a hierarchical judicial tribunal since, as Titorelli intimates to K, whereas the court itself is completely impervious to proof of innocence, this attitude changes completely when one moves into the consulting rooms, the corridors, or even into the painter's studio[22]—and in the office next door, behind the door, and so on horizontally ad infinitum. In the opening scene of *The Trial*, K finds himself immediately indicted as guilty, though the pestering bureaucrats provide no indications whatsoever as to the nature of his crime, deferring this latter disclosure to certain higher authorities before whom he is subsequently to appear.[23] Yet, despite being countenanced with a massive tribunal, the very substance of his supposed contravention of any sort of transcendent State Law swiftly founders into nothingness. The entire court session is a shambles. The examining magistrate mistakenly thinks K is a interior decorator and cannot control the uproarious mirth and fits of coughing in the assembly. As K makes clear to the court, he does not consider the court proceedings to be proceedings at all, regarding the court at downright pitiful.[24] Indeed, the investigating bureaucrats, far from following an effective trajectory toward substantive evidence to be found and brought against K, are instead seen to act in the manner of pernickety and scrupulous, though bumbling and obtuse, paralytics. These purlind men accept no evidence, and are instead only preoccupied with micro-events such as the happenings in the hallways, the noise heard behind the doors, whispers in the court, and off-scene murmurs. All of this further highlights the formal vacuity of the Law.

Moreover, far from K standing pensively before a series of hierarchical tribunals, what one finds in *The Trial* is his incessant migratory movement through an endless subterranean network of hallways and corridors connecting a vast plethora of offices. In these seemingly sempiternal peregrinations, he encounters inter alia women, all of whom are of a singularly libidinous bent: Elsa, Miss Burstner, the washerwoman, the bailiff's wife, the judge's lover, Leni, the lascivious little girls at Titorelli's abode, and so forth.

Thus, despite his position and preemptive indictment as guilty before a transcendent Law, quite at variance with the architectural embodiment and essential functioning of the latter, K finds himself sent along a seemingly interminable wayfaring nomadic trajectory along hallways and corridors leading him from office to office, in his journey engaging in a series of sporadic dalliances with highly equivocal women.

6. From the transcendent Law to the immanence of desire

The despotic astronomical state of architecture thus gives way, in Kafka, to endless peregrinations along the subterranean mazeworks of the second state of architecture that proliferates itself ever the more pervasively and virulently within the bowels of the latter. Here the infinite distance of the transcendent despotic Law and the perverse closeness of its bureaucratic functionaries yields to the continuousness of corridors and contiguousness of offices of the latter and the latter's erotic nuances.

The implication of this is that, on the one hand, it remains the case that the transcendence of the Law finds its embodiment and functioning in the first state of architecture. However, the insinuation of the second state—functioning as it does through the continuity and contiguousness of block-segments—also implies something else. For this inextricable imbrication of continuity and contiguity is precisely the nature which Deleuze and Guattari attribute to *desire itself*. That is to say, it pertains to desire in the molecular sense of desiring-production. Here, the continuity of the hallways is analogous to the continuity of the mutant flows emitted by the desiring-machines of the unconscious, whereas the discontinuity of the block-segments which still abides at the level of the second state of architecture is homologous to the breaks in the same flows through which desiring-machines are paradoxically connected. The contiguity of the blocks is isomorphic to the fact that the connections between the desiring-machines are local—recalling, of course, that at the level of the molar socius the connections between desiring-machines come to be global, causing them to be correlated at a distance. It is, moreover, the purely *immanent* nature of desire, as desiring-production, that is itself concretely embodied in the corresponding horizontal and earthly immanence of the hallways—as opposed to the transcendent tower of the paranoiac Law. This burgeoning subversion of the transcendent astronomical state of architecture in favor of the immanence of the earthly architectural state suggests that something very fundamental is at stake in Kafka's novels: that is to say, "*where one believed that there was the law, there is in fact desire and desire alone.*"[25]

In Pauline terms, one could say not only that the immanence of grace and the law of love must supercede the Statist Law, but that the latter in fact *presupposes* and is nothing outside of the former. Viewed from this aspect, Deleuzoguattarian desire appears not dissimilar to grace as conceived of in the register of the Apostle. Though the fact remains that the despotic social machine arrests desire, or equivalently *thwarts grace*, organizes lack in its midst, erects a transcendent paranoiac Law, fusing desire to the latter and to death; it is the latter molarity, the transcendent Law, which is in fact *secondary* to the immanent molecular agitations of desire, *grace*. As seen in the Prologue of the present text, molecular desire always already pervades and invests the entire social field of molar machines. The other inescapable manner in which the Badiousian Christian truth-event and Deleuzoguattarian molecular desire resemble one another resides in the fundamentally gratuitous quality that they share. Grace arrives in a singularly untimely manner without being due, that is, inter alia without being protended. Unlike the molar transcendent Law, it is by definition utterly refractory and oblivious to the hierarchy of means, ends and goals. Molecular desiring-machines are likewise bereft of goal, meaning nothing whatsoever and being solely what they make of themselves. That in which form and function coincide has no intrinsic meaning.

This ultimately means that justice itself is *not* Law, but in fact desire. Somewhere in his book on Job, Negri definitively contends that justice, far from being law, is indeed *generosity*. It is the generosity of the pure aneconomic gift that is Pauline grace—or the gracefully destinerrant nature of desiring-production and the diametrical opposition of its celerity vis-à-vis the turgid gravitas of the paranoiac Law and its unwieldy molar machinery. Like desiring-machines it is not on the side of necessity, but on that of chance. Titorelli portrays it as "blind fortune, a winged desire." It is "not a stable will but a moving desire."[26]

And it is precisely due to the aforementioned ubiquitous pervasion of the entire social field by desire that it is not only the bureaucratic caste that find themselves involved in the process and unfolding of justice. For in *The Trial* everyone is a veritable functionary of justice: the priest, the painter, the equivocal young ladies, and likewise the perverse little girls.[27] The erotic nature of justice is evinced from the fact that the accused are the most handsome and beautiful figures (as intimated to K by his advocate),[28] whilst justice's gratuitousness is manifestly evidenced in the childishly capricious actions and reasoning of the judges and in the fact that a mere impromptu joke is able to utterly derout the repression otherwise effected by the Law (as in the scene where K stands before a pitifully inept court).[29] The young women and ladies "simultaneously bring bliss to judges, lawyers, and accused, out of a single polyvocal desire. The whole of *The Trial* is overrun by a polyvocality of desire that gives it its erotic force."[30] The functionaries of justice are those who are attracted, and even propelled, by offense. For instance, K's advocate intimates to him that his secretary and nursemaid Leni finds almost all accused men attractive;[31] whereas warder Franz lets K know that the legal authorities, far from searching for guilt in the population at large, in accordance with the dictates of the Law, are instead *drawn* to guilt.[32]

The essential point here is this: whereas, from our discussion of the despotic regime, it seemed as though the locus of power resided in the transcendent position of the despot and his appropriation of all productive organs to himself, the despotic State's purportedly transcendent Law is seen to in fact be only a *surface effect* of the immanent peregrinations of desire investing the entire social field. In this connection, K notices in *The Trial* that it is not the tribunal that deals with the real problems of desire, power and justice. Rather, these latter appertain to the molecular agitations of desire animating the hallways of the congress, the happenings behind the scenes of meetings, the wings, the back doors and side chambers.[33]

The fact that the transcendent Law is merely the surface effect of the immanent peregrinations of desire is equivalent to the Lacanian contention that the big Other (*le grand Autre*), in this case as source of Statist power, *does not exist*. That is to say, as Žižek maintains, the big Other is in no sense a Thing-in-itself, but exists only as sustained by the machinations of the collectively of individual subjects. It is not of the order of the pre-symbolic material Real which exists independently of human subjectivity. Rather, it exists only as the pure, retroactive and virtual pressuposition of the subject.[34]

This nonexistence, or at least farcical nature, of the big Other as a fiction supporting the power structure of the despotic State apparatus and the Law is brought into clear evidence by K's barbed response to his lawyer, not in the Kafka's original novel, but in Orson Welles' somewhat divergent film version of *The Trial*. Not long before the end, in the scene in the cathedral, when the lawyer advises K to defend himself against the nonexistent charges made against him by presenting himself to his accusers and detractors as a victim of a diabolical plot hatched at the behest of obscure and malignant Statist powers; K immediately jibs at this notion, not merely pointedly refusing to assume the role of victim, but countering that the conspiracy of power itself resides in this sort of insinuation and imputation of victimhood.[35] He, thus, both refuses to be lured into admissions of guilt and into being victimized.

The reason why justice in the despotic regime founders into unrepresentibility is *not* fundamentally because of the supposed infinite transcendence of the Law, but rather because justice, as Deleuze and Guattari maintain, is *desire itself.*[36] Desire, in its fundamental molecular agitation, skirts around the arena of representation that transcendence—as born of the despotic molar social machine—forever attempts to foist upon the field of immance proper to it. Indeed, K—as a man of molecular desire—constantly circumnavigates all attempts at his being represented by any transcendent instance. For instance, having somewhat reluctantly taken the services of the advocate Huld at the behest of his uncle, he flagrantly risks jeopardizing his relationship with the latter through his dalliance with the latter's secretary and nursemaid, Leni.[37] Ultimately, he goes as far as simply ditching the advocate, doing so on the basis of his stated contention that having formal legal counsel had simply burdened him and that he would be better off without it.[38] He categorically refuses to let himself be represented and instead finds justice by incessantly perambulating from one room to the next in accordance with where he is taken by his desire.[39]

Moreover, unlike the Nietzschean man of bad conscience, or the Hegelian anguished soul, or the Pauline subject mortified by the indwelling of sin, we see that K is nowise swayed by any form of a priori guilt that the transcendent Law's machinations would serve to foist upon him. In his thoughts to himself, the rejection of guilt stands as the most urgent step. Instead of considering his case a matter of crime, guilt and punishment, he views it as a business deal attended by the usual dangers that need to be warded off. Instead of dwelling upon guilt, he resolved to focus on how to plot an advantageous course of action.[40] From the outset, he tells the warden Franz in his appartment that the entire matter raised against him cannot be particularly important, in that he stands accused of something for which there is no underpinning guilt.[41] When Frau Grubach opines to him that his supposed arrest, far from being bad like the arrest of a thief, is instead "something scholarly"; he retorts that not even this much can be said, since he views his arrest as "nothing at all."[42] Later, he publically declares before the court that his arrest has no greater meaning for him than being assailed by a random group of thugs in the street, and that it was vexatious primarily on the account of disturbing his early morning sleep.[43] In the scene in the cathedral, when the priest lets him know that his guilt is an established fact, K refuses to be suborned into bad conscience, retorting: "But I am not guilty. It's a mistake. How can a human being ever be guilty?"[44]

Instead of being a guilt-ridden masochist, he is an audacious, somewhat capricious, and even lewd figure, realizing as he does the derivative status of the Law in relation to the molecular peregrinations of desire.[45] This does not, however, mean that his actions and comportment are merely whimsical and willy-nilly. For, though he is in some sense playing truant, he does notwithstanding—as shall be increasingly seen—have a technique. As, of course, does Kafka. He moves endlessly through continuous passageways and contiguous offices, in so doing opening onto the unlimited field of immance of desire—for, as observed, it is precisely desire which functions through such co-functioning of the continuous and the contiguous.

7. The Sadeian movement: indefinite postponement

It is at this point that the third possibility intimated to K by Titorelli comes into effect: the constant deferral of legal procedures into the office next door, and then into the following one, ad nauseam. This ultimately implies that K leaves the arena of ostensible acquittal and enters into a movement wherein the Law is *indefinitely postponed.* As Titorelli avers, the technique of indefinite postponement, or "prolongation," consists in permanently maintaining all juridical proceedings in their initial stages. The advantage of this is that, in so doing, the defendant is safeguarded from the eventuality of a conviction almost as effectively as he would be had he actually been entirely free. Indefinite postponement serves to shield the subject from both the shock of sudden arrest and, likewise, from the vexation of striving to obtain an ostensive acquital.[46]

Superficial acquittal corresponds to the first state of architecture: it is infinite, limited and discontinuous. Its infinite character stems directly from the fact that it follows the circular distribution of offices around the central tower. The trajectory traced along a circle, of course, in that the circle has no specific beginning or end, may be described as infinite. But it is limited and discontinuous "because the point of accusation approaches and recedes in relation to this circulation, 'swinging backwards and forwards with greater or smaller oscillations, longer or shorter delays': opposed fluxes, opposed poles, opposed periods of innocence, *guilt*, freedom, and a new arrest."[47] Each respective period of innocence or guilt is clearly limited in duration, whereas the oscillation between these two poles is what accounts for the discontinuity.

In contradistinction to this, the indefinite postponement into which K enters partakes of the second state of architecture. It is finite, unlimited, and continuous.

It is finite in that all transcendence has been eschewed and "because it works by means of segments." On the other hand, the temporal delay that it foments is unlimited and continuous, since it entails an incessant paratactic adding of mutually contiguous segments, "operating piece by piece in order to always push the limit further back."[48]

This incessant pushing back of the limit is precisely homologous to Zeno's paradox of Achilles and the tortoise, wherein Achilles is unable to catch up exactly with the tortoise: he is only able to either partake in a temporally infinite asymptotic approach toward the slow animal, or overshoot it. Sexualizing matters a little, Alenka Zupančič delineates these two alternatives countenanced by Achilles as his two faces: the "Sadeian movement" and the "Don Juanian movement," respectively. In the former, the object of desire is conceived as a whole and, through the Sadeian fantasy of the immortality of the body to be enjoyed (though also quite likely tortured or even dismembered), approached *ad infinitum*. The object of desire is approached "bit by bit" and the whole distance is never covered, one more effort always having to be made, with the greatest pleasure attained precisely through the indefinite postponement of pleasure.[49] Though this Sadeian movement does not describe the manner in which K enjoys his women, it is homologous with the form of temporality he enters in his incessant peregrinations along the second state of architecture and the manner in which the molecular agitations unleashed by the continuity and contiguity of the corridors and the hallways through which he moves serve to undermine the transcendent molarities of the astronomical state of architecture by way of a piece by piece assemblage of segments.

8. The proliferation of series along a line of escape

In tandem with this burgeoning of a line of flight animating the continuity of contiguous segments within the very bowels of the State apparatus, one finds what Deleuze and Guattari term a "proliferation of series." In the bureaucratic apparatus found in *The Trial*, there is from the outset a proliferation of trios and also duos. Bureaucratic triangles are composed of two subordinate bureaucrats who always emanate from a superordinate über-bureaucrat "for whom they function as right arm and left arm."[50]

The two subordinate elements of any trio, constitute a structural novelty in the despotic regime, implying the emergence within its bowels of the primary elements of an entirely distinct regime: the *passional*, or *postsignifying*, regime. The subordinate pair of bureaucrats form a duo entailing not only a doubling of subjects but also the "doubling of the subject, as subject of enunciation and as subject of the statement"[51]; that is to say, the birth of the *binary* subject—one of the crucial features of the emerging passional regime.[52]

Now, in Kafka's novels, owing to K's experimentation with the legal machinery of the State, the terms of the bureaucratic trios and duos come to distribute themselves on and take flight along a line of escape in relation to the various contiguous segments such as those of the police, the lawyer, and the judge. In the same movement whereby they relinquish their double or triangular form, they begin to appear less and less as hierarchical representatives of the transcendent Law and—homologously to the manner in which the desiring-machines of the molecular unconscious function and are interrelated—increasingly assume the role of "connective cogs of an assemblage of justice, each cog corresponding to a position of desire, all the cogs and all the positions communicating with each other through successive continuities."[53] These vicissitudes are manifestly exemplified in the first interrogation scene of K in which the tribunal sheds its triangular form, with the two opposite sides or assembled galleries of the court over which the judge presides realigning themselves along a single continuous line—in so doing bringing into play "corrupt warders and fatuous supervisors and examining magistrates, of whom the best that can be said is that they are humble officials, but also … a judiciary of the highest rank with its inevitable vast retinue of servants, secretaries, police officers and other assistants, perhaps even executioners …"[54]

In tandem with this dismantling of triangles and duos and the realignment of their elements along a line of flight there forthwith and henceforth transpires an increasing usurpation of the hierarchical structures of the astronomical state of architecture in favor of the contiguity of offices of the second state of architecture. This transformation bears very strongly upon the precise nature of power and the relations between repressors and repressed in the despotic machine. As Deleuze and Guattari argue, it is not the case that power is *primus inter pares* that of the infinite transcendence of the despotic State apparatus in relation to a vast populace of slaves and accused milling and groveling down below. The transcendent Law ushers in the illusion that power is pyramidal, emanating from an infinitely removed apex. In truth, however, as in the case of molecular desire, power is segmentary and linear, proceeding as it does by way of contiguity, and not by way of height and farawayness.[55] Each of the segments "is power, *a* power as well as a figure of desire."[56] One must not, however, adduce from this co-implication of desire and power in the machinic assemblage of the second state of architecture that there is *primus inter pares* a desire for power—such as, for example, a sadistic despotic desire to repress or a masochistic and slavish desire to be repressed. This would be putting things the wrong way around. For, rather than there being a primordial sadistic desire for power, it is in fact "power itself that is desire." For desire, as molecular desiring-production and thus as a machinic assemblage, "is precisely one with the gears and the components of the machine, one with the power of the machine."[57] Therefore, the desire to repress or to be repressed, along with the divisions between repressors and repressed and of oppressors and oppressed "flow out of each state of the machine, and not vice versa."[58]

9. Toward a Kafkan ethics

A positive ethics that rails against the trappings of despotism, as may be extrapolated from Deleuze and Guattari's reading of Kafka, entails an audacious experimentation and incessant migration. The first step in this affirmative ethics is the abolition of any form of a priori guilt, in conjunction with a progressive dismantling of all transcendent pretensions of the paranoid Law of the despot. As observed, this is strongly homologous to St. Paul's deactivation (Agamben's *katargeō*) of the Law and its transmogrified preservation as an immanent trans-literal law of the spirit. This latter deconstruction would, moreover, entail a concomitant dismantling of any notion of power as fundamentally and primordially the possession of the despot and his bureaucrats, and as the latter's pure and simple repression of his subjects. Though the foregoing are in a sense the case, they are so only in a derivative sense and as the mere superfices of the more subterranean and fundamental molecular agitations of desire. As observed, the intrinsic functioning of desire thus conceived is that of a continuity and contiguity of segments along a line of flight or deterritorialization.

Now, since it is the molar machines (and we specifically have the socius in mind) that trap desire into systems of representation, K—in order to follow such a line a escape—must necessarily refuse to be represented, that is, abstain from legal representation. He does so by side-stepping all tribunals. For justice is not to be found in the transcendent Law, but in the office next door, in the continuity of passageways and in the contiguity of offices; not in bureaucratic triangles and duos, but in the dissolution of these and their distribution as cogs, as positions of desire, along a segmental line of escape opening onto the unlimited field of desire which is that of justice itself. K's process of experimentation, opening onto the field of immanence wherein desire and justice are identical, may thus be termed a "*method of segmentary acceleration or proliferation*" which "*connects the finite, the contiguous, the continuous, and the unlimited.*"[59] This technique precipitates an increasing collapse of the preponderance of the astronomical state of architecture over the subject and desire in favor of a proclivity toward the earthly and more rhizomatic configuration of the second state of architecture and the desiring migration of K through its passageways and offices.

However, if taken to the absolute limit of its molecular agitation, this movement of deterritorialization shall immediately entail a dissolution of all architectural sway over desire. This is so because architecture—comprised as it is of enclosures and forms—necessarily entails the existence of molarities serving to repress the flux of desire in its originary molecular destinerrancy. In this sense, it may be said that desire is fundamentally heterogeneous to architecture and, if pushed to the point an absolute deterritorialization, shall sweep away all forms of content in the movement of an unlimited metamorphosis of bodies. In the process, molecular desire's line of flight would be divested of all spatial extension and thereby become purely *intensive*—portending as it would the recrudescence and liberation of the BwO of desire from its incarceration by the despotic socius in the immediate wake of the latter's demise.

The latter becoming-intense would, moreover, entail a transformation of the implicit death sentence sempiternally passed on the subject by the despotic Damocles' sword into a line of flight wherein death is indefinitely postponed. Thus, corresponding to the indefinite postponement entailed in Titorelli's third theoretical possibility, there is at once a indefinite postponement of execution. In addition to the binary subject associated with the bureacratic duos seen earlier, we see here another trait of that new regime, the passional regime, that is now seen disgorging itself from beneath the teetering despotic regime. That is to say, *existence under reprieve.*[60] And, as observed afore, the herein implied absolute deterritorialization leads directly into the desert of the BwO at the limit of any socius—the true province of molecular desire as foreground production. This is where the dissident from the despotic regime, the scapegoat, shall arrive; choosing the goat's anus over the frontal face of the despot.

Yet, until such an eventuality can be reached, one must be reminded that every deterritorialization brings with itself a corresponding reterritorialization. So, on the one hand, the massive movement of deterritorialization anterior to the despotic megamachine's co-optation of the primitive territorialities of the savage socius as bricks on the new full body—draws a non-segmental line of flight. On the other hand, however, the attendent force of reterritorialization serves to relocalize desire in a given segment of the line. The mutual imbrication of these two movements implies that, desire itself will at once function in two coexisting states. On the one hand, it will be bound up with a particular segment, such as a particular office. On the other hand, it will take flight on the whole line, cavorting toward the unlimited realm of the immanence of molecular desire or justice.[61]

These two states of desire correspond at the same time to two different states of law; that is, the paranoiac transcendental Law of the despotic regime and the immanent "schizo-law" which functions like justice as molecular desire and subverts the paranoiac Law from within.[62] The co-implication of these two states of law entails at once the coexistence of two movements. The first movement "captures desire within great diabolical assemblages, sweeping along in almost the same movement servants and victims, chiefs and subalterns, and only bringing about a massive deterritorialization of man by also reterritoralizing him, whether in an office, a prison, a cemetery (paranoiac law)."[63] The second movement "makes desire take flight through all the assemblages, rub up against all the segments without settling down in any of them, and carry always further the innocence of a power of deterritorialization that is the same thing as escape (the schizo-law)."[64]

The poetic figuration of the paranoiac state of law deployed by Kafka, its form of content, is the bent head pressed up against the ceiling in submission to the despot and the bricks, or block-segments, cemented in molar architectural formations; whereas that pertaining to the schizoid state of law is the straightening head bursting through the roof of the courtroom and the brick hurled at and smashing through one of its windows.[65]

10. The Don Juanian movement: anti-conjugality

We have noted that the reterritorialization imposed by the paranoiac Law tends to render any line of escape a *segmentary* one. The question, of course, is: how does one de-segment this line? By way of an initial answer, we may note that Deleuze and Guattari allude to techniques of segmental acceleration such that, if taken to its very limit, this acceleration shall lead to an untrammeled and *non*-segmentary line of flight sweeping us away in a vertiginous liberated surge into the desert of the BwO of desire where the passional xenophile alone dares to venture. This segmental acceleration concerns itself *inter alia* with so-called *extraordinary series.*[66]

In the proliferation of series, the emergent series arose from a realignment of terms along a line of flight—such as that of the bureaucratic trios and duos mentioned earlier. Certain of these series, however, are comprised of special terms which, standing as they do at the end of a given series and the start of another series, serve to augment and proliferate the number of connections made by ordinary series in the immanent field of desiring-production suffusing the social field itself.[67] The first type of these "extraordinary series" are the young women. Each of these women is linked to a particular segment, marking its end and leading K onto a new segment. In *The Trial*, the banking segment is marked by K's girlfriend Elsa; the segment corresponding to the subordinate judicial officials, that of the bailiff, and that of the judge are all marked by the washerwoman; whereas the segment of the lawyers is marked by Leni.[68]

The young women do not merely stand at the turning points between one series and another. They, in addition, stand at the crossroads of a number of segments, which they serve to connect whilst at the same time distributing an extraordinary series through and precipitating a line of flight along all the proliferating chains assembled of the same linked segments. It is in this way that the young women reveal and establish the contiguity of segments which, though faraway from each other—in the terminology pertaining to the second and subterranean state of architecture—are in fact contiguous, by inter alia finding the location of service doors and such like apertures of thoroughfare.[69]

Crucially, much as do the connections between organs-partial objects at the level of desiring-production on the BwO, the encounters with these young women are brought about through chance, their very nature being evanescent and fleeting. It is the sheer capriciousness of the women that sends K off immediately onto a new segment. The chance-ridden contingency of these encounters and the chanciness of the young women themselves, for Deleuze and Guattari, implies that these extraordinary feminine terms serve to open K's peregrinations onto the unlimited field of immanence of desire, of justice. After all, these young women and justice resemble one another in that they are both devoid of principles.[70] In the scene in the cathedral, the priest, who claims to belong to the court, avers to K: "The court asks nothing of you. It receives you when you come and its releases you when you go."[71]

Most crucially, these renegade sisters, maids and whores, are radically anti-conjugal; that is to say, at variance with the triadic conjugality of the incestuous royal triangle despot-sister-mother.[72] This anti-conjugality is evinced by the fact that K is either deserted by the young women he encounters or himself deserts them. For example, when Frieda accuses K of infidelity to her, this latter transgression arises from the fact that he has by now already moved onto the new segment associated with Olga, whose arrival she in fact caused in tandem with her termination of her own segment.[73]

This anti-conjugality of K is homologous to the second face of Achilles as delineated by Zupančič: the "Don Juanian movement" of failing to catch up to the tortoise by outrunning it. Whereas through his entrance into the mode of temporality of the second state of architecture (that of indefinite postponement) K had assumed a mode of desire animated by a "bit by bit" approach toward the whole object of desire whose enjoyment stemmed from the infinite delay of pleasure itself, the more prodigious connectivity facilitated by the extraordinary series of the equivocal young women now allows him to completely enjoy each one of the women. Each of the two modes of desiring, however, has its distinct drawback. In the case of the Sadeian movement of indefinite postponement, K assembles segments, or pieces, which can never make up a *whole*, or One. With the Don Juanian movement, on the other hand, though he does enjoy each of the series of women "one by one," he can never say that he has enjoyed them *all*.[74]

Notwithstanding, what K accomplishes, as had Don Juan before him, is an unthinkably scandalous subversion of the Law itself. He effects the universal distribution of the very Lacanian *objet petit a*, the Platonic *agalma* within the subject, on account of which the desire and love of the other is provoked—something whose bestowal must by definition be exclusive. Here, it is not the case that K, as Zupančič similarly maintains of Don Juan, pursues his erotic dalliances with the vast array of women with whom he is countenanced in order to partake of a variegated menu. Rather, he seduces the women precisely because of what they have in common: that they are women, that is, without regard of their symbolic roles within Statist representation. The allocation of symbolic roles to women is the patriarchal mechanism used by the vast Statist machinery to repress the fact that, as Lacan infamously maintains, "woman does not exist." This representational machinery then permits the subject to enjoy woman solely in her specific designations as typist, washerwoman, nursemaid, and so on. Don Juan and K, on the other hand, call State representation's bluff that woman purportedly exists by suddenly demanding to enjoy woman *as woman*, that is, enjoyment of woman as the fundamental lack that she is outside her symbolic representations. In so doing, they undermine the panoply of symbolic representations by which the State represses the fact that woman does not exist and, by demanding the very "substance" of woman, show that she does not exist.[75]

Immersion in the extraordinary series of the young women allows K to become Don Juan, as he switches from the Sadean movement of the metonymy of desire (and its infinite approach toward the "true" and eternally elusive object of desire) to what Freud and Lacan term the death-drive (*Todestrieb*). With the drive, as opposed to striving to fill its constitutive lack, the subject instead takes lack itself as object. K and Don Juan, thus, attain satisfaction without attaining their aim. It is not, therefore, what was lacking in a particular woman that causes K to seek another. Rather, each woman does provide satisfaction and, on account of this very fact, causes him to move onto the next one, and so on, in an endless movement of circulation around the void.[76]

11. Beyond Don Juan: indexical homosexuality and homosexual effusion

The series of the young women precipitates and enables a marked segmental acceleration in K's peregrinations through the second state of architecture. There is, however, another series whose segmental acceleration is even more acute than that of the sisters, maids and whores. This is the series of the artist, Titorelli—a manifestly homosexual one. In this connection, Deleuze and Guattari aver that whereas the young women vouchsafe and spur on K's deterritorialization and flight from one segment to the next, the artist ensures a more "shifting and continuous line of flight" which "overflows all the segments and sweeps up all the connections."[77] Whilst the women are situated at the preponderant points of connection of the machinic assemblage of desire and justice, the artist "reunites all these points, arranges them in his own specific machine which extends across the whole field of immanence, and even anticipates it."[78]

Certain passages in *The Trial* have even been deleted by Kafka on account of their well-nigh pornographic explicitness. We have here the tight-fitting clothes of Arthur, Jeremy and Barnabas, and those of the subalterns in *The Castle*. The two police inspectors in *The Trial* wear skin-tight clothes with an assortment of kinky buckles, belts and other appurtenances, whereas the executioner who whips the inspectors in the stuffy and sultry room wherein K fortuitously finds them is sheathed in a short leather garment which leaves much of his chest exposed—rendering this latter scene strikingly reminiscent of something Edmund White would have beheld on countless occasions on his numerous libidinous sojourns to *The Mineshaft* gay club in New York, at which a vast array of sadomasochistic rituals were on display.[80] Shades here of what Christopher Hitchens recalls as former British Labour MP Tom Driberg's impromptu dalliances with members of the armed forces in the "cottages" of Leicester Square, Paddington Station, Clapham Common, St. James' Park;[81] and likewise his trawling of the Via Veneto for blue-collar workers with wing-man Gore Vidal.[82] "How he escaped gaol is a mystery," Paul Johnson once averred of Driberg.[83]

Nonetheless, though homosexual, the foregoing constellation of bureaucratic doubles represents, for Deleuze and Guattari, a lesser segmental acceleration and connection of contiguities in the immanent field of desire than that precipitated by the extraordinary series of the homosexual artist and K's homosexuality with the latter. They function merely as "homosexual indexes," whereas the artist's "Homosexual effusion has another finality that is no more than pointed to by these indexes."[84] Whether the authors would have adjudged the deterritorializations of Driberg and Vidal as anything more than indexical remains unknowable. Notwithstanding, they were certainly not as deleterious as the endlessly querulous and brawling homosexual dalliance of Rimbaud and Verlaine, whose deterritorializations found themselves definitively halted with Verlaine reterritorialized into a hard labor camp for having shot Rimbaud in the hand.[85] Or as self-defeating and harmful to the face and cranium as the homosexual deterritorializations of Hart Crane, who found himself suicidally reterritorialized at the bottom of the sea the very day after being beaten up by the last sailor upon whom he had made advances. Crane maintained that the bottom of the sea is cruel.[86] Or as embarrassing and shameful as the purported or probable homosexual deterritorializations between T.E. Lawrence and the Bey of Deraa, where the former's lingering guilt at having enjoyed being flogged by the latter to a greater degree than his conscience would permit extended itself into Lawrence's subsequent self-imposed series of flagellations at the hands of the naïve young Scotsman John Bruce and ultimately into his countervailing and tragic reterritorilization in death through what was not necessarily a fatal motorcycle accident.[87] Or as needlessly visceral and gruesome as the final denouement to the homosexuality between Yukio Mishima and Waseda University student Masakatsu Morita whose deterritorializations, owing to a conjunction of excessive patriotic fervor for the Emperor and Mishima's incorrigibly morbid and interminable thanatophilia, foundered into the cemetery when both where reterritorialized in death through near-simultaneous *seppuku*.[88] Lastly, one has the homosexuality of Brunetto Latini, who found himself reterritorialized for all eternity into the seventh circle of Hell.[89]

12. Beyond homosexual effusion: the bachelor

Notwithstanding, segmental acceleration can be increased even further. It is here that we move from the homosexual series of the artist to that of the bachelor. The authors maintain the bachelor to be a much more formidable and more intense state of desire than both incestuous desire and homosexual desire. The line of flight of the bachelor is distinct from that precipitated by the extraordinary series of the bureaucratic homosexual doubles, that of the young women and girls, and even that of the homosexual artist. For it now becomes properly-speaking *intensive*.[90]

In so doing, the bachelor engages in a journey in intensity in which, in his very solitariness and utter refractoriness to all family and conjugality, he is able all the more to invest the social field—producing intensive quantities directly on the socius itself: "The highest desire desires both to be alone and to be connected to all the machines of desire. A machine that is all the more social and collective insofar as it is solitary, a bachelor, and that, tracing the line of escape, is equivalent in itself to a community whose conditions haven't yet been established."[91] It is not so much that the bachelor flees the world. Rather, "he grasps it and makes it take flight on a continuous and artistic line."[92] The solitary, anti-conjugal and celibate figure of Saint Simeon of Stylites, who lived into the fifth century, exemplifies to some degree this investment of the social field by the bachelor. Though a fringe figure perambulating the countryside reciting the Beatitudes, people would flock around to admire his routine of standing in one position for ages without flinching, and crowd around to hear his murmured prayers. To increase his solitude, he dwelt permanently for the next thirty-seven years atop a sixty foot pillar. Yet the crowds that he drew only increased exponentially.[93] Sartre's favorite bachelor is, of course, Jean Genet. He maintains of this peculiar being that "He is a monster, the exception to the rule, the improbable, in fact the impossible. He is an undesirable, born not of man but of an incubus. He cannot found a family. Thus, he has not only been exiled from the fields, woods and springs, he has been excluded *from his own nature.*"[94] Sartre continues to harp on about this utterly anti-conjugal desiring nature of Genet thus: "Genet takes his stand on absolute particularity; he will derive only from himself. This upheaval affects even his sexuality. His contorted desire, sharp and nervous, derives only from itself."[95] Like that of the inherently ateleogolical desiring-machines, his desire is "sterile, rootless and aimless" and replete with "a fierce demand for its autonomy and singularity, in defiance of the rules, in defiance of nature, in defiance of life, in defiance of the species and in defiance of society."[96] Yet, the seemingly pure solipism of Genet's mode of desiring is shown to be fundamentally connected to the social field when Sartre avers that "Through a thousand human figures it is he himself that Genet is going to pursue, but himself as Other," in a ramified process that is "greater than his temporary incarnation and greater than Genet."[97]

Much, however, as in the case of the failed homosexual deterritorializations mentioned before our vignette on Genet, there likewise exists an extensive transhistorical litany of botched bachelor—or, for that matter, bachelorette—machines. Some of these are celibate, some not. Even the bachelor, as a state of desire, has its pitfalls and weaknesses: such as the Oedipal temptation to hermitry or the desire to commit suicide.[98] The first is evinced in the sixth century

Saint John the Silent who, having erected a celibate bachelor machine, immured himself in a private cell in a monastery in Jerusalem, walling over the door with bricks, and not speaking to anyone whomsoever for the next seven decades—indeed, until death.[99] The second weakness shall be seen soon enough. A third weakness, one may add, would be succumbing to incarceration or execution at the hands of the State.

The Adamite sect which originated in second-century North Africa abolished marriage in favor of nude prelapsarian promiscuity. Conjugality was seen as the origin of sin and misery. These deterritorializations were however to be countered in devastating fashion when, in the fifteenth century the State reterritorialized seventy members of Peter Kanis' Adamite community on the Nerarka River in Bohemia in death by burning them at the stake, whilst all remaining Adamites were wiped out by coordinated military action throughout Europe by the middle of the nineteenth century.[100]

From their emergence in the twelfth century, all-female communes known as the Beguines began to proliferate across Europe. The deterritorializations precipitated by their rejection of marriage in favor of polyamory were counteracted by inquisitorial reterritorializations whereby countless thousands of Beguines were reterritorialized into the grave for heresy, starting with the mystic Marguerite Porete in the early fourteenth century.[101]

The celibate anti-conjugal deterritorializations of the second-century Sicilian saint Agatha foundered when her refusal to marry Roman official Quintian brought about her brutal reterritorialization in a brothel, a prison, and finally in the grave after her mammaries had been removed.[102] Inundated by a vision of the recently martyred St. Agatha, the Sicilian-born Saint Lucia likewise attempted the erection of a celibate bachelor machine. Its deterritorializations were likewise counteracted when her refusal to contract an arranged marriage brought about her savage reterritorialization in a brothel and eventually in the cemetery after being repeatedly stabbed to death.[103]

As in the case of the Adamites and Beguines, the Khlysts, a Russian sect dating from the first half of the seventeenth century, so as to nullify the power of sin and hasten the internalized Second Coming of Christ into each votary's individual soul, likewise eschewed conjugality in favor of orgiastic self-flagellation and group sin. Whatever the deterritorializations herein effected were eventually brutally counteracted by the Soviets, who reterritorialized the Khlysts in death.[104] Shortly prior to this, the Khlyst Rasputin was likewise reterritorialized into death by marginal members of the royal family by being poisoned, shot four times, clubbed, castrated, drowned in a river, and cremated.[105]

If the Khlysts erected an aborted non-celibate bachelor machine, then the Skoptsy sect founded by the Russian peasant Kondratii Selivanov in the second half of the eighteenth century engineered an aborted celibate bachelor machine. Rebelling against not merely all forms of conjugality but against all form of eroticism, Selivanov held out auto-emasculation as the fastest route to salvation. Whatever the deterritorializations effected by Selivanov's bachelor machine, they were forthwith and repeatedly to be countervailed in the ensuing decades and centuries. For castrating thirteen men, Selivanov was reterritorialized in Siberia, then subsequently in an insane asylum, and eventually in a monastery. Starting in the final few decades of the nineteenth century, the State reterritorialized thousands of Skoptics onto their identity documents by classifying them as eunuchs, and later by parading them through streets dressed in female attire and dunce caps. In the first few decades of the twentieth century, the State reterritorialized the Skoptics into the courtroom, prison, and even death.[106]

After Marshall Applewhite's dabblings in indexical homosexuality had caused him to be temporarily reterritorialized into a psychiatric ward, in the manner of a Skoptic, Applewhite auto-emasculated in 1985 as a prelude to the erection of an infernal celibate bachelor machine. Whatever deterritorializations may have been set off by Applewhite were however at length savagely countervailed on March 26, 1997, when, so as to purportedly join Jesus on a spacecraft trailing in the fumes of the rapidly approaching Hale-Bopp comet, Applewhite and thirty-eight members of his Heaven's Gate UFO cult reterritorialized themselves onto death by dining on pudding and apple sauce laced with a fatal dose of phenobarbital chased down with a shot of imported vodka. Eight other members had likewise botched their attempted erection of celibate bachelor machines. Like Applewhite, they too had become skoptics.[107]

13. Suicide: the self-willed damnation of Don Juan

From the Lacanian perspective, the problem with these catastrophic suicides is that they only serve to preserve the consistency of the Statist big Other or, equivalently, the despotic reterritorializing sway of the astronomical state of architecture under which the subject coauthors its own death. This form of suicide, as Zupančič maintains apropos of Kant, follows the logic of an "infinite purification" in which the subject divests itself of all pathological objects of desire, to the point of ultimately sacrificing itself. At the same time, the Statist big Other only grows more imperious, sadistically demanding the subject to sacrifice more and more of itself, until such time as it literally commits suicide in its masochistic striving to destroy the extimate pathological element within itself.[108]

Therefore, what the subject must commit in order to accomplish an authentic subversive ethical act is a different type of suicide: not that of sacrificing its life on the altar of the Statist big Other, but rather that of "killing" itself through and *in the Other*. This entails an act of annihilation of that which, in the Statist symbolic order of the Other, shackles the subject to his particular ideologically interpellated identity. In Kant, this type of suicide necessitates a certain kind of *regicide*: not the corporeal destruction of the despot, but his *formal execution*: the destruction of the despot's symbolic body (or what Kantorowicz avers apropos of the Norman Anonymous to be his theo-political "body politic"), and therefore the fracturing of the very symbolic order of the Statist big Other itself. This act is *indeed* a suicide, in that the subject, or "the people," is/are only such through the despotic Statist symbolic order that is here to suffer the countersignifying vectors of insurrection. Since it is only the despot in his symbolic mandate who gives the subject its existence as standing before an absolutely vacuous yet pragmatically effectual Law, by undermining this same symbolic mandate, the subject effectively commits "suicide."[109] This act of the subject belongs *stricto sensu* to the order of Kantian "diabolical evil," in the sense that it effects a subversion of the purely formal and vacuous moral Law by the same moral Law itself, in which the distinction between "good" and "evil" falls into abeyance—much as, in Deleuze and Guattari's reading of Kafka, the immanence of desire functioning through the second state of architecture is exposed by K to be the very self-subverting essence of the Law, of which the latter is a mere superficial effect. Otherwise put, in Kafka's *The Trial*, K's molecular ingenuity effects the self-annulment of the transcendental appurtenances of the Law at the behest of its deeper element: the immanence of desire suffusing the entire social field. In this endless process of deconstruction, by way of incessantly skirting the arena of Statist representation itself, K may be seen as committing a variant of continuous symbolic suicide.

This symbolic self-annihilation is wrenched up to the most extreme pitch of exigency in the case of Don Juan. The manner in which the epic seducer comports himself in relation to the Statist big Other of the Law, at first, seems to entail a massive concession to the latter's imperious transcendence. For, far from viewing the absolute Other as nonexistent, Don Juan insists that God *does* in fact exist. What is scandalous about his position is, however, that it does not stop him from living in *complete disregard* of the Deity.[110] This is what moves Zupančič to aver that his position is *stricto sensu* an instance of Kantian "diabolical evil": by simultaneously maintaining God in existence and holding him in contempt, Don Juan subverts the moral Law through the moral Law itself. He does not oppose the despotic Law from a "negative" vantage-point on account of particular pathological motives. Rather, he does so *directly* and without regard for self-interest—even to the point of flirting with the machinery of death. And it is precisely in this that he accomplishes a pure ethical act, in which the "highest good" is indistinguishable from "pure evil."[111]

It is this comportment that Don Juan maintains until the bitter end. Unlike the pusillanimous Hegelian slave whose trembling before the prospect of biological annihilation underpins his espousal of a series of mendacious and self-contradictory ideologies, he refuses to back down before the threat of real or symbolic death with which he is threatened by the Absolute Master: Death. Indeed, in the finale of the entire play he effects what Zupančič terms a "hystericization of the Beyond, of the Other, God." This hysteria assumes the form of a vast procession of emissaries of Heaven who plead with him in variegated offers of repentance, and resembles the position of the hysterical subject firing hopeless imprecations at the Other to identify his desire: the Lacanian question "*Che vuoi?*" This divine hysteria is in effect the terminal desuetude of Heaven's masterly position, whose impotence is in clear evidence with its hysterical outburst of rolling thunder and lightning.[112] In apocalyptic scenes, the earth opens, Don Juan is swallowed up by the yawning abyss, engulfed in searing flames, and condemned to an eternity of torment whose sheer intensity of both physical and psychological pain is immeasurably more excruciating than anything that can possibly be imagined, let alone conceived.

14. The collective liberation of all imbeciles

What has emerged into view on account of K's ingenuous molecularizing deterritorializations, is the properly emancipatory and revolutionary comportment to be adopted in relation to the despotic Law and, more specifically, in relation to the Lacanian "big Other" of Statist ideological power. This comportment at once concerns itself with the subject's ideological interpellation by the big Statist Other and its entailed mode of dwelling in discursive language. In this connection, Žižek alludes to what he sees as two opposed styles of stupidity. The first is that of the idiot. The idiot may very well be exceedingly intelligent, such as one finds in the case of the autistic scientific genius, but is breathtakingly oblivious to the context within which enunciations are made. This places the idiot in a solitary position completely outside the big Statist Other. On the other hand, in perfect contradistinction to the idiot, the moron dwells completely within the discourse of the big Other—yet does so in a very stupid manner. For, though entirely privy to the multifarious contextual rules and cues of discursive language, he is utterly incapable of grasping the ideological distortions arising from causal factors heterogeneous to the order of institutional language that serve to interpellate and determine his subject position as what it is.[113] He is, thus, a covert Gadamerian, in the sense of believing that only prejudices exist and that these may be progressively unearthed and eroded through the dialogical encounter of two opposing sets of discursive formations. What he entirely forgets is the Habermasian insight that this thick substrate of tacit prejudicial assumptions underpinning overt statements, owing to its inevitable ideological distortion by properly non-discursive factors, such as the mode of production itself of a specific social machine, are themselves scrambled—herein requiring analysis to step outside the linguistic confines of *Verstehen* and move to the more scientific level of *Erklarung*. Main-stream liberals may be seen as the exemplar *par excellence* of the moronic relation to language and the State. Though the liberal moron is not autistic in the same manner as the idiot, one may aver that he remains strongly tinged by a different sort of autism: the collective autism wrought by the ideological interpellation of the Statist big Other which, on account of the liberal's singular inability to establish a more robustly curmudgeonly ironic pathos of distance between himself and the ideological formations of mainstream discourse, serves to mire the liberal subject in endless refrains of self-righteous yet insipid sanctimonious pablum from whose tepid inanity he is scarcely able to liberate himself. In this sense, in Kafka's *The Trial*, the bumbling yet pernickety and meticulously scrupulous bureaucratic perverts would certainly be idiots; whereas K's uncle, in his endless sanctimonious prodding of K to take his case seriously, is without doubt a moron.

This brings us to what Žižek posits as an intermediate position between that of the idiot and that of the moron: the imbecile. The imbecile is fundamentally marked by moderate to severe retardation and, as the science of phrenology has rendered pellucid since the nineteenth century with its analysis of the skulls of homicidal necrophiles in relation to their brain function, a definite innate proclivity to criminality. The imbecile, while cleverer than the autistic and intellectually precocious scientific idiot, whose intelligence quotient ranges from zero to twenty-five, remains more stupid than the liberal moron, whose intelligence quotient ranges from fifty-one to seventy-five. His intelligence quotient is, therefore, between twenty-six and fifty.[114]

Our overriding concern is, however, the imbecile's subject position in relation to institutional discursive language and the Lacanian Statist big Other of the transcendent despotic Law. In the same manner in which Nietzsche fails not to scrutinize who it precisely is that makes the ethical evaluations of nobility, baseness or evil, we must first inquire as to the exact identity of the imbecile. *Who is* the imbecile? Erasing the prefix "im" from the word "imbecile," Žižek arrives at "becile," which he etymologically traces to the latin *baculum*, meaning inter alia "walking stick." The imbecile is, therefore, someone who walks around without the aid of a stick. In Lacanese, the imbecile is a subject who is fully aware of the machinations of the Statist big Other, but is in the habit of cultivating an ironic distance with regard to it. He mistrusts it, being fully aware that it *sensu stricto* does not exist; that it is inconsistent, or barred. The imbecile, for this reason, occupies a strange interstitial position between the autistic idiot and the differently autistic liberal moron.[115]

On this account, if the pernickety little bureaucratic perverts in Kafka's *The Trial* are truly idiots and K's uncle is a complete moron, then what subject position does K himself occupy? What is his designated type of stupidity? Is he an imbecile? True, he completely mistrusts the despotic Law of the Statist big Other, realizing its chimerical and derivative status in relation to the immanence of molecular desiring-production. Yet, Žižek swiftly moves to complicate the whole matter of imbecility. For, much as no one in *The Trial* knows the content of the transcendent Law, no one knows precisely what the "im" in "imbecile" negates. Indeed, what would a "becile" be? Žižek contends that a "becile," far from being the simple negation of an imbecile, is the same as an imbecile—except that he redoubles the imbecile's lack of any substantial basis in the big Statist Other as a lack internal to the Other itself. That is to say, the "becile" is an imbecile whose imbecility is at once the imbecility of the transcendent Statist Law.[116]

In other worlds, K, as a "becilic" imbecile, is precisely he who knows full well that the despot is an imbecile. This, of course, implies that the despot's hauntingly absent speech- and existence-purloining deity is also an imbecile. K is thus a true revolutionary who, in the same manner in which the Christian subject's declaration of the Resurrection-event immanentizes the transcendence of God the Father as the filiation of all emergent son-subjects; by exposing the despot as an imbecile, precipitates the generalized demise of all idiots and morons and the incipient universal proliferation and ensuing loving collective labor of liberation from the despotic Statist machinery of all known imbeciles.

Notes

1. Kafka 1999, 140–167, 235–249.
2. Deleuze and Guattari 1986, 43, 44.
3. Kafka 2015, 126.
4. Ibid. 168.
5. Žižek 2013, 687, 688.
6. Ibid. 72.
7. Ibid. 75.
8. Ibid. 76, 77.
9. Ibid. 72, 77.
10. Kafka 2015, 126–127.
11. Ibid. 122, 123.
12. Ibid. 126–128.
13. Deleuze and Guattari 1986, 51.
14. Ibid. 73.
15. Kafka 2015, 66.
16. Ibid. 131.
17. Ibid. 75.
18. Ibid. 77.
19. Ibid.
20. Kafka 2015, 171, 172.
21. Ibid. 174, 175.
22. Ibid. 120, 121.
23. Ibid. 9, 10.
24. Ibid. 32.
25. Deleuze and Guattari 1986, 49.
26. Ibid.
27. Ibid.
28. Kafka 2015, 147.
29. Ibid. 32.
30. Deleuze and Guattari 1986, 49.
31. Kafka 2015, 147.
32. Ibid. 5.
33. Deleuze and Guattari 1986, 50.
34. Žižek 2013, 185.
35. Ibid. 687.
36. Deleuze and Guattari 1986, 50.
37. Kafka 2015, 88.
38. Ibid. 148, 149.
39. Deleuze and Guattari 1986, 50.

40. Kafka 2015, 101.
41. Ibid. 9,
42. Ibid. 16.
43. Ibid. 34, 35.
44. Ibid. 68, 69.
45. Deleuze and Guattari 1986, 46.
46. Kafka 2015, 128.
47. Deleuze and Guattari 1986, 52.
48. Ibid.
49. Zupančič 2011, 106, 107.
50. Deleuze and Guattari 1986, 53, 54.
51. Ibid. 52.
52. Deleuze and Guattari 1988, 127–129.
53. Deleuze and Guattari 1986, 55.
54. Kafka 2015, 36.
55. Deleuze and Guattari 1986, 55, 56.
56. Ibid. 56.
57. Ibid.
58. Ibid. 56, 57.
59. Ibid. 58.
60. Deleuze and Guattari 1988, 123–125.
61. Deleuze and Guattari 1986, 59.
62. Ibid.
63. Ibid. 60.
64. Ibid.
65. Ibid. 59, 61.
66. Ibid. 63.
67. Ibid.
68. Ibid.
69. Ibid. 63, 64.
70. Ibid. 64.
71. Kafka 2015, 177.
72. Deleuze and Guattari 1986, 64.
73. Ibid. 68.
74. Zupančič 2011, 107.
75. Ibid. 128, 131, 132.
76. Ibid. 136.
77. Deleuze and Guattari 1986, 69.
78. Ibid.
79. Ibid. 68.
80. Ibid.

81. Hitchens 2011, 412, 413.
82. Hitchens 2010, 153.
83. Johnson 2010, 99.
84. Deleuze and Guattari 1986, 68, 69.
85. Largo 2008, 224, 225.
86. Ibid. 67, 68.
87. Lawrence 2008, 452–454; Wilson 1989, 217–220.
88. Nathan 1975, 271–273, 278–280.
89. Dante 2003, 205–209, Canto XV.
90. Deleuze and Guattari 1986, 70, 71.
91. Ibid. 71.
92. Ibid.
93. Largo 2010, 459–460.
94. Sartre 1963, 81.
95. Ibid. 82.
96. Ibid.
97. Ibid. 86.
98. Deleuze and Guattari 1986, 70, 71.
99. Largo 2010, 276.
100. Ibid. 6, 7.
101. Ibid. 67.
102. Ibid. 9.
103. Ibid. 323, 324.
104. Ibid. 295.
105. Ibid. 422.
106. Ibid. 461, 462.
107. Ibid. 231, 232.
108. Zupančič 2011, 83, 84.
109. Ibid. 84, 85.
110. Ibid. 123.
111. Ibid. 90, 91.
112. Ibid. 126, 127.
113. Žižek 2013, 1, 2.
114. Ibid.
115. Ibid. 2.
116. Ibid. 3.

11 Demise of the Despot

1. Toward regicide

We have seen that the purported transcendence of the Law, which fuses desire to the desire of the despot (desire as the desire for the desire of the Other and the Other's desire, as the early Lacan would say) and to an ever-present threat of death that keeps the subject enslaved to the sovereign, gives way to something completely different, something which finds itself already burgeoning within the very bowels of the State apparatus. That is to say, the ostensive point of origination of power and the transcendental Law's force of indictment as emanating from the despotic apex of the imperial pyramid is glimpsed by K as nothing but a rather superficial effect of an immanence of desire ubiquituously and interstitially pervading the entire social body. This means that desire—and likewise power—is, in the first instance, in the infrastructure—and only secondarily in the superstructure.

The first step in K's dismantling of despotism is the abolition of any *a priori* sense of guilt before the Law. Here, his audacity moves him away from a static prostration before all hierarchical tribunals onto a line of flight through the immanence of passageways and faraway-though-contiguous offices, the transition from office to office being animated by the extraordinary series of equivocal women who serve also to connect his peregrinations to the vast network of segments comprising the entire social body—a body pervaded by the molecular agitations of desire itself.

This anticonjugal segmental acceleration finds itself still further exacerbated by K's participation in the more intense states of desire precipitated by his homosexuality with the artist Titorelli—a segmentary acceleration furthermore exacerbated via his animation by the bachelor machine which renders his journey and line of flight properly speaking intensive.

But how does K's intimation to us of all this bode for the supposed omnipotence of the despot and his bureaucratic perverts? The despot, the former bronze-eyed artist who—like the Hegelian master—had purportedly ascended to the level of a hitherto unprecedented omnipotence, owing to K's sober molecular ingenuity, is now reduced to the level of *comedy*. Even the quietistic Gadamer puts this quite well: though deceived into a belief in his absolute mastery, the master "no longer has the genuine self-consciousness of a master but that of a servant—a phenomenon that includes, we know, the most comical forms of anxious obedience to the servant. For us it is clear that such a master is no master. But is it clear for the master? Is he not comical precisely because he feels himself to be a master, yet in truth is afraid?"[1]

Though Gadamer is correct, the master or despot is even more ridiculous than this. Here, Deleuze and Guattari refer to the historial vicissitudes of the despotic State formations as a "long chronicle that will carry the assassinated, dismembered, dis-organized, filed-down body of the despot into the latrines of the city"—vicissitudes wherein the despot is ultimately revealed to be "nothing more than this 'dead rat's ass suspended from the ceiling of the sky.' The organs begin by detaching themselves from the despotic body, the organs of the citizens rise up against the tyrant."[2] The authors have in mind here the grisly end of the despot Heliogabalus as narrated by Antonin Artaud. Hunted by his own praetorian guards all over his palace whilst sensing "death on all sides" in a "fear-crazed" "nameless panic," the doomed tyrant is gutted in the latrines amid blood and excrement in "a scene from the meat-stall," "an ancient tableau of the slaughterhouse," his carcass reduced to a "lamentable lump of flesh already drained of blood," dragged through the town, flayed, and disposed into the Tiber "without inscription and without tomb," dying the death of an abject animal and a coward.[3] Coronation, as Ambrose Bierce maintains in his dictionary, is "The ceremony of investing a sovereign with the outward and visible signs of his divine right to be blown skyhigh with a dynamite bomb."[4]

This all being as it is, a markedly more thorough meditation on the mode of demise of the despot is to be found in Shakespeare's *Richard II*. In the tragedy, the king's overly burdensome escalation of taxation upon his own land, serves to unwittingly precipitate massive movements of deterritorialization within the then-roiling bowels of the State machinery. The burgeonings of these movements are first intimated by Northumberland—among the first privy to the inauspicious advance of "eight tall ships, three-thousand men of war" at the behest of the banished Henry Bolingbroke, Duke of Hereford.[5] That the pervasiveness of these massive movements of deterritorialization are rapidly compounding themselves is made all the more evident when word evinces that Lord Northumberland himself, his son, Henry Percy, the lords of Ross, Beaumond, and Willoughby, the Earl of Worcester, and much of the commons, have also posthaste joined the swelling revolt on the side of the seditious Bolingbroke.

The hopelessness of the king's situation is made clear by Bushy's pessimistic prognostication that "little office/The hateful commons will perform for us,/ Except like curs to tear us all to pieces"[6]—and Green's contention that York's strivings to thwart Bolingbroke's advance are potentially as futile as "numbr'ing sands, and drinking oceans dry."[7] From here on in, the gradual demise of the full body of the despot sets in like an incipient cancer that shall in time metastatize. In his reading of the play, Kantorowicz describes what he sees as the progressive fall into desuetude of the king's body natural—and likewise of his body politic. One observes the king's body natural and its perishability being foregrounded over the latter numinous body when Richard morbidly intones:[8]

> Let's talk of graves, of worms, and epitaphs,
> Make dusk our paper, and with rainy eyes
> Write sorrow on the bosom of earth.
> Let's choose executors and talk of wills:
> And yet not so, for what can we bequeath
> Save our deposed bodies to the ground?
> Our lands, our lives, and all are Bolingbroke's,
> And nothing can we call our own, but death
> And that small model of the barren earth
> Which serves as paste and cover to our bones. (*Richard II*, III.ii.145–154)

Kantorowicz now proceeds to go further, contending that—in addition to the foregrounding of the king's mortality over his immortality—kingship itself is glimpsed as essentially and inherently prone to death and as that which, in the end, is synonymous with death and nothing but death.[9] We witness here the interminable procession of slain kings being conjured up by Richard:

> For God's sake let us sit upon the ground
> And tell sad stories of the death of kings:
> How some have been deposed, some slain in war,
> Some haunted by the ghosts they have deposed,
> Some poisoned by their wives, some sleeping killed,
> All murdered—for within the hollow crown
> That rounds the mortal temples of a king
> Keeps Death his court, and there the antic sits,
> Scoffing his state and grinning at his pomp,
> Allowing him a breath, a little scene,
> To monarchize, be feared, and kill with looks,
> Infusing him with self and vain conceit,
> As if this flesh which walls about our life
> Were brass impregnable; and, humoured thus,
> Comes at the last, and with a little pin
> Bores thorough his castle wall, and farewell king! (*Richard II*, III.ii.155–170)

For Kantorowicz, the king that according to *inter alia* Plowden's *Reports* never dies, gives way to the king that always dies and cannot but die, revealing in tandem—quite in contradistinction to the solid and much vaunted monarch armed with diverse prerogatives—the feeble and enfeebled human nature of the king's body natural:[10]

> Cover your heads, and mock not flesh and blood
> With solemn reverence; throw away respect,
> Tradition, form, and ceremonious duty;
> For you have but mistook me all this while:
> I live with bread like you, feel want,
> Taste grief, need friends—subjected thus,
> How can you say to me, I am a king? (*Richard II*, III.ii.171–177)

Richard will, however, descend to a still lower level of abjectness. Very promptly one observes no longer merely the process of diremption of the king's two bodies, but ever the more so that of the denigration of his body *politic*, that is, the latter's increasing reduction to his body natural (and nothing beside) as the occluded truth and essence of the former numinous despotic full body.[11] And this Richard bespeaks in all its unadorned misery:

> I'll give my jewels for a set of beads;
> My gorgeous palace for a hermitage;

May gay apparel for an almsman's gown;
My figured goblets for a dish of wood;
My scepter for a palmer's walking-staff;
My subjects for a pair of carved saints;
And my large kingdom for a little grave,
A little, little grave, an obscure grave. (*Richard II*, III.iii.146–153)

A new and still greater level of abjectness shall, however, be reached. Northumberland orders the king to come down to the base court of Flint Castle to meet the insurrectionary Bolingbroke, to which demand Richard retorts in the following jocular tones:

Down, down I come, like glist'ring Phaethon,
Wanting the manage of unruly jades.
In the base court! Base court where kings grow base,
To come at traitors' calls, and do them grace:
In the base court, come down: down court, down king,
For night owls shriek where mounting larks should sing. (*Richard II*, III.iii.177–182)

This is the language of a fool. Richard is here a fool in two modalities: fool of kingship and fool of himself. He is a fool playing king and a king playing the fool. Here, at once, he is a mere semblance of both the kingship which he nonetheless possesses at most nominally and of the manhood of which he is well-nigh bereft. Not only is there now, as there assuredly is, merely a clear diremption between kingly body natural and monarchic body politic. For *both* are in a state of terminal desuetude. He is below both: less than king and less than man.[12]

Rock-bottom this may seem, yet it is merely fecund with and presaging of worse. Let us recall how Kantorowicz's Norman Anonymous characterized the king's body politic as the king's participation in the divine, as the fact of the king's *being* divine, as his being God—though only by grace: Christ is God eternally and by nature, the Christian ruler being God by consecration (through a life-long, though temporary, deification). Richard's abjectness thus becomes that of God *in the wretchedness of man*. As *christomimētēs*, the king now comes to resemble Christ at *his* most abject: as God humbled by his creatures.[13] Richard's Golgotha and Golgotha as such is approaching, as Carlisle now darkly foretells it:

Peace shall go sleep with Turks and infidels,
And, in this seat of peace, tumultuous wars
Shall kin and kin, and kind with kind, confound;
Disorder, horror, fear, and mutiny
Shall here inhabit, and this land be called
The field of Golgotha and dead men's skulls. (*Richard II*, IV.i.139–144)

And so does Richard proceed posthaste to identify himself unequivocally and sans prevarication with Christ delivered up by treachery to ravenous wolves:

> … Yet I well remember
> The favours of these men: were they not mine?
> Did they not sometime cry "All hail!" to me?
> So Judas did to Christ: but he in twelve
> Found truth in all but one; I, in twelve thousand, none. (*Richard II*, IV.i.167–171)

Soon hereupon he is forced to de-frock himself, to unking himself by performing the ceremony of kingly consecration in reverse—yet with all due solemnity. He continues in his role of *christomimētēs*, seeing his former subjects arrayed before him as so many Pilates, so many traitors, and Bolingbroke as the Lancastrian incarnation of Pilate himself:

> Nay, all of you that stand and look upon me,
> Whilst that my wretchedness doth baith myself,
> Though some of you, with Pilate, wash your hands,
> Showing an outward pity: yet you Pilates
> Have here delivered me to my sour cross,
> And water cannot wash away your sin. (*Richard II*, IV.i.236–241)

Richard remains the *Deus absconditus,* the concealed God, the betrayed hidden God, midst multivariegated Judases, assorted Pilates, and the blood-thirsty rabble.

Yet, immediately hereupon he shall, by a curious reversal, sink lower, lower still than these traitors to the despot's kingship. For in betraying his own kingly body as body politic by fearfully consenting to unking himself (a well-nigh sacrilegious traversy), Richard emerges (as he so himself sees it) still more treacherous than all these traitors combined—as *primus inter pares* the traitor *par excellence*:[14]

> Nay, if I turn mine eyes upon myself,
> I find myself a traitor with the rest;
> For I have given here my soul's consent
> T'undeck the pompous body of a king;
> Made glory base, and sovereignty a slave,
> Proud majesty a subject, state a peasant. (*Richard II*, IV.i.246–251)

Though, still and yet lower shall Richard sink. Here follows the scene where he requests that a mirror be brought to him. Gazing upon his countenance as reflected in the glass surface, Richard asks himself the crucial treble-question:

> ... Was this face the face
> That every day under his household roof
> Did keep ten thousand men? Was this the face
> That, like the sun, did make beholders wink?
> Was this the face that faced so many follies,
> And was at last outfaced by Bolingbroke? (*Richard II*, IV.i.280–285)

And indeed:

> A brittle glory shineth in this face,
> As brittle as the glory is the face,
> [*Throws glass down.*]
> For there it is, cracked in a hundred slivers.
> Mark, silent king, the moral of this sport:
> How soon my sorrow hath destroyed my face. (*Richard II*, IV.i.286–290)

As Kantorowicz has it, Richard's features as "reflected by the looking-glass betray that he is stripped of every possibility of a second or super-body—of the pompous body politic of a king, of the God-likeness of the Lord's deputy elect, of the follies of the fool, and even of the most human griefs residing in inner man. The splintering mirror means, or is, the breaking apart of any possible duality. All those facets are reduced to one: to the banal face and insignificant *physis* of a miserable man, a *physis* now void of any metaphysic whatsoever."[15]

This is the *demise* of Richard. The despotic full body of King Richard II, the monarch's body politic, is as such destroyed. And, in being reduced to less than man, that is, to less than *bios,* to perhaps less than Aristotle's *zoē* (as the "natural sweetness" of animal subsistence), to mere *physis*, to a raw matter kenotically evacuated of all semantic content; is Richard, formerly the sovereign despot presiding over bare life, through the most bewildering reversal twixt polar opposites, reduced to precisely the bare life over which he had presided. Richard's complete and utter demise, as intimated by the mirror-scene, entails at such short notice the reduction of the former Leviathan to *homo sacer*—to life that may not be sacrificed but which may be killed with impunity. And indeed, Richard, less than a man, that meaningless piece of bare life, finds himself conveyed to the Tower and—without ceremony—reduced to a lamentable bloodless lump of flesh similar to Heliogabalus.

In this connection, it is interesting to note that Žižek argues in favor of a homology between the State annihilation of *homines sacri* as effected by sovereign power and the killing spree associated with revolutionary terror, contending that both are instances of Benjaminian divine violence.[16] What is effected in Richard's tragic and ridiculous reversal of fate is therefore a peculiar transmogrification of divine violence as co-opted by the State—the displacement of its source of emanation (of those whom it delegates as its purported "agents") and of those whom it annihilates (its "victims"). This reversal is that by which the excess of the Law, the Law as the recrudescence of the impossible Real, of the obscenity of *das Ding*; turns against the sovereign in the form of Bolingbroke's vertiginous line of military abolition—whose final vicissitude is the destruction of Richard in his abject prison cell by an untimely murderer. Though Richard was condemned to defrock himself, it was not the force of divine violence *per se* that had effected this pitiful abnegation, since condemnation itself is not of the order of divine violence. Indeed, by analogy with what Robespierre uttered apropos of revolutionary terror, divine violence does *not* condemn kings; rather, in a "blind" extra-moral (though not necessarily immoral) enactment of immediate justice, it simply casts them into the ontological void.[17]

We have been speaking of the *homosacralization* of the despot. We have likewise noted in an earlier chapter that *homo sacer* was inherently subject to various Deleuzoguattarian becomings-animal. Now, if one co-opts some of Derrida's dwellings upon this matter, it comes to appear that there is in fact a double becoming here. For not only is there a becoming-animal of *homo sacer*, but at once a coextensive becoming-animal of the despot himself—or at once of the despot's deity. In one of his first glosses on this matter, Derrida alludes to the German god Wotan—or Odin. The very essence of Wotan is that of Sovereign King. Yet, sitting on the throne, he is flanked by two wolves. They are his coat of arms, insignia and living heraldry of sovereign majesty. It is to them that he gives everything which anyone hands him to eat. He himself does not eat, but merely drinks. Moreover, Odin Wotan was able to convert himself into a bird, a fish, or into a serpent. These are the becomings-beast or becomings-animal besetting this despot.[18]

This becoming-animal of both the despot and of the subjugated subject Derrida likewise finds in Grotius and Hobbes, who reduce the originary human community to an animal community and citizens to beasts and the sovereign to what Plato in the *Republic* opines to be a wolf-tyrant—one who, for Grotius and Hobbes, is always capable of devouring those whom he commands like cattle. In a similar vein, from Rousseau, Derrida borrows the image of the despotic sovereign as one who keeps his cattle solely with a view to devouring it, "*so as to* devour it savagely and gluttonously, tearing at it with his teeth ..."[19]

This recurring refrain of the coextensive becoming-beast of both the sovereign and the subject likewise emerges again in greater depth in Derrida's dwellings upon Carl Schmitt and Giorgio Agamben. As we already know, for the latter two, sovereignty *primus inter pares* resides in the sovereign's capacity to suspend the positive order to law. This entails, moreover, a certain exceptional power of the sovereign (*sovrano*) to place himself above right—and thus a right to "non-right." (Incidentally, what one may espy in the foregoing transcendence of the sovereign is his propensity to move beyond the human into the realm of something approaching divine omnipotence—a divine omnipotence which then comes to be seen as grounding the very principle of sovereignty in its self-posited sacred and theological origin. Is this not simply the direct filiation of despot with deity of the Deleuzoguattarian despotic regime?) Notwithstanding, our overriding point is that this arbitrary suspension or rupture of right, which situates him above and at a distance from the law, is that which tends to render the sovereign the most brutal beast of all—a beast who utterly scorns the law.[20]

Now, it is in this extra-legality that Derrida locates the commonality between the Agambenian sovereign (*sovrano*) and the beast. They both embody being-outside-the-law: the sovereign, in the sense that his is a being-above-the-law, as the originator and guarantor of laws which he notwithstanding need not respect; the beast, in the sense of being the criminal violator of the law.[21] So, even though the modes of being-outside-the-law of the sovereign, criminal and beast are seemingly mutually heterogeneous; there remains between these three figures an "obscure and fascinating complicity, or even a worrying mutual attraction, a worrying familiarity, an *unheimlich*, uncanny reciprocal haunting."[22] For this reason, it is not surprising that heads of State are often given animal nicknames. Mustapha Kemal Ataturk was called the "grey wolf"—alluding to the "blue wolf," Genghis Khan.[23]

What Derrida observes is that in this strange superposition of the sovereign and the beast one may detect a "sort of hypnotic fascination or irresistible hallucination," a "coupling, an ontological, onto-zoo-anthropo-theologico-political copulation," producing the apparition of "the face of the beast under the features of the sovereign."[24] Yet, he will move still further in this line of argumentation. For he proceeds to posit that the very force, power, absolute potency and indeed essence of sovereignty may very well and in the last instance quite simply reside in *devouring*, that is, devourment through the mouth, the teeth, the tongue, and so forth—all to the end of the sovereign incorporating his cowering subaltern into himself. (Was this fundamentally incorporating function of the despot not already intimated apopros of the Deleuzoguattarian despotic regime—especially through Elias Canetti?) Yet, at the same time, Derrida observes that the site of interiorizing devourment—the mouth, the maw, the teeth, throat, glottis, and tongue—is likewise the exact same region as that of the articulation of speech and language. Thus, as likewise observed regarding the semiotic attributed to the despotic regime by Deleuze and Guattari, we see here this devouring mouth to at once be the voice from above from whence arose the master signifier that overcoded the entire biunivocalized signifying chain.[25] We thus, once again, return to the notion of the mutual participation of sovereign and beast: the locus of devourment is at the same time that which projects the voice, "the *topos* of the *porte-voix* [megaphone, literally 'voice-carrier']," that is, "the place of *vociferation*."[26] We thus have the existential propinquity of devourment and vociferation, but in two modalities: vociferating devourment and devouring vociferation. The former interiorizes what is proffered: it eats. The latter exteriorizes what is interiorized: it speaks. However, as regards the former, it may interiorize by way of not merely eating, but by way of listening. For the devouring apparatus is also in very close contiguity to the auricular and likewise the audiovisual organs. Thus, devouring vociferation may at once be the pure assailment of the ears, that is, listening. In this regard, it is true that King Midas was afflicted by Apollo with ass's ears for preferring the latter's rival in a musical contest. Similarly, in *Tristan and Isolde*, King Mark acquires horse's ears. These are exemplars of the specifically auricular becomings-animal of the despot.[27] Ears can make one deaf. One may thus yet again see that the sovereign is beset by the same becomings-beast as is *homo sacer*—and to the point of some sort of obscene zoophilia or even bestiality.[28]

The overriding point to garner from our Derridean musings is thus that the homosaceralization of the sovereign, of Richard II for instance, which enables him to be killed like a dog—is that which, through the already pervasive becomings-beast which almost define the sovereignty of the sovereign, inheres within the vicissitudes of the despot from the start. The despot is always already potentially *homo sacer* from the very outset.

And he will die by the force of Benjaminian divine violence.

2. Exodus

(i) The death drive

Now, libidinously speaking, Benjaminian divine violence partakes of the drive: the latter's sempiternal undeadness and excessive life.[29] Here, if desire arises from the diremption between (what becomes) its object-cause and its object, i.e. exists insofar as it *lacks* its object (of which it is then the metonymy), then drive has lack *as* its object. Drive circulates monotonously around the void. It is of the order of the Freudian death drive (*Todestrieb*); i.e. as, paradoxically, the tendency of life to *persist* beyond death—as the living dead, as the undead, the spectral, the *immortal*.

Now, the centrifugal vector launching those who desert the territorial enclosure of the despotic regime beyond its ambit, is a line of flight precipitated by and partaking of the displacement of divine violence's angle of emanation and target of destruction. Hence, it is these men who shall likewise partake of the "living-on" characteristic of the death drive. They are the perpetual survivors, those whose existence is under reprieve, who persist and subsist amid the ambit of those whose existence is merely spectral.

(ii) The war machine

Here a group of dissendents takes flight from the despotic regime. These men choose the goat's anus over the frontal face of the despot and flee from the latter regime along a line flight or absolute deterritorialization.

The distinction between relative and absolute deterritorialization resides *inter alia* in the fact that relative deterritorialization is of the province of the State apparatus, whereas absolute deterritorialization is on the side of the *war machine*. We have seen that the despotic State apparatus is erected on the basis of a massive—though relative—movement of deterritorialization sweeping up the territorialities of the primitive regime. But relative deterritorialization continues to appertain to the machinations of the State in that the crucial preoccupation of every State apparatus is to render all movement a relative movement transpiring within a striated space defined by extensive coordinates. To this end, not only does it seek to thwart all nomadic movement as the absolute movement of a body taking place within a smooth space defined by intensive ordinates, but likewise tasks itself with closely restricting speed, and measuring all the relative movements of all subjects and objects within its ambit. The State apparatus decomposes all movement so as to recompose and strictly regulate it. Movement must cease to be the movement of a moving body within smooth space, and rather be that of a moved corporeal entity going from point to point in striated space.[30]

Now, on to the war machine. When Deleuze and Guattari refer to the war machine, they do not in the most fundamental sense conceive of it as a military or paramilitary organization—these latter arise only when the State *co-opts* the war machine for its own ends. In fact, taken in and of itself, the war machine does *not* actually have war as its object. Rather, its business is the "emission of quanta of deterritorialization, the passage of mutant flows," and so forth.[31] In its essence, the war machine is fundamentally irreducible to the State apparatus. It is radically exterior to the State's sovereignty and anterior to all Statist law. Its point of origination is elsewhere. It is akin to a pure and immeasurable multiplicity. It is a pack and an unforeseeable irruption of the ephemeral—the incalculable power of metamorphosis. It is a terribly fierce and rabid violence agains the very fabric of sovereignty. It is a celerity totally at variance and utterly bent against turgid gravity. It abides in secrecy against the transparent prurience of the public realm. It is power—in the sense of *puissance*—against sovereignty, *puissance* against *pouvoir.*[32] In fact, the war machine—conceived of in this its primordial and essential sense—exists solely in its own metamorphoses. It *is* solely its own metamorphoses.[33]

The pure power of metamorphosis, which is the essence of the nomadic war machine, is that of a movement of deterritorialization that has become *absolute*. The war machine operates within a smooth, unstriated space and limitlessly suffuses this space. It is carried not by migrants but by nomads who cavort at an absolute speed, to the extent that they become synonymous with speed itself. The war machine recrudesces each time seditious or properly insurrectionary activity surges up and directs itself against the State. One has in mind anarcho-syndicalism, guerrilla warfare, or out-and-out revolution. Each time such violence trains its sights upon the State, the war machine is revived coextensively with the renascence of a categorically novel nomadic potential, commencing within a reconstituted smooth space dissipating the striations of the despotic regime.[34]

(iii) Kengir, May 1954

As Solzhenitsyn relates it, in May 1954, a seditious subterranean line of absolute deterritorialization began to manifest its concrete ramifications from within the very bowels of the despotic regime of the Kengir labor camp, assuming the form of a well-orchestrated mutiny by camp inmates. This incipient line of flight had in fact been unwittingly precipitated by the Soviet Gulag authorities themselves, with their pointless shootings of innocents and in their having most imprudently poured thieves into the camp like "petrol fumes into an overheated atmosphere."[35]

The thieves, up until this point in history, had always been the steadfast and ineluctable class allies of the bureaucratic despots; functioning *inter alia* as the latters' agents of preemptive vengeance against the subjugated politicals. But at this point in time in Kengir, the thieves, in an unprecedented confounding of their prior allegiance, entered into alliance with the politicals.[36]

This political-cum-criminal movement of deterritorialization initially assumed the form of a tinkling of broken glass, as the thieves shot apart the lamps around the boundary zone with slingshots, rapidly proceeding hereupon to swarm all over Camp Division No. 2, their shrill bandit whistles rending the air as they broke down the service yard gate with a beam and smashed into the women's camp with a section of railway line. With the politicals having broken up bunks to arm themselves and carrying shields made of boards and mattresses, the coruscating line of flight swept forth the latter through a barricade, into the yard, and into the women's camp, freeing the latter and uniting the entire camp, with the exception of the jail.[37]

The authorities however swiftly bricked up the breaches that had been made, re-marked prohibited zones, replaced broken lamps, and posted sentries at the ends of walls with orders to fire. This, nonetheless, only precipitated a renewed wave of ramifications of the line of flight emblazoning its destruction upon and piloting the most pervasive of deterritorializations that would soon seem to render smooth much of the despotic striations of the camp. The whistles soon began to shrill through the camp again, the Second and Third Camp Divisions calling to each other like hooligans on a spree, as the zeks shattered the lamps yet another time, and legions of insurgents battered at the barbed wire and new fence posts with long tables, burrowed under the walls separating their divisions by frenetically scooping out earth using *only kitchen knives and mess tins,* thereupon smashing the boundary fences, battering holes in the walls, bursting through into the prison area, dispersing the guards as the freed prisoners wrecked the interrogation offices, as all inmates proceeded to rip off their number patches.[38]

In the wake of this liberating surge of pervasively ramified insurrection by the zeks, piloted by inexorable movements of deterritorialization, the scene emerging in the wake of these planomenic peregrinations is best described by Solzhenitsyn: "Posts with broken lamps sprawled against the wire fences. Even without their help the zeks moved freely from zone to zone by the trenches dug under the wires. Many of them took their street clothes from the storerooms and put them on. Some of the lads crammed fur hats on their heads; shortly there would be embroidered shirts, and on the Central Asian bright-coloured robes and turbans. The grey-black camp would be a blaze of colour."[39]

Notwithstanding, we do know from the historical record that the mutineers at the Kengir camp—though having fomented the most stout-hearted, largest and most successful uprising in the history of the Soviet servile labor system—did not prevail for longer than forty days. Moreover, though certainly piloted by considerable subterranean movements of deterritorialization, the revolt failed to precipitate the actual flight of a fraction of the populace centrifugally out from the strictures of the despotic regime itself. Were the mutineers truly convinced that the goat's anus was indeed preferable to Stalin's cockroach-whiskered countenance? Did they choose the goat's anus and make it their own? Probably not.

For Canetti, a revolutionary situation is that of a crowd discharging itself by way of a "collective deliverance" from its stings of command.[40] As we have seen earlier, these stings are hard crystals of resentment which form and remain embedded within the memory of those who carry out commands emanating from yonder. By way of a curious reversal, however, it is precisely these same stings which—in the movement of formation of an insurrectionary "reversal crowd"—function as the latter movement's driving force. By corollary of and tributary to this burgeoning of the reversal crowd, there emerge in its wake much smaller guerrilla entities known as "baiting crowds," which act to hunt down and kill selected members of the hegemonic caste—with or without trial. The relation between the baiting crowds and the reversal crowd is reducible to the fact that the "successive baiting crowds run their course on the surface while the waves of reversal rise slowly from the depths."[41]

It thus soon becomes pellucid that in the Kengir mutiny, if a reversal crowd had at all formed, the force of this latter was soon very rapidly attenuated and thereupon ignominiously extinguished. Nor may we be certain that the machinations of the thieves qualified as the actions of baiting crowds. What is certain is that the movement of deterritorialization animating the seditious forays of the Kengir mutineers was *not* absolute—did not reach the plane of consistency, the BwO of desire. Though in excess of the coefficient of deterritorialization constitutive of the signs of the despotic regime, they were still relative movements of deterritorialization. For a liberating flight from the despotic regime to be effected, this must proceed with the gait of a movement of deterritorialization that is *absolute.*

(iv) Moses

Such a vertiginous movement nonetheless did occur in history: at about the time of the insurrection by the votaries of the older cult of Amun against the heretical and newfangled proto-monotheism of Akhenaten—between the seventeenth and ninteenth dynasties of the Egyptian empire—a small and ragged band of Hebrews, cavorted inexorably by a tremendous movement of absolute deterritorialization, did indeed take flight from the Egyptian despotic regime, fleeing into the nomadic or smooth space of the desert. It is these men who, as Debray avers, transmogrified their persecution into secession and their political defeat into a moral victory. Their God was "the celestial alibi for terrestrial courage," who cast "the keys of the Kingdom to the other side of the mountain into order to force" the subjugated to "make their way through the pass and go and look elsewhere at whatever cost."[42] Through their God, they turned their non-egalitarian subjection into national liberation.[43]

It is across the vertiginous lunar expanses of an arid desert space that a new regime, although already beginning to burgeon its shoots within the very bowels of the imperialistic despotic regime, shall truly take shape. This is a nomadic regime whose impact upon human history shall forever reverberate its ubiquitous tendrils. It is the perpetually migrant regime of the Jews and their Judaism—and of the multivariegated progeny of the latter. And certainly—quite in contradistinction to the gaudy plumage with which the Kengir insurgents bedaubed themselves upon momentarily claiming the camp as their own—in penetrating the forbidding vastness of desert wastes to encounter the forbidding countenance of the living God, the passional Hebrews shall pass into monochrome.

Notes

1. Gadamer 1996, 162.
2. Deleuze and Guattari 2004, 229, 230.
3. Artaud 2003, 117–119.
4. Bierce 2011, 464, 465.
5. *Richard II*, II.i.285.
6. *Richard II*, II.ii.136–138.
7. *Richard II*, II.ii.145.
8. Kantorowicz 1957, 30.
9. Ibid.
10. Ibid.
11. Ibid. 32.
12. Ibid. 33.
13. Ibid. 34, 35.
14. Ibid. 38.
15. Ibid. 40.
16. Žižek 2008d, 198, 199.
17. Ibid. 202.
18. Derrida 2009, 10.
19. Ibid. 11, 12.
20. Ibid. 16, 17.
21. Ibid. 17.
22. Ibid.
23. Ibid.

24. Derrida 2009, 18.
25. Ibid. 23.
26. Ibid.
27. Ibid.
28. Ibid. 33.
29. Žižek 2008d, 198.
30. Deleuze and Guattari 1988, 385, 386.
31. Ibid. 230.
32. Ibid. 352.
33. Ibid. 360.
34. Ibid. 386.
35. Solzhenitsyn 2007, Vol. 3, 293–301.
36. Ibid.
37. Ibid.
38. Ibid.
39. Ibid. 301.
40. Canetti 1978, 58–60.
41. Ibid.
42. Debray 2004, 41.
43. Ibid. 71.

Part Three

The Passional Regime

12

The Vertiginous Foray into the Passional Regime

1. The art of monotheistic desertion and the Great Theophany

Régis Debray observes how God, before revealing himself to the Hebrews, drives his elect into a stony world, a wasteland for peregrinating caravans and antisocial fringe-figures; a hostile milieu roamed by plunderers, livestock thieves and assorted reprobates; a geography of grey interstitial zones and a subdesert no-man's-land between dry sands and fertile plains; from whence, midst the aridity of a steppe pocked-marked with strange boulders, twixt vertiginous cliffs of red granite, in a mineralized *umwelt* of brusque canyons, "limestone cliffs limned with geological phantasmagoria," and "at the abrupt peak of a *djebel* cut through with fissures and jutting rocks"—the *great theophany* transpires.[1] It is within the ambit of such forbidding environs and their seemingly sempiternal privations that the discreet and emaciated—these halfbreeds of the margins—the destinerrant, itinerant, migrant and nomadic, undergo what becomes a "period of abnegation and testing" wherein "the Desert serves as a refrain for dissidence, for monotheistic *desertion*."[2] This desertion is a flight from the hypocrisy of the city with its gilded filth, institutions and vampirical system of taxation.[3] The contempt of these xenophiles, as is that of Yahweh—the call from elsewhere, is that felt for the territorial despotic machine and all the noxious effluences of its sedentary narcissism and obscene proximities.[4]

2. The etiolation of personality and the solitudinous upsurge of Rosenzweigian "Character"

This contempt—as Richard Cohen maintains apropos of Franz Rosenzweig—is likewise that which the newfangled xenophile feels for "the insubstantiality of personality," for the lamentable "inadequacy of selfhood qua personality," its woeful "over attachment to the world" and the blithe and vacuous complacency whereby it contents itself "to wear the masks of the world, to play a role on the world's stage"—as the mere superficial ape that it is.[5]

Stéphane Mosès delineates the personality as "an accident of individuality" (that is, a mere sociological appurtenance of man as member of his own natural species) that serves to designate "a human as claimant to an originality among the other representatives of the human species," as that which "defines him as part of the social world," as merely a particular case subsumed under a more comprehensive—though admittedly vacuous—generality.[6]

What in the foregoing is meant by "world" may be elucidated by recourse to Badiou's notion of "appearing." Succinctly, a world is that in which beings appear. More precisely, "*We call 'appearing', that which*," of a being, "*is caught in a situated relational network (a world), such that*" this being "*comes to being-there, or to the status of being-in-a-world* [*étant-dans-un-monde*]. *It is then possible to say that this being is more or less different from another being belonging to the same world.*"[7] One may therefore maintain that "there is a 'world' to the extent that it is possible to identify a configuration of multiple-beings who appear 'there' and of (transcendentally regulated) relations between these beings. A world is ontologically assignable by that which appears, and logically assignable by the relations between apparents."[8] Badiou maintains that, though appearing is nowise reducible to pure Being in itself, it nonetheless remains something which beings must endure in order for their Being to be guaranteed. That is, beings must always manifest themselves in a world. This necessary appearance of beings in a world is regulated by the so-called logic, transcendental organization, or transcendental measure, of a world. This transcendental organization is that which serves to fix the mobile singularity of the said being's being-there in the world. Most crucially, however, the identity conferred upon a being qua its appearing in a world, much as in the case of the mere third-person existence of the personality, is strictly *relational*.[9] That is to say, the logic of a world effectuates the appearing of a being in the said world "by affecting a being with a variable degree of identity (and consequently of difference) *to the other beings of the same world* [my emphasis]."[10] The logic of a world, thus, by its very nature *ab initio* bears upon the identity or difference of more than one being in the said world.[11] "For *what is measured or evaluated by the transcendental organization of a world is in fact the degree of intensity of a difference of appearance of two beings in this world and not an intensity of appearance considered 'in itself'.*"[12] Therefore, a being as it appears in a world, qua this appearing, may in no sense be considered an ipseity. Appearing in a world is nowise that by which a being has its being in itself. Indeed, as an appearing, a being can be said merely to exist. And existence, as Badiou contends, falls short of being a category of Being and is thus only a category of appearing. Existence is nowise absolute or in-itself. It is only relative to a world, or the world, in which beings appear.[13]

However, as already observed, a being is in no sense reducible to its appearing or its existence in the world. That is to say, it always implicitly retains an ineluctable residuum that remains outside of and stubbornly resists the world, a hard kernel recalcitrant to and in excess of the logic of mere worldly existence. It is this excess of pure inexistent Being that may now serve as the basis of a subject or Rosenzweigian Self categorically at variance with the mere third-person existence of the personality. Indeed, the vertiginous movement of disaffection whereby the Jewish xenophile forever condemns the facile and fickle obsequiousness bred under the auspices of the despotic regime to the abyss of an inglorious past is that in whose wake a Self of this kind arises from the cinders of the older simpering jackanapes.

Bernard-Henri Lévy may be seen as hinting in the direction of such a Self when he speaks of the incipience of a "taste for life stripped of what usually furnished and, perhaps, encumbers it: habits, conventions, urbanity, money, that whole part of life that makes life secure—and that, suddenly, shatters into bits …"[14] For the foregoing revolutionary movement of monotheistic desertion is one which, as Richard Cohen maintains, "closes the self more firmly upon itself, ties the knot of selfhood tighter, seals the self more hermetically, and thus breaks with the apelike role-playing of personality."[15]

What emerges is a form of subjectivity marked by "a greater detachment from the rest of the world owing to a harder, firmer, and nearly impenetrable attachment to one's own unique created being."[16] This remarkable new Self, unprecedented in its obdurate cohesion, is a singularly *heroic* one, "a being so fully grounded in its own unique resources, in its own unique self, that it can make a stand in defiance of the whole world, unto death."[17]

In this sense, the aforementioned Self is strongly homologous to the Badousian subject. For the latter arises as an aside, an exception to the world. It has no place to be in the world and yet makes its incision in the latter. It is that which at once has no place to be according to the law of the world, and which subtracts itself from the world.[18]

The Self being delineated here entails what Rosenzweig refers to as the *Character*. The Character, in turn, is the fundamental inner kernel of that which Rosenzweig more broadly terms the "meta-ethical" Self. Rosenzweig delineates the refractory stance of the Self in relation to personality. The Self, as anchored in Character, is a solipsistic self-enclosed hard kernel that is in no sense part of "the world," "humanity," or "the children of earth." It is completely unrelated to the personality, as it has no role to play within society as a member of the human genus or through participation in the "common Good." It is absolutely alone and therefore cannot be counted as one among other men. It is, instead, the elemental and primordial essence of man. It is tautologically, persistently, and without difference, repeatedly Adam himself.[19]

The personality, the most *superficial* determination of the human, is defined in its essence by its relations with others. It refers to man as a gregarious and political herd animal; that is to say, as an obsequious dwarf. Character, in vast contradistinction, is strictly non-relational; it is a relation that the Self bears unto its own inner-most self. It is thus inwardly directed and centripetal. It is man conceived in the most extreme solipsism of icy solitude. This nowise, however, implies an autism on the part of a creature stripped to the bareness of Character. It is the personality that is autistic—or at least blithely collusive with the collective autism of the vapid world of masks. The upshot is that the vertiginous movement of radical disaffection that is the Jewish Exodus, entailing as it does the upsurge of Character from beneath the mendacious veils of personality, effects a definitive break from the political as such. The Character, as sundered or isolated from the smug cloak of personality, Rosenzweig terms the *daimon*.[20]

In the language of the Zohar, the movement from the superficiality of personality to the more consummate ipseity of Character may be viewed as entailing a transition from the *nefesh* to the *ru'ah*. As Scholem relates, these two latter terms designate, in ascending order of excellence, the first two parts of the soul—the third, and still higher, part of which is the *neshamah*. This tripartite division of the soul originally arose from Jewish neoplatonists such as Abraham ibn Ezra and Abraham bar Hiyya. The *nefesh*, while not exactly the personality, does seem to underpin it. It is the first element to be found in every man and the source of at once his animal vitality (*hiyyut*) and the entirety of his psychophysical functions. It is equally the property of all human beings. Now, since the personality and its ensconcement within the glibness of mere third-person relations necessarily entails a degree of spiritual development barely in advance of the prosaic dictates of vitality and its mendacious social appurtenances, the non-transcendence of personality and its superfices is at once the soul's failure to elevate itself to degrees higher than the *nefesh*. For the other two parts of the soul are indeed nowise congenital but indeed postnatal increments acquired solely through a spiritual awakening whereby man sheds his former simpering, platitudinous, hackneyed, trite, inane, insipid and moth-eaten parochialism. It is only thus that the *ru'ah*, or *anima*, is roused.[21]

Character's fundamental disaffection from, and detached comportment in relation to, the world is precisely homologous to Schelling's delineation of the modality of the subject as it appears anterior to its role as a medium of language. This, as Žižek argues, is the "night of the subject," "the infinite lack of being," defined in its very essence as a contraction through which the subject negates every being outside of itself. At the same time, it accords with Hegel's determination of madness as a profound gesture of withdrawal from the world, the self-enclosure of the soul, and its severance of all links with the outside—except not, as Hegel would have it, as an atavistic regression to the mere "animal soul," but rather as absolute negativity, as a nothingness that nihilates in Being, as the Freudo-Lacanian death-driven (*Todestrieb*) subject abiding in the surreal inter-space between nature and culture, in which the subject is not yet a being of language though its instincts have already been thoroughly denaturalized, perverted, and derailed from their natural integument.[22]

Character's uprootedness from the world may also be seen from the perspective of the Lacanian conception of the Freudian unconscious as a disembodied rational machine, whose trajectories are wholly refractory to the demands placed upon the subject by the world in which it finds itself embedded. In this sense, the unconscious is that which derails the subject's Heideggerian "being-in-the-world." Much as in the case of Character, the unconscious, as the psychic automaton which disregards the dictates and promptings of the Freudian reality principle, is precisely that within the subject which renders it fundamentally irreducible to its life-world context.[23]

3. Character as unworldly defiance and unconditional will

Character does not defer to the facile world of vain conceit and whimsical unction. Character is *obstinate*. It is *defiant*. This obstinacy consists in the Self's intransigent will to be itself, to unswervingly persevere in its own singularity. For Rosenzweig, Character is in some sense not susceptible to being annihilated—since it simply continues as pure unadulterated defiance. It is at base a defiance which forever continues to be defiance. It is relentlessness itself, and just goes on seemingly ad infinitum.[24]

Rosenzweig hereupon moves to elaborate the defiance inherent to Character in terms of will. He maintains that, as far as Character is concerned, its will is an entirely free will which fundamentally wills its own will. Character's will wills its own Character. Character wills Character. Its will is a defiant will. Otherwise, it would be nothing at all. It is defiance, pure and simple. It is a defiance completely condensed within the Character—to the point of being well-nigh identical with it. This is how the meta-ethical Self is configured.[25]

As Mosès maintains apropos of Rosenzweig, the obstinacy of the will is likewise a recrudescence of man's originary affirmation of Being as at once identical with the affirmation of himself as primary autonomy, as an originary independence in which his singularity seamlessly coincides with his very Being—a coincidence in which man's imperturbable ipseity is *each time and repeatedly Being itself.* It is in *this* latter sense that man, as Character, is identical in nature to the Godhead—insofar as the latter, the Schellingian Abyss of Freedom, is first and foremost *unconditional will.*[26] For it is not a personality, but a Self. Crucially, what Rosenzweig avers here rests on the point that man is created in the image of God. Again, man is Adam—and, for this reason, is not fundamentally a being of the world. He is in the world but not of the world, since the world is pure finitude as such whereas the Character is infinite.[27]

This unconditional nature of Character's will and its radical heterogeneity in relation to the world is homologous to the stubborn attachment which, as Žižek argues, the Hegelian subject must bear toward a certain exclusive idiosyncratic content in order to extricate itself from the particular ethnic totality in which it is immersed. It is this inexorable cleaving of the subject to some "pathological" feature, devoid as it is of inner necessity and made at the risk of losing its entire substantial content, which effects its break from the world. This often trifling particular feature, though completely indifferent in itself, gives body to the subject's arbitrary yet unconditional will. The same contingent and idiosyncratic feature is a metonymy of the void of nothingness, or of the abyss of the impossible Thing (*das Ding*), what Lacan following Kant terms a "negative magnitude": the excessive and obsessive attachment to which renders the subject indifferent to all other objects. The subject's unconditional and unflagging clinging to its idiosyncratic object is precisely the concrete form of its willing nothingness, in which excessive fixation on a particular object and wholesale disengagement from the entire world coincide and serve as recto and verso of the same movement of the Freudian death-drive (*Tobestrieb*). The death-drive is here the pure excess of the will over any self-contented and worldly attachment, which allows the subject to disentangle itself from the tepidly platitudinous parochialism of quotidianity, so as to risk its very life to the ends of a higher Cause.[28]

However, though man's *daimon* is indeed an unconditional will, in another respect, man differs sharply from God: for whilst the latter is infinite power, man is merely finite. Therefore, though unconditional, human will is not so much freedom for action—but like the will of God—the freedom for willing. It is not free power, but rather free will. For, in contrastiction to God, man's free power is utterly denied from the very outset.[29]

Notwithstanding, it is the obstinate affimation of Character—as entailed in man's being at bottom unconditional will—that, for Rosenzweig, properly-speaking consistutes the "selfhood" (*die Selbstheit*) of the meta-ethical Self.[30] This is a Self that, being radically heterogenous to the relational order of the quotidian, likewise escapes the very parochial and glib ethical order of the ancient city—and indeed that of the world itself. In this regard, Mosès is right to conclude that, as the *metaethical* reality that it indeed is, the Self "is not the object of ethics but its subject, its source"—and as such, far from being amenable to ethical subsumption, is *beyond the moral world*.[31] More precisely, as Rosenzweig puts it, what occurs—in a quasi-Kierkegaardian fashion—is a pushing back into the netherground of the world of the Ethical as the universal. It flies in the face of and cocks a snook at the human genus, society, peoples, and the State apparatus. For the meta-ethical Self, unlike the personality, simply does not live in the world as such. It does not breathe this sort of rather rancid and suffocating air. It only breathes itself. It is in contempt of or flagrantly indifferent to the laws of the world. For its laws are not those of the world. Its ethics is not that of the world. It simply has its own ethos. Its ethos is self-identically its own. In relation to this latter, it views the world as merely tedious and inane.[32]

4. Transmogrification of political defeat into moral triumph

But let us return to the Deleuzoguattarian Judaic line of flight itself, to the post-signifying regime as a collective unfolding. Millenia latter—as Régis Debray maintains—this latter becomes that of the primordial Christians, the "second Israel" who, pursuing the "movement of deterritorialization" and having divested themselves of any "fleshy attachment" to the ground, destinerrantly inhabit a space that is "centrifugal, dynamic, afocal, the sum of its vanishing points."[33] The draconian territorial machine, the despotic regime, had—before the emergence of this unique anthropomorphic shape of desertion—attributed a "negative sign" to any line of flight exceeding the coefficient of deterritorialization of the signs of the signifying regime, sending a scapegoat scampering down the incipient line. However, in the vertiginous centrifugal movement of self-positing of this categorically new breed of dissident, a coterie of signs breaks off from the circular semiotic system of the despotic regime and suddenly starts to meander off by itself, running off on a straight line, as if blown into a very tight and narrow corridor like a projectile hurtling through a vacuum tube.[34] Here, unlike in the case of the despotic regime, this line of absolute deterritorialization acquires a positive sign and is taken over by an entire people who find in it their singular destiny and reason for being.[35]

As Debray has it, entailed in this complete reversal of the sign of a former indictment for the crime of insufficient parochialism is the transformation of persecution into secession, of political defeat into moral victory; a moral triumph in which God—the author of the postsignifying ruse—is "the vitamin of the sacrificial victim, the celestial alibi for terrestrial courage, casting the keys of the Kingdom on the other side of the mountain in order to force those who live there to make their way through the pass and go and look elsewhere, at whatever cost."[36] It is an involution turning evil into good "by reversing an anti-egalitarian instrument of subjection (of peasants by administrators) into a tool of national liberation (of a people in relation to empires)"; in which a "factor of segregation in an opulent society was revealed to be, after the reversal, a factor of cohesion for a band of 'Habiru', as the turbulent, direputable bands from the Middle East were known."[37] It was a transvaluation turning necessity into virtue and misfortune into a fount of pride.[38]

As Eric Santer maintains, the Revelation on Mount Sinai is a Badousian "truth-event" whereby a certain "undeadness" that had hitherto "functioned as a support of ideological captivation," as the "surplus cause" of an entire people's "relational surrender" and "passionate attachment to ideological formations," of the same subjugated group's beholdenness to "various forms of *idolatry*"; through a curious reversal, now comes to be transmogrified into that which will henceforth serve as the *sine qua non* of the effective suspension and interruption of this foregoing captivation.[39]

These are the modalities and corollaries of the positive transvaluation of what was formerly but the ignominious line of the scapegoat. Indeed, we are now in a different regime: the Deleuzoguattarian post-signifying, subjective, or *passional* regime. On the one hand, we have the great paranoic Pharoah. On the other, there is the passional Hebrew. For the Jewish people, a constellation of signs detaches itself from the great Pharaonic network and scuttles along a line of flight into the desert. The Jews now move along this most deterritorialized line—that of the scapegoat, with the exception that they change its sign. From being the negative line of the scapegoat it is henceforth transformed into the positive line of Jewish subjectivity, of their passion, grievance or proceeding, in which they voluntarily elect to be their own scapegoat.[40]

Instead of the despotic regime's irradiating circular network of signs, whose coefficient of deterritorialization is defined by movements of (stratic) relative deterritorialization and where signs reterritorialize on the central and visibly radiant face of the despot marking the limit of the signs' deterritorialization (by supplying them with a signified but also recharging them by reimparting more signifier); we now have a strictly linear and temporal succession of finite proceedings.[41] The line of the scapegoat, tracing a centrifugal vector escaping the despotic regime, unlike the movement of relative deterritorialization defining the coefficient of deterritorialization of the despotic sign, now finds itself piloted by movements of *absolute* deterritorialization.[42]

It is this line of absolute deterritorialization, this line of flight, which the Judaic passional regime makes its own and along which it comes to cavort in an apparently interminable linear temporal succession of proceeding and grievance whose segments start, run their course, founder, and then once more resume in a potentially endless series of refoundings. Jewish history itself is such a long mobile nomadic line punctuated by an incessant series of disasters and consequent reconstructions, the former and latter only conveying sense if conceived conjointly. For the entire history of the Temple, of the Ark of the Covenant, the construction of the House of Solomon, and its reconstruction under Darius, only has meaning in relation to renewed proceedings of destruction—such as those of Nebuchadnezzar and Titus.[43]

This linear successional line, constantly emerging from and returning to the ashes in phoenix-like manner, is merely the most elementary delineation of this regime of signs, the passional or postsignifying regime—to which one may ascribe a multitude of idiosyncratic characteristics. It is to these salient traits of the regime that we presently turn.

Notes

1. Debray 2004, 38, 39.
2. Ibid.

3. Ibid.
4. Ibid. 84.
5. Cohen 1994, 50.
6. Mosès 1992, 70.
7. Badiou 2009, 118.
8. Ibid. 305.
9. Ibid. 118.
10. Ibid.
11. Ibid. 123.
12. Ibid.
13. Ibid. 208.
14. Lévy 2004, 140.
15. Cohen 1994, 50.
16. Ibid. 51.
17. Ibid.
18. Badiou 2009, 45.
19. Rosenzweig 2005, 77.
20. Cohen 1994, 56.
21. Scholem 1978, 155.
22. Žižek 1999, 34–36.
23. Ibid. 62, 63.
24. Rosenzweig 2005, 77.
25. Ibid. 78.
26. Mosès 1992, 69, 70.
27. Rosenzweig 2005, 78.
28. Žižek 1999, 104, 107, 108.
29. Rosenzweig 2005, 75.
30. Mosès 1992, 69, 70.
31. Ibid. 72.
32. Rosenzweig 2005, 81, 82.
33. Debray 2004, 134, 135.
34. Deleuze and Guattari 1988, 121.
35. Ibid.
36. Debray 2004, 41.
37. Ibid. 74.
38. Ibid. 85.
39. Santner 2001, 116.
40. Deleuze and Guattari 1988, 122.
41. Ibid. 120.
42. Ibid. 122.
43. Ibid.

13 Essential Features of the Passional Regime

1. Foundation via a non-ideational exposure to an exterior

As opposed to the despotic regime, which is defined by an "insidious onset" and a "hidden center bearing witness to endogenous forces organized around an idea"; the Deleuzoguattarian passional regime is launched by "a decisive external occurrence, by a relation with the outside" that is expressed as an emotion rather than as an idea, and much more so by striving effort or action than by the imagination.[1] Had this occurrence been subject to anterior protention or anticipation, saddling the latter in an ideational form would have been entirely possible—at least hypothetically. The strictly external and non-ideational nature of this occurrence thus stems from its being structurally always too early in arriving for it to be adequately couched in conceptual form. This entails the inescapable failure of symbolization of the occurrence's purported content.

Yet the symbolic intracability of the outside's implacable force may also be attributed to the latter's purely quantitative excess of pressure—and, analogously, to the excessive character of its solicitation. It is in these two latter ways that Freud would regard the Moses-event as the *terminus a quo* of the Jewish passional regime.[2]

Crucially, the upshot of the effective collusion of these foregoing characteristics is that of *trauma*. This is so because the non-ideational exposure to an exterior that arrives too quickly, too forcefully, and which harbors within itself an overbearing existential imperative, cannot but be *stricto sensu* traumatic. That is to say, trauma is generated by what may be termed an "excess of demand," a *Zuviel von Anspruch*; which latter, as Eric Santner aptly illustrates, entails "a *demand for work* placed on a system" which is at the same time characterized by "a *too much of address*, by an excess immanent to an address that resists metabolization, that is symbolically 'indigestible.'"[3] Trauma, therefore, comes to be possible when "a 'too much of address' persists beyond what can be *translated* into a demand for work, a task to be discharged, *something we can do* (or, for that matter, refuse to do, feel guilty for not doing, and so on). More precisely, trauma ensues when such a 'remainder' precipitates a breakdown of the very operation of translation, leaving the mind flooded by excitation."[4] This breakdown of translation/symbolization heralds the effective dereliction of the idea in its very incipience, a nascent conceptualization whose formative failure now comes to be filled instead with a heightened emotional tonality.

Mosès may be seen as expounding on the foregoing when he avers that, in the event of Revelation, subjectivity is inundated with something that cannot be objectivized and which therefore can only remain unidentifiable, that Revelation is indeed paradoxical since it is an event of personal experience that experience itself can nowise contain. Neither is Revelation a theme given up to thought. One must therefore conclude that the "exteriority of that which is revealed in the experience of Revelation is thus out of the reach of both experience and thought."[5] As we have seen above, the "too much of address" inherent to the event of Revelation implies that this former entails precisely the said paradox of its being an experience in excess of the very bounds of experience. Thus, it cannot but be traumatic.

For Rosenzweig, as we shall repeatedly observe, the traumatic event of Revelation is, moreover and fundamentally, the "intrusion—we might even say violent and disruptive intrusion—into the order of human life of a divine interpellation the content of which is" no less than an intransigent "demand for love."[6]

This last sally into Rosenzweigian territory, in fact, brings us back to the meta-ethical Self—which latter we must now scrutinize further. Rosenzweig poignantly observes how the Self emerges within man with the most unexpected and thieving brutality. The Self springs up in man like some kind of uninvited gendarme who assumes possession of all man's domestic goods. Until this momentous day, man is merely a piece of the world even before he is a consciousness. The untimely incursion of the Self robs him in one single stroke of all the goods and riches that he claimed as his own. Suddenly man has become very poor indeed. For he now only possesses and knows himself. In fact, no one at all knows him any more, since there simply isn't anyone apart from him. Unlike the personality, which is a consummately political animal, the Self is the quintessentially lonely man. He is alone in the most hard and arid sense of the term.[7]

Man does not know—and indeed cannot know—this dread day, for the irruption of the outside is far too untimely and excessive. Clearly, such a swift intrusion can only leave man bereft of what had hitherto been his overly fastidious and neurotic attachment to the world. All that remains of a man thus robbed, is the vertiginous destitution that is his experience of himself as at once an abyssal freedom and the most icy solitude. But this experience of himself as the abyssal and vertiginous freedom that he is by way of his *daimon* is simultaneously *the experience of a call from without.*[8]

Though man—as the meta-ethical Self—is in essence the said self-enclosed, self-grounding ipseity and at once the infinite will that is his being; his awareness and experience of being such (from whence stems the gesture of self-positing by which Character bursts through the mendacious veils of personality) arises solely from a divine *solicitation*, through an awakening by an Absolute that transcends the world. Man is in essence this self-grounding ipseity; yet it is only through the call from Beyond that he is henceforth able to recognize this fact and that his self-positing *daimon* may rouse itself from out of its former parochial dormancy. Solely through Revelation is man's authentic meta-ethical Self able to effect a recrudescence from beneath the garish cloak of the world where he had hitherto lived a mere facile third-person existence.[9]

Only now is he truly alive—at least in the sense attributed to living by Badiou. Since, for the latter, living is something quite other than mere existence, something very different to the mere relational coexistence of apparents in the world. For the possibility of living comes into being solely through an abrupt and violent solicitation of the world and of the personality by its exposure to an exteriority, or truth-event, refractory to the logic of the world. "It is not a world, as given in the logic of its appearing," avers Badiou, "which induces the possibility of living—at least not if life is something other than existence. The induction of such a possiblitiy depends on that which acts in the world as the trace of the fulgurating disposition that has befallen the world. That is, the trace of a vanished event."[10]

As observed, effected here is the upsurge, from man's own nullity, of the meta-ethical Self: through the unification of Character—that is, his particularity as enduring essence—with the sheer obstinate defiance of infinite will.[11] Yet, as this forged unity, the meta-ethical Self immediately betrays itself as conflicted and internally riven from the very outset. In contradistinction to God, in whom an unconditional will finds itself seamlessly conjoined with infinite power; the meta-ethical Self is the product of a conjunction of a (likewise unconditional) will with a degree of power and capacity to act that nonetheless remains quite *limited.* Man is after all a finite being. So, whereas man's capacity for free will is as vertiginously boundless as that of God himself; in the peculiar ramshackle that is his (man's) meta-ethical Self, this limitless will finds itself forcibly incarcerated within and frustrated by the constraining enclosure of his comparatively meagre capacity to act.[12] In this connection, apropos of the Rosenzweigian meta-ethical (elemental) Self, Benjamin Pollock observes that "unlike divine freedom, which expressed itself as 'infinite power' because it was united with the unconditional being of the elemental God, human freedom is unconditional and finite all at once. It does not express 'free power, but rather free will.' The human being can will unconditionally, but her freedom is inherently limited in the realization of this will. One might suggest that the coupling of unconditionality and finitude in the freedom of the human self makes for an uneasy marriage. To will unconditionally, but to remain wholly limited with respect to the realization of what is willed, marks an internal conflict within the elemental self which it cannot resolve closed up within itself."[13]

Now, as observed earlier, the meta-ethical Self arose in the first place from the *sine qua non* of the divine call—that of Revelation, an event averred by Santner to be properly-speaking traumatic: a violent intrusion characterized by a "too much of address," an excessively intransigent and inexorable imperative whose very excessiveness entailed the inherent impossibility of its being translated into a clearly definable "demand for work" wherein the said imperative would otherwise have found its consummation. Put more succinctly, the excessive character of the unconditional divine imperative necessarily precluces any possibility of its ideational appropriation—let alone its enactment.

But this unconditional nature of the divine imperative is likewise that of the divine will. It is, at the same time, the unconditionality of the human will whose upsurge from parochial slumber follows solely in the wake of its unlimited exposure to the same infinite and unconditional divine will. It is as though the infinite free will of the divine is transitive, or at least redoubles itself—in Revelation—as the infinitude of the human will. In Schelling's *Freiheitschrift* this convergence of human and divine wills well-nigh reaches the point of coincidence when he contends that human freedom in the most radical sense entails the recrudescence of the categorically anterior Abyss of Divine Freedom in the very midst of the intramundane.

Notwithstanding, the effect is that the excessiveness and inexorability of the divine demand is that which the incipient meta-ethical Self makes its own and comes to impossibly demand of itself in spite of being oblivious of its ideational content. The resultant hyper-excitation of the psychic system, its over-exalted internal atmosphere, the hypertrophy of its affective emotional state, can thus only appear as an implacable *compulsiveness.* These haemorraghing emotional storms of what could be said to outwardly resemble the writhing madness of a deity trapped within a claustrophic and unwieldy hominid shell—the fateful corollary of the absurd betrothal of an infinite and unconditional will to a vastly more restricted solipsistic ipseity far inferior to the latter's intractable demands, evince the meta-ethical Self as one in desperate need of assuagement.[14]

Furthering the latter's internal torment is the fact that, being essentially closed in on itself, the elemental meta-ethical Self is severed from the very world it inhabits.[15] It is here that the subject has entered into a mute realm entirely separate and foreign to the arena of quotidianity: a crepuscular interstitial zone in which it is entirely isolated from other wills, the landscapes of personality, the world, and God. Through the intrusion of its *daimon*, the meta-ethical Self directs all its defiant violence inward upon its own Character, beclouding the environs and itself with an all-pervading Schellingian and Hegelian "night of the world." Shrouded in this impermeable and opaque dusk, it knows not itself nor the world.

It has entered what Žižek terms the "zero-level," or pure void, of subjectivity: the absolutely unendurable abyss in which fully constituted objects and subjects dissolve and yield before a miasmal swarming of spectral partial objects, of exemplifications of the Lacanian *lamella*, of the undead, immortal, libido-as-object that is self-identically the Freudo-Lacanian death-drive (*Todestrieb*). This paronomastic modality of the subject is merely one "degree" removed from the pre-subjective noumenal Real. It is a return to, or recrudescence of, the impossible moment of upsurge of subjectivity in the form of a withdrawal, a recession from the light of phenomenality and its differential degrees of appearing in a world in accordance with its transcendental regime, which assumes the form of a contraction through which constituted reality yields before the surreal phantasmagoria of *membra disjecta*, of organs devoid of body functioning as stand-ins for the dread immortality of the object-libido in its semipernally stuck fixity and eternal psychotic circulation around the void. In Kantian terms, this crepuscular and vanishing mode of subjectivity is that of the pre-synthetic imagination gone mad, psychotic, and utterly demented—the impossible domain of absolutely purified and kenotically evacuated transcendental freedom and spontaneity. In this domain, the subject finds itself vertiginously suspended between direct unadorned animality, on the one hand, and the subordination of its freedom to the moral Law, on the other. In this gathering murk, far from being able to galvanize itself in the direction of forward-directed action in the world or attunement to the promptings of the Deity, the Self is submitted to the horrifying experience of dread organs without body, of itself as the Lacanian fragmented body, of its *membra disjecta* floating around in a nauseating and disgusting miasmal drift, the "night of the world," the "valley of the shadow of death," the hell-on-earth of Kundry in Wagner's *Parsifal* who, having committed the blasphemy of laughing at Christ dying on the Cross, is condemned to an eternity of wandering as an abject, multiply befouled, purely excremental self entirely rejected by the world, abandoned by God, and subject to involuntary, psychotic automatisms, spasmodic jabberings, slavering aphasia, dyslexia, *and the absolute horror of never being able to die.*[16]

In order to assuage its own roiling anxiety and the frustated groping of its hopeless and mute ignorance, the Self needs to make a crucial move: it must turn to its Neighbor. Of this, more quite a bit later. For now, however, we move onto the next essential feature of the passional regime.

2. The face-off: concealing-revealing countenances

In the move from the despotic to the passional regime, faciality undergoes a very profound transformation.[17] We no longer have the radiant public frontal face of the despot, which fuels interpretation and marks the limit of the deterritorialization of the signs in the system. Rather, God himself averts his countenance, for—as in the tremendous theophany on Mount Sinai—it must be seen by no one. At the same time the subject—seized as he is by a horrific fear of God—in turn, himself averts his face. In the process, the frontal view of the radiant face of the despot is replaced by mutually averted faces set in profile.[18] Moreover, unlike in the despotic regime, faciality, far from blocking or indicting the line of flight from an imperial network by marking the limit of deterritorialization of the sign which may not be breached, itself organizes it, attributing to it a *positive* sign. And it is now faciality itself that organizes the line of flight, in a sullen and morose face-off of two countenances that "become gaunt and turn away in profile."[19] More precisely, however, in the passional regime, countenances simultaneously conceal and reveal themselves to one another, wedding a line of flight, of absolute deterritorialization which incessantly both draws them together and drives them apart.[20] It is this face-off between concealing-revealing countenances which, drawing the line of flight or deterritorialization, sets off the segmentary line of passion constitutive of (or that is) the passional regime.

(i) Osarsiph and Mutemenet

In Thomas Mann's *Joseph and His Brothers*, invoking the occult, Potiphar's treacherous consort Mutemenet—abased, and with her heart sick unto death from the pangs of renunciation and loss, having foresworn the soul of her beloved Hebrew lad Osarsiph (Joseph)—having "abandoned any claim to your eyes," so as to possess him but only *in corpore*; laments that his "eyes will be closed and dead when we embrace" when she at last contemplates herself "delighting in the soulless bliss of his sweet corpse."[21] Yet, she desperately—yet vainly—hopes that "the dead boy may yet open his eyes to grant me the gaze of his soul—that by some means or other the strictures of magic may be deceived and undone."[22] This very blinding and occlusion of Joseph's gaze by Mutemenet's embrace, that she is only able to live the revelation of his soma with his eyes wide shut to her gaze, entails the effective aversion of their countenances in the very movement of their mutual disclosure—a unitary movement of concealing-revealing.

(ii) The world of Carlos Fuentes

In Fuentes' *Terra Nostra*, La Señora and the miniaturist monk Brother Julian—with pupils dilated by the effects of belladonna—stare at each other unseeing, although the former does see the convulsions of blood and war and of bed and gallows and the figure of a fecund youth multiplied infinitely lying on the shore in the pupils of the latter, whilst the latter sees the event of evanescent carnality in the eyes of the former.[23]

Elsewhere in the same novel, upon seeing the Lady of the Butterflies for the second time, the pilgrim to the New World beholds a veritable stranger who is then herself but also another and whose cruel eyes, despite the pilgrims manifold entreaties, fail to return his gaze.[24]

In another scene in Fuentes' magisterial opus, whilst supplicating himself repeatedly to the Divinity in the crypt and thereupon turning his imploring acts of devotion to the painting of Christ without a halo from Orvieto, El Señor sees that the naked figures—whose backs had until the present faced the viewer—turn to manifest their faces, whilst the Christ-figure who had hitherto faced the beholder, proceeds—to El Señor's tremulous aghastment—to avert his countenance.[25]

(iii) God, Christ and man

In Žižek's Hegelian reading of Christianity, Christ is the index of a triple mutual aversion: that of man and God from each other, that of man from man, and that of God from himself. In fact, Christ does not merely stand for these aversions, he *is* this triple aversion, the gap between man and God, that between man and man, and that between God and God—outside of which alienations neither God nor man would have been able to attain self-consciousnes.[26] The mutual aversion of the countenances of man and God, in the form of the face-off between Job and God, becomes (by way of a parallax shift) that between God and God, in the guise of Christ and God the Father, wherein Christ is the second Job.[27]

Yet, these aversions, these concealments of countenances, are precisely those through which the respective countenances effect their mutual reciprocal disclosure. This is the case, in that God discloses himself to man by way of introjecting into himself—in the form of the diremption between God and Christ—the hiatus that separates man from himself. Indeed, we "are one with God when God is no longer with himself, but abandons himself, 'internalizes' the radical distance which separates us from Him."[28] Here, our "radical experience of separation from God is the very feature which unites us with Him," to the extent that only "when I experience the infinite pain of separation from God do I share an experience with God Himself (Christ on the Cross)."[29]

3. Secrecy

The transition from the signifying to the postsignifying regime likewise entails a decisive movement from despotic deception to passional secrecy. Deleuze and Guattari aver that the subjective linear proceeding permits the subject to hold a secret, but solely as a sort of remainder capable of inaugurating a new linear proceeding; whereas this passional subject—marked by betrayal and treason—may be viewed as arising from the great nomads and their secrecy.[30]

Secrecy, as a salient characteristic of the present regime, may moreover be seen as a necessary corollary stemming from the traumatic character of the non-ideational exposure to radical exteriority that founds the regime to begin with. That is to say, as Žižek maintains, in that the original violent founding gesture—as we have seen earlier—is disavowed by the ensuing tradition following in its wake and in that this disavowed traumatic kernel nonetheless persists as the spectral dimension sustaining the said symbolic tradition as such; the obverse of this tradition is precisely this "secret history of traumatic fantasies transmitted 'between the lines', through the lacks in, and distortions of, the explicit symbolic tradition."[31]

(i) Joseph and his brothers

This secrecy is the secrecy of Dan, the son of Rachel's handmaiden Bilhah, who—when one of the several assembled Ishmaelite traders inquires as to the identity of the boy whose robe they had just pulled and whose face they had revealed—replies that he is a mere slave, an offspring of one of no standing, an utterly base menial, almost a kennel boy, who had to be punish for incessant thieving, lies, blasphemy, obstreperousness, contumacy, fornication, and countless other forms of moral corruption.[32] He was indeed Joseph, son of Jacob.

This secrecy is, however, likewise that of Joseph himself. For, far from moving to scupper Dan's dissimulation, he simply says nothing at all, merely sitting in his place with meek, downcast eyes, like a lamb being sheared without a single bleat.[33] When his brothers hereupon pass the remainder of the veil Jacob had gifted to Joseph through the blood of the lamb just slaughtered to consecrate the sale of Joseph to the Ishmaelites; despite being furtively concealed from the gaze of the desert merchants, this act is performed quite brazenly before Joseph's eyes—for the brothers know they can trust in the certainty of his silence: that he would nowise reveal their formerly attempted murder of his person nor his true identity to the Ishmaelites, as he in actual fact never does.[34]

Indeed, that *they*—Joseph's brothers—would say naught of having either abandoned Joseph or sold him—this secrecy—is also forthwith firmly cemented by a "fierce" oath, indeed a "gruesome" oath, sworn at the instigation of Levi by all the ten brothers. For, according to the oath, he who would fail to be entirely mute about what they were planning to do to Joseph, that is to say about "it," and dare to with even a blink, or a wink, or the slighest twitch, hint at "it," insinuate "it" or intimate "it," would instantly and forever become a "whore." Sin's daughter would emasculate him. He would become like a "mule," earning his wages by soliciting strangers in the streets, and be "driven from land to land, never knowing where to bed his whore's head, unable either to live or die, with life and death vomiting upon him in disgust for eon upon eon."[35]

(ii) Joseph and Jacob

Though not present at the swearing, nor compelled to entirely foreswear the disclosure of this dread secret, Joseph himself does keep it. This of course entails the upkeep of the appearance of his being dead to his father. And indeed, whilst ensconced as steward to his Egyptian master Potiphar, though he has ample and unimpaired opportunity to contact Jacob and emphatically controvert the latter's invincible faith that his son is no more, he does not. He acts like a calf that does not bellow to let his mother know upon which field it dwells. This cow—that is to say, Jacob—was thus to live in constant and unbearable expectation, having no choice but to believe the calf to have been mutilated and dead.[36]

(iii) Joseph, Mutemenet and Potiphar

During the precarious period when Mutemenet is haplessly in pursuit of Joseph's affections, with whispers she strives hopelessly to indelibly insinuate that the secret which they share with one another is one from which he is unable at any time to escape, both of them being purportedly irrevocably lost in it. Thus does she strive to interminably bind him to herself in this deepest of all secrets.[37] And, though it is hardly the case that her shameless lasciviousness of intent is a secret as such, the fact that Mutemenet's titular husband Potiphar is privy to the secret but does not disclose his being privy even to her, effectively keeps the secret quite secret for at least the duration over which Mutemenet thinks he is not thereunto privy.[38]

4. Necessary betrayal and fulfilment only in betrayal

The Deleuzoguattarian passional regime is a regime of universal betrayal, in which the true man never stops betraying God and God himself interminably betrays man.[39] Each betrayal is instigated precisely by the double turning away of countenances upon a line of flight that we mentioned earlier. In fact, it is Jesus who universalizes this system of betrayal by betraying the God of the Jews, being himself betrayed by the same God, betrayed by Judas, by betraying the Jews and in turn being betrayed by them.[40] The line of betrayals is endless, and extends far beyond the realm of the history of Judaism. Deleuze and Guattari mention the face-off of Richard III and Lady Anne in Shakespeare; the betrayals of companions, kings, indigenous peoples, and other terrestrial explorers by the greater discovering pioneers of Christendom; and Luther as traitor to "all things and all people; his personal relation with the Devil resulting in betrayal, through good deeds as well as bad."[41]

Adopting Žižek's line of reasoning, betrayal, as a salient characteristic of the postsignifying regime, may also be seen as an inexorable effluence of the disavowal entailed within the originary violent and traumatic gesture through which the regime founds itself. The betrayal spoken of here is symbolic—or, more precisely, is that which accounts for an occlusion at the level of the symbolic. The originary violent founding gesture (and its disavowal) precipitates *inter alia* the non-symbolization of the traumatic kernel by the Judaic tradition. This non-symbolization is the form assumed by the secrecy ensuing in the interminable aftermath of the founding violence that is here disavowed; that is to say, it is how the secret is kept. The postsignifying regime's sempiternal secrecy thus follows in the wake of its originary betrayal. Yet, paraxodically, this betrayal (as reflected negatively at the level of non-symbolization) shows itself to be that by which the passional regime maintains fidelity to the violent founding Event at the level of the phantasmatic.[42] Symbolic betrayal is thus at once phantasmatic fidelity.

(i) Jacob, Laban, Rachel, Leah, Joseph, Dudu, Mutemenet, and God

It is only in Jacob's betrayal of Rachel by unknowningly bedding Leah on his wedding night and by his subsequent consenting to Laban's gift of Zilpah and Bilhah as handmaidens (thus making a total of four "wives") and in Laban's betrayal of Jacob (by secretly slipping Leah in place of Rachel on the night of Jacob's marital consummation), that Jacob fulfils his promised consummation of love via matrimony to Rachel: since without taking on the other three as spouses, Laban would have almost certainly denied Rachel to him and they would never have been united in any form whatsoever.[43]

Yet, the line of betrayals shall continue unabated. Jacob betrays Rachel by consciously siring Ruben with Leah.[44] God then betrays Jacob in that, seeing Jacob's betrayal of Leah (his hatred of her and his well-nigh *idolatrous* love of Rachel—a form of privileging which by rights ought to belong to the Elohim alone), he renders Leah most fertile whilst causing Rachel's womb to be as barren as the desert itself.[45] Hereupon, Jacob in thought betrays God by placing the blame for Rachel's barrenness entirely on him.[46] Rachel betrays her original promise of monogamous beholdenness to Jacob by designating Bilhah as handmaiden, indeed leading Jacob directly to her by the hand.[47]

There is likewise Rachel's betrayal of Leah in her (Rachel's) daily taking of Leah's husband Jacob—but this only by virtue of Leah's much earlier betrayal of Rachel through her (Leah's) knowingly consenting to unite carnally with Jacob on what ought to have been Rachel's first nuptial night, and having Ruben poured into her—a child that should have by rights been born to her sister.[48]

Subsequently, Rachel betrays Jacob through her temporal sexual surrender of him to Leah for one night—but, due to her herein entailed acquisition of the fertility-enhancing *dudaim* from Leah, Rachel's becomes a betrayal geared toward the fulfilment of her and Jacob's mutual love in the form of the augmentation of her prospects of bearing a child by him.[49]

Rachel then proceeds, when Jacob and she are amid preparations for their secret escape from Laban's domains, to betray Laban and his household deities, the teraphim, by stealing the said idols—a betrayal, nonetheless, ultimatedly geared toward (and along the broader existential arc of) her gradual turn toward the Elohim, and thereby toward the further fulfilment of her love for Jacob through conversion to his God.[50]

Jacob's betrayal of Laban by way of his clandestine abdication and theft-tinctured elopement from the latter's employ and lands with Rachel and other of Laban's cherished items, as indeed Laban's numerous anterior betrayals of Jacob, come at length to function as inexorable antecedants to Jacob and Laban's convenant whereby both are mutually reconciled before God and above all to God.[51]

Yet, most swiftly amid the wake of these convoluted vicissitudes God comes to betray Jacob by way of not forestalling Rachel's death amid the pained agonies of moribund child-labor by the roadside—a betrayal in chastisement for Jacob's idolatrous love for her, and one in turn at once precipitative of Jacob's betrayal of God by way of his resentful and obstinately hateful indictment of the Elohim as sole reprehensible agent of Rachel's demise and of Jacob's subsequent doubly idolatrous feelings for Joseph: ungodly feelings for which God would later again punish Jacob through a renewed and still more heart-rending betrayal of the latter.[52]

Hereupon, there is Jacob's betrayal of his ten other sons in his blantant and unconcealed favoritism toward Joseph (he was truly willing to give "ten for one"), leading irrevocably to their subsequent betrayal of Jacob via their embittered severance of all ties with and abdication from their father's house.[53]

Later, with Joseph languishing in an abandoned well, Ruben betrays (though ultimately only at the level of intent—for his plan is thwarted) his brothers in contriving to secretly rescue Joseph so as to win the esteem of Jacob and his formerly lost rights as firstborn whilst moreover exposing his siblings' murderous fratricidal bent before their father.[54] This betrayal is quickly hereupon reciprocated in kind by his brothers, through their perfect nondisclosure of having in fact sold Joseph to Ishmaelite merchants—Ruben simply not having to know.[55]

Later, with Joseph having made his way along a line of flight at the behest of the Ishmaelites into Egypt and being at length ensconced in the house of Potiphar, the contumelious little dwarf Dudu—not without a fair trace of calumny—betrays Joseph to Mutemenet as some kind of "Habiru dandy," as a "puppy of the wretched desert"; portraying the favor shown him in Potiphar's house as a scandal, and as a malevolent "cancer"; contending moreover that he had been bought much too dearly from utterly base and ignoble wandering desert peddlers, who had stolen him from a dry well, from a "pit of punishment"; and that instead of sending the "foreign lout" to drudge in the fields, owing to the gormlessness of Shepses-Bes and Mont-kaw, the "gallows bird" had been left to loiter in the house and dine Potiphar's ears with "wily quibbles" that defamed Amun and blasphemed against the most exalted powers of the sun, and words that had inebriated Potiphar's senses and culpably bewitched him to the extent that he raised him up to be his own personal attendant and reader, whilst Mont-kaw treated this "mangy Asiatic" like a son in an Egyptian house residence.[56]

Yet, paradoxically, it is precisely through and only through this betrayal of Dudu's that the early seeds, of what will in time become a seething and unbridled infatuation and pulpitating lust for Joseph searing Mutemenet's brain, and the conditions of its possible consummation, were implanted; for it was whenever Dudu had complained to her of him that she thereupon came to look upon him with a heightened interest, and subsequently with more than simply interest.[57]

At length, however, it is precisely the gradual burgeoning of the poisonous ivy of this incipient lust within her riven heart that—filling her with dread foreboding—moves Mutemenet to betray Joseph to Potiphar, to bemoan before her titular husband the Habiru slave's monstrous, vexatious, and most untoward ascent within the courtier's house.[58]

The cascade of multiply ramified betrayals now gains momentum as Dudu proceeds to betray (though not without a fairly generous admixture of distortions and lies) Mutemenet and Joseph's purported sultriness to Potiphar, as—he claims—is evinced by their secret dalliances and meetings, billets-doux, "cooing and hot lisping."[59]

At last, having been spurned by the Habiru slave once too frequently, Mutemenet herself betrays Joseph by proceeding to calumniate the latter with the falsely-wrought charge of his having perforce attempted to make carnal acquaintance with her person, his mistress.[60]

After suffering the three years of incarceration of his second pit, as the man raised up from this mire and extolled to the position of Grand Vizier, Joseph finds himself imbroiled in practices wherein he comes to perpetrate his own betrayals. His marriage to Asenath, "this daughter of the sun," brings him into such proximity with the temple of Amun-Re that he draws his income from the benefice of an idol, officiating as priest in the burning of incense before the falcon Horakhte—an unalloyed betrayal of the God of his father Jacob and of his forefathers.[61] He proceeds to indeed doubly betray Jacob in calling his firstborn Manasseh, which name means "God has made me forget all my bonds and my father's house."[62]

When Joseph's ten brothers sally forth into Egypt during the time of the seven years of famine prefigured by the thin cows and the blighted grains, they find themselves betrayed by God, who here at last punishes their much earlier bloodthirsty betrayal of Joseph in the darkness of the abandoned well with forcible incarceration, as Judah holds forth: "Why did He not snort in his rage back then, and instead let lifetimes pass, let the repast of chastisement grow cold, until He prepared this meal for us?"[63]

But it was indeed to the ends of the ultimate and greater consecration of filial and paternal bonds and such that the implicit propensities inhering within Joseph would at end be fulfilled in the person of Joseph the Provider that the betrayals perpetrated by the brothers and by God ultimately point. As Joseph eventually makes clear to Judah and the other brothers, it is the kiss that the latter gave him before the Minaeans or Ishmaelites that he now returns—indicating hereby that their transgression against him has been blotted out. For he is certainly not angry with his brothers for having sold him to a great and opulent land in which he occupies a position second only to that of Pharaoh, furthermore maintaining that their betrayal of his person was in fact absolutely necessary and that it was not really them who had done it. For God had done it. It was El Shaddai who had separated him from his father's house very early on, and made him a stranger in accordance with his inscrutible plan. God had carefuly orchestrated even the most deleterious vicissitudes of Joseph's life precisely so as to enable him to provide for both his brothers and father but, through an astonishing rescue, also for the needs of Israel along with other nations in a time of dire famine.[64]

In a later conversation with his now-ailing father Jacob, Joseph resumes and elaborates on this same thread—of how God's betrayals (by way of stern chastisements meted out for man's own betrayals of the Divine) are merely necessary moments along a broader existential arc, such that the latter may reach its *telos* in the bringing to efflorescence and fulfilment of the greater events already implicit to the seemingly disparate vicissitudes of what would otherwise remain an interminable and blind sequence of senseless vengeance. For, as he pontificates, if God metes out punishment, this punishment carries within itself some kind of very serious moral and existential purpose, and serves ultimately to advance a greater and more immense event. Though God had admittedly seized Jacob and Joseph very roughly indeed and rendered them apart from one another to the extent that the son had verily died to the father, he did this only so as to send the son to Egypt in order to save his father and his brothers by providing for them and all their house in a time of famine—precisely so as to reunite the entire family with itself. This, Joseph concludes, is because God's very passion is providence, and his anger farseeing goodness.[65]

(ii) God, Christ, Judas, and man

As Žižek maintains, in an act of betrayal, God abandons man by separating him from himself, but only to the end that mankind may love him. This diremption, in turn, is introjected back into God himself as his own *self*-abandonment, but only such that the abyss of Godhead can give birth to Christ as reconciler of God with God, God with man, and man with man.[66]

God's self-abandonment is the ultimate and most pivotal betrayal of all. G.K. Chesterton contends that when the earth trembled and the sun was cast from the heavens that one vertiginous day on Golgotha, it was not at the time of the Crucifixion, but in the moment of the God-Man's anguished cry from the Cross—a cry that proclaimed that God was forsaken by God himself, the point in history at which God betrayed himself in the astounding and ridiculous moment of his own atheism, Christ's own commission of the mortal sin of infidelity—*with regard to his own Godhead.*[67]

Within another (though related) ambit, in the case of Christ and of Judas, Žižek suggests that—not only was it the case that Jesus needed the betrayal of Judas in order to redeem the human race—but that the inauspicious words of Christ to Judas at the Last Supper constituted a secret (though no less positive) *injunction* for Judas to betray him, and (still beyond this) became the opening salvo in the irreversible betrayal of Judas by God—the price of salvation being Judas' readiness to voluntarily lose his soul and suffer eternal damnation. Judas' betrayal of Christ, therefore, emerges here as an act of unsurpassable fidelity by man to the Divine mission of the Son, the *sine qua non* of Christ's transition from the universal concept "Jesus the Christ-Messiah" that he was prior to Golgotha to the unique singular "Jesus Christ" that he became by way of Crucifixion—in a movement of Hegelian *Aufhebung* wherein, through the murder of the particularity (God in the wretchedness of man), the eviction of Judas from the closed circle of the original twelve and the consequent enablement of St. Paul's subsequent inclusion into this circle, the universal Church was able to establish itself.[68]

In Žižek's reading, the moments of betrayal-diremption and fulfilment-reconciliation—moments hitherto cleaved by temporal delay, find themselves in a fundamental sense telescoped to the point of coincidence in the case of the Fall: it is not so much (or at least not *primus inter pares*) that God first betrays man by pushing him into sin so as to thereby forge the need for the fallen creature's redemption in the protended future act of himself assuming the role of Redeemer; rather, instead of being "followed by the Redemption, the Fall is *identical* to Redemption, it is 'in itself' already Redemption."[69] The Fall is *ab initio* the extrusion of a vertiginous freedom; it is "*strictu sensu* identical with the dimension from which we fall, that is, it is the very movement of the Fall that creates, opens up, what is lost in it."[70] In this sense, it is not primordially the case that what is lost in Adam is subsequently restorted in Christ (although this is what does happen in a more derivative and prosaic sense); but rather, Adam and Christ are the same, relating to one another as "negation" and "negation of negation," with Adam as Christ "in itself" and Christ's Redemption itself—far from being the straighforward "negation" of the Fall effected by Adam—the Fall's *accomplishment*.[71] O Felix culpa!

Betrayal in the passional regime, thus, takes upon the figure of a necessary and inevitable movement internal to the satisfaction of ethical injunctions, and is the inextricable concomitant of the face-off of concealing-revealing countenances drawing the line of flight or deterritorialization which sets the passional line in motion in the first place. Otherwise, it is nothing at all.

5. Subjectification: the doubling of subjects and the subject as double

In his meditation on Kierkegaard's reading of the binding of Isaac, Derrida avers that the self has within itself a "witness that others cannot see, and who is therefore *at the same time other than me and more intimate with me than myself* ..."[72] This, moreover, implies that I maintain a "secret relationship with myself ..."[73] Therefore, it becomes simultaneously both the case that God calls me and that I call myself, for I call myself solely on the condition that "God is in me, he is the absolute 'me' or 'self,' he is that structure of invisible interiority that is called, in Kierkegaard's sense, *subjectivity* [my emphasis]."[74]

Crucially, we see here that the subjectivity of the "I" presupposes *a double*: that is to say, there must necessarily be *both* God and myself for my own subjectivity to emerge as such and for me, in the enclosure of my own interiority, in secret, to be able to refer to or call myself. Therefore, in that the subjectivity of the self qua subjectivity is necessarily formed around a hidden God, this subject is hereby itself rendered properly-speaking *double* (in the sense that an "internal" "me" finds itself inevitably lodged within "myself"). More succinctly put: in order to be itself, subjectivity *ab initio* entails, on the one hand, a doubling of subjects (in the form of the couple God-I) and, on the other hand, that the subject is itself doubled (as inheres in the fact of the "I" being inhabited by God-as-"me"). That is to say: *the doubling of subjects and the subject as double.*

On the one hand, at the level of the subjective couple (although in this case a strongly assymetrical one), I betray and likewise maintain fidelity to God along the very same vector. At the level of the double subject, the "I" finds itself the more attuned to its ipseity through the "me" lodged into it by God himself. The "I," therefore, can hereby be said to spring back into the "me," in terms of being determined by the latter. Translated into Lacanese, the "I" is the subject of the enunciation whilst the subject of the statement corresponds to the "me." The fact that the ipseity of the "I" arises heteronomously from the "me" thrust into the midst of the former by God means that the "I" (as the subject of enunciation) *recoils* into the "me" (the subject of the statement).

Something very similar is afoot in the Deleuzoguattarian postsignifying semiotic. Formerly, in the despotic regime, we had the radiant face of the despot serving as the "center of signifiance" whereat the signifier found itself supplied with a signified, though was also recharged with more signifier. In contradistinction, in the passional regime, this center of signifiance gives way to a "point of subjectification" constituting the "point of departure" of the passional line.[75] This point of subjectification is precisely the "locus" of the face-off between concealing-revealing countenances glimpsed earlier (regarding the change in faciality from despotic to passional regimes). From the point of subjectification there issues a subject of enunciation and a subject of the statement "in a determinate relation to the first subject."[76] More precisely, the subject of enunciation arises from the point of subjectification "as a function of a mental reality" determined at the same point. From this subject of enunciation there issues forth a subject of the statement, that is, a subject fundamentally bound to statements in conformity with a "dominant reality" (of which the aforementioned "mental reality" is a part—even when it appears to purportedly oppose it).[77] The subject of the enunciation, thus, "recoils" into the subject of the statement as the "imprint of the process of enunciation in the statement"—to the extent that the subject of the statement in turn resupplies subject of enunciation for a further proceeding.[78]

(i) The God-man subjective double

This recoiling of subjects is discernible in the relation between Abraham and God where, as Thomas Mann has it, though God's potent attributes were certainly a definite reality outside the subjectivity of the great primordial patriarch, they were at once within him and an effluence of himself—to the extent that sometimes his own soul interpenetrated and melted into them and could no longer be distinguished from them.[79]

The mutual recoiling of God and man finds itself generalized in Meister Eckhart. For the latter, beyond the fact that God begets man but only truly becomes God through man; far more disconcertingly, it is indeed man himself who begets God, that God is naught at all outside man—this nothingness being the abyss of Godhead anterior to God in which the very difference between God and man is annihilated.[80] The mutual recoiling of God and man is the relatedness of God and man through the Otherness (Godhead) in God that functions as ex-centric center of man and of God alike, decentering man with regard to God and likewise decentering God in relation to his own Godhead.[81]

(ii) The point of subjectification

Concerning the point of subjectification as locus of this procees, this latter may in fact be effectively served by a multitude of diverse things. In the case of Judaism, the averted countenance of God is the point of subjectification from whence Moses issues forth as the subject of enunciation, constituted as it is on the basis of the tablets of God that replace the divine face; whereas the Jewish people constitute the subject of the statement, for the sake of betrayal but likewise for a new and blessed land, entering as they do a repeatedly renewed covenant.[82]

In fact, this point of subjectification—God's averted countenance—attains a unprecedented ubiquity. For Régis Debray, it insinuates itself "wherever a Jew prays and observes the Torah," consisting as it does in the "permanence and ubiquitousness of a God *accessible* wherever a *yad* followed the lines, a *sofar* was unrolled from right to left, or the Name was uttered in the dark."[83] The propensity for the miniature conferred by God upon man through the person of Noah enabled Judaism to multiply the averted face of God as point of subjectification through the interposition of a veritable slew of surrogates, such as phylacteries (*tefillin*), little quadrangular boxes in leather containing biblical passages written in miniature black ink and attached to straps worn by the pious on the forehead and left arm; and *ostracas,* which are the poor man's tablets of the Law.[84]

In general, regardless of what it actually is, the point of subjectification must merely satisfy the conditions of a double mutual turning away, betrayal, and—as shall be discussed soon enough—existence under reprieve.[85]

(iii) Syntagmatic axis of consciousness and paradigmatic axis of passion

Deleuze and Guattari elaborate further on this subjectification arising from the double turning: the doubling of subjects into the subject of enunciation and the subject of the statement, and the recoiling of one into the other is simply *consciousness*—more precisely consciousness as "passion." This is the syntagmatic axis of the line of subjectification. But to this axis there moreover corresponds a paradigmatic axis of subjectification: that of love as passion, "love-passion." There are thus two doubles occupying the twin axis of subjectification: a "celibate" doubled consciousness on the syntagmatic axis and, on the paradigmatic axis, a "passional love couple" that no longer requires consciousness or even reason.[86] As Deleuze and Guattari maintain, the line of subjectification is completely occupied by two different kinds of figures of a single double: "the syntagmatic figure of consciousness, or the consciousness-related double, relating to form (Self=Self [Moi=Moi]); and the paradigmatic figure of the couple, or the passional double, relating to substance (Man=Woman; here, the double is immediately the difference between the sexes)."[87]

(iv) Jacob, Rachel, and Joseph

Mann himself unearths two doubles—which are in turn themselves doubled. First there is the double Joseph-Rachel, evinced by the fact that Jacob had never known how to clearly distinguish between Joseph's and Rachel's eyes—since, for him, they were one and the same, and he had "kissed away the tears of impatience from beneath them." In death these two sets of eyes merged into one single pair of eyes, whereas the two beloved figures themselves fused into one unitary hermaphroditic image, "the object of a yearning that was itself suprasexual, both male and female, like all highest things, like God himself."[88] This would be heretical for both Christians and Muslims, but not necessarily so for many Jews.

This yearning of Jacob's was, however, an integral part of himself and he a part of it. For he likewise partook of this same nature—a conclusion with which he long felt quite emotionally comfortable. Very strangely, ever since Rachel's death he had in fact been both father and mother to Joseph, in this likewise assuming her role in their relationship—with maternal love in fact predomining. His equating of Joseph with Rachel, moreover, found its counterpart in his very own identification of himself with his deceased wife.[89] This nature of Jacob as a geminate being is the subjective double that is consciousness itself. That is to say, the consciousness of Jacob is that of Jacob insofar as he is the geminate being born of the introjection of Joseph-Rachel, where the redundancy Jacob-Jacob is at once the redundancy Joseph-Joseph and that of Rachel-Rachel implicit to the double Joseph-Rachel. The introjection of Joseph-Rachel to render up the syntagmatic figure of consciousness Jacob-Jacob, in that it is the corollary of the recoiling into one another of the subjects Jacob-Jacob and Joseph-Rachel, implies that the paradigmatic passional double here is Jacob-Rachel (in which, of course, Rachel may be interchanged with Joseph). This is how the doubling of subjects and the fact of the subject being double figures in the case of Jacob and Rachel (Joseph).

(v) The black hole of stochastic death

Crucially, one must be mindful that this process of subjectification is inherently suffused by a slew of glitches and the potential for vertiginous catastrophe. The doubling at the level of consciousness bears within itself the inexorble tendency that the doubled subject comes to be, at the terminus of this movement and at the highest pitch of intensity, incarcerated in the white wall of its own consciousness. At the same time, the doubling at the level of the passional love couple tends ineluctably toward the annihilation of both subjects in a common black hole of passion, of stochastic death, in which each of the two are abolished.

"She is vast," writes Bataille, "she is distant like that darkness in which she has trouble breathing, and she is so truly the vastness of the universe in her cries, her silences are so truly the emptiness of death, that I embrace her inasmuch as anguish and fever throw me into a place of death, which is the absence of bounds to the universe."[90] The black hole of stochastic death consists in the fact that, at the "most intense moment of fusion, the pure blaze of light, like a sudden flash, illuminates the immense field of possibility, on which these lovers are subtilized, annihilated, submissive in their excitement to a rarefaction which they desired."[91] In this pure blaze of incandescence, the subjects are burned by the sun, given over to the disintegration of the abyss, to ashes, and to the unfathomable night of the Schopenhauerian noumenal Will.

(vi) The Badiousian obscure amorous subject and death

The doubled subject of the Deleuzoguattarian post-signifying regime resembles a similar entity discussed by Badiou. Under the heading of "amorous subjects," Badiou avers that the event of amorous encounter precipitates the upsurge of "a scene of the Two" in which the male and the female species are bound around something in which they both participate. He maintains, however, that this common "object" is thoroughly enigmatic in the sense that, while its existence is vigorously affirmed, no one knows precisely what it is. It is this unknownness of the common object that is responsible for the strictly contingent nature of the amorous encounter. Notwithstanding, the enigmatic object, through the amorous encounter, constitutes "a bi-sexed body," or *couple.*[92] Badiou maintains that what is produced in this fashion is "an *enchanted existence* in which the truth of the Two is fulfilled in an asocial fashion."[93]

What Badiou maintains to be a degenerate form of the amorous couple arises when the contingent nature of the amorous encounter and likewise the unknownness of the common object are both denied. The contingency of the amorous encounter gives way to its necessity, whereas the unknownness of the common object yields before the lust for absolute knowledge. With this having come to pass, the amorous couple is replaced by the "obscure subject of love," as the bi-sexed body of the amorous encounter gives way to a single fictive and fusional body.[94] The obscure subject submits love to a "fatal ordeal," for it "demands an integrated originary destiny and consequently can only see a future for love in the chronic extortion of a detailed allegiance, a perpetual confession."[95] It sees love to be of its very nature *destined*. Moving beyond simple conjugality, it swiftly "institutes a deadly possessive reciprocity."[96] In the name of its fictive fusional body, it extinguishes the enchanted nature of the amorous couple and plunges into night. Badiou cites *Tristan and Isolde* as Wagner's poem par excellence on the obscure amorous subject.[97] The poem exemplifies the fact that "the very being of the amorous encounter" inevitably "carries an ineluctable dose of nothingness."[98] It reveals "the void that lies beneath the being-there" of existents as evinced by "the very romantic call to the fusion of lovers under the sign of death" itself.[99]

In Act III of *Tristan*, as Žižek avers, Tristan's state is one of desperation in that, being without Isolde, he is condemned to life without her, to the pain of eternal longing, and—above all—to the inability to *die*. For this reason, he anxiously and with tremendous pining awaits her arrival so as to be able to die with her, so that the single fusional body of the obscure amorous subject precipitates itself into the black hole of stochastic death.[100] As Edmond Jabès maintains, "when I ceased to be in order to become indefinitely *the other* in her desire to destroy me, love drew its strength from death and needed the sacrifice of my face in order to blossom."[101]

(vii) Lévy: haematology, haematomania

This latter self-destruction of the subjective double—and at once that of the double subject thereto correlated—is often tantamount to martyrdom or straightforward suicide. To be sure, the suicide hereby effected may well be indirect. The fact, notwithstanding, remains that—when the fusional nature of the subjective double reaches a certain point—this subjective double henceforth feels that it is only able to maintain fidelity to itself through martyrdom. This martyrdom may well be that of love. It may likewise also assume a markedly theological aspect. It may be purely political. Regardless, once a certain threshold of resonant intensity of the subjective double has been attained, the subject often feels compelled by an inexorable force to consummate himself through martyrdom and martyrdom alone. This point of no return being reached, the subject will henceforth seek itself in its own blood. Or in its own ashes. Such a subject is inescapably headed for the black hole of its own stochastic death. This penchant instilling itself within the obscure amorous subject—collective or otherwise—Bernard-Henri Lévy terms a "drunknennes of bloodshed," or better, a "haematology," or a "haematomania."[102]

Such a subject has moved beyond the point of merely staking his life for the defence of ideals of love, religion, politics, or art to which he is beholden with the utmost of conviction. For he is no longer merely gambling his life. Rather, his unswerving resolve is "to die, really die, without the shadow of hesitation, or of a doubt, or of a regret, or of a choice."[103] For such men it is no longer a matter of risking death through love of life. Indeed, they "don't risk it, they run to it, because in reality they love nothing but death."[104] It is not the case that such men are bereft of choice and are thereby compelled to risk death so as to avoid death, that they prefer a merely possible death to an absolutely certain death. For they do have a choice and could indeed not die. Yet, in spite of this staying of the Damocles' sword that constantly avails itself, it is "freely, with sovereign power, in full and absolute awareness, that they make the choice of *viva la muerte.*"[105]

These subjects headed for the black hole of stochastic death are thus of the order of fanatics and kamikazes possessed of "the cult of death for death's sake, pain for pain's sake," who notwithstanding "walk to martyrdom, and wish for their own execution, full of hope, with joy in their hearts, in a state of jubilation matched only by the desolation of their victims"—if there are indeed any.[106] Their's remains a "sacrificial dementia, self-abnegation, expiatory masochism," whose concern and destiny of pining is terminal immolation in the black hole of stochastic death and that alone.[107] "His main concern is to die."[108]

(viii) Eagleton: the self-immolatory freedom of the void

To these evocations of Lévy, we may add those of Terry Eagleton. One may divide Eagleton's ruminations into two parts or stages. The first aspect of his argument approximates the Deleuzoguattarian notion of the subjective double and its correlated double subject—as intrinsic to the post-signifying regime. On the other hand, in what we here designate the second stage of the same discussion, it is apparent that this quasi-Deleuzoguattarian level unearths and yields itself up to what appears a subterranean Schellingian substrate: that of the vertiginous abyss of freedom, that is, the Godhead.

Eagleton's first angle of attack is as follows. He presents human subjectivity as inherently part of a double. Firstly, as regards the originary interdependence of man and God, he maintains that it is "by being dependent on him that we came into our own."[109] Furthermore, God is "the unsearchable power at the core of the self which allowed us to be what we were."[110] Thus, not only is the self externally dependent upon God, but God also internally inhabits the self as the self. As seen in a similar vein in relation to Derrida, this, in Deleuzoguattarian terms, is tantamount to delineating man and God as the two sides of a subjective double in which God as subject of enunciation leaves his imprint on man as subject of the statement.

But this level of the Deleuzoguattarian subjective double swiftly reveals itself to be inhabited by the more vertiginous infractions of the Schellingian Abyss of the Godhead itself. The Schellingian Godhead is the primordial Abyss of Freedom. Eagleton's disquisition exhibits a subtle movement beginning with God *as subject*—this latter forming a component of the man-God subjective double typical of the post-signifying regime. Swiftly enough, however, Eagleton's argument moves toward God as the unfathomable Godhead. Such a movement is one from a pure disembodied form (God as subject) toward self-differentiating difference in itself (the Godhead as such). This all means to say that what first appears as a resonance and mutual recoiling of two subjects into themselves, inevitably reveals itself to be predicated upon a singularly allusive abyss. It is this unquantifiable abyss of freedom that the God-man subjective double reverts to—as a deeper instance of itself.

But let us follow Eagleton in this vein. Eagleton's contention is that the subject's imbrication with God as subject is at once the subject's sharing in God's unbounded liberty; that is to say, the unbridled participation of the subject in the infinite freedom of God—*not* as subject—but *as Godhead.* For the Godhead is always already and immediately freedom *itself.* The intimacy and formidable mutual resonance of the man-God subjective double thus reveals itself—at a more subterranean and vertiginous level—to likewise be the participation of man in the unconditional and boundless freedom of the primordial abyss. This primordial abyss is the Godhead.

Eagleton maintains that the foregoing was particularly the case in the pre-modern period, when God "was the answer to the question of where our freedom came from, in the sense that we became the free individuals we were by sharing in his own boundless liberty. This participation was traditionally known as grace."[111] The transition from God as the subject who imprints himself into man to both God and man as abyssal freedoms follows, since to "claim that God made us in his own image and likeness was to say that where we were most like him was in our autonomy."[112] God was "the ground of our freedom, not the obstacle to it. Being a 'creature' of the Almighty meant being dependent on his life for our own, and the life of God was nothing but freedom. This kind of dependency was therefore the reverse of slavery, as St Paul is eager to point out. Where we were most self-determining, there we were most truly his."[113]

So, this ineluctable and tautly drawn dependence of the subject on God—which in the Deleuzoguattarian postsignifying regime figures as the recoiling of man as subject of the statement into God as subject of the enunciation—is at once that through which man is a boundless freedom by way of his unmediated participation in the abyss of the Godhead. "Nothing could go further down than God, who was a bottomless abyss of being. To take your stand on him," Eagleton continues, "was thus to take your stand on infinity. Because this ground of our being consisted of pure freedom, we could be securely anchored while still feeling as though we were walking on air."[114]

With the modern epoch, the Godhead as the vertiginous source of man's freedom gives way to man as the unfathomable and purely self-determining abyss of his own freedom. As Eagleton puts it, "it is now not God but humanity which is the eternal author of itself, conjuring itself up out of its own unsearchable depths without visible means of support."[115] This transformation, though seemingly an upheaval, is however merely superficial. For modern freedom as pure human self-determination is merely a secular variant of the abyssal Schellingian Godhead. The latter has merely returned as the anarchistic impulse, whereas the man-God subjective double has been replaced with the subjective double comprising man and the deified entity Man.[116]

The issue at hand, notwithstanding, is that the abyss of freedom is latent to both of these two subjective doubles. Both stand upon an infinite precipice, for the freedom upon which they rest is absolute. This absolute freedom, as Eagleton avers, is what Hegel terms "the freedom of the void," that is, a form of liberty that is thoroughly shot through with "the sheer terror of the negative that contains nothing positive, nothing that fills it with a content."[117] For this reason, absolute freedom is a purely negative form of liberty and thus a "fury of destruction."[118] It can thus only feel itself fully alive in an act of such fury. The absolute freedom harbored by the passional subjective double is thus inherently nihilistic. The act or object which accords with its furious nihilism is therefore a kind of void. In that it absolutely refuses the positivity of all content, content that would limit it and render it relative, this form of freedom is pregnant with and portends the absolute negativity of death. Since its very own existence is likewise a barrier to the sheer

voracity of its absoluteness, this kind of freedom is bound to ultimately consume itself. The passional subjective double comes to be disgusted with bounds that it eventually cannot even tolerate itself. It thus ends up incinerating itself in the black hole of its own absolute negativity.[119] The subject dies of his own inexorably lethal will, the ferocious force of his own resolution, a will which invariably becomes a foreign body eating away or incinerating the flesh.[120] At once priest and victim, the subject presides over his own dismemberment in an final denouement of well-nigh ritual self-immolation.[121]

(ix) Jacob, Rachel, Joseph and Mutemenet

Apropos of Jacob and Rachel, Mann refers to their marital union as at once the moment of death wherein the feast of the solstice for the moon celebrates the lunar entity's most supernal height and plenitude only to inevitably turn its face once more toward the sun, into which celestial body it must now sink. Thus, Jacob was to obtain carnal knowledge of the woman he loved and, at the very same time, begin to die—and gradually do so by slowly dissolving his vital substance into his sons and surrendering his actual person to death.[122] This inevitability of being given over to death would at the same time overtake Rachel. She would die by the roadside when giving birth to Benjamin.[123]

The foreboding of the potentially looming black hole of stochastic death grows increasingly acute in tandem with Mutemenet's infatuation with Joseph. Again athwart their game board, Mutemenet and Joseph alike feel their hearts squeezed by death, Mutemenet's head falling forward onto his arm and exuding the steamy breath of her fevered stammers and hopeless lisps, as she embarks upon her increasingly desperate erotic and moribund pleas and exhortations. Here she acknowledges the cul de sac and absolute dead end into which she feels both herself and her beloved to have been condemned. The entire game is blocked. Their only recourse, she laments, is utter defeat for both: "my love is hot and high and eternal, so that now we can die together and descend into the night of desperate bliss."[124] And thus does she proceed to beg him that "we may die each in the other, for I can no longer bear that we live as two, you there, I here."[125]

The proximity of the black hole of passion is exacerbated all the more when, having her repeated lascivious entreaties spurned by Joseph with unfailing and invariable monotony, Mutemenet angrily resorts to threatening him with a terrifying death at the hands of the crocodile—where he would lie bound up defenseless with the voracious and ravenous beast approaching in its seething hunger, rolling up on its prey with its moist belly, beginning to devour firstly the thighs and shoulders, "so that your wild shrieks mix with its grunts of greedy pleasure, for no one hears or wants to hear you in your abandonment."[126]

But it is likewise Mutemenet who is ever the more inexorably feeling herself being sucked into the black hole of stochastic death, as she bitterly laments: he is her plight, her death, he has hurt her so fatally that "my red blood spills only in my anguish for him, and I shall die if he does not stanch it." Her "love for his beauty has slashed my heart and I am bleeding to death."[127]

(x) The Mad Lady, El Señor, Juan Agrippa, La Señora

In Fuentes's *Terra Nostra,* rabidly desirous of death, a desire animated by her inexorable and incorrigible obsession with and attachment to her dead husband (the once-philandering father of El Señor) with whom she pines to be reunited precisely through death; the Mad Lady almost perishes in the black hole of stochastic death she had hitherto so hankered for when her dead husband's maddened hounds, perhaps smelling the sweat and flesh of their deceased master in that of the moribund Mad Lady, viciously mutilate her limbs—during which time she utters not a single scream of pain but instead begs God that she may patiently suffer martyrdom right there and then so that, until the very end of her wicked and ridiculous days, she is left a quadriplegic invalid pushed around on a small cart by the grotesque dwarf Barbarica.[128]

Notwithstanding, the Mad Lady soon enough suffers terminal annihilation in the black hole of passion when, plummeting vertiginously from the high carved niche of El Señor's chapel, she smashes head-first into the granite floor far below, her mutilated old cadaver trampled by an oblivious rabble of a crowd as by a herd of rabid horses—the crooked remains of her crushed skull and vacantly staring eyes resembling no more than a bundle of discarded rinds and peels of rancid meat.[129]

In another scene from Fuentes' masterwork, when Juan Agrippa (one of the three identical youths found washed up on the sea shore by El Señor's halberdiers) finds himself gazing into the turbid depths of a black marble mirror held to his face by La Señora, out of the growing love he bears his own reflection burgeons a countervailing hatred by which he incarnates himself in a form that rapidly begins to increasingly assume the aspect of a woman, until such time as the form—by now supine upon the paving stones of a court yard (much as La Señora had once been for thirty-three and a half elemental days)—sighs its last sigh: the deathly sigh of both Juan and La Señora, two souls inhabiting the same body, both dying at exactly the same moment in the same dark black mirror.[130]

In the following chapter, we shall continue to investigate this dire vicissitude to which the Deleuzoguattarian double subject is prone, this absolute and irrevocable auto-destruction in its own self-precipitated black hole of stochastic death. Indeed, we shall do so at considerable length, and by delving into a rather massive expanse of history.

For now, however, we move to the next of the several povital features of the passional regime.

6. Segmentarity of the regime

Another property of the passional regime has been insinuating itself. As Debray avers: "The monotheist storm, streaked with lightning, seems to have been an extremely slow rise to power, marked by breakthroughs and relapses, a circumstance betrayed, in its way, by the dilated and repetitive temporality of the founding narrative."[131] Owing to its perpetually being haunted by the possibility or inevitability of betrayal, the passional line periodically runs aground and must constantly be recommenced. Hence, one may speak here of an absolutely fundamental segmentarity in which one proceeding must end, and its termination necessarily be marked, before another one is able to ensue.[132] Owing to the constant foundering and subsequent resurgence of the passional line, the postsignifying semiotic comes to form a broken line. Paraphrasing Emil Cioran and substituting the collective subject for the "violent" individual self, the passional line may be conceived as one along which a people perishes by the very identity and name which it assumes, yet—being forever beholden to its own suffering—persists in repeatedly and forever rising again from its own cinders.[133]

(i) Joseph and his brothers

This sort of line, with its commencements and recommencements, traces itself through Mann's story of Joseph and his brothers. We find in the narrative a string of non-identical repetitions of prior happenings proliferating themselves. Jacob's flight from the seething rage of his brother Esau effects the termination of the segment wherein he lived under the auspices of his father Isaac and mother Rebekah, this termination being effected by Jacob and Rebekah's betrayal of Esau's rightful claim as firstborn to the inheritance that would otherwise have been vouchsafed to him by Isaac. Likewise, the segment that the latter flight by Jacob ossifies into, itself founders upon Jacob and Rachel's betrayal of Laban twenty-five years later, again launching Jacob along a renewed line of flight with Laban's countersignifying line of abolition in tow.[134]

Again, a new segment of the regime is inaugurated when Jacob's son Joseph sets forth on a journey to his brothers at his father's behest, in a manner analogous to Jacob's own youthful sallying forth at the instigation of Rebekah—the betrayal in the later departure being Jacob's betrayal of his sons by Leah and her handmaidens through the favoritism he shows Joseph.[135] The foundering of this latter segment, by way of the betrayal of Joseph through his abandonment in a dry well by his brothers, leads to the refounding of the regime when—sold into Egyptian captivity by the same Ishmaelite merchants to whom his brothers had initially sold him—Joseph is made the slave of the rotund courtier Potiphar. Here, the years of labor performed by Joseph for Potiphar are a repetition of the labor done by his father Jacob for his (Jacob's) accursed uncle Laban, as is the fact that the founding of both of these segments stemmed from lines of flight precipitated by fraternal strife: that of the bloodthristy pursuit of Jacob by Esau and that of Joseph by a ten-fold rabid pack of Esaus, respectively.[136]

Notwithstanding, this segment itself comes to founder in its own mire when, at the behest of Potiphar's consort Mutemenet's betrayal of Joseph to Potiphar that he (Joseph) had desired carnal knowledge of his mistress (Mutemenet), Joseph is sent to languish in prison.[137] Here, Mutemenet's exhibition of Joseph's loincloth as an infallible token of his guilt (not the case) in the matter repeats Judah's showing of Joseph's bloodied garments to Jacob as a token (falsely) indicating Joseph's being gored by a wild animal.[138] Likewise, the three years spent by Joseph in prison (his second pit) repeats in non-identical form the three days he had spent in the well (his first pit) where his brothers had abandoned him to die.[139]

This, of course, is the regime in a state of desuetude—yet a temporal demise from whence it shall refound itself by way of Joseph's becoming Grand Visier of Pharaoh, a segment itself brought to term with the eventual joint departure from Egypt of Jacob, Joseph and the brothers.

Thomas Mann now brings to view the broader ambit of this postsignifying line in its segmentary. For indeed, in the final departure from Egypt, Jacob the old man was a repetition of Jacob the youth departing from Beersheba for Naharaim. He was the grown man Jacob, who with his wives and flocks of animals had left Haran behind after having sojourned there for twenty-five long years. But he was not only one self-identical person, in whom at various fundamental stages of his life departure rediscovered itself again and again. For he was likewise Isaac journeying to Abimelek and to Gerar in the land of the Philistines. Still further back in time than this, he experiences the return of the absolutely primordial departure: that of the wandering Abraham leaving Chaldean Ur. Yet even this was not really the primordial departure, but merely a comparatively mundane earthly reflection and imitation of a heavenly and divine wanderlust: that of the moon in its circuit, of Bel-Charran (the Lord of the Way), passing through his various fixed points.[140]

(ii) The Puritan line to the New Jerusalem

These non-identical repetitions internal to the history of Judaism, the Jewish people, and the narrative of the Torah, extend themselves far further afield in the Hebrew religion's subsequent monotheistic progeny.

Debray describes the Puritan line of flight as that of the "New Ancients," whose escape from "an England of nooses and burnings at the stake" to America was another crossing of the Red Sea by itinerant nomads—who confronted the forbidding wilderness as had their ancestors in Sinai, falling on their knees upon reaching Cape Cod and, upon arriving in Connecticut, withdrawing into "Canaan" to thank God for their first harvest.[141]

Much as Judaism commenced with a new beginning of God in history, so too was the American *novus ordo saeculorum* an excorcism of old European ghosts in the wake of the incipient nascence of "One Nation under God." The North American land ordained by the Divinity for the erection of the New Jerusalem, quite like the old, was decreed innocent and virginal; whilst the superfluity of the Indians replicated the near-negligible status of the Jebusites in pre-Davidic Jerusalem. The "shining city upon a hill" is Zion, whilst the Hebrew doctrine of the chosen people finds its homology in the notion of "Manifest Destiny." The millenarianism of the Book of Daniel and that of the Apocalypse of Saint John has attained vigorous recrudescences in pronouncements associated with not only the Bush Administration, whilst Armageddon itself appeared to have become a concrete reality as smoke engulfed Manhattan one fateful September morning. The American Declaration of Independence of 1776 draws its inspiration directly from the Book of Deuteronomy with God according his progeny their fundamental freedoms as a sovereign gift: "The Americans' Providence-God issues from a Moses illuminated by the fires of his camp and democratized by the technology of mechanical reproduction."[142]

(iii) Miscegenation with countersignifying and despotic regimes

With this refounding of the Jewish postsignifying regime in the form of the then-fledgling American nation being as it may, another reason for the segmentarity of the passional regime stems from the fact that regimes of signs only concretely exist in admixtures with other regimes of signs. More specifically, the passional regime is *inter alia* mixed with the countersignifying regime of the ancient nomads, since the Hebrews have not only a nomadic past, but likewise a continuing relationship with the "nomadic numerical organization that inspired them, and their own particular becoming-nomad; their line of deterritorialization owed much to the military line of nomadic destruction)."[143] This nomadic countersignifying regime is characterized by a line of "abolition" that turns back against empires, cuts right across them and annihilates them, or otherwise conquers them and fuses with them so as to form a mixed semiotic.[144]

The countersignifying regime, thus, is an military line that tends to annihilate the despotic, or signifying, regime. And it is with this latter regime that the passional regime likewise finds itself mixed. For the postsignifying regime has a very close and essential relation to the signifying semiotic. The Hebrews, for example, always had a nostaglia for the reestablishment of an imperial society of their own. They wanted to rebuild their Temple in an absolutely durable form. Here, David, Solomon and Zechariach are germane. They wanted to erect the spiral of the Tower of Babel and thereby find the face of God once more. And they wanted to do this so as to bring their wandering to an end, to overcome and surcease their diaspora, which itself exists solely as a function of an ideal gathering.[145]

This tendency inherent to the postsignifying regime to coalesce and ossify into a signifying regime of its own, arises *inter alia* from the ubiquitous and quotidian threat posed by rebellious factions, jealous subfactions, and anarchical gangs, whose tribal dislocations and generalized centrifugal pressures abrasively afflict the fabric of desert societies; herein precipitating and bringing to summons inundations of violent sentiments calling for the emergence of the suturing forces of cohesion—principle among these being God himself.[146]

We may therefore posit the existence of two countervailing, though mutually imbricated and entangled tendencies: the first centrifugal and nomadic, that of the postsignifying regime as if purged of signifying centralizations; the second centripetal and sedentary, that of the signifying regime as if voided of all postsignifying destinerrancies. Debray terms the former propensity the Moses complex: "Moses is the man of movement, not establishment. He resists settlement like a temptation. Better not to arrive; the most important thing is to tear oneself loose."[147]

The transition from the first tendency to the second entails the shift from "an ambulatory non-place, the Ark, to an obsessive site, Jerusalem; from a mystique of wandering to a strategy of occupation," wherein the mobile ark gives way to the immutable Holy of Holies—"the third chamber, the most recessed in the Sanctuary, in which only the high priest enters, once a year for Yom Kippur."[148] This process is one in which, quite paradoxically, a God decentered by the Letter becomes a source of centralization; during the course of which—it may be added—the vastly deterritorialized Judaic God is transmogrified back into the far more parochial territorial deity associated with despotic regimes, or rather (in what would effectively be an incipient henotheistic system), preserved more or less intact whilst nonetheless suffering circumvallation by a growing panoply of demigods (as exemplified by the dread recrudescence of Baal worship in Davidic Jerusalem).[149]

Notwithstanding, across the ambit of this transition, the tent (that "commanding centre of the tribes in march formation") sheltering the Ark of the Covenant, acquires a greater physical obduracy in forming itself as a temple. This movement is one in which it is as "though an epidemic content, the sacred sign, were transmitting to its container its own *aura*, in a cascade of virulent metonymies, but in reverse, the whole for the part: the Ark for the scroll, the tabernacle for the Ark, Solomon's Temple for the Tabernacle, and Eretz Israel, the land of Israel, for the second Temple (rebuilt lavishly by Herod, and set ablaze by Titus in 70)."[150]

With the sacred Hebrew Document ossified here into the Hyper-monument, the latter erected center proceeds to project outwards its irradiating circular network of signs in a potentially incessant metastasis of concentric circles extending from the "untouchable nucleus" of an "umbilicus in the middle of the sands" in which each antecedant circle "sacralizes" its successor: the scroll, the Temple, the City of the Temple, Eretz Israel. Monotheistic space, avers Debray, "is 'a circle with extendable and contingent limbs, constructed from a necessary point', Jerusalem, in which the periphery is defined as a function of the centre."[151] In this way, the forty years of wandering of Moses are cast into counterpoint with the forty years of rule of David—thirty-three of these in Jerusalem. This latter is the David complex.[152]

(iv) Imbrication of passional and despotic regimes in Christianity

This supervention of a David complex upon a Moses complex repeats itself later in history in several forms. For instance, the transmogrification of the earlier apostolic Christian postsignifying semiotic into a quasi-territorial signifying regime of its own entailed *inter alia* the bureaucratization of the functions of what had previously been the itinerant apostle into that of the sedentary bishop—in a movement of ontologization whereby hierarchies of functions became hierarchies of perfection, and supernatural graces in turn "became the decrees of attribution for prerogatives of rank, linked to careers."[153]

The standardization, generalization, and rendering official of this incipient system under Emperor Theodosius in 395 A.D. was in effect a reterritorialization at whose behest bishops where firmly affixed to dioceses, as were subsequently priests to parishes; the pastor of each territorial cell wielding powers at once political, administrative and spiritual. For example, in 420 A.D. in Hippo, Saint Augustine was commander, judge, recipient of taxes, garrison commander and police commissioner of the community. The bureaucratization of grace entailed here the minimal systemic closure and hierarchicalization requisite for the constitution of a body politic analogous to the one delineated in its essentials in Edmund Plowden's *Reports* and in the writings of Kantorowicz's Norman Anonymous.[154]

At the end of the seventh century, the preponderance of the signifying over the postsignifying semiotic within what was essentially a Christianity subsisting and functioning as a doubly-miscegenated despotic-passional regime, was thrust to the fore in the despot-deity filiation of Byzantine Emperor Justinian II with God. In this state-formation the sovereign Caesar-Pope, as royal tax-collector on behalf of Christ himself, certainly made the question of rendering unto Caesar as opposed to God well-nigh superfluous.[155]

Even long centuries after this, and in the ambit of a completely different time and place, the despot-deity filiation still subsists (albeit clandestinely) in the United States where, despite the separation of Church and State implicit to the First Amendment, far from the latter inducing the evacuation of religion from politics, the effect has been the exact opposite: "Thus the legislator in Washington is there in order to call Him on High to shower His grace over all the earth, until the fulfilment of time."[156]

(v) Canetti, despotic-passional regimes, and stagnant crowds

The recoalescence of the postsignifying regime and its segmentary line as an ideal gathering before the Divine countenance may in fact be seen as an instance of what Canetti terms a *stagnant crowd,* the most singularly impressive examplar of which is the "Standing on Arafat" at the climax of the Hajj—where hundreds of thousands of pilgrims, bare-headed and robed in white, circumvallate "The Mount of Mercy," an exposed hill rising from the middle of the arid plain. From the time when the sun burns with its greatest intensity until it sets much later, amid uninterrupted hours of searing heat; the arrayed circular ranks stand well-nigh stock still, some Muslims sobbing with exaltation, some beating their chests, many fainting from the terrible heat, whilst listening to the centrally ensconced imam's continuous glorification of God. "Hundreds of thousands of human beings in a state of mounting excitement are kept there on that plain. They stand before God, and, whatever happens to them, may not abandon their station."[157]

The most extreme case of stagnant recoalescence of the passional regime before the visage of God—and one in which the moribundity of the despotic regime foregrounds itself in its limit case—is, however, that of all the dead that have ever died. "All who have ever lived belong there, and there are so many of them that they cannot be counted. The earth between them is their destiny and, though they lie there separately, they are felt to be close to each other. They will lie there for an eternity, until the Day of the Last Judgment. Their life stagnates until the moment of resurrection, and this moment coincides with that of their assembly together before God, who will judge them. Nothing happens in between. As a crowd they lie there; as a crowd, they rise again."[158]

Conceived as a grand and all-encompassing tomb of all those who have ever died, the replete aggregate of all the dead of all ages and all peoples embedded in the crust of the earth constitutes a despotic regime of metaphysical subalterns in which *all* subjects are reduced to potential *homines sacri,* and in which *all* these subjects, these bare and quasi-bestial natures, are killed with impunity by the Damocles' sword of death that falls on *all* working organs. For *all* organs have been far too contumacious to the despotic purview, and thus *all* had to die in the wake of a hyper-Stalinist purge of cosmic proportions.

Notwithstanding, the gargantuan stagnant crowd lives on. What is the nature of this reprieve?

7. Existence under reprieve

As observed earlier, "existence under reprieve" is inexorably implicated within the double movement of mutual aversion along the line of flight on which subjectification occurs. The paradigmatic exemplar here is Cain. He turns away from God. And God turns away from him. Cain already embarks upon a vertiginous line of deterritorialization and, in cavorting along it, is in fact protected by a sign that allows him to escape death. This is the mark of Cain. In all of this, the Jewish God effectively invented the reprieve, existence under reprieve—a kind of *indefinite* postponement. He likewise created the positive character of alliance, of the covenant—that entirely novel relation with the Deity, in which the subject remains alive. Jonah betrays God, yet the latter betrayal nonetheless and quite paradoxically allows him to live, temporarily protected as he is by the tree of Cain. The subject of the postsignifying regime suffers something even worse than death or exile, wandering as he does on a bizarrely positive line of separation or deterritorialization, in whose vertiginous ambit the result ceases to be murder or abrupt death but in fact survival, reprieve and unlimited postponement—a kind of dreadful immortality.[159]

(i) Rosenzweig and the immortality of Character

In a strikingly similar vein, as regards Character in all its impervious obduracy, Rosenzweig contends that the incipient hero, who surges up in tandem with the egress of the latter impenetrable kernel, does not strictly speaking die. Death is for the tragic Self only that which relates to the temporal and radical contingency of man as a mere individual. Character, on the other hand, as melted into the heroic Self, is "immortal." For Rosenzweig, immortality is in fact the ultimate thing for which the Self longs. The personality does not crave and demand its own eternity. The meta-ethical Self does. Admittedly, the personality does strive for some kind of eternity. Still, it is of a very paltry sort: that of mere third-person relationships with other people. The meta-ethical Self, on the other hand, has no relationships with other individuals whatsoever. It engages in no such entanglements, and instead remains sempiternally itself. For it is, taken in and of itself, utterly aware of being eternal. And this its immortality is precisely its inability to die.[160]

In point of fact, Character—the immortal espied by Rosenzweig—surges up to behold man directingly in the eye especially when the latter, still beholden to his most vapid mode of mere third-person existence (that of the personality), suddenly faces the imminent threat of *death*. Character's imperviousness to death extrudes itself most acutely precisely in the very teeth of the Reaper's scythe. For when stumbling upon his Character, man as mere personality likewise encounters the sheer and naked givenness of created Being—as isolated, solitudinous, individual, and creaturely Being: nowise creating, not really free, but created as a pure given.[161]

Now, in that—as world-historical progress and development of the given—history is the arena of the humanly fabricated whilst, on the other hand, the Character represents the inalienable, unchangeable hard core or kernel of selfhood completely impervious to history; one may conclude that Character is so utterly isolated from history that it is for all intents and purposes *timeless*.[162] It is in this sense that Character may be described to be immortal: at precisely the vertiginous moment when man as facile third-person existent is confronted with the very real imminence of a death that threatens immediate perdition, the egress of Character as the sempiternal in man provides the paradoxical ground for a survival beyond the scope of mere personality alone.

(ii) Badiou and the immortality of the subject of a truth-event

The immorality of Rosenzweigian Character is an aspect likewise shared by the Badiousian subject, that is, the subject that rises up in the wake of a truth-event.[163] A truth-event, Badiou maintains, is "that by which 'we', of the human species, are commited to a trans-specific procedure, a procedure which opens us to the possi-bililty of being Immortals."[164] Averring that the subject's relation to truths is absolute, Badiou maintains that, if "we call 'Immortal' that which attains absolutely to some truth," then "'we', of the human species, have the power to be Immortals."[165] As observed earlier, the subject arises from that within Being which is not presently subject to the logic of appearing. The subject is immortal in the sense that, exactly like existence, death does not appertain to Being as such. It pertains only to apparents, to appearing. It is an affection of appearing. Death is not inexistence *tout court*. It is merely the reduction of the existent to an existence that is nil relative to the world in which it appears. Death is not an internal necessity transpiring at the level of Being. Rather, what comes to pass with death is an exterior change in the function of appearing that is contingently imposed upon the dying being. A being, Badiou maintains, cannot be said to be mortal in the sense that it is internally necessary for it to die. Instead, death is merely possible for it and transpires when a sudden, abrupt, and unprecedented change befalls its function of appearing. This sudden shift then minimizes the being's degree of existence in relation to the world. Dying, much as is existing, is a mode of being-there in a world. It applies only to apparents. It does not pertain to Being in and of itself.[166]

(iii) Heidegger and the imperishability of Dasein *as* Dasein

This immortality of Rosenzweigian Character and the Badiousian subject likewise finds strong homologies in the relation of Heideggerian *Dasein* to its own death. In Heidegger, Derrida distinguishes three ways of ending: *Ableben* (demising), *Sterben* (dying), and *Verenden* (perishing). *Dasein* as *Dasein* "can demise (in the medico-legal sense), when it is declared dead after its so-called biological or physiological death has been certified according to conventionally accredited criteria."[167]

Dasein is primarily determined in its existential structure as being-toward-death (*Sein-zum-Tode*) and, in this sense, is capable of demising. Animals, in contradistinction, are not determined existentially by a relation toward their own death, and therefore cannot demise. Note that *Dasein* as *Dasein* is capable of demising under the condition that it dies *(stirbt)*, and die it can. However, what *Dasein* is incapable of is perishing. To be sure, *Dasein* is capable of perishing as an animal or a biological, chemical, physical entity. It cannot, however, perish as *Dasein*. It is with regard to this third way of dying that *Dasein* can be said to be "immortal," or at least imperishable. More precisely, Derrida argues that *Dasein* is "immortal": "in its originary being-to-death, if by 'immortal' one understands 'without end' in the sense of *verenden*. Even if it dies *(stirbt)* and even if it *ends (endet)*, it never 'kicks the bucket' *(verendet nie)*. *Dasein*, *Dasein* as such, does not know any end in the sense of *verenden*. At least from this angle and as *Dasein*, I am, if not immortal, then at least imperishable: I do not end, I never end, I know that I will not come to an end."[168] Thus, even though its fundamental existential structure of being-toward-death implies that *Dasein* does demise and die, *Dasein* as *Dasein* is imperishable. This is the precise sense of a living-on which Derrida locates in Heidegger. There is therefore a sort of "existence under reprieve" in the imperishability of *Dasein*.

(iv) Lévy and the aversion of death by animal cunning

Bernard-Henri Lévy—speaking within the context of the perils of being trapped within a warlike scenario—maintains that what comes to the fore in the teeth of death's imminence is something which differs from, on the one hand, the "body" and its tendency to recoil in the face of death and, on the other hand, the "soul" and its desire to trick the body into facing up to death. This something—which he avers to be a qualitatively distinct feeling for existence or another experince of the self—is the *animal*. It is one's "untamed but intimate animal. This animal that resides in me, more inseparable than my shadow ..."[169]

The upsurgence of one's own untamed intimate animal ushers in an emotional tonality that is at once a feeling of terror and one of rebellion occasioned not so much only by the prospect of possible nothingness but more so by that of death. Yet not so much death *per se*—but rather death transmogrified. "Death, not as the truth, or the limit, of life, but as its hard kernel. A taste for playing with this death. A taste for outwitting it. A taste," as Lévy goes on to maintain with increasing force and licence, "for playing not with, but *against*, one's own death, and maybe, too, for averting it. Portrait of the adventurer as gambler. A gambler of outwitting death. Of the art of cheating one's death, of entering its game and triumphing over it."[170]

(v) The Freudian death-drive and immortality

The Rosenzweigian meta-ethical kernel of human existence—that is, Character—may be seen as entailing a strange surplus of *insistence* over existence.[171] The surplus that insists comes to the fore still more markedly in the Self's encounter with the Other, wherein is evinced the Self's "*ex-citation* by the other's internal alienness," an excitation which Santner describes as effecting "a kind of *unconscious transmission* that is neither simply enlivening nor simply deadening but rather, if I might put it that way, *undeadening*"; and that is such in the precise sense that "it produces in us an internal alienness that has a peculiar sort of vitality and yet *belongs to no form of life*"—an uncanny vitality that Santner expressly identifies with the Freudian death-drive (*Todestrieb*).[172] Here, of course, we must read the Freudian death-drive in the non-naïve sense: that of its being the tendency of nonorganic life to persist beyond the mundane homeostases of the pleasure principle. The death-drive is, therefore, not a principle of racidal finitude, but rather one of a curious and frightening immortality—that of something like Character, or the *daimon*, functioning as the imperishable "acephalic" drive that transcends the sphere of mere worldly existence.

Yet, here the subject, in being granted this particular sort of immortality, enters into a thankless and horrifying existential zone. It has entered what we have already referred to as the "zero-level," or pure void, of subjectivity: the absolutely unendurable abyss in which fully constituted objects and subjects dissolve and yield before a miasmal swarming of spectral partial objects, of exemplifications of the Lacanian *lamella*, of the undead, immortal, libido-as-object that is self-identically the Freudo-Lacanian death-drive (*Todestrieb*). This paronomastic modality of the subject is merely one "degree" removed from the pre-subjective noumenal Real. It is a return to, or recrudescence of, the impossible moment of upsurge of subjectivity in the form of a withdrawal, a recession from the light of phenomenality and its differential degrees of appearing in a world in accordance with its transcendental regime, which assumes the form of a contraction through which constituted reality yields before the surreal phantasmagoria of *membra disjecta*, of organs devoid of body functioning as stand-ins for the dread immortality of the object-libido in its semipernally stuck fixity and eternal psychotic circulation around the void. In Kantian terms, this crepuscular and vanishing mode of subjectivity is that of the pre-synthetic imagination gone mad, psychotic, and utterly demented—the impossible domain of absolutely purified and kenotically evacuated transcendental freedom and spontaneity. In this domain, the subject finds itself vertiginously suspended between direct unadorned animality, on the one hand, and the subordination of its freedom to the moral Law, on the other. In this gathering murk, far from being able to galvanize itself in the direction of forward-directed action in the world or attunement to the promptings of the Deity, the Self is submitted to the horrifying experience of dread organs without body, of itself as the Lacanian fragmented body, of its *membra disjecta* floating around in a nauseating and disgusting miasmal drift, the "night of the world," the "valley of the shadow of death," the hell-on-earth of Kundry in Wagner's *Parsifal* who, having committed the blasphemy of laughing at Christ dying on the Cross, is condemned to an eternity of wandering as an abject, multiply befouled, purely excremental self entirely rejected by the world, abandoned by God, and subject to involuntary, psychotic automatisms, spasmodic jabberings, slavering aphasia, dyslexia, *and the horror of never being able to die.*

(vi) Jewish survival

The tendency of the subject to survive in spite, and even because, of the perpetual imminence of death's threat may likewise be seen as the irrevocable corollary of a constant conjunction of two of the postsignifying regime's other characteristics: betrayal and secrecy. Here, as Žižek argues, it is precisely the phantasmatic fidelity sustained by the foregoing two features that "enabled the Jews to persist and survive for thousands of years without land or a common instututional tradition. In short, the Jews did not *give up the ghost;* they survived all their ordeals precisely because they refused to *give up their ghost,* to cut off the link to their secret, disavowed tradition."[173] In other words, phantasmatic fidelity to the symbolically disavowed original founding gesture (the Jews' refusal to give up *their* ghost), spawns the postsignifying regime's characteristic longevity, its living-on (the Jews' not giving up *the* ghost), its existence and that of its subjectivities as inherently an *existence under reprieve.*

It is precisely on multiple aspects of the astonishing history of seemingly improbable and well-nigh inexplicable Jewish survival that chapter 15 shall dwell—and do so at quite considerable length.

8. Aphasia of the prophet and discernment of the *puissances* of the future

The paradigmatic figure of the despotic regime is the priest who administers the radiant open face of the despot, adding the axis of interpretation or interpretance to the axis of signifiance of the despot or deity. The priest is the despot's principal bureaucrat, who interprets the words of the despot, therefrom engaging in loc-quatious exegeses and incessant perorations. In the passional regime, this couple is replaced by the averted countenances of the Jewish God and the prophet.

(i) Aphasia

In contradistinction to the priest, the prophet has no idea how to speak. Instead, words are simply shoved into his mouth by God. He does not so much expectorate words. He ingests them. This, Deleuze and Guattari term "semiophagy."[174] Rosenzweig may be viewed as expounding on the foregoing semiophagy when he maintains that the prophet is not fundamentally a mediator between man and God. He does not receive Revelation so as to then transmit it. Quite contrarily, God's voice comes from him without mediation. God speaks directly out of the prophet, to the extent that the prophet's voice is self-identically that of God. It is not so much that he allows God to speak. Rather, at the very moment when he opens his mouth, it is already God who is speaking. In this sense, God has verily assumed possession of the prophet's lips.[175]

This, of course, means that the prophet is in a certain fundamental sense an *aphasiac*. He eats his words, or they are paltry and perhaps even soundless mouthings, like the mute cry of a man with an atrophied larynx, or the utterances of a stammerer—like Robert Graves' Claudius inquiring as to his fate and that of Rome before the Sibyl Amalthea: "Sib … Sib … Sib … Sib … Sib …"[176]

(ii) The aphasia of Abraham, Jacob and Mutemenet

Mann, though conceeding that the great Partriarch could teach much about God, avers forcefully that Abraham never knew how to speak about God in the same eloquent manner in which other preachers could hold forth on diverse stories of their nationalistic or territorial deities. This is so because there simply were no stories to tell about God. So, Abram or Abraham, would merely posit the bare existence of God without reference to attendant circumstances or narratives. In fact, he would simply utter the word "God" and no more.[177]

Having received the blood-stiffened veiled, the garment of his son Joseph, and the token of his purported (though indeed fallacious) death at the claws of a wild beast; Jacob is seized in every muscle and fiber by a sudden stiffness so insurmountable that it turns his entire body into a stone. Any attempt at bending a single joint would shatter it. Notwithstanding, this human pillar of salt gradually dissolves into a man of sorrows, defenseless against his utter misery, his voice toneless, its muteness the muteness of the mute token.[178] In this dolorous and seemingly evacuated state, upon any and every hearing of the name of God, Jacob's lips would close tightly, uttering not a single intelligible syllable, as he would begin to scrape at the boils that he felt had broken out on his skin.[179]

Later on in Mann's great novel, Potiphar's consort, Mutemenet, dreams that she is sitting on the dais in the hall of blue columns, eating dinner in the midst of a silence so acute that all four diners are bereft of speech and making almost not a single sound with their utensils, in an atmosphere that was hereby rendered well-nigh hermetic. In this mute oneiric world Mutemenet, having accidentally cut her hand with a bronze knife whilst segmenting a pomegranate and, greatly ashamed (the effluence of blood here doubtless standing for her hopeless infatuation with the Abiru slave Osarsiph (Joseph)); in cowering embarassment, attempts to withhold from view the ruby red blood staining her petal-white attire. Forthwith, however, she beholds old Huya standing up and threatening her with a gilt thighbone, his mouth alternately opening and closing, berating the shamefaced Mutemenet in entirely inaudible words.[180]

This aphasia very soon afflicts Mutemenet herself: from her hapless fumbling amid little fits of shivering before Joseph in the garden, to her stammering, pained elisions, and infantilized speech whilst playing board games with the latter, to her split-tongue lisping upon his arm.[181]

(iii) Discernment of the puissances *of the future*

Yet, though afflicted with aphasia, the prophet and perhaps others caught up in the ambit of the postsignifying regime—harbor a certain advantage over the verbally dexterous and perorating priest of the signifying regime. The priest or seer seems to work with a "text" to which he attaches an interpretation. He suffers from the disease of interpretosis. In contradistinction, unlike the seer or priest, the prophet interprets nothing at all, and instead "anticipates and detects the powers *(puissances)* of the future rather than applying past and present powers *(pouvoirs)*."[182] Here, *pouvoir* pertains to a bureacratic form of power which is the result of a passing or dimming of the evanescent event of force (*puissance*) wrought via the ossification of its forceful dynamism in more or less stable structures.

(iv) The puissances *of the future in Abraham, Isaac and Jacob*

Mann avers that Abraham's suffering, his inner discomfort, torment, internal unrest, urge to wander, and historically novel experience of God, was fecund with (harbored the seeds of) the future—a future that he was, through the impregnation of many souls and the bringing forth of proselytes as countless as the sand upon the sea, destined to shape.[183]

When Jacob comes to Hebron to the oracle tree planted and sanctified by Abraham, there Isaac lays dying—yet speaks to all present in a kind of dreadful "prophetic confusion," about the blood of a ram that should be considered that of the true son shed for the atonement of all. Wide-eyed and with vacant stare, he babbles in strange ancient poetic words that a god will be slain. In this "rambling death rattle," the old man proclaims that a certain "it" was slaughtered, a father and an animal instead of a man and a son, and that "we ate of it." Then, with no one daring even to look upon him, the old man prophesies: "But truly, I say unto you the man and the son will be slaughtered instead of the animal and in place of God, you will eat of it."[184] Clearly, the prefiguration here is of the sacrifice of the God-man—the Crucifixion of Christ—that would occur far later, and of the Eucharist: the immortal and indestructible remainder insisting within being well beyond the ignominious spectacle of human and divine death on Golgotha.

Later, Mann tells of Jacob's dream in which he sees himself as if still fleeing from home, or alternatively fleeing yet another time, as if needing to ride out into the desert again. In this dream, there trots before him some kind of peculiar dog-headed beast with extended tail. This bizarre creature suddenly turns around to look at him and laugh, then lopes forth chicaning between rocks and scraggy bushes and disappears, only to reappear and look around again.[185] Suddenly, the creature disappears once more—only to be sitting on a boulder, facing Jacob with its little maleficent eyes, idiotic head, all the time exuding the pungent and acrid smell of a jackal. This creature is a kind of hybrid of boy and dog.[186] Hereupon, the broad mouth of the muzzle opens as the hideous composite creature begins to speak in a throaty and belabored voice, telling of how he was conceived by mistake, that Nephthys was not supposed to have been his mother, that she was not the right one, that the darkness of night was to blame, that a "majestic god" was in pursuit of Isis, his sister-wife, and in the blindness of night accidentally stumbled upon Nephthys, the sister of "the Red One." This majestic god made love to her, thinking her to be his wife, and out of this unseeing nocturnal coupling, Nephthys became pregnant with himself—that is, the repulsive dog-boy—whereas the god's true wife ought to have been.[187]

Here, the dog-boy intimates how the majestic god (Jacob), in pursuit of Isis (Rachel) in his bed chamber, will unknowingly have carnal knowledge of Nephthys (Leah), the sister (daughter) of the Red One (Laban), and that Nephthys shall conceive the dog-boy himself (Ruben). This is the form in which Jacob receives the enigmatic portent from this monstrous oneiric figure of the cruel and most crooked marital vicissitudes that shall beset him under the roof of his wicked master and uncle Laban.

Later on, the role of prophet comes to be played by the soothsayer Rimut who discerns in his divinations that two emerging rings of oil, one large, the other small—indicate that Rachel will give birth to a boy; whilst the larger ring's origination from the east and subsequent coming to stand-still—stand for the mother's return to health.[188] Rimut divines further that the child will go down into the pit, that is, the well into which Jacob's son Joseph would be cast by his brothers, and yet live; that "it would be as grain, which bears no fruit unless it dies."[189] This the soothsayer contends is rendered pellucid by the fact that when he had poured water into the oil, it had initially separated, but then had reunited with a marked sheen along the edge pointing toward the sun—this latter portending the raising up of the head from out of death.[190]

As Mann relates it, when Joseph does indeed go into the abandoned well and Jacob is deceived by the blood-stained garment that Joseph is dead; in his mad rants and delirious vituperations, Jacob inadvertently prophecies Joseph's rising up from the pit, his restoration to life: "I will sire him once again! Ought it not be possible to sire him anew, just as he was, the very same Joseph, and in that way, then, to lead him back from below? After all, I am still here, from whence he came; is he to be lost them? As long as I am, I cannot give him up for lost. I will awaken him anew and in siring him restore his image on earth."[191]

And, in a certain sense, this is what comes to pass. Much later, the years having dragged on, his head having been raised up by God from his second pit (that of prison in Egypt), Joseph—as Grand Visier of the Land of the Two Kingdoms—himself returns to prophecy when he interprets Pharaoh's dream of the lean cows devouring the fat cows and the blighted ears of grain the golden ones as portents of an ensuing seven years of plenty to be followed by a subsequent seven years of scarcity.[192]

Morever, Joseph's much earlier premonition that his brothers would one day bow down low before him—precisely the sort of youthful arrogance that had stoked up their ire and precipitated him into the abandoned old well in the first place—comes true in the remarkable scene when the brothers, wizened by the passage of years, suddenly find themselves standing before Joseph's throne in the Hall of the Provider—or, in fact, lie with their foreheads pressed to the floor—and stand up halfway to bend low with upraised palms and murmur in chorus, "You are as Pharaoh. Your servants are without blame before you."[193]

The following two chapters shall provide rather extensive and headlong transhistorical fugues on two of the above eight characteristics of the passional regime:

firstly, the postsignifying subjective double's proclivity toward self-annihilation in its own black hole of passion and, secondly, the astonishing survival of the Jewish collective subject in the very face of the Damocles' sword of its own imminent annihilation—and, beyond this, the universalization of this survival to the survival of what Alain Badiou terms generic humanity.

Notes

1. Deleuze and Guattari 1988, 120.
2. Santner 2001, 32–33.
3. Ibid.
4. Ibid.
5. Mosès 1992, 102, 103.
6. Santner 2001, 65.
7. Rosenzweig 2005, 80.
8. Ibid.
9. Pollock 2009, 100, 101.
10. Badiou 2009, 507.
11. Pollock 2009, 159, 160.
12. Ibid. 192.
13. Ibid. 193.
14. Ibid.
15. Ibid.
16. Žižek 1999, 51, 52.
17. Deleuze and Guattari 1988, 123.
18. Ibid.
19. Ibid. 124.
20. Ibid. 131.
21. Mann 2005, 1006, 1007.
22. Ibid.
23. Fuentes 2003, 148.
24. Ibid. 431.
25. Ibid. 155.
26. Žižek 2003, 24; Žižek 2009, 75.
27. Žižek 2003, 88.
28. Ibid. 91.
29. Ibid.
30. Deleuze and Guattarri 1988, 124, 126.
31. Žižek 2008b, vii, viii.
32. Mann 2005, 493.

33. Ibid. 494.
34. Ibid.
35. Ibid. 511.
36. Ibid. 779.
37. Ibid. 951.
38. Ibid. 965, 966.
39. Deleuze and Guattari 1988, 123.
40. Ibid. 124.
41. Ibid. 125, 126.
42. Žižek 2008b, 90.
43. Mann 2005, 252–254.
44. Ibid. 255.
45. Ibid. 255–257.
46. Ibid. 258.
47. Ibid. 259.
48. Ibid. 266.
49. Ibid. 267.
50. Ibid. 292, 293.
51. Ibid. 302.
52. Ibid. 303, 304.
53. Ibid. 425.
54. Ibid. 485.
55. Ibid. 507.
56. Ibid. 769.
57. Ibid. 773, 774, 777, 778.
58. Ibid. 853, 854.
59. Ibid. 968, 969.
60. Ibid. 1028, 1031.
61. Ibid. 1239, 1240.
62. Ibid. 1246.
63. Ibid. 1322.
64. Ibid. 1378, 1379.
65. Ibid. 1424, 1425.
66. Žižek 2009, 40.
67. Žižek 2003, 14.
68. Ibid. 15–18.
69. Ibid. 118.
70. Ibid.
71. Ibid. 87, 88.
72. Derrida 1996b, 109.
73. Ibid.

74. Ibid.
75. Deleuze and Guattari 1988, 127.
76. Ibid.
77. Ibid.
78. Ibid. 129.
79. Mann 2005, 346.
80. Žižek 2009, 33.
81. Ibid. 34.
82. Deleuze and Guattari 1988, 128.
83. Debray 2004, 85.
84. Ibid. 90, 91.
85. Deleuze and Guattari 1988, 129.
86. Ibid. 131.
87. Ibid. 132.
88. Mann 2005, 527, 528.
89. Ibid.
90. Bataille 1991, Vols. 2 and 3, 113–119.
91. Ibid.
92. Badiou 2009, 73, 74.
93. Ibid. 74.
94. Ibid.
95. Ibid.
96. Ibid.
97. Ibid.
98. Ibid. 368.
99. Ibid.
100. Žižek 2008a, 395.
101. Jabès 1991, Vol. 2, 240, 241.
102. Lévy 2004, 175.
103. Ibid. 178.
104. Ibid.
105. Ibid. 179.
106. Ibid.
107. Ibid. 180.
108. Ibid. 181.
109. Eagleton 2005, 69.
110. Ibid.
111. Ibid. 68.
112. Ibid. 68, 69.
113. Ibid. 69.
114. Ibid.

115. Ibid. 70.
116. Ibid.
117. Cited in Eagleton 2005, 71.
118. Cited in Ibid.
119. Eagleton 2005, 71, 72.
120. Ibid. 89.
121. Ibid. 96.
122. Mann 2005, 230.
123. Ibid. 303.
124. Ibid. 943, 944.
125. Ibid.
126. Ibid. 980, 981.
127. Ibid. 993.
128. Fuentes 2003, 182–185.
129. Ibid. 344, 345.
130. Ibid. 280.
131. Debray 2004, 32.
132. Deleuze and Guattari 1988, 128.
133. Cioran 2013a, 34.
134. Mann 2005, 293.
135. Ibid. 429.
136. Ibid. 669, 670.
137. Ibid. 1028, 1038.
138. Ibid. 1031.
139. Ibid. 1062.
140. Ibid. 1408.
141. Debray 2004, 202.
142. Ibid. 202, 203.
143. Deleuze and Guattari 1988, 122.
144. Ibid. 118.
145. Ibid. 123.
146. Debray 2004, 42, 43.
147. Ibid. 92–95.
148. Ibid.
149. Ibid.
150. Ibid.
151. Ibid.
152. Ibid.
153. Ibid. 146–148.
154. Ibid.
155. Ibid. 177–178.

156. Ibid. 204.
157. Canetti 1978, 37, 38.
158. Ibid. 39.
159. Deleuze and Guattari 1988, 123–125.
160. Rosenzweig 2005, 87, 88.
161. Cohen 1994, 51.
162. Ibid.
163. Badiou 2009, 49.
164. Ibid. 71.
165. Ibid.
166. Ibid. 269, 270.
167. Derrida 1993, 37, 38.
168. Ibid. 40.
169. Lévy 2004, 138.
170. Ibid. 139.
171. Santner 2001, 80, 81.
172. Ibid. 36, 37.
173. Žižek 2008b, 90.
174. Deleuze and Guattari 1988, 124.
175. Rosenzweig 2005, 192.
176. Graves 1989, 7.
177. Mann 2005, 349.
178. Ibid. 513.
179. Ibid. 518.
180. Ibid. 830–832.
181. Ibid. 897, 943, 945.
182. Deleuze & Guattari 1988, 124.
183. Mann 2005, 7.
184. Ibid. 147, 148.
185. Ibid.
186. Ibid.
187. Ibid. 231–233.
188. Ibid. 274.
189. Ibid. 275.
190. Ibid.
191. Ibid. 528, 529.
192. Ibid. 1171.
193. Ibid. 1326.

14

Transhistorical Sadeo-Deleuzian Fugue, 2. The Black Hole of Stochastic Death

In Richard Wagner's *Der Fliegende Holländer*, the main character's overriding motivation is to escape from existence itself. He longs for death as the only possible way of attaining equanimity, since for him peace is synonymous with pure, unadulterated nonexistence. This is not so because this or that aspect of existence is reason for despair. Rather, for him, existence *as such* is intolerable. Thus, the death for which he so desperately pines is in no sense conceived as a transition to a happier state. Quite contrarily, it is death as *complete* oblivion, *total* and *utter* annihilation. At the end of the opera, this much sought after redemption arrives in the form of a final release from the necessity of having to live at all. And it is made possible for him through the self-sacrificial love of a woman who is quite willing to share non-existence with him, by dying for him and with him, such that they both undergo terminal and irrevocable destruction.[1]

History itself furnishes us with ample instances wherein the resonance of the subjective double—as equivalently that of the double subject—climbs to an intensity whose sheer pitch inevitably destroys the subject in its own black hole of stochastic death.

And thus here doth follow, if one may so put it, a transhistorical Sadeo-Deleuzian fugue of obscure and moribund amorous subjects—spanning all the way from the legendary lapsarian duo in the primordial garden right up to the gigantic inferno of September 11. Whether they believe in a posthumous existence or, as in the case of the Flying Dutchman, do not; from a certain point within their existence onwards, their principle aim is to die.

c.4004 B.C. (Adam and Eve)

As Elie Wiesel observes, when Eve nibbled on a small piece of the forbidden fruit, she for the first time ever genuinely and deeply understood that there was a direct and inescapable relation between her person, Adam, and death.[2]

> ... In the day we eat
> Of this fair fruit, our doom is, we shall die. (*Paradise Lost*, IX.762–763)

Thus did Eve glimpse the common black hole of stochastic death toward which she and Adam, as subjective double, were jointly hurtling. For, instead of desisting and repenting, she opted to make use of the serpent's stratagems on her husband and draw him into the same deadly trap. Knowing the price of disobeying God, still she resolved to cajole Adam into becoming her accomplice. Since she would have to pay the ultimate price, she owed it to herself to drag Adam along into their joint black hole of stochastic death. The very thought that her husband might survive was intolerable to her. Since she had to die, she would not die alone:[3]

> ... Confirmed then I resolve,
> Adam shall share with me in bliss or woe:
> So dearly I love him, that with him all deaths
> I could endure, without him live no life. (*Paradise Lost*, IX.830–833)

Rejected by God, Adam himself drew closer to Eve. Never before had the two been so united.[4] That we have here a Badiousian obscure amorous subject, whose fusional character is contracted under the sign of death is rendered pellucid by Adam's words:

> ... Some cursèd fraud
> Of Enemy hath beguiled thee, yet unknown,
> And me with thee hath ruined, for with thee
> Certain my resolution is to die;
> How can I live without thee, how forgo
> Thy sweet convérse and love so dearly joined,
> To live again in these wild woods forlorn?
> Should God create another Eve, and I
> Another rib afford, yet loss of thee
> Would never from my heart; no no, I feel
> The link of nature draw me: flesh of flesh,

> Bone of my bone thou art, and from thy state
> Mine never shall be parted, bliss or woe. (*Paradise Lost,* IX.904–916)

With the internal resonance of the subjective double comprising Adam and Eve rising to a vertiginous intensity, this obscure amorous subject cavorted all the more rapidly toward death. Adam:

> However I with thee have fixed my lot,
> Certain to undergo like doom; if death
> Consort with thee, death is to me as life;
> So forcible within my heart I feel
> The bond of nature draw me to my own,
> My own in thee, for what thou art is mine;
> Our state cannot be severed, we are one,
> One flesh; to lose thee were to lose myself. (*Paradise Lost*, IX.952–959)

Eve:

> On my experience, Adam, freely taste,
> And fear of death deliver to the winds.
> So saying, she embraced him, and for joy
> Tenderly wept, much won that he his love
> Had so enobled, as of choice to incur
> Divine displeasure for her sake, or death. (*Paradise Lost,* IX.988–993)

After they had sinned, Eve proposes a joint lovers' suicide pact:

> Let us seek Death, or he not found, supply
> With our own hands his office on ourselves;
> Why stand we longer shivering under fears,
> That show no end but death, and have the power,
> Of many ways to die the shortest choosing,
> Destruction with destruction to destroy. (*Paradise Lost*, X.1001–1006)

Adam, however, cautioning that committing suicide would immediately cause eternal damnation, dissuaded such intentions and halted their implementation. Notwithstanding, due to their initial trangession, they had already reliquished their former immortality, and would—as a consequence and eventually—die anyway.

c.1078 B.C. (Samson)

According to Arthur Droge and James Tabor, the Hebrew Bible is saturated with the idea and replete with instances in which God sends death, or determines the moment of death, to the individual moribund. One finds, however, a blurring of the distinction between the autothanatophile requesting that God take his life, God himself determining the time and manner of death of the subject, and the moribund dying by his own hand. Notwithstanding, what remains consistent is that the subject's death is co-determined by God and the subject. Droge and Tabor relate this co-determination of death to the Socratic tradition of receiving the divine "sign," or "necessity," indicating to the subject that death is to be his fate or lot. Here, though the individual either wills to die or voluntarily cooperates in his own death, his death is at least partly brought about by God's machinations. The death is then interpreted either as heroic sacrifice, cooperative capital punishment, or simply a noble exit from the quagmire of existence.[5]

Nonetheless, the ambiguity which stubbornly insists is that regarding who actually is the ultimate author of the subject's death. For these acts of self-destruction are at once and the same time seen as acts of God. The request that God take one's life nowise precludes the taking of one's own life or allowing it to be taken by an external human agent, with the understanding that such is God's will.[6]

Moreover, for Josephus, voluntarily undertaken death likewise brings freedom and, when chosen in the face of absolutely hopeless and dire circumstances, is wholly justified and admirable. In more than one historical Jewish case, he extols such a form of death as born of loyalty to God.[7]

Following in the above mentioned Socratic tradition, he does however maintain that the proper necessity, or signal, be given by God. Again, an inextricable co-determination of the Jewish subject's self-willed death by God and by the subject is morally necessary. That is to say, the mutual resonance of something akin to the Deleuzoguattarian passional subjective double, with God as subject of the enunciation, and the man who wilt perish, as subject of the statement, is requisite for the ethical soundness of the subject's self-immolation. Otherwise suicide or "ownhandedness" (*autocheria*) would be a grotesque act of impiety.[8] "For it is from Him that we have received our being, and it is to Him that we should leave the decision to take it away."[9]

It is not that Josephus uses the term "ownhandedness" in any pejorative or disparaging sense. For him the term, in and of itself, is entirely neutral. It is only "laying mad hands upon onself," a form of self-destruction devoid of divine sanction and determination, that he proscribes and indicts as impious.[10]

Samson the Nazarite, enamored of a hag—the flicke Delilah—of the valley of Sorek, found his heart smitten to a blindness that soon presaged his own corporeal destruction. Gradually robbed of all probity by a rabidly metastatizing and idiotic love, upon being pressed by his beloved trollop's mendacious words with an incessantly reiterated daily relentlessness whose duress vexed his soul unto death, Samson eventually precipitously and callowly revealed the secret of his absurdly hypertrophied bodily strength. Having so cretinously borne his fluttering heart, posthaste was he betrayed by his deceitful harridan to the Philistines and forthwith sheared of his undulating locks the very night he assuaged his turbulent lust upon her moth-eaten loins. Thus reduced in physical prowess to the feeble indigence of an ordinary man, his eyes gouged out, was he bound with fetters of brass and hurled into a Gazan prison-house to grind (Judges 16: 4–22). There in that dank dungeon the black hole of his own terminal annihilation darkly forboded. In words put in his mouth by Milton, his intimation to his father Manoa was thus:

> *Samson.* All otherwise to me my thoughts portend,
> That these dark orbs no more shall treat with light,
> Nor th' other light of life continue long,
> But yield to double darkness nigh at hand:
> So much I feel my genial spirits droop,
> My hopes all flat, nature within me seems
> In all her functions weary of herself;
> My race of glory run, and race of shame,
> And I shall shortly be with them that rest. (*Samson Agonistes,* 590–598)

Before long, his deliverance into the hands of the Philistines was celebrated with sacrifices to the god Dagon, from whence—midst their noisomely invidious ribaldry—the revelers brought Samson out of gaol for purposes of sport. At the same time, the resonance of the subjective double Samson-Yahweh was escalating itself to a pitch the feverishness of which presaged its immiment and terminal self-immolation with the implacability of fate. Having already settled himself between the main two supporting pillars of the entire accursed building, he beseeched God one last time to grant him a final burst of imponderable strength such that he may avenge the Philistines for his twice smitten eyes—and likewise kill himself (Judges 16: 23–29).

At this point, as Droge and Tabor maintain it, God involves himself directly in Samson's self-willing annihilation, for he must grant him the sheer absurd quantity of strength necessary to collapse the entire structure upon himself and as a consequence, be crushed to death. The choice is made by Samson, God concurs, and jointly they effect Samson's self-immolation.[11] This is likewise the view of the Talmud and the Targum, which maintain that God answered his prayer, allowing him the strength required to destroy himself. Nowhere is his act condemned, whereas later Jewish interpretation understands it to be a wholy redemptive act of self-sacrifice.[12]

Taking hold of the two middle pillars, one with his left hand and the other with his right, and praying God that he may thus die with his enemies, Samson forcibly uproots the pillars and brings the entire unwieldy edifice down upon himself and the three-thousand corybantic Philistines (Judges 16:30). Milton:

> *Chorus.* O dearly-bought revenge, yet glorious!
> Living or dying thou hast fulfilled
> The work for which thou wast foretold
> To Israel, and now li'st victorious
> Among thy slain self-killed,
> Not willingly, but tangled in the fold
> Of dire Necessity, whose law in death conjoined
> Thee with thy slaughtered foes in number more
> Than all thy life had slain before. (*Samson Agonistes,* 1660–1668)

c.1007 B.C. (Saul)

According a midrash mentioned by Rabbi Joshua of Sikhnin uttered in the name of Rabbi Levi, God showed Moses every future generation and its judges, every future generation and its kings, every future generation and its sages, every future generation and its leaders, every future generation and its teachers of Mishnah, every future generation and its officers, every future generation and its communal heads, every future generation and its extortionists, every future generation and its brigands, and every future generation and its prophets. He likewise showed Moses King Saul and his sons marching into futile battle, knowing full well that he will certainly die, and falling by the sword.[13]

According to Droge and Tabor, the Hebrew Bible adjudges Saul's voluntary death as tragic, though ultimately an honorable and noble act. Fearing the dishonor, shame, and humiliation of falling into enemy hands, he and his armor-bearer decide to effectively commit suicide.[14] To be sure, the interpretation of Saul's death in the book of Chronicles maintains that it originated from the machinations of Yahweh himself: "So Saul died for his unfaithfulness; he was unfaithful to Yahweh in that he did not keep the command of Yahweh, and also consulted a medium, seeking guidance, and did not seek guidance from Yahweh. Therefore Yahweh slew him, and turned the kingdom over to David the son of Jesse" (1 Chron 10.13–14). Though this is undoubtedly the case, and Yahweh remains directly responsible for Saul's death, this death could nowise have come to pass had it not come by Saul's own hand.[15]

In the Midrashic commentary *Genesis Rabbah*, when the prohibition of the shedding of one's own blood in Genesis 9.5 is discussed, Saul is specifically exempted from such a charge and numbered among the martyrs. In a later Midrash collection, he is in fact instructed by the prophet Samuel, the heeding of whose injunction would allow his death to bring atonement and eternal life for himself.[16] In much the same vein, in the Midrashic *Leviticus Rabbah*, Samuel—brought forth in a śeance—intimates to Saul that his acceptance of the divinely decreed death sentence that hangs over him, and which is to be effected in the form of his own self-annihilation, will guarantee his exaltation in the world hereafter.[17]

In the actual biblical narrative of these tragic vicissitudes, at the time when the Israelites were engaged in battle with the Philistines and King Saul's men were fleeing in droves before and falling down slain at the hands of the intrepid foe's countersignifying line of military abolition, Saul's sons Jonathan, Abinadab, and Malchishua likewise died. Midst the pitched frenzy of a losing battle and having been wounded by the arrows of archers, with the resonance of the subjective double Saul-Yahweh and that of Saul-Samuel reaching a fevered intensity, did Saul—pining now for abrupt and terminal immolation in the nihilating substance of the Godhead—beseech his armor-bearer to slay him with his sword, lest the uncircumcised Philistines thrust him through and abuse him themselves. Though the fearful armor-bearer refused the command, the white heat of consciousness of his master as doubled subject on the point of proverbial self-incineration was hypertrophied to the point that Saul himself perforce seized a sword and, violently thrusting himself through, posthaste destroyed himself in the black hole of stochastic death. Immediately following him into the black hole of passion was the disobedient armor-bearer who unhesitantingly buried his weapon in his own innards, such that Saul, his three sons, and his armor-bearer were all annihilated on the same day—and their beheaded cadavers thereupon nailed by the Philistines to the wall of Beth-shan and burnt by the Israelites themselves at Jabesh where their bones henceforth lay buried (1 Samuel 31).

As Elie Wiesel maintains, Saul went to his death knowing that he was going to die. He faced death with his eyes open. It was ultimately Saul who killed himself. He was at once executioner and victim, mortal creature and Angel of Death. He brought his tragedy upon himself, and was his own worst enemy. He commited suicide.[18]

c.975 B.C. (Ahithophel)

In addition to that of Saul, God himself would likewise engineer the course of events and circumstances antecedant upon and effecting the self-immolation of the hapless Ahithophel.[19] To this same end, he would likewise addle his mind.

Among other commentators, the final suicide of Ahithophel is given a positive moral appraisal by Josephus: "Therefore, he said, it would be better for him to remove himself from the world in a free and noble spirit than surrender himself to David to be punished for having in all ways helped Absalom against him."[20] According to certain of the rabbinical material found in the Mishnah, Talmud, and Midrash, it is averred that, though Ahithophel's death is ultimately determined by David's curse—and therefore ultimately by God himself, his self-inflicted death taken in isolation received no censure.[21]

As we shall see presently, during the fateful period when Absalom was mounting a burgeoning conspiracy against his father David, the latter's counselor Ahithophel the Gilonite—in a vile act of seditious fickleness—defected to the side of Absalom. As David ascended Mount Olivet barefoot with those still faithful to his kingship—all weeping and proceeding with their heads covered—upon being made privy to Ahithophel's treachery, he was moved to beseech Yahweh that Ahithophel's counsel to Absalom and co-conspirators henceforth be turned to folly (2 Samuel 15). Herewith was the inevitability of Ahithophel's annihilation in the black hole of stochastic death presaged. For henceforth, with the resonance of Ahithophel as doubled subject extolling itself to the pitched intensity of a consciousness proceeding toward self-immolation in a bi-subjective passion oblivious to consciousness and reason alike, the traitor was in his newfangled madness divinely induced to counsel ever the more foolishly. Thus, so as to strengthen the hand of Absalom's supporters, did he urge the latter to make free with his father's concubines in the sight of all Israel, such that all be made privy to his being abhorred of David. This harebrained counsel was indeed followed by Absalom (2 Samuel 16).

Nowise did the divinely ordained dementia of Ahithophel cease here. Indeed, forthwith did he now urge upon Absalom a well-nigh suicidally idiotic plan to himself sally forth against the king that very night at the head of twelve-thousand men of his own selection—so as to cast fear into the heart of a startled, weary and enfeebled David, a man henceforth to be deserted by his own brethren and thereupon smitten by Ahithophel. The folly of Ahithophel's counsel to Absalom was, unbeknownst to Ahithophel, revealed to Absalom by Hushai the Archite. Not only was the latter's advice secretly heeded in lieu of Ahithophel's, but Absalom moreover henceforth recognized the terminal desuetude into which Ahithophel's perspicuity had fallen—and that this atrophy of the latter's faculties was divinely caused. Soon would Yahweh's maddening of Ahithophel, the hypertrophied resonance of the two subjects and that of Ahithophel as the subjective redundancy of consciousness in the whiteness of its heat, spell Ahithophel's demise. For, turning to the priests Zadok and Abiathar, Hushai set in train a course of action to betray the substance of both his and Ahithophel's seditious schemes to the king—precipitating as it did vicissitudes whereby David at last arose and by the light of morning passed over the Jordan with the mass of his people. At this point, the corybantic resonance of Athithophel as consciousness posthaste dissolved in its own infernal lividness, forthwith yielding to a subjective double—Athithophel-Yahweh—bereft of all but the black hole of passion into which Ahithophel was swiftly to disappear. For, upon now knowing that his counsel had verily been jettisoned, did Ahithophel dash headlong upon his ass until he reached his home. Here, he posthaste set his household in order and thereupon immediately hanged himself (2 Samuel 17).

399 B.C. (Socrates)

In Plato's *Apologia*, Socrates clearly intimates his proclivity toward his own death when he avers during his trial that: "Death might be the ultimate blessing enjoyed by mankind, for all that anyone knows; but people fear death because they are quite sure that it is the ultimate disaster."[22] Soon after this, he makes privy to the court that, should the court acquit him on condition that he abandon his practice of philosophy, he would rather continue the pursuit of wisdom even if it led to certain execution, maintaining that he hardly considers being put to death in any sense a calamity. He in fact maintains that, given the offer of such a conditional discharge, he would unabashedly persist in behaving in the same way as he had heretofore, "even if I had to die a thousand deaths."[23] During the course of this his discourse that he would persue such a path, he also adds that he would do so since he must obey "the god" in preference to his fellow Athenian citizens and because it is the command of this god.[24] This god is the Delphic Oracle, which had dubbed Socrates the wisest of men.

What one may therefore espy here is a Deleuzoguattarian passional subjective double comprising the Delphic Oracle as the subject of the enunciation and Socrates as the subject of the statement, wherein the mutual resonance of the two subjects has climbed to a pitch sufficiently intense to portend the terminal annihilation of Socrates in the black hole of stochastic death. When the jury votes in favor of Socrates' conviction and the prosecutor Meletus proposes the death penalty, Socrates admits to feeling no bitterness to the jury.[25] Though he is within his rights to propose an alternative sentence, such as exile, he professes to not be so enamored of life and so demented by the fear of death. Going into exile and keeping his thoughts to himself whilst there would, he contends, entail "disobeying the god" whose command is that Socrates philosophise publically even though this would inevitably lead to his execution.[26]

When he is fairly soon hereafter actually sentenced to death, he makes it clear to the court that he still prefers the manner in which he has defended himself, and the death he will suffer as a consequence, to a conventional defense that would have led to his survival.[27]

Later in his lengthy and meandering concluding speech to the court he again alludes to the "god," or "divine sign," that had guided his prior actions and manner of defense alike. He contends that the divine sign never thwarted him in any sense whatsoever when he came out of his house in the morning, nor when he entered the courtroom, nor at any point during his speech. It had not objected to a single word or action of his during the trial. He avers that this indicates that the outcome of the trial—that is, his death—is wholly advantageous to him. Once more does one see here the moribund mode of recoiling of the Delphic Oracle as subject of the enunciation and Socrates as subject of the statement. Still guided by the god, Socrates forthwith goes on to further expound upon his autothanatophilia. He maintains that death is an eventuality wholly to be desired, in that it is one of two possible things, each of which is entirely salubrious. Death is either the complete cessation of existence, or it is a process whereby the soul is translated from the body to another location. In the first case, if death is nonexistence, it is at once total unconsciousness, and thus akin to dreamless sleep, which Socrates regards as the most blessed state. If, on the other hand, death is the moving of the soul into another and superior world as maintained by the adherents of the Orphic and Eleusinian mystery cults, then it should likewise be coveted. In such an event, Socrates asseverates that he would be ready to die "a hundred times over."[28]

This is clearly what he is pining for in the *Crito*. Told by his friend Crito, who is visiting him in prison, that he would have to die the day after the return of the ship which carried an annual pilgrimage from Athens to the sanctuary of Apollo on the island of Delos in commemoration of Theseus' liberation of Athens from Cretan rule, Socrates prophesies that the ship will arrive the following day. He knows this, for in a dream the day before he saw a beautiful woman dressed in white robes approaching him, calling him by name, and saying: "Socrates, on the third day you will come/to the fertile land of Phthia." This is a saying attributed to Achilles wherein the latter describes the distance from Troy to his home, and which Socrates takes to mean that his soul will end its exile on earth in two days' time.[29]

In the *Phaedo*, in which the incarcerated Socrates is moving closer and closer toward death, he expounds the deeper esoteric underpinnings of his pining for the terminal demise of his corporeal form. What is revealed is a practice of philosophy that approaches autothanatophilia. Socrates intimates to Crito and Simmias that philosophy, if properly practiced, is nothing other than a preparation for the moment of death. The philosopher, therefore, above all lives his entire life in anticipation of his own death. When Simmias interjects with the old hit at philosophers that they are more dead than alive, Socrates replies that Simmias' joke is quite apt, going on to explain in what sense he considers this to be the case. Firstly, he defines dying as the separation of the body from the soul, and the state of being dead as one in which the body and soul, having become divorced, forthwith exist separately. Next, he contends that the philosopher is concerned not with the things of the body but with those of the soul, and therefore that his prerogative is to strive to increasingly divorce the soul from its association with the body. Socrates explains that it is in this sense that the common man lampoons the philosopher as more dead than alive: that he who is indifferent to the physical pleasures of the body already has one foot in the grave. The philosopher must likewise disassociate his soul from his body, in that the latter's physical senses are a hindrance to his acquisition of wisdom and knowledge. For when the soul tries to pursue and perceive reality and truth with the aid of the body, it opens itself to the deceptions and distractions of the latter.[30]

Socrates complains that "as long as we have our bodies, and as long as our souls are contaminated by this evil influence, we shall never reach our final objective, which, as we all agree, is Truth."[31] Therefore, "It has become abundantly clear to us that if we are ever to obtain pure knowledge of anything, then we must get rid of the body and pursue our search for pure Reality with the pure soul. Only then, it seems, will we obtain what we desire, the Wisdom whose lovers we profess to be; but logically, this can only be possible after we are dead, and not while we are still alive."[32] Socrates, of course, has already defined death as the complete separation of body and soul.

However, the pure and universal Truth which may only be attained after death is not attainable by simple mundane decease and naught besides. For Socrates maintains that, in order to posthumously attain the Truth, the philosopher must already be purifying himself by reducing the propinquity of his body and soul to the bare minimum. That is to say, he must spend his life training himself to live in a way which resembles death as faithfully as possible.[33] This ascetic practice, which for Socrates is the essence of philosophy, is what he likewise terms "practicing death."[34] Failure to do so will result in the contamination of the soul by physical nature and, upon decease, its being debarred from attaining the pure and uncontaminated Truth.[35]

Still worse, such a failure ensures that the soul is "cobbled and patched with pieces of the physical world" and has thus been turned into "a sort of hybrid specimen."[36] Such a hybridized soul will be dragged earthward by the heavy material accoutrement with which it is weighed down. It will "grovel in the graveyard among the tombstones, it will become one of the shadow-like apparitions, the visible manifestations of such souls which have indeed been seen in those places."[37] "They continue to wander aimlessly about, until desire for the physical world, which still clings to them, imprisons them in a physical body again."[38] Their particular form of reincarnation then mirrors their manner of bodily infatuation prior to their intramundane decease. Thus, drunkards, gluttons and rapists return as donkeys; robbers and intimidators as wolves, hawks and buzzards, and so forth.[39]

Above all, those recalcitrant to the practice of philosophy as thanatopraxis and thanatophilia, those who do not die their lives, will not be able to die in the fully-fledged Socratic sense. Instead, they survive in the form of hideous specters, monstrous spiritual-corporeal hybrids in whom the terminal annihilation of the body remains unaccomplished. Such overly hedonistic types, therefore, fail to even accomplish death in the mundane sense. Bearing as they do something akin to the mark of Cain, they tarry haplessly within a bewildered and destinerrant immortality. Only the Socratic thanatophile is able to attain his desired corporeal immolation.

As his death neared ever the more inexorably, Socrates composed songs. Swans, he averred, sing their most beautiful melodies when realizing that death is close at hand, knowing that their souls are journeying forth toward an eternal and indivisible communion with the immaterial divine nature of the god whom they serve. They are the sacred birds of Apollo, the same god whom Socrates serves and in whom he is imminently to suffer irreversible immolation.[40]

The poison gradually spread coldness through his body. When it reached his waist, he uncovered it and spoke his last words:

> *Socrates*: Crito, we owe a cock to Asclepius. Make sure that it is paid.

Like the sick man who sleeps in the temple of Asclepius, he expects that the sleep of death will heal him of the corporeal pollutions of life. A tremor runs through his body. The covering from his face is removed. His eyes stare into space. Crito closes his mouth and eyes.[41]

c.160–167 B.C. (The Maccabees)

Philo of Alexandra, as Shmuel Shepkaru intimates apropos of the latter, to the end that Jewish law not be controverted by insalubrious action, was a stern and implacable advocate of passive death in the name of Judaism.[42] "Jews," he maintained, "would willingly endure to die not once but a thousand times, if it were possible, rather than allow any of the prohibited actions to be commited."[43] Likewise an unswerving advocate of Jewish martyrdom in the face of a holy cause, Josephus was an ardent supporter of the noble act of voluntary death. For him the latter was an inextricable component of even ancient Judaism. "It is an instinct with every Jew," Josephus avers, "from the day of his birth, to regard them [the scriptures] as the decrees of God, to abide by them, and, if need be, cheerfully to die for them. Time and again ere now the sight has been witnessed of prisoners enduring torture and death in every form in the theaters rather than utter a single word against the law and the allied documents."[44]

Of the kinds of death toward which the devout Jew ought to lean, Josephus regards a passively suffered death as nobler. That is to say, he opts for that variety of prospective demise in which the sheer possibility of death is a priori rendered more likely. That is, the particular brand of moribund heroism of the Jew here extolled is nowise "that easiest of deaths, on the battlefield, but death accompanied by physical torture, which is thought to be the hardest of all."[45] Josephus was thus a champion not of the struggle in which possible death was merely to be hazarded as the inexorable adjunct in a higher ideal's defense. Rather, the road toward martyrdom was to be made more direct and to assume the form of the pursuit of a death that was excruciating and, above all, *certain*.

This increased urgency on the part of the individual Jewish subject to willingly cavort toward his own death may be discerned in the later portions of the Hebrew Bible.[46] In Isaiah, for instance, one finds the suffering servant of Yahweh who is willingly "brought as a lamb to the slaughter," (Isa. 53.7) is "cut out of the land of the living," making "his soul an offering for sin," (Isa. 53.10) pouring "out his soul unto death" (Isa. 53.12). Divinely vindicated, he chooses death in the face of persecution and evil.

The extolment of the nobility of self-willed death gains in levity and urgency with the emergence and burgeoning, starting in the vicinity of three centuries before Christ, of ideas of apocalyptic eschatology and the promise of the immortality of the soul. Beginning with Daniel, complete vindication at the final judgment is promised for those who undergo voluntary death. As in Isaiah, one finds in Daniel and later texts the notion of the an elect group, or remant, who, by enduring persecution and suffering to the point of death, ensure their immortality.[47] It is during the period beginning in the second century before Christ, that concrete instances begin to emerge of Jews willingly submitting to torture and death rather than violating religious prohibitions and practices.[48] Pining for a posthumous reward, these individuals are often themselves very eager to die, in several cases ending up directly killing themselves.[49]

The transformation in the relation to death by Jews in this their period under Hellenistic rule was thus that, whereas voluntary death may under certain circumstances have hitherto been considered noble, heroic, or the sole manner of avoiding shame—now death became a means of acquiring life. For the body is ever more seen as a prison of the immortal soul, belonging fundamentally to the world beyond the terminal demise of corporeality. Moving into this rarified atmosphere, thus, necessitated the destruction of the body in death—a death henceforth pursued with far greater urgency and venom than heretofore.[50]

During the early stages of the Maccabean uprising against the Hellenizing measures of Antiochus Epiphanes, one-thousand recalcitrants and followers of the seditious priest Mattathias—at the beckoning of the latter's cajoling cries that resounded throughout the city—fled from Jerusalem into the wilderness with their children, wives and cattle so as to practice their religion without let, hindrance or affliction. However, with this flight of the dissidents being at length intimated to Antiochus' servants, expeditionary forces caught up with and encamped against the xenophiles. In spite of the entreaties made by the Greeks to the Jewish dissidents to save themselves from harrassment and mortal peril by submission to the will of Antiochus Epiphanes and the rendering of sacrifice and fealty to the pantheon of pagan deities, the resonance of the collective subjectivity of the dissidents with that of Yahweh was such that—reflected as it was at the level of each doubled dissident subject as a consciousness whose internal resonance climbed to the heights of its own well-nigh self-immolating white heat—the black of hole of passion beckoned for all. For the Maccabean recalcitrants flatly refused not only the entreaty but also to militarily defend themselves on the sabbath and thereupon—at the highest pitch of consciousness's infernal self-immolating incandescence—without as much as casting a stone upon their mortal foes, were forthwith slain to the very last man, woman, and child, and beast of burden, so that the Maccabees-Yahweh passional couple hereby precipitated itself into the terminal annihilation of its own black hole of stochastic death (1 Maccabees 2).

Somewhat later in the Maccabean saga when it was unlawful for a man to keep sabbath days and feasts, or even to profess himself a Jew; when the temple in Jerusalem found itself replete with Gentile riot and revelling, with harlots dallying in holy places and other diverse profanities; when Jews were compelled under pain of death to eat of pagan sacrifices, to go in procession to Bacchus carrying ivy; when circumcised children were cast headlong from the city walls; when those who had sought the refuge of nearby caves to secretly keep the sabbath, upon being discovered by Philip of Macedon, were forthwith burnt to cinders; and when all these maddening calamities were thus joined in the inexorable momentum of their own gothic death march—then did Eleazar the elderly scribe likewise die. The pitched resonance of the subjective double Eleazar-Yahweh was here itself approaching a hypertrophied state whence the black hole of stochastic death and this said black hole of passion alone was solely and ineluctably presaged. Constrained with his mouth forcibly opened to eat swine's flesh, rather that stooping in acquiescence to besmirch himself with such hideous abomination, did the elderly scribe with great violence spit out the rancid carrion and "came of his own accord to the torment, as it behoved them to come, that are resolute to stand out against such things, as are not lawful for love of life to be tasted" (2 Maccabees 6.19–21).

Even though those in charge of the wicked feast were old acquiantances of the scribe and, that he be spared the eventuality of death, besought him to himself bring lawfully edible meat and to thereby feign the consumption of swine flesh whilst in fact merely feasting upon its simulacrum—the sheer pitch of resonance of the subjective double Eleazar-Yahweh and the attendant scorching incandescence of consciousness of Eleazar as double subject were at such a stratospheric corybantic levity that he nonetheless bade them to immediately send him to his grave where alone he could suffer terminal immolation in the divine substance. "For it becometh not our age," said he,

> … in any wise to dissemble, whereby many young persons might think that Eleazar, being fourscore years old and ten, were now gone to a strange religion, and so they through my hypocrisy, and desire to live a little time and a moment longer, should be deceived by me, and I get a stain to my old age, and make it abominable. For though for the present time I should be delivered from the punishment of men: yet should I not escape the hand of the Almighty, neither alive, nor dead. Wherefore now manfully changing this life, I will show myself such a one as my age requireth, and leave a notable example to such as be young to die willingly and courageously for the honourable and holy laws. (2 Maccabees 6.24–28)

Immediately hereupon, he was tortured to death (2 Maccebees 6.29–31).

During this torrid time of Antiochus Epiphanes' draconian Hellenizing measures others too were likewise compelled to eat of swine's flesh, amongst them seven brethren and their mother. In spite of being cajoled to the wicked deed with tormenting scourges and whips, the resonance of their subjectivities with that of Yahweh was already at such a pitch that—having well-nigh despensed with consciousness and reason—solely portended the black hole of stochastic death to the active exclusion of all else. There was an at once individual and collective lust for passively endured death at the determistic hand of an external agent. All seven sons and their mother would suffer annihilation in the black hole of stochastic death.

The Amoraic midrash *Lamentations Rabbah (Ekhah Rabbah)* elaborates on this moribund and doggedly thanatophilous drama by *inter alia* placing the following sentiments into the mouth of the mother. She fervidly enjoins her seventh and last moribund son to "go and say to Abraham your father, 'You bound one altar while I have bound seven altars'."[51] With livid boldness, and a determination to supercede the aborted sacrifice of Isaac, she prevails upon the her last living son to tell the ancient patriarch that "My mother has said as follows: 'Let not your spirit soar with the thought that I have built an altar and I have offered up my son Isaac. Behold our mother has built seven altars and has offered up seven sons in one day! Yours was a test but mine an actuality!'"[52]

This rabbinic text, thus, intimates that wilfully dying for the Jewish Law was to be sourced in an obligatory act in the eyes of the moribunds who had here so forcefully and spontaneously taken upon themselves the desire for martyrdom. In a manner quasi-derivative from that in which God had decreed the death of Isaac, similarly does the *Lamentations Rabbah* theologically enjoin the martyrdom of the seven sons in the book of Maccabees. The difference is, of course, that what was formerly a death-ridden compliance to a vertiginous command emanating from yonder, now finds itself interiorized in the form of an injunction endogenous to the moribund subject itself. What was hitherto an exogenous command to hypothetically align oneself with possible death has been transmogrified into an immanent command to die with a certitude that is henceforth courted with the full force of fate itself. They want to die. They want to die passively, under the inexorable force of an externally supervening agent. This death comes, for all intents, to be self-willed. What would have otherwise been an unwillingly succumbed death imposed from yonder, is now a self-willed death. Formerly acquiescing to death, they now will what was formerly imprinted with the hammer of fate. To be succint, they want to die. Death comes to be their own desire. The *Akedah* becomes synonymous with obligatory martyrdom for the Law.[53]

In a markedly similar vein, the *Sefer Yosippon* likewise presents the martyrdom of the mother and her seven sons as a sacrifice to God. This sacrifice moreover elevates the humble family to the rank of holiness. The heroine, for example, becomes "the holy mother" (*qedosha*). In fact, not only is it the fact that holiness is attained through self-immolatory human sacrifice. The reverse is likewise the case: a preexisting holiness on the part of the would-be martyrs is what leads them to sacrifice themselves.[54]

Returning to this specific Maccabean auto-immolatory catastrophe itself, we observe how the first of the brethren vociferously declared his readiness to die, a resolve that nowise alluded him after his tongue and extremities were severed from his body in the very sight of his mother and the rest of his brethren. Such was the scorching incandescence of consciousness of the first brother as double subject that nowise did he protest or perforce strive to save himself from the infernal temperatures of the giant pan upon which he was forthwith fried and upon which the subjective double brother-Yahweh destroyed itself in the black hole of passion whilst his mother and fellow brothers exorted each other thus:

> The Lord God looketh upon us and in truth hath comfort in us, as Moses in his song, which witnessed to their faces, declared, saying, "And he shall be comforted in his servants." (2 Maccabees 7.6)

With the first brother thus fried to death; the second brother, in spite of his being made a mocking-stock, in spite of the skin atop his head being pulled off by the hair, and in spite of open entreaties to save himself by the wicked consumption of unlawful meat; unswervingly opted for the death that he desired. With the corybantic resonance of the subjective double brother-Yahweh at an extreme and that of the brother as doubled subject hypertophied to the point of a consciousness whose white-hot incandescence was matched only by the infernal agonies of the pan in whose incinerating convulsions he was to destroy himself, thus did the brother—at his very last gasp and pining as he was for immediate immolation in the divine substance—cry:

> Thou like a fury takest us out of this present life, but the King of the world shall raise us up, who have died for his laws, unto everlasting life. (2 Maccabees 7.9)

After the third brother—also pining with frenzied impatience for instant immolation in the divine substance, likewise mocked, likewise amputated of tongue, likewise maimed in his members—was himself cooked to death in the infernal pan; with the corybantic resonance of the fourth brother's subjectivity in conjunction with that of Yahweh at the level of the passional couple climbing to the infernal zenith of hypertrophied consciousness; was the tormented, mangled, and maimed fourth brother moved to hold forth as follows:

> It is good, being put to death by men, to look for hope from God to be raised up again by him: as for thee, thou shalt have no resurrection to life. (2 Maccabees 7.13)

This being said, he too was annihilated in the incinerating pan's black hole of stochastic death. With the fifth brother likewise destroyed, the resonance of the sixth brother's consciousness as doubled subject was now itself at an apotheosis of infernal incandescence so stratospherically exalted that the prospect of death—far from being that of an impending tragedy—appeared to him as the object of well-nigh rabid pining, so that he addressed his executioners thus:

> Be not deceived without cause: for we suffer these things for ourselves, having sinned against our God: therefore marvellous things are done unto us. But think not thou, that takest in hand to strive against God, that thou shalt escape unpunished. (2 Maccabees 7.18)

And forthwith did he die, and after him the seventh brother, and the mother (2 Maccabees 17.24–42).

33 A.D. (Jesus Christ)

> My God, My God, why hast thou forsaken me?
> why art thou so far from helping me,
> and from the words of my roaring?
> O my God, I cry in the daytime, but thou hearest not;
> and in the night season, and am not silent. (Ps. 22.1–3)

With these words, the twenty-second Psalm of David portends the very similar words of Christ uttered from the ignominy of his gibbet on Golgotha hundreds of distant years hence. Graphically foretold, if not presaged, is the very death and manner of death which would then visit itself upon the Incarnate Word. For only a little later on, it is again the Crucified who would appear to be speaking through the pen of the Psalmist:

> They gaped upon me with their mouths,
> as a ravening and a roaring lion.
> I am poured out like water,
> and all my bones are out of joint:
> my heart is like wax,
> it is melted in the midst of my bowels.
> My strength is dried up like a potsherd;
> and my tongue cleaveth to my jaws;
> and thou hast brought me into the dust of death.
> For dogs have compassed me:
> the assembly of the wicked have enclosed me:
> they pierced my hands and my feet.
> I may tell all my bones;

They look and stare upon me.
They part my garments among them,
and cast lots upon my vesture. (Ps. 22.12–18)

Yet, the death on the Cross was in its planning stages very long before even this. In Milton, shortly after the original fall of man, the Son speaks to the Father:

Behold me then, me for him, life for life
I offer, on me let thine anger fall;
Account me man; I for his sake will leave
Thy bosom, and this glory next to thee
Freely put off, and for him lastly die
Well pleased, on me let Death wreck all his rage;
Under his gloomy power shall I not long
Lie vanquished; thou hast given me to possess
Life in myself for ever, by thee I live,
Though now to Death I yield, and am his due
All that of me can die … (*Paradise Lost,* III.236–246)

Crucially, this death, as per the dictates of Divine Providence, is necessarily one that is self-willed. "If God had to suffer and die," as Jack Miles avers, "then God had to inflict suffering and death upon himself."[55] He avers, moreover, that Christ is "a suicide by metaphysical definition," that, as God Incarnate, only he is able to take his own life away.[56] It was, therefore, neither the Jews nor the Romans who killed him. He killed himself.[57] Becoming human, he subjects himself to a death he could have completely avoided, absolutely refusing to defend himself against his persecutors.[58]

This is likewise the view of Tertullian, who described the death of Jesus as a form of voluntary martyrdom, and also that of Origen who forcefully asseverated that the Lord gave up his spirit quite freely since it would have been impossible that the Deity should be at the mercy of the flesh to the very end.[59] Even Augustine of Hippo, a steadfast opponent of self-willed death, would claim that Christ's "spirit did not leave his body against his will, but because he willed it to happen and he willed when and how it happened."[60]

As Droge and Tabor have it, the excruciating death on Golgotha was in no sense a calamitous mistake, or the execution of a failed apocalyptic prophet or that of a merely eccentric and off-beat preacher. Rather, it was Jesus' mission long prior to assuming the form of flesh. Hence the emphasis in all the Gospels on his clear prior knowledge of his eventual fate and on his unswerving, and loving, acceptance thereof. Droge and Tabor go on to maintain that in Luke, his agenda entailed a deliberate provocation of the Roman and Jewish authorities to bring about his arrest, and that he in fact actively baited them to execute him. Likewise, the early Christians nowise imagined his death as the contingent and more or less prosaic execution of a convicted criminal but as a properly divinely ordained event, predicted by God's prophets and provoked into actuallity by Jesus himself. He completely intended to die and embraced his death with the full compass of his infinite will. According to the Gospel of Mark, in addition to his death being ordained by God, it was furthermore predicted by Jesus himself. He deliberately accepted his own death as a necessary completion of God's plan, and accepted it only at the precise point when he had signified his readiness to embrace the full brunt of his Crucifixion. This point is given still further ballast by Droge and Tabor's observations that, moments prior to his actual death, Jesus reclines his head prior to actually dying: *klinas ten kephalen*, or "having reclined his head." For crucifixion victims, the reverse is invariably the case. The voluntary nature of his death is driven forth most emphatically in the Gospel according to John. For, in relation to the precise point of Jesus' death and the passing away of his spirit, Mark (15.37, 39) and Luke (23.46) use the verb *ekpneo* ("expire"), Matthew (27.50) has it as *aphiemi* ("give up"), whereas John is by far closest to the mark as regards indicating the voluntary nature of Christ's death in his use of *paradidomi to pneuma*, or "gave up his spirit."[61]

In this connection, Droge and Tabor quote D. Moody Smith's asseveration that "Jesus goes to the cross of his own volition and by his own decision. He decides when the hour for his departure in death has arrived, or rather he alone knows when the Father has decreed his hour has come. He lays down his own life; no one takes it from him (10.18). In John, Jesus' death is his glorification, not his humiliation."[62]

Jürgen Moltmann's polemic in favor of the overwhelmingly voluntary nature of Christ's ultimate sacrifice and death follows in the same vein. For him, the suffering and death of the Lord could nowise have been the mere "passive suffering of an uncomprehended fate, like a miscarriage, an illness, a plague, a premature death, or suffering from the deep rooted evil of other people, as social suffering, and suffering from the society which humiliates them."[63] "Jesus did not suffer passively," he continues, "from the world in which he lived, but incited it against himself by his message and the life he lived. Nor did his crucifixion in Jerusalem come upon him as the act of an evil destiny, so that one could speak of a heroic failure, as heroes have often failed and yet remained heroes to posterity. According to the gospels, Jesus himself set out for Jerusalem and actively took the expected suffering upon himself."[64] Moltmann goes on to aver that "he provoked the hostility of the guardians of the law," that he "incited the devout against him and was cast out into the godlessness of Golgotha. The more the mysticism of the cross recognizes this, the less it can accept Jesus as an example of patience and submission to fate."[65] Jesus' death was entirely his own will. And that of the Father.

Indeed, as Miles asseverates, Jesus' life itself appears to be overwhelmingly about his death. For the itinerary of the Lord appears to have one and only one appointment within it; that is, his fixed, obligatory, and predestined hour of execution in Jerusalem. In contrast, all else in his life seems random and optional.[66] Moreover, the fact that Jesus goes to his death in a priestly garment suggests that he was not merely the lamb who was to be sacrificed at the Passover, but likewise indeed the priest performing this same sacrifice upon himself. He is "A priest who is his own sacrificial lamb, a lamb who is his own sacrificing priest, a father who is own son, an Isaac who is his own Abraham with the dagger in his own hand …"[67]

The first more public and tangible indication of his long standing resolve to undertake and fully consummate his own self-ordained death appears when certain of the scribes and Pharisees ask of Jesus a sign of the divine origin of his preaching, to which he forcefully retorts:

> An evil and adulterous generation seeketh after a sign, and there shall no sign be given it, but the sign of the prophet Jonas. For as Jonas was three days and three nights in the whale's belly: so shall the Son of man be three days and three nights in the heart of the earth. (Matt. 12.38–40)

After the time when Jesus asks his disciplines whom men say he is and Peter proclaims in response that he is the Christ, the Son of the living God, he begins to intimate to his disciples that he is to suffer as a result of the actions of the elders, chief priests and scribes, be killed, and raised again on the third day (Matt. 16.13, 16, 21). When Peter begins to rebuke him on this account, saying "Be it far from thee, Lord: this shall not be unto thee," Jesus turns to him and chides him: "Get thee behind me, Satan, thou art an offence unto me: for thou savourest not the things that be of God, but those that be of men" (Matt. 16.22–23). To obstruct the accomplishment of this death is blind and meddling human autism.

After taking Peter, James, and his brother John to the high mountain where he is transfigured, he again foretells his inevitable self-willed death (Matt. 17.1–2). Jesus: "Tell the vision to no man, until the Son of man be risen again from the dead" (Matt. 17.9). Speaking then of the already accomplished death of John the Baptist, referred to by his disciples as Elias the precursor of the Messiah, he declares to them:

> Elias truly shall first come, and restore all things. But I say unto you, that Elias is come already, and they knew him not, but have done unto him whatsoever they listed. Likewise shall also the Son of man suffer of them. (Matt. 17.11–13)

This point he rams home again when he and his disciples abide in Galilee:

> The Son of man shall be betrayed into the hands of men: and they shall kill him, and the third day he shall be raised again. (Matt. 17.22–23)

Jesus, as Droge and Tabor moreover maintain, does not abide in Galilee for much longer, where he would only passively wait for the mere insipid possibility of being arrested. Quite in contradistinction, he ups the ante to accelerate the speed and increase the inevitability with which his apprehension by the authorities would come to pass. He concertedly heads off for the far more dangerous cauldron of Jerusalem. So doing, he deliberately sets out on a course that was to ensure his arrest and death to be inevitable. This unequivocal move further indicates the absolute degree to which he is commited to effect his death at the divinely appointed time.[68]

In the Johannine account, Jesus, utterly in control of all happenings, is the agent *primus inter pares* orchestrating all events leading him toward his Cross.[69] Not too long after this his last sojourn in Galilee, going up with his disciples to Jerusalem, he again repeats his prior declaration of the certainty of his impending and inexorable death (Matt. 20.17):

> Behold, we go up to Jerusalem, and the Son of man shall be betrayed unto the chief priests and unto the scribes, and they shall condemn him to death, and shall deliver him to the Gentiles to mock, and to scourge, and to crucify him: and the third day he shall rise again. (Matt. 20.18–19)

In one of Anne Catherine Emmerich's numerous visions, at about this same time, when Jesus and some of his Apostles and disciples were walking about Mount Olivet, the valley of Josaphat and Mount Calvary, he tells them that this very day he would give them his Body and Blood, his whole self, all that he was. His entire soul languished and pined with love and desire for the moment when he would give himself to man in death.[70]

Still later, after sermonizing to his disciples on the vicissitudes of the Eschaton, his determined march toward the black hole of stochastic death remains unflagging (Matt. 26.1):

> Ye know that after two days is the feast of the passover, and the Son of man is betrayed to be crucified. (Matt. 26.2)

Soon after this, at the house of Simon the leper in Bethany, when his disciples are indignant at a woman treating him with precious ointments from an alabaster box, contending that it could be sold dearly for funds to be given to the poor,

Jesus' retort again intimates the impending death that he must and desires to accomplish (Matt. 26.6–9):

> Why trouble ye the woman? for she hath wrought a good work upon me. For ye have the poor always with you, but me ye have not always. For in that she hath poured this ointment on my body, she did it for my burial. (Matt. 26.10–12)

On the evening of the passover, when the lamb was being immolated, he declared that the figure of the Pachal lamb was very soon about to attain its definitive consummation. He would be this consummate victim. The lamb was but a figure. He himself would be the true Lamb.[71] On this same evening, the most enduring symbolic figuration of the redemptive death that he must soon die is given:

> And as they were eating, Jesus took bread, and blessed it, and broke it, and gave it to the disciples, and said, "Take, eat, this is my body". And he took the cup, and gave thanks, and gave it to them, saying, "Drink ye all of it: for this is my blood of the new testament, which is shed for many for the remission of sins." (Matt. 26.26–28)

He would give his body up to ignominious death, a death in which his blood would be shed.

But this death, far from merely being the death of the corporeally embodied man Jesus, would verily, likewise and at once, be the death of God himself. Miles: "The Lamb who is the great I AM is that same God turned into a sacrificial animal."[72] In this connection, the second stanza of the Lutheran passion hymn "O Traurigkeit, O Herzeleid" by Johannes Rist reads:

> O great woe!
> God himself lies dead.
> On the cross he has died;
> And thus he has gained for us
> By love the kingdom of heaven.[73]

Analogous sentiments are expressed at the end of William Blake's *Jerusalem* where Jesus, appearing to Albion, says "unless I die thou canst not live." When Albion asks whether man can exist without "Mysterious Offerings of Self for Another," Jesus replies:

> Wouldest thou love one who never died
> For thee, or ever die for one who had not died for thee?
> And if God dieth not for Man & giveth not himself
> Eternally for Man, Man could not exist; for Man is Love
> As God is Love; every kindness to another is a little Death
> In the Divine Image, nor can Man exist but by Brotherhood. (*Jerusalem*, 96.23–28)

Already in Luther, one finds the expression "the death of God." Whereas for Zwingli, Christ suffers and dies solely according to his manhood whilst God's sovereignty remains unaffected by his assumption of flesh, for Luther, in that the person of Christ is determined by the divine person, the divine person also suffers and dies in the suffering and death of Christ.[74] Luther: "In his nature God cannot die. But now that God and man are united in one person, when the man dies, that is rightly called the death of God, for he is one thing or one person with God."[75]

Likewise, in Catholic theology, Karl Rahner understands the death of Jesus as the death of God. Far from maintaining that the death on Golgotha nowise affects God himself, Rahner contends that the death of Jesus in fact expresses God, that it is a statement of God about himself. Hans Urs von Balthasar likewise associates the paschal mystery with the omnious formula "the death of God." He traces the death of the crucified Christ back to the inner mystery in God himself. Karl Barth moves in the same direction, averring that it is the divinity of Jesus that is revealed precisely in his humiliation, thus drawing the harshness of the cross into the very concept of God. Speaking consistently of "God in Christ," Barth speaks in well-nigh theopaschite terms of God's suffering and involvement in the cross of the Son. He constantly stresses that God was in Christ, that God humbled himself, and that God himself was on the cross.[76]

Jürgen Moltmann avers that the expression *Ecce homo!*, which he takes to be a confession of faith recognizing God's humanity in the dehumanized Christ on the cross, at once and the same time says *Ecce deus!*, or Behold God on the cross! Humiliation to the point of death is essential to God's nature.[77] Moltmann: "When the crucified Jesus is called the 'image of the living God', the meaning is that *this* is God, and God is like *this*. God is not greater than he is in this humiliation. God is not more glorious than he is in this self-surrender. God is not more powerful than he is in this helplessnes. God is not more divine than he is in this humanity."[78] "The Christ event on the cross," he may be seen as concluding, "is a God event."[79] As Paul Althaus has it, "the full and undiminished deity of God is to be found in the complete helplessness, in the final agony of the crucified Jesus, at the point where no 'divine nature' is to be seen."[80] For "God himself really enters into the suffering of the Son and in so doing is and remains completely God."[81]

Miles alludes to an arresting image illustrative of this coterminous suffering of God and his Son on the Cross. This is Masaccio's *The Holy Trinity with the Virgin and Saint John*, in which the Father elevates the body of the Son above the altar as the priest in the Mass elevates the sacramental body of Christ. Miles maintains that the silent caption of the painting is that of God the Father, and not the Son, intoning: "This is my Body." By this Miles means to intimate that the suffering and death of the Father and of the Son are one and the same.[82]

In like vein, as Hegel likewise maintains, the Crucifixion is not merely the death of a specific man, not merely a solely natural human death. For this death must not be "represented merely as the death of *this individual*, the death of this empirically existing individual. Heretics have interpreted it like that, but what it means is rather that *God* has died, that *God himself is dead.* God has died: this is negation, which is accordingly a moment of the divine nature, of God himself."[83]

Among the heretics whom Hegel doubtless has in mind are *inter alia* the left-wing Origenists, who maintained that the Son was strictly subordinate to the Father, effectively denying the divinity of Jesus. Of these, it is especially the Arians who denied that God himself could have possibly suffered on the Cross. Hegel's position would certainly be closer to that of the right-wing Origenists who, so as to preserve the divinity of Jesus and the fact that it was verily God himself who suffered death on the Cross, maintained the complete identity of the Father and the Son, averring that the Father, Son, and Holy Spirit are nowise ontologically different but merely modalities of the same underlying substance. This position is termed Modalistic Monarchianism, whose chief representative was Sabellius. For Sabellius, the three divine persons were merely different names for the same reality. Strongly averse to this position, the left-wing Origenists branded it Patripassionism, maintaining the notion that God the Father suffers death with Jesus to be blasphemous in and of itself. For them, the foregoing entailed a vicious contamination of the unsullied eternity of God by the vicissitudes of temporality. The Son, therefore, had to be a mere creature of God, and in no sense God himself. In a letter to Eusebius, Arius confesses that the Son is neither unbegotted nor formed out of the divine substratum, that he was constituted by God, and is nowise a part of God. The conclusion that Mark C. Taylor draws here is that Arius unwittingly denies the Incarnation. For Arius, God does not become flesh and suffer death. His position is that such vicissitudes would vitiate his divinity. Most crucially for our present purposes, the heretical Arian position holds that God himself does not die on the Cross. A mere, albeit noble, creature dies on the ignominious gibbet. No more.[84]

Vasts apart, the right-wing Origenist position, that of the Alexandrians, chief among whom was Athanasius, vehemently asserted the divinity of Christ and the implied fact that it was truly God himself who died on the Cross. To this end, Athanasius avered the Father and the Son to be *homoouios*—of the same substance. If this is not the case, then God himself does not die at Golgotha. But, as Hegel knows, he does thus die.[85]

But the Alexandrians, those right-wing Origenists, would likewise themselves stand condemned by Hegel as heretics. For whereas the left-wing Origenist theologians of Antioch denied the divinity of the Son, the right-wingers denied his humanity. Apollinarius was the preeminent representative of the, equally flawed, position of these right-wingers, averring in a manner oddly analogous to that of the left-wingers that the corporeal embodiment of God would necessarily obviate his divinity. His position was, thus, that—as in the case of the left-wing Origenists—Christ could not have been God and man at the same time. To resolve the quandary into which he had driven himself, Apollinarius contended that the humanity of Christ could only be partially present in him. The co-presence of the divine and the human in Christ was, for him, not possible. More specifically, Apollinarius avers Christ to be God only in his mind, whilst his body is wholly human. He is God incorporeally and human corporeally. Such a position came to be labelled Monophysitism. As with Arius, Mark C. Taylor correctly avers that Apollinarius' position is tantamount to rejecting the Incarnation.[86]

The debate continued in the form of Nestorius' averment that Mary was merely the mother of Jesus' humanity and not his divinity. For him, she does not give birth to Christ as the divine Word or the Logos. She merely bears a man, exalted all be he amongst men. Whereas Apollinarius denied the humanity of Christ, Nestorius denied his divinity. Neither could fathom that Jesus was at once God and man. The rectifier was Gregory of Nazianzus, who fully affirmed that Jesus was truly both man and God.[87] It was, thus, God who died on the Cross—at once as God and as man.

Penetrating at an acute angle through the mire of this turgid debate, Hegel avers that the Son is at once one with and likewise other than the Father. Each of the two sees himself in the other and can only become and be himself through the other.[88] The Incarnation and Crucifixion are nothing if they are not truly the death of God himself. This death, however, God survives. He survives it as spirit.

Thomas Atlizer proceeds as far as Hegel in this vein, maintaining the sacrifice of Calvary to be the absolute self-negation and absolute self-emptying of the Godhead itself that transpires in the very depths of consciousness.[89] For him, the Crucifixion is absolute death in and of itself.[90] It is the self-annihilation of God in Christ. The atoning death of Christ is the sacrifice of the fallen and empty body of God.[91] Altizer posits the Christian name of God as refering to a process of absolute self-negation in which God is revealed in Christ as the God who dies freely for man. Christ is the triumphant epiphany of God in the sense of the latter's absolute act of self-sacrifice.[92] This death of God in Christ is at once a historical and a cosmic event.[93] Still more precisely, according to Žižek, what dies on the

Cross is not merely Jesus as God's representative-incarnation, but likewise the substantial, transcendent God of the beyond.[94] That is to say, after the death on Golgotha, there is neither Father and Son, but only the Holy Spirit.[95]

In this divine auto-immolation, love and death have attained a propinquity so acute that they have reached the searing and vanishing point of pure coincidence. They are conmingled in the sense that, as Hegel avers, "at the same time this death is to this extent the highest love."[96] This death is the abyss in which the coincidence of love and death, at once the supernal summit of love's consumation, is likewise that of the unification of God and man in the death of the God-man Christ. For "It is precisely love that is the consciousness of the identity of the divine and the human, and this finitization is carried to its extreme, to death. Thus here we find an envisagement of the unity of the divine and the human at its absolute peak, the highest intuition of love. For love consists in giving up one's personality, all that is one's own, etc. It is a self-conscious activity, the supreme surrender of oneself in the other, even in this most extrinsic other-being of death, the death of the absolute representative of the limits of life. The death of Christ is the vision of this love itself—not love merely for or on behalf of others, but precisely *divinity* in this universal identity with other-being, death. The monstrous unification of these absolute extremes is love itself—this is the speculative intuition."[97]

This absolute love is thus the strange and truly aberrant union of God and man in the death of a singular individual, the Son of God, who is simulantaneously God and man. For, as Hegel continues, "The speculative aspect is that the Son goes to death as the divine," as "he who is on his own, the absolute love."[98] Since, "although it is a natural death" and thus the death of a particular man, "this is the death of God ..."[99]

This most exalted love, which by its very nature finds itself limitlessly shot through by and consummated in the death of Christ, is that by which God denigrates himself to the lowest possible debasement of man and through which man is at the same time transfigured and exalted to a hitherto unprecedented proximity to God: "Death accomplishes the process whereby the divine idea has divested itself, divested itself unto the bitter anguish of death and the shame of a criminal, and thereby human finitude is transformed into the highest—that is the highest love. That is the deepest anguish, this the highest love; in anguish love is contained."[100]

In this death in which the coincidence of God and man attains a heretofore unthought identity and levity, nowise may it be said that human nature is diluted, let alone forestalled or jettisoned. For, in vast contradistinction, it undergoes unprecedented efflorescence and the fullest conceivable consummation: "This triumph is not an abstraction from human nature or a putting off of it; rather it is the preservation of it precisely in death itself and in the highest love."[101] Indeed, what comes to pass "involves the exaltation of human nature to heaven, where the Son of Man sits at the right hand of the Father, and the identity and the glory of the divine and human nature appear to the spiritual eye in the highest possible way."[102] For the Incarnation and Crucifixion are "the going forth of the divine idea into the uttermost cleavage, even to the opposite pole of the anguish of death, which is itself the absolute reversal, the highest love, containing the negation of the negative within itself and being in this way the absolute reconciliation, the sublation of the prior antithesis between humanity and God. The end is presented as a resolution into glory, the festive assumption of humanity in the divine idea."[103]

Yet the death of Christ is not merely the apotheosis of man. It is likewise, and conversely, the sine qua non of God's sojourning with man. It is that by which God's estrangement from man, and consequently his self-estragement, is terminated. For only through this death "does God come to be at peace with himself, only then is spirituality posited. God is the true God, spirit, because he is not merely Father, and hence closed up within himself, but because he is Son, because he becomes the other and sublates this other. This negation is intuited as a moment of the divine nature in which all are reconciled. Set against God there are finite human beings; humanity, the finite, is posited in death itself as a moment of God, and death is what reconciles. Death is love itself; in it absolute love is envisaged. The identity of the divine and the human means that God is at home with himself in humanity, in the finite, and in its death this finitude is itself a determination of God. Through death God has reconciled the world and reconciles himself eternally with himself. This coming back again is his return to himself, and through it he is *spirit*."[104]

The conjunction of God and man that transpires in the Incarnation clearly entails the bringing into existence of a Deleuzoguattarian subjective double, or passional love-couple, a mutual recoiling of two subjects whose most extreme pitch of reciprocal resonance coincides with the stochastic death of the double itself in the form of the Crucifixion of the God-man Christ at Golgotha. That God and man do conjointly comprise such a subjective double may readily be adduced from Hegel's theological musings on the nature of love. Here, a mutual recoiling of two distinct, yet utterly inseparable, subjects swiftly emerges into view: "For love is a distinguishing of two, who nevertheless are absolutely not distinguished for each other. The consciousness or feeling of the identity of the two—to be outside of myself and in the other—this is love. I have my self-consciousness not in myself but in the other. I am satisfied and have peace with myself only in this other—and I *am* only because I have peace with myself; if I did not have it, then I would be a contradiction that falls to pieces. This other, because it likewise exists outside itself, has its self-consciousness only in me, and both the other and I are only this consciousness of being-outside-ourselves and of our identity; we are only this intuition, feeling, and knowledge of our unity."[105]

The sort of subject that this particular Hegelian definition of love calls to mind is the Badiousian obscure amorous subject, or rather a conjunction of two such obscure subjects. As already observed earlier, for Badiou, two amorous subjects comprise something highly homologous to a Deleuzian passional subjective double, their stubbornly obdurate conjunction being underpinned and maintained by a single fictive fusional body. The obscure subjective double comes into existence under the auspices of the single integrated destiny upon which the subjects embark, a tautly bound trajectory whose deathly reciprocity hovers under the sign of death. For Hegel, in particular, "God is the one who as living spirit distinguishes himself from himself, posits an other"—that is to say, man—"and in this other remains identical with himself, has in this other his identity with himself."[106]

It does, however, appear that the God-man subjective couplet, though at this provisional level abiding by the contours of the Badiousian obscure subject, harbors within itself aspects and tendencies that breach the latter's closed confines. For "the precise determinacy of it is not that this incongruity of the two sides disappears for consciousness. The incongruity is there, it resides in spirituality. Spirit is the process of self-differentiating, the positing of distinctions."[107] Most crucially, though the Hegelian God-man subjective double proves itself to fully partake of the self-destructive thanatic aspect of the generic Badiousian obscure amorous subject, it at once contains within itself entirely contrary proclivities. For infinite spirit literally survives its own death. Its death, in fact, shows itself to appear merely as a fulgurating and evanescent moment of spirit. Here, the finitude assumed by God in the Incarnation nowise implies that the death that God suffers in Christ is something that the latter cannot transcend. Succintly, God survives the death implied within Incarnation. "The other-being, the finitude, the weakness, the frailty of human nature is not to do any harm to that divine unity which forms the substance of reconciliation."[108] For the otherness—that is to say, the Son—in which God reflects himself is not purely human, and thus nowise finitude *tout court.* Indeed, this "other is also God and has the entire fullness of the divine nature within itself. The character of otherness in no way detracts from the fact that this other is the Son of God and therefore God."[109]

We must, however, be clear as to what exactly dies and what precisely survives in the Crucifixion. For we have been at once speaking of both the death of God and God's survival of his own death. As already partially intimated a little earlier, according to Žižek, it is not merely Jesus as the terrestrial representative of God, but indeed the transcendent God of the beyond that likewise dies. Both the Father and the Son die. This is the form which the death of God assumes. God's survival, on the other hand, assumes the guise of a sublation of the opposition between Father and Son. In this sublation, God recrudesces as the Holy Spirit; that is, more precisely, as the Spirit of the community of Christian believers. In this death that God survives, God passes from being a transcendent and substantial reality into a virtuality that exists as the presupposition of the acting individuals of the Christian community.[110] In Žižek's formulation, it is "*the divine Substance itself (God as a Thing-in-Itself) which is sublated: negated (what dies on the Cross is the substantial figure of the transcendent God), but simultaneously maintained in the transubstantiated form of the Holy Spirit, the community of believers which exists only as the virtual presupposition of the activity of finite individuals.*"[111] A more particular form in which God survives the Crucifixion, the immortal remainder which his Roman executioners could never destroy, as Žižek maintains, is Christ's flesh and blood as consummed by Christian believers in the Eucharist.[112] Here, the transubstantiation endemic to the Eucharist repeats, or is the same as, the transubstantiation involved in the death and survival of God on the Cross.

But let us presently return to the death of God in the God-man Christ—a death survived by God as spirit. In Gethsemane, the forebodings of the demise of the mortal form that Christ had assumed would soon be weighing down upon his soul: it began to be sorrowful and exceedingly heavy (Matt. 26.36–37). To Peter and the two sons of Zebedee he was moved to say: "My soul is exceedingly sorrowful, even unto death: tarry ye here, and watch with me" (Matt. 26.38). He beheld sufferings surrounding him and drawing nearer and nearer from all sides, under the forms of frightful figures. These pursued him deep into the grotto to which he had retired to pray, sweating and trembling. They were images of the all the sins of mankind, for which he was to die. Overwhelmed by these enormities and deformities, he fell on his face and took upon himself the payment of this the most awful debt. He did so out of thanatic love for God and man.[113] Anguished and trembling in terror at the hideous, grotesque and innumerable crimes of man, he shudderingly prayed to God in these words:[114]

> O my Father, if it be possible, let this cup pass from me: nevertheless not as I will, but as thou wilt. (Matt. 26.39)

His will and that of his Father, due to the omnipotence of divine love, remained indissolubly bound. Tortured, shuddering with cold sweat, pallid, and crushed to the earth by the sins, wickedness, vices, and ingratitude of mankind, he felt the horror of his impending expiatory sufferings and death hurtling toward him under the form of hideous specters. He was prostrate in the grotto, his face to the ground, as he prayed to his Eternal Father. The precise details of his Passion, the torments he was to endure for the satisfaction of Divine Justice and due to the concupiscence of all mankind were shown to him in a series of ghastly visions. Horror overwhelmed his soul and a bloody sweat issued from the pores of his body.[115] Again he prayed:

> O my Father, if this cup may not pass away from me, except I drink it, thy will be done. (Matt. 26.42)

And he prayed thus a third time in the same words, each time surmounting the human fear of suffering and death welling up within him, so as not to be diverted from the full consummation of his divine mission to die for the sins of man (Matt. 26.44).

Much as when God had cast a deep sleep upon Adam to open his side and make Eve from one of his ribs; likewise was Jesus, the second Adam, pleased to be overtaken by the sleep of death on the cross and for his side to be opened, so that the second Eve, the Church, might issue from it. For he was desirous of offering his blood.[116] Upon beholding in visions how his death would liberate the souls trapped in Limbo and how fountains of salvation and sanctification would issue without surcease from his redemptive self-sacrifice, he was deeply consoled. So loved he his creatures that he would accept with joy all his impending sufferings and ignominous death to redeem but one of them. Every part and every most minute detail of these sufferings were shown to him in the most lucid light. He accepted them all voluntarily, for the love of God and of man.[117]

Returning to his disciples by the light of the moon, his face pale and bloody, he told them that in merely an hour's time he would be seized, led before a tribunal, maltreated, outraged, scourged, and suffer a most cruel and excruciating death.[118]

When he is not long hereafter betrayed with a kiss to a great multitude with swords and staves, and Peter smites the ear of a servant of the high priest so as to defend him from their clutches, Jesus upbraids his action as constituting an obstacle to the redemptive death that he must and desires to suffer (Matt. 26.51–53). "But how then shall the scriptures be fulfilled, that thus it must be?" (Matt. 26.54).

He was arrested and dragged barefoot along rough roads, over sharp stones, thistles, briars, fragments of rock and thorns, and through thick mires, by ropes attached to a collar and belt studded with iron points whilst being incessantly insulted, blasphemed against, and struck with knotted chords and pointed sticks, to the reechoing sound of sobs and lamentations from onlookers, like a dumb animal led to the slaughter.[119] As he entered Jerusalem, hatred and fury burst all around in the form of flames, together with feelings of horror and grief and the bleating of lambs to be sacrificed the following day in the Temple. But he was the one Lamb willingly embracing his own death. The sky was dark, gloomy, and threatening. The moon was red and covered with livid spots, as if dreading the imminent death of its Creator. It appeared as if the gates of Hell were flung open, as thousands of evil spirits busied themselves in all parts.[120]

Received at the court of Caiaphas by the groans and hisses of a mob, he was insulted by the high priest.[121] When accused on the basis of false witnesses that he had claimed he could destroy the temple of God and rebuild it in three days, he abstains from defending himself, despite being repeatedly struck by the archers (Matt. 26.67).[122] When the high priest adjures him to say whether or not he is the Christ, the Son of God, he fearlessly admits as much, widely opening himself to the charge of blasphemy, the penalty for which is death, as a yawning abyss crawling with vile devils opened beneath Caiaphas and a number of little black objects from Hell, like dogs with claws, entered into the bodies of the bystanders, suffusing them with hatred for Christ (Matt. 26.67)[123] "They answered and said, 'He is guilty of death'" (Matt. 26.66). Immediatley, the very scum of the people surrounded him like a swarm of wasps, heaping insults upon him, as the archers spat on him, struck him with their fists, wounded him with sharp sticks, ran needles into his body, and covered him in mud and spittle.[124]

Shut up in a small vaulted prison and tied to a pillar, he prayed to his Father, fervently offering up the expiatory sacrifice that he would suffer. As a faint ray penetrating the vent-hole of his cell signaled the dawning of the day of his Passion, a day for which he had ardently pined since the day of his birth on earth, he raised his fettered hands to his Father in thanksgiving.[125]

Colorless, haggard, and with bleeding face, he was dragged with ropes through a vicious rabble hollering curses in scurrilous language, received blows and buffets, and was led to the Palace of Pontius Pilate.[126] Asked by Pilate whether he be the king of the Jews, he openly admits to this potentially treasonous charge. When accused by the chief priests and elders in Pilate's midst, he again willingly fails to defend himself, and unequivocally wills himself into the hands of his butchers (Matt. 27.11–14, 24–25). For Droge and Tabor, Pilate's supposed agency here is merely contingent, effectively that of a pawn within the far greater ambit of the divine chess game. The Roman governor is no more than an accomplice. For

Jesus himself is the one who will very soon effectively hand down unto himself the death sentence that he and his Father had conceived before all eternity, along with its precise time and manner. That Pilate is ostensibly and bureaucratically a mechanism within this greater ambit is merely incidental.[127] Jesus' passion would, nonetheless, continue in all its harrowing intensity.

He writhed like a worm under hissing wooden whips to the alternate thunderous and defeaning clamor of the mob and the plaintive bleating of the Paschal lambs emanating from the Temple, his body riddled with black, blue, and red marks, as his blood stained the earth. Thorny sticks replete with knots and splinters tore his flesh to pieces; scourges composed of chains covered with iron hooks penetrated to his bones, tearing off large hulks of flesh at every blow, until his body was but one wound.[128] A crown of thorns was thrust into his head, as he was derided, spat upon, buffeted, and struck on the head with a reed so violently that his eyes filled with blood.[129] Presented to the crowd by Pilate as a bloodied pulp, the cry "Crucify him! crucify him!" resounded defeaningly from the seething rabble. Upon uttering the frightful curse "His blood be upon us, and upon our children" in words assuming the form of black flames, a gloomy sky covered with clouds the color of blood issuing fiery swords and darts lowered over the vociferating mob, encompassing them on all sides with darkness, as myriad devils whispered in their ears and entered their mouths.[130]

Approaching the cross, Jesus knelt down, embraced it with his arms, and kissed it three times, addressing a fervid prayer of thanksgiving to his eternal Father for the bloody expiatory sacrifice he was to offer on this august altar. With swollen and bleeding feet, a body covered with wounds and blood, fainting from exhuastion and loss of blood, a disfugured and bloody face, he commenced his headlong march toward the black hole of stochastic death—ignominious on earth and glorious in heaven. All the way the tumultuous rabble assailed him with opprobrius language, whilst slaves pelted him with mud and filth, and children with sharp stones. On each of the seven times that he collapsed under the overbearing mass of the cross, he was mocked and struck viciously.[131]

Spreadeagled upon the cross at Golgotha, the place of the skull, he was hammered to his ignominious gibbet with nails the thickness of a thumb and heads the size of a crown piece that drove with utmost brutality through his hands and feet, as demons and evil spirits under the forms of toads, serpents, sharp-clawed dragons, and venonous incests darkened the air and crept into the mouths and hearts of his executioners. He was raised upon the cross as the Temple resounded with the blast of trumpets announcing the sacrifice of the Paschal Lamb and the bellowing derisive cries of the uproarious rabble. The cross fell into a hole in the rock prepared for it, driving the sharp points of the crown of thorns still deeper into his head, as blood gushed from his hands and feet. A dead silence ensued, as all Hell shook with terror and rage. His countenance was disfigured, his chest torn, exposing muscles and bone, his limbs distented almost to the point of dislocation, his body a single wound, a corpse ripe for internment.[132]

The sun was obscured by a massive dark yellow body ringed by fire, as the sky's darkness grew progressively more dense, the stars cast a lurid red light, every sound ceased, and a thick fog penetrated everything. Jesus hung amid stillness on the cross, forsaken by God (and thus by his own self) and man, his soul suffused by indescribable bitterness, grief, darkness, gloom, wretchedness, dereliction, dreary solitude, and anguish. At three o'clock, the dead silence was pierced by his agonized cry:[133]

> "My God, my God, why hast thou forsaken me?"[134]

His death-struggle now commenced, cold sweat in his every limb. Raising his head, and crying out in a loud voice that pierced the surrounding silence, he spake thus:[135]

> "It is consummated. Father, into thy hands I commend my spirit."[136]

He bowed down his head and died.[137] All of nature wore a garb of mourning.[138]

73 A.D. (The Jews of Masada)

When the Judaeo-Roman war was nearing its final denouement and all of Judaea had been reduced to near-total impotence, only one fortress was still holding out: Masada. Occupied by the Sicarii under the command of one Eleazar—descendant of the infamous Galilaean rebel Judas, the fortress sat upon a massive rock broken off on all sides by deep ravines from whose bottom rose sheer cliffs so vertiginous that the summit was accessible only via an agonizing march along a narrow path, the slightest slip upon which meant instant death in the unfathomable abyss directly below.[139] Notwithstanding, before long the dauntless Romans under Flavius Silva were mounting their seige. Having swarmed up the White Cliff rocky projection like termites, forged a colossal earthen platform, a pier of massive stones and a huge tower clad with iron plates from whence huge stones pelted and drove off Jewish defenders; and after having bludgeoned one defensive wall with a massive ram and incinerated another; the Romans were finally poised for assaulting the fortress the next morning. Having beheld to his horror how a hitherto favorable northern wind had rapidly swung south and thus—as if through divine intervention—steered the

Roman's flaming volleys onto the very wall that his men had constructed in the face of such encroachment, Eleazar resolved that God's will had turned against the Jews and had moreover marked his people for terminal destruction.[140]

The eventual mass suicide of the Jews of Masada that would inevitably follow is regarded by the document *Sefer Yosippon* as a burnt offering aimed at pleasing God. "They gathered their wives, sons, and daughters to slay them to the ground. They will be considered a burnt offering (*qorban olah*) before God, because for His Name they went, not to be killed before the Romans."[141] Their wilful self-destruction would thus coincide with sacrifice to the Deity. The same text likewise views this self-sacrifice as a pure act of mutual and reciprocal love between God and the collective Israelite subject, wherein God grants out of his boundless love for his people the opportunity to suffer the death of martyrdom whilst Israel accepts the said divinely bestowed gift out of love for him. The final act of self-willed collective terminal annihilation in the black hole of stochastic death exemplifies an absolute love for God, a love that of its very nature exceeds and transcends the fear of death.[142] Needless to say, this love likewise of its very nature *courts* death.

The same sentiments are to be found in the *Scroll of Ahimaatz*, where the *qorban* (sacrifice) symbol dominates the notion of martyrdom. Here, self-fulfilment can only be attained through self-sacrifice, the latter being the highest spiritual experience. Through martyrdom as *qorbanot* the Jew expresses his love of God. This spiritual love is averred to be stronger than life and superior to death. It is an act of love in which the relationship between God and the Jewish people appears at its most essential.[143]

Ephraim of Bonn's *Sefer Zakhirah* continues in the same vein, forcefully contending that the blood of the pious ones is shed like water that ferments God's purple garment (*purpurin*).[144] Comparing Israel to a dove, the author maintains that "Whereas all other birds quiver when being slaughtered, the dove does not, but rather stretches out its neck, so none offers his soul for the Blessed Holy One except Israel, as it is written, 'For Thy sake are we killed all the day'."[145] Here, whereas other nations as seen as merely stumbling upon a perfuntory death in which they perish in humiliation and without cause or reason, the collective Israelite subject willingly and honorably undertakes self-immolation for the divine.[146] What they hereby attain is *qiddush ha-Shem*, or the sanctification of the Holy Name.[147]

Finally, for Philo of Alexandria, the mass self-slaughter of Masada is a sacrifice that allows the collective Sicarii subject to enter the world beyond in a purified condition.[148]

Returning presently to the Masada saga, with the resonance of the subjective double Eleazar-Yahweh climbing to a threshold whose sheer pitch portended the rapidly looming black hole of passion, Eleazar eschewed all notions of escape or heroic endeavor and instead resolved himself on death: his own and that of all. Gathering together the most valiant of his comrades, he urged thus:[149]

> And I think it is God who has given us this privilege, that we can die nobly and as free men, unlike others who were unexpectedly defeated. In our case it is evident that daybreak will end our resistance, but we are free to choose an honourable death with our loved ones....
>
> ... we ought perhaps to have read the mind of God and realized that His once beloved Jewish race had been sentenced to extinction. For if He had remained gracious or only slightly indignant with us, He would not have shut His eyes to the destruction of so many thousands or allowed His most holy City to be burnt to the ground by our enemies .... God Himself without a doubt has taken away all hope of survival ...
>
> For these things let us pay the penalty not to our bitterest enemies, the Romans, but to God—by our own hands.... One thing only let us space—our store of food: it will bear witness when we are dead to the fact that we perished, not through want but because, as we resolved at the beginning, we chose death rather than slavery.[150]

Hereupon, with the hearts of the more heroic of Eleazar's hearers enraptured by his stirring exhortation to martyrdom, the resonance of the collective subject constituted of these newfangled harbingers of self-immolation with that of Yahweh was already beginning to soar to a pitch at which—redounding upon itself as the white heat of a consciousness nearing dissolution—the black hole of passion forboded with the inexorability of fate. However, far from oblivious to the feelings of repugnance and cowardice evinced by the tearfully exchanged glances, furtive intimations of pity, and wincing countenances of those whose spirits quailed before his words; and lest those already enflamed with an appetite for suicide might in their turn be unmanned by the more cowardly; he roused himself and—with renewed ardor and hard-eyed bitterness—held forth with utter relentlessness on the immortality of the soul, the consummate folly of an overextended servile existence, and the unequivocal misery of life itself:[151]

> ... Ever since primitive man began to think, the words of our ancestors and of the gods, supported by the actions and spirit of our forefathers, have constantly impressed on us that *life* is the calamity for man, not death. Death gives freedom to our souls and lets them depart to their own pure home ...
>
> ... Why, pray, should we fear death if we love to repose in sleep? ...
>
> It might be expected that we ... would be an example ... of readiness to die.... let us look to those Indians who profess to practice philosophy. They are men of true courage who, regarding this life as a kind of service to nature, undergo it with reluctance and hasten to release their souls from their bodies ... they consign their bodies to the flames, that the soul may be as pure as possible when it is separated from the body ... Well then!

> Are we not ashamed to show a poorer spirit than Indians, and by our want of courage to bring the Law of our fathers, the envy of all the world, into utter contempt? Even if from the very first we had been taught the contrary belief, that life is indeed the greatest good of mankind and death a disaster, the situation is such, that we should still be called upon to bear it with a stout heart, for God's will and sheer necessity doom us to death....
>
> ... Who is such an enemy of his country, who so unmanly and so wedded to life as not to be sorry he is still alive today? If only we had all died before seeing the Sacred City utterly destroyed by enemy hands, the Holy Sanctuary so impiously uprooted! But since an honourable ambition deluded us into thinking that perhaps we should succeed in avenging her of her enemies, and now all hope has fled, abandoning us to our fate, let us at once choose death with honour and do the kindest thing we can for ourselves, our wives and children, while it is still possible to show ourselves any kindness. After all we were born to die, we and those we brought into the world: this even the luckiest must face. But outrage, slavery, and the sight of our wives led away to shame with our children—these are not evils to which man is subject by the laws of nature: men undergo them through their own cowardice if they have a chance to forestall them by death and will not take it. We were very proud of our courage, so we revolted from Rome: now in the final stages they have offered to space our lives and we have turned the offer down. ... Let us die unenslaved by our enemies, and leave this world as free men in company with our wives and children. That is what the Law ordains, that is what our wives and children demand of us, the necessity God has laid on us, the opposite of what the Romans wish—they are anxious none of us should die before the town is captured. So let us deny the enemy their hoped-for pleasure at our expense, and without more ado leave them to be dumbfounded by our death and awed by our courage.[152]

Eleazar would have long continued his paean to noble self-immolation. However, with the listeners' hearts bestirred to raptures by his rousing disquisition, the resonance of the subjective double Sicarii-Yahweh was by this stage already soaring to so vertiginous an intensity that even those who had afore flinched, teared and felt revulsion in the face of the orator's initial exhortation to mass suicide; now craved naught beside self-destruction in the black hole of stochastic death. Indeed, with the resonance of each double subject exacerbated to the point of a consciousness at the very point of self-incineration in its own searing incandescence; the slightest delay in the enactment of death, the least moment endured within the ambit of life, each most minute and trifling prolongation of existence, could now only be experienced with the infinite agony of a hapless victim trapped in an inferno of his own making. For, having abruptly cut off Eleazar's rambling exhortation; hastening with uncontrollable zeal, dashing off as if possessed, harrying themselves with unswervingly dogged determination, enflamed by and enamored of the purpose instilled within them through the diatribe now so infernally roiling their teetering and moribund minds; their last lingering kisses and desperate embraces upon their wives, their sons, and their daughters shiveringly being impressed;

forthwith did these now rabidly slavering men—their eyes streaming with caustic tears—move to immediately murder their own families with a pulverizing violence hitherto unbeknown to progenitors. With these beastly and demonic deeds having been committed with a hideous syphilitic madness verging upon apoplexy, the already proto-cosmic heights of resonance of the familial subjective doubles forthwith jolted to the very throne of God himself the Hell-bound auto-ramifications of the murderous double subjects to a corybantic apotheosis through which the ultimate demise of the said familial subjective doubles in the black hole of passion was posthaste to follow with an inevitability inseparable from out and out determinism. For indeed, utterly henceforth incapable of enduring the unequivocal horror of what they had perpetrated against their own flesh and blood, yet compulsively mindful that the eventuality of their outliving their dead families was tantamount to unilateral moral collapse; these now-bereft men, laying down beside their deceased wives and children and impatiently exposing their throats, unflinchingly suffered voluntary extinction at the behest of ten chosen angels of death. These last ten, incandescent with the self-convulsing auto-ramifications of their own executioner-Yahweh subjective doubles and rabidly pining for the swiftest possible terminal immolation in the divine substance; instantly annihilated themselves with such maddening quickness that, upon there being but one, this very last man destroyed himself much as had Simon of Scythopolis: with the full brunt of his strength he mercilessly drove his sword through his own body and thereby, as had all of the rest of the Sicarii, fell precipitously to terminal destruction in the black hole of stochastic death like a sack of rancid excrement slapping the earth with a sickening thud.[153]

March 7, 203 A.D. (Vivia Perpetua)

In certain forms of early Christian martyrdom, Farhad Khosrokhavar maintains, death ceases to be the mere indirect corollary of a trangression of State law, but instead becomes an end in itself. It suddenly transmogrifies itself into an open ambition to die per se, a burning desire to leave the vale of tears that is this world. A formerly unwilling death in the service of religious ideals becomes one which is fervidly desired in and of itself. Heretofore a defense of faith through self-sacrifice, martyrdom henceforth proceeds to outstrip its initial moorings and pursues death quite independently of what were its prior legitimate circumstances. The fascination with death comes to be divorced from its former religious logic and goals. Death ceases to be merely hazarded to consummate a sacred cause.

Now the martyr is swept up by a passionate love for death, a desire for self-annihilation. He verily thirsts for death (*tes maruriasr epithumia*). He despairs of living in the world and pines to leave it through a relentless quest for suicide at the hands of an external agent.[154]

This dialectical reversal from a death to which the believer was merely reluctantly resigned to one that emerges as the object of ferverent yearning and active courtship, at once produces an inversion of affects. The former fear and gloom occasioned by the prospect of terminal demise gives way to an inexplicable and radiant joy, equanimity, and even laughter and exultation in the face of imminent execution.[155]

As Droge and Tabor relate apropos of an early text that recorded a plethora of Christian martyrdoms at the hands of the Romans, there was not even the smallest indication that these Christians either hesitated in the face of their impending death or were in any way inexorably compelled by their executioners. Instead, they embraced death quite willingly and in a manner that struck Roman authorities as utterly reckless.[156] Lucian, a pagan of the second century after Christ, once wrote of them that, "The poor wretches have convinced themselves, first and foremost, that they are going to be immortal and live forever, in consequence of which they despise death and even willingly give themselves over to arrest."[157] One overwhelmingly dominant paradigm that they assumed was, not surprisingly, that of the Christ himself. For in a wilful interpretation, the Jesus of these early and beleaguerred moribunds was thought by them to have exhorted them to their deaths. "If any man would come after me," said the Lord, "let him deny himself and take up his cross and follow me. For whoever would save his life will lose it; and whosoever loses his life for my sake and the gospel's will save it" (Mark 8.34–35). They would act upon these words in the most concrete and literal sense imaginable. The representation of Christ's death would excercise a well-nigh inexorable force on early Christian martyrs. Behind each of their martyrdoms lay the self-willed sacrificed of the Incarnate Word himself.[158]

As Alfred Alvarez maintains, the deadly persecution of the early Christians was less directly religious and political than a creation and perversion of their own seeking. They blankly refused to save themselves by making the token gestures to State religion or by escaping during the interval between judgment and execution. They responded to clemency with provocation. Droves of men, women and children were beheaded, burned alive, flung from cliffs, roasted on gridirons and hacked to pieces, gratuitously and of their own free will. Tertullian even explicitly forbade his flock to flee persecution.[159] He, as Shepkaru relates, was convinced that "The sole key to unlock Paradise is your own life's blood"; with the implication that those who emulated the sacrifice of Jesus were thus "assuredly, efficacious in the sight of Christ."[160] For Tertullian, those stung by the evil of the intramundane and quotidian world and who consequently willingly maintained their subsistence within the soiled ambit of existence could not but, at least partwise, be in complicity with the diabolical. They were, in a strong sense, sustaining evil. Continued existence in this world was, for him, evil. To cure this, he contended that the "baptism of blood," produces the "*Antidote for the Scorpion's Sting.*"[161]

Life was at best insignificant and at worst unequivocally evil, and invariably replete with incessant tempation to sin. Death was impatiently sought as a release from this fetid quagmire. Given this, life for many a primitive Christian was intolerable from the very outset and irrespective of conditions. For him, death was a liberation. Bliss was a mere knife stroke away. Suicide was redemption. Much as baptism had purged original sin, martyrdom would wipe away all subsequent transgressions. His pining and sheer lust for death bordered on madness. John Donne went as far as conceding that the early Christian martyr was so hungry and ravenous for his own death that he appeared to be suffering from a disease. Some were baptized solely in order to be burnt. Children vexed executioners so as to be thrown into the fire. The point of culmination of this rabidly moribund trend was the insanity of the Donatists whose lust for their own annihilation was so extreme that the Church ultimately declared them heretics.[162]

This moribund predilection may also be seen as stemming indirectly from St. Paul. In his letter to the church at Philippi, he writes: "For to me to live is Christ and to die is gain. But if I live in the flesh, this is the fruit of my labour: yet what I shall choose I want not. For I am in the strait betwixt two, having a desire to depart, and to be with Christ, which is far better." (Phil.1.21–23). Droge and Tabor aver that, though St. Paul concedes that laboring with Chist in life is "more necessary" (*anangkaioteron*), dying in order to intimately be united with Christ for all eternity is for him "much more better" (*pollo mallon kreisson*). Though he accepts the former, the object of his pining is certainly the latter. For him, death is gain since it necessarily implies consummate and interminable union with Christ. For this reason, he "lusts after death" (*epithymia eis to analysai*).[163]

Ambrose's exposition of these passages from Philippians is that St.Paul "does not flee the servitude of life like a slave, and yet like wise men he does embrace the gain of death."[164] He asserts with the Apostle that the Christian must remain at the station that God has assigned him until such time as God issues the command to depart. Then, the Christian is to embrace the gain of death.[165] Commenting on the same text in his *Biathanatos*, John Donne avers of St. Paul that "At first he says he did not know which to wish for, life or death. Therefore, unless some circumstance inclinces or averts us, life and death are generally equal to our nature. Then, after much perplexity, he made up his mind, and he desired to be released and to be with Christ. Therefore, a holy man may wish it."[166]

This theme is taken up again in 2 Corinthians 5.1–8: "For we know that if our earthly house of this tabernacle were dissolved, we have a building of God, a house not made with hand, eternal in the heavens. For in this we groan, earnestly desiring to be clothed upon with our house which is from heaven: if so be that being clothed we shall not be found naked. For we that are in this tabernacle do groan, being burdened, not for that we would be unclothed, but clothed upon, that mortality might be swallowed up of life. Now he that hath wrought us for the selfsame thing is God, who also hath given unto us the earnest of the Spirit. Therefore we are always confident, knowing that, whilst we are at home in the body, we are absent from the Lord (for we walk by faith, not by sight). We are confident, I say, and willing rather to be absent from the body, and to be present with the Lord."

Christians after the death of St. Paul would decidedly up the ante as regards the urgency with which one clamored for the black hole of one's own annihilation. Some came to regard the mere vicarious reenactment of Jesus' death and resurrection in baptism as insufficient, contending that salvation could only be attained through an actual and voluntary reenactment of Christ's death.[167] Of the partisans of this view was Ignatius of Antioch, for whom a burning desire for *imitatio mortis Christi* was a necessary antecedant for becoming a man and a true disciple of Christ. Such a voluntary death would bring the moribund immediate salvation.[168]

Evarestus, chronicler of Polycarp of Smyrna's martyrdom, was another of the same view. He did, however, somewhat qualify the sort of death that was here being sought. Much as in the case of Socrates who justified voluntary death only in the event of having received a divine signal (the *anangke*), for Evarestus, only a death carried out at the instigation and under the direction of God could be considered blessed and noble. For the Christian to kill himself quite willy-nilly and outside of divine sanction would, thus, entail an usurpation of divine privilege. Such a suicide would not constitute a reenactment of Christ's death on Golgotha.[169] Rather, it would decidedly lead to eternal damnation.

Much like Evarestus, Clement of Alexandria firmly maintained that mere wilful voluntary death hardly constituted veritable martyrdom in the absence of divine instigation. He did, however, greatly extol the sheer salvific efficacy of a divinely guided voluntary death. For even in the case of one who had heretofore neglected to confess the Faith during his life, the willingness to nonetheless suffer death would guarantee his certain salvation. Though previously neglectful, he who would confess Christ before a tribunal and not deny him when being tortured to death would instantly be transformed through dying the death of a martyr.[170]

As regards the cajoling into motion of the headlong march toward the black hole of stochastic death, Tertullian would go beyond Clement's position in terms of radicality. For Clement, though voluntary death ensured salvation, it was not the sole road toward ultimate union with Christ. In vast contradistinction, for Tertullian, martyrdom was the *only* certain way of escaping eternal damnation. Contending that the worldly and corporeal existence of human beings is a prison, he nonetheless maintains that God has provided an escape route from the said carceral mire. Deliverance from this prison, however, cannot be attained by mere vicarious identification with Christ's sacrifice through baptism. A second baptism is required: not the baptism of water, but that of *blood*. He avers that God, seeing the overwhelming danger that men would fall into filth and sin after baptism, devised a secondary means of assuagement and redemption: a baptism of blood. That is to say, voluntary martyrdom. Since men well-nigh inevitably fall into sin and iniquity after baptism, the baptism of blood would be an inevitable necessity for one to save oneself from eternal damnation. Tertullian goes so far as to contend that God commands it. In a rather striking turn, he also connects the duty and necessity of voluntary martyrdom to God's proscription of idolatry. It is the commandment "Thou shalt have no Gods but me" (Exod. 20.3) that necessitates martyrdom. The necessity for martyrdom arises from the prohibition on idolatry. To free himself from the sin of idolatry, the Christian must move in the direction of voluntary martyrdom.[171] Averring life itself to be the primary form of death, Tertullian even regards martyrdom as a divine medicine: God "has chosen to contend with a disease and to do good by imitating the malady: to destroy death by death, to dissipate killing by killing, to dispel tortures by tortures, to disperse in a vapour punishments by punishments, to bestow life by withdrawing it, to aid the flesh by injuring it, to preserve the soul by snatching it away."[172]

Another crucial aspect in which Tertullian differs from Clement, and thus considerably accelerates the drive toward the destruction of the body, is in the fact that he makes no distinction between authentic and inauthentic martyrdom. He nowise condemns any sort of a headlong rush into death. There is no need to wait upon a special divine indication that voluntary martyrdom can or must be pursued. For God desires the martyrdom of the Christian subject, and has ordained it as the sole means of saving the soul.[173]

Tertullian is equalled in his stridency and vehement advocation of martyrdom by Origen. In a red hot outburst to a council of bishops in Caesarea in the middle of the third century, his seething polemic was thus: "Bring wild beasts, bring crosses, bring fire, bring tortures. I know that as soon as I die, I come forth from the body, I rest with Christ. Therefore let us struggle, therefore let us wrestle, let us groan being in the body, not as if we shall again be in the tombs of the body, because we shall be set free from it, and shall change our body to one which is more spiritual. Destined as we are to be with Christ, how we groan while we are in the body."[174] As with Tertullian, voluntary martyrdom is the sole road to salvation and union with Christ. It is a necessity and a duty. Sins commited after baptism can only be forgiven by the baptism of blood.[175]

In this connection, Origen's injunction—much as that of Socrates—is that the true Christian tries always and as much as possible to keep his soul separate from his body. Consistently maintaining this posture, the Christian can nowise be apprehensive in the face of death. For the latter would merely entail the final and irrevocable removal of the soul from the body for which he had always longed.[176] Origen alludes to the 116th Psalm: "What shall I give back to the Lord for all his bounty to me?" (Ps. 116.12). The answer is in the very next verse: "I will take the cup of salvation and call on the name of the Lord" (Ps. 116.13). This cup of salvation he associates with voluntary martyrdom through Jesus' question to his disciples: "Are you able to drink the cup that I am to drink" (Matt. 20.22). Here Droge and Tabor contend that this cup is, for Origen's true Christian, his own reenactment of Golgotha.[177]

A prime example of such a reenactment was the Carthiginian noblewoman Vivia Perpetua. Though she would not meet Origen, she would certainly come under the very tangible influence and tutelage of Tertullian. That the subjective double comprised of herself and her Lord had swiftly attained a pitch of resonance sufficiently acute to portend her own annihilation in the black hole of stochastic death is evinced by words uttered to her by Tertullian not long after her conversion. "And you, you, young lady," he said pointing an accusing finger at her, "deserve to die for every one of those sins." "Do you suppose that any animal's blood, any old goat or bull, can actually take the place of your own blood, the blood that a perfect God's justice calls out for?" She twisted her clasped hands and acquiesced in silence. Suddenly, she was submerged by a vision of Jeremiah preaching, his words surrounding her as if speaking God into her ear as she felt blood dripping from her hands. As Tertullian held forth on the Gospels, Perpetua beheld Jesus beaten, bound, spat upon, nailed, pierced, bloodied, and lying dead in the tomb. "The sacrifice of blood was made," he intoned, "but once for all. The death of Christ bore the sins of many. Now this is how you, a sinner as we know, can enter into His convenant, and eternal life."[178]

All the while, the tautness of the Perpetua-Christ subjective double was rendering itself still more obdurate as she increasingly assumed the fundamental condition of an obscure amorous subject. "Scripture," Tertullian advised her, "calls Him your husband, Perpetua. Marriage is the best way we can understand this, this intimacy of the spirit. A husband should know his wife completely: her character, her emotions, her soul..." She would forthwith pray to God with fervid and repeated assiduity to draw her near to Christ. She resolved to have eyes solely for him. As she walked along the corridor, she felt as though Christ was walking silently and invisibly alongside her. He was man, God, and a husband to her. She trembled from his nearness. During communal prayer he was right in front of her, watching her without reservation.[179]

This amorous union, as is proper to an obscure subject, had however been contracted under the sign of death. For soon Perpetua would find herself visited by a lucid dream whose comsummate purport was to be revealed several years later. She dreamt that she was one of a group of men and women ordered by soldiers to lie with their stomachs to the ground in the open forum in Carthage. When the captain's spear plunged with dull resounding thuds into two nearby supine men's backs and that of a prostrate woman, she felt no fear in her heart and lay in a death wait with greater inner peace than in her own bed. This sense of peace only increased with the proximity of the executioners footfalls:[180]

> *I felt the descending spear puncture my back, piercing through my chest until it hit the pavement beneath me and ground against the stone. The Roman pulled it out; in place of pain, an intense and beautiful warmth flooded out from my wound, pervading my body. Exquisite, perfect peace was all that existed, and I wondered if I could possibly be dead. Was death this absence of pain, this holy peace and safetly? I raised my body up slightly to look down at my chest. Yes, there was a pool of blood pouring from the wound in the center of my breast.*[181]

She awoke unable to move, the Holy Spirit having suffused the room so completely that not a crevice remained free of it.[182]

The more certain it became that she would inevitably be put to death for being a Christian, the greater was her pining for terminal immolation in the substance of the Godhead. Such a death would return her to the subjectless and objectless abyss of freedom:

> What an honour to be chosen to share in Christ's sufferings unto death! What is death in the face of an eternal relationship with the Creator of Life? A mere five-mile marker on the journey from Carthage to the ends of the earth.... I was free. Free from the empire. Free from the civilized, Roman world. Free from earth and life. Free.[183]

The more lucidly her own death appeared before her the more fervid and irremissible was her lust for it. To the extent that she experienced her own death as a fait accompli:

> I was dead. I was twice dead, I realized. Once then, once tonight. *What life do I have left that is my own? I am dead, Christ. This life is Yours now.* Perhaps death in the arena would not be such a hard thing. If I was dead now, the last blow would just be the end of what was already begun. A flash of blood and sand and peace reflected through my soul.[184]

As her unswerving pursuit of martyrdom gathered further inexorable force, she immersed herself repeatedly in meditations upon the Crucifixion. As she touched little cross shapes with her trembling fingers, images of Christ on his igmonious gibbet flashed before her. She thought of blood. Her fingertip rested on the ruby pool of blood at the front of the cross and burned. For hours on end she found herself transfigured by the swirling awe of the Divine sacrifice.[185] She felt herself to be dead and her life to be hidden with Christ in God, for whom her love grew the more intense as the certainty of her looming martyrdom burgeoned.[186]

After finally being baptized she felt herself to have tasted the manifest presence of God and, pining for his physical touch, she begged him to remove her from this world: "*Take me soon!* I now pray. *I want to be with You, no longer worshipping You from afar.*"[187] When hereupon receiving Holy Communion, the blood of the Redeemer rolling down her throat burned her inwardly much as had the fire of baptism.[188]

Being at length arrested with five others and conveyed by soldiers to the dungeon at the governor's mansion through a carnivaled mob that pelted them with rotten pomegranates and stones and from which rose the execrations "Christians!" "Scum!" "Atheists!"—she met her agony with childlike enthusiasm and untrammeled glee.[189] "*Lord, have mercy,*" she said to herself. "*Let me die. Let me die.*"[190]

Whilst in gaol, Perpetua—amid the slew of apparitions that assailed her—was inunadated by a vision of herself climbing up a tremendously high and exceptionally narrow bronze ladder festooned with treacherous swords, spears, spikes, and daggers, reaching all the way to the Empyrean and to the Godhead beyond.[191] When beseeched by her visiting father to save herself through the most trite of means, she retorted:

> Ah, Father! I grieve, not for myself, but over your gray hairs—that you alone of all my family will not rejoice in my passion! On that scaffold, when I am called to trial, whatever God wills shall happen. For I know that we are not placed in our own power, but that of God.[192]

When taken to be tried where the Senate convened, each step from the dungeon across the forum yard was for her one joy-suffused step nearer to terminal annihilation in the Divine substance.[193] "I wanted martyrdom; yes, I wanted it."[194] In the manner of a Muslim at the point of death, she said to herself "*I want to give back everything to You, Lord. Life came from You, and life belongs to You.*"[195] When brought again before a sizeable rabblement in the forum and repeatedly beseeched by the governor Hilarianus to have pity on her father's grey head and on her infant son and to offer sacrifice for the good of the emperors, despite seeing her—likewise beseeching—father thrown to the ground and beaten, she instead chose to die the death that her theophanic epiphanies portended. And thus was she transferred to the military prison with others likewise condemned to await their death at the hands of beasts.[196]

She entered the arena with four other moribunds, cursed and taunted by the enraged and gross noise of thirty-thousand ravenous eyes coalesced to engorge themselves on blood and death. She was whipped, stripped naked, incarcerated in a net and tossed and crushed to the ground by an apoplectic heifer.[197] When she was at last to be executed with others "she herself took the trembling hand of the young gladiator and guided it to her throat. It was as if so great a woman, feared as she was by the unclean spirit, could not else have been slain had she not herself so willed it."[198]

October 10, 680 A.D. (Imam Husayn)

When Imam Husayn set out from Mecca with his family and seventy-two armed men for Iraq in September 680, the resonance of the subjective double comprising these men, on the one hand, and that of their God, on the other, had already reached a pitch of intensity that portended its ultimate and terminal annihilation in the black hole of stochastic death.[199] A quixotic and ill-fated quest for Sunnis, Husayn's march to his own death is regarded by the Shi'a as a most noble act of self-sacrifice made in a state of higher consciousness and under divine guidance. The long journey from Mecca to Iraq would be his Gethsemane.[200] Kerbala would be his Golgotha. The destiny toward which Husayn was marching, and that was marching toward him, was *shahadat*—martyrdom.[201]

For Ali Shari'ati, Husayn was a conscious actor in his own death.[202] Shari'ati speaks of two kinds of martyrs. *Mujahaheed*, such as Hamzeh, fight to defeat the infidel and risk vertiginous death in the process. Husayn, on the other hand, was a *martyr of martyrdom* and as such entertained not the slightest intention of victory but instead rose to embrace his own death in a state of perfect lucidity. He chose to be bestroyed and to die. His purpose was self-denegation for a sacred cause. He knew perfectly well that he would perish.[203] Failure to die would be failure *tout court*.

As 'Abd al-Razzaq al-Muqarram avers, the lofty status accorded to the Imams from the progeny of Muhammad by God could only have been attained through their own annihilation of themselves in submission to the commands of the Almighty. To this end, they would willingly succumb to destiny and certain death. It is even averred that they had foreknowledge of the very moment of their deaths.[204] "Their acceptance of tragedy," contends Shaikh Yousuf al-Bahrani, "their being killed by the sword or by poison, and their acceptance of the affliction to which they are exposed at the hands of their oppressive foes, despite their knowledge of it and ability to avoid it, is due to their knowledge that it would cause the most Praised and Exalted One to be pleased with them, and that it was chosen for them by Him and was mandated upon them so that they would be closer to His holy Self...."[205]

Even prior to his departure from Mecca, the pitch of resonance of the Husayn-God subjective double had forged the Imam as an obscure amorous subject. He would submit divine love to a fatal ordeal and a deadly possessive reciprocity arising out of the demand for an integrated originary destiny. For when his mother Umm Salamah begged him not to go, intimating that the Messenger of Allah himself had informed her that Husayn would be killed in Iraq on a tract of land called Kerbala and that she even possessed a specimen of his grave's soil in a bottle given to her by the Prophet, the Imam replied:[206]

> Mother! If I do not die today, I will tomorrow, and if not tomorrow, then the day after. By Allah! There is no avoiding death. And I even know the day when I will be killed, and the time when I will be killed, and the grave in which I will be buried just as I know you, and I look at it just as I look at you. If you wish, mother, I can show you my grave and those of my companions.[207]

Having done so, he gave her a little of the soil to keep in a bottle, telling her that once she saw it boiling in blood she would know that he had been slain. On the afternoon of the tenth day of Muharram—the day of his death—the soil in both bottles boiled in blood.[208]

In his very last sermon prior to leaving Mecca, Husayn intoned thus:[209]

> Death is inscribed on Adam's children like a necklace on a girl's neck. My passion to be reunited with my ancestors is like that of Jacob for Joseph, and the demise I shall soon meet is better for me. I see my limbs being torn by speeding steeds in the desert between

> al-Nawawis and Kerbala', so they shall fill, through my death, hollow bellies and starved pouches. There is no avoiding a day recorded by the Pen. Whatever pleases Allah also pleases us, we Ahl al-Bayt.[210]

Though lucidly aware of the perils awaiting him at his destination, Husayn nonetheless set out despite the plethora of warnings from myriad sources that began even prior to his departure.[211] When one of his cousins warned him that the fickle Kufans would desert him in his struggle against the might of the caliph Yazīd and his governor Ubaydallah much as they had refused his father Ali's call to arms against then-Caliph Mu'āwiya and his governor Ziyad not twenty years since, Husayn retorted that "what is fated is fated, and will happen whether I heed you or not." When a messenger informed him just one day's journey outside of Mecca that the Iraqi's swords belonged to Yazīd, Husayn merely registered the warning and kept going. The next day, in response to a letter from the governor of Mecca guaranteeing safe conduct and protection should he abort his well-nigh suicidal mission and return to Mecca, Husayn replied that "The best guarantee of safe conduct is that of God." When confronted with the dire news that the twelve-thousand Kufans who had pledged to fight alongside him had been reduced to only four-thousand, then to three-hundred, and then to a mere handful, Husayn still refused to turn back, intoning: "Man journeys in darkness, and his destiny journeys toward him."[212] When counselled at Qadisiya by al-Hurr to swear fealty to the caliph or at least turn back to Mecca, Husayn again refused, retorting that with the world corrupted by apostates like Yazīd and Ubaydallah, "I can only see life with such oppressors as tribulation, and death as martyrdom."[213]

Travelling north along a desert bluff overlooking the immense flat valley formed by the Tigris and the Euphrates with al-Hurr in tow, Husayn at length encamped just below the bluff in sight of a branch of the Euphrates in an area called Kerbala, or "trial and tribulation."[214] It is here that he declared: "Here shall we camp and here will our blood be spilled and our graves be dug! My grandfather the Messenger of Allah had told me so."[215] It was the first day of the month of Muharram. On the third day of Muharram the encampment was surrounded by a four-hundred-strong army under the notoriously ruthless general al-Shimr. As Hazelton has it, "there was to be no escape. Nor did Husayn want any. Now that he had reached his final destination, he and all those with him would pass from the time-bound realm of history to the timeless one of heroes and saints."[216] Husayn and his collective *Ahl al-Bayt* subject was poised to disappear into the black hole of terminal annihilation.

One by one they would wilfully cast themselves into the void. Having married Husayn's daughter in the beleaguered encampment, his nephew al-Qasim immediately made for al-Shimr's lines and was killed. Husayn's half-brother Abbas charged the enemy lines, one man against a plethora, until both his arms were cut off and a sword pierced both his goatside water bottle and his heart, so that the water ran red with his blood as it spilled out onto the sandy soil. Preferring death in combat to death of thirst, Husayn's eldest son Ali Akbar ran up against al-Shimr's men and was rapidly slaughtered. Husayn's infant son was butchered with an arrow when he held him up approaching their lines begging for water for the children, its blood pouring through his fingers onto the ground.[217]

On the eve of their final day—*Ashura,* the tenth day of Muharram—when Husayn beseeched the remaining men to leave him to his fate, they all refused to avoid certain death. "We will fight with you until you reach your destiny," one proclaimed. Another: "By God, if I knew that I was to be burned alive and my ashes scattered, and then revived to have it done to me again a thousand times, I still would never leave you. How then could I leave when what I now face is a matter of dying only once?"[218]

Husayn hereupon uttered the Islamic phrase spoken in the face of death: "We belong to God, and to God we shall return." That long last night was a night of prayer and preparation for death.[219] In a dream in the predawn of the same night al-Husayn was confronted by a vision of dogs charging at him and mauling him, the most fierce among them being a spotted one, and the one who would kill him being leprous. His blood would be carried to the heavens in a green glass vase.[220] He took off his chain mail and donned a simple white seamless robe—a burial shroud. He anointed himself with myrrh melted in a bowl and his men with perfume. They all knew they were being anointed as were corpses—for death.[221]

As the sun inexorably rose and gained height on the tenth day of Muharram, the last of the seventy-two warriors went out to meet their deaths. After this, Husayn came out upon a steed, one man against four-thousand. Thrown to the ground by an arrow to the shoulder, he was butchered with thirty-three knife and sword thrusts, his corpse trampled again and again by steeds into the dust. The seventy-two blood-drenched, lacerated, macerated and crushed headless corpses were left in the desert to be fed on by hyenas and wolves.[222] Three long days and nights of pitched darkness followed, and it rained blood.[223] His head was hacked off, together with those of all his dead warriors, speared on a lance, carried to Ubaydallah's palace whose walls dripped with blood and flung before him.[224]

March 26, 922 A.D. (Mansur al-Hallāj)

In the *anal'l-Haqq* episode, Mansur al-Hallāj suddenly declared to his friend Junayd one day: "I am the Truth"—meaning "My 'I' is God" or "God is wholly within me."[225] Hallāj felt that he had surrendered his heart to the *Nutq*, or vital utterance of the inmost divine creative liberty, to the extent that his own speech became this same pure, inspired, divine utterance—a conviction that of its very nature compelled him to declare it openly.[226]

That the associated resonance of the subjective double Hallāj-God had already by this point attained to a pitch whose intensity portended annihilation in the black hole of stochastic death is evinced in Junayd's stern retort to Hallāj's dangerously blasphemous asseveration: "No, it is by means of the Truth that you are! What gibbet will you stain with your blood!"[227] Elsewhere Hallāj again declares: "if you do not know God, recognize Him in His sign; I am His sign, and I am the Truth, and I have never stopped being true to Truth. And even if I am killed, or hung on the gibbet, or have my hands and feet cut off, I will not retract."[228]

In reflecting upon what Badiou would here certainly describe as a fusional form of love with Hallāj as an obscure amorous subject, Louis Massignon refers to Ibn Dawud's musings on the melancholy of love. Love is considered at once the cause and the effect "of the inflammation of the blood and the bile that leads to increasing atrabiliousness; the greater the atrabiliousness, the more possessed one becomes by melancholy; so much so that love is a disease for which doctors have no cure."[229] According to Arabic poetry, love passes through the stages of admiration (*istihsan*), attraction (*mawadda*), attachment (*mahabba*), familiarity (*khulla*), inclination (*hawa*), desire (*'ishq*), enslavement (*tayim*), and delirium (*walah*). It is the third-last of these that Ibn Dawud takes as the unsatisfied bitterness of desire in the she-camel: a stage leading to a final frenzy of passion (*walah*), and the inexorable sign of demoniacal obsession (*waswas*) which his friend Niftawayh avers leads to mental slavery (*tatayyum*). From this point onwards love, that great malady of the soul, after progressively weakening the body and isolating the mind in evermore solipsistic thought, inevitably leads to fatal aggravation of mental illness, outright madness, and ultimately death.[230]

And it is from mental slavery to delirum and death that Hallāj would move. In his *The Death of Hallāj*, Herbert Mason intimates the foregoing with the following words that he puts into the mouth of Ibn Dawud himself:

> ...We shall see him
> Die for his presumption....

> He knows directly, so he claims, the mind
> Of God ...[231]

And indeed, soon enough, Hallāj's mere risking of his life for the sake of public fidelity to his having attained deifying amorous union (*ittihad*) with the divine gives way to a henceforth active pursuit of death. As related by Abu Musa 'Imran-b-Musa, when a Basrian beseeched him to pray for the cure of his brother, Hallāj replied: "I will pray for him, but on condition that you continue to speak evil of me, even more, that you accuse me of impiety, and that you agitate to get me executed."[232] On New Year's Day in Nihawand, when asked when *Nayruz* (New Year) will come for him, Hallāj answered: "The day when I will be exposed to view on the gibbet, when I will die in this way very close to God."[233]

Meanwhile, the resonance of the double subject Hallāj-God had climbed to such a plateau of intensity that, by the time of his last *Hajj*, Hallāj's will was perfectly united with that of God, in everything and everywhere directly interpreting the divine will. To the extent that, on this last pilgrimage, in the public prayer (*khutba*) that he performed at Mount 'Arafat before assembled pilgrims on *9 Dhu'l-Hijja* 295/907, he offerred thanksgiving (*shukr*) to God by beseeching God to do so himself:[234] "Come therefore into me to give thanks to Yourself, that is the true thanksgiving! there is no other."[235]

At the same time, Hallāj, as the obscure amorous subject locked in deathly reciprocity with his beloved, continued to pine for terminal self-immolation in the black hole of stochastic death. During his last period of preaching in Baghdad, he withdrew at night to pray in a secluded corner of the Quraysh cemetery, called *maqabir al-Shuhada'*, or the tomb of the martyrs. According to an account of Qadi ibn Haddad, Hallāj was espied one night weeping at the tomb of Ibn Hanbal, praying:[236]

> O You, who have intoxicated me with Your love, and let me wander through the esplanade of Your nearness, You are the Only One, in the loneliness of Eternity, You are the Only One to witness You, from the height of "the throne of truth" ...
>
> I beseech You, out of respect for this sacred possession that You bestow upon me and for the even higher degrees I am asking from You, do not return me to myself, after having robbed me of myself; do not show me my soul now that You have robbed it from me; increase the number of my enemies in Your cities, and the number of those among Your faithful who clamour for my death![237]

After glimpsing the interloper and berating himself as a novice in the ways of God, he utterred three cries and fell down with blood coming out of his mouth.[238]

In one of his last public discourses (*Khutab*) in the markets (*Aswaq*) of Baghdad, Ibn Marduya relates that a sorrowful Hallāj, after coming to a halt before the entrance of the 'Attab mosque in the Suq al-Qati'a and having bewailed his plight to the people, went on as follows:[239]

> As for me, there is no longer a veil between Him and me, not even a wink, my time of resting, so that my humanity may perish in his divinity, while my body is consumed in the flames of His omnipotence: in order that there remains behind neither trace, nor vestige of it—neither face, nor description![240]

At another time, Ahmad ibn al-Qasim al-Zahid likewise beheld Hallāj holding forth desperately in the Baghdad Suq, and thereupon launching into a fervid recitation of the following beseeching lamentation to God:[241]

> I have embraced, with my whole being, all Your love, O my Holiness! You have manifested Yourself so much that it seems to me that there is only You in me!
> I examine my heart amidst all that is not You,
> I do not see any estrangement between them and me, and only familiarity between You and me!
> Alas! here am I, in the prison of life, reunited
> With the whole human race … Ah take me with You, outside of the prison![242]

Very soon, able to tolerate the continuation of his own existence less and less, Hallāj resorted to outbursts of public impiety to rile the masses and the authorities into murdering him. Husayn ibn Hamdan relates hearing Hallāj hollering in the Baghdad Suq that he had set sail on the sea, that his boat was smashed and that he will die in the confession of the Cross, not caring to go to Mecca or Medina. Later Hallāj, pointing a finger at himself, answers an enquiring Hamdan that the aim of his impious remarks was:[243] "That someone will kill this cursed one!"[244]

Upping the ante still further, Hallāj entered the mosque of Al-Mansur, beseeching the massive gathered crowd thus:[245]

> You must understand that God has made the spilling of *my blood lawful* for you (*Abaha lakum dami*): therefore kill me![246]

With the crowd weeping and a great figure among the Sufi's enquiring of him how they would possibly kill a man who prays, fasts, and recites the Qur'an, Hallāj again implores:[247]

> O Shayk, the thing that prevents you from shedding my blood has nothing to do with prayer, fasting, or reading the Qur'an! Kill me then, and you will be rewarded for it, and I will gain peace by it, for you will have fought for the faith and I, will have died a martyr.[248]

Later on, when the same Shayk followed him to his house and enquired of the meaning of his upsetting public utterances, Hallāj retorted:[249] "My dear son, there is no more pressing business for the Muslims than my execution."[250]

At last, finally having cajoled the authorities into destroying him, Hallāj was indicted for spreading news of miracles (*ifshd' al-karamat*), usurpation of the power of God (*da'wat al-rububiya*), and the thesis of divine love (*zandaqa*).[251] He was captured, ferried by camel to Baghdad, paraded there, and hung up alive for four consecutive days on a gibbet were he was cursed, slapped and mocked by children. Hereafter followed an imprisonment of in excess of eight years during which he prayed a thousand *rak'a* daily in spite of being shackled from his neck down to his heels with thirteen chains.[252]

On the morning of his execution he was conveyed upon a pack mule from prison to a gibbett on the West Bank of the Tigris, had his clothes torn off, was flagellated with a thousand lashes until he fainted, had his hands and feet amputated, and was hoisted upon the gibbet overnight as screams rent the air, looters set fire to shops, and Baghdad convulsed with rioting. The next morning he was taken down, beheaded, doused with oil, incinerated, his ashes cast into the river, and his head hung on the Caliph's palace gate in public view.[253]

October 14, 1092 A.D. (Hassan-i Sabbah's Assassins)

Whereas Hallāj was an obscure amorous subject whose beholdenness to the single fusional body Hallāj-God submitted love to a fatal ordeal destining him for the blackhole of stochastic death, Hasan-i Sabbah actively set out to *create* obscure subjects of love well-nigh ex nihilo.

This much is evinced by the words which Vladimir Bartol puts into his mouth in his novel *Alamut.* Holding forth to Abu Ali, Sayyiduna (Sabbah) maintains: "I need believers who will long for death so much that they won't be afraid of anything. In fact, they'll have to be in love with death. I want them to chase after it, seek it out, beg it to have mercy on them, as though it were a hard and unwilling maiden."[254] To this end, Sayyiduna endeavors nothing short of ushering into existence an entirely new type of human being, wishing to "sneak into the workshop of Allah himself, and since the man is old and feeble, take over his work. Compete with him in artistry. Take the clay in my hands. And then truly create a new human being."[255] This new man would be the *fida'i.* "His distinctive trait will be an insane desire for death," a desire to be attained "through his utter faith—what faith!—his firm knowledge that the joys of paradise will be waiting for him after death."[256]

But he would not only use his knowledge of the human soul to manufacture these new beings. He would also reengineer man's relation to his body. With the aid of hashish and this-worldly houris, the sensual experience of paradise would be simulated. Having awoken, the fedayeen would not merely regret having been taken out of paradise. For, as Sayyiduna continues, "the poison of the hashash will be at work in their bodies, awakening an irrepressible desire to enjoy it again. That desire will be inseparable from their assumptions of heavenly bliss. In their mind's eye they'll see their beloved girls and virtually die longing for them. The erotic humours will regenerate in their systems and awaken new passions verging on madness. Eventually this condition will become unbearable. Their fantasies, their stories and visions will infect their surroundings completely. Their churning blood will blot out their reason. They'll no longer reflect, they'll no longer make judgments, they'll just pine away with desire. We'll provide them with comfort. And when the time comes, we'll give them their assignment and promise them that paradise will be open to them if they carry it out and perish. They'll look for death and they'll die with a blissful smile on their lips ..."[257]

One such obscure subject of love wrought by Sayyiduna was ibn Tahir. Narcotically precipitated into paradise, face to face with the houri Miriam, ibn Tahir recalled the far-off longing for some unknown thing that had gripped him the previous evening when he lay in the open air outside Alamut, gazing at the stars. Looking into her eyes, he felt that night's longing to have been a portent of her, that he had been waiting for her and for no one else, that her image was the perfect incarnation of some idea he had always carried within his breast—that she had always been *destined* for him.[258] His beholdenness to the single fictive fusional body effected by the intense resonance of the double subject Tahir-Miriam would already forbode a fatal destiny for ibn Tahir. Staring at her mesmerized, he felt as reverent as before a sacrament, moving on to say: "Yes, I'm crazy, I'm insane. This instant Sayyiduna and the martyr Ali mean as much to me as the emperor of China. I could dislodge Allah from his throne and put you in his place." "I don't care. Let me be in heaven or in hell. As long as you're with me, my Shirin, my heavenly Shirin." Soon his body was proving too fragile a vessel for the sheer violence of the passion assailing him, as he began to tremble, gradually succumb to weakness, proceed slowly to lose consciousness, grab his head, fall backwards, go temporarily blind, pass out, and eventually come to back at Alamut.[259]

In the ensuing weeks after his return to intramundane quotidian reality, ibn Tahir was rapidly passing through the final pathological phases attributed to the melancholy of fusional love by ibn Dawud. His desire (*'ishq*), like the bitterness of an unsatisfied she-camel, had plunged him into the mire of demoniacal obsession (*waswas*) and mental slavery (*tatayyum*). He escaped into solitude to be alone with his thoughts of Miriam. Assailed by sleeplessness, he felt she was omnipresent. He was teetering on the brink of delirium (*walah*), for frequently he was afraid that he might well go mad.[260]

Indeed, he was verily going mad and this madness was pregnant with the seeds of death. Already portended by the deathly obsessiveness of fusional love implicit to the rapidly climbing pitch of intensity of the Tahir-Miriam subjective double, ibn Tahir's fevered poetic effusions would now betray a delirious transmogrification toward an explicitly declaimed desire to actively pursue his own destruction. Ibn Tahir:

> Why won't you show the way that leads me
> Back to Miriam, O Sayyiduna,
> Kind uniter, cruel divider?
> If it takes death to buy my passage back
> To join her, say the word,
> And I'll leap from the highest rock.
> My smile will testify how much I love her.
> Or do I need to shove a knife into my heart
> To live beside my Miriam forever?
> Command! Perhaps I need to leap through fire
> And join the Daevas? Just no more waiting,
> Splitting me from paradise like Adam!
> Send me back to Miriam! Take me to her
> Before cruel longing rips my heart in two.

And it is toward his own self-willed destruction that he would forthwith march. For it was that very evening, his cheeks sunken and eyes aglow with the feverish, doleful fire of amorous sickness, that Sayyudina summoned him to send him to the black hole of his own annihilation.[261] Outfitted in the black garb of a Sunni seminarian and playing the role of al-Ghazālī's student Othman, under the pretence of delivering a letter from the university of Baghdad, he was to insinuate himself into the presence of and assassinate the Seljuq grand vizier Nizam al-Mulk, and in the process get himself killed.[262]

As he entered the tent of the vizier, with the poison in his body beginning to take effect, he felt as though he could already hear Miriam's voice and the murmur of paradise all around him. Whilst being wrestled to the ground and beaten and kicked by a mass of men after shoving his dagger into the old man's throat, he glimpsed Miriam's face through the blood streaming over his eyes. Eagerly awaiting the fatal blow, he exclaimed: "Kill me. I'm the sacrificial animal of Our Master."[263]

March 16, 1244 A.D. (Esclarmonde de Foix)

In his novel *The Fire and the Light*, Glen Craney alludes to one of the "The Secret Sayings of the Saviour": an admonition to take leave of the flesh and the bitter bondage of existing at all.[264] This passage toward the necessary dissolution and termination of the curse of corporeal embodiment was to be followed by the Cathar, Esclarmonde de Foix. Her aim, above all, was to die.

This utter annihilation would evidently require the coming into existence of something approaching the Badousian obscure subject. The forging of the obscure amorous subject comprising the Cathar Deity of Light, as subject of enunciation, and Esclarmonde, as subject of the statement, suddenly commenced with a riveting theophany in the Languedoc in March 1198 in which she was inundated by a shimmering orb of gold radiating with centrifugal illumination, its swirling growing more brilliant as it spun toward her, causing her to collapse in a spasm as pulsations coursed down her spine and distant words dimly visited her as if submerged in water. The golden orb split into two serpents, the black one of the two subduing its white adversary before choking on its conquest as a white dove emerged from its death mouth and transmogrified into the original orb. As a balm of calming ecstasy suddenly dissolved into fear, she felt that she was sinking into a hidden chamber of her own flesh. A voice intoned, "Follow this Light, even unto the mouth of the Serpent." She came to consciousness strapped to a stretcher.[265]

This intertwinement of the subjective double comprising Esclarmonde and the Cathar Deity would soon increment itself in its inner tautness and, so doing, commence to portend its slow march toward the black hole of its own terminal annihilation—or at least that of Esclarmonde de Foix. On one subsequent frigid evening in Gascony in October 1201, whilst kneeling before the Blessed Virgin and begging to know why she was still maintaining corporeal embodiment on an earth that she took for Hell itself, she was inundated by the same swirling orb of light that had incapacitated her earlier in Lavaur. The earlier Voice had returned, acting directly upon her body.[266]

Later on night in August 1205, after having carried a daughter who had died at birth up the great mountain of Montsegur and buried her in a shallow grave, she courted self-willed death for the first time when looking down at the welcoming darkness beyond cliffs that fell off sharply below her. She felt it a blessed relief to rid herself of this world through imminent suicide. Yet upon releasing her hold from the boulder separating her from death, she was held back by a hoary old man. "Preparation for death takes a lifetime," he averred. "Let me die!" she retorted. "To seek release from this world is the mark of wisdom," he continued,

"But only if accomplished in the manner that allows one to avoid returning to the flesh." He was Guilbert de Castres, a Cathar bishop. Her march toward the black hole of stochastic death, though it would have to eschew imminent completion, would nonetheless continue.[267]

For Cathars, an absolute death in the Light was the sine qua non of escape from the vileness of matter itself. The world had to be fled, as it was created not by God but by the evil Demiurge, by the Devil himself. The first more tangible step portending Esclarmonde's flight from the filth of corporeal embodiment was her covert initiation into Catharism in Foix in April 1206. She was led blindfolded down a long descending corridor, administered a bitter concoction of belladonna and wolfsbane, lowered onto a dank surface, over which a heavy slab was slid, effectively immuring her in a sarcophagus. Hours passed.[268] This was a simulation of the exile that the material world embodied, a visceral illustration that in this world one is trapped in the rotting coffin of one's vile physical sheath. This temporarily incarcerating coffin was doubtless likewise and at once an intimation of the final coffin in which one would be buried, a portending figuration and pre-image of the absolute death that one ultimately longed to accomplish. After this her initiation, Esclarmonde took the vow of a perfecta. As Castres intimated to her somewhat later on, she was buried to experience the true nature of death.[269] With the same bishop ministering, she vowed never to abandon her faith for fear of water, fire, or any other manner of death. This was her baptism of the Spirit.[270] The mutual recoiling of the subjectivity of Esclarmonde and that of the God of Light had now leaped in bounds in its obduracy and internal resonance. The knights and ladies in the audience filed up to her, offering the traditional Cathar blessing: "Help me, Perfecta, to make a good death."[271]

By June 1211, Esclarmonde had intensified her quest for terminal self-annihilation in the Light, through inter alia the study of kabbala. Together with hundreds of other perfects and perfectas assembled on the vertiginous terraces of Montsegur, she spiritually pursued a death more conclusive and absolute than prosaic physical death, though ultimately necessarily entailing it. A death that courted the possibility of a return to a renewed life of physical embodiment would scarely suffice. Taking up and staring for hours at a fistful of dirt and winnowing its essence, Esclarmonde dwelt upon how to escape the realm of earth, water, and air that is this grotesque creation and illusion of the Demiurge.[272]

On a cold morning in May 1243 at Montsegur, she deliberately caused a pot of melted sheep lard in the chapel to explode with a crackling flash so as to awaken the women of the community from sleep and to the necessity of preparing for death. Upbraided for her morbid fascination with the end of days, she retorted that these days were nearer than thought. Holding the flame emanating from a tallow candle in front of her face, she warned them that "The Light's approach at death will be more disorienting than the fright you've just undergone." "The entirety of our lives is but a preparation for this moment of passing," she added. Soon after this she rammed home her thanatic agenda, maintaining that "Our existence is not unlike spending years in total darkness only to have our eyes assaulted by the midday light at the trice of death." The soul had to be disjoined from the corporeal vileness of the body: "We can also relinquish the fear of separating our spirit from the body. This path offers the only true escape from this world." She enjoined meditation and asceticism as indispensible techniques in the preparation for accomplishing the proper sort of death that was desired. It was necessary to gradually raise one's spiritual temperature so as to meet the Light at the yonder side of death. Otherwise one would reincarnate, vitiating the absoluteness of death. In the thanatic passage from the flesh to absolute death in the Light, the fear of death had to be crushed by deadening the spirit to the desires for subsistence within the ambit of life instilled by the mendacity of the Demiurge. Else, death would only be relative and self-annihilation usurped by the stubborn return of corporeal existence.[273] Death would fail to damn up the dreaded recrudescence of the Lévinasian *il y a*, and its reconstitution of a consciousness that had to be destroyed under all conditions.

Esclarmonde then went on to enumerate the phases to pass through in order to attain this sort of irrevocable and definitive death. The soul would have to gradually divest itself of its demonic adherence to all material elements. It would need to accomplish the Shedding of the Earth, in whose approach the channels of the spiritual body would experience the loss of their wind; followed by the Draining of the Waters, in which a hazy gauze would spread over one's vision, as the veins released their fluid; leading unto the Consummation of the Flame, wherein the disintegration of corporeal sensation would usher in the merging of the sense of coldness with that of heat; finally proceeding to self-dissolution in the Light itself, harbingered by a red firestorm whose sparks would dance in the eyes like fireflies and render breathing impossible. This final stage would be that of a consummate, final, conclusive, and irreversible death, immune to the corporeal vicissitudes of reincarnation.[274]

On a fading afternoon in December 1243, when two-thousand marauding crusaders inched closer and closer to the Cathar stronghold upon the vertiginous terraces of Montsegur like slavering wolves and Basque mercenaries spawned on the scarp like dragon's teeth, Esclarmonde stood upon a perch near the cliffs where she had attempted suicide years earlier, a grey cloud shadowing her in darkness. She prayed to the Light and decided to undertake the Endura: the fearsome

Cathar sacrament, or ritual suicide, whereby one starves oneself to death so as to be terminally annihilated in the Light.[275]

She lay for days soaked in cold sweat on a stone slab without moving, cacooned in a bluish mist by pallid lunar light and entombed in a physical sheath that had surrendered all resistance. Visceral spasms and tormenting thirst had long given way to a dull, fragmented awareness. The icy numbness of her limbs increased. A purplish motling portended death. Her vision tunneled into darkness as the black hole of stochastic death drew near.[276]

Her shrouded corpse was buried in the dusky grotto of a subterrranean necropolis beside hundreds of neatly arranged skulls.[277]

As the crusaders closed in the day after Esclarmonde's burial, the other Cathars took the Consolamentum, the final sacrament whereby they vowed to forfeit their earthly life and never deny their faith, even in the face of fire and death.[278] In the final denouement, they resembled skulls on black-swathed stilts clustering in an oscillation of misery, their faces white as curdled whey, their protruding Adam's apples moiling above the fleshless hollows of their throats, splayed in muck with lips scabbed and limbs gangreous with frostbite. Shackled to a long chain, over two-hundred of them were dragged across rocks and bramble, crumpling in a pitiful bloodied heap. They were communally burned at the stake in a churning conflagration, their flesh melting like snow as they turned black in the voracious flames whilst their evermore frantic recitations of the Cathar Pater Noster rose above the fire wind.[279]

May 30, 1431 A.D. (Jehanne d'Arc)

Jehanne d'Arc first heard the voices of God's saints at the age of twelve when in the garden behind her parents' house in Domrémy. This forging of the Deleuzoguattarian subjective double comprising God as subject of enunciation and Jehanne as subject of the statement was already at this stage pregnant with propulsive vectors urging its henceforth inexorable surge toward its own, albeit indirect, self-annihilation in the black hole of stochastic death. For at this same time, Jehanne likewise thought of the saint that she loved most, Catherine. Above all, she was jealous of her, for Catherine had literally died for her love of God. Catherine had voluntarily died for love of God under the Emperor Maxentius, her head severed with an ax. She was for Jehanne the ideal prototype of a Badiousian obscure amorous subject long since headed for and consummating her divine love in the irreversible *terminus ad quem* of death. Enamored of the fulgurating image of the blade flashing and the pale face spinning through the air, she wished the same bravery and purity upon herself. Her love of God was a consuming fever. She would fervidly follow in the footsteps of Catherine in dying for the same Deity. Every evening, when the bells in Domrémy rang out for Compline, she ran to be with him, her heart pounding. God consumed all of her thoughts. He was all she wanted. When her mother spoke to her of Heaven as God's abode, Jehanne, with hungry eyes and heart swelling up, complained why she could not get there immediately.[280]

In a subsequent theophany, the mutual recoiling of God and Jehanne as the two poles of a single subjective double exacerbated itself in its internally strung tautness. Sitting in the garden with her face turned to the sun, a very loud and deep masculine voice rolled through her thunderously, setting her blood on fire. To her terror, weeping, and joy, it proclaimed her the Maid of France. She was blinded by the light, expecting to be ripped away from the earth, as the voice spread inside her entire soma. This is what she had prayed for, all that she desired. This line of flight along which the face of God and that of Jehanne arrayed themselves would, however, lead to the black hole of her own destruction. For when the presence dissipated, she immediately wanted to die.[281] This latter would prove a premonition of the ultimate death of the obscure amorous subject that the now-consolidated subjective double had ushered in.

Henceforth she would pray incessantly, falling on her knees and lifting her face to the sky every time the church bells resounded. In another theophany, the voices of Michael the Archangel, his deep lion's voice lighting up her bones and causing her to break apart with joy; Saint Catherine, a cool water flute transfixing her with certainty; and that of Saint Margaret, with her wildfire eyes, ensuring Jehanne of the inexorability of a mission she was bound to undertake; would further propel her on the course toward war and inevitable martyrdom.[282]

Finally, whilst kneeling with her eyes closed and face lifted to heaven, in a torrent of light, Michael would intimate to her in a deeply divine voice, and in words floating through her like underwater sounds, that she was to drive the English from France. This countersignifying line of abolition along which she was to cavort would likewise spell her own abolition in the black hole of stochastic death. She vomited.[283] It was the Archangel who would again catapult her ultimately self-immolatory line of abolition out of Domrémy onto Chinon, Orléans, and beyond.[284] Upon leaving Domrémy in a wagon, beyond the red eye of the sinking sun she bethought she espied the paradise to which her saints would lead her to dissolution in the Godhead, if only she fulfilled their holy military schemes.[285]

As she rode in Vaucouleurs to the cheers of "God bless the Maid of Lorraine!" the thought of God in her stood up her every hair and made her feel as if she were painted with light. Gallivanting from Vaucouleurs with the familiar theophanic voices goading her on she could sense the Godhead burgeoning inside her, spreading like a secret plant shackled to her in symbiosis, its roots fusing with her bones, its tendrils sprouting in her fingers, holy rivers coursing through her veins, ancient caves of sentience yawning open inside her skull, as the love of God suffused her. Her theological love was that of a new bride, blind as a mole, heedlessly charging toward the black hole of her ultimate terminal destruction.[286] Kneeling by a river in a small cove with her face skyward, the Archangel's voice again melted her bones in a mutual recoiling of divine and human subjects. She would be dead in two years, the voice intimated. The voice would intimate this again after the liberation of Orléans. Lying exhausted from sobbing after riding the same evening in frigid blue dusk, the voice again rose up inside of her, intimating that this death would indeed come to pass. Closing in on Chinon, she felt that love stood naked before her in all the ugliness, cruelty, and blood that it required. Death lurked behind the sky and amongst the flowers.[287]

Inching closer to Orléans, where the French would war with the English, Jehanne received communion with thousands of other men in a damp field, the love of Christ rising up through her when she swallowed his Body and Blood and the image of his anatomy hung up on the ignominious gibbet in the desert heat etching itself in her mind as she felt the sacred blood surging through her soul.[288] This would likewise be a dark prefiguration of the Christ-like sacrifice that she herself was destined to make in her last hours. In one of her very last dreams she would indeed even find herself oneirically willingly accepting Christ as her executioner.[289]

The mutual recoiling of God and Jehanne in what Badiou would describe as the single fictive fusional body that this obscure amorous subject had long since ushered into existence would further gain in obduracy and thanatic proclivity for Jehanne when, recovering from an injury sustained in a failed assault on Paris and sitting with closed eyes and face raised plate-like to the sky in the nave of a small limestone chapel, the Archangel's face gazed down upon her's yet again. With declarations of love, exhortations to abide in God's trust, and willingly accept her providentially determined tragic fate, it was intimated to her that her end was nigh, that the Burgundinians would imprison her. She willingly accepted this her looming death, but wanted it to be painless and immediate. This her death she would attain, but not without bodily agony. Outside Compiègne she was captured, and eventually sold to the English.[290]

She was dragged to the tower of Rouen amid the hisses and ferocious shrieks of a teeming corybantic rabble of faces contorted by hatred, baying for her to be publically flayed, torn limb from limb, and eaten by dogs. Standing with her wrists and ankles tightly chained before churchmen in the royal chapel at Rouen Castle, she refused to disclose her theophanies even under pain of death. For God had sworn her to secrecy. Despite a raging inferno of abuse, she would defer to no one save God. Pressed for days on end, she refused to disclose the secrets that inextricably bound her to her God, which airtight secrecy and which inexorable bond presaged her certain annihilation. Taken down to the dungeon and threatened with stretching rack, pincers, cat-o'-nine-tails, brazier of red coals, and the spiked Catherine Wheel, she remained unmoved. In a theophany, Catherine admonished her to suffer her torments, beckoning her to pursue the death destined for her and the final consummate union with God thereunto yonder.[291]

Declared a heretic, schismatic, and limb of Satan severed from the Church, she was dragged at dusk and in chains to a great wooden stake through a cold, raw wind, and an enormous crowd of thousands chanting "Burn the witch! Burn the witch!", throwing stones, screaming, and roaring in an ecstasy of horror and delight. Faced with the agony of fire, she recanted.[292]

But this abjuration would halt her inexorable march to the black hole of her own terminal incineration only momentarily. Soon the theophanic voices returned, violently upbraiding her for damning her soul to save her life, for betraying God for fear of fire.[293]

The tautness of the obscure divine and human subjective double would soon, however, reconstitute itself. In her last dream she runs to the market square, where a stake has been erected. Christ is her benevolent executioner. He helps her onto the platform, from whence she beholds thousands and thousands of devout onlookers. When she is upon the stake, Christ ignites the flames to the sound of an admiring sigh from the crowd who suddenly kneel down and bow their heads in solemn prayer. Looking up at a sky made of stained glass, she sees the three saints of her recurring theophanies. As the flames burn her feet, Jehanne hears the voice of God declaring her imminent canonization.[294]

Her body was burnt three times, reduced to ash, and discarded into the river.[295]

1555 A.D. (Protestant martyrs in England)

We enter the time during which Henry VIII reigned in England. After having afore abjured his protestant beliefs before the authorities, for two whole years was

Thomas Bilney vehemently pierced by sorrow and remorse—to the extent that, with the resonance of the subjective double Bilney-Christ already hereby having climbed to a well-nigh unbearable intensity, the black hole of stochastic death itself presaged. For now he resolved to die for the open confession of that which he had thence renounced and forthwith made haste for Norfolk—and, alluding to the commencement of Christ's passion, said he was headed for Jerusalem. Pining for terminal immolation in the divine substance, he risked his continued existence by brazenly preaching the reformed creed—such that, having gifted two of William Tindall's heretical tomes to the anchoress in Norwich, he was arrested and thrown into prison to await a writ authorizing his being burnt at the stake. Despite the stern warnings of friars and doctors sent to his cell that he would certainly suffer eternal damnation if he died in his prevailing opinions, Bilney was unmoved. Thus, in consonance with his earnest wishes, he was degraded and thence handed to the secular authorities for imminent execution. Finally, he was led to the low valley outside the Norwich city gates called the Lollards' Pit, apparelled in a layman's gown with his sleeves hanging and his hair piteously dishevelled, and—with the scorching white heat of consciousness as double subject now quite as infernal as the fire for which he was imminently bound—spake he thus to the onlooking rabblement:[296]

> Good people, I am come hither to die, and born I was to live under that condition, naturally to die again ...[297]

And ere long, having knelt before the execution apparatus in quiet prayer, he was chained to the stake and thence scorched by wind-tossed flames until the chain upon which his withered form now bowed was smote out with a halberd, causing his lame body to fall to the bottom of the fire where it was immediately incinerated in the black hole of stochastic death.[298]

During the bloody times when Queen Mary reigned in England—despite already having been imprisoned in Northampton for railing against papistry and likewise in spite of the present regime's unequivocal prohibition against such preaching—stubbornly did Laurence Sanders persist in his protestant proselytism. Already was the resonance of the Sanders-Christ subjective double at a pitch of intensity comfortably sufficient for the black hole of passion, of stochastic death, to insinuate itself with dark inevitability. For—having been brought before the bishop in London for refusing to heed the advice of a councillor to the Crown against preaching there—upon being arraigned before the said bishop and asked by the latter to write down his (doubtless heretical) views on transubstantiation, thus did the renegade preacher explode:[299]

> My Lord, ye do seek my blood, and ye shall have it. I pray God that ye may be so baptized in it that ye may thereafter loathe blood-sucking and become a better man.[300]

Far from content with confessing his having preached the Word against the Royal proclamation to the contrary and taunting the bishop with the latter's own earlier written contentions that the Queen was born a bastard, Sanders went quite as far as to lay open his heart's wish for terminal immolation in the Godhead. Upon which this "frenzied fool" was thrown into prison to languish in well-nigh hermetic silence for a year and a quarter, released, degraded by the same bishop, and thence crammed into a common gaol in Coventry to await execution. On the fateful day he was led barefoot and in an old gown and shirt outside the city to be burnt. At such a vertiginous pitch of inner resonance was the livid heat of consciousness of Sanders as double subject that, though given a last chance to recant, he instead made posthaste for the fire and, after falling to the ground in fevered prayer, embraced and passionately kissed the stake at which he was imminently to suffer incineration in the black hole of stochastic death.[301]

Now, during the reign of Edward VI, the heretic John Hooper commenced his return to England from Switzerland where he had long resided. Already at this early stage, the resonance of the subjective double Hooper-Christ had climbed to an intensity of auto-ramification that manifestly portended the black hole of passion. For, with the germs of an erstwhile lust for terminal annihilation in the divine substance in advanced ferment, and having bade his Zurich friend Bullinger farewell and assured the latter of his faithful correspondence, notwithstanding was Hooper moved to forewarn that:[302]

> … the last news of all I shall not be able to write; for there where I shall take most pains, there shall you hear of me to be burned to ashes.[303]

Yet Hooper's headlong passage toward the black hole of stochastic death was to swiftly gain in resoluteness. For, though Edward had died and the reign of terror of Queen Mary had already begun and the attentions of the papists were already upon him, still did he remain in London where he was—asseverating that, though much earlier he had fled the Damocles' sword of the regime,[304]

> … now, because I am called to this vocation, I am thoroughly persuaded to tarry, and to live and die with my sheep.[305]

And so, after being arraigned before a series of commissions, deprived of his bishoprics, and—upon repeatedly refusing to recant his heresies—degraded, was he thence condemned to die in Gloucester before the very people over whom he was pastor and at the very elm tree where he had preached. After having absorbed himself in solitary prayer for the most part of two days and nights, with the day of death being arrived, he walked in silence on a lowering and cold morning amid a mournful rabblement numbering several thousand toward the stake—at which he prayed inwardly whilst the people wept. With the well-nigh infinitely ramified inward resonance of Hooper as double subject reaching the sheer livid incandescence of a consciousness liquifying in its own molten heat, was he at last subjected to devouring wind-swept flames and—after three-quarters of an hour of infernal torment in which his nether parts were burnt, mouth blackened, lips shrunk to the gums, bowels fallen from the abdomen, with fat, water and blood dripping from his carcass—totally and utterly destroyed in the black hole of stochastic death.[306]

In the time of the reign of Queen Mary in Tudor England, Rowland Taylor was commanded by letter to appear before the fierce and cruel Chancellor Stephen Gardiner for his repeated obstruction of and cavilling at the reintroduction of papist rites in Hadley church. The resonance of the subjective double Taylor-Christ was already soaring to a pitch of intensity so vertiginous that the black hole of stochastic death portended itself with well-nigh irreversible force. For, in spite of beseechings by his friends that arraignment before Chancellor Stephen Gardiner could only mean imprisonment and death and despite the urgings and counsel of his servant John Hull that he flee from these dangers forthwith, he paid no heed to these men and to Hull in particular:[307]

> Remember the good Shepherd Christ which died for his flock. Him must I follow.[308]

And when he was before Gardiner and the latter was berating him as a knave, a traitor, a heretic, a villian who dare not behold his countenance, and a foolish and arrogant lunatic; did he continue to further court the black hole of passional death by not merely railing against the purported lies, superstitions, and blasphemies of the papists—but indeed by unsinuating that Gardiner had denied Christ and was in danger of eternal damnation. Wherepon was he immediately taken away and locked up in Kings' Bench prison, where he tarried for almost two years. After which time, appearing now before a commission of bishops and unequivocally refusing to recant his heresy, was he condemned to suffer death. Having been degraded, he was driven to Hadley with a hood over his head and thence through the town itself—whose streets were aflood with the weeping and lamentations of his mourners. At last reaching his Golgotha at Aldam Common and rending off his hood to reveal a head clipped like that of a fool, with the resonance of the

Taylor-Christ paradigmatic subjective double now at a well-nigh volanic frenzy and that of the syntagmatic doubled subject or consciousness of Taylor himself burning in the scorched incandescence of its own heat, he held forth thus:[309]

> Good people, I have taught you nothing but God's holy word, and those lessons that I have taken out of God's holy book: I am come hither to seal it with my blood.[310]

Upon this he was instantly bludgeoned with a yeoman's waster, kneeled down to pray, went to the stake, kissed it, set himself in the pitch-barrel at the stake, was bound in chains, smote in the face with a faggot by an earless subversive, struck on the lips for murmuring a psalm in English, and finally set ablaze with his hands folded—before a drunkard knocked Taylor's brains out with a halberd and his corpse fell into the fire and was thus forthwith incinerated in the black hole of stochastic death.[311]

November 26, 1872 A.D. (Kirillov from Dostoyevsky's *Demons*)

In Dostoyevsky's *Demons*, the provincial clerk and revolutionary member of the group of five, Sergey Vasilyevich Liputin, avers that the civil engineer Aleksey Nilych Kirillov was in the midst of writing an article on the reasons for the increase in incidence of suicide in Russia and on the general factors promoting or retarding the spread of suicide in society. Liputin went on that Kirillov rejected morality as such and instead adhered to the principle of general destruction—presumably and *primus inter pares* including self-destruction—in the name of purposes that were ultimately considered good.[312] As Kirillov claims somewhat later in relation to this purported article of his, he was solely searching for the reasons why people generally did not dare kill themselves, averring that the number of suicides were very few indeed. He moves on to contend that apart from this paucity of self-immolations, even those who did in fact kill themselves did so for the wrong reasons. They commited suicide out of either great sorrow, from sheer anger, or out of complete madness. All of these motivations for suicide were, therefore, well-nigh spontaneous self-destructions coming to pass—largely or solely—within the evanescently roiling cauldron of fleeting passion.[313]

Kirillov does, however, proceed to contrast these suicides with what he sees as their antipode: the far more deliberate self-murders of those who are given to much introspective thought and who kill themselves after much of such thought has been given. Though maintaining that many such deliberated suicides—suicides "for good reason" as he has it—have indeed been commited, these nonetheless remain far too few to his taste. They remain too few due to the stubborn persistence of prevailing prejudice, he maintains, adding that in the absence of this prejudice *all* would commit suicide. This prejudice, Kirillov asseverates, is that man continues to cling to life—rather than fleeing from it—through his obsessive attachment to pain and fear. That is to say, man has need of these latter as—strange to say—inducements to his continued subsistence within life. The example which he provides is that the vast majority of men would obdurately fail to accede to death at the behest of being swiftly squashed by a massive stone as large as a mountain, even though this death would be entirely bereft of pain.[314]

After this rapsodic discursus, Kirillov launches into what he contends to be a human freedom that is finally untrammeled by encumbrances of the foregoing sort. He avers that full freedom can only arise when the matter of continuing to live or not makes no difference, and that this must be the goal for all people. His interlocutor responds that, in such a case, no one at all would even want to be alive. Kirillov flatly agrees. His interlocutor returns that man is fearful of death because he *loves life* and that this is entirely *ordained by nature*. Kirillov, however, completely rips into this statement: "That's vile, and that's the basis of the whole deception!" He continues to pour on the vitriol that life is just pain, that it is fear, and that the human species is bitterly unhappy. Everything, he contends, is pain and fear. He goes even further. Man in fact loves life for this precise reason, because he has been made in this way. Man has, moreover, received life only in exchange for pain and fear. And herein resides his entire self-deception. Kirillov, notwithstanding, strongly believes that there can de facto emerge a "new man," who is proud and happy. This new man, he contends, is one who does not give a fig about whether he is alive or dead. Still more strikingly—and indeed, absurdly—this "new man" who has triumphed over fear and pain, *will become God*. "And that other God will no longer be." When asked whether "that other God" does not exist, Kirillov spontaneously agrees. But then, he immediately asserts that this other God does exist.[315] The dialogue continues as follows, with Kirillov averring that:

> "… God is the pain of the fear of death. Whoever conquers pain and fear will himself become God.… Then history will be divided into two parts: from the gorilla to the annihilation of God, and from the annihilation of God to …"
>
> "To the gorilla?"
>
> "… to the change of the earth and of man, physically. Man will be God and will change physically. And the world will change, and deeds will change, and thoughts, and all feelings. What do you think, will man change physically then?"

> "If it doesn't matter whether one lives or doesn't live, then everyone will kill himself, and that's perhaps what the change will consist of."
>
> "It doesn't matter.... Everyone who wants the main freedom must dare kill himself. ... There's no freedom beyond that; that's all there is, and there's nothing beyond it. Whoever dares to kill himself is God. Now everyone can do it so that there will be no God, and there will be nothing. But no one has yet done it, not even once."
>
> "There have been millions of suicides."
>
> "But never for that reason, always with fear and not for that purpose. Not to kill fear. Whoever kills himself only for the purpose of killing fear will immediately become God."[316]

There is much to unpack in this rabid *folie à deux*. When Kirillov broaches the notion and eventuality of man being animated by full freedom, his interlocutor contends that the upshot of such an unbridled freedom would be that perhaps no one at all will want to live. Kirillov, of course, flatly agrees: "Everyone who wants the main freedom must dare kill himself." This would seem to imply that such an untrammeled freedom is most kindred to the Hegelian "freedom of the void" that we traced somewhat earlier through Eagleton in connection with suicide bombers and other voluntary martyrs. As thence observed, this latter freedom's terrorism is so acute that, in addition to not being able to abide external prosaic encumbrances to itself, it swiftly comes not even to be capable of abiding itself—and thus inevitably leads to the self-immolation of him whom it so rabidly animates.

Earlier, Kirillov had contended that man fears death because of his love of life, for the reason that this has been so ordained by nature itself. This death is, as seen, the self-willed death inexorably entailed within the Hegelian-Kirillovian "freedom of the void." What stands before this absolute freedom is the love of life specifically *as ordained by nature*. More precisely, nature forestalls this vertiginous freedom through its carceral organic embrace. Nature is, thus, itself that which stands as the obstacle par excellence barring the way of the deathly consummation foreboded by the said and most abyssal freedom. Nature must, therefore, evaporate before the sheer vertiginous and strictly anti-organic ungrounding that Kirillovian freedom entails. This Kirillovian freedom is, therefore, none other than the most shocking and bottomless freedom of the Schellingian abyss of freedom that is quite simply the Godhead—a freedom bereft of both object and subject, and in whose vertiginous ambit all of organic nature instantly dissolves as in an acidic solution.

Kirillov, as mentioned, however, made it very clear from the outset that he who does not care whether he lives or does not live, will be the "new man" and that he will immediately become God. He who dares kill himself is God. Here we may aver that if man becomes the vertiginous abyss of his own freedom, as entailed in his desire to die (an inevitable actual death) in the macerating vertigo of this foregoing abyss, then—contends Kirillov—he will be God.

Kirillov has clearly attained the absolute summit of madness. This is surely poppycock. Yet, it may be made more sensible if one recurrs to the Schellingian distinction between God as subject and the Godhead as the vertiginous, bottomless, and absolutely groundless abyss of freedom. On Schellingian terms, if man commits suicide he will not become God as subject. He will return to and be immolated in the fathomless abyss of the Godhead, and in this sense undergo a sort of zero-level deification. Yet, for Schelling, man is in some way already God even prior to suicide. Though not God as a divine subject, as a "creature of the center" man is God by sharing in the Godhead. That is to say, the abyss of freedom that is God at his most fundamental is likewise man at his most fundamental. For, as observed much earlier in connection with Rosenzweig, both God and man are at base an unconditional and infinite will. The difference is that God is likewise infinite power, whilst man is merely finite in this latter sense. It is this conjunction in man of an infinite will with finite power that causes him to be so internally conflicted. For his will, though as infinite as that of God, finds itself stymied and incarcerated within the closed confines of his organic nature.

It is this encumbering organic nature that stands before man being God in the full bore sense. According to this logic, Kirillov is superficially correct in saying that man will himself become God by commiting suicide. He is ultimately wrong, however, for the attempt to conjoin an infinite will with infinite power in a finite organic nature, far from deifying the same nature, merely destroys it.

However, from the above quixotic discourse, it seems difficult to gauge whether Kirillov believes in God or not. Though averring that God does and doesn't exist, he maintains that man, by commiting suicide, himself can become God and cause God not to exist. Yet, a little later in the very same conversation he reveals an unending and lifelong preoccupation with God. In fact, he has only been able to think about God and nothing else. And this incessant thinking about God has been cause for relentless self-torture.[317]

> "… I can't think about something else, my whole life I've thought about one thing. I've been tormented by God my whole life."[318]

God is, thus, for Kirillov a toxic introject that he repeatedly interiorizes. We may, therefore, posit here the very long-standing existence of a Deleuzoguattarian passional subjective double comprising God as the subject of the enunciation and

Kirillov as the subject of the statement. In this arrangement, the sheer extent of Kirillov's obsession with God necessarily entails an ever tighter recoiling of the two subjects into one another, and a rise in their mutual resonance sufficiently acute and irresistable to forebode the terminal immolation of Kirillov in the black hole of stochastic death with an inexorable determinacy close to that of fate.

For the suicide that he advocates for the entire human race was one that he, above all, himself desires to commit. Though he had long contemplated killing himself in order to likewise—and perhaps, above all—kill the poisonous and metastatizing introjection of God within himself, his portended suicide would assume the outward form of agreeing to kill himself and at once taking the responsibility for the murder of a certain Ivan Pavlovich Shatov. Though what Kirillov is ultimately aiming at in the suicide that he has long since been planning for himself is that this suicide must in a most fundamental sense be absolutely pure. That is to say, it must necessarily be an unadulterated striving for and toward the Schellingian abyss of freedom in and of itself. It must, therefore, be unencumbered by any and all relations to intervening ends and means as possible. It must be suicide perpetrated out of the singular purity of a will that is in itself, for itself, and utterly itself. It must be the logical consummation of the most pure, absolute and infinite will. Thus, if this suicide is subordinated to the dictates of supervening projects within whose scheme it would find its measure, then it would not be the proper and unsullied Kirillovian suicide that Kirillov is seeking. This much Kirillov renders pellucid in his dialogue with Pyotr Stepanovich Verkhovensky later in Dostoyevsky's dark narrative. Here, whereas Verkhovensky is portraying Kirillov's agreed upon suicide as a contractual obligation, Kirillov, though he certainly desires above all to commit suicide, is utterly recalcitrant to the venomous notion that his self-immolation fundamentally arises out of any contract whatsoever—or even that it can be strictly aligned with one. At best, he conceives of it as pleasantly coinciding with the contract to which he agrees out of pure pragmatic indifference. Kirillov aims to commit suicide in and of itself. The contract with which it is aligned, thus, remains a matter of pure happenstance for him. He is at pains for it to be so. So, when Verkhovensky tells Kirillov that his suicide has been agreed upon, Kirillov replies:

> "What agreement?"
> "What do you mean, 'what agreement'?" Pyotr Stepanovich took alarm, even became frightened.
> "It's not an agreement and not an obligation, I didn't bind myself in any way; that's a mistake on your part."
> "Listen here, what *are* you doing?" Pytor Stepanovich now jumped up.
> "My own will."

> "And that is?"
> "The same as before."
> "But how is that to be understood? Does it mean that you're of the same mind as before?"
> "It does. Except there's no agreement and never was, and I didn't bind myself in any way. There was only my own free will, and now there's only my own free will."[319]

Later in their conversation, Kirillov again stamps forth his determination to kill himself out of his own pure and unsullied will. Contingent reasons are to be brutally swept aside, to the point of their absolute and terminal annulment. He wants to kill himself simply because he wants to kill himself.[320]

Much later, he intimates to a rather bewildered Shatov the strange epiphanies that he has been having, epiphanies that are ecstatic and in some sense quasi-theophanies. These latter, he maintains, last but a few seconds at a time, and in them one feels a sense of eternity that is unbearable for man. They are so unendurable that man must change physically or otherwise die. Yet, in them, one has a sense of nature as it was when God was creating the world, suffused by a joyous emotional tonality higher than love. In the five-odd seconds of the theophanies, Kirillov maintains that he lives an entire lifetime. But it is not only this, since for these theophanies he avers that he will give his whole life. However, if any given theophany would perchance last longer than five seconds, man's soul would be unable to endure it and would have to vanish.[321]

Shatov, of course, enquires as to whether Kirillov doesn't indeed have epilepsy. Whether epileptic fits or not, the principle features of these epiphanies is that they are unearthly and endurable for only a few short seconds by man both in his present finite organic form and likewise by his soul, beyond which death necessarily occurs. In these few evanescent moments, however, the soul is elevated to the levity of God in contemplation and lives its entire lifetime. Kirillov likewise avers that he would give his life for these brief ephiphanies.

Is not then what is intimated in these Kirillovian epiphanies the Schellingian abyss of freedom, or Godhead, and the premonition of its future recrudescence? Much as the overextension of a Kirillovian epiphany leads to death, likewise is all of organic nature incinerated in the abyssal Godhead as in a furnace. Much as in the eternity of the Godhead all of time is compressed into a single evanescent and incessantly self-annulling instant that fails to extend itself into any duration whatsoever, so is Kirillov's entire life compressed into and lived in one of his epiphanies. And it is toward the abyss of the Godhead that Kirillov is marching. For he wants to die—by his own hand.

Yet, the force impelling Kirillov toward his own self-destruction is the toxic introject that is God. We wager to say that the toxicity of God as introject arises from the fact that Kirillov is not certain whether God exists or not. Kirillov wants to kill himself because of the incessant torment herein implied. Perhaps, more precisely, the fact is that he, in the deeper recesses of his psyche, suspects that God *does* indeed exist and that this possible existence of God threatens his freedom. These awful thoughts have irreversibly ruined his life and he therefore wants to annihilate himself so as to finally and irrevocably cease the relentless and utterly excruciating torture that is none other than the entirety of his life. This much appears to be borne out by Kirillov's vacillation between affirming and denying God in another and still later quixotic dialogue that he has with Verkhovensky. Here, Kirillov begins, not so oddly, by asserting that God does indeed exist. He will, of course, almost immediately swing to the very opposite pole—and, bringing thesis and antithesis together, desperately declare that:[322]

> "Don't you really understand that a man with two such ideas can't go on living?"
> "Must shoot himself, you mean?"
> "Don't you really understand that he can shoot himself just because of that one thing? . . ."[323]

At this point in time Kirillov will kill himself within about an hour or so and Verkhovensky is thus already looking upon him as a semi-human entity. Their conversation comes to be less marked by musings on Kirillov's impending suicide as a means of removing within himself God as a toxic introject. Instead, Kirillov returns to the tenure of his earlier discourse about his projected self-destruction, that is, to his suicide as specifically arising out of the ultimately and inevitably self-immolatory nature of his Hegelian "freedom of the void." So absolute and terroristic is the freedom that he desires, that the only act worthy of it is suicide. For suicide is in and of itself, indeed, the most absolutely free act in the sense that prosaic appurtenances are stripped away and sloughed off to the greatest humanly possible extent. Suicide is the free act *par excellence.* Whereas from the other of his perspectives Kirillov desires to kill himself in order to annihilate the cancerous introjected divine subject within himself, here Kirillov wants to hurl himself into terminal and irreversible nonexistence in order to partake of the untrammeled freedom of God as Godhead. Self-will is what he proclaims this fathomless freedom of the void to be. This is why he desires to be God. Kirillov:

> "You ape, you're yesing me so as to get the better of me. Shut up, you can't understand anything. If there is no God, then I am God."
> "You know, I never was able to understand this particular point of yours: why should *you* be God?"
> "If God exists, then all will is his, and I can't escape his will. If he does not exist, then all will is mine, and I am obliged to proclaim self-will."

> "Self-will? And why are you obliged?"
> "Because all will has become mine. Can it really be that no one on this entire planet, once having put an end to God and having developed a belief in self-will, will dare to proclaim self-will, in the fullest possible sense? … I want to proclaim self-will. Even if I'm the only one, I'll do it."
> "So go ahead and do it."
> "I am obliged to shoot myself because my self-will in the fullest possible sense is for me to kill myself."
> "But look, you're not the only one to kill yourself; there are many suicides."
> "With good reason. But to do it without any reason, and solely for self-will—I'm the only one."[324]

At this exact point Verkhovensky interjects, averring to Kirillov that if he—that is, Verkhovensky—would be in Kirillov's place then, so as to enact unadulterated self-will, he would do so by shooting someone else instead of himself. So saying, Verkhovensky hints that he would ideally desire to employ Kirillov as an assassin—to make him useful. Kirillov, however, rapidly exhibits his recalcitrance to such a prospect. For such an enactment of freedom would inevitably sully the purity and absoluteness of his terroristic self-will by ensconcing and immuring it within the strictures of a means-ends network. To kill someone else for someone else's benefit would vitiate and diminish the absoluteness of the self-will toward which Kirillov is striving, and by which he is animated. To be sure, the Hegelian "freedom of the void" does bear within itself the imminent propensity to the murder of one's fellow man. But its full consummation and ultimate *telos* is that of annihilating oneself. This abyssal and terroristic freedom is ultimately in-itself and for-itself solely in suicide. The killing of another, though implied within the broad sweep and logic of Hegelian abyssal freedom, nonetheless remains a mere middling way station on its moribund trajectory. What one must ultimately do is, therefore, to kill oneself. One *must* commit suicide. Kirillov's reply:[325]

> "To kill someone else would be the lowest point of my self-will, and that's where you reveal who you are completely. I'm not you: I want the highest point, and I'll kill myself."[326]

Immediately after this, however, Kirillov reverts to displaying his other reason for killing himself: to rid himself of the toxic introject that is God, that most poisonously interiorized subject. Though he will now rapidly re-declare his purportedly untrammeled atheism, one has learned to disbelieve this to be the case. For what seems to lurk beneath Kirillov's atheism is a hypertrophied subterranean belief in God. Despite himself, he secretly believes in God. To the extent that his vehement rejection of God is the predominant form of his being tethered to the God that torments and harrows him to the marrow. This, of course, is not to say that God necessary exists or does not exist. I do not know whether God exists or not. Yet, I—like Kirillov—hate this question. I've perpetually been in fear of Hell. And this will never end. What, merely, is being said is that Kirillov's oblique and repressed hints seem to suggest that he does indeed believe in God—against his own will, in fact. And he hates this. This is why God, as divine subject, is for Kirillov a toxic introject. Despite all protestations, God existing or not, one must again and repeatedly affirm the existence of the double subject comprised of Kirillov and God as the two subject's within the Kirillov-God subjective double. The point is that Kirillov cannot rid himself of God, other than by suicide. And this suicide he will commit with absolute certainty. He has to. For he needs to rid himself of God. First, the atheism part. Kirillov:

> "For me there is no higher idea than the non-existence of God.... Man has done nothing but invent God in order to live without killing himself; that's the essense of world history to this point...."[327]

As seen, Kirillov has proclaimed the absolute necessity of suicide due to the unequivocal nonexistence of God. But this is far from all. For Kirillov is an unswervingly bizarre character and, as afore asseverated, this his purported atheism is—it is here contended—not the case at all. Since God is his predominant toxic introject, Kirillov desires to commit suicide not because he believes that God does not exist, but because he suspects that God might, may well, or really does exist. Or that God, because he vacillates between existence and nonexistence, has recoiled so cancerously into the subject Kirillov that Kirillov must kill this interiorized God by killing himself. Indeed, very soon after having declared the necessity of suicide on the basis of the nonexistence of God, Kirillov forthwith—and not so inadvertently—betrays a taut adherence to Christ:

> "... That is what *He* said."
> And in a fever of ecstasy he pointed to an icon of the Saviour before which a lamp was burning. Pyotr Stepanovich flew into a rage.
> "And so you still do believe in *Him* and you've lit the lamp. Isn't that an instance of 'just in case'?"
> Kirillov remained silent.
> "You know what, in my opinion, your belief is even stronger than a priest's."[328]

Kirllov may protest, but in truth he seems to truly believe in God. This God is the subject that is inexorably drawn into the persisting and persistent Kirillov-God subjective double—whose mutual resonance has long since reached a pitch the intensity of which portends his suicide. Yet, whether he believes in God or not, God is wholly within him and his atheism ultimately shows itself to be a mere modality of his desire to expunge the God inexorably and irrevocably lodged within his soul. This is the nature of Kirillov's proclaimed atheism. He believes in God so painfully that atheism is his sole mode of evacuating himself of a divine presence and inherence that torments him to the point of delirium. He cannot live with the introjected God in whom he, despite himself, believes—and must thus kill himself since God is far too indelibly embedded in his heart. Kirillov's atheism is thus a reaction to his subterranean theism. He wants—and certainly will—kill himself because he believes in God. His atheism is here the mere means of his evacuation of God—and at once of himself. In an ecstasy, he cries:

> "… I don't understand how an atheist could reach this point knowing that God does not exist and not kill himself immediately. To realize that God does not exist, and not to realize at the same time that you yourself have become God is absurd. Otherwise you'd certainly kill yourself…."[329]

The motivation for this necessary removal of God as toxic introject through the suicide of Kirillov now immediately comes to the fore. The indelible marks of God on the soul must be eradicated through suicide, for only such a suicide would be the supernal—and only fitting—enactment of self-will. Kirillov again:

> "Now I am still only God against my will, and I'm unhappy, because I'm *obliged* to proclaim self-will…. I'm dreadfully unhappy because I'm dreadfully afraid…. But I am proclaiming self-will, and I am *obliged to believe* [my emphasis] that I don't believe…. I have been looking for the attribute of my godhood for three years, and I have found it: the attribute of my godhood is Self-Will! This is the only way I can show, in its main aspect, my independence and my terrible new freedom. For it is very terrible. I am killing myself in order to show my independence and my new terrible freedom."[330]

With a pallid face and an unbearably heavy look in his eyes, he feverishly demands a pen from Verkhovensky, who thinks he is about to fall to the floor. Forthwith he proceeds to declare in writing that he has killed Shatov and that he will kill himself with a revolver that very day, not out of fear or repentence, but since he had long since intended to annihilate himself. Concluding the document with the words "*Liberté, égalité, fraternité ou la mort!*" and signing himself as "*Gentil-homme-séminariste russe et citoyen du monde civilisé!*" whilst bursting out in malicious laughter, he jumps up from the sofa, snatchs the revolvers from the window sill and runs into the next room, shutting the door tightly behind him. After ten minutes has elapsed, Verkhovensky listens for a sound at the door of the room. When he suddenly opens the door, Kirillov rushs at him, growling like an animal. Verkhovensky slamms the door shut immediately, cocks his revolver, ready to shoot Kirillov should he come out. Soon, deciding instead to pursue Kirillov, Verkhovensky kicks the door open with maximum force, revolver raised. Beholding no one in the room, he calls Kirillov's name softly, and then again slightly more loudly. No reply. Suddenly he beholds the protuberances of an utterly motionless figure standing erect between the wardrobe and the corner, with arms akimbo and head pressed tightly against the wall. Despite a burst of fury by Verkhovensky, this figure remains unmoved, as if made of stone or wax, its face unnaturally pallid, its black eyes completely fixed on some point in space. Verkhovensky scrutinizes the face from various points, illuminating it with his candle, imagining that a mocking smile has crossed its lips.[331]

Suddenly, this motionless figure of Kirillov knocks down Verkhovensky's candle and bites his finger whilst Verkhovensky strikes Kirillov's head as hard as possible with his revolver in the ensuing darkness and rushes headlong out of the house. About to exit through the entryway, Verkhovensky hears a shot, pauses for five minutes, and then hurries back up to the room. There lay the body of Kirillov, an entry wound in the right temple, the top left part of the skull shattered, and splashes of blood and brain visible, the revolver in his outstretched hand. He had finally become God.[332]

February 28, 1936 A.D. (Mishima's Shinji Takeyama and his wife Reiko)

Already in the wedding photograph of Lieutenant Shinji Takeyama and his wife Reiko one could discern that their countenances fixed with equal clarity upon the deaths which lay before them. Indeed, this subjective double's proclivity toward terminal annihilation in the black hole of stochastic death was vouchsafed and presaged with the implacability of deterministic rigidity as soon as the first night of their marriage. Sitting erect on the floor with his sword laid before him, Shinji made it very clear to his wife that a soldier's death may come to pass at any moment and that she must be unswervingly resolved to accept it. Immediately, she rose, pulled out a dagger from a nearby cabinet and, returning to her place, wordlessly laid the dagger on the mat before her in silent and irrevocable acquiescence. She too was fated to die.[333]

The couple were the recto and verso of the single fictive fusional body sustaining the obscure amorous subjective couplet that they comprised under the dull and foreboding sign of death. As Yukio Mishima narrates in his story "Patriotism," Shinji would daily think of his wife in the rest periods between training, whilst Reiko immersed herself in her husband's image all day long. He was the sun around which her entire world revolved.[334] The marks of the husband, as the subject of enunciation, engraved themselves with fateful indelibility in the inseparable wife, as the subject of the statement. These subjects would mutually recoil into one another ever more tightly as their moribund drama unfolded and drove forth to its irremissible denouement.

When the bugle sounded from the nearby Lord Privy Seal Saito's house on the morning of February 26, Shinji disappeared into the crepuscular dawn, only to return on the evening of the twenty-eighth. In the face of her departing husband, Reiko glimpsed the clear determination to die. If he failed to return, she resolved likewise to die. In his absence she begazed the great solar principle that her husband embodied, pining to be hurtled into it to her own destruction. Her body, she felt, could easily dissolve and be transformed into a mere fraction of his thought. Death hovered incessantly in her mind, and she knew that her husband's tortured thoughts and anguished feelings were cavorting her toward it.[335]

By the evening of the twenty-eighth, when Shinji returned, the resonance of the Reiko-Shinji subjective double had climbed to so soaring a pitch that its annihilation in the black of stochastic death was henceforth an irreversible and absolute certainty. Each of the lieutenant's words was firmly rooted in death. Gazing into his wife's eyes, he had a sense of hitherto unknown lucidity in the silence that rested between the two. For the first time he experienced true peace of mind. He knew that his wife clearly divined his moribund proclivities. With his gaze fixed firmly and unwaveringly on his wife's eyes, he declared he would ritually disembowel himself that very night. Unflinchingly, she immediately expressed her resolve to follow him into death, mesmerizing her co-moribund with her eyes. Overabundant joy welled up in their hearts. The thanatophilia that the lieutenant had instilled in his wife's heart at the very outset would now very soon attain concrete efflorescence in the rapidly looming abyss of their terminal demise. She felt as though she had returned to the night of her marriage. She looked upon her rapidly impending self-destruction as an opening onto a limitless expanse, a swiftly burgeoning recrudescence of the Schellingian abyss of freedom in which her organic being was to be incinerated as in a crematorium.[336]

Whilst shaving, death filled the lieutenant's body with warmth and pleasurable anticipation. He was absolutely certain that there was not the slightest trace of impurity in their resolution upon commiting double-suicide.[337] When the two moribund countenances looked upon death in each other's eyes they felt themselves "safe once more behind steel walls which none could destroy, encased in an impenetrable armour of Beauty and Truth."[338] The lieutenant shaved himself with the utmost fastidiousness. For this would be his death face. He felt the association of his radiant face and death to be suffused by a compelling elegance.[339] Already, his face had "half departed from the lieutenant's personal possession and had become the bust above a dead soldier's memorial."[340] When he closed his eyes everything was swaddled in utter blackness. He no longer seemed to himself a living, sentient being.[341] Soon he would not be.

When the lieutenant looked into his wife's face he knew that it would be the last face that he would ever see in the entire world and the last face of his wife, before both faces would dissolve in the searing void of their own terminal annihilation.[342] He "scrutinized the face minutely, with the eyes of a traveler bidding farewell to splendid vistas which he will never revisit."[343] Poised upon making love to his wife for the last time, death and the wild ecstasy of the senses seemed to conmingle and overlap, "almost as if the object of this bodily desire was death itself."[344] This peculiar miscegenation of love and death suffused his soul with a freedom the soaring and untrammeled nature of which he had never experienced before.[345] Death impregnated the single fictive fusion body supporting the two entwined obscure amorous subjects with singular erotic force. They "felt as if the still-unknown agonies of death had tempered their senses to the keenness of red-hot steel. The agonies they could not yet feel, the distant pains of death, had refined their awareness of pleasure."[346] Begazing his wife's countenance, a radiant death face conjured itself in his mind. His lips pressed against the white throat where her blade would soon strike. It reddened faintly in a premonition of blood.[347] In his armpits "a sweet and melancholy odour emanated from the growth of hair, and in the sweetness of this odor was contained, somehow, the essence of young death."[348] Gazing at his stomach, a foreboding of its imminent disembowelment transfixed her. The gentle touch of her tears upon his stomach galvanized his courage in the face of rapidly impending self-destruction. After this their last erotic fusion, the countenance of death peered down from the ceiling. Death was rapidly encroaching upon them, edging in ever the nearer. They would soon reach out and seize it themselves.[349]

With their blades, Shinji and Reiko stood before the god shelf, praying silently. They put out all the downstairs lights and ascended to the eight-mat room. The lieutenant, sitting erect, laid his sword before him. His wife sat facing him, a mat's width away. They gazed intently into each other's eyes across the mat which lay between them. The sword before the lieutenant reminded his wife of their first night, when the Shinji-Reiko subjective double commenced its headlong march toward the nullity of its own self-willed annihilation. For the lieutenant the death that he would die would simultaneously be a lonely death on the battlefield and a death died under the steady and loving gaze of his wife. The realization of this impossible union suffused him with unutterable sweetness.[350] Having every smallest moment of his death scrutinized by her eyes was for him "like being borne to death on a gentle, fragrant breeze."[351] Her face was the embodiment of all that he loved in life and for which he would very soon commit gruesome suicide. The image of her soon-to-be dead husband, upon whom she intently gazed, was for her a vision of unprecedented grandeur. The severe brows and tightly closed lips that contemplated their own self-intended death were for her the absolute summit of masculine beauty.[352]

Fixing his wife with an intense, hawkish stare and mustering the full brunt of his strength, the lieutenant buried five or six inches of naked point into the left side of his stomach with a piercing cry, as "a fearful and excruciating pain came welling up as if the ground had split open to disgorge a boiling stream of molten rock."[353] She beheld a "deathly pallor fall across his face, like an abruptly lowered curtain."[354] His eyes were those of a small animal.[355] She felt that he "had already become a man in a separate world, a man whose whole being had been resolved into pain, a prisoner in a cage of pain where no hand could reach out to him."[356] She "began to feel as if someone had raised a cruel wall of glass high between herself and her husband."[357] So tightly wrought were the two subjects of their subjective double that ever "since her marriage her husband's existence had been her own existence, and every breath of his had been a breath drawn by herself. But now, while her husband's existence in pain was a vivid reality, Reiko could find in this grief of hers no certain proof at all of her own existence."[358] In order to acquire this confirmation of her existence she would, as per the dictates of the obscure amorous subject and its incorrigible deathly reciprocity, herself become a harrowed cauldron of pain and fuse with her husband in joint annihilation on the other side of death itself.

The mat before the lieutenant was drenched crimson in splattered blood. He was stricken by fits of vomiting and cried hoarsely as his abruptly heaving stomach retched its entrails. His head drooped, his eyes were narrow slits. Soaked in blood to the knees, he was a crumpled and listless semi-being, his head drooping, a raw smell suffusing the room, as his wife gazed in lurid fascination at the tide of blood advancing in her direction. His face was that of the dead. The eyes were hollow, the skin parched, the cheeks and lips the color of dried mud. His corpse lay with its face submerged in a sea of blood, its eyes wide open, a metal point protuding from its neck. His wife bestowed a last kiss on the bloodless lips.[359]

She would very soon follow him into the abyss. She would close the terrifying gulf that her dying husband's pain had opened by rendering his pain a part of her own experience. She would enter the void in which her husband had already been consumed in a final fusion of obscure amorous subjects in death. What was heretofore merely faintly intimated to her in her husband's facial agony, she would now savor with her own flesh—and in its terminal desuetude. She plunged the dagger into her throat. Blood flooded in a searing gush into her mouth. Everything before her eyes reddened. Shinji and Reiko were dead.[360]

November 25, 1970 A.D. (Yukio Mishima)

Kimitake Hiraoka—later to assume the name Yukio Mishima—was taken away from his mother on the fifteenth day of his life by his grandmother Natsu, who guarded him jealously, fiercely, and hysterically in her darkened sickroom for twelve long years. In her insanely possessive selfishness she effectively caused him to share in her physical pain, humiliation, and despair. Knowing that any form of truculence would cause her misery, Kimitake unfailingly deferred to Natsu and concealed his anxiety from his mother and others. This performance would, however, come at an enormous emotional cost to himself.[361] With neuralgia and worsening sciatica, stomach ulcers and kidney disease, Natsu would cry and tear her hair at night, imploring him to comfort her, occasionally threatening to slit her throat. This was torture for him.[362] Mishima later averred that "everything in the room would be permeated with her grieved, monotonic moaning."[363] When "the spasms," he went on, "continued for an entire day, or at times for nights on end, a more distinct sign would appear: 'disease' would accupy the house as if it were the owner."[364] Natsu was, thus, for Kimitake his earliest and most stubbornly inextricable toxic introject. We observe here a Deleuzoguattarian passional subjective double comprising Natsu, as subject of enunciation, and Kimitake, as subject of the statement. The tautness of the recoiling of these two subjects and their mutual resonance would already at this early stage portend Mishima's ultimate, though much later, self-annihilation in the black hole of stochastic death. For even after Natsu physically died, she remained inside him as a malignant disease-carrying parasite that could only ultimately be dislodged through the simultaneous destruction of the host.

Mishima was obsessed with death, particularly his own, from a very early age. The odor of the sweat of soldiers marching, mixed with the sea breeze, would intoxicate him as he imagined the various ways in which they would die and likewise how he would die.[365] He immersed himself in literature in whose narratives a beautiful youth is stabbed to death and decapitated whilst kissing the rose gifted to him by his beloved; others in which a prince is torn to pieces by a dragon, shot full of poison and devoured by a giant spider, and flung into a pit lined with thousands of deadly knives. This all made him pine for "Death and Night and Blood."[366]

In a scene described in *Confessions of a Mask*, when Omi, a classmate in school to whom Mishima was sexually attracted, was issuing intimidating challenges to others, Mishima felt an impulse within himself that yearned for suicide, an impulse even stronger than any instinct for self-preservation.[367] Mishima's descriptions of Omi's face, with its "haughty cheekbones," resembled that of another man to whom he would be strongly sexually drawn later in life: Masakatsu Morita. In his morbid fantasy of the death of Omi, he imagined how the latter would be stripped naked and impaled with an arrow in his chest and another in his armpit in the same Saint Sebastian pose he himself would assume in a later morbid martyrdom photograph. Morita, on the other hand, was the man with whom he would eventually commit suicide in a violent blood-soaked denouement of homosexual beauty and death.[368]

The Second World War would profoundly intensify and hasten Mishima's increasingly headlong march toward the black hole of stochastic death. With American bombs raining down, Mishima and his fellow schoolmates believed that conscription and almost certain death awaited them after graduation. Mishima pined for it.[369] During the months in 1945 when he worked at an aeroplane factory, his fantasies of beauty, death, and destiny possessed him entirely. He fancied himself a genius destined for early death, as a member of beauty's *kamikaze* squad.[370] When the fiery air raids on Tokyo commenced in March 1945, they furnished him with incendiary fuel for an exaltation of death as a glorious and privileged destiny.[371] When a rumour spread that the enemy would soon make a landing, "again, even more than before, I found myself deeply immersed in a desire for death. It was in death that I had discovered my real 'life's' aim."[372] "I was hoping," he continues, "that during that month the Americans would surely land at S Bay and we would all be sent out as a student army to die to the last man …"[373]

After the war, he continued in his thanatophile proclivities, indirectly protraying himself in his *Kyoko's House*, in the character Osamu: a narcissistic youth who is only able to extract a proof of his existence in a masochistic pain that anticipates his own death—a death through which the consummate verification of his existence would finally bring itself to bear.[374] Here existence finds itself irremissibly bound wih Mishima's recurring conjunction of beauty and death: "You must realize that a man's determination to become a beautiful person is very different from the same desire in a woman; in a man it is always the desire for *death*."[375] Death, or the moment on the cusp of death, is the only reality. To live, exist, and participate in reality, is thus self-identically to die.[376] Shortly after the publication of the same book in 1959, he expressed the wish to play in a gangster film, wear a leather jacket, and die at the end.[377]

What troubled Mishima was that the post-war lull of continuous peace, with its paucity of dangers to life and limb, had vastly diminished his ability to feel that he was truly a sentient being. This lament, which was to be his leitmotif until the bitter end, he first sounded in August 1962 in an article entitled "These Seventeen Years of *Warlessness*."[378] Here he maintained that, though the three-hundred years of Tokugawa peace was tedious, during that time "the samurai class," nonetheless "maintained an artificial consciousness of peril which it seems to have employed as spiritual hygiene."[379] Then, what *Bushido* demanded was that the warrior bear the prospect of his own death ever before him, to the extent of seizing the very first opportunity to die.[380] With Mishima's own sense of the stubborn and seemingly interminable indwelling of the diseased introject that was his grandmother, death could not but be tantamount to spiritual hygiene. "My literary Wanderings," an autobiographical essay of early 1963, continues the thanatic agenda: "What remains then is the concept of death, present, momentary, instant to instant death. It seems likely that to me this is the only truly enticing, truly vivid, truly erotic concept."[381]

Having reached the point of not being able to feel alive in a world bereft of peril, he gave himself increasingly to fantasies of an agonizing death by the sword. In January 1965, he enacted this fantasy by acting the role of the lieutenant who ritually disembowels himself in the film version of his own story "Patriotism." Renamed "The Rite of Love and Death," the feature had only one sound track, the "Liebestod" from Wagner's *Tristan und Isolde*—that perfect sonic marriage of beauty and death.[382]

Later, in "The Voice of the Hero Spirits," he again laments the "smiling full-bellied peace" in which "Running blood is sullied and clogged with peace/ The gushing river of blood has dried up."[383] The text presents a scene in which young officers salute the emperor with their swords, looking up into his august face, whilst from his white horse he instructs them to die immediately. Unhesitatingly, they open their jackets, shout the emperor's praises, and drive their bloodied swords deep into their sides, feeling no pain, as the generalissimo stands on the snow dyed scarlet by their blood, saluting them as their bodies move toward decease and behold a miraculous tear on his majestic cheek as death descends upon them as the quintessence of bliss.[384]

On April 9, 1967, as a propaedeutic to patriotic death of a similar sort, he secretly enlisted in the Army Self-Defence Force, subjecting himself to forty-six days of basic training.[385] The main *raison d'etre* behind his enlistment was as an initial step toward transforming himself into a samurai. For this preparation was always meant to be a preparation of nothing short of his own eventual death.[386] "The samurai's profession is the business of death. No matter how peaceful the age in which he lives, death is the basis of all his action. The moment he fears and avoids death he is no longer a samurai."[387] This his thanatic words were merely a paraphrase of the *Hagakure*, an eighteenth century *bushido* treatise by Jocho Yamamoto, wherein the samurai is enjoined to know that "the essence of *bushido* is to die," meaning that "the samurai when faced with a choice between life and death chooses death."[388] "So long as you have chosen death," it continues, "it does not matter whether you have died 'in vain'—death cannot be to your discredit."[389] On February 26, 1968, Mishima and eleven students signed a blood oath. This fantasy, which would eventually turn into a full bore reality in less than three years, was something he nourished since his childhood.[390] It would, above all, be the consummation of the headlong thanatic vector spawned of the Natsu-Kimitake subjective double's mutual recoiling forged in the fetid cauldron of the disease-ridden room of his infancy and childhood. He would, at length, evacuate this long-introjected and eternally abiding malignancy in the manner of a samurai. The guise under which this suicide would occur would be formally presaged by his November 3, 1968, formation of a forty-strong private army of co-moribunds that would ostensibly publically commit itself to heroically dying for the emperor. This would henceforth be the Tate no Kai, or Shield Society.[391]

More voracious, deeper, and subterranean reasons for his headlong march toward the black hole of his ultimate terminal self-immolation would, however, be evinced in his subsequent work *Sun and Steel*. Here again returns the older and persistent leitmotif of his only feeling alive and real in the desired moment of death, or his inexorable movement thereunto.[392] Again the constant conjunction of beauty and death recrudesces. The homosexual beauty of his long-cultivated musculature would at once be the preparation for his death at the behest of the same beauty and homosexual muscular strength. His assidiously cultivated musculature, though a movement toward and terminus ad quem of a longed-for beauty, would only reach its properly preordained *telos* in a death in which this same homosexual musculature would be the victim of itself own death. The haitus twixt his seeing of his muscular beauty and the proof of its existence through the sine qua non of death had to eventually close in upon itself and be rendered null: The subtle contradiction between self-awareness and existence began to trouble me. I reasoned that if one wants to identify seeing and existing, the nature of self-awareness should be made as centripetal as possible. If only one can direct the eye of self-awareness so intently toward the interior and the self that self-awareness forgets the outer forms of existence, then one can 'exist' ... Let us picture a single, healthy apple ... The apple certainly exists, but to the core this existence as yet seems inadequate; if words cannot endorse it, then the only way to endorse it is with the eyes. Indeed, for the core the only sure mode of existence is to exist and to see at the same time. There is only one method of solving this contradiction. It is for a knife to be plunged deep into the apple so that it is split open and the core is exposed to light.[393] The proof of existence, thus, existed only in the annihilation of the same existence craving such proof—at the very zenith of existence's homosexual beauty. This tormenting lacuna between beauty and death, presaged by beauty, could only be closed through immersion in a revolutionary and moribund group.

This group would be the Tate no Kai that Mishima had already forged. It is this group that would faciliate the antecedant elimination of individuality requisite for his ultimate self-immolation.[394] "The tragic quality of the group was also necessary, the quality that constantly raised the group out of the abandon and torpor into which it was prone to lapse, leading it to an ever-mounting shared suffering and so to death which was the ultimate suffering."[395] The pining proclivity toward death that had beckoned him since youth was now, at last, opened to consummation by the group. It was through the group that he was enabled to embark upon the final fatal step. The group was his access to death.[396]

Through immersion in the group, from the middle of 1968 to the end of 1969, the mode of death that he sought was that of *kirijini*—that by which the samurai, fighting in defense of his lord against forces that he knows will definitely overwhelm him, dies with sword in hand. In Mishima's fantasy, he pined to be killed defending the emperor in a conflagration against the Left.[397] Rehearsing what would be his impending death, he played the terrorist Shinbei Tanaka in a film in which, after effecting much carnage with his clan, he inexplicably commits *seppuku*.[398]

By this time, he had found his principle co-moribund: a Waseda University student named Masakatsu Morita, who pined to die for the emperor and incessantly spoke of his own death with relish. His desire for death was hardly less fervid than that of Mishima. In their last years they were together frequently, like two fiancées. Mishima would intimate to friends that he had pledged to kill himself for the emperor and that Morita pledged to die for his sake, adding that it was Morita who would kill him. Rumor had it that Mishima and Morita were embarked upon a homosexual lovers' suicide, Mishima having found the lover of his life with whom he would die a violent warrior's death, in a manner nowise recalcitrant to the orthodox *bushido* code. Certainly, he was strongly sexually attracted to Morita—an attraction he pined to consummate by having Morita sever his head.[399]

In a series of final artistic turns, Mishima was slowly winding down to the final tragic denouement in which the taut internal resonsance of the subjective double comprising himself and his grandmother, and the more recently forged double formed of himself and the thanatophilous Morita, would find terminal dissolution in the black hole of stochastic death. About two months prior to his self-annihilation, Mishima posed in a series of photographs called "Death of a Man." They included images of him drowning in mud, with a hatchet in his brain, crushed beneath the wheels of a cement truck, and as Saint Sebastian with arms roped above his head and arrows lodged into his armpit and flank. In another photograph, he is naked on the floor with a short sword buried in his abdomen, as a man stands behind him ready to behead him with a long sword. These images were included as part of a photographic retrospective of himself that he mounted just over three weeks prior to his death. The exhibit represented his entire life as emptying into the final work of his own self-destruction.[400] "The murderer knows," he had written at the age of eighteen, referring to himself, "that only by being murdered can he be completed, realized."[401]

Less than two weeks prior to his death, he met with other members of the Tate no Kai, to approve the manifesto he had drafted.[402] Its concluding few lines read: "Rise with us and, for righteousness and honour, die with us."[403] "We will restore Japan to her true form," it continues later, "and in the restoration, die."[404] "Are none of you willing to die by hurling yourselves against the constitution that has torn the bones and heart from that which we love? If you are there, let us stand and die together."[405]

Five members of the Tate no Kai had resolved to die: Mishima, Morita, Chibi-Koga, Furu-Koga, and Ogawa. Two days prior to their death, on November 23 and 24, they convened in a room in the Palace Hotel, holding eight complete rehearsals of the death they would very soon deal to themselves. They wrote the customary thirty-one syllable farewell *tanka* poem, which Japanese soldiers wrote when going out to meet their deaths during the Second World War.[406]

The quintet arrived at the Ichigaya headquarters of the Army Self-Defense Force at 10:50 am of November 25, 1970, proceeding to tie up Commandant Masuda in his second-floor office. By twelve noon Mishima began to address a multitude of eight-hundred suddenly-assembled soldiers from a high balcony,

urging them to rise up against the Constitution and die with him for the emperor. Jeered and ridiculed as they were, he and Morita exclaimed three times "*Tennō Heika banzi*" (Long live his imperial Majesty!), whereupon they abruptly withdrew, Mishima immediately unbuttoning his uniform jacket as he stepped back in from the balcony. Sitting on the floor facing the balcony with Morita standing with sword raised above his head, he grunted and drove the blade of his short sword into his left side, drawing it slowly across his abdomen whilst slumping forward. After two failed blows by Morita, Furu-Koga severed Mishima's head with a third stroke. Morita drove the short sword into his abdomen. Furu-Koga beheaded him with a single stroke.[407]

November 18, 1978 A.D. (Rev. Jim Jones and Peoples Temple)

From as early as his 1973 Los Angeles sermon, Reverend Jim Jones declared himself to be God. He announced himself a living alternative to the creator or Sky God, claiming for himself all the powerful acts falsely attributed to the latter. Unlike this "Buzzard God," Jones declared himself a visible and embodied Deity, the "actual, personal, present-tense God in a body." He was a deliverer who—like the gnostic savior emanating from the fullness of divine light—would make a heaven out of the hell created by the Sky God. He was an embodied God in the sense of being the incarnation and living manifestation of the real God. And this real God was Principle, Love, and Socialism. For Divine Principle was love itself, and love was Socialism. Jones was thus the "God personification of Socialism." He was likewise Jesus Christ, both as God and as man, in the sense of being the living embodiment of socialist revolution for justice. Though manifestly white, in a 1972 sermon in Redwood Valley, he claimed to have been miraculously conceived and born black when his mother, desirous of bearing a savior to liberate humanity, had contacted a higher vibration of cosmic black consciousness. His other incarnations had been Moses and Lenin.[408] "You can call me an egomaniac, megalomaniac, or whatever you wish, with a messianic complex," he said. "I don't have a complex, honey, I happen to know I'm the Messiah."[409] "Because I am freedom. I am peace. I am justice.... I AM GOD!!!!"[410] But if his listeners did not want to call him God, then they could simply call him "Daddy-O".[411] His listeners did, however, believe that Jones was God. "I know that Pastor Jim Jones is God Almighty himself," a member of a crowd of over one-thousand once cried.[412]

What one thus espies here is a subjective double composed of the terms Jim Jones and God, with God of course being merely the quixotic and self-aggrandizing auto-deification of Jones himself—by himself.

Entangled with the foregoing, there is, however, another subjective double involved in the vicissitudes of the Jones saga. For Reverend Jones, so as to accomplish his divine socialist revolution, proposed to make the votaries of his Peoples Temple church themselves gods. Every one of these socialists was an embodiment, reproduction, son, or daughter of Divine Socialism.[413] "I am personifying God," Jones revealed, "so that you may know your rightful inheritance.... I have come to show you that ye are gods because I am God."[414] His faithful were to effect their own deification either by reproducing an image of Jim Jones in themselves, or by attuning themselves to Jim Jones, or by a process of imitation in which they were to think like Jim Jones, talk like Jim Jones, and act like Jim Jones in order to absorb his superhuman power.[415] "I'm Daddy God," he said, "and you're my baby gods, and I want you to be just like me."[416] They were to become gods, Principle, by being crucified with Christ, and with the revolution.[417] Inextricable from this auto-deification would be the elimination on the part of each Peoples Temple adherent of his desire for property.[418] "That's what I have come to show you," Jones intimated to his faithful in one sermon, "that you can be God, that all men can be God, if they eliminate the urge to gain, accumulate," if they obliterate their, "capitalizing, aggrandizing spirits."[419]

This other implicated and imbricated subjective double thus brought into propinquity the deified Jones, on the one hand, and the collective Peoples Temple or "Jones-ite" subject, on the other. In that the latter would by its very design be molded in the image of Jim Jones, the divine Jones would serve as subject of the enunciation whilst the Jonesites were the subject of the statement—with the one recoiling into the other through the indelibility of Rev. Jones' enduring marks of polemical and demagogic enunciation.

As pertains to the obscure amorous subject, this collective Jones-Jonesite subjective double would before long attain a pitch of resonance sufficient to forebode its own terminal destruction in the black hole of stochastic death. Already as far back as 1972, Jones sermonized that the role of the Peoples Temple in the drama of human history would very likely play itself out on the gallows of martyrdom.[420] "Now, you might not like the kind of history we make," he warned. "We may be swingin' through the sky on a rope, but we're gonna make history."[421] In the same year, in a sermon in Redwood Valley, referring to John Brown's Christ-like sacrifice in protest of slavery, Jones polemicized to boisterous applause that people's faith in money, racism, and America could be shaked solely "if we will be willing to go to the gallows for what we believe."[422]

In like manner, the resonance of the subjective double comprised of Jones and the deified Jones had long since reached a plateau of intensity sufficing to portend his own march toward terminal annihilation. "Maybe one day I'll have to lay down my life," he sermonized in California, "but I'm not afraid to lay down my life for my bethren."[423] In a January 1973 sermon in San Francisco, reflecting upon the possibility of triumphant social struggle through sacrificial death, Jones assured his votaries that "If I could die on a cross and save all of you people from some real or imaginary sin, I would say, 'Get me the cross, and put it in the ground and nail my hands as fast as you can,' cause I'd be glad to save you with one act."[424] Later in Jonestown, Guyana, he reminded his assembled community that he had desired death every day of his life, ever since as a child he had seen a dog die.[425] "I'd have found the first fuckin' bomb, you heard me five years ago, I'd have found out where it was going to be and parked my ass under it."[426] He foresaw himself on a podium going down in a fusillade of hot lead.[427]

Such a death, Jim Jones termed revolutionary death. In his California sermons, Jones announced that, in the face of insuperable racial and capitalist oppression, such a death would be more than desirable. "If I can't do anything about it," he shouted, "we're gonna take 'em on full-square and die in the process!"[428] In fact, only a death suffered in struggle against and at the hands of the forces of oppression was regarded as fully human. "Dying comes to all," Jones maintained, "but that dying cannot be noble unless it's a revolutionary death."[429]

Yet, revolutionary death would not merely entail the willingness of Jim Jones, or any other single Peoples Temple sectarian, to die for the sake of the whole community. It likewise, and ever more, came to entail the propensity of the entire collective Jonesite subject to suffer death for any single votary.[430] Thus, even ostensibly trifling exogenous perturbations to the fictive fusional body of the collective Jones-Jonesite subjective double would potentially precipitate the latter into a headlong march toward its own self-destruction. For, as Jones declared in the early 1970s, "if they try to stop one of us, they'll have to kill all of us."[431] "If they wouldn't take one of our people," he promised, "we'd just draw a circle around us and say, 'You'll just have to kill us all!'"[432]

Yet, as time passed and the sheer pitch of resonance of the Jones-Jonesite collective subjective double steadily rose, the propensity for revolutionary communal death in the face of external oppression would gradually give way to an ever more fierce proclivity to revolutionary suicide. Collective suicide was first broached within the Peoples Temple inner circle during 1973, and first acted out on January 1, 1976, in the form of a "suicide drill" wherein thirty inner circle members were given a glass of wine which they later were informed was poisoned and would kill them within an hour. Though the ritual was a mere farce, most members clearly exhibited a willingness to die.[433] In fact, no one rose to question the proceedings, to challenge its logic, or to call for an antidote. Some even considered it a beautiful sight to behold.[434]

Termed "white nights" by Jim Jones, collective suicide drills would become a permanent feature of the Peoples Temple, increasing in frequency and soon coming to encompass the entire community when it moved to Jonestown, Guyana. They functioned as a centripetal force serving to consolidate ever the more tautly the obduracy and purity of the fictive fusional body of the collective obscure Jonesite subject in the face of external defiling influences.[435] These suicide rituals would start during the day and last well into the night. Soon they were being called about every two weeks.[436]

One of the main sources of defilement was the potential fragmentation of the collective Jonesite subject felt in the face of defection. The initial collective suicide rehearsal of 1973 took place in the face of a defection by eight votaries, whilst subsequent white nights were enacted in response to the activities of a group of dissidents called the Concerned Relatives.[437] The latter, Jones described as the "contrary relatives and cantankerous sons of bitches that we call our blood kin."[438] In September 1977, with the Concerned Relatives fight for the custody of one John Victor Stoen coming to a head, preparations for collective suicide were made when all Jonestown babies were given sleeping pills and plans were made to pack all Jonesites into a warehouse that was to be burnt down.[439]

When the Concerned Revatives petitioned Congress, alleging human rights violations at Jonestown, on April 11, 1978, Jones called a white night the very following day.[440] "The only fuck I want," Jones declared before the assembled community, "is the orgasm of the great fuck in the grave."[441] Averring that "this could be your night of death," he called up previous white night participants to publically express their willingness to die: "How do you feel about it, you might die tonight?"[442] They were willing.[443]

Another white night was called when inner leadership member, Deborah Blakey, defected on May 13, 1978. On this day, Jones stated he was so distraught by the betrayal that he had in fact died. The collective Jonesite subject primed itself for a plunge into the black hole of stochastic death. But it was not yet their time to go.[444]

However, when Jones learned in earlier November 1978 of an impending visit to Jonestown by a congressional delegation led by Congressman Leo Ryan, the momentum propelling the collective Jonesite subject into the black hole of its own self-willed annihilation was now inexorable.[445] After fourteen community members had defected with the departing delegation on November 17, five of which were killed and nine wounded in a retaliatory assault by the Jonestown "Red Brigades," the final white night was called. At this point, the perturbation wrought upon the single fictive fusional Jonesite body could only be assuaged by a single act of self-immolatory redemptive violence.[446] The Jonestown community would now only exist as a single fusional revolutionary body in death.[447] The Eschaton, the much anticipated Apocalypse, the final culmination of cosmic time in the dissolution of timelessness, was nigh.[448] It was to be attained through a communion with death itself.[449]

Hysterical testimonials thanking Reverend Jones for giving the community life, love and death rang out as Jones himself moved among and expressed his love for a community convulsing, gagging, vomiting, bleeding from the nostrils, foaming from the mouth and wailing in contortions of pain from potassium cyanide-laced drink. Well under way by 6 pm on November 18, the final holocaust had devoured over nine-hundred lives by 11 pm that same evening, Jim Jones lying dead with a gaping exit wound in the temple, his brains and skull blown away, everyone save a miniscule remnant dead.[450]

1980–1981 A.D. (Khomeini's "Basiji" or child soldiers)

In Iran, a particularly vehement—and indeed well-nigh maccabre—emphasis is placed on the mourning of Imam Husayn, felled with his followers at Kerbala in 680 by the infinitely superior forces of the incumbent caliph Yazīd. Husayn's martyrdom is commemorated on Ashura, the tenth day of the month of Muharram, during which passion plays known as the Taazieh re-enact Husayn's death whilst in massive street processions women in the ubiquitous black chador wail, men beat themselves bloody with chains, and even the young flail themselves and fall prostrate in grief.[451]

During Iran's war with Iraq, Ayatollah Ruhollah Khomeini, then ruler of Iran until his death in 1989, assidiously encouraged such events. In a speech to fellow Shi'a clergy of June 20, 1982, Khomeini declared: "These mourning sessions have developed young men and youths who voluntarily go to the war fronts seeking martyrdom and feel unhappy if they don't achieve it. These Ashura mourning gatherings develop such mothers who urge their sons to go to the war fronts and if they do not return, the mothers wish they had more sons to send or say we have other sons to send to the war fronts."[452]

Such fervent communal religious practices and Khomeini's inexorable exortations served to engineer a gargantuan collective Shi'a subject, wrought from entire swathes of the Iranian populace, whose sheer pitch of resonance with God-as-subject posthaste soared to such an intensity that tens of thousands of young volunteers, the *Basiji*, found themselves pining with rabid fury for terminal immolation in the black hole of stochastic death. One particular Iranian youth who had been martyred by throwing himself under an Iraqi tank at Khorramshar in November '80 whilst detonating a grenade was the skinny, innocent-looking Mohammad Hosein Fahmideh—lionized by Khomeini as "our leader."[453] Khomeini: "Our leader is that twelve-year-old child who threw himself with his little heart against the enemy. He is worth more than a hundred pens and a hundred tongues."[454]

From the very outset of the war the Khomeini regime incessantly and vehemently cajoled the populace into voluntarily dying whilst fighting the eternal Sunni enemy and rank infidel Saddam Hussein. Khomeini: "Can anyone who believes in the world beyond be afraid? We must thank God if He confers on us the honour of dying in the Holy Battle. Let us thrust our way into the ranks of the martyrs in our hordes … if we have been afraid, this means that we don't believe in the world beyond."[455] Khomeini's indefatiguably reiterated mantra was: "Every land is Karbala; every month is Muharram; every day is Ashura."[456]

This rabid cajolery to martyrdom by Khomeini garnered mass response from teeming hordes of young Iranians since, for them, other modes of self-realization were stymied, well-nigh entirely cast out of existence. As Khosrokhavar maintains, violent self-willed death is the upshot of a decision made in the face of the impossibility of self-realization by other means. When the desire for self-assertion meets with its own impossibility, self-destruction is presaged. Self-assertion and resignation come to be increasingly conflated, inducing an inexorable aspiration toward self-realization and self-destruction in one unitary, coterminous, fell swoop. This latter confluence implies that the sole mode of individuation of the subject thus ensnared is in death itself. And this death is sought with the utmost urgency.[457] Through martyrdom, the subject's incapacity to assert himself in life transmogrifies itself into a formidable proclivity to self-assertion in death—to the extent that death itself becomes his sole and preponderant mode of individuation.[458]

Khosrokhavar, notwithstanding, distinguishes between two different kinds of individuation through death in martyrdom. The first, an optimistic kind of individuation, comes to pass when the subject verily risks death but nonetheless retains a positive self-image irrespective of happenstance. Though possible death is courted with exemplary aggression, such a subject does not unequivocally intend to die as such. Death is nowise the primary goal. Secondly, however, as a dark counterpart to the foregoing, there exists a deeply pessimistic kind of individuation via death. Here the subject is no longer concerned with life at all. He wants to die.[459]

Though derived from the first variant, this form of martyrdom remains profoundly distinct both anthropologically and in its affects. The movement involved is one from martyrdom to what Khosrokhavar terms "martyropathy." In the former type of individuation by martyrdom, the subject hazards life and limb so as to ultimately transfigure life itself. In the latter, on the other hand, life is a priori occluded and the overriding ambition is simply to die. It is no longer a case of sacrificing life for the sake of an ideal higher than life by which life may be elevated beyond its hitherto preponderant baseness. Instead, the pursuit of a transfigured life via the detour of possible death in the teeth of a frustrated ideal yields before an unequivocally deathly logic. Conventional martyrdom is inverted and all utopian idealism evaporates in the face of a rabid, univocally dystopian, and apocalyptic desire to take leave of existence itself. Heretofore a merely contingent risk to be taken, death now becomes a burning desire, an unquenchable thirst. The desire for a life redeemed by possible death mutates into the desire for death in and of itself. Death is now the only aim. What ensues is a fascination, a luxuriation, and a quest for happiness in and solely through death. Despair turns life into death and death henceforth comes to be apprehended as the ultimate form of plenitude. Death becomes an obsession, a voluptuous fascination, the only assuagement, the sole and greatest good, the only object of desire. Outside of death, there is nothing.[460]

And as martyropathy turns life into death, consumption becomes self-destruction. The voluptuous moribund literally consumes himself. His suicidal martyrdom becomes a form of homophagy, an ingestion of his own flesh.[461]

In this autophagious necrophilia the martyropath devours himself to, moreover, escape from the sullying effects of life itself. Continued subsistence within life is seen as a form of impurity. All attachments to this world offend his morbid convictions. He must purify himself of this nauseating gunk, and his urgent ablution can only be attained through death.[462] His blood must leave his body and be spilled, his ego must leave the world. Both must separate from and consummate their rupture with the impure confusion of life.[463]

Post-Shah Iran rendered the pursuit of the absolute a morbid neo-mysticism in which persistence within being became a matter of insanity. The sole site of promethean self-realization, the purgatorial creation of the self, was the infinitely dark void of death. Consumption was incarcerated in a deadly form of the sacred, in a neo-ascetic auto-necrophilia.[464]

Self-annihilation was inevitable for the Iranian martyropath. His body was contaminated. To be purified, it had to rid itself of its blood. And just as much as his blood was a burden to his body, his sins were a burden to his soul. Much as his body had to be drained of its blood, to redeem itself of its sins, his soul had to divorce itself from his body. This is how the martyropath sought his salvation. By an evacuation of its blood, the body cast off its vile burden and became pure since it was lighter. This, in turn, enabled the soul to disemburden itself from the indwelling sin with which it was weighed down, and thus to liberate itself in death.[465]

The youth enlisting under the banner of Khomeini's *Basiji* was a moribund of this stripe. His desire to live was null. His aim, above all, was to die.[466] To inveigle this most willing of moribunds, recruiters trawled schools countrywide, instilling young children with a lust for their own deaths. Robert Fisk recounts beholding these recruited youths pining for terminal immolation in the Godhead at Friday prayers at the University of Tehran. He describes them as "these miniature soldiers—every bit as young and carefree of life and death as those I had met in the trenches outside Dezful. The inscription on the red bands round the little boys' heads was quite uncompromising. "Yes, Khomeini, we are ready," it said. And the would-be martyrs, identically dressed in yellow jogging suits, banged their small fists against their chests with all the other worshippers, in time to the chants. This cerebral drumbeat—at least ten-thousand hands clapping bodies every four seconds—pulsed out across the nation, as it did every Friday over the airwaves of Iranian radio and television.[467]

Christoph Reuter tells of Lebanese journalist Muhammad Salam, witness to an attack launched by one of Khomeini's suicide battalions. It occured on the Iraq-Iran front at the start of '84 near to the village of al-Usair, a modest distance from the Tigris Bridge. Salam beheld an Iraqi marksman behind a heavy machine gun, his body securely wedged in, his hand on the trigger. Reuter: "And then, Salam says, they arrived. Or, rather, you could hear them first: a high, buzzing sound, as if a swarm of locusts were approaching. The sound swelled: thousands of human mouths coming closer and closer, all of them roaring: 'Ya Karbala! Ya Hussein! Ya Khomeini!' as they came. As a human wave, they emerged from the trenches and dugouts, from behind ramparts and hillocks; thousands, tens of thousands of them, coming closer and closer. And almost all of them were children, youths, some of them holding Kalashnikovs with difficulty, others just with clenched fists. Every now and then, you could see an older man among them, egging them on."[468]

Meanwhile, the "marksman with the machine gun started shooting. And he shot, and shot. And he simply didn't stop: 'And he shot these children, in the same way you'd shoot at a row of empty bottles; he just mowed them down—and these were nothing but children. And they kept running; they were climbing over the dead, jumping over them, falling over them; and the man with the machine gun just kept shooting. What's more, from above, helicopters were bombing the Iranians, who were just running as if nothing could stop them'."[469]

Salam reached the front of another such suicide offensive further to the south on the marshy terrain at the edge of the Shatt al-Arab. "Corpses had lain everywhere: on firm ground, floating in the still waters of the bomb craters, and half-covered by mud, as if those following after them had trampled over their bodies as they ran. Many of them, their heads shaven, wore red headbands or scarves, and every one of them had a key around his neck—this, so they had been told, would open the gate to Paradise once they had died a martyr's death."[470]

Despite the scorching heat, the unforgiving sands, the omnipresence of malaria emanating from the dank marshes, and the abject paucity of water and blood supplies for the wounded, the atmosphere at the battle fronts was suffused with a well-nigh other-wordly piety, selflessness, and—above all—lust on the part of the youths for their own death. To the extent that, according to the words of the few survivors, those who died claimed not only to know in advance that they would indeed die but also experienced dreams the night prior to their deaths revealing the place and time at which they would die.[471]

For indeed, those "who die as martyrs are totally convinced that nothing is more useful to the Mahdi's return, to his dawning empire as a whole, than their sacrificial deaths that clear the way for it. From the blood of the martyrs arises salvation, the tree of true Islam, not from the depths of the enemy."[472]

This dawning empire of the Mahdi, however, posthaste turned out to be an empire of the dead. Fisk observes that the cult of mass martyrdom evinced an entire nation masochistically pining in headlong fashion for its own dissolution in the black hole of stochastic death. The entire collective Iranian subject had become a nation "in which women were to be dressed in perpetual mourning, in which death was an achievement, in which children could reach their most heroic attainment only in self-sacrifice," to the extent that "the country was neutering itself, moving into a black experience that found its spiritual parallel in the mass slaughter of Cambodia rather than on the ancient battlefield of Kerbala."[473]

As martyrdom becomes ever the more ubiquitous, Khosrokhavar avers, it forms a veritable ghostly community in death, or necro-community, in which the incorrigible yearning for death burgeons. The ego comes to flourish in a plurality of mortiferous desires whose only consummation is in death. Through a malignant form of contagion, the sheer number of auto-necrophilous moribunds pining for and inevitably meeting their own sacred annihilation escalates exponentially and inexorably.[474] A coterie of individual subjects hellbent on their own self-destruction give way to a rapidly metastatizing collective martyropathological subject marching in million-fold mass towards the black hole of stochastic death.

Visiting the cemetery of Imam Zadeh Ali Akbar on the cold slopes of the Alborz mountains at Chasar, Fisk saw that "bulldozers had dug deep into the icy graveyard and there was now fresh ground—two football pitches in length—for the next crop of martyrs."[475] The thin, dark-faced cemetery keeper maintained that, though the graves of ordinary people were destroyed after thirty years, the bodies of the martyrs would lie in the ground for more than a thousand years. Yet, the cemetery at Chasar harbored merely the war dead of one small suburb of north Tehran. Across the entire country, the arrayed corpses increased to three-quarters of a million, or perhaps far more. In the Behesh-i-Zahra cemetery outside Tehran, the martyrs lay in their tens of thousands.[476]

April 19, 1993 A.D. (David Koresh and the Branch Davidians)

While once praying in a field and beholding the sky, Vernon Howell found himself suddenly enshrouded and encompased by a presence that zeroed in on him, causing him to tremble with mortal fear. He felt that he was being watched from every angle, confronted by a being and with no place to run, at once terrified and awed. A voice overruled all his perception banks, emblazoning a perfect picture in his mind, and reviewing to him all the liminal experiences of his entire life. The voice complained that it had loved him for nineteen years and that for nineteen years he had turned his back on it and rejected it.[477]

On another occasion, also whilst praying, he felt as if he were being thrust up an elevator shaft, after the ascent seeing two gigantic walls, one inscribed with the word "law," the other with the word "prophecy." Hereupon he saw God himself, who had in one hand a book, while holding out the other to Howell, who reached forward for it. Though not described further, Newport avers that Howell considered this his "anointing" as a Christ.[478]

We espy here a subjective double comprising Howell and God, with God as the subject of enunciation and Howell as the subject of the statement, and the mutual recoiling of the two as the indelible marks of enunciation of the former upon the latter.

The resonance of this subjective double was soon to rise to a still further pitch. For Howell further claimed in a January 28, 1984, letter to George Roden that he was the angel of Revelation 14.17, a verse which reads, "And another angel came out of the temple which is in heaven, he also having a sharp sickle." In Branch Davidian theology, this figure will gather the one-hundred and forty-four-thousand in the end times and lead them to Mount Zion. Later on, during a 1985 trip to Israel, after being made privy to the secrets of the seven seals by seven angelic beings on the same mountain, Howell was also claiming to be the angel of Revelation 10.7, of whom it is written that "when he shall begin to sound, the mystery of God should be finished, as he hath declared to his servants the prophets." As such, Howell was the one chosen by God to lead the remnant people and to reveal to the world the secrets of the end times.[479]

This same epiphany also made known to him that he was the antitypical David—he who had come to rule the remnant in the end-time.[480] He would furthermore claim to be the very appearance of Christ as intimated by the mainstream Seventh-Day Adventist sanctuary doctrine, according to which Christ will come to the earth prior to his second coming proper to chastise the unfaithful shepherds in the church. He was also Christ in the sense of being the figure in Psalm 2 who would take off the "bands" that seal the book of Revelation.[481]

This internal resonance of the subjective double Howell-God would, however, soon forebode its own eventual terminal annihilation in the black hole of stochastic death. For even in a relatively early tape, "The Bird" (August 1987), Howell unequivocally intimated that he would be put to death, in accordance with the prophecy of Psalm 2, by the King of the North for his future polygamous activities and his claim to be the Son of God.[482] At another time, referring to how Moses struck the rock twice in Numbers 20.11 and to Paul's assertion in 1 Corinthians 10.4 that the rock struck by Moses was Christ, he averred that Christ would be killed twice, once by the Jews and once by the Gentiles. Since Vernon Howell (by now rechristened David Koresh) was Christ, his being slain by the Gentiles was inevitable.[483]

The slaying of Koresh would occur in the end times wherein the present age would pass by means of a violent rite of passage. As regards this final denouement, Seventh-Day Adventist, Davidian, and Branch Davidian traditions view America itself as the "lamb-like" beast of Revelation 13 which nonetheless "speaks like a dragon" and will seek to destroy God's remnant and enforce false worship—a drama destined to play itself out on American soil (Rev. 13.11).[484]

Thus, when on a cold Texan Sunday morning on February 28, 1993, two cattle trailers thundered up to the Branch Davidian compound on Mount Carmel, the awaited beast had finally arrived. And the eighty heavily armed, black-uniformed ATF officers pouring out of the backs of these trucks were none other than the locusts erupting from the bottomless pit of Revelation 9. The Beast had come to make war on the Saints (Rev. 13.7).[485] This was the event eagerly awaited by the inhabitants, for even on the prior day the community was already bracing itself for the coming of "the end." This end, Thibodeau believed, would very likely involve individual and communal death.[486]

In fact, even before this and as early as the second half of 1992, ominous portents suggestive of fate driving Koresh's Branch Davidian community to the cataclysm foretold in Revelation were beginning to suffuse the air around Mount Carmel. Koresh even intimated that the compound would at length be surrounded by tanks.[487] That the prophecies of Revelation would concretely involve the annihilation of the Waco community he foresaw even after the Gulf War. "And fear not them which kill the body, but are not able to kill the soul," he said with a stark look in his eyes. Those last months prior to the collective self-immolation of the community, he spoke of "quickening": the bringing of the soul into harmony with the divine.[488]

Indeed, on the evening of that very February day when the tanks rolled in, Koresh declared that Psalm 2 was being fulfilled, that the leagues of Satan had come to break his bands asunder. He reiterated Moses striking the rock twice as a sign of Christ having to die twice, thereupon ominously intimating that people would die, that the fifth seal had been entered into and that the souls of the faithful would have to go under the altar (Rev. 6.9–11), that the sixth seal and its darkening of the sun, moon, and stars as prophecied in Joel 2 was nigh, and that he was going back home to God. He expected to die.[489] And his death would come, as long since he had prophecied, at the hands of the King of the North—or the United States.[490]

As the ATF siege wore on, with the resonance of the Koresh-Branch Davidians collective subjective double, and that comprising the Branch Davidians and God, rising to feverish pitch—this latter now marched all the more inexorably toward the black hole of its own self-willed destruction. For, so as to fulfil their divinely ordained destiny, they began to speak all the more about "going back en masse" in an act of mass suicide.[491] They had gone into the siege with the expectation of communal death.[492] From its first day, or not much thereafter, they had resolved to bring on the Apocalypse.[493]

And their apocalyptic role, as the remnant of God, was to pass through a cleansing fire—a fire of their own making and prepared long in advance.[494] For they knew that they had been long since divinely primed for an infernal death.[495] This death was an irrevovable part of their eschatological scheme.[496]

Indeed, from the very beginning of the Davidian movement forerunning the Branch Davidians, founder Victor Houteff was adamant that the faithful would have to inexorably pass through a cleansing eschatological fire.[497] At the Eschaton—prior to the establishment of the kingdom and the onset of the millenium—the remnant of the Church of God, or the Seventh-day Adventists, would themselves have to be cleansed in a violent this-worldly rite of passage. For Houteff, this is the time when the Lord would come "with fire" to rebuke "with flames," and "by fire" to plead with all flesh. Relevant is Matthew 3.11–12 where John the Baptist announces the coming of one who "shall baptize you with the Holy Ghost, and with fire," and who "will thoroughly purge his floor, and gather his wheat into the garner," and "burn up the chaff with unquenchable fire." For Houteff, this prophecy refers to the Christian rite of baptism with water as the type, or prefiguration, the anti-type of which was to be an end-time baptism with fire wherein God's faithful would be refined as silver in a furnace.[498] "For, behold, the Lord will come with fire, and with his chariots like a whirlwind, to render his anger with fury, and his rebuke with flames of fire. For fire and by his sword will the Lord plead with all flesh: and the slain of the Lord shall be many" (Isa. 66.15–16).

The Branch Davidians drew strongly on this apocalyptic imagery, Lois Roden adding that the baptism by eschatological fire would visit itself specifically on Mount Carmel. Steve Schneider and Koresh, brutally uping the ante, insisted that the people of God would literally die through this cleansing fire—not in a spiritual death but in a real corporeal *earthly* death. To this end, Koresh openly courted a confrontation with the authorities such they he would die and such that his community would be incinerated.[499] The ever more searing pitch of resonance of the collective Branch Davidian subject with that of God would inexorably resolve itself through a terminal *auto da fé* whereby, as Newport has it, in "that second or two before death there would be a moment that would give the opportunity for life-giving faith. In the few seconds between knowing for sure that death was about to embrace the believer and that death actually coming there would be the opportunity for absolute pure faith. As the believer looks down the barrel of the gun, sees the tank tracks, or feels the heat of the flames upon the skin, he or she can now rely totally on God and on God alone."[500]

The apocalypse arrived in the form of a pall of black smoke shot through with orange and yellow flames, as methylene chloride dust, propane and methyl chloride vapors, mingled with kerosene, acetone and ethanol, to ignite an inferno approaching cremation temperatures.[501] In the final holocaust, children and parents coughed, choked, screamed and vomited like souls in the lower cirles of Hell as they suffocated while buried alive, their bones crushed, their bodies incinerated, in scenes reminescent of the end of the world.[502]

July 1, 1993 A.D. ("The Giants of al-Qassam" of the first Intifada)

In the grainy video footage that comprises the film *The Giants of al-Qassam—imlaq*, one may behold three living martyrs uttering what they believe to be the last words that they will ever pronounce. The first *shadid* (martyr), Mahir "Hamza" Abu-Surur, on occasion interrupted by the ill-rehearsed interjections of his dispatcher Muhammad al-Rushdi, speaks thus:[503]

> It is the day of encounter. It is tomorrow. Thursday, insha'Allah, is the day of leaving … leaving this world….
>
> It is the day of meeting the Lord of the Worlds and approaching and bearing witness to the Messenger….
>
> To all of these, we will, insha'Allah, present our spirits and make our blood cheap for the sake of Allah….
>
> We expect that our blood will flow cheaply in the service of our homeland, with the hand of Allah the one and only, insha'Allah….
>
> Of course, the greatest expectation is martyrdom, and this is what we have been aiming for since we were knee high to a grasshopper.[504]

Observing the manner in which Abu-Surur "fondles his words, caresses them in his mouth before releasing them," Oliver and Steinberg are moved to infer herein "a hedonic relation to language and death—language in the face of death."[505] For the latter, he epitomizes the so-called "happy death" touted in the innumerable vituperations of Hamas during the time of the first intifada—boisterously auto-necrophilous slogans such as "I will die smiling in order that my religion live," "Death has become sweet, become sweet/Revolution against the strangers," and "O Muslim, say 'Allahu akbar!' with joy, and let's march forward, carrying our coffins."[506] The thought of blood was on everyone's lips. There was blood everywhere. The blood and wounds of martyrs were on well-nigh ubiquitous display whilst people dipped their hands in it and left bloody handprints on walls. Blood was said to flow from everywhere, streaming over walls, forming oceans, soaking the land, which was often alluded to as haemorraghing like a festering wound. The streets were washed with blood, "haennaed" with it.[507] The blood of martyrs was said to light up the way, make haenna on the hands of the living, flow across the land, covering it like a consummate libation, irrigating the soil. "The revolution is a tree irrigated by the blood of the pure"; "My homeland, I irrigate the soil of your borders"; "Palestine is a green tree whose thirst can only be quenched with the blood of martyrs"; "When the blood of martyrs irrigates/the land then roses appear"; "The land, O my comrades, is thirsty, the blood of the martyrs will irrigate it/I listen to the sound of guns, in which the poor seek help//What am I? I am a fida'i//I will fight until the last drop, for the land is calling me/Then make from my body torches and from my love, bombs for the homeland"; "Purify your land—mix its sand with blood." Waterfalls of blood, rivers of blood, volcanoes of blood, seas of blood. Life itself was suspended in a bloody emulsion. "The debt of our blood is on you"; "Blood demands blood … martyr demands martyr." Blood in Palestine was cheap: "The blood of our hero-martyr is a spring." Muzaffar a-Nawwab's much-quoted *Night Songs*: "My homeland taught me that the letters of history are not real without blood." The chanted introduction (*mawwal*) of one Hamas anthem bays: "Flay my bones with a whip; put my neck under the knife," continuing with, "Break, break my bones and shed my blood/I'm a Muslim, and Allah named me," whilst another song declares, "We are in love with the colour of blood."[508]

Clearly envinced here is the existence of a collective Palestinian subject whose degree of resonance with God-as-subject has long since foreboded this subjective double's proclivity for terminal anihilation in the black hole of stochastic death. For during the uprising the land was saturated with the ubiquitous mantra that all had to meticulously prepare themselves for the imminent eventuality of self-sacrifice, that it behooved one to repeatedly and fastidiously prepare for death—to the extent that "natural" and "accidental" death swiftly came to be vilified and demonized as downright ignoble. Hence, the PFLP imperative: "Beware of natural death. Don't die except in a shower of bullets."[509] Indeed, in the *Giants* video itself, al-Rushdi himself inveighs with utter sarcastic disdain against such ignoble forms of pseudo-death:

> And the truth, O brother, as everybody knows, is that there is the one … you know, who falls off his donkey and dies, and there is—May Allah postpone it—the one whose donkey tramples him, and he dies, and there is the one who is hit by a car and dies, and there is the one who dies by heart attack, and there is the one who falls off the roof of his house, and there is … the one whom—I take refuge in Allah—a collaborator kills, and he

> dies, but what a difference there is between one death and another, and like Dr. 'Abdullah 'Azzam said, "Truly, there is only one death, so let it be on the path of Allah."[510]

During the intifada, there was not one place that one could turn without being solicited by a call to death. The injunctions to self-destruction were ubiquitous: "Palestine is a green tree whose thirst can only be quenched with the blood of the martyrs"; "When the land calls us, we sacrifice ourselves proudly"; "The blood of the martyrs is calling us"; "The call of truth summoned its sons … so respond to the voice of the summons. Purify your land, purify it … dye its sand with blood"; "Till death, our hearts are with Palestine … Because of it, death has become sport for us. And if death calls us to protect it … we will run to meet death."[511]

In this connection, Franco de Masi compares the unthinkable lust for self-immolation of suicide bombers to the self-annihilatory proclivities of anorexics, borderlines, and psychotics. In such cases, the anxiety felt before death likewise disappears, yielding to a well-nigh inexplicable attraction to the destruction of the self. The suicide bomber in some sense resembles an anorexic patient who, unware of her own impending death, believes that she will be transformed into pure spirit.[512]

Referring to Speckhard, de Masi posits the source of the suicide bomber's etiolated death anxiety as dissociation; that is, to a disruption in formerly integrated functions of consciousness, memory, identity, and perception, wrought by the brunt of andecedent trauma. In more extreme cases, this dereliction is so severe that the event or reality of death comes to be isolated from the remainder of life. The entailed emotional stupefaction may induce in the sufferer a sensation of himself as already dead. It is in such a stupefied and deadened subject that death's anxious foreboding falls into sempiternal desuetude, instead occasioning the aforementioned contrary states of excitement and pleasure. In the face of his own self-willed annihilation, the suicide bomber enters into a veritable state of grace in which the survival instinct is nullified and death itself is seen as a relief, an escape from the incessant roiling cauldron of suffering, pain and impotence. All the cells in his mind, de Masi maintains, appear dead with the exception of those desirous of liberation through death. By ruining emotional perception, dissociation obliterates any empathy that the suicide bomber may entertain for his own dying self. It is thus an affective *felo de se* anterior to, yet portending, his eventual destruction of his own body. He never emerges from this moribund state and maintains himself in it until such time as he is finally and irrevocably dead.[513]

This imperviousness in the face of his own death of the suicide bomber wrought by antecedent trauma and dissociation, contends Ruth Stein, is *inter alia* sustained and even reinforced by a hypertrophied immersion in the vision of God—resulting in a depersonalized trance state and a restriction of consciousness wherein death anxiety gives way to a sustained mood of euphoria. The latter assumes the form of an ecstatic love of God, in which the ideal fusion the subject feels that he has—or longs imminently to attain—with the Deity induces him to believe his portended self-destruction to be divinely willed.[514] It is here that one espies the confluence of fantasies of fusion with the Deity and redemptive violence—in the form of an inexorable pining for martyrdom through which the suicide bomber longs to consummate his narcissistic vision of unsullied purity by way of a terminal and self-willed immolation in the Godhead.[515] And this eternal purification of the subject in the paradisiacal void through the immediate absolution of all sins, as Luca Ricolfi finds from Andoni and Hassan's profiling of the suicide bombers, "is very, very near, right in front of our eyes. It lies beneath the thumb. On the other side of the detonator."[516]

Yet the terminal annihilation of the suicide bomber in the vastness of the Godhead, in that it is born of his phantasmatic fusion with God, is also that by which he longs for a form of deification. It is as if the subjects in the suicide bomber-God subjective double, at the extreme height of the latter couple's intense resonance, tend to become interchangeable upon their submersion in the Godhead's vertiginous abyss. For as Ali Shari'ati has it, by immolating himself, man, who is made from dirt but is possessed of a divine soul, casts the base half of himself into the inferno of faith and in so doing becomes wholly divine.[517] Man is both a being made of mud and a divine soul, as Khosrokhavar paraphrases.[518] As Shari'ati himself has it: "Martyrdom is an act whereby a man suddenly, and in a revolutionary mode, flings his vile being (the diabolical half) into the fire of love and faith and thus becomes completely good."[519] That is to say, he becomes completely divine.

Yet, this God into whom the suicide bomber dissolves cannot be identified with the severe but just God of Abraham. For De Masi describes the God of the suicide bomber as a *cannibal God*, one who effects the simultaneous deification and annihilation of the moribund subject by *devouring* him.[520] He is the Schellingian abyss of vertiginous freedom.

One of those thus cannibalised was Abu-Surur. Long had he tormented his mother with incessant tirades about how he despised the *dunya* ("this world"—in contradistinction to *al-akhira*, or the world to come) and was impatient for the Last Days.[521] He prayed punctilliously from the age of seven, never skipping a prayer; read the Qur'an assidiously, even memorizing sections of it; went to mosque regularly; fasted on Tuesdays and Thursdays; prayed the *tarawiyah* prayer during Ramadan; forbade certain songs as *haram*; watched Hamas videos all the time; and listened to the polemicist Ahmad Didat. Upon finding out from television that someone had been martyred, Abu-Surur would exclaim, "O Lord, make me like you!"—and immediately performed *wadhu* (ritual ablutions) whilst saying, "I declare there is no god except Allah, and Muhammad is the Messenger of Allah." Pining for *shahadat* (martyrdom), he would disconsolately declare, "O Lord, what world is this?" Instead of working, he caroused with the *shabab* all day, constantly flirting with death by throwing stones at Israeli soldiers, to the extent that they had once followed him all the way home and left a bullet hole above the couch as a sign of their former presence. As an adolescent he spent three months in prison in the Negev desert in southern Israel and upon release hardly ate anything and lived long indolent hours doing nothing, just sitting.[522]

A photograph of Abu-Surur after he had become a member of the *mutaradun*, or the "hunted," reveals a man with a dark and slightly patchy beard, staring outward, blankly. "There is a whiff of fear that exudes from this last photo, the kind of blank concentration one sometimes sees in the dead and the dying."[523]

By this time, the height of resonance of the subjectivity of Mahir Abu-Surur with that of God had acceded to a threshold so stratospheric that his decision to embark on a martyrdom operation and terminal immolation in the incinerating substance of the Godhead was to follow with the ineluctability of fate. Muhammad al-Hindi and Salam 'Uthman were his accomplices in an explosive death that was deigned to destroy all three. In order to gain more mystical familiarity with the vicissitudes of death, the three—together with their dispatcher Rushdi—spent ten uninterrupted days in a cave near to Hebron so as to immerse themselves in meditations upon the Torment of the Tomb that follows death, a tomb that after forty days becomes either one of the gardens of Paradise, or one of the pits of Hell.[524]

This eerie time between one's death and one's entrance into the afterworld is that of the interrogation of the dead by the angels Munkar and Nakir. Blue-black in hue, possessed of thunderous voices, lightning-bolts for eyes, and iron whips—in the descriptions of ibn Taymiyyah—these monstrous twins come to sit by the dead inside their tombs after the last footsteps of the mourners have faded away. Surrendering themselves to the claustraphobic darkness of their cave, the three were merely repeating certain rituals that Hamas typically foisted upon their would-be martyrs, such as mock burials, forcible incarceration in dark chambers, and being interred in an empty grave whilst reciting Qur'anic verses associated with the tomb's torment before being permitted to reascend into the world of the living.[525]

Al-Hindi himself had pined for terminal immolation in the Godhead ever since he had met the youth Yasir an-Nimruti during his fifteen months of detention in Ansar desert prison. Al-Hindi and an-Nimruti themselves came to form a subjective double bent upon the black hole of stochastic death. Once, he rebuked a youth who had just turned thirty thus, "Review your faith, O my brother, for you have passed thirty years and have not met Allah as a martyr." Indeed, he was very soon doubting his own self for he had failed thusfar to martyr himself, in spite of the double death vow he had made with Yasir an-Nimruti.[526] In the words of Mahir's brother Suhail al-Hindi,

> I still remember your calling the people beseechingly until tears ran down your cheeks such was the fervour with which you plead with your Lord, May He be exalted, to bless you wih martyrdom and join you with your brothers who proceeded you, for they attained martyrdom, while you were doubting your faith because you had not joined them.[527]

Later, having been refused admittance to the Battalions by its leadership, he announced frustratedly,

> No one in the world will be able to stop me from meeting Allah as a worthy martyr whose wonders are mentioned in the verses of Allah and the hadith of the Messenger of Allah, May Allah bless Him and grant Him salvation.[528]

Eventually, the agony of pining of al-Hindi and Abu-Surur was to be slaked. Having spent the said ten days in their cave at Hebron, they travelled with 'Uthman to Jerusalem where they slept their last night, awoke, performed *wadhu*, prayed two rak'atain, preparing for death. On July 1, 1993, at around 6:45 in the morning, Abu-Surur, 'Uthman, and al-Hindi boarded Egged bus number 25, on French Hill. Abu-Surur proceeds to have a scuffle with the bus driver David Yom-Tov whom he shots in the thigh, whereupon the double-bus jackknifes in the middle of the street as people fall out of the doors onto the street screaming whilst Salah 'Uthman is shot in the head and Abu-Surur and al-Hindi flee and move on to hijack a Ramot Eskol teacher's Renault 5 on Midbar Sinai Street in the neighborhood of Givat Hamivtar which takes them along Hebron Road past the Old City to the military checkpoint at Gilo Junction, at which point the car is fired upon by soldiers, loses control, collides with a tree, ignites, and blows up as Abu-Surur and al-Hindi—with the internal resonance of their doubled subjectivities soaring to the scorching heat of consciousness on the very point of self-dissolution—are incinerated in the black hole of stochastic death.[529]

September 11, 2001 A.D.

Near the start of Ramadan in December 1999, Muhammad el-Amir Atta, Ziad Jarrah, Marwan al-Shehhi, and Omar—all members of al-Qaeda's Hamburg cell—foregathered at the House of Ghamdi in Kandahar. It was here that Osama bin Laden invited them to join his ranks and accept suicide assignments. Persuasion was not required. Long since had the resonance of their subjectivities with that of God reached a threshold presaging their self-willed annihilation in the black hole of stochastic death. For the eventuality of becoming martyrs in a holy war was something that they had already been discussing among themselves for years. Some had pined for such a death for very long indeed. They had debated the morality of martyrdom, concurring that dying for the sake of terminal absorption in the divine substance differentiated what they lusted for from the mere anomic suicide of the infidel. There had been periods back in Harburg when they poured over such issues every day for hours on end. Omar and al-Shehhi, in particular, were unswervingly resolved upon their own demise. Omar's resignation to his own pending *ishtishadat* (self-martyrdom) arose from his oneiric acceptance of the word of God and his well-nigh mystically exalted belief in the intrinsic goodness of Islam. Shehhi's eagerness for death on the path of God had been enflamed the moment he arrived in Harburg. For him martyrdom, far from being a case of morbid fate, was an eventuality to be embraced with joy.[530]

As regards Muhammad Atta, he already knew—or this is at least how Martin Amis has it—"that the first thing he would be asked was whether he was prepared to die. But the Sheikh was smiling, almost with eyes of love, when he said it. 'The question isn't necessary,' he began. 'I see the answer in your face'."[531] The black whole of stochastic death loomed large for the lot. For the September 11 hijackers as a whole were, according to Amis, a "peer group piously competitive about suicide," a collective subject so aggressively enamoured of the imminent eventuality of their own auto-immolation that it could verily be said of them that "death was not death—and life was not life either."[532]

These men were now bent upon immolating themselves by flying passenger aeroplanes into the twin towers of the World Trade Centre. On the night prior to the day of their portended terminal demise they were to pour over the instructions drawn up by one Abdul Aziz al-Omari. They were *inter alia* exorted thus:

> … Vow to accept death, renew admonition …
>
> Read the surahs of Repentance and The Spoils. Contemplate their meaning and the bounties God has prepared and established for the martyrs.…

> Remind your base self to listen and obey this night ... Train your base self, make it understand, convince it, and goad it on to this end....
>
> Purify your heart and cleanse it of stains. Forget and be oblivious to that thing called the world. For, the time for playing has passed, and the time has arrived for the rendezvous with the eternal Truth....
>
> Let your breast be filled with gladness, for there is nothing between you and your wedding but mere seconds. Thereby will begin a happy and contented life and immortal blessing with the prophets, the true ones and the righteous martyrs....
>
> Then fix your gaze, such that you fall into tribulations, you will know how to behave, how to stand firm, how to say "We are, verily, from God and to him we shall return"....
>
> Then recite the words of God, "You were wishing for death before you encountered it, then you saw it, and are looking for it." And you wanted it.[533]

It is especially the injunctions to repentence, subjugation of the base self, and the purification of the heart, that are here of the most acute paramountcy. For, as Stephen Holmes renders pellucid, it was the consciousness of internal defilement that constituted the kernel of the highjackers' psychological state and the overriding driving impetus steering their hellbent march toward the black hole of their own self-destruction. They felt morally tainted, spiritually gangrened, by a society that could not fail to seduce them—the West. They were riven by stress-riddled self-contempt and self-blame. Atta, for one, was a man flagellating himself. He felt contaminated, repulsed, and humiliated. His sexually tormented will was in permanent evidence. This vile stain on the living martyrs' souls was a cancer on the threshold of metastatis that had perforce to be erased despite all lethal consequences to life and limb. The inevitable ramification was a rabid devotion to self-mortification in which self-inflicted death emerged as an intrinsic good.[534]

In order to effect the removal of the western gangrene that was causing them to slowly decay from within, the nineteen men embarked upon what Holmes sees as a concurrent performance of the Greater Jihad and the Lesser Jihad, so as to slay the infidel whilst simultaneously incinerating the nauseating traces the same had deposited in their souls in a cataclysmic inferno of furious self-immolation. If the first drop of blood shed in martyrdom would not wash away their sins, then suicide terrorism would at least permanently extinguish their debilitating and sickening sense of guilt.[535]

Yet, as Roy avers, in the case of the hijackers, it was the Greater Jihad waged against the base self that preponderated over the Lesser variant. The sacrifice of the self-immolatory attackers' own lives was possessed of a mystical dimension and was, above all, an act of supreme *devotion*. Concurring with this tenure of argument, Makiya and Mneimneh add that the hijackers' viewed their pending martyrdom as an end in itself, being imbroiled in a peculiar and absolutely deadly personal act of *worship* above all intended to please God.[536] Indeed, in a recorded pre-attack message, one of the moribunds had uttered a striking supplication: "Oh Allah, take from our blood today until you are satisfied."[537] We thus find the subjective double consisting of the hijackers, on the one hand, and God, on the other, to be held together by an act of singular devotion limitlessly shot through with inexorable suicidal proclivities.

It, however, also appears to be held together by *fear*. For in the instruction manual distributed to the attackers, the latter were reminded not only that fear is *a form of worship* but that fearing anyone or anything, including death, apart from God was nothing less than a mortal sin that would damn them to an eternity in Hell.[538] Eternal damnation.

Yet, though in fear and trembling before the Deity, the attackers were at the same time besotted with the thought of making other people fear *them*—thus arrogating to themselves the blasphemous desire to be godlike or perhaps even to be God. They thus embraced two mutually opposed self-images: at once omnipotent and impotent, both thunder gods and despicable sinners, simultaneously occupying both positions in a subjective double comprising their own deified selves and themselves as reduced to utter self-flagellatory misery and abjection, in a single fictive fusional body cavorting headlong towards infernal death.[539] And having thus arrogated divinity to themselves, these rabid moribunds were perhaps ultimately quite godless and destined not for God but for Hell—or at least for the unsullied void of nonexistence.

As Amis avers, the sheer unadulterated purity of Muhammad Atta's self-immolatory nihilism in particular, meant that his specific personal pining for terminal immolation did not assume the mediate form of a longing for paradise. Paradise was frivolous, bereft of the inspiring arctic coldness of ascetic auto-evacuation. "He didn't expect paradise. What he expected was oblivion."[540] Indeed, one of his utterances to the imam he visited in a hospital in Portland, Maine, the day before September 11, seemed to indicate that his desire was expressly to go to Hell:

> It's from the traditions. The Prophet said: "Whoever kills himself with a blade will be tormented with that blade in the fires of Hell … He who throws himself off a mountain and kills himself will throw himself downward into the fires of Hell for ever and ever … Whoever kills himself in any way in this world will be tormented in that way in Hell."[541]

Yet, subsequently, in his last telephone call to Ziad Jarrah he mentions having obtained a certain "holy water" from the convalescing imam, water "from the Oasis." "It absolves you of what the imam called the 'enormity', the atrocious crime, Ziad, of the self-felony."[542] Atta goes on, in addition averring that Ziad will go to Hell:

> The water is within you and preserves you for God. It's a new technique—it began in Palestine. Your hell, Ziad, will burn with jet fuel for eternity. And eternity never ends—it never even begins.[543]

Notwithstanding, according to Amis, it appears that all of Atta's invocations of the fires of Hell were merely thinly veiled intimations of death as the consummate and blessedly neutral nullity of pure non-Being—and naught beside:

> He didn't believe in the Devil, as an active force, but he did believe in death. Death, at certain times, stopped moving at its even pace and broke into a hungry, lumbering run. Here was the primordial secret. No longer closely guarded—no longer well kept. Killing was divine delight. And your suicide was just a part of the contribution made—the massive contribution to death. All your frigidities and futilities were rewritten, becoming swollen with meaning. This was what was possible when you turned the tides of life around, when you ran with the beasts, when you flew with the flies.[544]

The final reckoning was upon Atta and his co-moribunds. The black hole of stochastic death, long presaged by the maddening resonance of the collective jihadi subject with that of the Deity, was imminently to find its rabid consummation in the lividly infernal white heat of consciousness approaching molten auto-immolation in the rapidly vanishing epicenter of "ferocious fires of more than 1300 degrees" in which "far-flung hopes, dreams, and memories" disappeared from the earth "in the premature demise of 2 797 suddenly, but completely, vanquished lives."[545] In the same terminal morass, Muhammad Atta was annihilated in the singular void of his eternal longing, dying in a scorchingly evanescent instant of "physical torment—a panic attack in every nerve, a riot of atoms" italicizing the "last shinings of his brain."[546] Disappearing in the same flitting instant was a life—his own—that "he had hated, and had wished away," absenting itself in the face of his "helpless grief," "imperturbable in its beauty and its power. Even as his flesh fried and his blood boiled, there was life, kissing its finger-tips."[547]

"Then it echoed out, and ended."[548]

Notes

1. Magee 2001, 178.
2. Wiesel 2005, 22.
3. Ibid. 22, 23.
4. Ibid. 28.
5. Droge and Tabor 1992, 63.

6. Ibid. 67.
7. Ibid. 90.
8. Ibid. 94.
9. Josephus 1970, 3.371.
10. Droge and Tabor 1992, 96.
11. Ibid. 55.
12. Ibid. 100.
13. Bialik and Ravnitzky (eds.) 1992, VI.77, 78.
14. Droge and Tabor 1992, 54.
15. Ibid. 55.
16. Ibid. 100.
17. Ibid. 102.
18. Wiesel 1981, 94, 95.
19. Droge and Tabor 1992, 56.
20. Josephus 2006, 7.229.
21. Droge and Tabor 1992, 99, 100.
22. Plato 1972, 45.
23. Ibid. 47.
24. Ibid. 46.
25. Ibid. 55.
26. Ibid. 58.
27. Ibid. 59, 60.
28. Ibid. 62–64.
29. Ibid. 70, 71.
30. Ibid 103–107.
31. Ibid. 107.
32. Ibid. 108.
33. Ibid. 108–110.
34. Ibid. 135.
35. Ibid. 109.
36. Ibid. 135, 136.
37. Ibid. 136.
38. Ibid.
39. Ibid. 136, 137.
40. Ibid. 140, 142.
41. Ibid. 197.
42. Shepkaru 2006, 36.
43. Cited in Shepkaru 2006, 36.
44. Cited in Ibid.
45. Cited in Ibid. 36, 37.
46. Droge and Tabor 1992, 69.

47. Ibid. 72.
48. Ibid. 69.
49. Ibid. 75.
50. Ibid. 85.
51. Cited in Shepkaru 2006, 71.
52. Cited in Ibid. 72.
53. Shepkaru 2006, 72, 73.
54. Ibid. 119.
55. Miles 2002, 5.
56. Ibid. 169.
57. Ibid. 180.
58. Ibid. 188, 189.
59. Droge and Tabor 1992, 114.
60. Cited in Ibid.
61. Droge & Tabor 1992, 115–119.
62. Cited in Ibid. 119.
63. Moltmann 1974, 51.
64. Ibid.
65. Ibid.
66. Miles 2002, 41.
67. Ibid. 234.
68. Droge and Tabor 1992, 117.
69. Ibid. 118.
70. Emmerich 2007, 75, 76.
71. Ibid. 77, 78.
72. Miles 2002, 5.
73. Cited in Hegel 1998, 125.
74. Moltmann 1974, 233.
75. Cited in Ibid. 234.
76. Moltmann 1974, 201–203.
77. Ibid. 205.
78. Ibid.
79. Ibid.
80. Cited in Ibid. 206.
81. Cited in Ibid.
82. Miles 2002, 10.
83. Hegel 1998, 219.
84. Taylor 2009, 146, 147.
85. Ibid. 147.
86. Taylor 2009, 148, 149.
87. Ibid. 150.

88. Ibid. 159.
89. Altizer 2003, 18.
90. Ibid. 68, 69.
91. Ibid. 77.
92. Altizer 1967, 88.
93. Ibid. 103.
94. Žižek 2009, 29.
95. Ibid. 33.
96. Hegel 1998, 125.
97. Ibid.
98. Ibid.
99. Ibid. 127.
100. Ibid. 131.
101. Ibid. 131, 132.
102. Ibid. 132.
103. Ibid. 100.
104. Ibid. 219, 220.
105. Ibid. 276.
106. Ibid. 311.
107. Ibid.
108. Ibid.
109. Ibid. 311, 312.
110. Žižek 2009, 60, 61.
111. Ibid. 61.
112. Ibid. 289.
113. Emmerich 2007, 99, 100.
114. Ibid. 102.
115. Ibid. 102–106.
116. Ibid. 107, 108.
117. Ibid. 117–119.
118. Ibid. 116.
119. Ibid. 130, 133, 136, 137.
120. Ibid. 141–144.
121. Ibid. 151.
122. Ibid.
123. Ibid. 156.
124. Ibid. 158.
125. Ibid. 167–169.
126. Ibid. 176–178.
127. Droge and Tabor 1992, 118.
128. Emmerich 2007, 206–209.

129. Ibid. 221.
130. Ibid. 223–226.
131. Ibid. 233–246.
132. Ibid. 251–260.
133. Ibid. 263–269.
134. Ibid. 269.
135. Ibid. 270, 271.
136. Ibid. 271.
137. Ibid. 272.
138. Ibid. 281.
139. Josephus 1970, 7.247–275.
140. Ibid. 7.304–318.
141. Cited in Shepkaru 2006, 120.
142. Shepkaru 2006, 120.
143. Ibid. 127.
144. Ibid. 222, 223.
145. Cited in Ibid. 223.
146. Shepkaru 2006, 223.
147. Ibid. 230.
148. Droge and Tabor 1992, 97.
149. Josephus 1970, 7.318, 7.319.
150. Ibid. 7.323–336.
151. Ibid. 7.337–340.
152. Ibid. 7.341–388.
153. Ibid. 7.389–405.
154. Khosrokhavar 2005, 7–9.
155. Ibid. 8.
156. Droge and Tabor 1992, 2.
157. Cited in Ibid.
158. Droge and Tabor 1992, 119.
159. Alvarez 1972, 58, 59.
160. Cited in Shepkaru 2006, 95.
161. Cited in Ibid. 97.
162. Alvarez 1972, 59, 60.
163. Droge & Tabor 1992, 122.
164. Cited in Ibid. 124.
165. Droge and Tabor 1992, 124.
166. Cited in Ibid. 125.
167. Droge & Tabor 1992, 126.
168. Ibid. 131.
169. Ibid. 133, 134.

170. Ibid. 142.
171. Ibid. 144–146.
172. Cited in Ibid. 146.
173. Droge & Tabor 1992, 148.
174. Cited in Ibid. 149.
175. Droge & Tabor 1992, 149.
176. Ibid. 150.
177. Ibid. 151.
178. Peterson 2004, 65–69.
179. Ibid. 77, 78, 81.
180. Ibid. 75, 76.
181. Ibid. 76.
182. Ibid.
183. Ibid. 208.
184. Ibid. 241.
185. Ibid. 300, 301.
186. Ibid. 308, 309.
187. Ibid. 356.
188. Ibid. 357.
189. Ibid. 358.
190. Ibid. 361.
191. Ibid. 365, 366.
192. Ibid. 368.
193. Ibid. 369.
194. Ibid. 370.
195. Ibid.
196. Musurillo 1972, 109–117.
197. Peterson 2004, 391, 392.
198. Ibid. 394.
199. Hazelton 2009, 173.
200. Ibid. 178, 179.
201. Ibid. 181.
202. Khosrokhavar 2005, 37.
203. Ibid. 42, 43.
204. al-Muqarram, 49, 51.
205. Cited in Ibid. 51–52.
206. al-Muqarram, 117.
207. Ibid.
208. Ibid.
209. Ibid. 145.
210. Ibid.

211. Hazelton 2009, 174.
212. Ibid. 174–178.
213. Ibid. 180, 181.
214. Ibid. 184.
215. al-Muqarram, 164.
216. Hazelton 2009, 185.
217. Ibid. 185–187.
218. Ibid. 187.
219. Ibid.
220. al-Muqarram, 195.
221. Hazelton 2009, 187, 188.
222. Ibid. 190–192.
223. al-Muqarram, 262, 264.
224. Ibid. 264; Hazelton 2009, 190–192.
225. Cited in Massignon 1982, 12, 65.
226. Massignon 1982, 12.
227. Cited in Massignon 1982, 65.
228. Cited in Ibid. 66, 67.
229. Cited in Ibid. 158.
230. Massignon 1982, 158–160.
231. Mason 1991, 6.
232. Cited in Massignon 1982, 88.
233. Cited in Ibid. 91.
234. Massignon 1982, 134.
235. Cited in Ibid.
236. Massignon 1982, 137.
237. Cited in Ibid.
238. Massignon 1982, 138.
239. Ibid. 142.
240. Cited in Ibid. 142.
241. Massignon 1982, 143.
242. Cited in Ibid. 143, 144.
243. Massignon 1982, 144.
244. Cited in Ibid.
245. Massignon 1982, 146.
246. Cited in Ibid.
247. Massignon 1982, 146.
248. Cited in Ibid.
249. Massignon 1982, 146.
250. Cited in Ibid.
251. Massignon 1982, 147, 148.

252. Ibid. 215–220.
253. Mason 1991, 76–81.
254. Bartol 2007, 169, 170.
255. Ibid. 170.
256. Ibid. 171.
257. Ibid. 225.
258. Ibid. 234.
259. Ibid. 236, 237, 243.
260. Ibid. 270.
261. Ibid. 277.
262. Ibid. 280, 281.
263. Ibid. 288–290.
264. Craney 2008, 140.
265. Ibid. 61, 62.
266. Ibid. 109, 110.
267. Ibid. 145, 146.
268. Ibid. 149.
269. Ibid. 164.
270. Ibid. 154, 155.
271. Ibid. 157.
272. Ibid. 261, 262.
273. Ibid. 389–393.
274. Ibid. 394, 395.
275. Ibid. 412, 421, 426, 427.
276. Ibid. 434, 437.
277. Ibid. 452, 453.
278. Ibid. 457, 458.
279. Ibid. 461–467.
280. Cutter 2011, 4, 5.
281. Ibid. 6.
282. Ibid. 13, 14.
283. Ibid. 27.
284. Ibid. 42, 43.
285. Ibid. 46.
286. Ibid. 84–86.
287. Ibid. 91–94, 199.
288. Ibid. 159.
289. Ibid. 274, 275.
290. Ibid. 238, 240, 254.
291. Ibid. 260–266.
292. Ibid. 268, 269.

293. Ibid. 270, 271.
294. Ibid. 274, 275.
295. Ibid. 277.
296. Foxe 1965, 82–85.
297. Ibid. 85.
298. Ibid. 86, 87.
299. Ibid. 215, 216.
300. Ibid. 216.
301. Ibid. 216–218.
302. Ibid. 219, 220.
303. Ibid. 220.
304. Ibid. 220, 221.
305. Ibid. 221.
306. Ibid. 222–227.
307. Ibid. 229–231.
308. Ibid. 231.
309. Ibid. 232–240.
310. Ibid. 240.
311. Ibid. 240, 241.
312. Dostoyevsky 2008, 103, 104.
313. Ibid. 126, 127.
314. Ibid. 127.
315. Ibid. 127, 128.
316. Ibid. 128.
317. Ibid. 129.
318. Ibid.
319. Ibid. 414.
320. Ibid. 415.
321. Ibid. 653.
322. Ibid. 681, 682.
323. Ibid.
324. Ibid. 683.
325. Ibid. 683–684.
326. Ibid. 684.
327. Ibid.
328. Ibid.
329. Ibid. 685.
330. Ibid. 685, 686.
331. Ibid. 686–691.
332. Ibid. 691, 692.
333. Mishima 1966, 93–95.

334. Ibid. 95, 96.
335. Ibid. 96, 97.
336. Ibid. 99, 100.
337. Ibid. 102.
338. Ibid.
339. Ibid.
340. Ibid.
341. Ibid.
342. Ibid. 103.
343. Ibid.
344. Ibid.
345. Ibid.
346. Ibid. 104, 105.
347. Ibid. 105, 106.
348. Ibid. 106.
349. Ibid. 107, 108.
350. Ibid. 110, 111.
351. Ibid. 111.
352. Ibid.
353. Ibid. 112, 113.
354. Ibid. 113.
355. Ibid.
356. Ibid. 113, 114.
357. Ibid. 114.
358. Ibid.
359. Ibid. 114, 115, 116.
360. Ibid. 117, 118.
361. Nathan 1975, 8–12.
362. Ibid. 18, 19.
363. Cited in Nathan 1975, 20.
364. Cited in Ibid.
365. Stokes 2000, 44.
366. Ibid. 46.
367. Ibid. 58.
368. Ibid. 60.
369. Ibid. 67.
370. Nathan 1975, 53.
371. Ibid. 57.
372. Cited in Ibid.
373. Cited in Ibid.
374. Stokes 2000, 123, 124.

375. Cited in Nathan 1975, 167.
376. Nathan 1975, 168.
377. Ibid. 170, 171.
378. Ibid. 188, 189.
379. Cited in Ibid. 189.
380. Nathan 1975, 189.
381. Cited in Ibid. 190.
382. Nathan 1975, 198, 199.
383. Cited in Ibid. 211.
384. Nathan 1975, 212.
385. Ibid. 220.
386. Ibid. 223.
387. Cited in Ibid.
388. Cited in Ibid.
389. Cited in Ibid.
390. Nathan 1975, 225.
391. Ibid. 229.
392. Ibid. 236–237.
393. Cited in Stokes 2000, 204.
394. Nathan 1975, 239.
395. Cited in Ibid.
396. Nathan 1975, 240.
397. Ibid. 244, 245.
398. Ibid. 252, 253.
399. Stokes 2000, 264, 265.
400. Ibid. 189–191.
401. Cited in Nathan 1975, 270.
402. Nathan 1975, 270.
403. Cited in Ibid. 271.
404. Cited in Ibid.
405. Cited in Ibid.
406. Stokes 2000, 247, 248.
407. Ibid. 27–31.
408. Chidester 1988, 55–58, 60, 62.
409. Cited in Ibid. 61.
410. Cited in Reiterman and Jacobs 2008, 149.
411. Chidester 1988, 61.
412. Cited in Reiterman and Jacobs 2008, 212.
413. Chidester 1988, 62.
414. Cited in Ibid.
415. Chidester 1988, 62.

416. Cited in Ibid.
417. Chidester 1988, 62.
418. Ibid. 100.
419. Cited in Ibid.
420. Chidester 1988, 117.
421. Cited in Ibid.
422. Cited in Ibid. 119.
423. Cited in Ibid. 126.
424. Cited in Ibid.
425. Chidester 1988, 146.
426. Cited in Ibid. 147.
427. Reiterman and Jacobs 2008, 199.
428. Cited in Chidester 1988, 147.
429. Cited in Ibid.
430. Chidester 1988, 127.
431. Cited in Ibid.
432. Cited in Ibid.
433. Chidester 1988, 131.
434. Reiterman and Jacobs 2008, 295, 296.
435. Chidester 1988, 132, 137.
436. Reiterman and Jacobs 2008, 390.
437. Chidester 1988, 138, 139.
438. Cited in Ibid. 139.
439. Chidester 1988, 145.
440. Ibid. 145, 146.
441. Cited in Ibid. 146.
442. Cited in Ibid. 147.
443. Chidester 1988, 147.
444. Ibid. 150.
445. Ibid. 151.
446. Ibid. 153.
447. Ibid. 156.
448. Ibid. 155.
449. Ibid. 159.
450. Reiterman and Jacobs 2008, 559–566.
451. Davis 2003, 49.
452. Cited in Ibid.
453. Davis 2003, 49.
454. Cited in Ibid.
455. Cited in Reuter 2006, 44.
456. Cited in Ibid. 42.

457. Khosrokhavar 2005, 45.
458. Ibid. 49.
459. Ibid.
460. Ibid. 50, 60.
461. Ibid. 86.
462. Ibid. 92.
463. Ibid. 99.
464. Ibid. 87.
465. Ibid. 100.
466. Ibid. 50.
467. Fisk 2006, 283.
468. Reuter 2006, 34.
469. Ibid.
470. Ibid. 35.
471. Ibid. 47.
472. Cited in Ibid. 47.
473. Fisk 2006, 315.
474. Khosrokhavar 2005, 50, 91.
475. Fisk 2006, 315.
476. Ibid. 315, 316.
477. Newport 2006, 175, 176.
478. Ibid. 176.
479. Ibid. 180–185.
480. Ibid. 216.
481. Ibid. 225, 226.
482. Ibid. 204, 264.
483. Ibid. 226.
484. Newport 2006, 234.
485. Ibid. 235.
486. Ibid. 245.
487. Thibodeau 1999, 135.
488. Ibid. 137.
489. Newport 2006, 253, 254.
490. Ibid. 264.
491. Ibid. 262, 263.
492. Ibid. 264, 265.
493. Ibid. 278.
494. Ibid. 281, 294.
495. Ibid. 298.
496. Ibid. 301.
497. Ibid. 308.

498. Ibid. 310, 311, 313.
499. Newport 2006, 314, 315, 319.
500. Ibid. 318.
501. Thibodeau 1999, 264, 265.
502. Ibid. 273, 276, 284.
503. Oliver and Steinberg 2005, 119.
504. Cited in Ibid. 119, 120.
505. Oliver and Steinberg 2005, 121, 122.
506. Cited in Ibid. 122.
507. Oliver & Steinberg 2005, 90, 91.
508. Cited in Ibid. 90–93.
509. Cited in Ibid. 122.
510. Cited in Ibid. 156.
511. Cited in Ibid.
512. De Masi 2011, xxxvi.
513. Ibid. 78, 79.
514. Ibid. 14.
515. Ibid. 16.
516. Cited in Ricolfi 2005, 111.
517. De Masi 2011, 26, 27.
518. Khosrokhavar 2005, 44.
519. Cited in Ibid.
520. De Masi 2011, 93.
521. Oliver and Steinberg 2005, 120.
522. Ibid. 126, 127.
523. Oliver and Steinberg 2005, 125.
524. Ibid. 139.
525. Ibid. 74, 75.
526. Ibid. 149–151.
527. Cited in Ibid. 151.
528. Cited in Ibid.
529. Oliver and Steinberg 2005, 180, 181.
530. McDermott 2006, 178, 179.
531. Amis 2009, 98.
532. Ibid. 116–117.
533. Cited in McDermott 2006, 249, 250.
534. Holmes 2005, 153, 154.
535. Ibid.
536. Ibid. 155.
537. Cited in Ibid.
538. Holmes 2005, 155.

539. Ibid. 155, 156.
540. Amis 2009, 102.
541. Ibid. 110.
542. Ibid. 115.
543. Ibid. 116.
544. Ibid. 122.
545. McDermott 2006, 243, 244.
546. Amis 2009, 123.
547. Ibid. 124.
548. Ibid.

15

Transhistorical Fugue, 3. From Jewish Survival to the Survival of Humanity

1. Jewish Survival

Paul Johnson maintains the Jews to be the most tenacious people in all history.[1] If we read this tenacity as specifically relating to their astonishing perdurance and will to live and survive in the face of quite terrible odds, this remark is certainly nowise overly audacious. In this vein, Max I. Dimont compares the sheer longevity of the Jews to the notable absence of this quality in an entire gamut of peoples part-wise contemporaneous with them. All of these either underwent fragmentation and miscegination after their period of eminence—as in the case of the ancient Greeks and Romans, or—in the case of the Babylonians, Persians, Phoenicians, Hittites and Phoenicians—simply vanished from the historical record and are now completely nonexistent. In stark contrast to the foregoing, not only do the Jews possess a continuous living history of in excess of four-thousand years; they have moreover survived three-thousand of these years entirely bereft of a country of their own, whilst at the same time preserving their own unique identity even whilst living in the midst of a vast array of alien cultures. This means that they have not merely survived for this incredible stretch of time, but have survived *as Jews*. History has threatened the Jews with extinction, whether by physical annihilation or complete dilution of their strain through assimilation, by hurling six broad obstacles in their direction. Each one of these seemingly fatal Damocles' swords presented itself as a dire threat to Jewish survival. The Jews, notwithstanding, survived every one of them.[2]

They survived near-extinction in the pagan world of the Babylonians, Assyrians, Phoenicians, Egyptians, and Persians. Having returned to their country after their Babylonian exile, they survived both the intensive Hellenizing measures of the Greeks and the pulverizing military juggernaut of the Romans. From the time of their expulsion from Jerusalem in the sixth century B.C. by the Babylonians until their liberation from the ghettos in nineteenth century Europe, they survived all fissiparous tendencies threatening to undermine their cultural cohesion. This threat of assimilation they obviated through the creation of the Talmud, which served as a centripetal force of cohesion. From the seventh century onwards, they survived existence in the House of Ishmael, that is, life under Islamic dominion. They survived near-extinction in the Middle Ages, during which torrid time all vanquished nations found themselves converted to Christianity—with the exception, of course, of the Jews. In spite of a plethora of depredations, the Jews survived this awful twelve-hundred-year period, and emerged from it with renewed spiritual and cultural effervescence—to the extent that it did not take more than a single generation after the fall of the ghetto walls for the Jews to rise to prominence within the firmament of Western civilization at large. The final Damocles' sword listed by Dimont was the Modern Age itself, which ushered in nationalism, industrialism, communism, and fascism. The Jews overcame these threats as well.[3]

This seeming indestructability of the Jews has, in fact, confounded quite a few schools of history. For many a school of historical thought would maintain that, as in the case of an individual human being, any given civilization has only one life span, lasting perhaps five-hundred years, but certainly not longer than a thousand years. Yet, the Jews have lasted four-thousand years, and have had six distinctive cultures in six different alien civilizations.[4]

Simplifying somewhat, Hegel, Spengler, and Toynbee conceive each civilization as naturally passing through infancy, a childhood, adolescence, maturity, old age, and finally terminating in death—as if this latter was their fate from the very outset. For Spengler in particular, civilizations are, in fact, foredoomed to death. They pass through the spring of early origins, mature into the summer of their greatest physical achievement, grow into the autumn of great intellectual heights, decline into the winter of their civilization, and finally die and are erased from the historical record. Jewish history is fundamentally refractory to this pattern, as the Jews have far outflanked their purported life span and simply refuse to pass into obsolescence.[5]

Dimont likewise approaches this phenomenon from the vantage point of Abraham's original covenant with God. He considers it entirely irrelevant whether it was Abraham who projected his experience onto an imaginary Yahweh or whether a real Yahweh proposed it to him. The fact remains that after four-thousand years the pivotal idea of a covenant between the Jews and God still stubbornly abides and is sedulously remembered on a quotidian basis. Though many aspects of Jews and Judaism have been altered or modified during their subsequent four-thousand-year history, the idea of a covenant with God has remained absolutely constant. And it is this latter which in turn gave rise and continues to give rise to a will to survive as Jews, which has been the ultimate motor of Judaism. Without it, there could have been, and can be, no Judaism and no Jews. In fact, if this notion of the covenant falls away—that is, when the Jew, through a lack of inner compulsion, no longer wishes to retain his identity as a Jew, then nothing will stand between him and assimilation and final disappearance. This inner compulsion has, however, not fallen into desuetude, and thus neither have the Jews. The Jews have, in fact, successively designed new ideas to perpetuate this aim of survival: those of Abraham, the rite of circumcision, and the prohibition of human sacrifice—as allegorized in the *Akedah*, or binding of Isaac.[6] The laws of Moses, on the other hand, proved virile enough to vouchsafe the survival of the Jewish national entity for eight-hundred years until the prophets in turn renovated them.[7]

For Arthur A. Cohen, "Judaism is a tradition which has apostrophized the miracle of survival from its very beginnings, underscoring in each generation the sign of the Exodus and its symbolic adumbration in each historical searing …"[8] As Eliezer Berkovits maintains, the Jews from the outset staked their very existence on the original biblical proposition that life and the moral good were identical, as were death and evil. The Jewish God is likewise on the side of life. Playing in midrashic style on the Hebrew word *emumin* (faithful), *amen* (an affirmation) and *emunah* (faith, trust), Berkovits concludes that the faithful are those who answer with *amen* in complete trust (*emunah*) of God. They believe that God is the one who quickens the dead. He is the one who preserves the faithful. He is a redeemer and resurrector.[9] Moreover, it is precisely through their fear of his awesomeness that the Jews are enabled to survive among the nations. Conversely, the awesomeness of God is revealed in the survival of Israel. He reveals his presence in the survival of his people. He renders himself powerless through forebearance and long-suffering, yet precisely through this his lack of manifest power ensures the continued protection and survival of a powerless people.[10]

Jewish history, or faith history, is outside of what is termed power history. The latter is the history the nations. What Berkovits regards as miraculous is that faith history has not been removed from the face of the earth by the immeasurably superior material and military force that power history has almost incessantly arrayed against it. They are a small people whose very existence is forever beset by the forces of power history. Yet they mysteriously survive without recourse to power, and do so through God's hidden and power-divested guidance.[11] In Deleuzoguattarian terms, this powerlessness of God and of the Jews would not be an absence of power *tout court*. Rather, it indicates that divine and Jewish power is not *pouvoir*, or the bureaucratic power of the State. It is instead power as *puissance*, as in the *puissances* of the future sensed by the prophets. Here, as Irving (Yitzschak) Greenberg maintains, faith is not primordially a belief or an emotion, but rather "an ontological life force that reaffirms creation and life in the teeth of overwhelming death."[12]

We have seen that the miraculous longevity of the Jews is vouchsafed by their joining of their terrestrial destiny to that of God. However, the converse is likewise the case. For God's own destiny in human history is also linked to that of the Jews. His fortunes are joined to those of the Jews, as theirs are joined to his. As he uttered through the mouth of his prophet, "Ye are My witnesses, saith God." Rabbinical commentary on this is: "If you are My witnesses, I am God; if not, I am not God." However, since God necessarily is, so too must the witnesses be. This is precisely the guarantee of Jewish survival "until the end of days."[13]

Now, another transhistorical Sadeo-Deleuzian Fugue.

c.4115 B.C. (Cain)

The original sin of Adam in the garden of Eden would, by way of a contagion polluting the moral and ethical probity of all of his progeny, precipitated the first murder in history. So it is thought. This was to be the heinous and deleterious act of Cain upon his own kith and kin, his brother. God unto the first man:

> Adam, now open thine eyes, and first behold
> Th' effects which thy original crime hath wrought
> In some to spring from thee, who never touched
> Th' excepted tree, nor with the snake conspired,
> Nor sinned thy sin, yet from that sin derive
> Corrpution to bring forth more violent deeds. (*Paradise Lost*, XI.423–428)

This was the foretelling of the murder of Abel by Cain. When Cain had risen up against and slain his brother Abel after the firstlings of the latter's flock had found favor with God whereas his own agricultural produce had not and the voice of his brother's blood had cried to heaven from the ground, he was driven far from the face of the Lord to live as a fugitive and a vagabond on the earth. In this vagabond existence, the Damocles' sword of death would forever hang over him, for he complained fearfully that anyone who found him would kill him. Reassuring Cain that whosoever would slay him would reap vengeance sevenfold, God however, set a mark upon him protecting him from all of his would-be murderers—thus placing his continued existence under reprieve (Gen. 4.3–15). Cain, despite his heinous crime, would live.

In this connection, Alan Dershowitz maintains that in spite of the severity of his crime, God is rather soft on Cain. To be sure, being excluded from the clan could carry serious consequences in a society as the one in which Cain dwelt, since it returned the excluded person to the state of nature and exposed him to the elements as well as to animals. Even in early England, being denied the protection of the king's peace was dangerous. Notwithstanding, the mark imprinted upon Cain by God effectively ensconces him in God's "witness protection program."[14]

As Wiesel maintains of Adam and Eve, despite being faced with death, they decided to soldier on by giving life, by conferring a meaning upon it. For as long as Adam lived, even at a distant remove from paradise and from God alike, victory nowise belonged to death but to him. Though to begin is God's prerogative and privilege alone, it is given to man to begin again. Like Adam, he does so every time he chooses to defy death and side with the living.[15]

This would likewise be the case with Cain. He, in fact, enjoyed a strange immunity: no one has the right to imitate him by murdering him. In this, he was protected by God. But also by his very deed itself. Having perpetrated an act nobody before him had ever done, he became untouchable precisely on account of the newness and enormity of his crime.[16]

In José Saramago's novel on the primorial murderer, Cain's survival is taken to far greater lengths than merely these. He goes on to be the illicit lover of Noah's wife Lilith in a city in the region of Nod. As a result, the Damocles' sword of terminal annihilation is once more raised over his cranium. For Noah, privy to these goings on, plans to kill him. Suddenly, Cain is ambushed by three men armed with swords and knives in cahoots with a treacherous slave, one of whose swords is unsheathed and pointed at his chest. Suddenly, however, the sword turns into a cobra, which the horrified man shakes from his hand. The would-be assassin's accomplices flee, leaving him head down, weeping, tearing out his hair, and regretting a thousand times over ever having chosen the profession of a murderous bandit.[17]

From here on, Cain travels through time, exhibiting perfect timing to stop Abraham from murdering Isaac,[18] to witness the destruction of Sodom,[19] the giving of the Mosaic Law to the Israelites at Sinai and the massacre of the worshippers of the Golden Calf by the Levites,[20] the destruction of the Midianites,[21] Joshua's annihilation of Jericho,[22] the travails of Job,[23] eventually hopping onto Noah's Ark, whilst the mark on his forehead grows bigger and bigger, "like a black sun rising up above the horizon" of his eyes.[24]

c.2588 B.C. (Noah)

A midrash from the *Sefer ha-Aggadah* maintains that one day the angel Metatron sent emissaries to the angel Shamhaza who informed the latter that the Holy One will destroy the world by bringing a flood upon it. Shamhaza fell to weeping and grieving. This contention was buttressed when that same night, Hiva and Hayya, the sons of Shamhaza with the human maiden Ishatar, both had prophetic dreams. The first saw a stone on the ground, lying flat like a table, with its top incised with many lines of script. Suddenly, an angel came down from heaven with a chisel, with which he scraped and erased all the lines, with the exception of only four letters. The other dreamt of a large and beautiful orchard with various kinds of trees. However, in the orchard were angels who felled all the trees, leaving but one solitary tree with three branches. With his sons having in consternation intimated these dreams to their father, Shamhaza was all the more confirmed in his prognosis that the Holy One would bring devastation to the world.[25] The Damocles' sword of death was set over the entire human race.

Indeed, when God saw the wickedness of man throughout the earth and the evil that pervaded his heart, he swiftly came to regret that he had made him and decided to annihilate the entire species: "I will destroy man whom I have created from the face of the earth; both man, and beast, and the creeping thing, and the fowls of the air: for it repenteth me that I have made them" (Gen. 6.8). "The end of all flesh is come before me; for the earth is filled with violence through them; and behold, I will destroy them with the earth" (Gen. 6.14). "And behold, I, even I, do bring a flood of waters upon the earth, to destroy all flesh, wherein is the breath of life, from under heaven, and everything that is in the earth shall die" (Gen. 6.18).

In this connection, Robert Alter observes the words "All the wellsprings of the great deep burst and the casements of the heavens were opened" of Genesis 7.11 to be a direct reversal of the second day of creation, when a great vault was erected to divide the waters from above from the waters below. Biblical cosmogony and apocalypse are always the recto and verso of the same coin. The Flood abounds in verbal echoes of the Creation, in which what was made in six days is wiped out in forty. Moreover, the words "and the waters multiplied" from Genesis 7.17 attaches the very verb of proliferation employed in the Creation story for living creatures to the instrument of their destruction.[26]

"For yet seven days, and I will cause it to rain upon the earth forty days and forty nights: and every living substance that I have will I destroy from off the face of the earth" (Gen. 7.4). With the tremendous divine Damocles' sword falling on mortals below, all beasts, cattle, fowl, creeping things and men upon the face of the earth were put to death and destroyed by the flood waters which covered even the mountains (Gen. 7.21–23). According to another midrash, Rabbi Yohanan maintains that every rain drop that God caused to come down upon the generation of the flood, he first brought to a boil in Gehenna.[27]

The Damocles' sword would, however, be obviated in the case of Noah and his three sons (the four remaining letters in Hiva's dream and the solitary tree with three branches in Hayya's dream). For, after the global carnage of all sentient life had come to pass and with the waters beginning to recede, his ark, bearing as it did his wife, sons, their wives and a multitude of beasts, found itself resting safely upon the mountains of Ararat (Gen. 8.4–5). Henceforth the divine Damocles' sword of death would likewise be lifted from above the necks of all living beings upon the earth: "I will not again curse the ground any more for man's sake; for the imagination of man's heart is evil from his youth; neither will I again smite any more everything, as I have done" (Gen. 8.21).

c.2191 B.C. (Isaac)

According to a midrash attributed to Rabbi Yose ben Zimra, following the feast given upon Isaac's having grown and been weaned, Satan speaks thus to God: "Master of the universe, out of the entire feast that this old man upon whom You bestowed fruit of the womb at the age of one-hundred—out of the entire feast he prepared, could he not have spared, say, one turtledove, one fledgling, as an offering to You?" To this, God replies: "Is it not true that Abraham prepared the feast in honour of his son? Still, if I say to him, 'Sacrifice your son to Me,' he will sacrifice him at once." Hereupon, Satan retorted: "Try him." God did, indeed, try him.[28]

The divine Damocles' sword would come to stand poised inauspiciously over the tender young life of Abraham's son Isaac. God commands the patriarch one day: "Take now thy son, thy only son Isaac, who thou lovest, and get thee into the land of Moriah: and offer him there for a burnt offering upon one of the mountains which I will tell thee of" (Gen. 22.2). Obeying the Lord, Abraham rose up early on the morning of the fateful day and, saddling his ass and taking Isaac and cleft wood for the burnt offering, made for the divinely appointed place where he would slay his son (Gen. 22.3–8). According to a midrash, along the way, Satan appears to Isaac, warning him: "O hapless son of a hapless mother! How many fasts did your mother fast, how many prayers did she utter until at last you were born! And now this old man has gone mad in his old age and is about to slit your throat."[29]

Indeed, as Alan Dershowitz maintains in this connection, even in Abraham's mind the imminent death of his son seemed inevitable. For nothing in God's past behavior led Abraham to believe that his son would survive. After all, this was the same God who destroyed the world in the flood and was prepared to annihilate the innocent with the guilty in Sodom. Why would such a God not also cajole one of his votaries to kill a single child?[30]

Abraham came to the place where the ghastly filicide was to be commited, built an altar for human sacrifice, laid wood upon it, bounded his son, and placed him upon the altar (Gen. 22.9). Isaac said: "Father, hurry, do the will of your Maker, burn me into a fine ash, then take the ash to my mother and leave it with her, and whenever she looks at it she will say 'This is my son, whom his father has slaughtered.' ..." Abraham's eyes were directed at Isaac's, and Isaac's at the heavens. Tears flowed from Abraham's eyes, until his entire body was all but afloat in them.[31]

However, upon stretching forth his knife with murder in his heart and about to slaughter his own child, suddenly the Damocles' sword of death was stayed by divine intervention (Gen. 22.3–10). A midrash maintains that Satan appeared and shoved Abraham's arm aside, so that the knife fell out of his hand. When he reached out attempting to seize the knife once more, his mouth fell wide open with weeping as a great cry of anguish erupted from his mouth. Then, with his eyes blinking frantically, he looked up to the divine Presence and pleaded in a rising voice. At that, God appeared above the angels and flung open the firmament. God said to the angel Michael: "Why are you standing still? Do not let Abraham go on!"[32] Isaac beholds the higher sanctuaries of the *merkava*, of creation, with entire rows of angels lamenting: *Yakhid skokhet veyakhid nishkat*—"Look at the slaughterer, he is alone and so is the one he is about to slaughter." Isaac had become the center of the universe. Wiesel: "He could not be allowed to die, not now, not like this. And die he would not."[33] And so the angel of the Lord called upon the would-be slaughterer from heaven, saying: "Abraham, Abraham," "Lay not thy hand upon the lad, neither do thou anything unto him: for now I know that thou fearest God, seeing thou hast not withheld thy son, thy only son from me" (Gen 22.11–12).

Reflecting upon the sheer distance separating Moriah and Golgotha, Wiesel emphasizes how the placing of the existence of Isaac under reprieve apostrophizes the Jewish insistence that man cannot use death as a means of glorifying God. No man has the right to sacrifice another, not even to God. Had he actually killed his son, Abraham would still have become the forefather of a people—but not the Jewish people. For the Jew, all truth must arise from life, and never from death. At the top of Moriah, not only is it the case that the living remains alive, but the *Akedah* moreover marks the end of an era of ritual murder.[34]

As Dershowitz has it, Moriah likewise marks the distance between the older Canaanite God, who allows sacrifices to go forward unhindered, and the Abrahamic God, who brings them to a halt.[35] Heretofore, parents would quite willingly immolate their children to many a deity. Though Abraham was willing to do the same, his hand was stayed—thereby signaling that the God of the Jews was different: not a God of death, but a God who ensures the existence of the subject despite overwhelming odds.[36]

Dershowitz, in what could be initially seen as a sudden volte-face, provides an analogous though alternative interpretation concerning the reason for Isaac's survival of the Damocles' sword looming over his proto-corpse. He refers to a contemporary Conversative Rabbi named Harold Schulweis who, in a rather strangely secularizing move, renders internal what had hitherto been a dialectic of externally opposed moments (Abraham's resolve to kill his son, as against God's sudden prohibition of the said act), to the extent that the angel who stays Abraham's hand is now his own moral conscience. Notwithstanding this internalization, what is foregrounded is still the same: the appearance of the God of Abraham as a moral God who would and could not will the death of an innocent.[37]

Herein, the future survival of the Jews is linked to that of Isaac, for they shall repeatedly undergo their own *Akedah*—and on each occasion survive as a collective subject. Isaac is beyond the reach of Satan: he defies death. After Moriah, Isaac would devote his life and his right to immortality to the defense of his people. In this connection, Wiesel recounts the following midrash: "At the end of time, say our sages, God will tell Abraham: Your children have sinned. And Abraham will reply: Let them die to sanctify Your name. Then God will turn to Jacob and say: Your children have sinned. And Jacob will reply: Let them die to sanctify Your name. Then God will speak to Isaac: Your children have sinned. And Isaac will answer: *My* children? Are they not also Yours? Yours as well?"[38]

According to Rabbi Yehoshua, Isaac was always going to survive the Damocles' sword of Mount Moriah: "*This ram had been living in paradise since the sixth day of creation, waiting to be called. He was destined from the very beginning to replace Isaac on the altar.*"[39]

c.2006 B.C. (Joseph)

Already when it was evident that his father Jacob, son of Isaac, loved his son Joseph more than all of his other sons, the Damocles sword of death was beginning to hang over Joseph's neck. For, because of Jacob's selective love, Joseph's brothers began to hate him. When Jacob had made his love of Joseph especially manifest in making him a multicolored coat, the other brothers ceased to speak peaceably to him. When Joseph intimated a dream to his brothers wherein he espied himself as a sheaf to which the sheaves representing his brethren would bow in obeisance, their ire was piqued all the more. When he blithely made them privy to a subsequent dream in which the sun and the moon and the eleven stars also made obeisance to him, the situation was compounded, for they came to envy and hate him still further. With this, the Damocles' sword of death, navigated by his seething brethren, moved even closer to the neck that he so casually exposed. For when they begazed his silhouette approaching them from afar at Dothan in Shechem, his brothers were already conspiring to kill him (Gen. 37.3–18): "Behold, this dreamer cometh. Come now therefore, and let us slay him, and cast him into some pit, and we will say, 'Some evil beast hath devoured him': and we shall see what will become of his dreams" (Gen. 37.19–20).

This their Damocles' sword was, however, temporarily stayed when Reuben delivered him out of their murderous hands: "Let us not kill him," "Shed no blood, but cast him into this pit that is in the wilderness, and lay no hand upon him." Thus did he desire to rid him from their thanatic hands and deliver him back to his father. Notwithstanding, the Damocles' sword still hovered, for they stripped him of his coat and cast him into an empty and waterless pit to die of starvation and dehydration (Gen. 37.21–25).

Wiesel avers that, at the crucial moment, while his brothers were binding him and throwing him to the ground, Joseph would have been confronted by a sudden searing and terrifying thought: perhaps his father had sent him there, since he wanted him to be killed. Perhaps, Jacob was seeking hereby to emulate Abraham by sacrificing his son. It is no coincidence that in the Shechem and the Moriah episodes, respectively, both Isaac and Joseph are designated by the same word: *naar*, or adolescent. Moreover, when called by God, Abraham had responded: *Hineni*, here I am; whereas Joseph, when sent by his father, replied with exactly the same word.[40]

Moreover, as regards the waterless pit, pivotal here would be Thomas Mann's referral to the word *bôr,* used by Joseph's homicidally inclined brothers, meaning at once both "well" and "prison."[41] This word was also very closely associated with the idea of the "lower" world, that is, with the realm of death. Here, the thought of prison and that of the underworld were exactly the same. They were synonyms for one another, especially inasmuch as a well was of itself conceived of as tantamount to being an entrance to the underworld and moreover that the rotund stone that usually fit over it even intimated death itself. For it covered its circle like the shadow of the dark moon. *Bôr* was, furthermore, the underwordly sheepfold, the Etura, the realm inhabited by the dead—where the son becomes Lord, shepherd, martyr, victim, and mutilated god. And it was Joseph, the semi-divine son, who had already been mutilated, his lip ripped open, his skin lacerated, his robe torn from his body and mutilated and soiled with goat's blood, that is, the blood that was to symbolize his own.[42]

In this sense, *bôr* conjures up the demise (*Ableben*) of the *Dasein*—that of Joseph—beholden to it. Joseph's brothers had not only cast him into the pit, but had in every single way bound him to the grave and absolutely precluded his rising again.[43] More pointedly: having been abandoned in his pit and in being dead to Jacob through his brothers' subsequent foul skulduggery, Joseph demises without biologically dying—though the said skulduggery ensures that it is *just as good as if he had indeed factually died.* Mann avers further that it was the three days of cruel pain that he had undergone in the well that had swayed him to affirm his brothers' trust in death, and that his decision and resolve not to disappoint them was all the more firm since it was not voluntary, but rather as automatic and logically necessary and inescapable as the silence of a dead man, a silence whose keeping became over time as obdurate and weighty as the earth covering the deceased.[44]

But, Mann and we know that Joseph does not die in the pit. For the Damocles' sword hanging over him comes soon to be stayed when he is conveyed into the hands of Ishamaelite traders from Gilead, who lifted him out of the pit and transported him to Egypt where he was sold to Potiphar, an officer of Pharaoh's, and captian of the guard (Gen. 37.26–28, 36).

We may refer this staying of the Damocles' sword to the fact that the underworld is replete with imagery of an elemental life suggestive of a living-on beyond the prosaic fact of death, the miasma of the undead excess of pullulating life (suggesting—though not identical with—life as interminable senescence along the diachronic axis of Blanchotian dying)—alluding to the immortality proper to *Dasein,* that *Dasein* does not perish (*Verendet*) as *Dasein.* Thus, the fact of Joseph's living-on as recorded by the biblical narrative may be read as an allegory of the imperishability of *Dasein* as *Dasein*, with Joseph's ensuing life beyond the pit standing for this same immortality of *Dasein* as itself.

The Damocles' sword of death would hover over Joseph once more when he is falsely accused by Potiphar's wife of making advances upon her. The punishment of death, notwithstanding, is commuted to imprisonment (Gen. 39.7–20). Upon interpreting Pharaoh's dreams he was raised out of this second pit and elevated to grand vizier of Egypt (Gen 41.25–45).

c.1576 B.C. (The Mosaic Exodus)

During their Egyptian captivity, the Damocles' sword was reared and ready to strike the Jews. For Pharaoh commanded the Hebrew midwives to slay all newly born Jewish sons. Yet their existence was placed under reprieve, as the midwives refused to do so, allowing the people to multiply liberally and to flourish (Exod. 1.15–17).

Referring to Exodus 1.19, wherein it is said that the Hebrew midwives are "hardy," Robert Alter associates the latter adjective with *hayot*—already used twice with regard to the newborn—maintaining that it is derived from the verb "to live."[45]

A midrash attributed to Rabbi Hana maintains that the Israelite women hid the infants in holes in their homes. Notwithstanding, the Egyptian despotic Damocles' sword continued to hang over the necks of male Hebrew infants, as Pharaoh charged all his people to cast all such newborns into the Nile. However, upon espying this deathly threat, the Holy One said to his ministering angels: Descend from my Presence and look at the children of My beloved Abraham, Isaac, and Jacob being thrown into the river. The angels immediately rushed headlong from the divine Presence and caught the children of Israel in their arms and set them upon rocks, out of which God brought forth nipples which suckled the children.[46] Other infants were hid in the fields. When the Egyptians became aware of these infants and sought to slay them, the infants were miraculously swallowed by the ground and, upon the departure of the Egyptian would-be murderers, burst forth from out of the ground like grass.[47]

Commenting on Exodus 1.22, wherein Pharaoh, despairing of cooperation from the Hebrew midwives in his genocidal project, now enlists the entire Egyptian population in a search-and-destroy operation whereby the Hebrews would be eradicated by cutting off all male progeny while the girls are raised for sexual exploitation and domestic service by the Egyptians by whom they would be rapidly assimilated; Alter notes that these plans of Pharaoh will once more be frustrated, as the future liberator of the Hebrews will indeed be gently placed (and not flung) into the Nile and eventually emerge to thwart Egypt.[48]

The future liberator is, indeed, Moses. The same Damocles' sword hung over his life, yet was stayed when his mother hid him for three months and then placed him on a river in an ark of bulrushes which conveyed him to the safety of Pharoah's daughter (Exod. 1–10).

As regards the foregoing, Alter alludes to the fact that the basket in which the infant Moses is conveyed is, in fact, called a *tevah*, or ark, the same word used to describe Noah's ark. Whereas in Genesis, a universal deluge nearly destroys the entire human race, here Pharaoh's decree to drown every Hebrew male infant threatens to destroy the people of Israel. As the ark in Genesis bears on the water the saving remnant of mankind, the child borne on the waters here will save his imperiled people. Moses is borne to safety upon water, which Pharaoh had bethought the very means to destroy all the Hebrew male children. Moreover, his floating among the reeds (*suf*) foreshadows the miraculous Hebrew triumph over the Egyptians that he will lead in the parting of the Sea of Reeds (*yam suf*). His very name, in that folk etymology relates it to the Hebrew verb *mashah*, "to draw out from water," and in that the active form of the verb used for the name is *mosheh*, "he who draws out," points—as Alter renders clear—Moses' future destiny of rescuing his people from the water of the Sea of Reeds.[49]

The Damocles' sword of death would, however, soon again hover over Moses. For when he slew an Egyptian taskmaster that was abusing a Hebrew slave, Pharaoh wanted his blood. One midrash maintains that Pharaoh besought a sword that had no equal and had him struck ten times on the neck, but that Moses' neck became like a marble pillar, and the sword did him no harm. According to Rabbi Yannai, after Moses was seized, taken to the scaffold, tied up, had the sword laid upon his neck, and was condemned to be beheaded; an angel disguised in the form of Moses descended from heaven and was seized by the executioners as Moses escaped. Hereupon, as Rabbi Joshua ben Levi maintains in a midrash, to further assist Moses' flight from the Egyptian Damocles' sword of death, the Holy One made some of Pharaoh's counsellors mute, some deaf, some blind and some lame, such that when Pharaoh asked them where Moses was, the mute ones could not speak, the deaf ones could not hear, the blind could not see, and the lame could not walk. Moses thus evaded death by fleeing into the land of Midian. Notwithstanding, the self-same Damocles' sword hanging over Moses was, moreover, hanging over the entire enslaved Hebrew populace (Exod. 1–2).[50]

Notwithstanding, appearing to Moses in the burning bush at Moriah, the living God vouchsafed to lift the Damocles' sword hanging over Moses' people, saying: "I have surely seen the affliction of my people which are in Egypt, and have heard their cry by reason of their taskmasters: for I know their sorrows; and to bring them up out of that land unto a good land and a large, unto a land flowing with milk and honey, unto the place of the Canaanites, and the Hittites, and the Amorites, and the Perizzites, and the Hivites, and the Jebusites. Now therefore behold, the cry of the children of Israel is come unto me: and I have also seen the oppression wherewith the Egyptians oppress them. Come now therefore, and I will send thee unto Pharaoh, that thou mayest bring forth my people the children of Israel out of Egypt" (Exod. 3.7–10). To assure Moses that this liberation would, at length, verily come to pass, God specifically assumed the form of a thornbush that was not consumed (Exod. 3.2).

A midrash avers that this was because Moses had thought to himself that Egypt might consume Israel. Therefore, the Holy One showed Moses a fire burning in a thornbush which latter was nevertheless not consumed, effectively saying to him: As the thornbush, though burning, is not consumed, so the Egyptians will be unable to consume Israel.[51] Rabbi Simeon ben Yohai contends in a midrash that God chose to reveal himself to Moses specifically in a thornbush, in that of all bushes, the thornbush is most apt to inflict harm. Whenever a bird flies into it, the bird cannot emerge unscathed, for many parts of its body will be ripped open. This thorn bush was precisely the Egyptian Damocles' sword hanging over the Hebrews. For the divine task of aiding anyone to emerge free from and unscathed out of slavery in Egypt was more delicate by far than out of any other slavery in the world. In this connection, according to Rabbi Yose, it is characteristic of a thornbush that when a man sticks his hand into it, he is not injured, as the sharp ends of its thorns are pointed downward. On the other hand, when he attempts to draw his hand out of the bush, the thorns fasten onto it. Likewise, when the Hebrews first entered Egypt, they were well received. However, desirous to leave, the Egyptians fastened onto them.[52]

As Alter maintains of Exodus 3.12, to the question about bringing the Israelites out of Egypt, God responds that the fire in the bush is the concrete token of the miraculous power Moses will exert as God's agent in rescuing his people.[53] In reference to the Ten Plagues, Alter maintains that the story of liberation from Egyptian bondage will be a process of violent oppression to be broken only by violent counterstrokes.[54]

Yet, prior to the foregoing, the Damocles' sword of death would again hang over Moses, as God himself would seek to slay him. The existence of Moses would, however, be placed under reprieve with his circumcision of his son (Exod 4.24). For Alter, circumcision here serves as an apotropaic device, to ward off the hostility of a dangerous deity by offering him a bloody scrap of the son's flesh, a sort of symbolic synecdoche of human sacrifice. This is a life-threatening trial and perilous rite that the hero must undergo before embarking on his mission proper. Notwithstanding, the deity that appears here on the threshold of Moses' return to Egypt is dark and dangerous, a potential killer of father or son. Blood, in the same double function it will serve in the Plagues narrative, is set starkly in the foreground: the blood of violent death, and blood as the apotropaic stuff that wards off death—the bloody foreskin of the son will be matched by the tenth plague by the blood smeared on the lintel to ward off the epidemic of death visiting the firstborn sons.[55] From existence under reprieve to existence under reprieve. And once again.

In the plagues narrative, when Pharaoh refused to let the Hebrews go, God smote Egypt by turning its waters into blood, such that the fish in the rivers died and the water was undrinkable (Exod. 7.14, 20–22). When Pharaoh again refused to let the Hebrews go, frogs were made to cover the land of Egypt (Exod. 7.23, 8.6). The next time Pharaoh hardened his heart, all the dust of the land was turned to lice (Exod. 8.15, 17). But Pharaoh's heart was again hardened. So, the Lord infested the entire land with grevious flies (Exod. 8.19, 24). The next time Pharaoh refused to let the Hebrews go, God afflicted all the cattle, horses, asses, camels, oxen and sheep of the land with death (Exod. 8.32, 9.3, 6). When Pharaoh again refused to liberate the Hebrews, the Lord afflicted all Egyptians with boils (Exod. 9.7, 10, 11). The next time Pharaoh refused to let the Hebrews go, God rained down hail and fire from the heavens, slaying all herbs, trees, men and beasts that chanced to be in the fields (Exod. 9.12, 23–25). Again would Pharaoh refuse to free the Hebrews, and God sent a plague of locusts that darkened the entire country and destroyed every herb of the land and the fruit of all trees (Exod. 9.34, 35, 10.13–15). The following time Pharaoh hardened his heart and refused to let the Hebrews go, God covered the entire land with thick darkness for three whole days (Exod. 10.20–23).

In this connection with the last-named plague, Alter maintains that the contrast between the light in Goshen, where the Israelites resided, and the terrifying darkness in the rest of Egypt sets the stage for the distinction between life for the Israelites and death for the Egyptians in the tenth and last plage.[56] For when Pharaoh hardened his heart yet again, the Lord killed all firstborn Egyptian children, even that of Pharaoh and his maidservants, those of prison inmates, and the firstborn of cattle and other beasts, such that there was not a single house in which there was not one dead (Exod. 10.27, 11.5, 6, 29, 30). Yet while this divine Damocles' sword was falling upon the entire land, the existence of all Hebrews was placed under reprieve, saved as they were by the blood they smeared on the lintel and side-posts of their houses (Exod. 12.23).

For Alter, this smeared blood serves to ward off evil spirits, the "Destroyer" (*mashhit*) of Exodus 12:13 being a particularly scary instance of such a spirit. Much as the blood of circumcision had saved Moses' life, so here the blood on the lintels wards off the sacrifice of Hebrew firstborns.[57] Referring to Exodus 12.48, Alter notes the resonance between the apotropaic blood of circumcision, the apotropaic blood of the lamb on the doorposts, and God's saving Israel from the bloodbath of Egypt.[58]

This time Pharaoh was induced to lift the despotic Egyptian Damocles' sword from over the Hebrews' heads. Six-hundred-thousand men, women and children marched out of the land with their flocks and herds along a line of flight, of absolute deterritorialization, into the desert, led by God in a pillar of a cloud by day and in a pillar of fire by night (Exod. 12.31, 32, 37–50, 13.21).

However, the despotic Damocles' sword of Pharaoh would again rear its head and threaten the Hebrews with death. A massive army of six-hundred chariots pursued the children of Israel and overtook them encamping by the sea, beside Pi-hahiroth, before Ball-zephon. Beholding the Egyptians marching after them, the children of Israel were overtaken by blood-curdling fear and cried out to God (Exod. 14.7–10). They complained to Moses: "Because there were no graves in Egypt, hast thou taken us away to die in the wilderness?" (Exod. 14.11).

But the Egyptian Damocles' sword would be stayed. For Moses spoke thus to his people: "Fear ye not, stand still, and see the salvation of the Lord, which he will show you today: for the Egyptians whom ye have seen today, ye shall see them again no more for ever. The Lord shall fight for you, and ye shall hold your peace" (Exod. 14.13–14). Hereupon, the pillar of cloud came between the Egyptians and the Israelites, becoming a cloud of darkness to the former and a light to the latter. Moses parted the sea and the children of Israel walked through it on dry ground, whereas the pursuing Egyptians were drowned en masse by the returning waters, all lying dead upon the seashore in the aftermath as the Hebrews were conveyed to safety (Exod. 14.19–31).

Concerning the parting of the waters in Exodus 14.22, Alter refers to Ilana Pardes who maintains the foregoing to be a passage from the saving of a man from the water in which he floated as an infant, and a genocidal decree in which water was to be the means of killing Hebrew babies, to water as the salvation and beginning of the birth of a nation.[59]

The school of Rabbi Ishmael, in a midrash, compares Israel prior to crossing the Red Sea to a dove who flees from a hawk and is about to enter a cleft in a rock when she suddenly comes upon a serpent nesting there. If she enters the cleft, the serpent will hiss at her threateningly. If she turns back, the hawk stands outside ready to seize her. In this imbroglio, the dove cries and beats her wings, so that the owner of the cote would come and deliver her. Similarly, the Israelites could not go down into the sea, as it had not yet split. Nor could they go back, as Pharaoh had already drawn near. They thus call upon the Lord and he delivers them.[60]

Another midrash compares the fleeing Israelites to a child who, when threatened by a wolf from the rear and brigands from the front, is taken into his father's.[61] Here, God's vouchsafing the children of Israel's safe passage across the sea is compared by Rabbi Hama bar Hanina to a shepherd leading his sheep across a river faced with a wolf bent on attacking his sheep. So as to save his sheep, the shepherd takes a large he-goat, throwing it to the wolf, saying: "Let him wrestle with this one until we cross the river, and then I will return to recover it." Similarly, with Israeli having departed from Egypt, Satan arises questioning God's moral probity in being ready to split the sea for what were heretofore mere idol-worshippers. In response, God turns the upright and non-idolatrous Job onto the Devil, conveys the Israelites to safety across the sea whilst Satan is busy with Job, and at some point thereafter saves Job.[62]

Ultimately, being forever saved from the Egyptian Damocles' sword of death, the Israelites burst out into triumphant song:

> I will sing unto the Lord, for he hath triumphed gloriously:
> the horse and his rider hath he thrown into the sea.
> The Lord is my strength and song, and he is become my salvation:
> he is my God, and I will prepare him a habitation;
> my father's God, and I will exalt him.
> The Lord is a man of war:
> The Lord is his name.
> Pharoah's chariots and his host hath he cast into the sea:
> his chosen captains also are drowned in the Red Sea.
> The depths have covered them:
> they sank into the bottom as a stone.
> Thy right hand, O Lord, is become glorious in power:
> thy right hand, O Lord, hath dashed in pieces the enemy.
> And in the greatness of thy excellence thou hast overthrown them that rose up against thee:
> thou sentest forth thy wrath, which consumed them as stubble.
> And with the blast of thy nostrils the waters were gathered together,
> the floods stood upright as a heap,

> and the depths were congealed in the heart of the sea.
> The enemy said, I will pursue, I will overtake, I will divide the spoil:
> my lust shall be satisfied upon them.
> I will draw my sword, my hand shall destroy them.
> Thou didst blow with they wind, the sea covered them:
> they sank as lead in the mighty waters.
> Who is like unto thee, O Lord, amongst the gods?
> who is like thee, glorious in holiness, fearful in praises, doing wonders?
> Thou stretchest out thy right hand,
> the earth swallowed them.
> Thou in thy mercy hast led forth the people which thou hast redeemed:
> thou hast guided them in strength unto thy holy habitation.
> The people shall hear, and be afraid:
> sorrow shall take hold on the habitations of Palestina.
> Then the dukes of Edom shall be amazed:
> the mighty men of Moab trembling shall take hold upon them:
> all the inhabitants of Canaan shall melt away.
> Fear and dread shall fall upon them;
> by the greatness of thy arm they shall be as still as stone;
> till thy people pass over, O Lord,
> till thy people pass over, which thou hast purchased.
> Thou shalt bring them in, and plant them in the mountain of thy inheritance,
> in the place, O Lord, which thou hast made for thee to dwell in,
> in the Sanctuary, O Lord, which they hands have established.
> The Lord shall reign for ever and ever. (Exod. 15.1–18)

As Elie Wiesel has it, were it not for the Mosaic revolution, the Jew would not be Jewish; that is, he would no longer exist.[63]

c.538 B.C. (The Babylonian Captivity)

Yet, a thousand years later the survival of the collective Jewish subject is again threatened. God makes privy to the prophet Jeremiah the looming Damocles' sword that he would wield against sinful Israel through the instrumentality of Neduchadnezzar and his frightful Babylonian war machine:

> … I will bring evil from the north,
> and a great destruction.
> The lion is come up from his thicket,
> and the destroyer of the Gentiles is on his way;
> he is gone forth from his place to make thy land desolate;
> and thy cities shall be laid waste, without an inhabitant.
> For this gird you with sackcloth, lament and howl:
> for the fierce anger of the Lord is not turned back from us. (Jer. 4.6–8)

This Damocles' sword would bring untold death and dereliction:

> And the carcasses of this people shall be meat for the fowls of the heaven, and for the beasts of the earth; and none shall fray them away.
>
> Then will I cause to cease from the cites of Judah, and from the streets of Jerusalem, the voice of mirth, and the voice of gladness, the voice of the bridegroom, and the voice of the bride: for the land shall be desolate.
>
> At that time, saith the Lord, they shall bring out the bones of the kings of Judah, and the bones of the princes, and the bones of the priests, and the bones of the prophets, and the bones of the inhabitants of Jerusalem out of their graves.
>
> And they shall spread them before the sun, and the moon, and all the host of heaven whom they have loved, and whom they have served, and after whom they have walked, and whom they have sought, and whom they have worshipped: they shall not be gathered, nor be buried; they shall be for dung upon the face of the earth. And death shall be chosen rather than life by all the residue of them that remain of this evil family, which remain in all the places whither I have driven them, saith the Lord of hosts. (Jer. 7.33, 34, 8.1–3)

Death would come to the Jews in a diverse number of ways, and widowhood and the bereavement of children would cover the land like a plague:

> And it shall come to pass, if they say unto thee, Whither shall we go forth? then thou shalt tell them, Thus saith the Lord; Such as are for death, to death; and such as are for the sword, to the sword; and such as are for the famine, to the famine; and such as are for the captivity, to the captivity.
>
> And I will appoint over them four kinds, saith the Lord: the sword to slay, and the dogs to tear, and the fowls of the heaven, and the beasts of the earth to devour and destroy.
>
> . . .
>
> Thou hast forsaken me, saith the Lord, thou art gone backward: therefore will I stretch out my hand against thee, and destroy thee; I am weary with repenting.
>
> And I will fan them with a fan in the gates of the land; I will bereave them of children, I will destroy my people, since they return not from their ways.
>
> Their widows are increased to me above the sand of the seas: I have brought upon them against the mother the young men a spoiler at noonsday: I have caused him to fall upon it suddenly, and terrors upon the city.
>
> She that hath borne seven languisheth: she hath given up the ghost; her sun is gone down while it was yet day: she hath been ashamed and confounded: and the residue of them will I deliver to the sword before their enemies, saith the Lord. (Jer. 15.2–3, 6–9)

Death would be ubiquitous and would fall on all:

> For thus saith the Lord concerning the sons and concerning the daughters that are born in this place, and concerning their mothers that bore them, and concerning their fathers that begot them in this land;

> They shall die of grievious deaths; they shall not be lamented; neither they be buried; but they shall be as dung upon the face of the earth: and they shall be consumed by the sword, and by famine; and their carcases shall be meat for the fowls of heaven, and for the beasts of the earth
>
> ...
>
> Both the great and the small shall die in this land: they shall not be buried, neither shall men lament for them, nor cut themselves, nor make themselves bald for them ... (Jer. 16.3–4, 6)

It is in this vein that, upon the ascension to the throne of Jehoiakim and the ensuing resumption of another relapse of the inhabitants of Judah into their former moral depravity, Jeremiah began to announce the doom, complete oblivion and utter annihilation of the entire nation. However, during one of his first public polemics at a festive gathering in the Temple where he caustically and vehemently denounced the cult of Neith and the burning of children, the Damocles' sword of death would soon threaten to fall upon Jeremiah himself. For he was seized by the idolatrous priests and false prophets there assembled who cried: "Thou shalt die, for thou hast prophesied that this temple will become like that of Shiloh" (Jer. 26. 8, 9).[64] He was, however, released from harm at the behest of several princes and especially his friend Ahikam.[65]

Shortly after Nebuchadnezzar's victory over Egyptian pharaoh Necho, Jeremiah foretold of the impending Damocles' sword that God would wield against Judah through the instrumentality of the Babylonian war machine. This eventuality would become the more acute when, upon being given the choice between subjection and annihilation by Nebuchadnezzar, Jehoiakim prevaricated, anticipating Egyptian aid or protection within the Temple the people had befouled with idolatrous perversities. This tempting of the rapidly looming Chaldean Damocles' sword was all the more in evidence when, upon being read a prophetic scroll of Jeremiah's foretelling doom in the absence to submission to Nebuchadnezzar delivered to the Temple by his disciple Baruch, Jehoiakim cast each page of the scroll into the fire.[66]

Jehoiakim would in fact rapidly wield his own Damocles' sword, ordering the arrest of Jeremiah and Baruch. The princes, however, took care to hide the two prophets in a secret place where they could not be found, thus saving their lives. Moreover, with Jehoiakim having finally obligated himself to the duties of vassalage to the Chaldeans, Jeremiah could leave his hiding place without fear of being killed by the king's henchmen. For he was protected not only by his prince's friends but also by Nebuchadnezzar himself who knew Jeremiah had prophesied a great career for him and had advised the people of Judah to submit to him.[67]

However, Egyptian seductions and Phoenicia's declaration of independence induced Jehoiakim into alliance with the former and into refusal to pay tribute. His son and successor Jehoiachin's continuation of his reckless and suicidal recalcitrance finally precipitated the raining down of the Chaldean Damocles' sword. Nebuchadnezzar captured all of Judah with the exception of a few cities in the south and laid siege upon Jerusalem, dethroning Jehoiachin and dragging him into exile with his entire family and over ten-thousand other inhabitants.[68]

The Babylonian Damocles' sword would at length fall with still greater rapacity upon Judah when Jehoiachin's successor and Nebuchadnezzar's vassal Zedekiah—at the urging of Egypt, other countries subjugated by the Babylonians, and his self-seeking nobles—himself refused to pay his tribute.[69] After easily capturing most of the fortified cities of Judah, Nebuchadnezzar renewed his siege of Jerusalem. This Damocles' sword was, however, stayed when the Chaldeans were compelled to postpone it due to a momentary Egyptian distraction. During this time, Judah again lapsed into its former depravity and Jeremiah, standing before the king and nobles, rebuked them and announced that the Chaldean Damocles' sword would promptly return and that Nebuchadnezzar would capture Jerusalem and that fire, the sword, pestilence and famine would rage in the city.[70]

Once more, Jeremiah's announcement of the looming Damocles' sword threatening to fall upon Judah brought into existence a corresponding Damocles' sword threatening to fall upon Jeremiah himself. For hereupon the nobles of Judah turned upon him with bloodthirsty hatred. The situation was compounded when he desired to leave Jerusalem to visit his native Anathoth, for herein he was suspected of wanting to defect to the Babylonians. He was thus treated by the nobles as a traitor and a spy, struck in the face and hurled into a cistern in a dungeon controlled by the sadist Jonathan. Yet, though he languished for many days in solitary confinement in a narrow and filthy cell (where he sobbed out a heartrending dirge from the third chapter of the Lamentations), his existence was notwithstanding still under reprieve—for the nobles dared not kill him after all.[71]

With Jeremiah being absolutely right, after routing the Egyptians and the wholesale defection of neighboring nations, the Chaldean Damocles' sword now hovered over Judah still more inauspiciously. They returned with fury to once more besiege Jerusalem. Secretly, in the inner chamber of his palace, Zedekiah inquired of Jeremiah as to the future outcome of the war. Jeremiah replied that defeat was certain. Though it seemed as if this would cause Zedekiah to kill Jeremiah for purported treachery, it in fact humbled the former and induced him to release his captive from Jonathan's prison, and place him under protection in his palace, in the court of Mattarah. Meanwhile, the Chaldean Damocles' sword was falling all the more violently and horribly upon Jerusalem from day to day. Jews were being slaughtered and felled by famine by the thousands—the streets of Jerusalem strewn with dead and unburied pestilential corpses decomposing in the scorching summer heat. Children expired in their mothers' arms, and mothers cooked and ate their own children to assuage their ravenous hunger.[72]

When in the midst of this carnage Jeremiah continued to urge surrender to Nebuchadnezzar as the only means of staying the Chaldean Damocles' sword of death, he himself came under the Damocles' sword of the nobles who again wished to murder him for discouraging the resistance of the defenders. He was cast into a slimy pit in the court of Mattarah to die of starvation:[73]

> Mine enemies chased me sore
> like a bird, without cause.
> They have cut off my life in the dungeon,
> and cast a stone upon me.
> Waters flowed over my head;
> then I said; I am cut off. (Lam. 3.52–54)

He would have undoubtedly perished there had it not been for an Ethiopian servant of the king, named Ebed-melech, who prevailed upon the king to remove the prophet from his dank pit. Thirty men raised him out of the hole with ropes. Jeremiah was not completely released but placed under guard in the court of Mattarah and protected by the king from the homicidal Damocles' sword of the nobles:[74]

> I called upon thy name, O Lord,
> out of the low dungeon.
> Thou hast heard my voice:
> hide not thy ear at my breathing, at my cry.
> Thou drewest near in the day that I called upon thee:
> thou saidst, Fear not.
> O Lord, thou hast pleaded the causes of my soul;
> thou hast redeemed my life. (Lam. 3.55–58)

Meanwhile, the Damocles' sword was raining down upon Jerusalem without surcease in the triple guise of famine, pestilence and the sword. The bread supply failed and, with the depletion and exhaustion of the defenders, the Chaldeans breached the wall—pouring into the city, raping women, killing youths, and slaughtering men. They threw the population into chains, massacred the priests and prophets in the Temple, plundered the treasures, and desecrated all holy places. Zedekiah's sons were murdered in front of him—after which he was blinded, chained and led captive to Babylon to die. Nebuchadnezzar ordered the commander of his guard, Nebuzaradan, to destroy the city without mercy and with bludgeoning homicidal violence. The city walls were razed to the ground and all houses, palaces and the temple were burned down.[75]

The greater part of the population of Judah perished by the sword, famine and pestilence; a smaller portion was led into captivity; and a tiny fraction remained in the country, surrounded by a plethora of bitter and ravenous enemy nations bent on their terminal annihilation.[76] This last remnant was carried away into captivity in a final punitive expedition by Nebuzaradan, leaving Judah a veritable, depopulated desert—an arid, bereft, and desiccated abode roamed by slavering wild beasts.[77]

The Damocles' sword which hung over Jeremiah was however stayed, as Nebuzaradan removed his chains and gave him liberty to go wherever he saw fit.[78] He would be left to lament the demise of his nation:

> The Lord hath swallowed up
> all the habitations of Jacob,
> and hath not pitied:
> he hath thrown down in his wrath
> the strong holds of the daughter of Judah;
> he hath brough them down to the ground:
> he hath polluted the kingdom
> and the princes thereof.
> He hath cut off in his fierce anger
> all the horn of Israel:
> he hath drawn back his right hand
> from before the enemy,
> and he burnt against Jacob like a flaming fire,
> which devoureth round about.
> He hath bent his bow like an enemy:
> he stood with his right hand as an adversary,
> and slew all that were pleasant to the eye
> in the tabernacle of the daugher of Zion:
> he poured out his fury like fire.
> The Lord was as an enemy:
> he hath swallowed up Israel,
> he hath swallowed up all her palaces:
> he hath destroyed his strong holds,
> and he hath increased in the daughter of Judah
> mourning and lamentation.
> And he hath violently taken away his tabernacle,
> as if it were of a garden:

he hath destroyed his places of the assembly:
the Lord hath caused the solemn feasts
and sabbaths to be forgotten in Zion,
and hath despised in the indignation of his anger
the king and the priest.
The Lord hath cast off his altar:
he hath abhorred his sanctuary,
he hath given up into the hand of the enemy
the walls of her palaces;
they have made a noise in the house of the Lord,
as in the day of a solemn feast.
The Lord hath purposed to destroy
the wall of the daughter of Zion:
he hath stretched out a line,
he hath not withdrawn his hand from destroying:
therefore he made the rampart and the wall to lament;
they languished together.
Her gates are sunk into the ground;
he hath destroyed and broken her bars:
her king and her princes are amont the Gentiles:
the law is no more;
her prophets also find no vision of the Lord.
The elders of the daughter of Zion
sit upon the ground and keep silence:
they have cast up dust upon their heads;
they have girded themselves with sackcloth:
the virgins of Jerusalem hand down their heads to the ground. (Lam. 2.2–10)

A midrash from the *Sefer ha-Aggadah* has God ordering Jeremiah to summon Abraham, Isaac, Jacob, and Moses from the dead. Standing on a bank of the river Jordan, Jeremiah calls upon Moses, the son of Amram, from his grave to behold how enemies have ripped apart his flock. Raised up from the earth, Moses immediately rends his garments of glory with which the Holy One had clothed him and, putting his hands on his head, walks weeping and moaning. Upon the destruction of the Temple, Abraham weeps before the Holy One, plucking his beard, pulling out the hair of his head, striking his own face, and rending his garments. With ashes on his head, he bestrides the site of the Temple, lamenting and crying.[79]

Another midrash has Jeremiah maintaining the vanquishing and near-annihilation of the collective Judaic subject as a divine chastisement, much like that inflicted by God upon Job. As were Job's sons and daughters taken away from him, so too were the sons and daughters of Jerusalem taken away. As were Job's gold and silver taken from him, likewise was the gold and silver of Jerusalem taken away. Much as Job was cast upon a dung heap, so too was Jerusalem reduced to a heap of dung. But, much as God had redeemed and placed the existence of Job under reprieve, so too would he comfort, resurrect and vouchsafe the resurrection and ultimate salvation of the Jews.[80]

Having been summoned from their graves, Abraham, Isaac, Jacob and Moses beseech God to have mercy upon the children of Israel, such that the collective Judaic subject—at the time faced with annihilation through being rendered politically and ethnically obsolete—be redeemed and saved from terminal desuetude. Abraham and Isaac petition the Holy One to save the children of Israel from the death of exile as he had saved Isaac from Abraham's dagger. Jacob petitions the Holy One to save Israel from the death of exile as he had saved Jacob from the murderous rage of Esau. In the same midrash, Moses says to Jeremiah: "You go before me, so that I may lead them back to the Land of Israel and see who dares to lay a hand upon them." When the exiles see Moses at the rivers of Babylon, they say to one another: "The son of Amram has come up from his grave to deliver us from the hand of our adversaries." Then a divine voice comes forth from the heavens, saying: "It is a decree from My Presence."[81]

Millions of Jews perished by the sword, by famine and by pestilence, or were carried to distant countries, to be perpetually lost to posterity. Only a miniscule remnant remained—a remnant that was itself threatened with terminal extinction. In this connection, Heinrich Graetz maintains that, according to the normal strictures of human calculation, the latter handful of exiles could not possibly have survived in the strange and hostile environment in which it was forced to abide. That the remnant ultimately did not in fact perish, but instead rose phoenix-like out of the dust, Graetz unequivocally regards views as entirely inexplicable.[82]

As Max Dimont maintains, according to the Spenglerian conception of history; with the death of Israel and Judah, the time was nigh for Jewish civilization to die. The spring of Jewish civilization had been ushered in by Moses, Joshua, and the Judges. Its summer stage was embodied by the attainments under the aegis of David and Solomon. Its autumn phase may be considered as commencing at some point after its diremption into the separate kingdoms of Israel and Judah. The winter stages for both of these would be the period of decline running up to their ultimate annihilation. In accordance with the foregoing conception of the history of human civilizations, that of the Jews seemed at this point to be at an end. And, certainly in the case of the kingdom of Israel, this proved to be the true; since, in the wake of the Assyrian death-wielding depredations, the Ten Tribes of Israel had completely vanished from the historical record, never to reappear again. Countenanced as they were with their Babylonian exile, it seemed as though the exact same fate awaited the Jews of Judah—and, therefore, the Jews as a whole.[83]

Such imminent obsolescence of the Jews—as a collective ethnic, racial, religious and linguistic subject—would have seemed an absolute certainty in light of what happened to other nations in pagan times. Such nations, when marching into captivity, almost invariably marched into terminal desuetude—if not physically, then certainly as a collective national subject. These vanquished nations embraced the idols and *Weltanschauung* of their conquerors alike, and would dilute their strain almost to the point of nullity through interbreeding. This is so because these pagan captives did not particularly care whether or not they survived as Hittites, Phoenicians, Syrians, Jebusites, and so forth. In a manner homologous to that of the Hegelian proto-slave, they were merely desirous of forestalling their own biological extinction—and quite willing to relinquish their religious and national identity to vouchsafe this lesser end. This was even the case with the kingdom of Israel. It would, however, *not* be the case with Judah. The reason for this is that something happened between the destruction of Israel by the Assyrians in 722. B.C. and the defeat of Judah in 586 B.C. This occurrence, or linked series of occurrences, enabled the Jews not merely to survive but moreover to regenerate themselves.[84]

It is these historical vicissitudes which imbued the future captives of Judah with an obdurate and implacable will to survive—and to do so *as Jews*. What gave them the will to retain their Jewishness despite every obstacle and threat may be traced back to a grand spiritual awakening, a sort of revivalist movement, between the fall of Israel and that of Judah. This prototypical monotheistic revival was not altogether unlike the far later Christian revivals that would occur on both sides of the Atlantic. We have in mind here, obviously, the original Protestant Reformation, as well as the uncompromising stance of the Covenanters in seventeenth century Scotland and Ulster, who came to Virginia and Pennsylvania with Presbyterians and other non-conformist groups; the Methodism of John Wesley and George Whitefield in eighteenth century England; the First Great Awakening in the American colonies in the 1730s and 1740s, which massively reshaped the Congrenational, Presbyterian, Dutch Reformed, Baptist and Methodist denominations; the Second Great Awakening of the first four decades of the nineteenth century in the United States; the Third Great Awakening in North America in the decades from the run up to the American Civil War onwards; and on to the various Pentecostal revival movements reaching far into the twentieth and twenty-first centuries. As in this slew of Christian revivals, during the intervening hiatus between the defeats of Israel and Judah, the very contours of individual Jewish character and the concept of Jewishness itself were rebored and burnished. That Judah was also facing the Damocles' sword of terminal obsolescence is cearly evinced from its bitter factionalism, internal strife, and the terrifying external threat by which the nation was beset. Jewish identity was, in fact, countenanced with a triple threat. Idolatry was poisoning religion, the oppression of the poor was undermining morality, and interbreeding was eroding racial purity. The point was approaching at which the unity of the Jewish collective subject had all but been eviscerated. Judah was primed for disappearance. In the Toynbeean sense, at least, Jewish civilization had in fact already come to an end. It had likewise perished at this stage for Spengler, as in his view the Jews were by now purportedly fossilized entities. Yet, the Jews continued to reemerge in ensuing history, refusing to fit into these frameworks.[85]

The manner in which they managed to forestall the onset of terminal obsolescence was by canonizing part of Holy Scripture. These vicissitudes were ultimately brought to fruition at the behest of King Josiah, who enjoined his High Priests to edit what are now termed the Jahwist and the Elohist documents, so as to collate them into Holy Writ. Though ultimately well-intentioned, there was more than a smidgen of casuistry in these machinations. For, after the revamped documents had been surreptitiously secreted in a secluded location of the Temple, Josiah suddenly proclaimed to boisterous fanfare that a long-lost book written by Moses under divine inspiration had perchance been unearthed in the Temple and would be read to the general public. This book is now known as the book of Deuteronomy. There is an alternative version of this event, perhaps equally casuistic: the text had, in fact, lain in the deep recesses of the Temple since as early as the reign of Solomon, and was fortuitously discovered during the course of the Temple's renovation. Whether casuistic or not, this sleight of hand and shrewd savvy by Josiah and his charges, served to unleash a well-nigh unprecedented surge of renewed patriotism and religion fervor throughout the nation. Riding the effervescent tide of this collective subjective upsurge, Josiah moved to eviscerate idolatry and ram through a barrage of sweeping social reforms.[86]

Along the same coruscating revivalist arc, Josiah likewise established a new form of power: charismatic power. Whereas political power *stricto sensu* rests on physicial and military force; charismatic power, bereft as it is of these means of enforcement, relies instead on the sanctity of office itself. The distance separating these two forms power may be seen as analogous to the distinction between the mode of power wielded by, on the one hand, a president and, on the other, the Pope. Charismatic power ultimately functions through the inculcation and cultivation of an inner discipline in the subject, and is thus a sort of pastoral power. It therefore galvanizes and renders more taut and impervious the hard inner kernel of what is here an almost Rosenzweigian meta-ethical Self. The subject yields to charismatic power on the basis of adherence to the dictates of an inner voice and within the horizon of an ideal of consummate freedom, and does so even in the very teeth of physical danger and probably death.[87]

Though initiated by Josiah, charismatic power and its renewal of subjectivity was itself refined and brought far closer to perfection by the Prophets; who at once served to maintain the purity of Judaism and—still more crucially—rendered it exportable to foreign soil. This purification and untethering of Jewish religious practice from the parochialism of spatio-temporal particularity is what enabled the Jews to survive their Babylonian captivity. After the annihilation of Israel at the hands of the Assyrians, Isaiah and Jeremiah were *primus inter pares* among those instrumental in driving their words into collective Jewish consciousness—to the extent that, by the time of the fall of Judah, the Prophets had already developed a finely-tuned exportable religion to enable the exiles to cling steadfast to their identity from the moment they trudged their way to captivity in Babylon and far beyond. For the Prophets' words had seared themselves into the Jews' racial memories, effectively becoming memes to be transmitted across the generations.[88]

The survival of the Jews as Jews in their Babylonian captivity was likewise effected at the behest of the Prophets' undermining of the value of sacrifice and the heretofore overly fastidious attachment to the sedentary Temple in Jerusalem, in favor of rendering morality superior to ritual—henceforth making Judaism more transcendent of the strictures of time and place. Through synagogue and prayer, the Jew was no longer tied to any specific priesthood, temple, or country. He could be in direct communication with God in any country—sans intermediaries. Formerly immobile and rigid, Judaism became an exportable commodity, resilient and clandestine. These reforms massively bolstered the Jews' resources for survival during the course of their captivity and later dispersion.[89]

This resistance of the Jews to wholesale extinction as a political and moral subject at the behest of the Babylonian Damocles' sword of death may be seen as intimated in a midrash attributed to Rabbi Joshua ben Korhah. The midrash maintains that when Nebuzaradan, the captian of Nebuchadnezzar's guard, had killed ninety-four myriads on one stone in Jerusalem and their blood had flowed and touched that of Zechariah—a priest and prophet murdered by the Israelites for constantly reproving their transgressions—he noticed that the blood of Zechariah kept seething and bubbling up. Thinking that he could still Zechariah's blood, Nebuzaradan slew the members of the Great and the Small Sanhedrin over the same blood. The blood was, however, nowise stilled. He then consecutively slew young priests, youths and maidens, schoolchildren, until he had slain ninety-four myriads altogether—and each time the blood of Zechariah was not stilled. Then he drew near the blood and said: "Zechariah, Zechariah, I have slain the best of them. Will it please you to have me slay them all?" Suddenly, the blood became entirely still. Instantly, Nebuzaradan was racked by remorse and fled away and became a convert to Judaism. Though many individual Jews could be and were killed, the *collective* Jewish subject as a whole would not perish.[90]

The particular, and somewhat counter-intuitive, nature of the Jews' survival of the Babylonian Damocles' sword may be gleaned from Paul Johnson's asseveration apropos the ostensibly bleak and dystopian fatalism of Jeremiah's vitriolic tirades that Judah was duly bound to receive its comeuppance at the behest of Nebuchadnezzar for its own backsliding foulness. Missed by his contemporaries was another message secreted in the wild prophet's rants: the annihilation of Judah as a State was irrelevant as the bond between the Jews and their God, being intangible and indestructible, could well endure beyond the confines of a mere petty nation-state.[91] This uncompromising soteriological stance was likewise that of Ezekiel. Much as with Jeremiah, Johnson contends that the profoundly anti-Statist message of "this weird and passionate man" was that empires and thrones were pettifogging entities and that the important entity was none other than man himself and his direct, albeit law-mediated, relation to God.[92] It was before the terrified eyes of this latter eccentric firebrand that the resurrection of the collective Jewish subject would announce itself, and do so in the form of a bizarre hallucinatory apparition rising from a heap of desiccated bones:[93]

> The hand of the Lord was upon me, and carried me out in the Spirit of the Lord, and set me down in the midst of the valley which was full of bones,
>
> And caused me to pass by them round about: and behold, there were very many in the open valley; and lo, they were very dry.
>
> And he said unto me, Son of Man, can these bones live? And I answered, O Lord God, thou knowest.
>
> Again he said unto me, Prophesy upon these bones, and say unto them, O ye dry bones, hear the word of the Lord.
>
> Thus saith the Lord God unto these bones; Behold, I will cause breath to enter into ye, and ye shall live:
>
> And I will lay sinews upon you, and will bring up flesh upon you, and cover you with skin, and put breath in you, and ye shall live; and ye shall know that I am the Lord.
>
> So I prophesied as I was commanded: and as I prophesied, there was a noise, and behold a shaking, and the bones came together, bone to his bone.
>
> And when I beheld, lo, the sinews and the flesh came up upon them, and the skin covered them above: but there was no breath in them.

> Then said he unto me, Prophesy unto the wind, prophesy, son of man, and say to the wind, Thus saith the Lord God; Come from the four winds, O breath, and breathe upon these slain, that they may live.
>
> So I prophesied as he commanded me, and the breath came into them, and they lived, and stood up upon their feet, and exceeding great army.
>
> Then he said unto me, Son of man, these bones are the whole house of Israel: behold, they say, Our bones are dried, and our hope is lost: we are cut off for our parts.
>
> Therefore prophesy and say unto them, Thus saith the Lord God; Behold, O my people, I will open your graves, and cause you to come up out of your graves, and bring you into the land of Israel.
>
> And ye shall know that I am the Lord, when I have opened your graves, O my people, and brought you up out of your graves,
>
> And shall put the spirit in you, and ye shall live, and I shall place you in your own land: then shall ye know that I the Lord have spoken it, and performed it, saith the Lord. (Ezek. 37.1–14)

According to a midrash, this future resurrection of the collective Jewish subject was already glimpsed by Isaac who, stretched out on the altar on Mount Moriah with his wrists and ankles bound, and thinking that he would soon die, received a vision of the Temple in Jerusalem being destroyed and thereupon rebuilt, that what was about to happen to him would happen to many other Jews down the centuries.[94] Yet, though the tangible reconstruction of the Temple would indeed come to pass, the deeper element in the preservation of the Jews as Jews is to be found in their exile itself. Here, it is through the written word that the identity of the Jews is vouchsafed through the scribes assiduous ordering, editing and rationalizing of formerly oral traditions and the accession of these to print. The scribes are elevated to a special caste and financed by a newly ascendant merchant class.[95] Deprived of their own State apparatus, the Jews forthwith came to endure as a nomocracy. Indeed, their now far more fastidious adherence to the Law and the rite of circumcision in exile served to define and distinguish them more starkly and solidly against the gray pagan mass in their midst. Moreover, their absorption and deployment of Babylonian astronomy further entrenched their religiously defined identity through the temporal punctuation of their calendar. Johnson goes as far as to maintain that, far from eroding their ethnicity, the Babylonian exile served as the enabling ground of the regular practice of their religion and maintenance of their strain. Quite strangely, in a Hegelian retroactive positing of presuppositions in which Jeremiah emerged as the first Jew and Ezekiel as the formulator of Judaism, exile served to connect rather than separate the Jews from their virtual past.[96]

Paradoxically, the adversity of exile served as a spur for Jewish survival, in that it chivvied the Jews not only into a more resolute adherence to their own defining principles and strategies of internal cohesion, but likewise enlivened their imaginations, inventiveness, lucidity of thought and conception, and zeal. Herein lay the essence of Jeremiah's perspicacity and surreal prescience: the well-nigh preternatural observation that, for the Jews, powerlessness at the level of *pouvoir* translated into immense strength at the level of *puissance*. Indeed, they appeared for all intents to be better off without a State of their own. Jeremiah actually comes close to the Marxist and Nietzschean contention that the State is evil in and of itself. But there was also another factor in Jewish survival: the Persians.[97] Cyrus would be the instrument for the purpose of redeeming Israel and of hastening its salvation. This redemption of the Jews would at the same time be the in-gathering of the scattered remnant of Judah and of Israel into Jerusalem. Coextensive herewith would be the transmogrification of the former heart of stone of the Jews into a heart of flesh:[98]

> And I will give them one heart, and I will put a new spirit within you: and I will take the stony heart out of their flesh, and will give them a heart of flesh, that they may walk in my statutes, and keep my ordinances, and do them: and they shall be my people, and I will be their God. (Ezek. 11.19, 20)

This imparting of a new heart to Judah and Israel signals at once the resurrection of the collective Jewish subject and its reprieve from its being mortified under the Babylonian Damocles' sword of death. The eunuchs of the royal family of Judah who had remained faithful to a pure version of Judaism in conjunction with Jehoiachin's grandson, Zerubbabel, obtained the freedom of the Jews from Cyrus. The latter granted the people of Judah the privilege to return to their former homes, and to reconstruct Jerusalem and the Temple from their charred embers.[99]

In the same month in which, eight or so centuries earlier, the Jews had escaped from out of Egypt; their descendants departed from Babylonia after a captivity of forty-nine years to return to and once more take possession of their long-lost and longed-for home of Jerusalem. Cyrus deployed a thousand horsemen to protect the Jews on their lengthy journey from the rabid attacks of marauding robber bands that infested the highways.[100]

Indeed, the Jews—as Dimont maintains—had not only prospered in their Babylonian exile and become more consolidated and resilient as a collective national subject; they had in fact multiplied their numbers. Whereas at the beginning of the exile there had been merely 125,000 Jews in the whole world, these numbers had grown to 150,000 Jews in Babylonia alone. About a fourth took advantage of Cyrus' edict and returned to Jerusalem. Here, they joined the small number of Jews who survived amid the debris and ruin of three devastating wars which had led to the exile in Babylonia in the first place.[101]

70 A.D. (The Roman destruction of Jerusalem)

A midrash has it that, already forty years before the destruction of the Temple by the Romans, the lot "For the Lord" did not come up in the high priest's right hand, nor did the crimson stand turn white, nor did the western-most lamp burn continually. Moreover, the doors of the Temple Hall would open by themselves, until Rabbi Johanan ben Zakkai reprimanded them: "Temple Hall, Temple Hall, why do you yourself keep sounding the alarm? I already know about you, that you are to be destroyed, for long ago Zechariah, a descendant of Iddo, prophesied to you, 'Open thy doors, O Lebanon, that the fire may devour the cedars.'"[102] Lebanon is, of course, another name for the Temple Hall, since the word *Levanon* means "whitener of sins."

As Simon Schama relates, in about 62 A.D., a strange prophet named Jesus began to be obstreperous at various Jewish pilgrimage festivals by expectorating lamentations of impending doom: "Woe, woe to Jerusalem!" He would continue these vitriolic tirades for a full seven years and five months, despite being flogged for his pervasion of ostensibly subversive rhetoric until "his bones were laid bare." We are, of course, not speaking about Jesus Christ.[103]

This was to prove true. From the spring of 68 A.D., Vespasian's Roman armies were already turning all parts of Judaea and the environs surrounding the capital into an arid and abject desert of sempiternal desolation. In March 70 A.D., the Roman Damocles' sword stood poised to fall upon Jerusalem itself, as Titus assembled an army of no less than eighty-thousand men and a greater number of siege engines was provided than in any previous war. As his furious assault unfolded, captives were crucified outside the city walls, sometimes as many as five-hundred a day, to terrify stubborn defenders with the grisly fate that was theirs. To prevent the besieged from leaving the city through secret passages, Titus surrounded the entire city with a gargantuan earthen wall more than four miles in circumference. This wall in itself was a Damocles' sword, in that it effected a suffocating blockade resulting in the ravages of famine. The houses and streets were filled with corpses which remained unburied. Living skeletons stalked the city. The ubiquitous agony of death drove people to surrender and suffer death by disembowelment. The Romans breached the walls and set the Temple on fire. All Judaeans who did not escape the burning structure were murdered. Titus entered the Holy of Holies and gloated over the sight before being driven back by the suffocating smoke.[104] The ghastly spectacle of the defunct sanctuary sent the Judaeans into wailing lamentations. The sea of flames, whose infernal roaring shook the very earth and sent sickening echoes throughout the surroundings, signaled to the inhabitants that all hope had been banished from the world.[105]

Defeated by despair and no longer desirous of surviving the Temple, many immediately committed suicide by throwing themselves headlong into the incinerating flames, annihilating themselves in the black hole of stochastic death. The Romans threw themselves upon the thousand men, women and children who had remained in the southern collonades, massacring them. The Temple burned to the ground, with only the foundations and a few ruins of the western wall remaining. These rose from the earth like giant specters. Ominously, the second Temple was reduced to ashes on the very same day as the first Temple fell. The walls were completely destroyed. The entire city was razed to the ground. Over a million people perished.[106] Graetz: "Zion again sat upon the scene of her desolation and wept; her sons had fallen; her daughters were led into ignominious slavery, or distributed among the Roman slavery to satisfy their lust. She was still more unhappy than at the time of the first destruction, since no prophet appeared to foretell the end of widowhood and sorrow."[107]

Over nine-hundred-thousand Jews were taken prisoner during the war. Of these, all suspected of being combatants were crucified on the spot. Of the remainder, seventeen-thousand died of starvation. Male captives over the age of seventeen were doomed to pine and wither away in the Egyptian mines as imperial slaves, and a still greater number of young Jews were given over to the provinces to die at the hands of wild beasts in the arenas of public depravity. Boys and female captives glutted the slave market, each sold for a trifle. Judaean youths were dragged in chains through Syria. They were devoured in cages by savage beasts in Caesarea, whilst twenty-five-hundred perished in forced gladiatorial contests against themselves, irrigating the sand of the arena with their blood. Seven-hundred Jewish captives, including Zealot leaders John of Gischala and Simon bar Giora, were dragged in chains along Titus' grotesque and obscene victory parade through Rome. Bar Giora was hurled down the Tarpeian rock. Seventeen-thousand men were put to the sword at the fortress of Machaerus, and the women and children sold into slavery. The incendiary Roman siege of the fortress of Masada induced over nine-hundred men, women and children to voluntarily annihilate themselves in the black hole of stochastic death. Six-hundred Zealots were captured and executed in Egypt, and the rest scattered throughout the country, gradually seized and tortured to death. The Zealots of Cyrenaica were seized and executed, their ringleader Jonathan dragged in chains with other prisoners to Rome, scourged and burned alive.[108]

The nation of Israel was no more. Judaea was now for the most part an arid, depopulated, and desiccated wasteland. Its cities and villages lay in melancholy ruin.[109] The national body was dismembered and suppurating from a thousand bloody wounds. Its very heart, the Temple, had been ripped out. It seemed as though this would be the last hour of the Judaean race, and that God himself had doomed the Jews to terminal destruction. This is, indeed, what many of the bedraggled and emaciated survivors felt.[110]

The hopelessness of the situation was poetically reinforced by the fact that the particular Jesus who had dementedly, though correctly, foretold the annihilation of the city was himself brought to naught by a stone fired by one of the Roman siege engines half-way through one of his frenzied lamentations. The seeming termination of Jewish existence is, in fact, still visible today. On the reverse side of the Western "Wailing" Wall there is a massive discombobulation of masonry dislodged by a Roman projectile. One of the stones contains a niche large enough for the shofar blower, whose trumpet would announce the Sabbath. This regulator of the systole and diastole of Jewish life had been silenced for good. In the "Burnt House" in the Jewish Quarter in the dusty frame of a glossily carbonized roof beam there is a faded photograph of the skeletal arm of a young girl found in the ruined basement, her outstretched finger-bones reaching out in desperation. In vain.[111]

Yet, something strange had happened before this lamentable scene had entered the historian's purview. Whilst the war was still raging, and in spite of the dangers of such an undertaking in a faction-splintered Jerusalem controlled by a Zealot party whose firebrands would murder anyone who attempted to leave Jerusalem, a sealed coffin redolent of a corpse in the final stages of decomposition is covertly smuggled out of city. Suddenly, in a preternatural scene, the body resurrects, uttering "*Vive Imperator*." The addressee of this utterance is Vespasian, whose troops are in the process of concertedly laying the city to ruin. Isaiah's much earlier prophecy that only a king could take Jerusalem, as emanating from the resurrected man's lips, would at length become true. Vespasian would become Roman emperor within a few short years. Yet, the prophecy was not granted him before the man in the coffin, the always living Yohanan ben Zakkai smuggled beyond the perimeter in a coffin packed with rancid meat, was granted a small wish by the Roman commander: to build a simple center of Jewish learning in the town of Jamnia (Jabneh).[112]

Schama maintains here that the Jews would survive the physical obliteration of Jerusalem, destruction of the Temple, desuetude of the priestly cult, and erosion of much of the Jewish settlements across Palestine and environs, owing to a very similar move to the one effectuated during the Babylonain exile: a move from the State to cathedocracy and the intensified proliferation of the written Word in the hermitic and hyper-intellectual underworld of the Talmud. With ben Zakkai, all trappings of Temple cult, sacrifices and pilgrimage, undergoing the effects of pervasive movements of deterritorialization, would be virtualized. Judaism floated free from its historical grave as chronological Jewish time ground to a halt and underwent transmutation into an eternal present that would forthwith incessantly be reanimated through the endless refrains of cultural memory. Even though Titus' grand and boisterous victory procession would with promiscuous ostentation parade the outward symbols of their purportedly powerful defeat of the Jews, inclusive of dragging rebel Simon bar Giora by a rope around the neck and torturing him to death in the Forum at its conclusion, the Romans were utterly unable to do away with the essence of Judaism as embodied in the Sifrei Torah. Although the swords and projectiles of the Romans had defeated the more improvised weapons of the Jews, the words of the Jews would far outstrip the swords of the Romans. These words would float free from their material embodiment, much as would the *nefesh*—although surely Schama here means *neshama* or at least *ru'ah*—from the body. The universalism of the Torah had made Judaism imperishable. Yohanan ben Zakkai had been right. The Romans would not endure. The Jews would.[113]

1492 A.D. (The Marranos under the Spanish Inquisition)

Moses Maimonides—reflecting upon the response of a rabbi to an enquirer as to whether, when forced at sword point, to choose either death or conversion—emphatically bethought ostensible conversion and consequent life by far the more prudent and noble path to pursue. What moved him unequivocally in this direction, as he rendered pellucid in his *Mishneh Torah*, was Leviticus 18.5, which enjoins Jews to live by the commandments and not die by them. This is bolstered by Deuteronomy 13.15: "I have this day set before you a blessing and a curse; life and death; therefore choose life."[114]

However, in time, even conversion would cease to deflect the numerous rapacious Damocles' swords pointed at Jewish necks. In the fifteenth century, rabbis who had refused to follow the path of ostensible conversion were already beginning to warn Jewish converts what was in store for them. Rabbi Yitzhak Arama: "You will find no rest among the gentiles, and your life will hang in the balance." Concerning the forcible converts, or *anusim*, he foresaw: "One third burnt by fire, one third flying hither to hide and those who remain living in deadly fear." Rabbi Yehuda ibn Verga conceptually divided the *anusim* into three distinct pairs of turtle doves: the first pair would remain in Spain, be fleeced of their property and either be slaughtered or burned; the second pair would likewise be dispossessed, yet would survive by fleeing the country when the nefarious depredations of the State apparatus loomed; whereas the third pair would make off with both their lives and their property by ensuring that they were the first to escape. Events would brutally confirm these sentiments. The Spanish Jew was no longer able to evade anti-Semitic hostility through conversion.[115] Things would, however, become still worse than even this. According to the Jewish author, Samuel Usque, in the latter stages of the fifteenth century, a veritable monster had emerged from out of Rome, whose hideous countenance or the mere mention of its name alone would send a shudder through all of Europe. This ravenous beast, whose voice alone killed more readily than the dread stare of the basilisk, was an unprecedented homophagiac and vampire. It would plunge whole swathes of humanity into an Egyptian midnight. Everything that it touched withered and died. With its poison, it would reduce all in sight to a parched desert entirely bereft of but a scintilla of life.[116] "Every misery, every mortal pang, conceived only by the most extravagant imagination of the poet; every horror that can thrill the heart of man to its lowest depths, these monsters in the garb of humility brought upon the Jews of the Hesperian Peninsula."[117] Spain was soon to be turned into one vast, smoke-enveloped, spiteful, vindictive, treacherous, necrophilous, snitching, petty-minded dungeon.[118]

Already during the lifetime of Don Henry IV, the Damocles' sword of death would hang poised over Spanish Jews. This looming sword was already presaged by Tomas de Torquemada's early extortion of a vow from the overly scrupulous Infanta Isabella to the effect that she set herself to the extirpation of heresy immediately upon assuming the throne. The movement toward the final and full wielding of the Inquisitorial Damocles' sword was accelerated by the Dominican prior Alfonso de Ojeda, who strained every nerve to set the Inquisition in motion against the Marranos (Jewish Christians). The papal nuncio in Spain, Nicolo Franco, supported his proposition for a tribunal to call these latter to account for their transgressions. The scheme would be given the assent of Don Ferdinand, eager to fill his coffers with the plunder of the accused. In November 1478, Pope Sixtus IV issued a bull empowering the Spanish sovereigns to appoint inquisitors from among the clergy, with full authority to sit in judgment on all heretics, apostates, and their patrons. The commission appointed by the sovereigns concluded that the Marranos were obstinate in their errors and irreclaimable. The tribunal of the Inquisition was appointed on September 17, 1480. As soon as it took up residence in the convent of St. Paul at Seville, on January 2, 1481, it issued an edict for officials to deliver up the Marranos and confiscate their goods. Soon Marranos who either avowed their old religion or made horrible confessions on the rack were incinerated in the black hole of stochastic death. Not only the hunted Marranos, but in fact every Spaniard was called upon by an edict of the inquisitors to become an informer. Everyone was bound to give a list of acquiantances guilty of Jewish heresy within three days, under pain of excommunication. Even the most seemingly insignificant actions were regarded as signs of aggravated heresy. *Autos da fé* started from January 6, 1481. Unhappy victims clad in the hideous, short and clinging San Benito were marched to the place of torment where unrepentent Jewish heretics were given over to the flames and repentant ones were first straggled. In fact, not even death itself afforded a safeguard against the Damocles' sword of the Inquisition. For the corpses of proselytes who had died in purported heresy were dug up from their graves and incinerated.[119]

The Damocles' sword of the Inquisition would be exacerbated on February 11, 1482, when Sixtus IV appointed Tomas de Torquemada among others as chief inquisitors. This brought a fresh influx of victims to the stake.[120] Sixtus IV hereupon issued a bull the logical consequence of which was that only one step remained before clergy of Jewish blood would be damned to the stake.[121]

Schama maintains that the phenomenon observed here was not so much an ultimate incarnation of pure Medieval madness, but rather an embryonic form of the modern police State. It was, one may add, a primitive form of Stalinism, with firebrand theologians as *intellegentsia*, inquisitors as *nomenklatura*, the general populace as one gray snitching mass of stool pigeons, and the Jews as enemies of State. Snitches were even to be found in monasteries and convents, as monks and nuns reported on their brothers and sisters in Christ for purportedly looking down when the Host was raised during Mass, or for stumbling over their prayers in the manner of temporary aphasiacs.[122]

In May 1483, Tomas de Torquemada was appointed inquisitor general. He was "the incarnation of the Holy Inquisition with all its devilish malice, its heartless severity, its bloodthirsty ferocity."[123] He filled his ranks with individuals ready at his word to perpetrate the most hideously necrophilous barbarities with the composure of homophagiacs. These men filled Spain with the putrefaction of the dungeon, the stench of corpses, and the crackling flames of burning

Jews. A blood-curdling cry of collective Jewish misery suffused the land like the scream of a thousand score lunatics in an echo chamber, as the arm of every man was paralysed to put a stop to the butchery. Those who tendered confessions of their relapse into Judaism within thirty days were spared death, the catch being that they were obliged to put their confessions into writing and betray all fellow offenders and those even suspected of Judaizing tendencies. If the confessions were deemed insincere or partial, the recalcitrants were incinerated—these latter presumably to thereupon suffer eternal damnation.[124]

The slaughter and incineration of Jews by the proto-Stalinist Spanish State apparatus likewise proceeded in accordance with the barbarous prescriptions of Alfonso de Espina, a firebrand Franciscan scribbler. His compiled volume *Fortalitium Fidei* listed a slew of infractions by which deviant *conversos* could be identified. These included slovenliness in the practice of Christianity, such as merely perfunctory Mass attendance, seldom mentioning the names of Jesus Christ and the Virgin Mary, avoidance in making the sign of the cross, working on Sundays, and abstaining from receiving the sacraments. The familiar, utterly absurd and indeed non-existent crime of host theft and desecration to the end of smearing its blood over matzo cakes, in addition to new ones such as "holding philosophical discussions" were also included as actions meriting bodily incineration.[125]

A Jew of the time, Isaac Arama, observed: "In these days the smoke of the martyr's pyre rises unceasingly to heaven in all the Spanish kingdoms and the isles. One-third of the Marranos have perished in the flames, another third wander homeless over the earth seeking where they may hide themselves, and the remainer live in perpetual terror of a trial."[126] The eleven tribunals would transform Spain into "a blazing Tophet."[127] Soon the Damocles' sword of the Inquisition, having poured out its wrath on Jewish Christians, would come to stretch itself over non-Christian Jews.[128] During the course of fourteen years as inquisitor-general, Torquemada burned at least two-thousand Jews.[129] His successor Deza, the archbishop of Seville, was still worse. His death-wielding instrument Diego Rodriguez Lucero, a hangman in Cordova, was insatiable for Hebrew blood. "Give me Jews to burn," was his constant cry. His thanatophilia degenerated into cannibalistic fury. The third chief inquisitor, Ximenes de Cisneros, continued the march of Marranos to the stake.[130]

It appeared as though the entire Jewish collective subject in Spain would ultimately be devoured by the raging inferno. Yet, some Marranos were already beginning to successfully flee the inquisitorial Damocles' sword somewhat before this moment. Many took refuge in the Muslim kingdom of Granada, in Portugal, Africa, Provence, or Italy.[131] Often a Jewish Christian proselyte condemned by the tribunal would succeed in escaping to Rome, where he purchased absolution from the papal throne.[132] The Jews of northern Spain, in Catalonia and Aragon, fled to neighboring Navarre and were granted shelter by people there who sternly resisted the Inquisition.[133]

Ultimately, in the spring of 1492, the Jews were expelled from the Pyrenean Peninsula. In a sense, this event was experienced as a third destruction of the Temple in Jerusalem and as a third exile threatening the Jews with the typical privations and desolations of fissiparity, starvation, exposure, shipwreck, and death. Part and parcel of Torquemada's drive to effect a complete and utter purification of all converts and their ineluctable reunification with the full body of the Catholic Church was the inexorable adjunct of a wholesale eviction of all Jews from the purview of the micro-Stalinist Spanish Inquisitorial State apparatus. Yet, as Schama fails not to observe in his perspicacity, this was none other than a strangely perverse backhanded complement to and unwitting affirmation of the sheer tenacity and dogged perdurance of Judaism. In a precisely homologous manner to what had transpired with both the Babylonian destruction of the kingdom of Judah and the Roman annihilation of Jerusalem, Torquemada, Deza and Cisneros could gingerly go ahead with the wielding of cold Statist power and unleash the pestilential rabble upon the Jews in the burning of books, incarceration and incineration of bodies. All of this could for all intents be destroyed. Yet, by the very nature of Judaism, there would always be an immortal, "undead"—one may even say "death-driven"—surplus that would unfailingly survive. It was as if, as Schama maintains with eloquence, from the decomposing or charred bodies of the Jews their indelible words "floated free from bodies and parchment, into the air itself like particles of fatal miasma."[134]

Despite, or rather because of, its cowering in trepidation under the Inquisitorial Damocle's sword of death, the collective Spanish Jewish subject strangely exalted itself in Phoenix-like fashion. For, as Graetz expresses it, "the very enormity of the misery they endured raised the dignity of the Sephardic Jews to a height bordering on pride. That they whom God's hand had smitten so heavily, so persistently, and who had undergone such unspeakable sorrow, must occupy a peculiar position, and belong to the specially elect, was the thought or the feeling existing more or less clearly in the breasts of the survivors. They looked upon their banishment from Spain as a third exile, and upon themselves as favourites of God, whom, because of His greater love for them, He had chastised the more severely. Contrary to expectation, a certain exaltation took possession of them, which did not, indeed, cause them to forget, but transfigured, their sufferings. As soon as they felt even slightly relieved from the burden of their boundless calamity, and were able to breathe, they rose with elastic force, and carried their heads high like princes."[135]

The Jews had an entire plethora of means whereby to flee Spain. Armies of ships from the country and from Genoa and Venice were ready in Spanish seaports to take the wanderers and convey them to havens of safety. King Ferdinand I of Naples made his realm free to the Jews of Aragon, Catalonia and Valencia. Many thousands landed in the Bay of Naples on 24th August 1492, with the native Jewish community defraying the costs of passage of those bereft of financial means. Jews were also accepted in Pisa.[136] The Greek islands of Corfu, Candia, and others became filled with Jews.[137] Egypt, and especially Cairo, became a home for Spanish Jewish fugitives.[138] Jerusalem and other Palestinian cities likewise welcomed Sephardic refugees. The number of Jewish families in Jerusalem alone swelled from barely seventy to fifteen-hundred within less than three decades, and the populace rose from its former destitution and barbaric coarseness to prosperity and moral and aesthetic refinement. The community was no longer harassed to death, and driven to despair or exile by rapacious, tyrannical and treacherous factions of immoral thugs.[139] The communities of Safet in Galilee and Damascus in Syria also burgeoned at a prodigious rate.[140] Jews flooded into Italy, where popes Alexander VI, Julius II, Leo X, and Clement VII were tolerant of them. Northern Italian cities received even pseudo-Christian Jews re-converted to Judaism, allowing them all the privileges of free intercourse. Venice, Rome, Padua and the port of Ancona received Jews without hindrance.[141] Asylum and citizenship was likewise extended to fugitive Jews in Poland and Lithuania. Swarms of Jewish families settled on the banks of the Vistula and the Dnieper. The German rabbinical school established itself in Poland, Lithuania, Ruthenia and Volhynia, spreading itself in all directions.[142] Sultan Bajazet II issued a command to all of his dominions to the effect that all hunted and harassed Jews arriving in Turkey be treated with generosity, and threatened with death anyone who deigned to oppress them. Chief rabbi, Moses Kapsali, levied a tax from rich native Jews for the liberation of those Spanish Jews who had come to Turkey as beggars or slaves.[143] Turkey would serve as the refuge and redemption of the greater part of the remnant of the three-hundred-thousand Sephardic exiles. Subequent sultans Selim I and Solyman I (Suleiman I) would grant the Jews the same liberies as enjoyed by Armenians and Greeks.[144] Once again, a passage away from imminent annihilation to continued existence and freedom had been forged for the Jews—this time not through a parted sea, but into vast oceanic expanses.

1656 A.D. (The Chmielnicki Massacres in Poland)

Bogdan Chmielnicki, the fiercesome Ukranian warlord, had long been vexed by Jewish tax farmers and suffered personal wrongs at the hands of Jews. His Damocles' sword would soon hang poised over Poland and especially over Polish Jews. The cruelty of Chmielnicki would combine with the murderous hatred of his neo-Assyrian and proto-Nazi Cossack butchers. Zaporogians and necrophilous Tartars put Polish troops to flight on May 18, 1648. These wild bands would then proceed eastward from the Dnieper, between Kiev and Pultava, plundering and murdering especially Jews who had failed to take flight. Regular troops of murderers called Haidamaks (the Tatar word for partisans), were formed under brutal leaders oblivious to human life, and morbidly luxuriated in the death-throes of their Polish and Jewish foes. Catholic and Jewish women were dragged along the ground by thongs slung around their necks. Dressed as Polish officers, Cossacks massacred the six-thousand Jews of the stronghold of Nemirov amid grisly tortures. One-thousand-five-hundred Jews who had taken refuge in the town of Tulczyn were tortured and executed before the eyes of Polish nobles. All inhabitants of Homel as well as fugitives were stripped outside the town, surrounded by Cossacks, with one-thousand-five-hundred suffering martyrdom instead of being forcibly baptized. The pusillanimity of the Poles before the Tartars under Tugai Bey in the second war (September 1648) against the Cossacks prepared a bloody fate for even those Jews who felt themselves far from the field of battle. The blood of slaughtered and mutilated Jews marked the vast tract of land from the southern part of the Ukraine to Lemberg by way of Dubno and Brody. In the town of Narol the Zaporogians butchered forty-five-thousand, twelve-thousand among them Jews. Among the corpses remained living women and children, who were forced to feed on human flesh. In Czremieniec, hundreds of Jewish children were slain, their corpses scornfully examined as Jews do with cattle, and thrown to the dogs. The failure of a Polish commission to accede to the conditions of Chmielnicki's proposed peace treaty of February 16, 1649, once more brought the Damocles' sword of death upon the Jews. For, deceived that a settlement had been reached, they only returned to their homes to find themselves surrounded and massacred by Cossacks. With the resumption of hostilities between Pollacks and Cossacks in the early part of the year 1651, Jews were again the first victims of Chmielnicki and the Zaporogians. After Chmielnicki broke a November 11, 1651, treaty signed with Polish king John Casimir, the Ukranian Damocles' sword would again fall upon the Jews in Poland. Allying himself with the Russians in the early part of 1654 and 1655, Chmielnicki now started to annihilate the communities formerly spared by Cossack swarms. The community of Vilna was completely depopulated by Russian slaughter and migration. To the Cossack and Russian Damocles' swords ravaging the Jews of Poland was added that of Charles X of Sweden, through which the communities of Great and Little Poland, from Posen to Cracow, were reduced to despair. Even the Poles would begin to ill use and murder the Jews. General Czarnicki, who hated the Jews, assailed them with utter barbarity, destroying synagogues, and tearing up scriptures. Poland was reduced to a bloody battlefield where Jews were savagely abused and slain by all. By the end, a quarter of a million Polish Jews had been slaughtered with the remnant having seemingly fallen into desuetude.[145]

Yet, the Jews would again survive as a collective subject through various subterfuges and forms of solidarity. From the outset of Chmielnicki's campaigns against Poland, hundreds of Jews underwent baptism in the Greek Orthodox Church, and pretended to be Christians, in order to save themselves. Those who fell into the hands of the Tartars were transported to the Crimea, and ransomed by Turkish Jews. Four notable Jewish communities escaped massacre by surrendering to the Tartars with all of their property. They were sold into Turkey, where their coreligionists ransomed them. The Constantinople Jewish community sent a deputy to Holland to collect money from the rich communities for the ransom of the captives.[146]

Much as after the expulsion of the Jews from Spain and Portugal, every place was filled with fugitive Sephardic Jews; so too during the Cossack-Polish war fugitive Polish Jews, haggard in appearance, with hollow eyes, having escaped the sword, the flames, hunger, and pestilence, were seeking shelter everywhere.[147] Much as after their expulsion from the Pyrenean Peninsula, Turkey had emerged as the destination par excellence where the collective Sephardic Jewish subject would evade the Inquisitorial and environmental Damocles' sword of death and indeed survive and flourish; so too would a certain country emerge as a safe haven for them in the wake of the Chmielnicki massacres, when the Jews of Poland were being trodden down, slaughtered, or driven through Europe like terrified quarry. This land of freedom would be England. As Poland was crumbling and England rising to preeminence, the latter would again peacefully admit Jews after a hiatus of more than three-and-a-half centuries. The obstacle to this admittance was, however, the still lingering prejudice that existed against the Jews even in England. But this obstacle was removed by one Manasseh ben Israel, the second or third rabbi at Amsterdam. The gradual release of the Jews from their thousand years of contempt and depreciation in European society and the consequent Damocles' sword which this dangled over the collective Jewish subject would begin with this man.[148]

When the king of England, Charles I, was executed in 1649 and the royalists routed by the philosemitic Puritans, Manasseh glimpsed an opportunity, as many Puritan luminaries sensed the imminence of the Second Coming yet believed, as per Daniel 12.7, that this would transpire only when the Jews were scattered to "the end of the Earth," or *Kezeh ha-Arez*—the Hebrew term for England. Latching onto these sentiments, Manasseh penned two books brimming with ideological ballast supporting this form of millenarianism and, in September 1655, traveled to London himself to present a petition to Lord Protector, Oliver Cromwell, requesting the repeal of centuries old laws forbidding Jewish entrance.[149]

The manner in which Manasseh's aspirations were at long last brought to fruition was, however, very peculiar—yet, at the same time very illustrative of the anti-Statist and extra-legal character of the passional regime itself. No agreement was drawn up according to which the prospective Jewish immigrants were granted special legal status. If this had indeed happened, they would have been relegated to second-class citizens and, with the restoration of the monarchy under Charles II in 1660, any deal would have been renegotiated on possibly much harsher terms. The idea was, therefore, to bypass the law so as not to raise anti-Semitic protestations. This entailed a fundamental secrecy.[150]

The matter was ultimately resolved through what Paul Johnson terms a "tacit conspiracy," marked by a odd event. The background was the Anglo-Spanish war. A certain Antonio Rodrigues Robles, a *marrano* legally, though actually a practicing Jew, was being proceeded against in court as a Spanish alien. Seeing this, twenty *marranos*, declaring themselves refugees, petitioned the court to practice their Judaism privately. The charges against Robles were quashed, the petition was seemingly granted, although the minutes for it were mysteriously lost. This would have entirely salubrious reverberations for all Jews bent on filtering into England.[151]

In fact, the matter of the special status of the Jews was purely and simply abandoned. There being no statute barring their entry, they entered.[152] Ultimately, it was through an almost bizarre act of omission that the Jews would accede to full citizenship in England and, much as in the manner of Catholics and other Nonconformists, could belong to the Church of England and have their religious idiosyncrasies recognized when swearing in court.[153] A 1732 judgment would extend legal protection to Jews against the various ridiculously superstitious libels at the behest of which they had been calumniated and persecuted for countless centuries. At length, they would establish themselves firmly in England and, over the ensuing centuries, forge an Anglo-Semitic alliance that would inexorably support international Jewry across the globe for a very long time.[154]

May 8, 1945 A.D. (The *Sho'ah*)

As Elhanan Wasserman notes, the prophets had long since prophesized that there will be a time of troubles for the people Israel such has never before beset a nation. Referring to the Goan of Vilna's *Even Shelemah*, he maintains that whilst the travails of exile were similar to pregnancy, the catastrophy that would assail the Jews would be like the pangs of birth. Whereas during the exile Jews countenanced with the Damocles' sword of death in one country could still survive by fleeing to another country, such as from Spain to Turkey, Poland and Holland; during the immeasurably more terrible future catastrophe they would be persued everywhere and have nowhere to escape. Whereas during the exile there had been an interval of time between one evil decree pronounced against the Jews and the next, during the prophesized national calamity evil decrees would "flow like a river" (Isa. 59.19) without interruption.[155] Every day would, moreover, be worse: "In the morning thou shalt say, Would it were even!" (Deut. 28.67).

In Ignaz Maybaum's formulation, Auschwitz would be the twentieth-century Calvary of the Jewish people. It would be the Golgotha of modern mankind, where the Cross would be replaced by the gas chamber.[156]

Arthur A. Cohen terms the death camps the *tremendum*, for they are the "monument of a meaningless inversion of life to an orgiastic celebration of death." And due to his dogged persistence, sheer unflagging endurance, and refusal to die, the Jew was the ideal prospective victim of the *tremendum*.[157] The *tremendum* would be an elaboration of the most harrowing Jewish fear: that the eternal people are actually not eternal, but in fact mortal. Indeed, the *Sho'ah* would come perilously close to writing *finis* to the Jewish mythos of indestructibility and their three millennia of endurance.[158]

Already in August of 1920, the Damocles' sword of death began to hoist itself over the collective Jewish subject. Hitler railed in favor of combating the "racial tuberculosis" besetting Europe through the extirpation of its "causal agent, the Jew."[159] In June 1922, Hitler proclaimed that the inexorable mission of the German State was to destroy Bolshevism, and together with it "our mortal enemy: the Jew."[160]

In the second volume of *Mein Kampf*, published on December 10, 1926, Hitler even deployed the necessity of this extermination of the Jews retrospectively, contending that if twelve to fifteen-thousand "Hebrew corrupters of the people" had been put under poison gas at the beginning of the First World War, then "the sacrifice of millions at the front would not have been in vain."[161] Soon the Damocles' sword would begin to actually fall on individual Jewish subjects. In

Berlin on January 1, 1930, eight Jews were killed by Stormtroopers, whereas for the following nine months Jews were molested in public and synagogue services were interrupted. During the September election campaign of that year, seventy-eight Jews were wounded by SA thugs. On the day of the opening of the Reichstag, the windows of Jewish department stores were broken. Nazi deputies walked to the Reichstag amid crowds chanting "*Deutschland erwache, Juda verrecke!*", "Germany awaken, death to Judah!" In 1931, fifty synagogues were desecrated, and several thousand tombstones defiled in more than a hundred Jewish cemeteries.[162]

From March 9, 1933, Jews began to be incarcerated in concentration camps, starting with Dachau. In so-termed *Eizeloperationen*, or "individual operations," all around Germany, Jewish shopkeepers, rabbis and communal leaders were beaten and arrested. The Star of David was painted with the slogan "Perish Judah!"[163] They were likewise being assaulted in the growing archipelago of camps, such as Esterwegen and Sachsenhausen.[164]

On April 26, the implicit death sentence pronounced upon the Jewish subject loomed still larger, with the usurpation of the *Geheime Staatspolizei*, or Secret State Police, by the Nazis. The Gestapo was forthwith given full authority to shadow, arrest, interrogate and intern, without reference to any other State authority. Law courts and due process of law were consigned to the past. The full thanatic despotic State apparatus, with the *Schutzstaffel* (SS) security service, its *Sicherheitsdienst* (SD) intelligence arm, and the Gestapo, was in place. Cowering under its sovereignty was the Jewish subject excluded from the juridical sphere of positive law, that is to say, the Jewish subject as *homo sacer*—or bare life that may be murdered with impunity.[165] In Dachau, Jews were already being killed for trifles.[166]

After the German annexation of Austria in March 1938, Jews were kicked and beaten in Vienna synagogues, made to eat grass in parks and trampled upon. In Frankfurt, caricatures of Jews hanging from gibbets were painted on Jewish shop windows. This was likewise the beginning of mass arrests during which scores of thousands of Jews were hurled into Dachau and the newly opened concentration camp of Buchenwald. From June 9, synagogues were burned down, followed by more and more arrests of thousands of Jews who came to suffer the "tortures of hell" in the "quarries of death," or camps, in which they died frequently. In Berlin, Jews suffered looting and violence, as obscene and bloodthirsty pictures of their being beheaded, hanged, tortured, and maimed littered the cityscape. They were subjected to systematic house-to-house searches, hunted like rats.[167]

With the murder of the German diplomat Ernst vom Rath by Hirsch Grynszpan on November 6, 1938, the Nazi Damocles' sword fell upon the Jews with yet greater force. In the *Reichskristallnacht* pogrom of November 9-10, a thousand synagogues were gutted and smashed up or demolished with explosives. Over seven-thousand Jewish shops had their windows shattered, fittings ripped out, and stock looted and hurled into the streets. Over ninety Jews were killed and countless women and even girls raped. Jews were dragged to kneel before their burning synagogues, driven barefoot over shards of broken glass, and herded out of hospitals with such force that they died. Women were forced to dance naked in basements, sometimes spat on by other women whilst tied to tables. The streets looked like an abattoir. Jews were packed in lorries like livestock headed for slaughter.[168] Thousands of Jews were driven to concentration camps in Sachsenhausen, Dachau, and Buchenwald, where they were beaten to death with spades, clubs and whips, their skulls smashed to pulp.[169]

Speaking in Berlin on January 30, 1939, Hitler declared that in the event of a European war—which he was, of course, planning—the result would be "the annihilation of the Jewish race in Europe."[170] From the time of the crossing of German forces into Poland in the early hours of Friday, September 1, Jews were immediately singled out for massacre by the *Einsatzgruppen* following in the rear of German fighting forces. They were forced into and incinerated in synagogues that were burned down, and herded into churches where they were massacred.[171] In the Warsaw ghetto, hundreds of Jews died every day of hunger and disease. Cadavers were excavated from mass graves by gangs. They were forced into cannibalism. They were raped and executed en masse. Horribly emaciated living corpses died by the thousand among filth of malnutrition and dehydration. Jews froze to death. They fell dead in the middle of the street.[172] Sick children with comatose eyes lay half-dead, covered in sores, breathing heavily with a rattle in their throats. Rats devoured rotting and decomposing corpses in houses haunted by starvation and typhus.[173]

From the very first hours of the German invasion of the Soviet Union on June 22, 1941, the Damocles' sword of death fell with still more rabid and concerted fury upon the Jews. The policy was the systematic destruction of entire Jewish communities in the conquered territories. Jews were butchered by hooligans with axes and clubs studded with nails and buried in pits. Newborn babies were bludgeoned to death at their mothers' breasts. Hundreds of Jews at a time were driven into barns and incinerated.[174]

In Dantesque scenes, Jewish women were pulled out of their homes and driven to killing sites outside city perimeters, completely disrobed, raped, struck and kicked, machined-gunned to death in the wastelands outside industrial suburbs, and buried in pits.[175] They were drowned in blood spilling in puddles out of clay pits by Soviet Prisoners of war co-opted by the SS.[176] Men, girls, women and old men were stripped naked and shot point blank at edge of the Babi Yar ravine outside Kiev. Children were thrown alive into the pit below. Many went insane prior to reaching the place of execution.[177] On roads corpses lay strewn with smashed noses, broken limbs, their tongues cut out, their eyes gouged.[178] As part of the "sardine packing" method, Jews were forced to walk into burial pits before being shot.[179] Victims were forced to lie on top of corpses already supine in pits while shooters fired into their heads. Children, torn from their mothers' hands, were thrown onto the piles of dead and dying or killed with rifle butts.[180] In Odessa, dozens of Jews were strung up from apartment block balconies. Thousands of mutilated bodies lay at their feet. It was a town of the hanged. Tens of thousands of Jews were driven to warehouses, packed inside, and annihilated with machine-gun and artillery fire or incinerated in petrol-doused infernos.[181] Jewish babies were ripped in half, grabbed by the feet, and their heads smashed on stones. Women had their breasts cut off. Entire families were buried alive or burned alive in massive bonfires.[182]

Things would get worse. Over several meetings between 7 and 8 December 1941, Hitler briefed the highest cadres of the Nazi Party of his plans to considerably step up the destruction of the Jews. Goebbels: "The world war is here, the destruction of the Jews must be the inevitable consequence."[183] In Berlin, on December 16, Hans Frank declared before the administration of the General Government: "We must put an end to the Jews." The intent was bloodthirsty and unequivocal: "they will disappear. They must go."[184] Frank added: "We must destroy the Jews wherever we encounter them and wherever it is possible, in order to preserve the entire structure of the Reich."[185]

These vicious sentiments would lead to the Wannsee Conference on January 20, 1942, at which the terminal annihilation of European Jewry would be decided. This would be the "Final Solution," at which the hitherto preponderant proclivity toward a protracted deportation programme aimed at the annihilation of Jews through forced labor and starvation in the occupied territories of a defeated Soviet Union swiftly gave way to the rabid desire to systematically destroy the Jews in death camps in the territory of the General Government well before the end of the war.[186]

This necrophilous intent was again rendered pellucid ten days hereafter in another vitriolic speech by Hitler at the Sportpalast in Berlin. "We are clear that the war can only end either with the extermination of the Aryan peoples or the disappearance of Jewry from Europe."[187] He continued, rabidly averring that "the result of this war will be the annihilation of Jewry." "And the hour will come when the most evil world-enemy of all time will have played out its role, at least for a thousand years."[188] As Martin Gilbert maintains, "The Jews were to disappear altogether. It was the Jews alone who were marked out to be destroyed in their entirety: every Jewish man, woman and child, so that there would be no future Jewish life in Europe."[189]

Upon the decision of a Berlin meeting presided over by Himmler on July 7, 1942, Jews would be subjected to death in yet another way: draconian medical experiments upon Jewish women. So as to undercut the propagation of the Jewish race, Jewish women would be sterilized in concentration camps and subjected to other vile experiments. The genitals of both men and women were subjected to high intensities of X-rays which caused radiation burns, peritonitis and fatal infections. The likelihood of death was further exacerbated by the violent extraction of the irradiated testicles and ovaries.[190] They were likewise killed with injections of phenol, petrol, chloroform and air.[191] As Olga Lengyel maintains, camp inmates were used as guinea pigs for malaria experiments. Some were simply killed so as to be dissected for the purpose of an autopsy. Those suffering from tuberculosis were injected with medicines that swiftly killed them, after which their lungs were removed for laboratory experiments. The sick and feeble were killed with benzine or petroleum injections to the heart.[192]

With the industrialized mass murder of Jews in death camps now well under way, this vastly fast-tracked necrophilia of the Nazi State apparatus was clearly reflected in Himmler's October 4, 1943, speech to his senior SS officers in Posen:[193] "I'm referring to the evacuation of the Jews, the extermination of the Jewish people. Most of you will know what it is like to see one-hundred corpses lying side by side or five-hundred or one-thousand of them."[194] He went on: "We had the moral right, we had the duty to our people, to destroy this people which wanted to destroy us."[195]

In the deployment of the precursor to the gas chambers, Jews were packed into vans, asphyxiated with carbon monoxide fumes fed through a pipe from the exhaust, and their cadavers bundled out and dragged to burial pits by Jewish prisoners in shackles. These latter were thereupon themselves flung into the pits and machine-gunned to death.[196] Women who gave birth were often immediately stretchered to death pits and their newborn infants drowned in latrines.[197] Starving women were even driven in desperation to cannibalism, in certain cases resorting to eating the body parts of their dead children.[198] Shortly before his deportation to Birkenau in April 1944, the Polish Jewish poet Yitzhak Katznelson wrote the poem "Song of the Murdered Jewish People", in which he relates a terrible dream wherein his people no longer existed and in which he wakes up crying, only to find that this was indeed true.[199]

Jews were conveyed to grisly death camps in cattle wagons in a state of acute dehydration and unbearable heat, sandwiched in the midst of the stench of disintegrating flesh, drooling, imbecile, live people with dead minds, and the moans of the dying and cries of pain and fear, like famished and thirsting animals in cages destined for slaughter. Many died in these box-cars after hours of suffocation. Those who broke out were shot and expired on the tracks, their corpses chopped up by ensuing trains. Arms, legs and miscellaneous body parts in various states of decomposition were scattered alone the spur lines.[200] Olga Lengyel remembers how "horror ran through the tightly packed mass of humans" in the cattle cars. Voices lifted themselves, murmuring the traditional mourners' chant over the dead and dying. "The cattle car had become an abbatoir. More and more prayers for the dead rose in the stifling atmosphere." In this "infernal trip" in the "death car," the "dead, the contagiously ill, those suffering from organic diseases, the parched, the famished, and the mad must all travel in this wooden gehenna." This "compact mass of animal humanity" was ever the more assailed by creeping "anxiety as though a sixth sense were warning us of impending disaster."[201]

Death stalked the camps without respite. People fell like flies, from sickness, overwork, starvation, exhaustion, bullets, and from the axe.[202] Lengyel describes that in the camps the Damocles' sword hovered over the inmate so pervasively that "the idea of death seeped into our blood. We would die, anyway, whatever happened. We would be gassed, we would be burned, we would be hanged, or we would be shot."[203] The "soil of the camp was thick with blood."[204] "I was frightened by the thought that tomorrow they would be dragging me through the block, a nameless dishonoured corpse, unmourned by anyone," writes Sara Nomberg-Przytyk. "I was thinking about my death, feeling its cold breath washing over me."[205] As Elie Wiesel writes in his own camp memoir, "Death enveloped me, it suffocated me. It stuck to me like glue. I felt I could touch it. The idea of dying, of ceasing to be, began to fascinate me. To no longer exist."[206] "All around me, appeared to be a dance of death. I was reeling. I was walking through a cemetery."[207] Corpses were strewn all over the place, bodies hung from barbed-wire fences. Starving, emaciated human skeletons uttering incoherent sounds stumbled about, collapsing and gasping their last breaths. The faces of the deceased were distorted with terror and awe, their eyes gaping wide open, their tongues protruding, skulls crushed, bodies eviscerated. Blood was everywhere.[208] Wiesel describes how "In one terrifying moment of lucidity, I thought of us as damned souls wandering through the void, souls condemned to wander through space until the end of time, seeking redemption, seeking oblivion, without any hope of finding either." The inmates had become *homines sacri*, mere bodies devoid of human spirit. "The child that I was, had been consumed by the flames. All that was left was a shape that resembled me. My soul had been invaded—and devoured—by a black flame." They were "withered trees in the heart of the desert," "a pack of leprous dogs clinging to life," "more dead than alive, poor creatures whose ghostly faces peeked out from layers of prisoner's clothes."[209] It seemed that, Jabès avers of the *Muselmann* of the *univers concentrationnaire*, "his brain was no longer irrigated by his blood, but by some kind of, probably yellow, liquid; sweat or urine?" "He no longer has a mouth, just a useless cavity where air was reluctant to venture. Voices drone in his ears and spread in a confused blur." "Flayed, on a path bordered with thorns, he would soon be only the awkward tracing of a thousand mortal wounds."[210] The Jewish camp inmate became, as Primo Levi maintains, "a man whose life or death can be lightly decided with no sense of human affinity, in the most fortunate of cases, on the basis of pure judgment of utility."[211] For Jean Améry, he would die the subhuman death of "an animal intended for slaughter."[212] Prior to this subzoological death, he would already have been reduced to "nothing more than an involucre, like the slough of certain insects which one finds on the banks of swamps, held by a thread to the stones and shaken by the wind."[213] The Jewish camp inmate, Améry continues, "did not live next door to but in the same room with death. Death was omnipresent."[214] He was "promised to death, already in the midst of life," his present survival being "a period of false grace that could be revoked at any second."[215] Through inhuman privations, he was reduced to a body and naught besides. This reduction of the Jew to a mere subsentient corporeal entity, to mere unadorned body, was at once his reduction to pure pain—and pure pain is a premonition of death.[216] He was reduced to mere flesh, from which point he could be "driven beyond the border of death into Nothingness."[217]

Jews died of dysentry and typhoid fever every day.[218] Their feet were covered with eczema, they suffered from chronic enteritis, diphtheria, scarlet fever, dysmenorrhea, oedema, phlegmon, panaris and furunculosis.[219] Naked and shorn women were beaten and whipped, and pushed onward like cattle with bayonets toward the gas chambers. Those unwilling to enter were stabbed and bloodied.[220] As Vasily Grossman describes it in *Life and Fate*, they "made their way into a wide corridor, lit by dim ceiling-lamps covered by ovals of thick glass. The muscular strength of smoothly curving concrete sucked in the stream of people."[221] "This wasn't how people moved. It wasn't even how the lowest form of animal life moved. It was a movement without sense or purpose, with no trace of a living will behind it. The stream of people flowed into the chamber; the people going in pushed the people already inside, the latter pushed their neighbours, and all these countless shoves and pushes with elbows, shoulders and stomachs gave rise to a form of movement identical in every respect to the streaming of molecules."[222] They moaned, shouted and convulsed in despairing, blood-curdling cries behind the closing doors of the death chambers.[223] As Shlomo Venezia describes it, crammed in so tightly against one another that the smallest and weakest suffocated, they gripped one another, suffocating with anguish, desperately fighting for air, climbing on top of one another until the last one died.[224] Thereafter their soiled and lifeless blue corpses were hurled into huge mass graves, flooding the latter with pools of their thick black blood.[225] Their corpses were dug up, drenched in petrol and burned on enormous flaming wooden pyres.[226] Their bones were ground up and the ashes scattered in nearby wooden areas or rivers.[227] Every day during the last summer of Hitler's despotic reign, twenty-thousand people were annihilated in Auschwitz. From the gargantuan masses of cadavers burning in the ravines next ot the crematoria "flames leaped upwards, and the sky was turned red by the gigantic fire." The empyrean was a "sea of human blood," "human life oozing out of existence everywhere."[228] As Paul Celan sings it in his *Todesfuge*, the Jews would "scoop out a grave in the sky where it's roomy to lie."[229] Nelly Sachs writes of the "the ingeniously devised habitations of death/When Israel's body drifted as smoke."[230] After being evacuated, what remained of the camps were large empty expanses stretching for kilometres, littered with tens of thousands of skulls and bones.[231]

The specter of defeat increasingly haunting the Nazi war machine from the autumn of 1943 would only intensify the murder of Jews so as to posthaste complete the "Final Solution."[232] Jews were subjected to grueling death marches, dragging their emaciated and exhausted bodies barefoot through snow, and being shot and killed by the tens of thousands for stumbling, falling or lagging behind, their bodies thrown into the nearest roadside ditches.[233] In shoreline death marches they were even murdered and hurled down cliffs into the sea, so that entire coasts were covered with piles of semi-clad and emaciated corpses.[234] Left behind in the barren wilderness of the camps was an almost continuous carpet of dead bodies soiled in excrement and piled up like twisted branches, dying gangrenous skeletons and silhouettes teeming with lice, naked and emaciated corpses in various stages of decomposition, some of whom had drowned in their own vomit.[235]

As Victor Frankl maintains of the almost lunar environs of the *univers concentrationnare* in which the Jew lived a semi-dead life barely distinguishable from non-existence, the disappearance of the last layers of subcutaneous fat through the ravages of semi-starvation turned his body into a mere skeleton thinly disguised by a diaphanous gossamer epidermis and tattered rags. Further exacerbating this inexorable process of emaciation was the fact that the body digested its own protein, shriveling the muscles down to barely more than tissue paper precariously stretched over a spider's web of filament-thin sinews. In this state, one after another member of this invisible necro-community fell to the ground and died. Every such spectral denizen could calculate with fairly rigorous accuracy when the pale ghost alongside him would fall onto the yonder side of nothingness and the time when he himself would faint and dissolve into the dust. During his search for lice, each man would see his own own naked body and say: "This body here, my body, is really a corpse already. What has become of me? I am but a small portion of a great mass of human flesh … of a mass behind barbed wire, crowded into a few earthen huts; a mass of which daily a certain portion begins to rot because it has become lifeless."[236]

In the grim, kenotically evacuated and ethically nil universe of the lager, the Jew dwelt within a world devoid of all value. This environment's innate tendency was to annul his will and nullify his dignity. He was reduced to an object whose sole purpose was to be exterminated. Indeed, the brazenly unnuanced and utterly prosaic utilitarian nihilism of the *univers concentrationnaire* did not even factor the fundamental status of the prisoner as living or dead. Instead, it was merely his prison number that counted, irrespective of whether or not it was attached to a living or a dead body, or to anything whatsoever. Under the influence of such a completely nihilistic and necrophilous terrain of semi-existence, man was himself subjected to the wholesale and terminal desuetude of all internal values that he had hitherto held as inextricable to his continued sense and consummate efflorescence of self. As Frankl observes, in the face of this unbearable pressure, if the inmate would fail to cultivate within himself an inner will to strive with the utmost assiduity to counteract and militate against the inherent tendency of camp existence to corrode and pervert his will and erode all sense of value within him, he would inevitably lose all sense of self and individuality—to the point of suffering terminal dissolution in the gray anonymity of massed captive and immiserated sub-humanity of which he was but an insignificant part. With such a dereliction of all inner volition having come to pass, he would ineluctably sink to the level of bare, unadorned animality, and hereupon pass into the undifferentiated void of pure nonexistence.[237]

The aspect of camp life which particularly inexorably inculcated a sense of unremitting unreality and existential nullity in the inmate was the absolute uncertainty of the term of incarceration. Not only was the term of imprisonment uncertain; it was, moreover, unlimited. This rendered the life of the inmate a so-termed "provisional existence." The herein entailed danger of this elision of man's future-anticipatory projection would, however, swiftly herald the onset of rapidly impending decay, beginning with generalized apathy, progressing to a complete desuetude of the will, and terminating solely in death. The inmate feels as though he is marching to his own funeral and, in fact, regards himself as already dead. Indeed, the confluence of the temporal unlimitedness of imprisonment with the narrow spatial circumscription of camp existence causes the prisoner to inwardly experience actual life outside the surreal enclosure of the lager as would a dead man looking upon life from another planet. Such a man, like the Nietzschean man of *ressentiment*, inevitably comes to be mired in the past. This incessant reminiscing and rumination, if left unhindered, would destroy all possibility of survival. The prisoner would sink into passive nihilism, terminal *anomie*, and in fact begin to deteriorate not only mentally, morally and ethically, but likewise *physically*. It is at this point that the prisoner would begin to truly die, and fairly soon actually die.[238]

However, in another sense, through what must inevitably count as the most massive degree of subjective destitution imaginable on earth and by way of possibly the greatest immorality ever perpetrated on earth, the subject at the same time has, via its subjection to the hideous privations of the *univers concentrationnaire*, entered the "zero-level," or pure void, of subjectivity: the absolutely unendurable abyss in which fully constituted objects and subjects dissolve and yield before a miasmal swarming of spectral partial objects, of exemplifications of the Lacanian *lamella*, of the undead, immortal, libido-as-object that is self-identically the Freudo-Lacanian death-drive (*Todestrieb*). This paronomastic modality of the subject is merely one "degree" removed from the pre-subjective noumenal Real. It is a return to, or recrudescence of, the impossible moment of upsurge of subjectivity in the form of a withdrawal, a recession from the light of phenomenality and its differential degrees of appearing in a world in accordance with its transcendental regime, which assumes the form of a contraction through which constituted reality yields before the surreal phantasmagoria of *membra disjecta*, of organs devoid of body functioning as stand-ins for the dread immortality of the object-libido in its semipernally stuck fixity and eternal psychotic circulation around the void. In Kantian terms, this crepuscular and vanishing mode of subjectivity is that of the pre-synthetic imagination gone mad, psychotic, and utterly demented— the impossible domain of absolutely purified and kenotically evacuated transcendental freedom and spontaneity. In this domain, the subject finds itself vertiginously suspended between direct unadorned animality, on the one hand, and the subordination of its freedom to the moral Law, on the other. In this gathering murk, far from being able to galvanize itself in the direction of forward-directed action in the world or attunement to the promptings of the Deity, the self is submitted to the horrifying experience of dread organs without body, of itself as the Lacanian fragmented body, of its *membra disjecta* floating around in a nauseating and disgusting miasmal drift, the "night of the world," the "valley of the shadow of death," the hell-on-earth of Kundry in Wagner's *Parsifal* who, having committed the blasphemy of laughing at Christ dying on the Cross, is condemned to an eternity of wandering as an abject, multiply befouled, purely excremental self entirely rejected by the world, abandoned by God, and subject to involuntary, psychotic automatisms, spasmodic jabberings, slavering aphasia, dyslexia, *and the absolute horror of never being unable to die.*

In this vein, and sometimes almost as fantastically, even when faced with existential circumstances still worse than the concentration camp universe, the Jews were still somehow able to conjure instances of almost dumbfounding individual survival of the Nazi Damocles' sword of death. A particularly harrowing course of events began to unfold starting on the first day of the Hebrew month of Ellul on the evening of August '42. Five-hundred Jewish families of the ghetto of the village of Zagrodski, near Pinsk, were bundled into a truck and spirited off to a hillock where they were being stripped, naked, shot dead, and hurled down into a trench with foul brutality. Among these moribunds headed for their terminal demise was Rivka Yosselevska and her family who were pushed along and undressed, save for the father who, upon refusing to remove his clothes was forcibly stripped naked and shot. Hereupon Rivka's mother was shot, after which her eighty-year-old grandmother was shot with babies in her arms, followed by her father's sister who likewise was shot cradling infants. Then her younger sister, upon pleading for her life in stark nakedness, was looked straight in the eye by a German and shot with her friend. After this, her second sister was shot. More, of course, were shot. Finally, after the German had ripped her child from her arms and immediately shot it, the Nazi Damocles' sword of imminent death was looking Rivka herself flush in the face in the form of a gun pointed right at her. She was about to be shot. The murderer grabbed her by the hair, turned her head around and shot her. Having been shot, she fell into the pit amongst the other bodies and felt nothing, despite having been shot.[239] The moment she did come to feel something after she had been shot, she, in her own words:

> ... felt a sort of heaviness and then I thought maybe I am not alive any more, but I felt something after I died. I thought I was dead, that this was the feeling which comes after death. Then I felt that I was choking; people falling over me. I tried to move and felt that I was alive and that I could rise. I was strangling. I heard the shots and I was praying for another bullet to put an end to my suffering, but I continued to move about.
>
> I felt that I was choking, strangling, but I tried to save myself, to find some air to breathe, and then I felt that I was climbing towards the top of the grave above the bodies.

> I rose, and I felt bodies pulling at me with their hands, biting at my legs, pulling me down, down. And yet with my last strength I came up on top of the grave, and when I did not know the place, so many bodies were laying all over, dead people; I wanted to see the end of this stretch of dead bodies, but I could not. It was impossible. They were lying all, all dying; suffering; not all of them dead, but in their last sufferings; naked; shot, but not dead. Children crying, "Mother", "Father"; I could not stand on my feet.[240]

In turned out that the Damocles' sword of death had been postponed, for Rivka—far from having passed onto the yonder side of the grave—discovered that she had merely been wounded in the head. Not only this. The Germans had indeed gone. Rivka, for herself, had managed to crawl out of her own grave. However, after the Germans had briefly reappeared to heap the corpses together and had thereafter left once more, the Damocles' sword of death loomed again for Rivka and the three other survivors she had stumbled upon in the thanatic quagmire of blood, death, and putrescence. It was not, however, that of the returning Nazis.[241] Rather, it was that which the coterie of four survivors desired forthwith to bring down upon themselves:

> I saw that they all left, and the four of us, we went on to the grave, praying to fall into the grave, even alive, envying those who were dead already and thinking what to do now. I was praying for death to come. I was praying for the grave to be opened and to swallow me alive. Blood was spurting from the grave in many places, like a well of water, and whenever I pass a spring now, I remember the blood which spurted from the ground, from that grave.
>
> I was digging with my fingernails, trying to join the dead in that grave. I dug with my fingernails, but the grave would not open. I did not have enough strength. I cried out to my mother, to my father, "Why did they not kill me? What was my sin? I have no one to go to. I saw them all being killed. Why was I spared? Why was I not killed?"[242]

The Damocles' sword still beckoned for Rivka. For she continued to lay streched out where she was amid the pestilence of the mass grave, not moving for whole days and nights—pining for death to overtake her. Even when, after three days, shepherds had driven their herd onto the field and began throwing stones at her, she did not move and was taken for a dead woman or a mad woman. Yet, eventually, when the imminence of certain death would otherwise have overtaken her, a farmer took pity on her, took her in, hid her, and fed her. She survived the war and went on to recount her ordeal in a Jerusalem court a full nineteen years later.[243]

Ultimately, however, sporadic invidivual survivals would not suffice. What was required was the survival of the *collective* Jewish subject in the face of its wholesale imminent annihilation by the Nazi Damocles' sword of death. The impetus for Jews not to give up the ghost was stoked by events such as the meeting on May 1, 1943, of a group of Jewish writers and poets in the Vilna ghetto, when the poet Hirsch Glik first sang a song that was soon to spread like wildfire in the ghettos and camps and among Jewish Partisans. The song enjoined the Jews to resist, or at least to survive.[244]

This bold defiance in the face of imminent demise was even in evidence at the very threshold of the gas chambers. On April 8. About to be gassed with countless others within the next few minutes, one Mosze Friedman, formerly Rabbi of Bayonne, grabs the lapels of a SS lieutenant, saying: "You common cruel murderers of mankind, do not think you will succeed in extinguishing our nation. The Jewish nation will live forever and will not disappear from the world's arena." He spoke in a strong lion's voice and with tremendous energy, and then with immense fervor cried out "Shema Israel!" "Hear! O Israel." Immediately, all those present with him cried the same words. An eye-witness averred that this impressive effusion confirmed the eternal spiritual power of the Jews.[245]

Their hope could not be crushed by either the squalor of death or the debilitating force of fear. On April 11, from her refuge in Holland Anne Frank wrote in her diary:[246] "Who has inflicted this upon us? Who has made us Jews different to all other people? Who has allowed us to suffer so terribly up till now? It is God that has made us as we are, but it will be God, too, who will raise us up again. If we bear all this suffering and if there are still Jews left, when it is over, then Jews, instead of being doomed, will be held up as an example. Who knows, it might even be our religion from which the world and all peoples learn good, and for that reason and that reason only do we have to suffer now. We can never become just Netherlanders, or just English, or representatives of any country for that matter, we will always remain Jews, but we want to, too."[247]

One of the ways in which the collective Jewish subject managed to maintain its existence under reprieve in the teeth of the Nazi State apparatus' Damocles' sword of death was through the assidious preservation of Jewish heritage. In the face of the Nazi swastika, which was bent on erasing the very memory of the Jewish people from the collective consciousness of mankind itself, the Jews ensured their cultural survival through the production of historical records for the attention of future generations. Such that the collective Jewish subject remain intact, individual victims and survivors of the Holocaust resorted to an unprecedentedly assidious recording of their near-death experiences for the posterity of the Jewish nation and of humanity as a whole. In this sense, they preserved the collective Jewish subject in a manner highly homologous to that effected almost two-thousand years ago by Rabbi Johanan ben Zakkai at the time of the near-annihilation of the

Jews during the Roman destruction of Jerusalem in 70 A.D., when the great rabbi ensured the continuation of Judaism through Vespasian's befuddled acquiescence to the seemingly miniscule request that the rabbi be permitted to found a school of Jewish learning (*jeshiva*) in the small town of Javneh in Palestine. Yet, it was through this ostensibly negligible bequest that ben Zakkai was able to save the Jewish heritage, and thereby the continued existence of the collective Jewish subject, from extinction through inter alia its terminal absorption into the grey mass of the gentile masses. This gesture was repeated almost two-thousand years later by Rabbi Nachum Yanchiker to his students at the Slabodka-Musar Yeshiva in the fateful year of 1941. He enjoined that the collective Jewish subject be preserved through the written word.[248] "And do as our Holy Sages have done—pour forth your words and cast them into letters. This is the greatest retribution which you can wreak on the wicked ones. Despite the raging wrath of our foes, the holy souls of your brothers and sisters will remain alive. These evil ones schemed to blot out their names from the face of the earth; but a man cannot destroy letters. For words have wings; they mount heavenly heights and they endure for eternity."[249]

We have mentioned Elhanan Wasserman's comparison of the prophesized Holocaust travails of the Jews to the pangs of birth. More precisely, they are the birth pangs of the Messiah (*hevlei mashiah*), meaning that implicit within these throes is a form of redemption. This redemption is the rescue of the collective Jewish subject from the Nazi Damocles' sword and its rebirth.[250] "For I the Lord change not. For ye, O sons of Jacob, are not consumed" (Mal. 3.6). The children of Israel will not be doomed. Though a Haman has been set up to persecute the Jews, the Deity rescues them: "And God seeketh that which is pursued" (Ecc. 3.15). He stands on the side of the persecuted. The harsher the persecution of the Jews, the greater the grounds for their rescue. In that Jewish persecution under the swastika was historically unprecedented, the birth of the ultimate redemption would result. Wasserman compares the survival of Israel to silver. As silver goes into fire and comes out intact, so does Israel enter under the dominion of other nations and likewise comes out whole.[251] Israel will be in the fire and come out whole: "The bush burned with fire, but the bush was not consumed" (Exod. 3.2). "When thou passest through the waters, I will be with thee.... When thou walkest through the fire, thou shalt not be burned. Neither shall the flame kindle upon thee" (Isa. 43.2).

In *Parshat Behukotai Sevel Yisrael Ugeulato*, Reuven Katz derives a model for the Jewish survival of the Holocaust from the *Akedah*, or "binding of Isaac." Of course, a move is here made from the individual to the collective. As opposed to the sacrifice of one man, one finds here the multitudes of Israel offered up as burnt offerings, their ashes piled up on the altar of the gargantuan *Akedah* made to the

Creator. These ashes then awaken the mercy of heaven over the atrocious situation of the Jews and serve to hasten their redemption and deliverance. Though it is the case that Isaac's death is averted at the last moment whereas the sacrifices of millions of individual Jews during the *Sho'ah* were fully consummated in death; the Holocaust does remain homologous to the *Akedah* in that the sacrifice of the *collective* Jewish subject, threatened as it was with wholesale extinction, was effectively averted. For, as Katz maintains it, the ashes piled up on the Holocaust altar served to raise the memory of the children of Israel before God.[252] They induced the Holy One to fulfil his promise: "Then will I remember My covenant with Jacob, and also My covenant with Isaac, and also My covenant with Abraham will I remember; and I will remember the land" (Lev. 26.42). The deliverance of the Jews from the Nazi Damocles' sword is compared to an owner of an ox immediately breaking its yoke and tearing up its harness. Analogously, when God beheld Israel almost exterminated from the world, he immediately broke its yoke and led the Jews to a life of freedom and independence.[253] Katz has here in mind the founding of the modern State of Israeli, according to the promise: "Unto thy seed will I give this land" (Gen. 12.7). In *Mei Merom: Mima'yenei Hayeshua*, Ya'akov Mosheh Harlap in fact sees the great *Akedah* of the *Sho'ah* as the necessary precondition of the said redemption of the collective Jewish subject through the termination of the diaspora. During the time of the onset of the Messiah, as he maintains it to be, the *Akedah* is actualized collectively, bringing about the rebirth of *Eretz Yisrael* in Palestine: "Exiles are not gathered in, except by the merit of Isaac."[254]

In *Kol Kore fun'm Lubavitscher Rabin*, Yosef Yitshak Schneersohn speaks in the same vein of the survival of the collective Jewish subject being vouchsafed by the blood of individual Jews. As he maintains, Jewish blood has been spilled across the entire European continent to enable the Jewish people to come to life again.[255] It is through its bleeding that Israel returns to life.[256] "In thy blood, live!" (Ezek. 16:6). "Blood fertilizes, perpetuates," writes Edmond Jabès. The Jews, for him, have a "will to live in death, as strong as that which quickens truth in its fight against the lie," a "will to grow, to love, to procreate within death ..."[257] Indeed, in one of her poems, Nelly Sachs enjoins the day of destruction to arrive so that, in being put to sleep in the earth and in the very midst of death, Jewish life shall be heard to spring up with irrepressible force.[258] In the same vein, though more expansively, Jabès effuses of the Jew: "Dead of wanting to live against life, alive by virtue of being lost in death's labyrinth, he comes into his own in survival, as if the beyond were his place."[259]

As with Wasserman and Harlap, in *Yelammedeinu Rabbeinu*, Schneersohn likewise refers to the indissoluble paradoxical imbrication of the thanatic Nazi

Damocles' sword and the Jews' continued subsistence within existence through a form of resurrection. He views the war and the Holocaust as the ineluctable propaedeutic to the complete redemption of Israel through the messiah in which Jews become more exalted in the life of the Torah. This renascence of the collective Jewish subject has the destructive agonies of the birth pangs of the messiah as inexorable antecedent. For before the great ram's horn awakens everyone to divine holiness, there must be a preparatory cleansing from sin. Hereafter, the collective Jewish subject resurrects through the return of estranged Jews to an emergent *Eretz Yisrael*, whereby its continual survival is vouchsafed.[260]

In *Kol Dodi Dofek*, Joseph Soloveitchik describes these birth pangs of the messiah as one of terror, horror and absolute divine self-concealment (*hester panim muhlat*), in whose dark night God suddenly comes into manifestation, knocking at the tent of this the sufferer of the convulsive pains of Hell. The result of these knocks is the exaltation of the tormented through the foundation of the State of Israel.[261]

Moreover, as Berkovits contends, it was in fact *only* through the establishment of the State of Israel that the Jewish people could have been saved from extinction from hopelessness.[262] What Soloveitchik terms *hester panim muhlat*, Berkovits refers to as the *Galut haShekhinah*, or exile of the Divine Presence. During the Holocaust, which Berkovits prefers to call *Hurban* (annihilation) rather than *Sho'ah*, the *Galut haShekhinah* reached its absolute nadir, it's most tragic intensification in all history—to the extent that life itself became absurd. Since Jewish existence is ineluctably bound up with the *Galut haShekhinah*, they too were dragged down into the depths of suffering and martyrdom almost to the point of terminal desuetude. Even the spark of *Ge'ulah* (redemption) hitherto present in every prior Jewish *Galut* (exile) was extinguished this time around. Therefore, due to the sheer unprecedented magnitude of the *Hurban* and the virulence of its Damocles' sword of death, the resurrection of the collective Jewish subject had to be as radical and revolutionary as the destruction itself. It had to assume the form of the restoration of Israeli's sovereignty in Palestine in the sight of all men. Only in this way were the "dry bones" of Israeli espied much earlier by Ezekiel resurrected from the dust.[263] This latter foundation is what Irving (Yitchak) Greenberg sees as the fundamental act of life of the Jews in the wake of the evil and death of Auschwitz.[264]

With regard the strange renascence of the collective Jewish subject through the shedding of its own blood, Yosef Achituv alludes to the startling image of the Holocaust used by Rabbi Zvi Yehudah. The latter refers to the *Sho'ah* as a divine healing operation performed on the body of the Jewish nation in order to cut it off from the bonds of exile in which it long lay mortified. Here, the separation of the single body composed of millions of Jews from its exile is performed by God in a singularly cruel way and against the Jews' will. Yet, it is only through this violent surgical operation that redemption is attained via the return of the exiles to the land of Israel. Through the traumatic convulsions of this act of severance, the holy people are revived. It is a sweetness appearing within bitterness, light emerging from darkness. Crucially, it is not that the Holocaust precedes redemption. Rather, it forms an integral part of it.[265] The Nazi Damocles' sword of death is employed in the manner of a scalpel surgically excising the tumor of exile sapping the vital strength of the collective Jewish subject. Through individual Jewish demise, the latter's collective existence is resurrected. Michael Wyschogrod even goes as far as to say that with the Holocaust Judaism, oddly enough, received a new lease of life. In fact, he avers, in the absence of the Holocaust no further justification for the further survival of Judaism could possibly have been found.[266]

As Eleizer Schweid maintains it, this resurrection is, in fact, so emphatic that the consciousness which united the survivors of the *Sho'ah* into a covenant of destiny marked by the establishment of the modern State of Israel was stronger in the two decades after the war than at any other prior point in Jewish history. It superseded all differences of ideas and conflicts of interest. For most Jews, the founding of the modern Jewish State was the conclusive victory over Nazi Germany.[267] In this connection, Simhah Elberg avers that the reason for the Jews' survival is to be located precisely in the fact that their history is exile history. He contends the history of exile to ultimately be more formidable than any Hitler. Hitler went into battle against exile history, bent on terminating this exile with the destruction of those who created it. However, the power of the Jews differs from that which characterizes nation States. For whereas powerful nations may make glorious history, in the case of the Jews it is the other way around: it is their glorious history which makes them into a powerful nation.[268] Jewish survival and Nazi demise was vouchsafed by the superiority of the latter form of national power over the former.

Apart from the incredible perdurance which exile history bequeaths upon a nation defined by it, another source of the seeming indestructibility of the Jews is to be found in Emil Fackenheim. He maintains that survival, for the Jews, is moreover a holy duty. This is not merely the injunction for Jews to affirm their mere *right* to continue to exist as Jews. Nor is it an injunction for the Jews to affirm the necessity of their continued existence as human beings in general. Rather, the "commanding Voice of Auschwitz" demands that Jews survive specifically *as Jews*, and because they are Jews. What renders this moral injunction to survival still more acute is, as Fackenheim maintains, the fact that, precisely and only in being Jews, the Jews represent all of humanity. Through a transmogrification of the thanatic logic of the *univers concentrationnaire*, the commanding Voice of Auschwitz commands that the Jew suspend the heretofore extant Jewish exaltation of martyrdom and descend from the cross. It orders the preponderance and preeminence of life over death. Life becomes more sacred than death, even if the latter is for the sanctification of the divine Name. For after the death camps, the one fundamental and supreme value to be fought for at all costs is life itself.[269] The Jew is *commanded* to continue to exist, to survive. He simply *must* continue to exist.

A fairly elaborate understanding of the Jewish survival of the Holocaust may be gleaned from the manner in which Jeffrey Alexander conceives the latter as a *traumatic* event. Alexander avers that, for a trauma to be experienced at the level of a collective—and not so much an individual—human subject, it must be culturally constructed. This cultural construction of trauma consists of the coding, weighting, and narration of the traumatic event. Firstly, the coding of the trauma is, more precisely, its being coded as *evil.* In this way, the trauma comes to be perceived and experienced as evil by the collective at the epistemological level.[270]

In addition, this evil must be placed in rigid binary opposition to that which is concomitantly, and by corollary, normatively constructed as good. In addition, so as to preclude the miscegenation of the culturally wrought good and evil, this division and separation must be rigidly enforced and policed.[271]

Secondly, once the traumatic event is coded as evil, it must be weighted. That is to say, it must be accorded a certain *degree* of evil. More specifically, the distinction alluded to here by Alexander is that which exists between ordinary evil and radical evil. Generally-speaking, ordinary evil may be termed mere wickedness; though in the case of radical evil, there has come to pass a perversion of the will itself—which may now be described as singularly malign. This is crucial, since these two forms of evil—differing fairly sharply as they do—do not have the same implications as regards, for instance, the urgency and scale of the remedial action that must be undertaken with respect to the victims of the traumatic event, on the one hand, and the degree of culpability and intensity of punishment to be accorded the perpetrators of the event, on the other. Thirdly, and finally, the trauma must be narrated. That is, the meaning of the trauma must be explicated and unfolded by delineating the evil and traumatizing actions, singling out the victims and perpetrators, and deciding upon the exact remedial actions to be undertaken in the wake of the trauma's eviscerations.[272]

From the late thirties onwards, in accordance with what Jeffrey Alexander terms the *progressive* narrative under which the Holocaust came to be constructed as a traumatic event, Nazism was already being coded as evil in opposition with the universalism and good of Anglo-American liberal democracy. This evil of Nazism was weighted as radical, in that it stood for violence and death on a hitherto unprecedented scale—and as directed by an incorrigibly malign will. As Abraham Lewin maintained from the testimony of escapee, David Nowodworski, "This is without a doubt the greatest crime ever committed in the whole of history … God!"[273]

The ethnic, racial, and religious hatred on account of which Nazism was codified as evil, in conjunction with the fact that its systematic discrimination, coercive force, and death-wielding machinations were directed inter alia and preeminently against the Jews, began to bear within itself the upshot that anti-Semitism itself was evil—and indeed radically evil. Inter alia, *Kristallnacht*—the rhetorically virulent and physically violent explosion of Nazi repression against the Jews on November 9 and 10 in 1938—was immediately seen by large swathes of the American public to represent the evil of Nazism. It made page one news in the *New York Times* for three consecutive days. At a November 14 press conference, President Roosevelt declared: "The news of the past couple of days from Germany has deeply shocked public opinion in the United States … I myself could scarcely believe that such things could occur in twentieth century civilization."[274]

Crucially, it was against the backdrop of the evil Nazi persecution of German Jews that democratic forces began to militate in historically unprecedented fashion against the anti-Semitism which in fact still existed within the United States itself. According to the logic of symbolic association and antinomy employed, in that Nazism was represented as absolute evil and in that the greatest enemies of the Nazis were the Jews, the Jews had to by implication stand on the side of the good and therefore on the side of American liberal democracy. In a startling and diametrical reversal of hundreds of years of history, overtly positive representations of Jews began to proliferate throughout both popular and highbrow cultural firmaments. In that Nazi Germany was now distinctly seen as the common enemy of both Jewry and of the United States, Jewish and American fates came increasingly to be seen as merging. Part of this convergence was the birth at this time of the phrase "Judeo-Christian tradition." Insofar as it was the latter composite that constituted the bedrock of Western democratic life being threatened by Hitler, it was the Jews in addition to the United States that had to be defended against nefarious and apocalyptic Nazi evil. These sentiments and resolve would intensify with the United States' declaration of war on Nazi Germany on December 11, 1941.[275]

With Nazi plans for the Final Solution and their actual implementation already having been publically documented by June 1942, in July of that year more than twenty-thousand people rallied in Madison Square Garden to protest against anti-Jewish Nazi depredations. The rally was organized by the American Jewish Congress, B'nai Brith and the Jewish Labor Committee. The audience heard messages from Roosevelt and Churchill pledging retribution following emphatic Allied Victory over the pestilence of Nazism. On October 7, the British and American governments announced the formation of a United Nations War Crimes Commission for the collection of evidence to be used in the trials of those guilty of atrocities against the Jews. A surge of public sentiment pouring into the press from politicians, labor unions, and clergy induced Roosevelt to renew his stentorian warning to the Nazis at a meeting with a Jewish delegation at the White House on December 8. Ten days later Anthony Eden read out a joint statement condemning Nazi war crimes in the House of Commons.[276]

As we shall see, what one may espy here is the swift emergence, already a few years before anti-Semitic Nazi violence had reached the gargantuan scale and intensified genocidal frenzy of the Final Solution agreed upon at the Wannsee Conference on January 20, 1942, whereby all of European Jewry would be slated for wholesale annihilation, of the basis upon which the Jews—as a collective subject—would be enabled to *survive* the Nazi Damocles' sword of the Holocaust in spite of the destruction of millions of individual members of the Jewish race. According to this the progressive narrative of Jeffrey Alexander, the trauma created by Nazi evil and Nazism itself would be overcome predominantly on account of the latter's overly anomalous and parochial historical status by comparison with the good of Western liberal democracy whose superiority over fascism stemmed from its far greater inherent potential for being universalized to the entire human race—to, as we shall observe vis-à-vis our pending discussion on Badiou, generic humanity. Though it profoundly threatened the continued existence of the collective Western liberal subject, this trauma and danger was merely liminal and—despite the certain impending deaths of millions—the same collective subject would attain social salvation through the elimination of Nazism.[277]

In that, according to the progressive narrative, the fates of the American collective liberal subject and the collective Jewish subject were seen as convergent, the salvation of the former would intrinsically entail the salvation of the latter. And, indeed, in the immediate postwar years the *salvation* of the collective Jewish subject from death via historical obsoleteness proceeded apace by way of not merely the de-Nazification of Germany but likewise the progressive purging of American society of its own festering pockets of anti-Semitism. The American non-

Jewish audience was now determined to *redeem* the Jews. Jewish groups engaged in vastly heightened political activism from 1945 to 1948. By 1950, these activities had effected a radical improvement in the practical and cultural position of American Jews. The ramifications of the progressive narrative of the anti-Nazi war would thrust anti-Semitism into an ancient and ignominious past.[278] Or, at least, this was the intention of some.

In fact, in that anti-Semitism was so inextricably symbolically associated with Nazi evil, post-war democratic tolerance came to be predominantly and specifically linked to tolerance of the Jews—with intolerance and anti-Semitism seeming, in certain walks, to be virtually indistinguishable. The obverse of this was the increasingly broad drive to protect and exalt the Jews. Indeed, due to the aforementioned fundamental intertwinement of American and Jewish paths, the efflorescence and *survival* of Israel came to be a sine qua non of the maintenance of the redemptive *telos* of the American progressive narrative itself. To this end, on 22 June 1945, President Truman tasked the US representative on the Intergovernmental Committee for Refugees, Earl G. Harrison, with pushing for a redemptive Jewish State. In a report, Harrison declared that the Jews in Germany and Austria must have the first claim upon the conscience of the American and British people. This entailed facilitating their emigration to Palestine. Harrison's report was rapidly endorsed by Truman, who instructed General Dwight Eisenhower, supreme commander in Europe, to act on it immediately. Copious relief funds poured in, leading to a massive efflorescence of Jewish life in Displaced Persons camps in US zones of occupation as they resumed control of their fate. A baby boom, accompanied by rituals of circumcision, retaliated against nefarious Nazi depredations that sought to wipe the Jews off the face of the Earth.[279]

What is crucial about the progressive narrative concerning the traumatic evil inflicted by the Nazis is that, though it concerned itself with the redemption of all those vanquished and obliterated by the latter—and above all the Jews; its teleology focused not so much on the Holocaust itself but rather on the dire need to purge post-World War Two society on either side of the Atlantic, and indeed the world, of all residues of the fascist virus.[280]

The genocidal assault upon of the Jews would, however, likewise fall under a different sort of narrative. Jeffrey Alexander terms this second narrative the *tragic* narrative. Here, the focus would fall *specifically* on the Nazi mass murder of the Jews as an object in its own right. This event would receive far greater symbolic weight than was the case with the progressive narrative.[281] In this new cultural construction of the Jewish mass killings as a collective trauma, the Holocaust would acquire far greater weight whilst at the same time becoming increasingly unhinged from Nazism and its variegated profanations per se. Rather than merely being regarded as the most emblematic typification of Nazism's nefarious machinations, the mass killings came eventually to be construed as not representative or typical of anything whatsoever. Instead, the Holocaust came to be seen as a completely unique and unprecedented event, as evil of such intensity, diabolical villainy, and colossal scale as had never previously occurred within the entire gamut of human history. Hitherto accorded a fairly minor and ancillary position in portraits of the Second World War, the Holocaust presently seems to occupy the entire picture—virtually to the exclusion of all else. In fact, now it is so ubiquitous that, on November 1, 2005, the United Nations General Assembly made January 27 International Holocaust Remembrance Day—a day which is not only commemorated universally, but moreover considered the worldwide marker of evil and crimes against humanity of the our epoch.[282]

In fact, the sheer gargantuan enormity and horror of the Holocaust led not only to its being set apart from any Badiousian world, or totality of bodies and languages, but above all rendered it virtually inexplicable, absolutely mysterious and incomprehensible. This led to the well-known injunctions against representing the Holocaust. Their contention echoes Wittgenstein's last proposition in the *Tractatus*, which maintains that one must be silent about that of which one cannot speak. Elie Wiesel, for one, maintains that there can be no novel about the Holocaust, as it would automatically neither be about the Holocaust nor be a novel. Similarly, Jorge Semprún avers that there may be no prosaic documentary about the Holocaust. Claude Lanzmann went as far as saying that if he were to stumble upon actual documentary footage of the murder of inmates of Auschwitz, he would immediately destroy it.[283]

In this connection, Marxist historian Isaac Deutscher maintained: "I doubt whether even in a thousand years people will understand Hitler, Auschwitz, Majdanek, and Treblinka better than we do now."[284] After learning of the experience intimated to him by the escapee David Nowodworski, Abraham Lewin declared: "From his words we put together a testimony of such stark anguish, so shattering, that it cannot be grasped and put into words." One Seidman, after hearing an analogous testimony from Yaakov Rabinowitz, wrote: "words fail; me … All I want to do is cry, cry, and cry."[285]

Geoffrey Hartman, for one, maintains that "something in the specific case of the Shoah remains dark at the heart of the event; and it leads to reflections that seem 'theological'."[286] Hartman refers specifically to the theological averment of Emil Fackenheim, who declared the Holocaust to be a veritable "caesura or rupture in historical time, the emergence of a *novum* that must be recognized as such, even when it cannot be understood."[287] It is this *unrepresentability* of the Holocaust that Hartman wishes to underline when he quotes Edmond Jabès' description of Auschwitz as "the indestructible memory of the void," Henri Razymow's "cry without a voice," or a survivor visiting the Warsaw ghetto shortly after its liquidation hearing the unidentified, ghostly sound of a "woman's voice calling from the rubble."[288] Likewise referring to the singularity and unrepresentability of the Holocaust, Claude Lanzmann avers the latter to be "above all unique in that it erects a ring of fire around itself."[289] Aharon Appelfeld contends that the Holocaust is of such a peculiar nature that its memory is subject to a perpetual curfew.[290]

It is this unequivocal apartness of sacred-evil from all other forms of evil that came to demand that the trauma in question no longer be designated as "mass murder" or even "genocide." For even these very emotive terms were felt to unacceptably normalize it and place it in too great propinquity to the mundane and the banal. It was therefore renamed "Holocaust," or "something burnt up"—except that this already existing common noun, from the early sixties onward, came henceforth to be employed in ordinary English usage as a *proper* noun. Its Hebrew equivalent, *Sho'ah*, was already being used by Israelis only several years after the Nazi mass murder. This shift in linguistic usage emphasized, and still emphasizes, the radically *singular* nature of the Holocaust. Henceforth a proper name, "Holocaust" would be a name naming the unnamable, a signifier without a signified, referring only to itself, indicative of the Holocaust as *sui generis*. This unrepresentability of the Holocaust is likewise the case at the visual level. Žižek alludes to the "cinematographic" techniques through which aspects of the Holocaust are "represented" in Semprún's *Le Grand Voyage*: details of images, objects and sounds shown in close-up in such a way that their excessive intrusion and over-proximity bursts through and undermines the narrative context in which they are set. These overbearing details are pure exposed fragments of the Lacanian Real which inherently resist symbolization, since the trauma of the *Sho'ah* is too shocking to be assumed. It cannot thus be symbolically inscribed or registered by any figure of the big Other.[291]

This *en nuce* is the tragic narrative that elaborated itself in the wake of the Jewish mass murders and through which the Holocaust truly became the "Holocaust." Notwithstanding, the main thrust and purpose of Jeffrey Alexander's discussion of the two narratives under which the Nazi mass slaughter of Jewry was constructed as a collective trauma resides in their subsequent *universalization* of the same traumatic event to all of humanity. That is, in the generalization of the redemption, or survival in the very teeth of foreboding historical obsoleteness, of the Jews as a collective subject to the survival of human civilization itself. We shall first proceed to show how the progressive narrative may be universalized in this sense and then demonstrate how a similar universalization transpires in the case of the tragic narrative.

2. Universalization of the Jewish survival of the *Sho'ah* to the survival of all humanity under Alexander's "progressive" narrative

Under Jeffrey Alexander's progressive narrative, the first movement toward the universalization of the socio-political survival of the collective Jewish subject to the survival of the entire human race occurred fairly shortly after the discovery of the prodigious network of concentration camps in which Jews had been massacred. It is this burgeoning evidence of the sheer gruesomeness and gargantuan scale of the Nazi atrocities perpetrated against Jewry that would write *finis* to the heretofore preponderant construal of these mass killings as mere symptoms of Nazi evil. For now, in a radical shift of perspective, they would be regarded as collectively comprising the very core and essence of the evil whose transcendence was the very *raison d'être* of the progressive narrative.[292] As Meyer Levin wrote concerning his visit to Ohrdruf, the first camp liberated by American soldiers: "it was as though we had penetrated at last to the center of the black heart, to the very crawling inside of the vicious heart."[293]

The move toward the universalization of the Jewish mass murder quickly gained pace when, within a few months of the initial revelations, they came to be framed within the public sphere by the term "genocide"—a jurisprudential innovation defined by the Polish Jewish jurist Raphael Lemkin as the deliberate and premediated effort to destroy, in whole or in part, an entire people on racial, ethnic, linguistic or religious grounds. This swiftly led to the next step in the universalization of the public representation of the mass murder of the Jews: the extension by the Nuremberg War Crimes Tribunal of charges made against twenty-three high-ranking Nazi officials from mere war crimes, conspiracy against peace, and the waging of aggressive war to a charge that had never before been made in all of human history: "crimes against *humanity*." The latter crime, much as in the case of the Holocaust under the tragic narrative, was considered entirely separately from the other crimes and wrought as a universalization of the charge of annihilation of the Jewish people.[294]

It is with the latter that the Nuremberg trials and their ramifications, in accordance with the progressive narrative, would serve as a lesson to *all* mankind in the construction of a more morally sound world—a world, above all, free of such death-wielding depredations. Here, the genocide of European Jewry, as part of the massive trauma of World War II, was seen as a birth trauma signaling an ensuing process of *redemption* and *reconstitution* of human—and not merely Jewish—civilization from its charred embers and scattered ashes.[295]

3. Universalization of the Jewish survival of the *Sho'ah* to the survival of all humanity under Alexander's "tragic" narrative

In vast contradistinction to the progressive narrative, the tragic narrative represented the Holocaust as nowise a birth trauma but indeed a death trauma. What we have here is effectively almost a dialectical reversal: the Holocaust is no longer a "beginning" but indeed an "end." Therefore, the *telos* is henceforth no longer redemption and life, but instead suffering and death. Moreover, in that the tragic narrative would place the Jewish mass murders as an event outside the world and history alike; through its detachment from the specifics of time and space, the Holocaust acquired a transcendental status. Jeffrey Alexander maintains from this that it erected a basis for psychological identification with its victims on a hitherto unprecedented scale.[296]

Though our argument will later shy away from Alexander's foregrounding of psychological identification, it is still germane to dwell upon this notion for the time being. In his *Poetics*, Aristotle defines tragedy as a "representation of an action of a superior kind—grand, and complete in itself—presented in embellished language, in distinct forms in different parts, performed by actors rather than told by a narrator, effecting, through pity and fear, the purification of such emotions" (*Poetics* II.6.1449b24-28). He maintains that it is through either the reversals in or discoveries of their fate by the protagonists, or both, that pity and fear are evoked in the beholder of tragedy. That which ultimately elicits the viewer's pity and fear is the suffering undergone by the protagonists in their tragic denouement. Here, Aristotle alludes to such occurrences as entail pain or destruction, in particular, extreme agony or people being murdered. Germane to our purposes is the distinction he makes between what he sees as the task of the historian as opposed to that of the poet when this latter works within the tragic form: whereas the former relates events which had actually happened, the latter does so for those which *would* happen in accordance with the dictates of "probability and necessity." More precisely, events that transpire with probability or necessity do so in the sense that they follow inexorably under condition that certain antecedent occurrences take place. Given that tragedians are likewise partly historians by portraying certain veridical historical personages, events which transpire in their dramas with conditional probability are, in addition, credible. From this, Aristotle adduces that tragic poetry moves beyond history's mere enumeration of facts and proceeds to the enunciation of universal truths. This universality concerns what would probably and necessarily appertain to individuals of a certain type more broad than, and at a remove from, the specific characters of a given historical drama. This transcendence of tragic poetry in relation to the specifics of its characters likewise arises from Aristotle's averment that tragedy represents not so much the character of its protagonists, but the web of actions within which they are caught. This means that the moral character traits of historical and other actors, while important, are decidedly ancillary in relation to the story which serves as the foundation and soul of tragedy. Still heightening tragedy's fundamentally universalizing tendencies is the transcendence of a tragedy's plot over any one of its specific performances, wherein the anguish-riddled episodes in the narrative alone ought to evoke pity and fear. Of these two emotions elicited by tragedy, pity arises in the spectator from the undeserved misfortune of the innocent, whereas fear wells up within the beholder when the victims of outrageous fortune in certain fundamental ways resemble ourselves (*Poetics* II.6.1450a16-III.13.1453a13).

The conjunction of tragedy's universal address with the pity and fear that it evokes in the spectator serves to mirror the tragedy of moribund protagonists as the tragedy that resides within us, or as the tragedy that we ourself are. For the fear attendant upon the pity aroused by the unjust fate suffered by tragedy's victims is what makes their tragedy potentially our own. This is what effects an ineluctable identification with the victims by the beholder. For the inexorable forces of darkness that spell their catastrophic demise likewise lay dormant within ourselves. Tragedy also, however, serves to purify and exalt us, heightening and sharpening our insight into the sheer depth of the catastrophe it represents and our moral probity in general. As Percy Bysshe Shelley has it in his *A Defense of Poetry*, "The tragedies of the Athenian poets are as mirrors in which the spectator beholds himself, under a thin disguise of circumstance, stripped of all but that ideal perfection and energy which every one feels to be the internal type of all that he loves, admires, and would become. The imagination is enlarged by a sympathy with pains and passions so mighty, that they distend in their conception the capacity of that by which they are conceived; the good affections are strengthened by pity, indignation, terror and sorrow; and an exalted calm is prolonged from the satiety of this high exercise of them into the tumult of familiar life: even crime is disarmed of half its horror and all its contagion by being represented as the fatal consequence of the unfathomable agencies of nature; error is thus divested of its wilfulness; men can no longer cherish it as the creation of choice."[297] Since the spectator is "stripped of all but that ideal perfection and energy which every one feels to be the internal type of all that he loves, admires, and would become," tragedy both elucidates our feelings and emotions and purifies our emotive and imaginative faculties. In that "an exalted calm is prolonged from the satiety of this high exercise of them into the tumult of familiar life," tragedy brings relief. The conjunction of psychic purification and relief is, indeed, *catharsis*. However, inasmuch as in tragedy the horror of crime, or rather evil, is "represented as the fatal consequence of the unfathomable agencies of nature" and human error, or rather misfortune, is "divested of its wilfulness"; unlike the progressive narrative, the tragic narrative cannot offer any immediately tangible or obvious redemption and means of avoiding the catastrophe it lays bare. Instead, much as the psychotic rotary movement of Schellingian eternal divine nature incessantly recurs to the auto-contraction of the Godhead, Athenian tragedy merely puts into circulation an unremitting and relentless mnemonic return to the primordial scene of the anterior catastrophic trauma. In this, the subject is inundated by the recrudescence of his "extimate" traumatic kernel, to the internal roiling horror that is himself.

This is especially the case for the Holocaust as a tragic drama, as its victims are the victims of the *sui generis* archetypical trauma—our pity and sympathy for them being excited all the more acutely. But what then of the indictment by Semprún and Lanzmann of prosaic documentary representations of the *Sho'ah*? Would not tragedy also fall under this interdict? No, since it is not literal. As we may recall from Aristotle, tragedy differs from history in that, through fictional aestheticization, it elevates factual historical action to the eternal plane of its "truth." Indeed, Semprún maintains that it is only the arts that can represent the Holocaust, contending that its "truth" can solely be borne out through aestheticization. In this connection, Žižek alludes to Lacan's thesis that truth has the structure of a fiction, this being especially true in the case of an overwhelmingly traumatic truth such as that of the *Sho'ah*. Here, tragedy is entrusted with indirectly representing the Holocaust through the cathartic sensation wrought by fiction's aestheticization of its macerating traumatic kernel. In so doing, fiction extracts from the symbolically elusive reality of the Holocaust its *inner* form, or eternal archetype. As Žižek maintains of Semprún's *Le Grand Voyage*, the subject immersed in this tragic work has his identity shattered and, being ripped out of the time of succession, forthwith moves within an eternal present in which, after passing through the nightmare of camp life, his perception is immeasurably broadened as he sees all three dimensions of time simultaneously. In Deleuzian terms, past, present and future are telescoped into a single synchronic "sheet of time," which in the case of the Holocaust is our collective memory of this catastrophe.[298]

This traumatic memory of the Holocaust, instilled in post-war generations by its transmogrification into tragic drama and retained within collective consciousness as an atemporal Deleuzian "sheet of time," renders it a less territorially bound, less temporally specific, and *more universal* drama. Through this, empirical socio-political reality has in Platonic fashion come to "participate" in the Idea of the Holocaust as the eternal archetype of traumatic human tragedy. Here, we may refer again to Žižek who, deontologizing Plato's Ideas, avers that they are thoroughly virtual and "insubstantial" entities generated by spatio-temporal material processes yet which subsist within them. Žižek now moves to link his deontologized variant of the Platonic Idea to Badiou's repetition of this gesture in his defense of the autonomy of the "immaterial" order of Truth, wherein the mere human individual or collective accedes to placing its life at the service of a transcendent truth, in so doing becoming a subject dedicated to an overriding emancipatory Cause.[299] In the case of the trauma drama of the Holocaust, this would swiftly become the dedication of myriad human collectives to intervene in diverse Holocaust-like catastrophes that threatened and still threaten to transpire in countless milieu across the globe. In the manner in which Shelley maintained that the tragedies of Aeschylus, Sophocles and Euripides intimate to us at a fundamental psychic and even visceral level the inexorable and sinister forces of elemental nature stirring within the midst of all men, so too did the Holocaust drama entrench the more deeply in contemporary human sensitivity the panoply of evil depredations threatening the existence of many a collective subject. As a transcendent and "insubstantial" Platonic Idea, the Holocaust symbol rendered privy that all human beings in every society are at once the potential victims and perpetrators of such humanitarian catastrophes, in the process morally binding generic humanity to forestalling and, if need be, thwarting such necrophilous depravity, so as *to place the existence of collective national, ethnic, and cultural subjects threatened with the Damocles' sword of terminal historical annihilation under reprieve*. Herein, the bathetic events of the Holocaust drama under the tragic narrative, formerly the sole preserve of the Jews, became the trauma of all mankind. The categorical imperative enjoining vast swathes of humanity, thus, became *the necessity of transforming the necessity of Jewish survival into the necessity of the survival of the entire human race.*

We have spoken of the universalization—via personalization—of the Holocaust as trauma drama within post-Holocaust consciousness as that of not merely victims but likewise that of perpetrators. Now, Alexander avers that the latter form of generalization gained particularly acute momentum with the 1961 trial of Adolf Eichmann in Jerusalem. Oddly then, in this connection, Prime Minister Ben-Gurion's express intention for the trial was ostensibly not to generalize away from the originating events, but indeed to return to them in their localized singularity. For him, the purpose of the trial was inter alia to acquaint the Israeli generation born after the Holocaust with its tragic minutiae. The paradoxical effect of the trial was, however, that, far from moving in a parochial direction, it inaugurated a universalization of Nazi radical evil—and by implication—radical evil as such. Relevant here is, of course, Hannah Arendt's insistence that the trial compelled global recognition of the "banality of evil."[300]

This ushers in a seeming contradiction. For banal evil is the disinterested evil perpetrated by what may be aptly termed the "desktop" mass murderer whose function is merely that of a gray, bloodless, bureaucratic cog within a vast network of identical gray bureaucratic cogs blasély contributing to the deaths of countless innocents through the effortlessly perfunctory signing of documents. Radical evil, on the other hand, is vicious, almost apoplectic or apocalyptic and mind-numbing hatred—to the point of being diabolical and necrophilous. It is that which—through the perversion of the will and the value-positing eye as such—goes as far as to cause good and evil to switch places. This circle, if in need of squaring, is quasi-successfully squared by Zygmunt Bauman's averment that the Holocaust was *both* uniquely evil within the historical expanse and, at the same time, completely normal for modernity. Bauman maintains here that the Holocaust may be seen as the unique and deleterious confrontation between, on the one hand, certain fundamental unresolved tensions that modernity itself had long since repressed and, on the other hand, modernity's own potent means of instrumental reason and pragmatic efficiency. Though this encounter was a very singular outcome of a unique confluence of circumstances, the actual factors themselves which coalesced in the encounter were very "normal" indeed for modernity.[301]

Bauman swiftly comes to universalize the Holocaust to all of humanity, maintaining that, given that the already extant historical reality of the *Sho'ah* and its sheer massiveness, we have since become tremulously aware that such an event can very easily happen again elsewhere, or indeed anywhere—that humanity is entirely capable of inflicting an enormity of this sort upon itself in the future, for "Auschwitz expands the universe of consciousness no less than landing on the moon."[302] Bauman goes on to expound this conjunction of the singularity and potential universality—by way of normalization—of the Holocaust, averring that, with the sole exception of the gargantuan moral odium of its goal, the set of activities it combined and orchestrated did not differ in any formal sense from all other carefully organized activities effectuated by the quotidian administrative and economic divisions of modern bureaucracies.[303]

Bauman continues his universalization of the Holocaust through his abiding contention that it was wholly consonant with the formalized ethical myopia characteristic of modern bureaucratic rationality and its intensely narrow pursuit of efficiency. Not only was the unprecedented scale of its mass murder inexorably dependent on the finely-tuned, firmly entrenched proficiency, meticulousness, precision, efficient command structure and informational flow, and perfectly synchronized co-ordination of autonomous actions endemic to modern bureaucratic culture; but, conversely and rather alarmingly, one is moved to the startling dystopian conclusion that "*the very idea of the* Endlösung *was an outcome of the bureaucratic culture.*"[304]

In this connection, Bauman maintains that it was the normal routine bureaucratic procedures, such as means-ends calculus and universal rule application, that formed the thick and fecund background against which the choice to effect the removal (*Entfernung*) of European Jews—specifically by way of physical extermination—was made. More precisely, the decision in favor of extermination arose naturally from the bureaucratic predilection to, in the face of changing circumstances, treat all importuning contingencies as "problems" to be resolved by the deployment of rational "solutions." Likewise instrumental in this regard was the ubiquitous and very normal bureaucratic tendency to displace goals—in our case, the shift from the mere deportation of a population to its physical annihilation in Auschwitz and other labor and death camps.[305]

The Holocaust was, above all, not some sort of antediluvian recrudescence of repressed pre-modern irrational fluxes resurfacing in all their barbarity. For it was, in fact, entirely proper to modernity, and indeed uniquely and specifically modern—meaning that it was something that categorically could not have taken place at any point during the long phase of humanity's pre-modernity.[306] The Holocaust, though radically singular in terms of its sheer level of odium, was normal for modernity, in the sense that it does not lie within the competency of modern instrumental rationality—taken in and of itself—to preclude such enormities. This is all the more the case, given that nothing exists in the rules of modern instrumental rationality which would serve to disqualify nefarious proto-fascist—or literally fascist—techniques of "social engineering" as improper or irrational.[307]

Indeed, such proclivity to social engineering may be seen as, in many ways, intrinsic to modernity's bureaucratic culture—in which one is almost enjoined to conceive society as an object to be administered, as comprising a set of problems to be solved, and as an otherwise raw and inchoate nature to be controlled, mastered, and re-forged. As Bauman maintains, it was precisely in this sort of context that the idea of concerted mass-scale bureaucratic and military action against European Jewry could be conceived and ultimately brought to its logical conclusion with the Final Solution.[308] But there is more. For, not only does the spirit of instrumental rationality as institutionalized in modern bureaucracies render Holocaust-type solutions to perceived social ills possible; it furnishes a framework within which they imperceptibly emerge as "reasonable," hereby making the eventuality of their adoption all the more likely.[309]

In this connection, one may argue that the absolute singularity and normality of the Holocaust are reconcilable and constantly conjoined in the sense that the way to Hell is initiated by radical evil and paved with banal evil. What is crucial, however, is that the said radical evil would in no sense be effectively radical if not supplemented by the inexorable adjunct of banal evil. Jeffrey Alexander maintains that the Eichmann trail served to universalize the Holocaust since, as a banally evil man, Eichmann could be *anyone*. As a mere obedient cog within the thanatic Nazi bureaucracy, he was not really very different to anyone else.[310] He was a cipher of generic humanity.

Indeed, as Hannah Arendt reports, half a dozen psychiatrists certified him as "normal," one of these maintaining that Eichmann's entire psychological outlook, his attitude to his spouse, offspring, parents, and siblings, was not merely normal but in fact commendable. Experts found him morally and legally sane, and opined that there was no trace of fanatical anti-Semitism of any kind in him.[311]

In fact, after the judgment was read out, Eichmann's defense counsel Dr. Servatius contended that Eichmann had merely carried out "acts of State," and that what he had done could be done in future by anyone, that this was a problem facing the whole civilized world.[312] As Arendt maintains, "The trouble with Eichmann was precisely that so many were like him, and that the many were neither perverted nor sadistic, that they were, and still are, terribly and terrifyingly normal."[313]

Bauman concurs wholeheartedly, underlining that most of the perpetrators of the Holocaust were very normal people indeed, who would freely pass through any psychiatric sieve, however dense. Moreover, the actual institutions responsible for the Holocaust were in no legitimate sense pathological or abnormal.[314] When these men took off their uniforms, they were not evil in any way. They behaved much like all of us in their quotidian and intra-mundane affairs.[315]

Himmler, for instance, expressed deep, and probably sincere, concern about maintaining the mental sanity and upholding the moral standards of his subordinates. He likewise expressed pride that, in his belief, both sanity and morality emerged unscathed. Contrary to what is generally maintained, SS leaders counted on organizational routine, not on individual zeal; on discipline, not ideological dedication.[316]

We may at this point summarize the precise manner in which singularity and normality, as characteristics of the Holocaust, are mutually imbricated. Bauman, in fact, maintains the Holocaust to be unique in two distinct ways. Firstly, its uniqueness *as a concrete historical instance of genocide* stems from the fact that it is modern. Secondly, its uniqueness *within the ambit of modernity itself* stems from the fact that it gathers together certain quite ordinary factors of modernity which otherwise remain separate. It is, on the other hand, normal since the factors it uniquely collates are—viewed in isolation—entirely normal for modernity.[317] It is, thus, a singularity arising within—and indeed already intrinsically presaged by—the inherent structures of modernity in its quotidianity.

This potential emergence of Holocaust-like propensities within modern quotidianity was poignantly adduced from Stanley Milgram's experiments of the sixties. These demonstrated that ordinary well-educated college students would follow the orders of professional authorities with scant resistance, to the point of gravely endangering the lives of innocents. The findings of Milgram's experiments profoundly undercut prevailing beliefs about the human capacity for empathy, revealing instead that the latter much-vaunted capacity was indeed far more limited than hitherto thought—to the point of the hominid as such suffering from what one may term an intrinsic moral autism.[318] Milgram's experiment swiftly came to form part of the moral and cultural folklore of its decade. Yet, far from wholly undercutting the human resolve striving to universalize the Jewish survival of the Holocaust to other collective human subjects' survival of their own portended extinction, the dystopian findings of both Eichmann's trial and Milgram's experiment instead alerted Western consciousness to the sheer extent of the difficulties involved in all future humanitarian undertakings—rendering privy to those undertaking such salvific interventions that filibustering naïvety would need to yield to painstaking and measured diligence.

As already observed, the personalization wrought by the tragic narrative of the Holocaust would bring about its universalization. This would move identification beyond the Jewish victims themselves and, in tandem, expand the circle of possible perpetrators beyond that of the Nazis. Thus, as the death-wielding machinations of the Vietnam War unfolded, it was now the American government itself which came to be viewed by tens of millions of oppositionists worldwide as a symbol of apocalyptic evil. This transmogrification in the moral status of the United States during these fateful decades was inter alia brought to bear by the various revolutionary impulses of the sixties, such as student and black power movements inside the States and guerrilla movements from without. Napalm bombs became gas pellets, the flaming jungles of Vietnam became gas chambers, whereas the US army came to be identified with the Wehrmacht and seen as perpetrating bloody genocide against a hapless Vietnamese populace. Bertrand Russell and Jean-Paul Sartre established a sort of counter-"War Crimes Tribunal" to apply the logic of the Nuremberg judgment to the United States. The focus fell especially intensely upon the 1968 My Lai massacre as a typification of nefarious American aggression and evil. The *universalization* herein wrought entailed the newfangled ability of nations other than the United States, antigovernment groups, and even historic enemies of the Jews to cast and produce the trauma drama and thus to compel identification and channel catharsis on a far less parochial scale than hitherto.[319]

In a curious reversal, Palestinian liberation movements would equate the Israeli government with the Nazis and Palestinians with, ironically, the Jewish victims of the Holocaust. This symbolism became all the more firmly entrenched following Ariel Sharon's complicity in the 1982 Sabra and Shatila massacres in Lebanon and by the release of documented reports of Palestinian torture and occasional death in Israeli prisons.[320]

We may surmise that the tragic narrative universalizes the Holocaust by way of a combination of symbolic extension and emotional identification. In so doing, it vastly broadens the reverberations of subsequent traumas reminiscent of the great *ur*-trauma that is the *Sho'ah*, in so doing vastly expanding their impact upon collective social consciousness at large. Jeffrey Alexander terms this the "engorgement of evil." The latter is an almost self-ramifying process whereby evil, in its fluid and labile miscegenation, pollutes everything it touches. We have seen how this process has functioned above. The "Holocaust" came to function as what Alexander terms a "bridging metaphor" to parse post-*Sho'ah* events as good or evil and to define inhumanity in the present era. This engorgement of evil was exacerbated when no longer merely perpetrators but likewise passive and complicit non-actors came to be represented as evil. Thus, according to post-Holocaust moral standards, States, groups and even individuals become normatively bound to intervene in a prospective or transpiring holocaust, to the end of saving lives. This is so in the sense that, as a crime against humanity, a "holocaust" is considered a threat to the continued existence of humanity itself. It is through the engorgement of the Holocaust as symbol of radical evil that *similar* death-wielding machinations were powerfully moralized and *prohibited*. This almost viral pollution has of itself vastly enlarged the circle of the culpable in post-Holocaust crimes, to the extent that guilt soon came to be conferred by way of simple propinquity, that is to say, by way of metonymic association. Due to this, moral crimes could more readily be imputed in that reference to obtuse exculpating circumstances, or lack of direct involvement, were no longer admissible.[321]

Though the foregoing may be the case, the manner in which the Holocaust as bridging metaphor came to function more typically and profoundly was through the device of analogy.[322] For instance, in the 1990s, the evil that seeped from the engorged Holocaust metaphor provided a compelling analogical framework for the nefarious evens transpiring in the Balkans. For, whether the events were conceived as falling under the category of a dictatorial purge, ethnic rampage, civil war, ethnic cleansing, or genocide, it was the engorged Holocaust symbol that served as the ultimate semiotic basis propelling American diplomatic and then American-European military intervention against Serbian ethnic violence. In this connection, Senator Joseph Lieberman declared that the atrocities perpetrated by Serbian forces were eerily reminiscent of the deleterious vicissitudes of the Second World War in Europe. Democratic presidential nominee, Bill Clinton, effectively contended that mugwumpery could hardly be a sustainable position in the face of the mass extermination of a people. Retreating forthwith from the polluting passivity of the Allies during the Second World War, he forcefully declared that, if elected, he would commence airstrikes against the Serbs so as to restore the basic conditions of humanity presently being eviscerated by Milošević. It was President George H.W. Bush's threat of precisely such military deployment that forced Serbia to sign the Dayton Accords and surcease what were extensively represented in American and European media as genocidal depredations. The exact same symbolic analogies motivated, justified and launched the allied bombing campaign when Serbian militias threatened to enter Kosovo. This humanitarian military intervention was represented as responding to the widely experienced horror that the trauma drama of the Holocaust was being reenacted directly under the civilized world's purview. In view of the systematic slaughter of innocents at the hands of the Serbs, Bill Clinton would go as far as to draw a direct analogy between Milošević and Hitler, averring that the Germans could scarcely have perpetrated the Holocaust without the demagoguery of the Führer. One the very same day, Joschka Fischer, foreign minister in the coalition "Red-Green" German government, defending the allied air campaign before a special congress of his Green Party, likewise drew an analogy between the Serbian evil and the Holocaust. His deputy foreign minister and party ally, Ludge Volmer, described Milošević's systematic ethnic cleansing policy as an instance of fascism. Believing the analogies wholly germane, about sixty percent of the Green

Party voted in support of Fischer's position. The analogizing would continue. Two weeks later, when the allied bombing campaign had not yet succeeded in bringing Milošević to heel, Clinton asked Holocaust survivor Elie Wiesel to make a tour of the Kosovar Albanians' refugee camps. The rationale for this move, as given by a spokesperson of the U.S. embassy in Macedonia, was that focus had been lost on the ultimate underlying reason for the ongoing military intervention—meaning that the proper analogy was not being made consistently enough. To remedy this, direct metonymic association would be required, and Wiesel would provide it by commending Washington's response in Kosovo as a considerable advance on its prevarication during the Holocaust. When the air war and the growing threat of an allied ground invasion had finally driven the marauding Serbian forces out of Kosovo, thus shattering the said nefarious Damocles' sword and enabling the survival of the collective Kosovar subject; the *New York Times* would directly link this august historical vicissitude to the Holocaust trauma, lauding the ultimate salvation of the collective Albanian subject as the fruit of the conscious resolve on the part of the America and Europe to stop any revival of genocide dead in its tracks.[323]

It is through the Holocaust as bridging metaphor deployed as an analogical frame that one may espy the preponderant means through which the *Jewish survival of the Holocaust universalized itself to the will to vouchsafe the survival of humanity as such.* This universality, always already implicit to the *sui generis* nature accorded the *Sho'ah* by the tragic narrative, would continue to burgeon much further afield. In the 1980s, for instance, the engorged and free-floating Holocaust symbol came to be analogically associated with the movement against nuclear power and nuclear testing, and with the ecological movements that emerged at this time. Here, politicians and intellectuals campaigned against the testing and deployment of nuclear weapons by alluding to the "nuclear Holocaust" that governments contumaciously bent on perpetuating such polices may well unleash. Activist's evocative depictions of the nuclear holocaust that could well result from such egregious activities were poignantly dramatized by dystopian and grisly images of "Auschwitz," the iconic representations of which were rapidly becoming a universal medium for portraying demented violence, abject human suffering, and pointless death. Similarly, the environmental movement advanced claims that industrial nations were committing ecological genocide against assorted species flora and fauna, thus subjecting the planet itself to the possibility of extermination.[324]

Admittedly, the extension of the survival of the collective Jewish subject of the Holocaust to that of generic humanity as such has had its massive snags. For, despite President Truman's recommendation that the United States draft a code of international criminal law to thwart similar occurrences around the world, and despite the "human rights" foreign policy of Jimmy Carter, and despite the nineteen UN treaties and covenants condemning genocide and exalting the new mandate for human rights, new international legal codes were never drafted. Nonetheless, over the same period, an increasingly ample body of "customary law" has been developed *to militate against nonintervention against sovereign States engaged in large-scale death-wielding depredations threatening the very existence of human collectives.*[325]

4. Universalization of the Jewish survival of the *Sho'ah* in terms of the Badiousian event

The radically singular and unrepresentable nature of the Holocaust likewise suggests certain similar features to the Badiousian conception of "historical situations" and the "event." For it is precisely the *sui generis* character of the last-mentioned that Badiou sees as the source of universal "truths" and of an "immortal" subject, and it is the latter which suggests existence under reprieve.

(i) Propaedeutic: Badiou's ontology

In order to understand the exact idiosyncratic meaning Badiou ascribes to these terms and how he deploys them, we must first present a brief disquisition on his ontology. For Badiou, Being as such is neither one nor multiple. It is merely what presents itself. The multiple is Being's regime of presentation. The one, which is not, is only an operational result of this presentation. Any presented multiplicity, Badiou terms a *situation*. The *structure* of a situation is what composes the multiplicity as a multiplicity of units, terms which belong to the situation. This count-as-one that is the situation's structure retroactively splits multiplicity into two: *consistent* multiplicity and *inconsistent* multiplicity.[326]

The former is a multiple of ones. The latter is the pure multiple, the multiple "in-itself," the multiple without any other predicate other than its own multiplicity, the multiple qua multiple.[327]

However, inconsistent multiplicity is not actually presented as such since all presentation falls under the structure's regime of the count-as-one. Therefore, from the standpoint of the situation, inconsistent multiplicity is *nothing*. The latter, Badiou notwithstanding maintains, is not non-Being, but rather a "being-nothing."[328]

The nothing is merely the name of unpresentation in presentation. It names the unpresentable of presentation that, moreover, distributes itself between purely inert inconsistent multiplicity and consistent multiplicity. It wanders throughout the situation in the form of a subtraction from the count, and is scattered all over it, both nowhere and everywhere. It likewise manifests that the situation is sutured to Being. Badiou terms this nothing the *void*. The void is the name of Being, inasmuch as the situation unpresents it in the mode of a non-one, as the errancy of the nothing.[329]

Yet, the void is the very Being from which any situation is woven.[330] The errant void is omnipresent in the situation in the manner in which the empty set is included in, is the subset of, any set in mathematics.[331]

Now, what all presentation of the multiple must guard against is that the void, which is the name of inconsistency in the situation under the law of the count-as-one, cannot, in itself, be presented. Such an eventuality would be the presentation of inconsistency as such and the consequent ruin of the oneness imposed by the structure of the situation. Badiou maintains that, in order to preclude such a catastrophic presentation of the void, the structure of the situation must be doubled by a metastructure. That is, the structure must itself be structured, the count-as-one itself counted as one, that the structure's presentation be reinforced by metastructure's re-presentation. The metastructure is likewise termed the *state of the situation*. Its mode of reinforcement is as follows. The first count, the structure, composes consistent multiples, that is, one-multiples, as terms (or elements) within the situation (set). The second count, the metastructure, takes one-multiples (elements) and, in turn, composes parts, or sub-multiples (subsets), out of them. The terms, or one-multiples, are said to *belong* to the situation, as elements belong to a set; whereas the parts, or sub-multiples, are said to be *included* in the situation, as subsets are included in a set. The latter sub-multiples, moreover, *belong* to the state of the situation, much as the subsets of a set belong to its power set (the set of all of the set's subsets). This belonging is effected by the metastructure's counting of these sub-multiples as ones, much as the structure had done for one-multiples. In the absence of this latter capturing action of the metastructure, the situation would have been flooded with an irremediable excess of sub-multiples. It would not have been able to count them as ones and thus render them consistent multiplicities. According to the theorem of the point of excess, these sub-multiples of a situation are in irremediable excess of the situation's one-multiples. One of these sub-multiples is the void, or empty set, itself. In that the empty set likewise belongs to the power set, the metastructure also counts the void as one—thus thwarting its disturbing the integrity of the structure of the situation with a catastrophic recrudescence of inconsistency. In that the entire original set itself also belongs to its power set, the metastructure counts the sub-multiple comprised of the entire totality of all the one-multiples counted by the structure. The completeness of the initial one-effect of the first count is therefore counted by the state in the form of its effective whole, establishing the reign of the one.[332]

Badiou avers that the degree of connection between the structure of the situation and its statist metastructure is, in fact, variable. This entails a typology of terms or one-multiples. A multiple counted as one by the structure of a situation is *presented* therein. If this multiple is also counted as one by the metastructure, or state of the situation, then it is *represented*. Such a multiple belongs to the situation (presentation) and is likewise included in the situation (representation). It is a term and a part. However, according to the theorem of the point of excess, there are included (represented) multiples which do not belong to (are not presented in) the situation. These multiples are parts but not terms. Finally, there are multiples that are presented yet not represented in the situation. These are terms but not parts. Badiou calls multiples that are both presented and represented *normal*, multiples that are represented but not presented *excrescent*, and terms that are presented but not represented *singular*. The reason why singular multiples are not represented, that is, why they are not sub-multiples or parts of the situation, is because some of the elements of which they are composed are not presented in the situation separately. A singular multiple therefore unifies multiples which are not themselves one-multiples of the situation. Badiou terms such a multiple "indecomposable." It thus belongs to the situation but is not included in it. It is hence not counted as one by the metastructure of the state.[333]

(ii) Badiou's ontology, socio-historical situations and the State

The foregoing general ontological structure Badiou likewise attributes to historico-social situations. Here the multiples can be either individuals or groups of individuals. The state of a historico-social situation is the State. Whichever of these individuals or groups are counted as ones and thus presented in the situation, the State, as metastructure of any historico-social situation, will re-present as one-compositions of multiple individuals or groups. Much as does the state of an ontological situation, the State does not recognize individuals or groups per se but rather collective subsets composed of the latter. In that the totality of these collective subsets is in immeasurable excess of the individuals or groups from which they are composed, the State itself is likewise in immeasurable excess of the situation over which it holds dominion. When the State does "deal with" a particular individual or clearly defined group counted as one by the structure of the situation, it does not do so directly with a personal "him", "her", of "they." Rather, the State considers the individual or idiosyncratic group as a subset. That is, it constructs a singleton set out of these latter. Individuals or idiosyncratic groups are, therefore, held by the State not to belong to society but merely to be included in it. As with the metastructure of an ontological situation, the State must count as one each of the compositions, or subsets, of individuals or groups of the situation, as their excess over these individuals or groups would otherwise serve as a hiatus for the recrudescence of the void—this time in the form of a Hobbesian war of all against all.[334]

(iii) The Badiousian singular multiplicity, the Nazi State and the Sho'ah

It is here that we shall begin to introduce the Badiousian conception of the event, or truth-event. In this connection, we must return to what were earlier termed singular multiplicities, that is, multiples that are presented by the situation but not represented by the state of the situation, multiples which belong to the situation but are not included in it. We recall that a multiple composed of elements, some of which are not presented (counted as one) separately within the situation, is not represented (counted as one) by the state of the situation. In a historico-social situation, a singular multiple would be a group, some of whose individual members are not presented in the situation. In that the State can only represent individuals that are already presented in the situation, these unpresented members will likewise not be represented by the State. Notwithstanding, though not separately presented in the situation, these individuals will still be indirectly presented in the situation by virtue of their belonging to their group. On the other hand, those members of the group who are presented in the situation the State will represent as singleton sets of the said members.[335]

Returning to the Holocaust, we may certainly venture to aver that, when the Jewish massacres were being perpetrated, the Jews formed such a singular multiplicity. Here one may divide the Jewish multiplicity counted as one by the German historico-social situation into two groups. Firstly, there were certainly those Jews who were presented within the situation and, on that account, represented as singleton sets of themselves. That is, being identity document holders, they were registered as citizens by the Nazi State. Of course, on account of being individually counted as one by the State, they could the more easily be corralled into ghettos and thence conveyed to their deaths in camps. Secondly, there were the *Muselmanner* of the concentration camps, the living dead, unpresented in the situation and, being no longer within the interest of the State, not represented by the latter's metastructure. In that the *Muselmann* is the most telling and ultimate figure of the Holocaust as an enigma, the epicenter of its inscrutable core, we may plausibly venture to draw here a parallel between his not being represented by the Nazi State (and indeed by any State) and the unrepresentability of the Holocaust itself in Jeffrey Alexander's tragic narrative.

But let us return to, not so much the figure of the *Muselmann*, but the Jews themselves as an ethnic group. There is another way in which the Jews, as a group, were merely presented by the Nazi situation, but not truly represented in any fully-fledged sense by the Nazi State apparatus. Here we may refer to Bauman's discursus on the so-termed "conceptual Jew"—who had meandered through European vistas even prior to the advent of the Third Reich. The conceptual Jew was viewed by the Nazi State as *visqueux*, in Sartean terms; that is, as "slimy." He was "the prototype and arch-pattern of all nonconformity, heterodoxy, anomaly and aberration. As an evidence of the mind-boggling, uncanny unreason of deviation," of "*chaos and devastation.*"[336] The Jews, collectively espied through the notion of the conceptual Jew, were on these terms "historically construed as the universal 'viscosity' of the Western world."[337] They were possessed of "a multi-dimensional unclarity" which imparted to them "*an extra cognitive incongruence*" entirely foreign even to other acutely marginalized groups.[338]

The Nazi State conflated the Jews with prevailing forms of turmoil and instability. From its vantage-point, they were "that glutinous substance which blurs the boundary between things which ought to be kept apart, which renders all hierarchical ladders slippery, melts all solids and profanes everything sacred."[339]

The *primus inter pares* dimension of endemic Jewish incongruence, as espied from the apex of various increasingly obdurate and idolatrous State apparati—inclusive of the hyper-idolatrous Nazi one, was that they were viewed as a "non-national element in a world of growing or existing nations."[340] Through their territorial dispersion and ubiquity, the Jews were both an international-nation and a non-national nation. They were constantly experienced by nation States as a relativity threatening the limits of both individual self-identity and communicational interest alike, which the criterion of nationality hankered through State metastructures to determine and fix with absolute and conclusive authority. Still worse: "*The Jews were not just unlike any other nation; they were also unlike any other foreigners.*" They were "flexible and adaptable; an empty vehicle, ready to be filled with whatever despicable load ..."[341]

The Jews had even been described as "an invisible cohesive web of slime fungus (plasmodium), existing since time immemorial and spread over the entire earth."[342] Finally, the Führer himself spoke of the Jews as bacilli, decomposing germs, or vermin. "The discovery of the Jewish virus," he opined to Himmler in 1942, "is one of the greatest revolutions that have taken place in the world. The battle in which we are engaged today is of the same sort as the battle waged, during the last century, by Pasteur and Koch."[343] Bauman: "*The world tightly packed with nations and nation-states abhorred the non-national void. Jews were in such a void: they were such a void.*"[344]

Right here, Bauman is speaking in well-nigh Badiousian terms. The Jews were the void, pure inconsistent multiplicity, repressed by the presentative structure of the Nazi situation and the representative metastructure of the Nazi State apparatus, and incessantly threatening to recrudesce and discombobulate this entire body politic and regime. As we shall see fairly soon, a certain sub-portion of the Jews verily were on the very edge of the void. The consequences of the latter shall likewise soon be excavated.

(iv) The collective of Muselmanner *as a Badiousian evental site*

We now move again from the Jews to Badiou. After this, we shall once more move from Badiou to the Jews. Now, from the notion of an abnormal or singular multiple, Badiou moves on to define what is meant by an *evental site.* This is an abnormal (singular) multiple, *none* of whose elements are presented in the situation. The singular multiple, or evental site, itself is presented—but nothing from which it is composed is presented. Under the Nazi State apparatus, the multiplicity of *Muselmanner* would form such an evental site—in that, though each individual *Muselmann* is unpresented, the said multiplicity is presented in the situation through the plethora of images of anonymous emaciated and moribund *Muselmanner.* Badiou maintains that an evental site, as the one-multiple that it is, is *on the edge of the void.* This is so because, from the perspective of the situation, it is composed solely of unpresented multiples. The evental site is, therefore, composed solely from what are, with respect to the situation, inconsistent multiplicities. Right beneath the evental site there is, thus, only the void. Crucially, the notion of an evental site is relative. Though a situation may contain evental sites, there is no evental situation. A situation with at least one evental site, Badiou terms *historical.*[345]

(v) Evental site and Muselmann *in Beckett's* Worstward Ho

Badiou espies in Beckett's *Worstward Ho*, the deployment of a literary method whereby beings (or "shades") presented within Being—that is, counted as one (being presented) within a situation, undergo a process of gradual "worsening" through which they ever the more closely approach the void. That is, they are made to more and more closely approximate the condition of being unpresented within the situation. Worsening entails inter alia operations by means of which the shades in Being are "iller said than said before [*être plus mal dit quer déjà dit*]."[346]

One of the shades presented within the situation is the head—which stands for thought. For Badiou, Beckett reduces thought to its two absolutely primordial constituents: the visible and saying. Since Beckett wishes to reduce seeing and saying to their utter bareness, that is, to an "ill seen ill said"; he reduces the presentation of the head to merely eyes and a brain oozing words: two holes on a brain.[347] Beckett: "No face. Skull and stare alone."[348] "In the skull all save the skull gone. The stare.... In the skull the skull alone seen. The staring eyes."[349] So as to reduce seeing to an ill seeing, Beckett describes the eyes as "Clenched eyes. Staring eyes. Clenched staring eyes."[350] As for saying, that is, for words, Beckett avers: "How almost they still ring. As somehow from some soft mind they ooze. From it in it ooze."[351]

But in what sense does Badiou maintain that the essence of saying is ill saying? Ill saying is opposed to "well saying." "Well saying" brings about an adequation between the saying and the said, that is, with "what is said." For Beckett, a saying that is adequate to the said suppresses saying. The saying must, therefore, be disjoined from the said in order not to be robbed of its essence.[352] Beckett: "Say for be said. Missaid. From now say for be missaid."[353] In order to attain a free saying, moreover, this saying must never be direct and must be a saying that leastens, that worsens. Through worsening, it must not, however, be brought to abolition.[354] Beckett: "Least never to be naught. Never to naught be brought. Never by naught be nulled. Unnullable least. Say that best worse. With leastening words say least best worse. For want of worser worst. Unlessenable least best worse."[355]

Is not this movement, through worsening, toward a pure saying absolutely disjoined from what is said, analogous to the Lévinasian Saying (*le dire*), which solicits the subject directly exposed to the face (*le visage*) of the Other and admonishing it, in a pre-ontological and non-thematic manner, not to overwhelm the Other's otherness? For, as Badiou avers, in a similar manner to which what is said robs saying of its essence; Lévinas maintains that ontological (discursive) language, or the Said (*le dit*), subordinates and hence betrays the Saying (*le dire*).[356]

Though *Worstward Ho* performs worsening exercises over the entire field of the shades, it is the worsening of the head that is the most pertinent for our present purposes. Beckett's worsening of the eyes, in particular, shall continue: "The eyes. Time to try worsen. Somehow try worsen. Unclench. Say staring open. All white and pupil. Dim white. White? No. All pupil. Dim black holes. Unwavering gaping. Be they so said. With worsening words."[357] In this foregoing passage, Beckett passes from "clenched staring" to "staring open." "Open" then gives way to white, white to black. One passes from clenched to open, from open to white, and from white to black. The outcome of this process of worsening is that, in place of "clenched staring," we will have "black holes"—and that, from now on, eyes will no longer even be described with the word "eyes." Beckett will merely mention two black holes—as blind seats of visibility. The eyes are abolished in favor of a pure seeing linked to a hole.[358] "Skull and lidless stare."[359] "To skull and stare alone.... Now way the fore alone. No dome. Temple to temple alone. Clamped to it and stare alone the stare."[360] "Two black holes. Dim black. In through skull to soft.... Two black holes in foreskull."[361]

What this worsening of the head to merely two black holes lodged in a skull calls to mind is Giorgio Agamben's notion of *homo sacer*. That is, *bare* life, life conceived—or reduced to—its raw functions, life distinguished by nothing other than its pure *naked* existence.[362] This bare life is aptly captured by Primo Levi's description of the living-dead, or *Muselmann*, of the Nazi concentration camps as "nothing more than an involucre, like the slough of certain insects which one finds on the banks of swamps, held by a thread to the stones and shaken by the wind."[363] This extreme reduction of the human is likewise espied in Jean Améry's description of the tortured (he was tortured by the Nazis). Torture transmogrifies the tortured into pure "flesh," implying that "the tortured person is only a body, and nothing else beside that."[364] For Agamben, that *homo sacer* is unadorned naked existence causes him to be bereft of the accoutrements that would otherwise bestow human dignity upon him—to the point that he teeters on the brink of the inhuman. He is indeed the extreme threshold between the human and the inhuman;[365] and likewise the threshold of indistinction and passage between man and animal.[366] This zone of indiscernibility between man and beast is detected by Deleuze in Francis Bacon, when the latter replaces the human head with that of an animal. He maintains this zone of indistinction to be "meat." Bacon reduces man to "a piece of meat."[367] Is this piece of meat not the *Muselmann*?

Moreover, does not the face and head in *Worstward Ho*, worsened almost to the point of disappearing, not bare some resemblance to the face of the Other as the Neighbor (*le prochain*) in Lévinas? After all, the Lévinasian face is less than phenomenal: "It escapes; it is the very collapse of phenomenality. Not because it is too brutal to appear, but because in a sense too weak, non-phenomenal because less than phenomenal."[368]

Similarly, Derrida refers to the "dissymmetry that exists between the divine regard that sees me, and myself, who doesn't see what is looking at me,"[369] maintaining that "God is himself absent, hidden and silent, separate, secret, at the moment he has to be obeyed."[370] God is this "Thing that meanwhile looks at us and sees us not see it even when it is there,"[371] a kind of "ghost [that] would have shown neither his face, nor his look, nor therefore his identity ..."[372]

Contending that Lévinas tends to elide the truly radical "inhuman" dimension of Otherness, that is, "the Otherness of the human being reduced to inhumanity, the Otherness exemplified by the terrifying figure of the *Muselmann*, the 'living dead' in the concentration camp" as the "zero-level neighbor"; Žižek maintains that it is precisely when confronted with the "blank wall, a lack of depth" that is the "faceless" face of the *Muselmann*, that one encounters the Other's ethical call "at its purest and most radical," that it is this utterly blank muteness of the *Muselmann* that makes us "aware of our responsibility toward the Other at its most traumatic."[373]

The well-nigh disappeared faceless and gazeless skull with two black holes for eyes that Beckett patiently engineers in *Worstward Ho* likewise echoes Bernard-Henry Lévy's descriptions of the various "damned" in many a war zone. He alludes to "faces without gazes; gazes without light; that look of broken birds, faces so shattered they seem to have lost that obscure radiance, that superhuman dignity a gaze usually offers."[374] As Elie Wiesel maintains of the denizens of the concentration camps, the *Muselmanner* were "poor creatures whose ghostly faces peeked out from layers of prisoner's clothes."[375] It seemed, as Edmond Jabès avers of the *Muselmann* of the *univers concentrationnaire*, that "his brain was no longer irrigated by his blood, but by some kind of, probably yellow, liquid; sweat or urine?" "He no longer has a mouth, just a useless cavity where air was reluctant to venture."[376]

The *Muselmann* of the *univers concentrationnaire* is, of course, the ultimate victim of the Holocaust. It is precisely in the *Muselmann*, however, that a large hiatus comes to appear between Jeffrey Alexander's tragic narrative and Lévinasian and Žižekian thought. We recall that Alexander maintained of tragedy that it causes *identification* between the beholder of the tragic drama and its victims, leading to *catharsis* in the former. However, as we saw in Žižek's contentions, the ultimate Holocaust victim—the *Muselmann*—nowise enables identification with his plight. Instead, he confronts one as a blank, mute and traumatic void. One does NOT identify with a *Muselmann* and attain catharsis by beholding his plight. For he is the ultimate Other, and utterly heterogenous and refractory to the self. The Other, as Eric Santner expresses it, is at once the Neighbor—but is such, oxymoronically, only to the extent that he is a stranger; that is, his is *the absolutely inassimilable outlandishness of the very nearest.*[377] Far from being a source of catharsis, the "faceless" gaze of the Other is rather the cause of the beholder's being ripped out of his own world, and of his world itself being ripped to shreds. Both world and subject in tatters and shreds. Nohow more shredded.

But let us return to *Worstward Ho.* Beckett's operations of worsening likewise pertain to the will, and move to the will's unlessenable least. This is the reduction of the will to a willing of the non-will, or a will that wills that there be no more willing, the will as non-will, the "longing that vain longing go."[378] In Beckett's words: "Dint of long longing lost to longing. Long vain longing. And longing still. Faintly longing for the least of longing. Unlessenable least of longing. Unstillable vain last of longing still.//...Longing go. Vain longing that vain longing go."[379] Is this worsening of the will in Beckett, to the point that it is virtually indistinguishable from a nullity of will, not likewise the production of the *Muselmann* of the Nazi concentration camps? After all, the will of the *Muselmann* is merely sufficient to support bare subsistence—if that. He is on the very cusp of a complete desuetude of the will, of Nietzschean passive nihilism. Yet, his still remains a bare minimum of the will.

(vi) From Beckett's production of the Muselmann *to the Badiousian event*

But let us continue with *Worstward Ho.* As worsening—contends Badiou—is exercised upon the shades, one approaches the edge of the void. One does so through the worsening of appearances, through a slow chipping away at phenomenality itself. However, although worsening steadily and patiently approaches the void, it cannot reach it. One must go worstward, but one may not go voidward [*Nous ne pouvons mettre cap sur le néant, seulment sur le pire*].[380] The head and the skull, for instance, cannot disappear.[381] "In the skull all save the skull gone."[382] "Unmoreable unlessable unworseable evermost almost void."[383] That worsening must approach the edge of the void but not reach it is further intimated by Beckett as follows: "Such last state. Latest state. Till somehow less in vain. Worse in vain. All gnawing to be naught. Never to be naught."[384] And finally: "At bounds of boundless void. Whence no farther. Best worst no farther. Nohow less. Nohow worse. Nohow naught. Nohow on."[385]

Through these Beckettian operations of worsening, sedulously and relentlessly prising apart and chipping away, bit by bit, piece by piece, element by element, atom by atom, at all aspects of phenomenal appearing in the world, to the point of reaching the very edge of the void itself, we have here entered the absolute "zero-level," or pure void, of subjectivity: the absolutely unendurable abyss in which fully constituted objects and subjects dissolve and yield before a miasmal swarming of spectral partial objects, of exemplifications of the Lacanian *lamella*, of the undead, immortal, libido-as-object that is self-identically the Freudo-Lacanian death-drive (*Todestrieb*). This paronomastic modality of the subject is merely one "degree" removed from the pre-subjective noumenal Real. It is a return to, or recrudescence of, the impossible moment of upsurge of subjectivity in the form of a withdrawal, a recession from the light of phenomenality and its differential degrees of appearing in a world in accordance with its transcendental regime, which assumes the form of a contraction through which constituted reality yields before the surreal phantasmagoria of *membra disjecta*, of organs devoid of body functioning as stand-ins for the dread immortality of the object-libido in its semipernally stuck fixity and eternal psychotic circulation around the void. In Kantian terms, this crepuscular and vanishing mode of subjectivity is that of the pre-synthetic imagination gone mad, psychotic, and utterly demented—the impossible domain of absolutely purified and kenotically evacuated transcendental freedom and spontaneity. In this domain, the subject finds itself vertiginously suspended between direct unadorned animality, on the one hand, and the subordination of its freedom to the moral Law, on the other. In this gathering murk, far from being able to galvanize itself in the direction of forward-directed action in the world or attunement to the promptings of the Deity, the Self is submitted to the horrifying experience of dread organs without body, of itself as the Lacanian fragmented body, of its *membra disjecta* floating around in a nauseating and disgusting miasmal drift, the "night of the world," the "valley of the shadow of death," the hell-on-earth of Kundry in Wagner's *Parsifal* who, having committed the blasphemy of laughing at Christ dying on the Cross, is condemned to an eternity of wandering as an abject, multiply befouled, purely excremental self entirely rejected by the world, abandoned by God, and subject to involuntary, psychotic automatisms, spasmodic jabberings, slavering aphasia, dyslexia, *and the absolute horror of never being able to die.*

But now, a Badiousian truth-event shall transpire. Beckett's operations of worsening, having reached the very edge of the void (and having literally produced the *Muselmann* of the *univers concentrationnaire*), constitute what Badiou terms an evental site. We now leave Beckett and finally move onto what Badiou defines as a truth-event as such. The event is what-is-not-being-*qua*-being.[386]

An evental site of a situation is that which opens up the possibility of the occurrence of an event. On the other hand, a site is only evental inasmuch as it is retroactively qualified as such by the occurrence of an event. Much as is the Holocaust in terms of Jeffrey Alexander's tragic narrative, an event is not of the order of presentation or of re-presentation. Formally, Badiou defines the event of an evental site as the one-multiple composed of the elements of its site and, in addition, *the event itself.* That is to say, the event is the multiple which both presents its entire site, and, by means of the pure signifier of itself immanent to its own multiple, presents the presentation of the one-multiple that it is. As with the Holocaust in Jeffrey Alexander's tragic narrative, a Badiousian event is self-referential. It refers solely to itself. This definition, moreover, renders it intrinsically impossible to determine whether it does or does not belong to its site. Badiou concludes here that the belonging of an event to the situation of its site is *undecidable* from the standpoint of the situation itself. That is, the signifier of the event is supernumerary to its site.[387]

Deciding that the event belongs to the situation is, therefore, of the order of a wager. To be recognized as an event requires an intervention. It requires inter alia that the event be *named.* The name of the event, avers Badiou, must be made out of an *unpresented element* of the evental site. Since the evental site, as a one-multiple, is by definition composed purely of inconsistent multiplicities, the foregoing means to say that the name of the truth-event must emerge from the void. Badiou terms this name a supernumerary name, in that inconsistent multiplicity is subtracted from the count which renders up consistent one-multiples. This intervention extracts a supernumerary signifier from the void. The name of the event chosen by the intervenor represents the void. Again, as with the name "Holocaust" in Jeffrey Alexander's tragic narrative, it is a representative without representation and sutures the event to the unpresentable. Much like the event itself, its supernumerary name belongs only to itself—and therefore, as in the case of the name "Holocaust" in the tragic narrative, it must be a name that is deployed as a *proper* noun. It is this name that now circulates within the situation. Once more, as is the Holocaust in the tragic narrative, for the State, this name is a pure enigma.[388] In fact, the truth-event itself—much like the Holocaust in its unrepresentability—is utterly incomprehensible according to the prevailing language and established knowledge of the situation.[389] As Claude Lanzmann maintains, it erects a ring of fire around itself.

Badiou maintains, moreover, that it is in the wake of an event that the individual is convoked to enter into the composition of a *subject* (the subject of a truth). This is the movement in which the individual is called upon to enable the passing of a truth. A truth, for Badiou, is the process whereby the subject relates to the situation from the perspective of the event. This relating is termed *fidelity*. To be faithful to an event, the subject must move within the situation that the event has supplemented in accordance with the event. In these terms, a truth is the process of a fidelity to an event. It is what this fidelity produces in the situation. The subject, on the other hand, is the bearer (*le support*) of a fidelity, the one who bears a process of truth. The subject of a truth is compelled by the event to invent a new way of being and acting in the situation. Truth is, thus, not merely a process but likewise an invention. Much as in the case of the event, that which issues from a truth-process is incommunicable (*ne se communique pas*).[390]

This incommunicability calls to mind the incomprehension with which many, or all, Holocaust survivor testimonies of life within the *univers concentrationnaire* are received. Moreover, the fact that the survivor is only an imperfect witness of the Holocaust—the perfect witness being the *Muselmann* who is deceased, still further underlines the sheer incommunicability of the Holocaust-event. In that it is radically heterogeneous thereunto, a truth punches a "hole" in prevailing knowledges. Notwithstanding, it is the sole known source of new knowledges. It *forces* knowledges. Though it is never communicable in and of itself, a truth is nonetheless implicated in fundamental reshapings of the forms and referents of communication.[391] This was certainly the case with the Holocaust as truth-event, in the manner in which it rendered human consciousness privy to the draconian propensities inherent to the efficient instrumental functioning of modern bureaucracies and in its being the ineluctable spur in the wake of which an extensive framework of international law and humanitarian institutions were progressively wrought and consolidated in the many decades following the war.

(vii) The Sho'ah *as Badiousian event and the universalization of Jewish survival*

Badiou contends that a truth-event, furthermore, convokes the individual to be "the immortal he was not yet."[392] Yet, it is not so much this individual who can become this immortal, but rather the subject into whose composition he enters which, on account of the fact that it exceeds him, comes to pass as immortal.[393]

This Badiousian subject is far closer to the Deleuzoguattarian collective subject—specifically as the latter's "group subject," as opposed to what they term a "subjected group." The immortality, and survival—the existence under reprieve from the nefarious Damocles' sword of death—herein implied, would therefore entail, not necessarily that of the individual human subject (although this may, in certain cases, be the case), but rather of particular ethnically, religiously, or socio-politically defined human collectives. In our specific analysis, this collective survival would be that of the Jews. We shall, however, swiftly come to universalize this survival beyond the pale of Jewry: that is, from the vanguard of humanity to humanity itself.

For one observes that—in Jeffrey Alexander's tragic narrative and in Badiou alike—it is precisely from the event (the tragic narrative's *sui generis* unrepresentable Holocaust-event, and Badiou's truth-event as such), in its singularity, that universalization transpires. As Badiou maintains, a truth, in its invention, is the only thing that applies to all humans.[394] This is so because it relates to the particularity of the situation solely from its void. As the multiple-of-nothing, the void neither excludes nor constrains anyone. As seen earlier, it is universally included in any situation. Thus, the fidelity that originates in an event is necessarily universally addressed.[395]

This may likewise be argued from the averment that a truth is inherently generic, that is to say, subtracted from all precise designations, and in excess of what the said designations can possibly discern.[396] The event is per se without object.[397] The Being of a truth emanating from an event is the generic multiple. A truth process is that which induces the recrudescence of inconsistent multiplicity from the unfathomable depths of consistent multiplicity. It makes rise to the surface that which is maximally subtracted from consistent multiplicity and the count-as-one which had heretofore repressed and dominated it. This repressed—like the skull with two black holes in *Worstward Ho*—is an indistinct multiple, sans clear contours, and bereft of any specific name. Badiou terms such a multiple, "any multiple whatsoever." Since, as seen earlier, any situation is woven from the null-multiple that is the void, its Being is its inconsistency. Therefore, a truth about Being shall be any-multiple-whatsoever, that is, a multiple that is purely presented, a minimal consistency without a nameable predicate which says nothing of the situation except for the inconsistency (inconsistent multiplicity) which is its very Being.[398]

In his later terminology, in which situations are termed "worlds" (materialist totalities of bodies and languages), Badiou maintains that a truth created in one world is actually valid for other worlds and virtually valid for all worlds. That is to say, a truth is immediately universal in the sense that it can be appropriated by other worlds, other cultures and other individuals than the worlds, culture or individuals that partook in the truth's upsurge and development, despite the difference in the bodies and languages constituting the materiality of these other worlds. A truth, therefore, functions in a transworldly fashion. It is available to all.[399] This, all the more, reinforces our abiding contention that Badiou's theoretical framework, in a manner somewhat homologous to that of Jeffrey Alexander's tragic narrative, is able to universalize the Holocaust-event to all of humanity, irrespective of cultural particularity.

That an event's address is universal and as such pertains to all men, whereas a truth convokes the immortal in the individual; means that a truth-event universally heralds the advent of the immortal in humanity as such. Speaking generically, Badiou maintains that truths are what render man "precisely something other than a victim, other than a being-for-death, and thus: *something other than a mortal being*."[400] "An immortal: this is what the *worst* [my emphasis] situations that can be inflicted upon Man show him to be, in so far as he distinguishes himself within the varied and rapacious flux of life."[401] Human rights, if they exist, do so only as "the rights of the Immortal, affirmed in their own right, or the rights of the Infinite, exercised over the contingency of suffering and death.... And we know that every human being is *capable* of being this immortal.... In each case, subjectivation is immortal, and makes Man."[402] Once more, we have the universalization of the survival by the Jewish collective subject of the Holocaust to the survival of humanity as such of diverse Holocaust-like vicissitudes.

5. Survival of various human collectives in the wake of the Holocaust-event

Let us return to the singular truth-event that we are presently considering: the Holocaust. The faithful truth-procedure that gradually unfolded in the wake of this truth-event was as follows. The Nuremberg judgment of September 30, 1946, that we have already alluded to earlier, signaled the universalization of this singular event with the unprecedented use of the term "crimes against humanity." Thus, the truth manufactured by the said event, which had transpired within one particular world—that whose State metastructure was the draconian and blasphemous Nazi State apparatus, was deemed virtually exportable to all human worlds; that is, to all of humanity, to generic humanity. Inextricable from this pronouncement was the promise that such crimes would henceforth be deterred by the inculcation within the psyche of would-be perpetrators of the tremulous realization that their actions would almost certainly be punished.

This much may be garnered from the words of presiding judge Justice Robert Jackson, who declared that crimes against humanity could only be forestalled if those bent on perpetrating them find themselves noticeably discouraged by the thought that the long arm of the law will, in due course and perhaps inevitably, overtake them. The universalization herein implied is pellucid, and arises naturally from the very definition of what constitutes a crime against humanity: an offense of such exceeding severity that it may be tried and punished by *any* court in *any* part of the world, on territory presided over by *any* Statist metastructure, and without regard to the offender's ethnicity, the identity of the victims, or the place of crime's commission. Such a crime, of its own inherent nature, immediately renders the jurisdiction of courts absolutely universal, that is, trans-worldly.

In that its overriding purpose was the prosecution of those deemed responsible for crimes against humanity, international criminal law was in effect the direct progeny of the truth-procedure emerging in the wake of the Nuremberg judgment. This is so because—and we can see here the universalization being wrought—the very fact that *any* member of the generic human species could even conceive, let alone be capable of and actually commit such enormities, immediately demeans *every* single member of the entire human race, irrespective of the particular Badiousian world—that is to say, totality of bodies and languages—they inhabit (Universal Declaration of Human Rights, Preamble). This crime is specifically defined as one which threatens, or renders actual, the termination of the existence of individual or collective human subjects across the globe. Particularly tightly circumscribed within its ambit are acts of genocide, mass murder, and terror, but likewise rapacious warfare and systematic torture (Nuremberg Charter, Article 6(3)). The perpetrators to whom it ascribes responsibility include not only political or military leaders, but also heads of State, assorted abettors, theocrats, ideologues, and even industrialists (Nuremberg Charter, Articles 6 and 7). To thwart the Damocles' sword of the foregoing, the truth-procedure commenced by the Nuremberg judgment assumed the form of a firm resolve to establish international institutions of justice powerful enough to undercut and even to surcease the impunity of these sadistic and necrophilous rogues. It would kick-start an ensuing evolution of international human rights law and endow the latter with the force necessary to compel municipal courts and an ever-burgeoning constellation of international tribunals to prosecute men of blood (Nuremberg Charter, Articles 10 and 11). The heretofore closed aperture of State sovereignty would be unlocked so as to deter, thwart and punish the depredations of political and military leaders (United Nations Charter, Article 2(7) and Article 42).

Apart from the bloodthirsty comportment of States, it was now likewise the hitherto prevalent impunity of individuals carrying out the orders of nefarious governments that was to fall into desuetude. For the Nuremburg judgment had decreed that international law forthwith concern itself and be empowered to punish these as well. In this connection, the move toward the universalization of the Holocaust-event—its intended exportation to all human worlds on the planet—is swiftly espied. Here, the Nuremberg Charter effectively ruled that all individuals have duties which far exceed the aegis of the particular States to which they are nationally obligated, especially in cases where these States authorize actions outside their competency. This competency is, of course, one stipulated by international law. That is to say, the moral probity of State actions would be measured against a *universal*. In accordance with this universal basis, all transgressing individuals without exception would be held liable for crimes perpetrated against the core dignity of the human species. It was through this deterrent that the existence of would-be victims would, ideally, be placed under reprieve. Article 8 of the Charter, for instance, in no sense absolves defendants acting in compliance of orders received from their superiors or governments. This would include the actions of both torturers and commanders following the malign dictates of their governments.

The truth-procedure whose principle concern was ensuring the survival of human collectives countenanced with the potentially death-wielding machinations of draconian States and homologous forms of rapacity would continue. In December 1946, the Nuremberg Charter and the overall reasoning of the Nuremberg Tribunal were unanimously confirmed by the United Nations General Assembly. Crimes against humanity, as expounded in the Nuremberg judgment and Nuremberg Charter, were very soon to acquire a privileged position in international law through the various conventions which ensued. According to these latter, every State was forthwith saddled with an *erga omnes* ("towards everyone" or "towards all") obligation to assist with the trial and punishment of this particular sort of crime, even in the absence of any evident connection between the States in question, on the one hand, and the crimes, perpetrators or victims, on the other. To this end, all States were endowed with so-termed "universal jurisdiction" in the prosecution of such matters (Geneva Convention 1, Article 2). Once again, we observe here the universalizing tendency inherent to a Badiousian truth-procedure in general, and in the specific truth-process here unfolding itself. For one finds here a movement from the far more parochially limited jurisdiction of States—each with its own idiosyncratic metastructure—over ordinary crimes, to the "universal jurisdiction" as pertains to crimes deemed against humanity. In the former, a—usually territorial—link is required in order to prosecute violations; in the latter, no link whatsoever other than the unadorned humanity of the victims or homosaceralized would-be victims is needed.

As observed, intervention upon the Holocaust-event spawned at Nuremberg the charge and notion of "crimes against humanity" for the first time. Article 6(c) of the Nuremberg Charter criminalized the authorization of torture or genocide by future State agents against their own populations. Such perpetrators could henceforth legitimately be punished by any court in the world. This unfolding truth-procedure would be stepped up and lead to the Universal Declaration of Human Rights announced by H.V. Evatt, president of the United Nations General Assembly, in Paris on December 10, 1948. Hereby, though arising from the relatively parochial ambit of the customary laws and constitutions of a limited number of States, human rights would transcend their domestic point of origination and emerge into clear view on the international—that is, trans-wordly—plane. Along the same trajectory, local State-bound protections against nefarious depredations would likewise be universalized to the international milieu (Universal Declaration of Human Rights, Preamble). A formerly localized reprieve of existence would be writ large to the level of the human species as such.

That the drafting of the UN Charter for human rights was verily the continuation of a truth-procedure emanating specifically from the singular event of the Holocaust is evinced by the fact that, for the drafting committee and for speakers in the General Assembly debate, it was explicitly the Holocaust-event which justified the adoption of such universal rights. Behind the adoption of most of the articles there is always some reference to the perversions of Nazism. Indeed, the Human Rights Commission had received detail reports on the prosecution evidence at the Nuremburg trials prior to drafting the Universal Declaration of Human Rights. Therefore, whilst compiling their drafts for the declaration, the members of the commission were constantly privy to the specific minutiae of the various Nazi homosaceralization and extermination projects, such as the purging of "useless eaters," the scorched earth policy of the *Einsatzgruppen*, the gas chambers, Mengele's experiments, "night and fog" decrees, and so forth. "Gazing" as they were at the multiplicity of *Muselmanner* constitutive of the Holocaust-event's eventual site, their acute awareness that these atrocities had been perpetrated in consonance with perfectly valid laws of the Nazi State's metastructure, for instance, moved the committee to discard the qualification "except in cases prescribed by law" from the phrasing of the universal right to life, liberty, and personal security encapsulated in Article 3.[403] The immortality of this truth and the generic character of the truth-process from whence it had arisen may clearly be adduced from the fact that it subsequently inspired in excess of two-hundred international treaties, conventions, declarations, and the bill of rights in nearly every national constitution adopted since the Second World War. This imperishability and universality is evinced from the fact that Article 28—which declares that all humans are entitled to a trans-worldly order facilitating the full realization of all rights and freedoms in the Universal Declaration—is still very much germane today. The truth embodied in this article was, in turn, itself the product of a truth-process unfolding from Chapter VII of the UN Charter, which had authorized and empowered the Security Council to intervene in all cases where human rights violations potentially threatened world peace, and did so by thwarting the death-wielding machinations of any sovereign State—hereby vouchsafing the survival of endangered populations. In that, in our day, such depredations are seen as diminishing the fundamental dignity of generic humanity and insofar as the UN Charter and the Universal Declaration of Human Rights are universally addressed, the legal mechanisms provided by these may now readily be triggered to undercut the murderous bent of many a draconian State apparatus.

This generic truth-procedure, geared to the ends of ensuring the continued survival of individuals and collectives, would continue with the Convention on the Prevention and Punishment of the Crime of Genocide. The latter was signed in 1948 one day after the Universal Declaration of Human Rights, entering into force in 1951. It required States to deter political mass murder or, more specifically, all acts intended to destroy, in whole or in part, a national, ethnic or racial group, committed by rulers, officials or private individuals (Article 2), by punishing perpetrators through international criminal tribunals with jurisdiction (Articles 5 and 6).

The Genocide Convention likewise continued the universalizing tendencies of the UN Charter and the Universal Declaration. This is the first time that crimes against humanity were unhinged from any temporal connection with such forms of belligerence. This may be adduced from the fact that Article 1 of the Convention declares genocide, whether committed during times of peace or of war, a crime under *international* law. That the very definition of genocide expounded by the Genocide Convention is verily a milestone along the broader trajectory unfolded by the truth-process following in the wake of the Holocaust-event may be clearly adduced from the fact that it reflected the then-prevailing preoccupation with the genocidal Nazi policy toward the Jews as unearthed at the Nuremberg trials. To stifle the multifarious depredations of agents bent on effecting the destruction of collectives, the United Nations Security Council is empowered by Article 8 of the Genocide Convention not only to suppress genocide but moreover to thwart its incitement. The Convention likewise authorizes the extradition of perpetrators of genocide without regard of their specific political motivations (Article 7).

Sixty years of painstaking fidelity to the truth-event first commenced in the immediate post-war years have led generic humanity to the point where today a firmly entrenched panoply of fundamental rights are accepted as universal. These include freedom from the death-wielding machinations of the State and other draconian practices that may likewise lead to death: torture, arbitrary arrest and unfair trial (Universal Declaration, Articles 5, 9, 10). A plethora of strategies have been developed in order to keep nefarious State apparati at a safe remove from their subjects. Failing this, imminent or actual State depredations against populations are to be met, as a matter of international law, by the routing of their militaries or paramilitaries. The right to intervene against nefarious State violations likewise extends to the malign actions of third parties, such as vigilantes and death squads. In addition to their suffering military chastisement at the hands of the international community, Article 8 of the Universal Declaration and Article 2(3) of the UN Covenant on Civil and Political Rights oblige offending States to implement effective reparations for their malevolence.

What Article 3 of the Universal Declaration most fundamentally guarantees is the right to life. This right is absolutely non-derogable in all human rights conventions. As the most primordial and base-level right, it serves as the preponderant prophylactic against Statist infractions such as summary execution without fair trial, the shooting of surrendering soldiers, and the liquidation of dissidents by death squads. Not only is military intervention against proto-genocidal or genocidal States enjoined under Chapter VII of the UN Charter, but the International Court of Justice (ICJ) is itself empowered to bring "provisional measures" under the Genocide Convention (Article 9).

Another manner in which the truths produced through fidelity to the Holocaust-event would serve to place the existence of threatened human collectives under reprieve is through the right of individuals to leave States in which their lives are endangered. Here, in accordance with the terms of the Universal Declaration, the fleeing subject is permitted refuge in the first safe country he reaches (Articles 13(2), 14(1)). Once more, the production of this truth through the Universal Declaration was a direct effluence of the truth-procedure arising from the Holocaust-event, for its promulgation aimed inter alia to shame States that had turned away Jewish refugees fleeing the Nazi Damocles' sword. The universal address of this truth would be more firmly entrenched in 1951 with the International Convention Relating to the Status of Refugees, ratified by one-hundred and forty-seven countries. This convention forbids any country to deport refugees to a State where it is known that they will be persecuted for ethnic and ideological reasons, especially if such individuals are likely to be maltreated or even killed at their would-be destination (Article 33(1)). Adding further ballast to all of this is the Convention against Torture and Other Cruel, Inhuman or Degrading Treatment and Punishment, entering into force in 1987. The latter prohibits deportation of individuals to States where they may be tortured, perhaps to death. In accordance with the convention, those who have successfully escaped the nefarious depredations of their home State must be given sanctuary if it is plausibly evident that returning would probably lead to subjugation or even death (Article 3(1)).

The Rome Statute of 1998 continues fidelity to the truth-procedure consequent upon the Holocaust-event. That this statute ultimately sources itself in the foregoing event is evident from its preamble; which explicitly recalls the "unimaginable atrocities" of the past century, vividly underscoring the threat such heinous crimes pose to the very lives of persons across the globe. The statute necessarily invites UN intervention under Chapter VII of the Nuremberg Charter in the face of the potentially death-wielding machinations of rogue States. The universality of its address is evinced from the fact that its wording unequivocally endorses trans-worldy jurisdiction over crimes against generic humanity (Article 5). So as to entrench the latter, the Rome Statute led directly to the establishment of the International Criminal Court in The Hague—the rationale behind whose creation recurred again to the endlessly reiterated refrain that such shocking crimes imminently threaten world peace and, even before and beyond this, always already undermine the well-being of those unfortunate enough to behold the suffering induced (Preamble).

The Rome Statute's fidelity to the post-Holocaust truth-procedure is clearly evident in Article 6, which confirms the truth produced by the 1948 Genocide Convention in its definition of genocide as the intentional destruction of a national, ethnic, or religious collective, whether in whole or in part. But it goes further, also factoring in more indirect means of the perpetration of this crime, including the systematic imposition on designated groups of conditions of life likely to cause death, such as the forcible eviction of children from a group. Included under Article 7's definition of crimes deemed to be against humanity, are not only murder, torture, rape, enslavement, but also false or unlawful imprisonment or deportation. The abhorrent quality of these acts resides in their commission comprising a strategy deployed within the more comprehensive ambit of systematic depredations pursuant to or in furtherance of counter-humanitarian State or organizational policy.

Article 8 of the Rome Statute further adumbrates those truths of the post-Holocaust truth-procedure produced by the 1949 Geneva Conventions. All of the breaches considered war crimes under the latter convention are covered by Articles 8(2)(a) and 8(2)(b) of the Rome Statute. These are, however, updated in a subsection which lists twenty-six violations as war crimes. Included here are assaults on UN peacekeepers; attacks known beforehand to lead to disproportionate loss of civilian life or severe damage to national environments excessive of military objectives; intentional attacks on non-military targets; use of asphyxiating or poisonous gases; and conscription of children under the age of fifteen years for active military purposes. Article 8(2)(c) extends jurisdiction over internecine conflicts in respect of all serious violations of common Article 3 of the 1949 Geneva Conventions, that is, inhumane attacks on civilians or sick or surrendered soldiers. This includes the subjection of the foregoing individuals to unfair trial.

The staying of the Damocles' sword of death countenancing collective subjects across the globe has likewise been facilitated by post-Nuremberg international law's adoption and entrenchment of diverse minority rights. This effluence of the post-Holocaust human rights truth-procedure can already be detected as early as 1946, with the UN's creation of the Sub-Commission on the Prevention of Discrimination and Protection of Minorities. This was stepped up in 1948, when the Convention on the Prevention and Punishment of Genocide outlawed the partial or complete destruction of national, ethnic, racial or religious collectives (Article 2). Minority rights were to be further consolidated and universalized through the declaration, in 1951, by the International Court of Justice of genocide as a crime under customary international law. The declaration upheld the right to continued existence of minorities in the face of Statist depredations bent on splintering or physically annihilating such groups. This protection was extended by international treaty through Article 27 of the Civil Covenant. Ultimately arising from the 1949 Geneva Conventions, the right to existence of minorities derives its enforcement mechanism from the pervasive notion and conviction that universal jurisdiction exists to punish crimes deemed against humanity. The truth-procedure has led to the establishment of tribunals in Rwanda and former Yugoslavia to bring perpetrators of genocide to trial. It is precisely in that international law threatens perpetrators with prosecution and the intervention of foreign militaries, that minorities are protected—at least in part—from homicidal actions such as ethnic cleansing. In the face of an actual breach of minority rights by nefarious State actors, afflicted minorities are given the right to secede.

Another manner in which international human rights law places the existence of the human subject under reprieve is through its laws against war crimes. This legal inscription was first made in the Nuremberg Charter and judgment, and in the ensuing war crimes trials in Germany and Japan. Through the Charter, the Nuremberg Tribunal was enabled to punish not only war crimes, but likewise crimes against peace, and—as already seen—crimes deemed against humanity; that is to say, death-wielding depredations aimed against civilians in pursuance of their belligerent objectives. As we have already seen, this truth-procedure was continued with the Geneva Conventions of 1949, which elaborated the principles of international law regarding the treatment of POWs, sick and wounded combatants, and civilians caught within the maelstrom of war. More specifically, Article 3 aims to protect all non-combatants caught in the crossfire of a civil war. Prohibited here are not only murder, torture, hostage-taking, and extra-judicial executions, but also military, police or guerrilla actions deliberately pursued so as to kill or maim civilians and prisoners of war.

The truth-procedure would continue and receive additional ballast with the 1977 Geneva Protocols, which would update the Geneva Conventions within a more modern context. The latter's Article 35 prohibits the deployment of such means and methods of waging war—including weapons, projectiles and materials—as would cause superfluous injury or spurious suffering. The Protocols likewise rule that, in order to minimize civilian casualties, bombardment must be confined solely to military targets, and moreover that particular vigilance be exercised in protecting refugees, women and children, and even journalists.

Presently, the viral omnipresence of images and videos of atrocities imminently accessible to an international public further narrows the scope for the flaunting of The Hague and Geneva Conventions by criminal parties or States, in that public outrage is now ever the more deployable as a means of pressurizing reluctant politicians to effect military intervention to the end of saving lives.

The precise conditions under which the UN is authorized to legitimately penetrate the sovereignty of States bent upon, or in the process of, unleashing dead-wielding machinations against a collective subject are delineated in certain select articles of the UN Charter. Article 2(7), for instance, allows the UN to intervene in matters essentially falling within the domestic jurisdiction of any State, except solely by recourse to enforcement measures under Chapter VII. However, according to Article 39, in applying the latter, the Security Council must first establish whether there exists a threat to or breach of peace, or act of aggression. If measures short of armed force are deemed insufficient, the Security Council may be empowered under Article 42 to take military action in order to restore international peace and security.

Nonetheless, the condemnation of a State by the Security Council for crimes against humanity does not suffice to activate the "just war" provisions of Chapter VII of the Charter. In addition, at least nine of the fifteen members must pass an actual resolution to employ military force, with no veto cast from the five permanent members (the US, China, Russia, France and Britain). Notwithstanding, there is an interesting loophole within the UN Charter. This is to be found in Article 51, which inter alia maintains that the UN may in no way thwart a militarily threatened sovereign nation's "inherent" right to self-defense, regardless of whether the Security Council decides to intervene or not. According to this, force as pursuant to a rule of a customary law which had arisen independently of the UN Charter, was not to be excluded. It also made room for something alluded to later by Václav Havel in a speech to the Canadian parliament six weeks into the NATO bombing campaign to liberate Kosovo from Milošević's nefarious depredations. Havel maintained that, though it had no direct mandate from the UN, the intervention was nowise refractory to international law but in fact acted by recourse to a law "higher" than that appertaining to sovereign States. Geoffrey Robertson maintains here that Havel was alluding to an evolving principle of humanitarian necessity, a kind of "Good Samaritan" principle writ large, according to which the exigency to surcease the massacre of innocents necessarily overrides the UN's own prohibition against meddling in the affairs of other States as expressed in Paragraph 4 of Article 2 of its Charter. This paragraph itself has a caveat: the kind of intervention it forbids is only that which is inconsistent with UN purposes. This would generally permit the deployment of force, aimed not at undermining the territorial integrity or political sovereignty of States, but in order to rescue collectives threatened with possible extermination by Statist actors or their proxies.[404] It is this sort of logic that was used in Israel's justification of its incursion into Uganda to rescue Israeli nationals held at Entebbe airport by hijackers colluding with the Ugandan government. A similar rationale was deployed in Tanzania's invasion of Uganda, which signaled *finis* to Idi Amin's reign of terror; the US's 1971 humanitarian intervention in Grenada; and India's 1971 military intervention against genocide in Bangladesh. After the first Gulf War—this time in accordance with UN Security Council acquiescence (Resolution 688)—the US, Britain and France invaded Iraq to establish protected enclaves for displaced Kurds in Northern Iraq faced with Saddam's death-wielding depredations.

(i) The survival of the Kosovars

The Damocles' sword of death first began to hover in spectral fashion over the 1.7 million Albanians in Kosovo as far back as 1989, when Slobodan Milošević imposed direct rule from Belgrade and presided over pro-Serb discrimination. In 1990, Milošević effectively ordered Muslims ranged against him to "surrender or die," announcing a "time for struggle." They were effectively reduced to *homines sacri*, to a collection of shades on the edge of the void. In March 1998, Serbian death-wielding machinations would commence in earnest, starting with a massacre in Kosovo's Drenica Valley. This precipitated a refugee exodus. In October 1998, four-hundred-thousand Albanians were driven from their homes by a bludgeoning Serb military offensive. More violence would follow by the year's end when Milošević's minions, flaunting the ceasefire agreement brokered by Richard Holbrook, reprised savagely against Albanian recalcitrants—shocking the positioned Organization for Security and Co-operation in Europe (OSCE) monitors. The Yugoslav army slaughtered hapless innocents. The world would see forty-five lifeless bodies picked through by television crews in the village of Racak in January 1999. It was this atrocity more than anything else that would signal the beginning of the reversal of fortunes for Milošević and the bedraggled Kosovars alike. For it convinced the US administration that NATO had no choice but to employ military force. Soon hereupon, Milošević's defiant refusal to let ICTY prosecutor Louise Arbour to enter Kosovo to gather forensic evidence for crimes against humanity would bring America's European allies to the same resolve. But the Serb slaughter of one-thousand-eight-hundred civilians in Kosovo was the last straw. Meanwhile, a number of resolutions under Chapter VII had already been adopted by the Security Council. An arms embargo and calls that the Serbs reconsider Kosovo's "political status," had already been wrought through Resolution 1160 of March 1998. Further resolutions in September and October demanded that Milošević surcease his depredations, threatening "further measures" in the event of his refusal of a political settlement to the ongoing imbroglio.[405]

In the end, it was the burgeoning realization that the headlong spate of killings and deportations were merely a part of a far more comprehensive and premeditated homosaceralization project by Milošević to completely purge Kosovo of its 1.7 million Albanians that chivvied NATO to intervene in a manner in which the Security Council would not. Further bolstering this resolve was NATO's March 23 report that, solely within the past three months, one-hundred-thousand Kosovars had been forcibly displaced—fomenting a burgeoning refugee crisis in neighboring States, clearly evincing that a mass deportation program was already well underway. In the face of this, the international community was verily empowered to intervene militarily, since mass deportation is considered a crime against humanity by Article 7(1)(d) of the Rome Statute. The article defines the latter as the forced displacement of persons from their lawful area of residence by eviction or other means of coercion. Yet, Milošević's would persist in his contumacy. In that it would thwart his ethnic cleansing of Kosovo, he rejected the Rambouillet peace proposal giving Kosovo autonomy and self-government under a continued NATO presence. Even when the NATO bombing campaign commenced on March 24, far from desisting or at least de-escalating, Milošević actually quadrupled his troops in Kosovo. Ten-thousand Kosovars were massacred, and over a million inhabitants put to flight. There was destruction, pillage, torture and rape.[406] Dead Albanian bodies killed by Serb forces were dug up from mass graves in Kosovo and reburied beneath a police station outside Belgrade, or dumped into the Danube.[407]

But Milošević would be stopped, and his Damocles' sword smashed to pieces. For two solid months, NATO would pulverize his economic infrastructure, bombing power grids, bridges, and even a television station and water purification plant—crushing enemy morale.[408] By way of an agreement signed June 9, the homicidal Serb forces would leave Kosovo. Even refugees from the beaches of Australia returned to their burned-out homes. Milošević and six of his senior commanders were indicted. The liberated province became a protectorate of the UN overseen by forty-thousand UNMIK troops, who would—in accordance with Resolution 1244 of June 10, 1999—gradually cede governmental functions to the Democratic Alliance of Ibrahim Rugova, elected November 17. By 2006, it was widely accepted that Kosovo was morally and legally entitled to secede from the State that had tried to exterminate its populace. On February 17, 2008, the country declared independence.[409]

(ii) The survival of the East Timorese

The East Timorese have at best cowered in sullen silence under the ever-poised Indonesian Damocles' sword for entire swathes of the last quarter century of the past millennium. The blood-soaked nature of Indonesian political affairs is already evinced in President Suharto's accession to power in a 1965 military coup, which led to the murder of a million citizens. Sporadic pockets of resistance to the draconian rule of the often-nefarious Indonesian State apparatus were struck down with tremendous ferocity. Tens of thousands died in these crackdowns. In 1983, the entire village of Kravas was wiped out. In 1991, there was a notorious massacre of protesters at Santa Cruz cemetery, followed by another massacre later the very same day as soldiers dragged wounded protesters from hospital beds and executed them. These and other atrocities were committed by the élite Kopassus force commanded at times by Lieutenant-Colonel Prabowo, Suharto's son-in-law. Much later, in 1998, though faced with Suharto's necrophilous depredations, the East Timorese stymied these by concerted mass action. In January 1999, the Indonesian despotic regime would be countenanced with the flight of working organs from its reterritorializing sway, when a prospective East Timorese independence referendum was anounced. Already in February of the same year, certain army leaders conspired to intimidate the population even prior to the vote, inter alia by arming militias in each of the thirteen provinces to wreak havoc in the event of the referendum swaying in the direction of independence. In collusion with the Indonesian military, these militias would brazenly bare down the Damocles' sword of death upon East Timorese appearing to favor independence in the months running up to the ballot. On April 6, sixty-seven people were gunned down or hacked to death in the Liquica church massacre. The terror aimed to instill a very clear message: voting for independence was tantamount to signing one's own death certificate. Casting such a ballot would mean certain death. Prospective voters were, thus, homosaceralized—reduced to almost completely etiolated shades on the edge of the void, floating just above the abyss of pure inconsistency. This meant that a majority vote in favor of independence would almost certainly result in a blood bath. This would indeed happen. The East Timorese voted overwhelmingly in favor of independence. When this was announced on September 4, the free-lance militias and sections of the Indonesian army immediately commenced their slaughter. The Falintil intelligentsia and its supporters were rounded up and murdered. Half-a-million East Timorese were forced to flee into the jungle in fear of their lives. Thirteen-thousand were forced at gunpoint across the border to West Timor. Fourteen Catholic priests and nuns were murdered in Dili. The head of the Protestant Church was murdered.[410]

The international community was riled. The United Nations General Assembly would deem the occupation a blatant breach of the 1960 Declaration on the Granting of Independence to Colonial Territories and Peoples, whose fourth paragraph condemns all military or other repressive measures aimed at thwarting the self-determination of a dependent people. In the face of this, five UN ambassadors were immediately expedited to Jakarta to ask the government's permission for an international force to enter East Timor. On the fifth day of the massacres, Kofi Anan said the UN could not simply stand by and watch the people of East Timor being killed. Meanwhile, however, in an absolutely relentless and headlong scorched earth campaign, the militias and supporting Kopassus troops were still in the process of wreaking livid-brained vengeance on those who had voted for independence, massacring countless innocents and dumping them in mass graves. They were zealously executing a thoroughly premeditated, highly organized and systematic policy of intimidation, mass murder, and displacement of an entire ethnic and national collective. They were massacring the East Timorese and transporting the bodies of their victims across the border and secretly disposing of them in West Timor.[411]

But, it is here that the hitherto moribund fate of the East Timorese would be irrevocably reversed. For the death toll was mounting so rapidly that, on September 12, President Clinton was moved to threaten President Bacharuddin Jusuf Habibie with the loss of billions of dollars in loans aid unless he allowed a UN peacekeeping force to be introduced. With Habibie having no choice, a 7,000-strong Australian-led force came ashore a week later, forcing the Indonesian soldiers to leave in surly silence and the makeshift militias to slink across the border to West Timor. Not long hereafter, the beleaguered East Timorese reemerged from their hiding places in the hills. The referendum result was ratified by a new Indonesian parliament. Habibie was succeeded as president by Abdurrahman Wahid, under whom East Timor was made a protectorate of the United Nations on October 25, 1999. In February 2000, after a historic meeting between Wahid and Fretelin leader Xanana Gusmaoa—who had spent long years incarcerated in a Jakarta cell under Suharto's brutal despotic regime—a plausibly congenial relationship was engineered between the United Nations Transitional Authority in East Timor (UNTAET) and the Indonesian government. Hereafter, the East Timorese would—with the aid of the United Nations—be primed for self-government under a new constitution. This would come in April 2002, when Gusmao was democratically as president. The East Timorese would endure.[412]

(iii) Reprieve from and survival of the death penalty

A crucial manner in which the universalization of the Holocaust-event, through fidelity to the truth-procedure unfolding in its wake, placed the existence of those cowering under the Damocles' sword of death under reprieve was by way of contributing to a gradual abolition of the death penalty in many countries. Prior to the time at which it was imposed upon Nazi war criminals at the Nuremberg judgment in 1946, no nation of significance had abolished this ultimate and most definitive punishment. However, within the subsequent half century that would elapse, most of the countries of the world would do so—both in law and in practice. The seeds for the unfolding and efflorescence of this effluence of the post-Holocaust truth-procedure are traceable to at least as far back as Article 6 of the International Covenant on Civil and Political Rights. For the Human Rights Committee, the article's wording certainly points toward the desirability of the abolition of the death penalty, as this would be the inextricable verso to the recto of the right to life. Article 6(6), in particular, declares that "Nothing in this article shall be invoked to delay or to prevent the abolition of capital punishment by any State Party to the present Covenant." The further fruit of this human rights procedure, concerning the reprieve placed on human existence, is that the imposition of the death penalty in times of peace has been forsworn by all Council of Europe members that have ratified the Sixth Protocol of the European Convention for the Protection of Human Rights and Fundamental Freedoms concerning the Abolition of the Death Penalty (Article 1)—which latter went into force on September 3, 1983. The right to life in general is affirmed in Article 3 of the Universal Declaration, whereas Article 6(2) of the Civil Covenant is somewhat more specific and prohibits the "arbitrary" deprivation of human life. Apart from those transpiring in defiance of due legal process, the foregoing likewise covers executions inflicted under the aegis of legal systems whose practices and structures are below international human rights standards. Factors such as these have, moreover, inclined domestic and international courts to increasingly hinder the imposition of the death penalty by rendering it more difficult to accomplish—inter alia by diminishing the number of classes of offenders who may be executed. Presently, there exist within international law a number of safeguards that limit the applicability of the death penalty in countries which still retain this punishment. This comportment of international law stems inter alia from certain prohibitions of the death penalty as contained in Article 3(d) of the Geneva Conventions. The article specifically prohibits executions carried out in the absence of a preceding judgment as pronounced by a regularly constituted court, vouchsafing the judicial guarantees deemed indispensable by all civilized persons. The foregoing limitations on the death penalty are strengthened by the so-termed ENOSOC safeguards. These consist inter alia of the irrevocable judicial guarantees initially expounded in Articles 6 and 14 of the Covenant and collected in the Safeguards Guaranteeing Protection of the Rights of Those Facing the Death Penalty, as adopted by the Economic and Social Council—endorsed by the United Nations General Assembly in 1984.

We now move on to the much more particular ways in which the life of individuals, threatened with death by the penal system of assorted States, may well be placed under reprieve by the anti-State infractions of international human rights law. Article 6(2) of the Covenant had already decreed that death may only be imposed for the most heinous crimes. The Human Rights Committee interprets this rule to imply that the death penalty should only be administered as a most exceptional measure. This applies even in the case of murderers. A milestone within the unfolding of this post-Holocaust truth-procedure aimed at the saving of lives was *Woodson v. North Carolina*—decided on July 2, 1976—at the conclusion of which a mandatory death penalty was struck down by Justice Stewart for certain categories of murder. His contention was that legal processes cannot treat all persons convicted of a specific offense as mere anonymous exemplars of a gray human mass, blindly to be submitted to the ultimate penalty of death. In lieu of this, each legal process must be acutely mindful of and factor in the charged individual's relevant character traits, record, and the specific circumstances of his offense. When considering the ultimate penalty, it must exercise compassion in the face of the variegated frailties of human nature (*Woodson v. North Carolina* 428 U.S. 280 (1976)). The immortality of the truths herein wrought was confirmed when the passages bearing these sentiments and convictions were approved by the Privy Council in 2002, when it struck down mandatory death penalties in the Commonwealth Caribbean on humanitarian grounds arising from mitigating factors (*Fox v. R (Saint Christopher and Nevis)* UKPC 13 (2002)). The universality of this truth may be adduced from the fact that international human rights law nowise considers all persons convicted of murder as deserving of death. It likewise forbids the imposition of the death penalty for the crimes of treason or sedition (American Convention on Human Rights, Article 4(4)). In this specific vein, politically motivated executions have time and again been condemned by the United Nations General Assembly. This bolsters the rather broadly held view that imposing the death penalty for infractions against the State short of murder flies in the face of international law. This is especially the case when such offenses are born out of ideology rather that arising from more ignoble motivations, such as greed or revenge.

International law makes several stipulations concerning requirements linked to the legal imposition of the death penalty. Inter alia, it insists that it may not be carried out unless preceded by fair and scrupulous trial procedures, during the course of every stage of which the defendant has been accorded sufficient legal resources, advice and assistance to realistically assert his innocence against the potentially death-wielding machinery of the State (Rome Statute, Article 67(1)(b)). Specifically as relating to *Robinson v. Jamaica*, the provision of such legal assistance for capital cases was deemed by the Human Rights Council to be "axiomatic." (*Robinson v. Jamaica*, Communication No. 223/1987, U.N. Doc CCPR/C/35/D/223/1987). As regards *Mbenge v. Zaire*, it would go as far as ruling any violation of due process rights in a capital case—as expounded in Article 14 of the Covenant—a breach of Article 6(2). It is on this basis that the State was obliged to withdraw the sentence of death in absentia passed on Mbenge, whose existence would duly be placed under reprieve (*Mbenge v. Zaire*, Communication No. 16/1977, U.N. Doc. CCPR/C/OP/2 at 76 (1990)). This reprieve of existence, as engineered by the efflorescing post-Holocaust truth-procedure, is likewise exemplified in the prohibition placed on the retrospective application of the death penalty. That is to say, the death penalty may not be imposed as punishment for a crime which did not carry the said penalty at the time of its commission (ENOSOC Safeguards, Article 2). A still more comprehensive protection of the right to continued existence in this ongoing truth-procedure has been brought to bear by the American Convention on Human Rights, which goes so far as to contend that States which have abolished the death penalty may not reintroduce it (Article 4(3)). There is, moreover, a nascent human rights rule which prohibits executions during any period in which the abolition of the death penalty is a real political prospect.

Provisions geared toward staving off the death-wielding machinations of substandard judicial processes abound. Among these, one finds ENOSOC safeguards which require, not only that evidence for convictions in capital cases be so convincing that it effectively shuts out any alternative explanation of the facts (Safeguard 4), but also that each such case be meticulously scrutinized by domestic courts for the possibility of error and that the defense retains the right to appeal to a higher court (Safeguard 6). Even beyond this, there is still room for saving the accused from death. For, at the appeals level, but a single dissenting judgment to the upholding of the conviction is viewed as sufficient cause for commuting the death sentence. The rationale behind this is that even a modicum of such dissent casts serious doubt upon the procedure used to establish guilt in the first place—and, moreover, this purported guilt itself.

This is, however, not all. For the International Covenant on Civil and Political Rights still makes provision for staying the Damocles' sword even in the event when the condemned prisoner has exhausted his appeals. Specifically, Article 6(4) accords him the right to seek commutation of sentence or even the possibility of pardon. Herein entailed is the provision granted for recourse to a forum for deciding whether the condemned truly does deserve to be executed. This forum typically assumes the form of a "mercy" committee, which advises the head of State or at least a senior government official minister. The final decision on the continued existence of the accused lies with the latter executive organ. In that this forum is strictly extra-legal, this life-or-death issue is now resolved not as a matter of law but as one of common humanity. Once again, the universalizing tendency inherent to a Badiousian truth-procedure is here palpable. For the recourse here made is one to something more universal than any law, however inclusive: it is an appeal to generic humanity. Considerations in this regard include: residual uncertainty about guilt, evidence that the prisoner may be mentally insane, and other mitigating factors rendering imposition of death excessive in relation of moral culpability.

In addition to this, the imposition of the death penalty is forbidden during the time when appeals and/or clemency petitions are pending. This rule is supported by General Assembly resolutions and the Article 4(6) of the American Convention, and is further elaborated in Article 8 of the Economic and Social Covenant safeguards. Moreover, a stay of execution must likewise be granted pending the determination of a point of international law. This position has been maintained unanimously by the International Court of Justice, resulting in the survival of many a condemned subject. In the wake of the ICJ's judgment in the 2001 *LaGrand Case*, aliens may now be granted a stay of execution if there exists a treaty between their home country and the country in which they have been charged. In such a scenario, the accused has the right to consular access on arrest and may obtain a provisional order from the ICJ, which must be obeyed by a domestic court prior to its handing down of judgment. This is, of course, if such an offender's home State swiftly dispatches lawyers to The Hague (*Germany v. United States of America*, General List 104).

Furthermore, imposition of the death penalty upon certain specific classes of individuals is simply forbidden by international law. Among these are pregnant women. This proscription may be found in all conventions and likewise in the domestic law of all State that still retain the death penalty. The exception is Iran. It is likewise forbidden to execute women with dependent infants—at least according to the Geneva Protocols. Some governments, in fact, go as far as to prohibit the execution of women altogether, regardless of the vileness of their crimes. It is likewise forbidden under international law to impose the death penalty on children who were under eighteen years of age at the time of the commission of their crimes (Geneva Protocols, Article 6(4)). That the retention of the juvenile death penalty, let alone its actual imposition, was an effective violation of an emerging norm of international law was confirmed by a 1986 ruling by the Inter-American Commission—which now likewise prohibits the execution of persons over the age of seventy. This aspect of the post-Holocaust truth-procedure gained further traction when, in 1989, this norm was brought to actuality in Article 37 of the Convention on the Rights of the Child—ratified by all countries with the exception of the United States and Somalia. The cruel and unusual nature of this punishment was recognized in 2005 by a majority in the US Supreme Court. This principle has by now acquired such universality that presently only Iran and the Congo still execute teenagers. Another category of offender for whom the death penalty generally does not apply is the mentally impaired individual. Most domestic legal systems presently balk at the perpetration of such an enormity. Not only has the US Supreme Court ruled the execution such persons unconstitutional; it moreover compels deployment of procedures determining the sanity of a prisoner prior to possible imposition of death.

Still another category of offender increasingly reprieved from execution is the prisoner who has been on death row for an excessive length of time. Deeming such incarceration to have reached the threshold at which the condemned has undergone cruel, degrading and inhuman suffering, the Privy Council (*Pratt and Morgan v. Attorney General for Jamaica*, Appeal 10 (1993)) and European Court of Human Rights alike have declared the execution of such a person unlawful. Such treatment of those countenanced with the penalty of death is, in fact, outlawed by Article 5 of the Universal Declaration and by all human rights conventions—for it exceeds the threshold of severity tolerated by human rights law. Here I undergo seemingly interminable years brooding and agonizing over my own impending death, the terminal annihilation of my possibility to be. I pine in dread over losing my vital substance. Hemorrhaging my mind are unremitting thoughts of being thrown headlong into nothingness. Unendingly, I contemplate with horror the imminent prospect of ceasing to exist. To make matters still worse, I vacillate ceaselessly between the hope of being allowed to live, on the one hand, and despair in the face of the futility of such a prospect, on the other. The relentless oscillation between these two diametrically opposed poles leaves me eternally on the precipice of outright madness. The environment in which I dwell is a sterile dungeon. Immured with the bowels, within the very intestines, of the State apparatus, I see other men who can at any moment be taken out to meet their death. I am flung mercilessly between alternating nerve-storms of annihilating panic and the suffocating and brackish mire of depression. This existence of the long-term death row prisoner therefore resembles that of the inhabitant of the *univers concentrationnaire* for whom death as terminal annihilation is ubiquitous and omnipresent. He is the subject cowering in terror upon the despotic Damocles' sword. This scenario is what the European Court described as a "death row phenomenon." The Privy Council has contended that the prospect of executing a man who has undergone years of such infernal torment must necessarily elicit an instinctive revulsion as regards the continued incarceration of a man in this seemingly interminable way as deeply inhuman. As seen, the appeal made here is to the vulnerability endemic to generic humanity. It thus partakes of the universalizing tendency that inheres within any Badiousian truth-procedure and to our post-Holocaust exemplar in particular. These truths would indeed be put into operation. In 1989, a European Court ruled that the United Kingdom was not permitted to extradite Jens Soering to the United States for a murder trial in Virginia, as the waiting time between conviction for a capital offense and execution in this State averaged between five and seven years (*Soering v. United Kingdom* 161 Eur. Ct. H.R. (ser. A) (1989)). In *Pratt and Morgan v. Jamaica*, the senior court of the Commonwealth ruled in 1993 that, within its jurisdiction, prisoners who had been incarcerated for more than five years could not be submitted to the death penalty (*Pratt and Morgan v. Attorney General for Jamaica*, Appeal 10 (1993)). This ruling led to the immediate commutation of the death sentences on one-hundred-and-sixty prisoners in Trinidad and Jamaica, and has since led to a plethora of other commutations in retentionist countries of the Commonwealth.

The great fall of the Damocles' sword onto the subject has at short notice also been deferred or annulled in the face of the likely prospect of a particularly gruesome mode of execution. In this regard, the United States listed the grotesque spectacle of dissidents hung up in a crowded football stadium as one of the almost visceral motivations for bringing down the Taliban government in Afghanistan. The proliferation of images and video footage of hangings, gassings and lethal injections has of itself fomented much opposition to the death penalty. The film in Spandau prison of participants from the foiled plot, led inter alia by Claus von Stauffenberg to assassinate Hitler and depose the Nazis, strung up and slowly suffocating to death from nooses made of piano wire convinced many a European statesman to do away with the death penalty altogether. In a similar vein, foresight that malfunctions in the death-dealing apparatus could well inflict excruciating pain has led to stays of execution in American courts. More specifically, in 2005, *Lancet* published evidence that death by lethal injection could very likely be accompanied by astronomical pain. In fact, a fair number of executions have been stayed on the basis of averments by anesthetists to the effect that facilitating deadly anesthesia would constitute a breach of the Hippocratic Oath. Gruesomeness of execution has been written a little larger to likewise include within itself gruesome death row conditions, such as seeing the execution of fellow inmates and subjection to a maddening tirade of death-defying last-minute reprieves. Meanwhile, the truth produced through the decision in *Pratt and Morgan v. Jamaica* was in short order exported to many a world outside its point of origination, to the extent that it has not merely saved the lives of over a thousand individuals in the Commonwealth Caribbean, but likewise been adopted by a number of countries in Africa. Recourse is now available to prisoners facing execution amid circumstances and conditions refractory to the salubrious dictates of international law. Failing to find relief in his national court, such an inmate can bring his qualms to the United Nations Human Rights Committee (UNHRC), on condition of course that his home country has signed the Optional Protocol, or to a regional human rights court, if there is one. The June 27, 2001, decision of the ICJ in the *LaGrand Case* now forms an authoritative basis for postponing a death row prisoner's sojourn with terminal annihilation during the period when his plea for leniency is investigated and, if held admissible, at the time of its adjudication. Even failing this, the prisoner may still be saved from the Damocles' sword of death by calling upon the United Nations Special Rapporteur on "summary and arbitrary" executions. If the prosecuting State is deemed to have flaunted the Covenant safeguards linked to the death penalty, then the Rapporteur is free to inundate the defaulting government with a urgent formal communiqué. Failing this, UN Secretary-General reserves the right to intervene on a "best endeavors" mandate. As an absolutely final resort, the Pope may intercede. In 1958, Chief Justice Earl Warren, maligning capital punishment as refractory to the constantly evolving moral standards of a maturing society, deployed aspects of the US constitution to erode its prevalence. Though Justice Antonin Scalia would counter this some thirty years later, Americans are growing increasingly weary of tinkering with the machinery of death. Fewer and fewer death sentences are being passed and fewer felons are being executed. Worldwide, there has been a gradual increase in the number of abolitionist nations or nations in which courts have made executions virtually impossible to carry out.[413]

Notes

1. Johnson 1988, 3.
2. Dimont 2004, 3–5.
3. Ibid. 5–7.
4. Ibid. 7, 8.
5. Ibid. 9, 10.
6. Ibid. 20, 21.
7. Ibid. 34.
8. Cohen 2007, 568.
9. Berkovits 2007, 466, 467.

10. Ibid. 471, 472.
11. Ibid. 473–475.
12. Greenberg 2007, 514.
13. Berkovits 2007, 488.
14. Dershowitz 2000, 50, 51.
15. Wiesel 2005, 31, 32.
16. Ibid. 55.
17. Saramago 2012, 51–54.
18. Ibid. 69.
19. Ibid. 86.
20. Ibid. 88–91.
21. Ibid. 94.
22. Ibid. 100.
23. ibid. 126–128.
24. Ibid. 141, 142.
25. Bialik and Ravnitzky 1992, II.113.
26. Alter 2004, 44.
27. Bialik and Ravnitzky 1992, II.124.
28. Ibid. III.45.
29. Ibid.
30. Dershowitz 2000, 106.
31. Bialik and Ravnitzky 1992, III.45.
32. Ibid.
33. Wiesel 2005, 90.
34. Ibid. 76.
35. Dershowitz 2000, 116.
36. Ibid. 125.
37. Ibid. 126.
38. Cited in Wiesel 2005, 96.
39. Cited in Ibid. 101.
40. Ibid. 165.
41. Mann 2005, 473.
42. Ibid. 473, 474.
43. Ibid. 484.
44. Ibid. 544.
45. Alter 2004, 310.
46. Bialik and Ravnitzky 1992, IV.9.
47. Ibid. IV.10.
48. Alter 2004, 311.
49. Ibid. 312–314.
50. Bialik and Ravnitzky 1992, IV.9, 10, 24.

51. Ibid. IV.31.
52. Ibid. IV.29.
53. Alter 2004, 320, 321.
54. Ibid. 327.
55. Ibid. 331, 332.
56. Ibid. 370.
57. Ibid. 378, 379.
58. Ibid. 384.
59. Ibid. 394.
60. Bialik and Ravnitzy 1992, IV.77.
61. Ibid. IV.80.
62. Ibid. IV.84.
63. Wiesel 2005, xi, xii.
64. Graetz 2012, Vol. 1, 182, 183.
65. Ibid. 183.
66. Ibid. 185–188.
67. Ibid. 188.
68. Ibid. 189, 190.
69. Ibid. 196.
70. Ibid. 200, 201.
71. Ibid. 201.
72. Ibid. 202, 203.
73. Ibid. 203.
74. Ibid.
75. Ibid. 204, 205.
76. Ibid. 206, 207.
77. Ibid. 214.
78. Ibid. 209.
79. Bialik and Ravnitzky 1992, VII.11, 12.
80. Ibid. VII.8.
81. Ibid. VII.12.
82. Graetz 2012, Vol. 1, 219, 220.
83. Dimont 2004, 52, 53.
84. Ibid. 53.
85. Ibid. 53, 54.
86. Ibid. 54, 55.
87. Ibid. 55, 56.
88. Ibid. 56, 57.
89. Ibid. 58.
90. Bialik and Ravnitzky 1992, VII.13.
91. Johnson 1988, 77.

92. Ibid. 81, 82.
93. Ibid. 81.
94. Wiesel 2005, 72.
95. Johnson 1988, 82, 83.
96. Ibid.
97. Ibid. 84.
98. Graetz 2012, Vol. 1, 239–241.
99. Ibid.
100. Ibid. 245, 246.
101. Dimont 2004, 60, 61.
102. Bialik and Ravnitzky 1992, X.1.
103. Schama 2014, 150.
104. Graetz 2012, 222–229.
105. Ibid. 229, 230.
106. Ibid. 230, 231.
107. Ibid. 231.
108. Ibid. 232–239.
109. Ibid. 245.
110. Ibid. 245, 246.
111. Schama 2014, 152, 153.
112. Ibid. 149, 150.
113. Ibid. 153, 154.
114. Schama 2014, 327, 328.
115. Johnson 1988, 224.
116. Graetz 2012, Vol. 4, 324, 325.
117. Cited in Ibid.
118. Graetz 2012, Vol. 4, 308, 309.
119. Ibid. 310–318.
120. Ibid. 319.
121. Ibid. 321.
122. Schama 2014, 405.
123. Graetz 2012, Vol. 4, 323.
124. Ibid. 323–328.
125. Johnson 1988, 225.
126. Cited in Graetz 2012, Vol. 4, 332.
127. Graetz 2012, Vol. 4, 332.
128. Ibid. 334.
129. Ibid. 349, 350.
130. Ibid. 483, 484.
131. Ibid. 318.
132. Ibid. 320.

133. Ibid. 357, 358.
134. Schama 2014, 407, 408.
135. Graetz 2012, Vol. 4, 386, 387.
136. Ibid. 358–360.
137. Ibid. 363, 364.
138. Ibid. 392.
139. Ibid. 396, 397.
140. Ibid. Vol. 4, 399, 400.
141. Ibid. 407, 408.
142. Ibid. 418–420.
143. Ibid. 364.
144. Ibid. 400.
145. Graetz 2012, Vol. 5, 7–15.
146. Ibid. 7, 8.
147. Ibid. 15, 16.
148. Ibid. 18, 19.
149. Johnson 1988, 276.
150. Ibid. 277.
151. Ibid. 278.
152. Ibid.
153. Ibid.
154. Ibid.
155. Wasserman 2007, 36.
156. Maybaum 2007, 406, 407.
157. Cohen 2007, 566.
158. Ibid. 571.
159. Cited in Kershaw 2009, 149.
160. Cited in Ibid. 150.
161. Cited in Ibid.
162. Gilbert 1986, 28–30.
163. Ibid. 32–34.
164. Ibid. 36.
165. Ibid. 37, 38.
166. Ibid. 40.
167. Ibid. 60–65.
168. Cesarani 2016, 184–191.
169. Ibid. 199–201.
170. Ibid. 221, 222.
171. Ibid. 244, 245.
172. Ibid. 346, 347.
173. Ibid. 348.

174. Ibid. 371.
175. Ibid. 402.
176. Ibid. 403.
177. Ibid. 404.
178. Ibid. 377.
179. Ibid. 390.
180. Ibid. 393.
181. Ibid. 410, 411.
182. Ibid. 412.
183. Cited in Cesarani 2016, 448.
184. Cited in Ibid.
185. Cited in Ibid.
186. Cited in Kershaw 2009, 696.
187. Cited in Ibid. 697.
188. Cited in Ibid. 697, 698.
189. Cited in Gilbert 1986, 824.
190. Cesarani 2016, 529.
191. Gilbert 1986, 582.
192. Lengyel 1947, 175, 176.
193. Cited in Kershaw 2009, 776.
194. Cited in Cesarani 2016, 665.
195. Cited in Kershaw 2009, 776.
196. Cesarani 2016, 461.
197. Ibid. 483.
198. Ibid. 488.
199. Gilbert 1986, 672.
200. Cesarani 2016, 505.
201. Lengyel 1947, 9–12.
202. Gilbert 1986, 596, 598.
203. Lengyel 1947, 157.
204. Ibid. 163.
205. Nomberg-Przytyk 1985, 22, 23.
206. Wiesel 2006, 86.
207. Ibid. 89.
208. Cesarani 2016, 506.
209. Wiesel 2006, 36–38.
210. Jabès 1991, Vol. 2, 270, 271.
211. Levi 2003, 33.
212. Améry 1980, 16.
213. Levi 2003, 48.
214. Améry 1980, 15.

215. Ibid.
216. Ibid. 33, 34.
217. Ibid. 35.
218. Gilbert 1986, 758, 759.
219. Lengyel 1947, 44, 46, 63, 83, 122.
220. Gilbert 1986, 415, 416.
221. Grossman 1995, 549, 550.
222. Ibid. 551.
223. Gilbert 1986, 416.
224. Venezia 2009, 65, 68.
225. Cesarani 2016, 641.
226. Ibid.
227. Höss 1996, 33, 45.
228. Nomberg-Przytyk 1985, 81.
229. Celan 2005, 46.
230. Sachs 2011, 21.
231. Gilbert 1986, 765.
232. Ibid. 627.
233. Ibid. 772, 773, 775.
234. Cesarani 2016, 752.
235. Ibid. 757–759.
236. Frankl 2008, 42.
237. Ibid. 60, 63.
238. Ibid. 78–82.
239. Gilbert 1986, 419–421.
240. Cited in Ibid. 421, 422.
241. Ibid. 422, 423.
242. Cited in Ibid. 423.
243. Gilbert 1986, 423, 424.
244. Ibid. 568.
245. Gilbert 1986, 667, 668.
246. Ibid. 668.
247. Cited in Ibid. 668.
248. Nomberg-Przytyk 1985, 164, 165.
249. Citied in Ibid. 165.
250. Wasserman 2007, 36.
251. Ibid. 37, 38.
252. Katz, Reuven 2007, 107, 108.
253. Ibid. 108.
254. Harlap 2007, 150.
255. Schneersohn 2007a, 175.

256. Schneersohn 2007b, 181.
257. Jabès 1991, Vol. 2, 241.
258. Sachs 2011, 55.
259. Jabès 1991, Vol.2, 298, 299.
260. Schneersohn 2007c, 186.
261. Soloveitchik 2007, 390.
262. Berkovits 2007, 484.
263. Ibid. 487, 488.
264. Greenberg 2007, 515.
265. Achituv 2007, 323, 324.
266. Wyschogrod 2007, 459.
267. Schweid 2007, 236.
268. Elberg 2007, 196.
269. Fackenheim 2007, 435, 436.
270. Alexander 2006, 31.
271. Ibid. 32.
272. Ibid. 33.
273. Cited in Cesarani 2016, 503.
274. Ibid. 214.
275. Alexander 2006, 37.
276. Cesarani 2016, 575, 578.
277. Ibid. 767–769.
278. Alexander 2006, 42–45.
279. Ibid. 47.
280. Ibid. 48.
281. Ibid.
282. Cesarani 2016, xxv.
283. Žižek 2013, 23.
284. Cited in Alexander 2006, 50.
285. Cited in Cesarani 2016, 503.
286. Hartman 1996, 4.
287. Cited in Ibid.
288. Cited in Ibid. 79.
289. Cited in Ibid. 84.
290. Hartman 1996, 89.
291. Žižek 2013, 28, 29.
292. Alexander 2006, 40.
293. Cited in Ibid. 40, 41.
294. Cesarani 2016, 784.
295. Alexander 2006, 52.
296. Ibid.

297. Aristotle 2013, 75.
298. Žižek 2013, 23–28.
299. Ibid. 36, 41.
300. Alexander 2006, 60, 61.
301. Bauman 1989, xiv.
302. Ibid. 11.
303. Ibid. 14.
304. Ibid. 14, 15.
305. Ibid. 17.
306. Ibid.
307. Ibid. 18.
308. Ibid.
309. Ibid.
310. Alexander 2006, 61.
311. Arendt 2006, 25, 26.
312. Ibid. 247.
313. Ibid. 276.
314. Bauman 1989, 19.
315. Ibid. 151.
316. Ibid. 20.
317. Ibid. 94.
318. Martinez 2016, 214–216.
319. Alexander 2006, 62–64.
320. Ibid. 66, 67.
321. Ibid. 67, 68.
322. Ibid. 69.
323. Ibid. 70–72.
324. Ibid. 70.
325. Ibid. 74.
326. Badiou 2005, 24, 25.
327. Ibid. 28.
328. Ibid. 52, 53.
329. Ibid. 55, 56.
330. Ibid. 59.
331. Ibid. 86.
332. Ibid. 93–98.
333. Ibid. 99.
334. Ibid. 105–110.
335. Ibid. 174.
336. Bauman 1989, 39.
337. Ibid. 40.

338. Ibid. 40, 41.
339. Ibid. 50.
340. Arendt quoted in Bauman 1989, 52.
341. Ibid. 52.
342. Cited in Ibid. 68.
343. Cited in Ibid. 71.
344. Bauman 1989, 53.
345. Badiou 2005, 174–177.
346. Badiou 2003a, 83.
347. Ibid. 85.
348. Beckett 2006, 477.
349. Ibid.
350. Ibid. 472.
351. Ibid. 480.
352. Badiou 2003a, 90.
353. Beckett 2006, 471.
354. Badiou 2003a, 93.
355. Beckett 2006, 479.
356. Lévinas 1981, 5, 6.
357. Beckett 2006, 478.
358. Badiou 2003a, 94–96.
359. Beckett 2006, 478.
360. Ibid. 480.
361. Ibid. 483.
362. Agamben 1998, 81–83.
363. Levi 2003, 48.
364. Améry 1980, 33.
365. Agamben 1999, 47.
366. Agamben 1998, 104, 105.
367. Deleuze 2003, 20–22.
368. Lévinas 1981, 88.
369. Derrida 1996b, 56, 57.
370. Ibid. 57.
371. Derrida 1996a, 6.
372. Ibid. 8.
373. Žižek 2006, 112, 113.
374. Lévy 2004, 207.
375. Wiesel 2006, 36–38.
376. Jabès 1991, Vol. 2, 270, 271.
377. Santner 2001, 9.
378. Badiou 2003a, 106.

379. Beckett 2006, 481.
380. Badiou 2003a, 98, 99.
381. Ibid. 102.
382. Beckett 2006, 477.
383. Ibid. 483.
384. Ibid. 484.
385. Ibid. 485.
386. Badiou 1999, 105.
387. Badiou 2005, 178–181.
388. Ibid. 201–208.
389. Badiou 2002, 43.
390. Ibid. 51.
391. Ibid. 70.
392. Ibid. 40.
393. Ibid. 43.
394. Ibid. 32.
395. Ibid. 73.
396. Badiou 1999, 80.
397. Ibid. 91.
398. Ibid. 106, 107.
399. Badiou 2011, 20, 21.
400. Badiou 2002, 11, 12.
401. Ibid. 12.
402. Ibid.
403. Robertson 2006, 34–37.
404. Ibid. 471–476.
405. Ibid. 478, 479.
406. Ibid. 480, 481.
407. Ibid. 491.
408. Ibid. 482.
409. Ibid. 482–486.
410. Ibid. 494–498.
411. Ibid. 498–500.
412. Ibid.
413. Ibid. 154–156.

16

From the Violence of Divine Love to the Redemption of the World

Rosenzweig, Lévinas, Derrida, Blanchot, Kierkegaard

1. The quandary of Character

We have seen in Rosenzweig how Character, emerging as it had from the traumatic event of Revelation, represented a considerable advance in comparison with the utterly facile third-person existence (that of the personality) from under whose mendacious—though now rent—veils it had extruded itself. The resultant tautly involuted meta-ethical Self was, however, also observed to be highly inadequate in its own right. For in spite of acceding to and assuming the kernel of its authenticity, its unique ipseity, the Self at the very same time found itself coldly abandoned to the arid wasteland of its own insurmountable solitude. Moreover, being itself a bizarre ramshackle of unconditional free will and its own vastly more limited mortal finitude, the meta-ethical Self showed itself to be a roiling cauldron of anxiety inwardly riven by the absurd marriage of incompossibles. Reduced by the trauma of divine solicitation to the authentic though entirely barren kernel of its inner-most quiddity, the Self dwells within a bewildering and bewildered darkness entirely its own. It knows not the world nor even itself.

In this sense, the Self—in the most raw nakedness of Character—bears a somewhat shocking resemblance to what Schelling described as the second divine will that blindly and vainly searches for eternity (the pure willing that is the Godhead). In this regard, is not the well-nigh psychotic roiling cauldron of the conflicted and angst-ridden meta-ethical Self as bare Character akin to the blind and unremitting rotary motion of Schellingian eternal divine nature? And is not the immurement of the Self within the impervious shell of Character (in the very movement of its accession to authenticity) not strikingly reminescent of the manner in which God's contraction of Being is at once his ensnarement in the incarcerating morass of his own self-begotten eternal divine nature?

It is here that the subject has entered into a mute realm entirely separate and foreign to the arena of quotidianity: a crepuscular interstitial zone in which it is entirely isolated from other wills, the landscapes of personality, the world, and God. Through the intrusion of its *daimon*, the meta-ethical Self directs all its defiant violence inward upon its own Character, beclouding the environs and itself with an all-pervading Schellingian and Hegelian "night of the world." Shrouded in this impermeable and opaque dusk, it knows not itself nor the world.

It has entered the absolute "zero-level," or pure void, of subjectivity: the absolutely unendurable abyss in which fully constituted objects and subjects dissolve and yield before a miasmal swarming of spectral partial objects, of exemplifications of the Lacanian *lamella*, of the undead, immortal, libido-as-object that is self-identically the Freudo-Lacanian death-drive (*Todestrieb*). This paronomastic modality of the subject is merely one "degree" removed from the pre-subjective noumenal Real. It is a return to, or recrudescence of, the impossible moment of upsurge of subjectivity in the form of a withdrawal, a recession from the light of phenomenality and its differential degrees of appearing in a world in accordance with its transcendental regime, which assumes the form of a contraction through which constituted reality yields before the surreal phantasmagoria of *membra disjecta*, of organs devoid of body functioning as stand-ins for the dread immortality of the object-libido in its semipernally stuck fixity and eternal psychotic circulation around the void. In Kantian terms, this crepuscular and vanishing mode of subjectivity is that of the pre-synthetic imagination gone mad, psychotic, and utterly demented—the impossible domain of absolutely purified and kenotically evacuated transcendental freedom and spontaneity. In this domain, the subject finds itself vertiginously suspended between direct unadorned animality, on the one hand, and the subordination of its freedom to the moral Law, on the other. In this gathering murk, far from being able to galvanize itself in the direction of forward-directed action in the world or attunement to the promptings of the Deity, the Self is submitted to the horrifying experience of dread organs without body, of itself as the Lacanian fragmented body, of its *membra disjecta* floating around in a nauseating and disgusting miasmal drift, the "night of the world," the "valley of the shadow of death," the hell-on-earth of Kundry in Wagner's *Parsifal* who, having committed the blasphemy of laughing at Christ dying on the Cross, is condemned to an eternity of wandering as an abject, multiply befouled, purely excremental self entirely rejected by the world, abandoned by God, and subject to involuntary, psychotic automatisms, spasmodic jabberings, slavering aphasia, dyslexia, *and the absolute horror of never being able to die.*

Certainly, the internally riven cauldron of eternal divine nature and that of Character—alike psychotic—are both the crepuscular dusk anterior to an abrupt dawning. In the case of eternal divine nature, this sudden irruption is the colossal evacuation of the latter and its maddening pulsation in its forcible repression into the immemorial past attendant upon the coterminous upsurge of God's accession to subjecthood. What will it be for the meta-ethical Self immured within the unyielding obduracy of its own Character?

2. The violence of divine love

The Self must escape its quandary. But it is able to do so, as Richard Cohen maintains, only at the behest of an "alterity encountered frontally in the intersubjective encounter" and possessed of the "power to open up the hard core, the hard nugget, the very fixity of character."[1] The opening up of the otherwise impermeable, impenetrable, and imperturbable obduracy of Character is effected by what one may term the "excessive significance" inherent to the intersubjective encounter—the encounter with the Other.

Crucially, as avers Lévinas, the relation to the Other is "not a participation in a third term—intermediate person, truth, dogma, work, profession, interest, dwelling, or meal; that is, it is not a communion. It is the fearful face-to-face situation of a relationship without intermediary, without mediations."[2] In this connection, Eric Santner maintains that the otherness of the Other consists not so much in his spatial exteriority with regard to the Self, but rather by virtue of his being *strange*; that he is a stranger not merely to the Self but even in relation to his own self. The Other, that bearer of his own internal alterity, as Santner avers, is at once the Lévinasian Neighbor (*le prochain*)—but is such, oxymoronically, only to the extent that he is a stranger; that is, that his is *the absolutely inassimilable outlandishness of the very nearest.*[3]

The excess of significance precipitated by the encounter with the Other emerges through and as *love*; that is to say, the otherwise implacable Character (the elemental Self) finds itself unfolded by the shock of its sudden experience of being loved.[4] In the process of Character's being thus turned inside-out by its stunning and shocking realization that it is loved, an entirely new self is portended by the intersubjective encounter. This is the "Soul" (*Seele*).[5]

The Self is loved by the Other whom it so vertiginously encounters. Yet, as Rosenzweig contends, what is experienced in love cannot be merely human. This is so because humans have no "reason" whatsoever to love one another, love being—especially from the purview of the beloved—entirely inexplicable. The love for the beloved must therefore be, in its most fundamental sense, *divine*.[6] Here, as Richard Cohen explicates, since it "cannot be made sense of in ontological or in human terms without denying its character of pure exteriority and unearned bestowal," man's being loved is an absolute and continuous miracle and thereby simply another aspect of Revelation.[7]

The crucial distinction to be drawn here is that between love conceived in the more prosaic and quotidian sense, and the infinitely more radical sense of love—that is, love as divine love. This distinction is Rosenzweig's. The former love is relational, love in the third person, love based on the Self's predicative being—the Self's knowable quiddity; whilst divine love is love as an eventful and vertiginous encounter with the Other in what Santner terms the Other's uncanny "death-driven singularity."[8]

Lévinas terms the relation to the Other *proximity*. In proximity, the self and the Other find themselves related in a manner that, however, nowise entails an overarching totality subsuming the former and the latter under a higher unity. Rather, the relation to the Other is a limitness exposure to the Other in which, such that the otherness of the Other be safeguarded, the Other must remain sempiternally inaccessible. As Lévinas maintains, proximity is in no sense fusion: rather, "it is contact with the other. To be in contact is neither to invest the other and annul his alterity, nor to suppress myself in the other. In contact itself the touching and the touched separate, as though the touched moved off, was always already other, did not have anything in common with me."[9]

Now, the fact that love emerges in the frontal encounter with the Other, this love being—as is all love—fundamentally divine, a modality of Revelation; implies that the intersubjective encounter is at once the encounter with God himself. In other words, the Deity is encountered in and through the alterity of the other person.[10]

3. *Le visage*

Here, the relation to the Other, a direct exposure, is immediately—for Lévinas—the "epiphany" of the face (*le visage*) of the Neighbor (*le prochain*). More precisely, the *illeity* "that bypasses the present" moment leaves "a trace by the very fact of bypassing the present. The trace lights up as the face of the neighbour."[11] In the same vein, but indeed elsewhere, Lévinas is more succinct—maintaining that the "dimension of the divine opens forth from the human face."[12] This does not, however, mean that God somehow incarnates himself in the Other, but rather that the Other is God's "face, in which he is *disincarnate* [my emphasis], is the manifestation of the height in which God is revealed."[13] For Rosenzweig, the violence of divine love that fulgurates in the face of the other person nowise entails some kind of obdurate and immovable form of his countenance. It is not of the hardness of a death mask. Rather, it is the infinite modulations of facial expressions, the incessant scintillation of light on the eternal features of the divine. Love fundamentally balks at rendering up a portrait of the beloved. This would merely serve to mortify the living face.[14]

Nowise is God's countenance somehow the Other's face itself, its outer surface, its epidermis and contours; rather, the face is conceived here as the *dimension* in which God is revealed. This "dimension," as observed, is that of "height." As Richard Cohen explains, height as referred to here is that "found nowhere else or nowhere more than in the human face, in the face-to-face"; a dimensionality "which is that of the divine" and at once "that of the human, what is human about the human face"; that is to say, "a *moral* dimension, the dimension of moral height, goodness"; for indeed "the face of the other manifests and is manifest in a moral height which is the dimension of G-d, the revelation of G-d."[15] Bernard-Henri Lévy intimates something quite similar when he conjures the human gaze as "the sign of the soul" and "the area of the body, above all others, where light filters through (from the mind … from beyond the mind …)."[16] "It's the infinite," he continues, "in reach of faces. Their divine portion. Their grandeur."[17]

Although Richard Cohen alludes to manifestation whilst Lévy likewise appears to be doing the same, we must not grasp the latter in the sense that attaches to phenomenality. For the face of the Other is *not* strictly-speaking manifest. It is—in point of fact—*less* than phenomenal. Lévinas: "It escapes; it is the very collapse of phenomenality. Not because it is too brutal to appear, but because in a sense too weak, non-phenomenal because less than phenomenal."[18] This is so because, if the face (*le visage*) of the Other were indeed of the phenomenal order it would thereby be reduced to a mere intentional object of a perceiving consciousness—in the process abdicating its otherness.

In this connection, it is interesting to take note of Žižek's criticism of the Lévinasian characterization of Otherness. Žižek contends that Lévinas tends to elide the truly radical "inhuman" dimension of Otherness, that he has a tendency to occlude "the Otherness of the human being reduced to inhumanity, the Otherness exemplified by the terrifying figure of the *Muselmann*, the 'living dead' in the concentration camp."[19] Taking the *Muselmann* as the ultimate figure of the Other as the Neighbor, Žižek continues his critique of Lévinasian ethics by claiming that when "we are confronted with a *Muselmann,* we precisely cannot discern in his face the trace of the abyss of the Other in his or her vulnerability, addressing us with the infinite call of our responsibility—what we get is a kind of blank wall, a lack of depth."[20]

It is this sort of "de-facement" that Lévy aims to invoke in his descriptions of those "damned" and bedraggled by the ravages of senseless wars in Sri Lanka, Angola, and Burundi. Here one may speak of "faces without gazes; gazes without light; that look of broken birds, faces so shattered they seem to have lost that obscure radiance, that superhuman dignity a gaze usually offers."[21] These faces, though they see other people and thus have eyes to see, eyes that are useful and that allow their faces and bodies to live and to survive; are not real eyes. For real eyes "aren't content with seeing; they look, they love, they dislike, they think, they question, they dream."[22] The eyes of the "damned," on the other hand, "seem to serve just to see" and are thus "eyes without gaze."[23]

So extreme is this harrowing figure of the living damned that Lévy christens him with the epithet of the *Wreck*. The Wreck, it is contended, "is beyond the known forms of misery, especially that of the Refugee. It is inaccessible to any idea of *Erlösung,* that is to say of 'redemption,' of 'deliverance'"—to the extent that "it is incapable of bearing witness and hence, literally, of martyrdom" itself.[24] It is an "anti-witness" or an "anti-martyr" and hereby "differs, whether you like it or not, from even the most afflicted deportees."[25] The Wreck—that is, the Other conceived in his most abject destitution—therefore, as such serves absolutely no direct function within the structure of Being.[26]

This all may well be the case. But is not Žižek's "amended" version of the Lévinasian conception of the relation to the Other not effectively fairly close to what Lévinas maintains from the outset? Indeed, Žižek himself does not appear to be contending that the *Muselmann* is somehow refractory to the category of the Lévinasian Neighbor (*le prochain*), but rather that he is the "zero-level neighbour"; and that it is precisely in the latter, in the "faceless" face of the *Muselmann*, that one encounters the Other's ethical call "at its purest and most radical," that it is the utterly blank muteness of the *Muselmann* that makes us "aware of our responsibility toward the Other as its most traumatic."[27]

In this connection, it is fairly intriguing to compare the foregoing conceptions of the Neighbor's face with the face, or faciality, as it appears in Deleuze and Guattari. The latter firstly distinguish between the face and the head, contenting that the head, even the human head, is by no means necessarily a face. In fact, the face must be *produced* and is produced only when the head (and indeed the entire body) has been deterritorialized and decoded and then overcoded by the face; that is to say, when it has been "facialized." The face, as conceived by Deleuze and Guattari, is neither animal nor *stricto sensu* human. This is so because the deterritorialization that attends it is an *absolute* deterritorialization and therefore removes the head from the organic stratum entirely, whether this stratum be specifically human or animal. It is, in fact, contended that there is something profounding *inhuman* about the face—although not in the sense that it only becomes inhuman beyond a certain threshold, as in tends to be in the case of a close-up, extreme magnification, or recondite expression. Rather, the face is inhuman *ab initio.* It is the inhuman in humans as such.[28] For the face is intrinsically a close-up, "with its inanimate white surfaces, its shining black holes, its emptiness and boredom. Bunker-face."[29] "*The face, what a horror.* It is naturally a lunar landscape, with its pores, planes, matts, bright colours, whiteness, and holes: there is no need for a close-up to make it inhuman; it is naturally a close-up, naturally inhuman, a monstrous hood."[30]

Moreover, the Deleuzoguattarian face is neither ancillary to a pre-existing subject nor may it be assimilated to the Sartrean, or the Lacanian, gaze. The converse is in fact true, for the subject only emerges with the face, whilst "*The gaze is but secondary in relation to the gazeless eyes, to the black hole of faciality.*"[31]

But there is more. For the movement of deterritorialization that separates the head from the organic stratum likewise connects it to the strata of signifiance and subjectification.[32] The production of the face is thus inextricably interwoven with the coexistence of both the signifying (or despotic regime) and the postsignifying (or passional) regime of subjectivity. Of course, as observed, the passional regime inevitably comes to be miscegenated with the despotic regime almost from the outset. The former and the latter, as Deleuze and Guattari maintain, are assemblages of power (*pouvoir*) that require the production of the face.[33] The face is here, more precisely, conceived as a *white wall/black hole* system, and that which at once provides the wall upon which signifiance inscribes its signs and the black hole into which subjectification lodges its consciousness and passion.[34]

The primitive regime, on the other hand, does not require the face, as its semiotic is at once nonsignifying and nonsubjective—and characterized by a polyvocality that operates through bodies. Here, presignifying codes, even masks, rather than making the head into a face, ensure its belonging to the body. At the same time, movements of deterritorialization in the primitive regime connect the body not with faciality, but with becomings-animal. Primitives have no face and need none.[35]

Furthermore, Deleuze and Guattari aver that the face is effectively a *politics*. The social production of the face and its overcoding of the head is the sine qua non that permits and ensures the almightiness of the signifier and at once the autonomy of the subject. The signifying and postsignifying semiotics never operate through mere bodies. Without the face, the signifier and the subject would never be installed.[36] We may thus term the inhumanity of the face a political inhumanity. And since this inhumanity is that constitutive of man as *homo historia*, it may be still better to term it *human inhumanity*. This inhumanity assumes *inter alia* the form of despotic and authoritarian control. For the face defines a zone of frequency or probability which neutralizes and precludes any expressions unamenable to appropriate despotic signifiances, whilst likewise forming loci of resonance that ensure and preordain the conformity of mental with dominant reality at the level of subjectivity.[37] What the face subsumes and places at the service of signifiance and subjectivity are faciality *traits*, clamping down and taking hold of any such trait attempting to escape its sovereign organization by blocking its line of flight.[38] The face is that through which facialized bodies are disciplined and becomings-animal hounded out: "You will be pinned to the white wall and stuffed in the black hole."[39] The face announces the statements of language, providing ballast for these latter in relation to the signifiers in progress and subjects concerned: "The face is a veritable megaphone."[40]

Let us surmise the features of the Deleuzoguattarian face. Initially, the latter appeared to share all the features of that mute blank wall that was the "faceless" face of the *Muselmann*. The face was averred to be an inhuman horror and lunar landscape with inanimate white surfaces, shining black holes, and gazeless eyes filled with emptiness and boredom. Ultimately, however, it exhibited itself as likewise a cog in a greater regime—despotic and authoritarian—of political repression, and by no means enigmatic, let alone mute. The face was indeed averred to be loud and brash. It therefore appears that the political inhumanity of the face, as conceived by Deleuze and Guattari, is not of the same order as the inhumanity of the *Muselman's* gazeless gaze. It is indeed human, all too human. For, as observed, the face is the inhumanity which ultimately and inevitably *supports* the edifying, luminous and most human gaze of man. In order to move behind the luminous human gaze, we must therefore likewise move behind the inhumanity of the Deleuzoguattarian face.

What then of the primitive head? Though initially seeming to concede that the primitive head was *stricto sensu* human in relation to the inhuman face, Deleuze and Guattari, in a strange volte-face, move to aver that the head of the primitive regime is likewise inhuman—since, though it is still a part of the body, this body is nonetheless already deterriorializated and subject to becomings-spiritual/animal.[41] One may therefore term primitive, or pre-facial, inhumanity *animal inhumanity*.

Now, a movement from the inhumanity of the face, and the all too human gaze that it supports, to the animal inhumanity of the head may, however, be effected. And it is, as Deleuze avers, brought to bear by Francis Bacon. Bacon, as a portraitist, is a painter not of faces, but of heads. More precisely, he undertakes the *dismantling* of the face so as to effect the emergence of the head from beneath it. The face, as the structured spatial organization or monstrous hood that conceals the head, now yields before the head as "a spirit in bodily form, a corporeal and vital breath, an animal spirit."[42] For the head is "the animal spirit of man: a pig-spirit, a buffalo-spirit, a dog-spirit, a bat-spirit …"[43] Under the pressure of Bacon's techniques of rubbing and brushing, the face is disorganized as the animal traits of the head recrudesce, deforming the body. The traits of animality are the spirits that haunt the wiped-off features, pulling at the head.[44]

Crucially, through the latter techniques, Bacon brings into effect a *zone of indiscernibility or undecidability* between man and animal in which man is swept up in a becoming-animal. This zone of indiscernibility between man and animal, avers Deleuze, is *meat*. Meat is, however, in no sense dead flesh; for it assumes all the hues of living flesh—and moreover retains all flesh's *sufferings*. It is *pain* and *vulnerability*, and as such is Bacon's principle and indeed only object of pity. This zone of indistinction between man and animal is one in which the man who suffers is a beast and the beast who suffers is a man.[45]

Now, since man as *homo historia* is the product of the face and its political inhumanity, whereas animality risides in the head recrudescing from under the face; would not Bacon's constitution of a zone of indistinction between man and beast at once be the intimation of the beastialized humanity occluded by the machinations of political inhumanity? A bestialized humanity whose gaze, formerly supported by the face now being dismantled, has etiolated and approaches the condition of being but a head subject to becomings-animal. With the desuetude of its facial megaphone and solicizing gaze alike, the sole intimation of this nonhuman humanity's suffering is its flesh, its meat.

And is this silently suffering meat not that of the *Muselmann*? For, as Agamben avers, the *Muselmann* is precisely "the moving threshold in which man passes into non-man," the paradoxical figure who "makes it forever impossible to distinguish between man and non-man," he who inhabits "the extreme threshold between the human and the inhuman."[46]

4. *Illeity*

So much for the Other as the Neighbor (*le prochain*). Now we move onto God himself. Lévinas' conception of God—which he consistently terms *illeity*—is many ways close to Kierkegaard's God. According to Žižek, in both cases God is an Absolute Other, thoroughly desubstantialized, beyond the order of Being, and as such "*nothing but* the mode of how we relate to him," or—still more precisely—this very relating itself.[47] God, desubstantialized and heterogeneous to the ontological order and thereby likewise to phenomenality, must therefore remain hidden.

Derrida is therefore right to posit a "dissymetry that exists between the divine regard that sees me, and myself, who doesn't see what is looking at me"; and—in the same vein—to maintain that "God is himself absent, hidden and silent, separate, secret, at the moment he has to be obeyed"; that—in fact—"if he were to speak to us all the time without any secrets, he wouldn't be the other, we would share a type of homogeneity."[48] In being merely the mode of our relating to him, God must remain invisible to me, even when right in front of me, whilst nonetheless still holding me under his gaze. In this sense, he would be like the narrator of Henri Barbusse's novel *Hell* who, from a rented hotel room, through a small hole in the ceiling is able to see the goings-on of the denizens of the appartment above without himself being seen.[49]

Such a Deity would thus likewise accord with a logic of "spectrality," as alluded to in Derrida's *Spectres of Marx*. That is to say, God would be this "Thing [that] meanwhile looks at us and sees us not see it even when it is there. A spectral asymmetry interrupts here all specularity."[50] God, on this account, is a kind of "ghost [that] would have shown neither his face, nor his look, nor therefore his identity."[51] In terms of Žižek's aforementioned reading of Lévinas, this "defacialized"/"defaced" God would accord precisely with the Lévinasian characterization of the divine countenance's apparition as an event beneath the ambit of phenomenality.

5. Diachrony, *visage, illeity*

To the face (*le visage*), as conceived by Lévinas, there corresponds a very specific form of temporality—one radically at variance with and incommensurable with the synchronic time of habit, of clocks, of chronology, of the intramundane and the quotidian, and of the phenomenon. That is to say, the temporality that attaches itself to, subtends, and—in a sense—*is* the face, is not that of Deleuze's first synthesis of time. For this temporality entails a past infinitely deeper than that of habitual memory. Of this latter, Lévinas writes: "Immemorial, unrepresentable, invisible, the past that bypasses the present, the pluperfect past, falls into a past that is a gratuitous lapse. It cannot be recuperated by reminescence not because of its remoteness, but because of its incommensurability with the present."[52]

Richard Cohen may be seen as elaborating on the foregoing when he maintains that the apparition of the face of the Other is the irruption of "the excessive dimensionality of time, 'diachrony,' irreversibility, the transcendence of an immemorial, irrecoverable past and an always surprising future"; whilst, on the other hand, "the face crystallizes a whole life," yields up "evidence of accumulated and accumulating vulnerabilities and powers, of experiences etched as character in lines and wrinkles on its skin."[53] The face is thus at once "presence and passage, living now and through all its times—a fitting image of the divine-human juncture, where time and eternity meet."[54] Here, even though "the face of the other person neither is nor is not G-d's face—for G-d has no real face, no spatial or conceptual face"; it may still be contended that "the face of the other person *nonetheless, enigmatically, paradoxically, disturbingly* 'appears' *as though* it were G-d's face."[55]

6. Judgment, sinfulness, shame, surrender and the birth of the Soul

We are now ready to return to the violence of divine love. The love entertained for the Self by the Other—and by implication, by God—must not, as Leora Batnitzky maintains apropos of Rosenzweig, be collapsed into a mere horizontal lyricism of mutual affirmation and understanding. Quite contrarily, the sheer alterity of the Other alone implies that the latter's mere "tautological" presence (the Other as beyond predicative Being) to the Self—of itself—*judges* the Self. This judgment of the Self is primarily that handed down by God. It is in this very movement whereby the violence of divine love passes judgment upon the meta-ethical Self, that the Soul can potentially be precipitated out of the forcibly enfolded imperviousness of its stony character. At the same time, far from spawning a symmetrical reciprocity, the divine judgment of the Self by alterity immediately finds itself transmogrified as the Self's own distressing self-judgment.

The Self comes to view its former self with *shame*. With shame does it reflect back upon its prior existence as steely Character, replete as it was with purblind solipsism, obstinacy, and the relentlessly defiant affirmation of its own arrogant independence. Shamed is the Self now by its ealier incorrigible persistence in thinking itself entirely self-sufficient. The Self's destinerrant and shameful robinsonade has come to an abrupt halt. The Soul, writhing disconsolate in the birth-pangs of its shame, brings itself to admit that it has *sinned*.[56]

This acknowledgement marks the beginning of the Self's opening to the love of the Other. Yet, in order for the Self to wholly open itself to the Other's love and in order for it to thereby love the Other in return, the Self must *intensify* its self-judgment. This intensification entails a movement from the Self's prior confession that it was hitherto a sinner to its further conceding that it *is still*—in point of fact—sinful. At the same time, the Self opening itself to the love of the Other must—in addition to this—pass beyond the said shame-riddled discomfort of self-judgment and thereby literally *surrender itself* to the Other's judgment. This is so since love is not—at least not in the first instance—affirming, but rather *judgmental*. Love is at once commanding and judgmental, for the Other's love demands of the Self that it fashion for itself a categorically new self—a loving Soul (*Seele*)—out of the former and present Self that it (the Other) deems repugnant. In fact, it is *only* through the deferral of judgment to the Other on the part of the sinful Self, in conjunction with the admission of its own present shamefulness, that the Self is able to transmogrify itself into a loving Soul. Only through the Self's *submission* to the Other's judgment is the Soul born—for the first time. It is only hereupon that the Other forthwith responds to the Soul's admission of its present sinfulness—by claiming it. God claims the Soul as his own.[57] "'I called you by your name. You are mine.'"[58]

The birth of the Rosenzweigian Soul may be seen as homologous to what the Zohar maintains to be the rousing of the third and highest part of the soul, that is, the *neshamah*, or *spiritus*. Having already risen above his merely vitalistic and vapid side by way of the transition from *nefesh* to *ru'ah*, with the further ascent to his *neshamah*, man now acquires an intuitive power that connects him with his Creator. According to the kabbalists of Gerona, the *neshamah* is a divine spark in man that emanates directly from the Godhead. The *nefesh*, on the other hand, had merely evolved from the four elements. Similarly, for Isaac Israeli, whereas the *nefesh* was still a common denominator between man and animal, the *neshamah* originates from the divine *Sefirot*, and more particularly from the *Sefirah Binah* or Divine Intelligence. It is a divine spark created from nothingness, but a nothingness from the realm of the Godhead itself. Certain of the Gerona kabbalists accorded an even higher source for the *neshamah*: the *Sefirah Hokhmah* or Divine Wisdom.[59]

For Lévinas, this abandonment of the Rosenzweigian Soul (*Seele*) to the Other falls under the rubric of "substitution." That is to say: the beholdenness of the Self to the Other is *so* extreme, its responsibility and obligation *so* great and *so* utterly unidirectional; that it is—in point of fact—*infinite*, and therefore unconditionally requires that the Self go as far as to stand in the place of the Other in *all* possible circumstances. What in Rosenzweig figures as the submission and abandonment of the Soul to the Other, and thereby to the divine; in Lévinas, amounts to the subject's being obligated to the Other *all the way*—even to the extent of being *responsible for the Other's death and for the Other's responsibility*.[60]

7. From defiance to humility, faithfulness, diachrony

The Soul's surrender to the Other midst the vicissitudes of divine love is also a movement whereby the defiant pride of the elemental Self transmogrifies itself into a different kind of pride—a pride which, purged of its prior contumacious obstreperousness, allows itself to be carried by the divine love of another; that is to say, defiance becomes *humility*.[61]

This newfangled humility is, however, nowise the dialectical the end opposite of pride; it is merely a different modality of the latter. As observed, pride's modality had hitherto been that of defiance. In this former modality, defiance and Character—*hybris* and *daimon*—found themselves conjoined in man, making him a mute Self, turned in exclusively upon himself. That is to say, pride in its former modality of defiance, in its endless eruptions as the defiant pride of free will, that had hitherto restricted the Character to the Self, now comes to be the very first aspect to intrude itself and effuse from the Self, yet:

> … no longer necessarily in the shape of passionate eruptions, every one of which, in its momentariness, reaches the highest threshold, but in the form of tranquil spreading.[62]

It thus becomes a pride that:

> … in its convulsive violence, would no longer distort the countenance of man, … and which, instead of transforming man till he is no longer recognizable, would, like calm water, spread around him from all sides and under him and carry him.… But this pride … in which man is silent and by which he lets himself be carried, is therefore the exact reverse of the defiance that constantly explodes anew. It is humility.[63]

With the transmogrification of the solitudinous Self's impervious obduracy into an incipient humble equanimity, what is granted unto the latter is none other than Being itself. The received Being must, however, prolong itself. This perdurance of the Being granted unto man in his experience of self-effacement before the countenance of the Other is acheived through *faithfulness.* Here, as regards the Soul born of submission to the violence of divine love, Rosenzweig avers that the new quality of faithfulness gives it the fortitude to live in a permanent love of God:

> The faithful belief of the beloved acquiesces to the love of the lover, bound to the moment, and reinforces it so far as to make it a lasting love. This is the counterpart of love: the faith of the beloved in the lover. The faith of the soul testifies, in its faithfulness, to the love of God, and it gives it permanent being.[64]

The Soul's faithfulness is, moreover, none other than man's *will to be* that hitherto, in the solitude and intransigence of Character, had made its appearance as defiant pride. Faithfulness is thus a *new modality* of man's will to be. More precisely, in the experience of Revelation as the encounter with the Other, the Soul affirms itself no longer as authochthonous Being but as *instituted* Being. Quite alarmingly, coextensively with the Self's surrender of its own tragical and elemental ipseity in its upsurge as Soul, *God himself* yields up his *own* Being unto man. That is to say, in the fulgurating and irruptive event of Revelation in the modality of intersubjective encounter, *God's very Being is reduced to its being affirmed by man.* This affirmation of God by the Soul is what Rosenzweig understands to be *faith.* Moreover, the foregoing adduces the startling—and somewhat heretical—conclusion that faith, far from consisting in the mere fact of believing that God exists, is more fundamentally the constantly renewed affirmation that *makes* God exist.[65]

We have observed that the face of the Other belongs to a diachronic, immemorial past, an indefinite recess of infinite depth. Now, it is in the Self's unconditional abandonment to the Other under the commanding and judgmental violence of divine love that this more ancient and primordial temporality extrudes itself in a well-nigh insurrectionary recrudescence that overturns the prosaic order of quotidian temporality. Here, the overturning of the Self's obdurate ipseity in the teeth of overexposure to the Other's excessive vulnerability and destitution is at once a disruption of Being's synchronous self-collection. Revealed in this fulguration is time as an irrecuperable past, a "created" past anterior to force's diremption into cause and effect at the behest of representational consciousness, and thus a past prior to the freedom (causal necessity's inexorable adjunct) attributable to the autonomous agent. (This immemorial past is, however, *not*—let us be quite certain—anterior to the Abyss of Freedom that is the Schellingian Godhead.) For it is precisely the autonomous agent, with its complacent self-collection in the punctual present that it claims as its own and in which it comes to presence, that is here being uprooted. In the face of the Other's overburgeoning privation, time—as that of created being—appears as that of an *extreme passivity* solicited out of the Self's circular ipseity, that is, out of the centripetal self-collecting from which the Self had formerly emerged as obdurate Character. It is the recrudescence of this infinitely recessive past of the Self's originary passivity that reveals the Self as in essence chosen by the Other, as for-the-Other, and not as the self-positing for-itself that the defiance and solipsistic blindness of bare Character had hitherto mollycoddled it into believing.[66]

For the past of the Self's primordial passivity is, as maintains Lévinas, much older and deeper than the sources whereby the Self, as obdurate Character, comes to autochthonously posit itself in the self-referential solitude of its own ipseity. Indeed, this infinite past of the Self's originary passivity is beyond the purview of the Self as consciousness. Yet, it is the mode of indwelling of the Other in the Self—for the Other has always already passed into the Self. More precisely, the infinite obligation introjected into the Self by the Other's sheer proximity—a Self subject to the Other, an obligation which is at once the divine election of the Self—is suffered by the Self as having always already passed into the Self, having already judged the Self and enjoined the Self's love.[67]

8. *Il y a*, dying

This radical anterior passivity that the Self undergoes in its exposure to the Other is barely distinguishable from that which elsewhere in Lévinas falls under the rubric of *il y a*. The *il y a* may be described as a shapeless nocturnal rumbling, the event of mere Being without beings, and of consciousness without a subject. It is from the *il y a* that one may view the Rosenzweigian Character, bursting as it does through the mendacious veils of personality, surging up via what Lévinas elsewhere terms a hypostasis—a hypostasis whereby the existent takes up its existence in an egress that is at once a withdrawal, a paradoxical double-movement foundational of the very instant in which the said incipient being comes into existence and attains and repeatedly reattains evanescent (though plenary) contact with Being.

The amorphous anteriority that is the *il y a* befits in a multitude of ways Régis Debray's effusive description of the desert of monotheistic desertion and its spectral denizens. This latter aridity of the soma and of the spirit, "is as magnetic as it is repellent. It is charged with ambivalence: the punishment of Adam and the salvation of Moses, a place of trial and a space of temptations, near the *Sheol,* the shadowy domain of the dead," avers Debray—for whom this arid interstitial zone of privation, desolation and abnegation, is "also the dust to which man will return *in fine.* In which demons and jackals and serpents loiter."[68] The spectral denizens of this infinitely forbidding wasteland of dread foreboding are those who, "through emaciation by hunger and thirst, becoming skeletal and diaphanous," hereby "become shadows in order to escape our theatre of shadows, outstripping the grave by becoming sepulchral before their time."[69] This vast liminal realm that blurs the dichotomy between life and death is traversed by those who are already dead: the living dead, the undead, *revenants,* specters of the dead, and those whose life is but spectral.

It is therefore likewise the crepuscular realm of the what Lévy terms the "living-dead of the South Sudan," of "those bodies that drag themselves along, those men with limp gestures, those shadows that, sometimes, seem already to be no longer alive except in a few details—their limbs numb, their lips blue, their eyes or smiles extinguished," well-nigh surrendered to desuetude.[70] The *il y a* is the realm of malignant spectral remanences, a domain such as war-torn Burundi where the dead "inhabit the living, harass them, hound them, torment them, leave them no rest," where the very nature of existence "is never to be finished with the dead," where the "proper function of the dead is tirelessly to persecute, to torture the survivors."[71] This miasmal Burundian abyss is one in which the living, "not content with being inhabited, to the point of hypnosis, by their dead," inhabit the very same world as do the dead—a world in which "the cemeteries are destroyed and the world itself is a cemetery."[72] "Transparent shades of the living … Resounding, confused commotion of the dead … One foot in the realm of the living for the dead, one foot in the realm of the dead for the living …"[73] This undeadness of the *il y a* is either the case of "a dead soul in a living body. Or a body here, forgotten—and a soul elsewhere, already damned. Or, finally, a body, just a body, but one that is dead in some places, living in others"—"faces that" seem "alive while their eyes" are "already dead" or "half-dead bodies where it's only the gaze" that survives or "those bodies in tatters where you had the feeling that the blood was circulating only in certain zones—the rest already" possessed of "the blue tint that is a sign that death has begun."[74]

It may likewise be seen that the *il y a* is the shadowy and interstitial realm of the *Muselmann,* since—as Agamben maintains—the latter marks the moving limit between life and death, in which life is no longer life and death is not death either. The *Muselmann* marks the impossibility of the experience of death as the most proper and insuperable possibility. In him the being of death is inaccessible and men do not die. He appears as the inscription of life in a dead area and of death in a living area.[75]

The *il y a*—this vertiginous anteriority to the self-collected and punctual subject—may also be approached via Blanchot's delineation of what he refers to as the second slope of literature: that of interminable dying. This second slope pertains to the silence and materiality of things as things anterior to the act of naming that supervenes upon them: it is "the concern for the reality of things, for their unknown, free, and silent existence ..."[76] The second slope is that of the event of Being in general divorced from and anterior to the beings which dominate it in the assumption of their existence as subjectivity: it is "my consciousness without me, the radiant passivity of mineral substances, the lucidity of the depths of torpor."[77]

It is an experience which consciousness has of darkness as the presence of absence, the density of the void, an uncanny swarming of points, an experience which is at once wakefulness and watchfulness (*la veille*). The experience of the *il y a* is that of consciousness without a subject. It is an anteriority to subjectivity, anterior to the subject. It is the "site" (which is a non-site) of what Blanchot terms "writing," and of a dying stronger than death. As in Lévinas' *il y a*, the second slope of literature moves within the ambit of diachronic time—a time sans punctuality, sans the coagulation of the evanescent instant into the obduracy of a synchronic present. There not being the punctuality of a present moment in the *il y a* onto which a self-collected being could launch itself as *inter alia* the moment of death which it could claim as its own, consciousness in the *il y a* is expropriated from its own present moment as well as from itself as such (consciousness "tears itself away from the meticulousness of an I") in what Blanchot terms the interminabity of dying wherein death as idiosyncratically that of the existent is impossible: "I am no longer capable of dying, my impending death horrifies me because I see it as it is: no longer death but the impossibility of dying."[78]

The temporality of the *il y a* is what Lévy—by way of analogy with utopia as "no place"—describes as *uchronia*, or "no time." It is a "dead time," a "nowhere-time," an "unsituated," "ahistorical," "abstract and semi-legendary temporality."[79] This "Burundian time," Lévy continues, "is a time that is not suspended but empty, truly empty, without content, mechanical, automatic, repetitive, homogeneous," and moreover "the time of *malheur*, the 'evil hour,' misfortune."[80] It is a "time without dates or reference-points," a time "where events supposedly" occur "only in the manner of shots fired with blanks, without reference-points, hence without echo."[81] This temporality that pertains to the axis of interminable dying is "like an exhaustion; a fatigue of dates; it's a humanity so profounding worn-out that time itself seems to have exhausted its resources and its momentum. A fixed time, yes; a sempiternity; a time without dates, without anything to hold onto, where horrors indefinably identical to each other keep piling up—a true time of the End of History."[82] An atemporal time of the end of history within the very ambit proper to history itself. It is a time of "black holes," a time that, "because it is immobile, is without memory"; and one that "because it is without memory," "erases first and foremost speech, complaint," and the "suffering of the people."[83] This "time without intermittence is the foretaste, not of paradise, but of hell."[84]

One may also speak of the interminability of dying in the *il y a* as a *grundstimmung* of existence anterior to and more fundamental than the futural movement of forward-looking projection in which Heideggerian *Dasein* launches itself in an arc of heroic masculine virility upon its death, that which in Heidegger is the reachable fulcrum of *Dasein's jemeinigkeit*, *eigentlichkeit* and assumption of Being. With Blanchotian interminable dying we are here, instead, within the unfathomable ambit of spectrality and of the sheer irremissibility of an existence that never ceases to haunt the existent, sempiternally returning to the hapless existent who cannot get rid of it. Consciousness in the *il y a* is not the tragically incessant return of the subject's full weight to itself as an unbearable quasi-mechanical torsion. It is the horror of there being no subject whilst consciousness uncannily persists regardless: an indestructible consciousness devoid of subjectivity. It is not the weight of eternal return of the self-mired subject to itself, but rather the timelessness of the crepuscular paronomastic *l'entretemps* in which the inescapable insomniac vigilance of consciousness bereft of subjectivity is that of the night itself upon consciousness, a consciousness incapable of any form of retreat, sleep, or respite whatsoever. Dread worse than tragedy, a fate worse than death itself.

Lévy espies the *il y a* in the thankless abyss of Burundi's war-ravaged zones. Here one is struck to the core by the "disappearance of laughter, of smiles, of gazes, all these traits that function, in principle, to individualize a subject."[85] It is "this radical and general desubjectification," "this night of non-subjects, this hell where all subjects are grey," "this indifferentiation," "this empire of the mass and the identical, of the anonymous and the intensely close," "this midnight of souls that seems, for eternity, the time of the damned."[86] Lévy even goes as far as to accord a different kind of humanity to the subjectless reprobates of the *il y a*. This latter "subhumanity"—nowise to be confused with the Nietzschean *untermensch*, yet quite distinct from the humanity "that is spontaneously named, the one for which having a name and a face is not just a right, but a fact, something self-evident"—is a humanity only insofar as it is that of "the subjectivity and, thus, the humanity, of the impossible face crossed out."[87] This pseudo-humanity is that of "these destitute people, these fallen people, these excess men, to whom it is forbidden to give shape and style to their lives."[88] In a more ostensively Deleuzoguattarian vein, the *il y a* as subhuman abyss is "the path of this becoming-animal of neo-hunamity."[89] It is the unfathomable void of the "animalization" and "liquidation of the subject-form as such."[90]

(i) Isaac's dying words

It is precisely this dreadfully nauseating abyss of eternal damnation that is elicited by Isaac's dying words in the hearts of Jacob and of all those present at his death. Of this episode, Mann maintains that the assembled felt as though their very innards had turned over within themselves, causing them to want to vomit out their guts. For in the dying man's utterances and countenance they had perceived the intimation of something absolutely primal, something that was in fact obscene and utterly ghastly. They sensed something sacred yet radically anterior to all sacredness, something lurking beneath all the various layers of civilization in the most repressed and pre-personal underworlds of their souls. In roiling up these depths, the dying Yitchak (Isaac) had fomented utter nausea. From distant and long since buried times there had arisen a most obscene specter. This was an animal that was, as it were, identical to God. This animal was in actual fact a ram, the Hebrew tribe's god and ancestor, out of which the tribe had arisen in the first place—and whose blood, in those primordial and obscene times, they had spilled and devoured in order to renew their tribe's kinship with the divine animal.[91]

This primordial ram that is at once God is precisely what Bataille terms the *sacred* (not to be confused with Agamben's *homo sacer*). The *sacred* is that part of nature which is denied (or disavowed) via prohibition and thereby cast outside the ambit of the profane life of the community.[92] In the process, however, this despised piece of bestiality is transfigured and effectively *deified*. It is henceforth divine.[93] The primordial foreclosure of what becomes here sacred and divine, however, does not preclude the denied animality from continuing to persist at the level of the spectral, the undead, the realm of ghosts. The antediluvian rescrudescence of the foreclosed bestiality roiled up by the dying Isaac in a nauseating experience of the *il y a* is precisely the Lacanian phantasmatic Real. It is likewise the Bataillean *sacred* in the movement of its being dredged up as "nature at the point where its effervescence closely joins life and death, where it is death gorging life with decomposed substance."[94] This "return of the natural given in the guise of the definitive collapse" which "goes against the plenitude of the world of efficacy," this elemental horror associated with the elusiveness of that which remains unassimilable to "the reassuring order"; is the untimely recrudescence of the "world of trembling."[95]

(ii) Jacob and Joseph

When Jacob renders asunder his garments to the point of shameless nakedness in the wake of the false indication received of Joseph's purported death, the shame that drives away all beholders of his act, argues Mann, is merely an attenuated outward appurtenance of the absolute horror and dread that burgeons when a repressed primal state bursts through the accumulated layers of civilization where it had hitherto endured only in the guise of superficial allegorical traces.[96] Jacob's frenzied act of anguished disrobing, a painful debasement to the state of the abject and utterly naked creature, effected a recrudescence of the very source of all dread—extruding the recessive depths and allowing them to well up at the surface. Had he bethought expressing the rank and obscene profundity of his misery by literally bleating like a ram, his beholders—family and servants alike—could scarce have been more nauseated to the very bowels of their being.[97]

Later in Mann's great novel, the dread elicited by Mutemenet's libidinal advances upon Joseph consisted precisely in the intimation and conjuring up of Egyptian religion's obsession with the beyond of death (a predilection irrevocably rejected by the Judaism of his fathers), an evocation of the nauseatingly formless realm of the undead, of monstrous metempsychoses, of the spectrality of the Ka's bare insistence in the anonymity of the *il y a,* and of the obscene motley of corpse-like gods lying rigidly in the death of their temple graves.[98] In Mutemenet's wooing, which disgusted him to the point of recoil, Joseph felt no less than the complete coincidence of death and obscenity united in sexual temptation. This temptation was that of Sheol, something that would have spelled complete destruction were he to yield and bare the nakedness of his body.[99]

Here, whereas it is certainly the case that the dread and utter revulsion elicited by the object of desire—with all its intimations of death as at once decay and the pullulation of nascent life, and this endless equivocation itself—is a *sine qua non* of the welling up and maintenance of desire. Mutemenet's desire was elicited by the disavowed bestiality associated with the base proto-human Habiru slave, Osarsiph (Joseph). It is this dread—and alike the horror of prohibition—which holds Mutement captive to her anguished desire. And, indeed, the greater the subjection to horror, the greater the desire elicited. However, the horror cannot be so great that the subject is unable to bear it.[100] Indeed, the sheer excess of horror roiled up in Joseph by Mutemenet destroys what could otherwise have been an irrepressible seething of lust transfiguring his soul in the Epicurean sty.

(iii) King Philip II of Spain

In Carlos Fuentes's *Terra Nostra*, the dread of the *il y a* breaks forth even more forcefully. El Señor (King Philip II of Spain) dreams that he is trapped inexorably in an arid stony valley in which rapacious birds of prey are swooping in jagged arcs. Without logical transition, he dreams that he is at once three different men—although each man is himself, all three are captives to the arid valley having no escape but the sky above.[101] After fleeing the countenance of the first man, he encounters that of the third man in which he now feels himself an ancient man clad in black and lying upon a massive flat rock with his face turned directly to the sun. This sun, however, neither illuminated his light-repelling waxen face, and through the facial apertures there crawled worms even less white than the ancient's epidermis. From this repulsive face, its ears, its lips, its nostrils, worms twisted and pullulated. Behind the eye-lids and behind the cornea of the eye, writhed a miasma of disgustingly terrifying eggs.[102]

It is especially the last-mentioned that epitomizes the dread of the *il y a* as precisely *not* that of the fear of death itself but in fact of the exact opposite: that of the *impossibility* of death. That is, the sheer dread of the lingering and inextricable sense of discomfort occasioned by the fact that what would appear at first glance to be death, death as completed, as having definitively come to pass, as a conclusive ontological closure—soon reveals itself to be merely another site of the inexorable return of pullulating life in all its viscous dread and gut-roiling nausea. It is, as Bataille has it, "that sticky movement that is the privileged object of our disgust" associated with "those unstable, fetid and lukewarm substances where life ferments ignobly"—the "procreative power of decay" wherein "nascent life is not distinct from the putrefaction of life which death is …"[103] It is the abandonment to a "silence without appeal"; "that shipwreck in the nauseous"—a seething forth of "abject nature and the purulence of anonymous, infinite life, which stretches forth like the night, which is death."[104]

In Lacanian terms, the *il y a* is the "zero-level," or pure void, of subjectivity: the absolutely unendurable abyss in which fully constituted objects and subjects dissolve and yield before a miasmal swarming of spectral partial objects, of exemplifications of the *lamella*, of the undead, immortal, libido-as-object that is self-identically the Freudo-Lacanian death-drive (*Todestrieb*). This paronomastic modality of the subject is merely one "degree" removed from the pre-subjective noumenal Real. It is a return to, or recrudescence of, the impossible moment of upsurge of subjectivity in the form of a withdrawal, a recession from the light of phenomenality and its differential degrees of appearing in a world in accordance with its transcendental regime, which assumes the form of a contraction through which constituted reality yields before the surreal phantasmagoria of *membra disjecta*, of organs devoid of body functioning as stand-ins for the dread immortality of the object-libido in its semipernally stuck fixity and eternal psychotic circulation around the void. In Kantian terms, this crepuscular and vanishing mode of subjectivity is that of the pre-synthetic imagination gone mad, psychotic, and utterly demented—the impossible domain of absolutely purified and kenotically evacuated transcendental freedom and spontaneity. In this domain, the subject finds itself vertiginously suspended between direct unadorned animality, on the one hand, and the subordination of its freedom to the moral Law, on the other. In this gathering murk, far from being able to galvanize itself in the direction of forward-directed action in the world or attunement to the promptings of the Deity, the self is submitted to the horrifying experience of dread organs without body, of itself as the Lacanian fragmented body, of its *membra disjecta* floating around in a nauseating and disgusting miasmal drift, the "night of the world," the "valley of the shadow of death," the hell-on-earth of Kundry in Wagner's *Parsifal* who, having committed the blasphemy of laughing at Christ dying on the Cross, is condemned to an eternity of wandering as an abject, multiply befouled, purely excremental self entirely rejected by the world, abandoned by God, and subject to involuntary, psychotic automatisms, spasmodic jabberings, slavering aphasia, dyslexia, *and the absolute horror of never being unable to die.*

9. *Sprechende Sprache*

But let us return to our earlier discussion on the violence of divine love irrupting in the limitless exposure to the Other. It is of paramount importance not to forget that the Self's being infinitely obligated to the Other, its self-judgment and consequent surrender to the Other; all follow in the immediate wake of the apparition of the face of the Other to the Self. In this connection, Rosenzweig maintains that the violence of divine love:

> … can be nothing other than the self-negation of a merely mute essence by a word uttered out loud, the opening up of something locked, of a silently reposing permanence by the movement of a blink of the eye.[105]

As in Lévinas, it is the *face* (*le visage*) of the Other which commands the Self at once not to murder the Other and also to love God. Yet, it does so not through prosaically spoken words, nor through its brute physionomy, but *through the pure resistance of naked vulnerability*. This means that the face of the Other signifies prior to the advent of spoken words, prior to language. It is the pure gaze of the Other, of itself, that at once judges and commands the Self.[106]

And it is in this latter sense that Revelation is lived by man as intersubjectivity. That is to say, the irruption of radical otherness into the midst of the Self is—as Stéphane Mosès has it—that of a moral *injunction*. This injunction, this solicitation by the bare gaze of the Other, being anterior to language *tout court*, is a commandment devoid of any positive content.[107] This commandment is the pre-discursive "speech" (*sprechende Sprache*) of the Other. But this speech of the Other, appearing as it does within the Self open to the Absolute as precisely the very lack of Absolute within the said Self, is at once experienced by the Self as its *own* speech. The Other's command/injunction transpires within the Soul as the very fact of the Soul speaking, the very fact of the latter addressing a "thou" that it experiences as an (internal) absence. This speech of the Soul, addressed to pure exteriority, is *prayer*. Like the commandment of the Other, the prayer of the Soul—in the originary sense intended here—is likewise devoid of all positive content. It is the bare fact of the Other being addressed. Yet, it is nonetheless significative.[108]

Rosenzweig calls the anterior command of divine love—which is at once the Soul's internal prayer—*sprechende Sprache* (speaking language), whilst Lévinas terms the same *le dire* (the Saying). Both of these terms refer to the action or praxis of speaking—as opposed to the actual words spoken. Both terms are anterior to the discursive formations of language as a system of significations. *Le dire* and *sprechende Sprache* concern not the cognitive/conceptual function but the ethical dimension of speaking; that is to say, the bare fact of the Other enjoining the Self and the latter's attendant responsiveness/apology to the Other.[109] In this sense, they are homologous to the truth-event as conceived by Badiou. For "a truth is trans-linguistic, insofar as the general form of thought that gives access to it is separable from every specifiable language."[110]

(Lévinasian *le dire* and Rosenzweig's *sprechende Sprache* may be seen as strongly homologous to what Blanchot quite uncomplicatedly terms *poetry*—which moves along the second slope of literature. This latter, in the idiosyncratic and profound sense that it attains in Blanchot, pertains to the wide-eyed insomniac vigilance to things themselves prior to their punctual entry into the order of transparency within the medium of language as a system of signs. Poetry, through its circumlocutious peregrinations, is a tangential evocation of things-in-themselves, an evocation born of the anterior nocturnal rumblings that are the *il y a*.)

For this reason, Rosenzweig is right to maintain that the pre-discursive command/injunction of the Other—that is, Revelation—belongs to the *imperative* mood. Discursive language as used in unequivocal descriptions, on the other hand, is of the *indicative* mood. Now, though it is true that the Other's command and the Soul's prayer are alike devoid of positive content (that is, non-descriptive), they are nonetheless the antecedent condition of discursive language/speech. It is in this sense that Leora Batnitzky views indicatives as *necessarily presupposing* imperatives. This is so because it is only possible to describe something on the basis of a prior commitment that in turn itself devolves from an anterior command. It is only the Self's commitment, as elaborated from an antecedent injunction, that is able to ground and orient the latter within its environs to the extent of providing an adequate basis from whence descriptions of states of affairs may proceed. Succinctly put: indicatives refer back to the imperative that lies beneath them, whilst descriptions presuppose the commitments that are their basis of orientation.[111] The ethical imperative inherent to the intersubjective encounter is hermeneutically anterior to the descriptive statements of the indicative and at once their condition of possibilty.[112]

The foregoing means that language (as an unspoken system of significations) depends upon the singular command/injunction of the Other evoking the Self's responsibility to respond; or, more succinctly, that which Lévinas terms the Said (*le dit*)—or what Rosenzweig calls unspoken language (*unausgesprochene Sprache*)—relies upon the anterior Saying (*le dire*) or, equivalently, Rosenzweig's *sprechende Sprache* for its very emergence.

(For Blanchot, Lévinasian *le dit* and Rosenzweigian *unausgesprochene Sprache* correspond to what he matter-of-factly refers to as *prose*. We have already delineated Blanchot's second slope of literature; that is, that of interminable dying. Prose, in contradistinction, transpires along literature's first slope. Here, things qua things, which hitherto shimmered in the effervescent and untimely crepuscularity evasive of the attentions of subjectivity and held by the *il y a* in the writhing inchoate and implacable soup of an endless expropriation, are appropriated by the rising and triumphant subject which now comes to claim things for itself as things-for-consciousness under the taxonomic matrix of language and naming: "literature is turned toward the movement of negation by which things are separated from themselves and destroyed in order to be known, subjugated, communicated."[113] The upsurgent subject, in this movement of appropriation, annexes things to itself, negates them, and thus kills them as things qua things. Death is thus present in this movement and moreover speaks in the subject which appropriates, which is an appropriation. Simon Critchley refers to the first slope of literature in Blanchot as "Hegel avec Sade," since it combines Hegelian incorporation of otherness with Sadean murder.)[114]

The very emergence of time as temporal succession—that is, what in Deleuze figured as the first synthesis of time (that of habit)—rests upon the infinitely ancient and deep past in which the face (*le visage*) of the Other makes its apparition and in which, by implication, the Saying (*le dire*) had afore always already solicited the self. For the Said (*le dit*) is of the synchronic order of representation and transpires in a punctual present contemporaneous with that in which the self collects itself in the economy of self-presence.

(It is for this reason—one might add—that the limitless immemorial past of the Saying (*le dire*) is homologous to the virtual past of the Deleuzian second synthesis of time (that of memory). Both entail an irrecuperable past that never was, and can never be, present—a past irrevocably untethered from the punctiliousness of any present moment. Much like the infinitely deep past of the ethical Saying, the Deleuzian virtual past also served as the inexorable errant condition of the present's passing. Crucially, that the Deleuzian second temporal synthesis was characterized as that of memory, whilst in that the Lévinasian deep diachronic past is termed immemorial; must not mislead one into positing this as a point of divergence between the two. For the sense of memory that attaches to the Deleuzian virtual past, as the gargantuan and infinitely receding (yet incessantly burgeoning) Bergsonian cone that it is, is that of matter in gestation; that is to say, it is a cosmic memory anterior to and transcendent of the molar subject. Therefore, as with the deep past of the Saying, the virtual past of the Deleuzian second temporal synthesis is *likewise* refractory to the reminiscing faculties of the human subject.)

However, in that it is not of the order of presence, the Saying (*le dire*) cannot appear directly or unequivocally within the Said (*le dit*). Rather, the Saying is traced in the Said only indirectly: as at once an overdetermination and a disruption of the Said, as its unsaying or resaying; a disruption that is moreover the undercutting and undermining of the egoist subject, whereby the latter is passively submitted to an undergoing that has always already stunned it with an intractable vulnerability it is nowise able to ground or recuperate.[115]

The divine command/injunction, given as it is in the purely imperative mood, is—as observed—devoid of anything resembling a positively delineated content. In that it is anterior to discursive speech/language, it cannot be unequivocally or directly stated or thematized. Indeed, it is moreover *not even of the order of knowledge*. Its recalcitrance to thematization stems, however, not from any inherent lack on its part—but rather from its inherently *excessive* character. That is to say, the divine command is possessed of a *too great exigency* for it to be circumscribable by reason alone.[116]

(We have thus returned to what Santner earlier described as the "too much of address" that inheres in the event of Revelation. One must, however, be careful not to conflate the two senses of Revelation found in Rosenzweig. Our use of the latter term in connection with the "non-ideational exposure to an exterior" referred to the traumatic event—the handing down of the Mosaic Law on Mount Sinai and the coterminous violent repression of the Jews' prior idolatrous past—through which the Judaic postsignifying regime was founded. This was alike the event whereby God forcibly removed the Self, the latter having hitherto led a facile and parochial existence as mere personality, from the mendacious and vapid world of third-person relations—in the process exposing the meta-ethical Self to the obdurate inner kernel of its own autochthonic authenticity (the Character). In contradistinction, the sense of Revelation being used in this present section refers to the irruption of divine love in the Self's direct frontal exposure to the face of the other person—the traumatic intersubjective encounter whereby the contumacious obstinacy and solipsistic defiance of the all-too-obdurate Character yields to the surrender, submission, and faithfulness of the nascent Soul to the Other. These two senses of Revelation nowise imply qualitatively different events. Rather, both are alike Revelation—except merely in different modalities.)

For Lévinas, the divine injunction that transpires in the limitless exposure to the Other is moreover not merely outside knowledge but also *beyond Being*; that is to say, beyond Being as presence. It is at once *further* and *closer* than presence. It is further because the Other remains transcendent of the self and Being alike, irreducible to the latter. It is closer because the divine command is far more exigent that whatever may be thrown up by self-presence, imposing as it inevitably does infinitely greater and more urgent demands than the self could possibly impose upon itself when subsisting purely upon its own resources. Richard Cohen avers that the command/injunction of divine love is that of goodness, and the "way of goodness is, in a word, *better* than the thematization of being, more glorious. This 'better' is emphatically more important and serious than being; its 'more' is immeasurable because its force lies outside the calculus of making-present. Not truer, of course, but better."[117]

10. Toward the Neighbor

As observed, in Revelation, by way of the violence of divine love, the defiance and solipsistic pride of Character finds itself inverted into an act of faith whereby the Self, by freely abandoning itself to and affirming its origin not in itself but in God, awakens to itself as the passivity anterior to self-presence that it—in essence—is. Herein, the anterior and pre-discursive speech of God (*sprechende Sprache*—or *le dire*) inhabiting the Soul as the injunction to divine love, is at once and alike the Soul's prayer addressed to the Absolute.

Yet, the prayerful dialogue of the Soul with God already burns with an exigency that exceeds itself:

> ... for the cry that the soul moans at the moment of the supreme and immediate fulfillment overflows the limits of the dialogue; it no longer comes out of the blissful calm of the receiving of love, but on the contrary rises, in a new uneasiness, from the depths of the soul still unknown to us, and, beyond the invisible but sensed nearness of the lover, keeps sobbing into the very twilight of infinity.[118]

This lugubrious yearning of the faithful Soul that arises in the very midst of the solitudinous dialogue of divine love is an incipient discomfort which proves even the faithful and speaking Soul to be incomplete. Though faith is born of the opening unto alterity of the hard and impervious kernel that is Character, it nonetheless fails to take up for itself even the least part of this *daimon*. For this reason, as Mosès explains, the man of pure faith, though penetrated by and receptive to the commandment of divine love, is—as was Character in all its obstinate solitude—still severed from the world.[119]

In fact, contends Rosenzweig, the prayerful Soul evinces itself to be *even more* bereft of the world than is the contumacious hero of tragedy: the meta-ethical Self. Indeed, for all the solitude and muteness inherent to the impervious stiffness of Character, the tragic hero is nonetheless *visible* to others as the spectacle that he is; whereas the intimate dialogue of the Soul with the divinity remains inescapably immured within its own hermetic arena. This solipsistic Soul communing with his God is the *mystic*. His flaw is that he regards the world as if it does not exist. For the mystic—much as for the Hegelian stoic slave—the world is a meaningless procession of facile appearances devoid of reality, a senseless ramshackle of fleeting contingencies. He is a man without world, lacking as he does the world as concrete lived reality. For these precise reasons, Rosenzweig heaps opprobrium upon the mystic. For his relation to the world is *fundamentally immoral*.[120]

Remaining as he does on his own level, the mystic thus represents a *cul de sac* in the movement from Revelation to Redemption. In point of fact, the mystic's stultifying passivity and complacent immobility runs the perpetual risk of regression. For the mystic is in danger of growing deaf to the solicitation of divine love that drew him out of the obstinate solipsism of Character in the first place. This is so because the possibility and impetus—and indeed the injunction—toward the Redemption of the world already inheres within Revelation from the very start. More precisely, the commandment to love issued in the traumatic event of Revelation is at once the injunction to love one's Neighbor.[121]

This inextricable mutual imbrication of the first two commandments is still more pronounced in Lévinas, where the violence of divine love is seen to irrupt in the very encounter with the Other whose mere gaze compels response through the resistance of naked vulnerability and destitution, and in the fact that the apparition of the face (*le visage*) marks the junction or zone of indiscernibility between the human and the divine. In Lévinas, the passage from the relation to the Other as God or *illeity* and the relation to the Other as the Neighbor already inheres in and is presaged by the former: "The illeity in the beyond-being is the fact that its coming toward me is a departure which lets me accomplish a movement toward a neighbour [*le prochain*]."[122]

So, the mere unreserved surrender to the passivity of essential dependence inherent to the Self's opening to divine love is far from sufficient. The Self must turn toward the Neighbor—and thus accomplish the first step on the path toward Redemption. This latter is only possible when the faithful Soul, whilst maintaining its abandonment to the enjoining violence of divine love, takes up the resources of Character that the impasse of mysticism had hitherto felt content to lay fallow. That is to say, the mystic must—while not abandoning fidelity to the divine—resurrect and subsume within himself the elementary reality of the tragic hero. This movement being effected, a new self arises: the *saint*.[123]

Conversely, we may view the saint as—in a certain sense—the converted tragic hero. Like the mystic, the hero-become-saint abandons himself to the commanding violence of divine love. Unlike in the case of the mystic, however, the Character does not lay fallow and thereby weigh the incipient Soul down with the leaden gravitas of its (the Character's) impervious obduracy. Instead, the Character is taken up by the hero-become-saint. It is, however, subsumed *not* on its own contumacious terms—but as a resource to be molded by the force of negation.[124] Within the elemental Self of the tragic hero, Character figured as the Self's dogged perseverence within Being; that is, it was what constituted the Self's ipseity as enduring essence. As Benjamin Pollock explains, in the transition from the tragic hero to the saint,

Character transmogrifies itself into a plastic force that repeatedly negates itself at its very base so as to wrestle itself anew from out of its own nullity. Formerly a *sine qua non* that indelibly marked human identity in its impervious and ossified fixity in the crepuscular dusk of arid solitude; Character, as it appears in the saint, emerges as a transformative force extinguishing and creating itself anew at every moment via self-negation.[125] As Rosenzweig contends, Character must engineer for itself:

> … an internal reversal from a once-and-for-all "affirmation" to the wrestling forth in ever new self-denial of its own origin, the Self closed on itself.[126]

Upon this reversal, the Character shall henceforth be that which:

> … at every moment seems to break forth in its newness and at every moment to come out of his own innerness with all the violence of the oriented will.[127]

It is precisely this newfangled plasticity of Character as possessed by the saint that, in breaking through the solipsistic confines of the tragic hero and the mystic alike, henceforth allows the Soul to *truly* surrender to the commanding violence of divine love by enacting the latter's outwardly oriented modality: *love of Neighbor*.[128] For the foregoing force of will that extrudes itself from the depths of the Soul and externalizes itself in sempiternal newness is that which must render complete the surrender required by the commandment to love God. This is, of course, nothing apart from the love of Neighbor. That is to say, love of Neighbor is that which at each and every moment surmounts, yet always must presuppose the foregoing wholesale surrender.[129]

One also observes how the fulfilment of the commandment of love—that fulgurates in the violence of divine love—in and through love of the Neighbor can only come to pass at the behest of the above newfangled plasticity of the now sempiternally recrudescing will of the meta-ethical Self. For it is not merely one singular act, but instead entails an entire series of acts, since love of Neighbor incessantly arises anew. It is itself always an absolutely new beginning, which does not allow itself to be dissuaded by disappointments and reverses. Quite contrarily, it absolutely needs dissappointments and reverses so as not to ossify and cease to well up in its freshness.[130]

Moreover, in a manner homologous to that whereby God finds himself uncannily gripped, in Revelation, by a spontaneous love for man; likewise is the obdurate Character of the tragic hero rendered supple in its entirely unexpected seizure by the incipient suddenness that is love of Neighbor.[131] As regards the latter, Rosenzweig contends that love of Neighbor entails an incessantly commencing and recommencing rupture of the solidity of Character, through the perpetually unforeseen irruption of the act of love. Indeed, it cannot be known or foretold in advance what this act of love consists in in any particular case. In fact, it must necessarily be unforeseen—for otherwise it would not be an act of love at all.[132]

11. *Le tiers,* distributive justice

The saint, then, is capable of Neighborly love. Yet, even this modality of love—if left to linger merely at its own level—is, much as was the worldlessness of the mystic in solipsistic communion with his Deity, unsustainable. As observed, the exigency for Redemption is already present within and inherent to Revelation. Love of the Neighbor, however, constitutes merely the first stage on the path toward the Redemption of the human race as a whole. Love, whether in the modality of the prayerful Soul's mystical communion with God or that of its outwardly-oriented variant (that is, Neighborly love), Rosenzweig describes as *lyrical*. Love—of its very nature—is, however, fleeting. The pure speech (*sprechende Sprache*)—as addresses and solicits in the violence of divine love bursting forth in the passively resistant vulnerability of the Other's bare countenance, in point of fact, negates itself upon being spoken. This is so because the reality of the spoken posthaste ceases to be present. The evanescence alluded to here is, more precisely, that which pertains to the Neighbor. It is, in point of fact, the *Neighbor* who is evanescent—and is so essentially. The fading and vanishing quality of the Neighbor is, in fact, from the very outset inherent to the manner in which the injunction to Neighborly love appears in the second commandment.[133] In this connection, Rosenzweig makes clear that:

> ... both in the holy language and in Greek, the word means the neighbour *at the precise moment of love* [my emphasis]; it matters little what he was before this moment of love and what he will be afterwards, in any case, at this moment, he is only the neighbour for me.[134]

It is for this precise reason that the exigency for Redemption contained within the Saying (*le dire*)—or equivalently in Rosenzweig's speaking language (*sprechende Sprache*)—inherently presages a movement beyond the flickle evanescence of the merely lyrical. This progression beyond the lyrical implies a movement toward the *dramatic*. That is to say, ethical exigency demands a necessary movement away from the intimacy of the Soul and the Neighbor in the form of an attendant movement toward a *third party*. Rosenzweig argues that this embrace of the Third

Party—the Neighbor of the Neighbor, the next nearest after the most near—is an immediate corollary arising from the Neighbor's inherent transience.[135] More precisely, the transigent nature of the Neighbor implies that the latter:

> … is therefore only a representative; he is not loved for himself, he is not loved for his beautiful eyes, but only because he is just *there* [my emphasis]. In his place—in this place that is for me the one neighbouring on me—there could just as well be another person….
>
> The neighbour is therefore, as just noted, only a place-keeper; oriented by way of substitution toward the one who is, each time in the fleeting moment of his present moment, his neighbour …[136]

The crucial upshot here is that Neighborly love:

> … is really oriented toward the embodiment of all those—men and things—that could at any moment take this place of its neighbour, in the last resort it applies to everything, it applies to the world.[137]

From the Neighbor, to the Third Party, and onward toward the entire world.

For Lévinas, the foregoing argument is approximately reducible to the asseveration that, in the assymetrical and unmediated relation to the Other, one inevitably finds that the Third Party (*le tiers*) is—in point of fact—somehow always already implicated. Lévinas: "The third party looks at me in the eyes of the Other—language is justice."[138] Indeed, the "third party is other than the neighbour, but also another neighbour, and also a neighbour of the Other, and not simply his fellow."[139]

Herein entailed is the transition from the imperative mood of the of the "I-Thou" toward the plural imperative of what Rosenzwieg terms the *cohortative.* So as to avoid succumbing to the suffocating and potentially regressive quagmire of the mystic, lyricism must dramatize itself. Here, from the flitting injunction of divine and Neighborly love in the intimacy of intersubjective encounter, one moves on to the ever-burgeoning cohortative of the community and its gradual Redemption of the world. As observed, with the violent irruption of divine love in the limitless exposure to the Other's countenance, the Self experiences the presence of the Absolute within itself as this latter's absence. Now, it is precisely this lacuna that the Other burns within the Self which instills the latter with the desire and impetus to strive to escape the lyrical dialogue of merely two voices and to thereby pursue its lack of concreteness elsewhere. This concreteness the Self attains in the polyphonic multiplicity of voices that is drama.[140]

In the very same vein, for Lévinas, the exigency to move beyond the immediate ambit of the self's exposure to the Other's countenance (*le visage*) is born of this very ambit itself. This is so since, in the singularity and vulnerability of the face, a still deeper obsession inheres in the self's infinite beholdenness and obligation to the Other; an obsession whose very pervasiveness renders the infinite ethical relation born of man's exposure to the Other inadequate to the moral desire with which it (this (non-)relation) itself burns. This means that it is not solely to the Other who faces that the self is ethically beholden. For the self is herein likewise beholden to Others bearing upon the Other and thereby—though mediatedly—upon the self. It is in this sense that the infinite beholdenness of the self to the Other opens out onto a responsibility to *all* Others. Lévinas: in "the proximity of the other, all the others than the other obsess me, and already this obsession cries out for justice, demands measure and knowing, is consciousness."[141]

Thus, the seemingly private and intimate exposure to the Other's ethical Saying (*le dire*) at once opens out onto what comes to appear as the practical demands of society; that is, with the demands of justice at the social and political level, with distributive justice and its far more prosaic order of rules, regulations, and—above all—law. This all, of course, involves the Said (*le dit*). "An event like mediation—synchronization, comparison, thematization—is the work of justice, an entry of the diachrony of proximity, of the signifyingness of the saying into the synchrony of the said, a 'fundamental historicity' in the sense of Merleau-Ponty. It is the necessary interruption of the Infinite being fixed in structures, community and totality. Synchronization is the act of consciousness which through representation and the said, institutes 'with the help of God,' the original locus of justice, a terrain common to me and the others where I am counted among them, that is, where subjectivity is a citizen with all the duties and rights, measured and measurable which the equilibriated ego involves, or equilibriating itself the concourse of duties and the concurrence of rights."[142]

If the initial transition from the elementary and solitudinous Self, solicited by the violence of divine love and its attendant love of Neighbor, was one from atheism to ethics; then the movement implied here is one from ethics to *justice.* The far broader terrain of justice unveils itself initially within the movement from the somewhat hermetic intimacy of Neighborly love toward what—as observed—Lévinas terms *le tiers* (the Third Party). From here on in, however, follows an incessant and ever-expanding trajectory galvanized by the inexorable inner exigency to ultimately encompass the whole world. In the process, the sheer excessiveness of the initial moral imperative fulgurating in the destitute countenance of the Other reveals itself to have been always and from the very start possessed of an inherent exigency for the Redemption—without reserve—of the entire human race.[143]

Here, though justice's exigency is already implicit in ethics and likewise presupposes the latter as its genetic condition, these two are not to be mindlessly conflated. For, whereas distributive justice entails the unequivocal assertion of equality before the law and the painstaking embodiment and establishment of the said equality through enduring institutions, ethics entails the fundamental *inequality* of the Other's strict priority over oneself. The call for the equal treatment of persons is nowise, as much historical liberalism tends blunderingly to assert, sourced in a preexisting human essence. Rather, it derives strictly from the irrevocable *inequality* inherent to the Self's infinite obligation before the moral height of the Other's destitution. Liberalism's erection of an apparatus of distributive justice upon the uncertain ground of a well-nigh arbitrarily posited primordial equality of persons, thus appears perilously close to the forcible subsumption of human singularity under the auspices of a single monolithic genus.[144]

But, as maintains Richard Cohen, just "as the ethical self is not the specification of a genus but the uniqueness and exigency of an irreplacable responsibility for the other, humanity for Lévinas is not the genus of which individual human beings are the specification. Humanity is precisely the exigency for justice which shines in the face [*le visage*] of the other person, for the other who faces is also the other of the others who do not face. Thus, in the face of the other, beyond its absolute transcendence, as it were, lies the transcendence of all humanity, of all others."[145] Indeed, the severance of this essential link between distributive justice and the singular ethical relation to the Other from whence the former initially devolves, and the attendant occlusion of humanity's essential transcendence at the behest of genus and its coterie of epigones, is—one is compelled to add—precisely that which inevitably renders justice properly-speaking *tyrannical.*[146]

However, this propensity to the tyranical is, sadly, already implicit to the ethical relation itself. For, in the very process of its centrifugal unfolding as what will in due time become distributive justice and indeed *law*, the ethical relation to the Other—such that, through the panoply of instances of the Third Party (*le tiers*), all Others meet with due response—necessarily comes to appear as a demand for a rectifying distantiation and partial disengagement on the part of the Self from its anterior infinite and immediate obligation to the Other. Thus, it seems—in a very palpable sense—that, in its modality as distributive justice, ethics inherently, necessarily, inexorably, inextricably and unavoidably—such that its essence be maintained—*implicitly calls for the very betrayal of this same essence.* This is certainly the case. For the formalization of the anterior ethical relation in the Said *(le dit)* as inscribed in the decrees and strictures of law, though implicitly enjoined by the primordial Saying (*le dire*) itself, all the same entails the emergence of a finite and unambiguous discursive formation at the behest of a truncation whereby the anterior ethical injunction's originary equivocity and excessiveness is—for all intents—*betrayed.* For Rosenzweig, this supervention of the Said onto the Saying does indeed entail a betrayal, in that *sprechende Sprache* (the Saying) is none other than the commandment to love—which latter, in contradistinction to:

> … every other commandment, at least when considered from the outside and as it were after the event, could have been just as well law, the commandment of love alone is absolutely incapable of being law; it can only be commandent. All other commandments can pour their content into the form of the law, this one alone refuses to be decanted, its content tolerates only the form of the commandment …[147]

And it is precisely the Said (or equivalently *unausgesprochene Sprache*) that perforce decants that which cannot—in its essence—be decanted. This movement can therefore only be a supervening violence which ruthlessly immures the commandment of love at the cost of the betrayal and distortion of the latter's singularity. Distributive justice necessarily and unavoidably betrays the originary ethical condition of its very emergence, an emergence and a betrayal already presaged and moreover enjoined by the ethical relation itself. It is as if the inevitable sequence of ensuing betrayals as *sine qua non* of justice was from the very outset enjoined by the violence of divine love itself. For it is indeed the case that, though enjoined by the Saying (*le dire*), the Said *(le dit)*—despite its being the Saying's only means of manifestation—irrevocably dissimulates the Saying through its incarceration of this latter within the radically heterogenous order of ontology. In other words, the Said forcibly ensconces the Saying into the order of Being: forcing the untimely, fundamentally anachronistic, sub-phenomenal, non-luminous, and enigmatic into the prosaic ontological order of presence, chronology, phenomenality, obscene radiance, and non-equivocation. But this is precisely the manner in which the Saying finds itself inexorably betrayed in the Said. Lévinas: "The correlation of the saying and the said, that is, the subordination of the saying to the said, to the linguistic system and to ontology, is the price that manifestation demands. In language qua said everything is conveyed before us, be it at the price of a betrayal."[148]

Lévinas terms this passing of the Saying into the Said—in which the Saying finds itself subordinated by the Said in the form of its (the Saying's) unambiguous thematization—the *refutation of skepticism.* Through the forcible synchronization of Saying and Said, skepticism is refuted. Skepticism, however, is hereby nowise ultimately defeated. For skepticism—as that which in essence "puts an interval between saying and the said"—though temporarily refutable, inevitably returns.[149] More precisely, the Saying "that betrays in its said the proximity it conveys before us still remains, as a saying, a proximity and a responsibility"; that is to say, the Saying's subsumption under the Said, though it "circumscribes the life of the approach" and moreover "measures obligations before the third party [*le tiers*] with justice and knowledge," nowise serves to irrevocably undo this life.[150] In that the Said fails to exhaust the Saying, there is always an unsaid in the Said which serves to "destroy the conjunction into which saying and said continually enter."[151] For Rosenzweig, this disruptive recrudesence of the Saying at the heart of the Said is none other than the violence of divine love that fulgurates in the evanescent gaze of the Neighbor and in the attendant lyricism of Neighborly dialogue:

> Love, apparently in order to throw things into confusion, breaks into this structured edifice, and ... soon separates a member here and there to give it its own life that threatens to explode what holds it all together.[152]

In Lévinas, this undermining of the Said at the behest of the Saying—that is, the unsaying of the Said—is wrought by the force of diachrony which *refuses* the synchrony wherein Saying and Said are correlated. This diachronic force of the recrudescent Saying as unsaying is that of skepticism. For Rosenzweig—as for Schelling before him, this diachrony that extrudes itself is none other than the recrudescence of eternity within the very ambit of the prosaic time of habit and chronology, the fulgurating resurgence of eternity from beneath the intramundane veils of the temporality of succession. Rosenzweig:

> Therefore all that is worldly in all time has its history: law and State, art and science, all that is visible; and it is only at the moment when in such an It of the world there resounds the echo of the wake-up call for God's Revelation to man that a bit of temporality dies the death of the resurrection of eternity.... In eternity there is silence.[153]

Yet, there shall ever be a renewed conjunction of Saying (*le dire*) and Said (*le dit*), a rehashed thematization of the anterior ethical relation to the Other wherein a new "truth of justice" is forged. Thus does discourse—in roughly the sense accorded to the term by Foucault—as the correlation of Saying and Said, afore evanescently interrupted by the diachronic moment of skepticism, resume. Indeed, every "contestation of this power of discourse is at once related and invested by discourse. It thus recommences as soon as one interrupts it"; for in "relating the interruption of the discourse or my being ravished into discourse I connect its thread"; with the result that the "interruptions of discourse founded again and recounted in the immanence of the said are conserved like knots in a thread itself tied again, the trace of diachrony that does not enter into the present, that refuses simultaneity."[154]

A broken line appears to be what is traced. In an initial movement, emerging as it does from out of the Saying; the Said unhesitantingly dominates the former, in its wake erecting a certain discursive "truth of justice" which functions coextensively with legal, juridical, social and other structures. Left to its own devices, this latter would develop itself unhindered in what would resemble a straight line. However, the inevitable recrudescence of the skeptical force of diachrony serves to effectively interrupt this unadulterated line; an interruption from whence the developmental line of the Said (*le dit*) nonetheless resumes in the guise of a newfangled discursive formation. Therefore, what emerges at the historico-political level is a kind of *Deleuzoguattarian segmentary passional regime*, marked by its own constant founderings and consequent recommencements, a regime in which the very functioning of distributive justice rests upon a betrayal of its own foundation in the anterior ethical relation to the Other, and wherein in turn distributive justice finds itself incessantly betrayed by the diachronic force of a recrudescing Saying (*le dire*) by which it is unsaid. We may thus speak of a regime of successive proceedings punctuated by structurally irremissible betrayals. The Saying is betrayed by the Said, whilst the Said is in turn betrayed by the Saying. And so on—ad infinitum. The violence of divine love bursting forth in the exposure to the destitute countenance of the Other, due to the structural inherence of the Third Party (*le tiers*), finds itself inevitably betrayed by the exigency of meting out distributive justice to all such parties. Yet, these same third parties are themselves in turn inevitably betrayed by the recrudescence of the Saying's diachrony which necessarily undercuts distributive justice's accordance of justice to all equals—or, at least, potentially so.

How does all this augur for justice? Justice for Lévinas, pithily stated, is the "ethical relation to the Other"; that is to say, ethics as pre-distributive justice. Yet, as readily observed, (pre-distributive) justice—inhabited as the ethical relation to the Other primordially is by an inherent Third Party (*le tiers*)—is from the very outset in peril of truncation by the subordination of Saying to Said. Otherwise put, justice as the relation to the Other is inevitably—and indeed immediately—betrayed by its very own exigencies. Yet, the call for distributive justice to be meted out to all third parties is, in turn, itself rendered inherently impossible by the inevitable return of the Saying and its disruptive diachronic force. Thus, justice—both in its "pure" originary (pre-distributive) sense as the relation to the Other and in its distributive modality—inevitably founders. The Said betrays the Saying, while in return the Saying betrays the Said in the unsaid.

Yet, being the originary locus of subjectivity/selfhood, the responsibility to the Other is irremissible. Far from being in its most primordial determination, as Sartre would have it, a *pour-soi* ("for-itself"); as observed, the subject/self is quite contrarily *pour l'autre* ("for-the-other"), bound to the Other and responsible for its deeds and misdeeds. However, in that the Third Party is likewise inherent to this same primordial locus of selfhood, the exigency of the Self is at once that which presages distributive justice. Hence, there must be both the Saying and the Said. But this is "impossible," as these two mutually exclude and betray each other. Thus, responsibility to the Other inevitably throws the subject into an irresolvable paradox, placing as it does the Self on a path or way *(poria)* which cannot be traversed and yet which must be traversed. That is to say, a "pathless" path. Thus, Rosenzweigian and Lévinasian ethics are fundamentally *aporetic*. It is an ethics entailing an utterly impossible task, the exigency for which nonetheless remains fully and interminably in force. The herein entailed aporia stems, at base, from a complex imbrication of two fundamentally heterogeneous orders of discourse: as again, that of the Saying and that of the Said. It may be added that, though Saying and Said are chiasmatically imbricated, they cannot hereby be considered simultaneous. They do, to be sure, occupy the same instant. This instant is, however, nowise a punctual and self-identical present. It is, rather, a paradoxical instant which—in order to accomodate the two incommensurable orders of the Saying and of the Said—must itself be *internally distended*. Such an instant is diachronic: it is, as Simon Critchley argues, "the punctual present falling out of phase with itself."[155]

It is precisely such a paradoxical instant which the ethical subject occupies—suspended as it is within the vertiginous chasm between Saying and Said: between the universality of law, the nonequivocation of the thematic, and the phenomenal, on the one hand, and the singular sub-apparition which founds and likewise suspends the latter, on the other.

A strikingly similar aporia obtrudes itself in Blanchot. As observed, the Lévinasian Saying was seen to be strongly homologous to the movement of poetry along the second slope of literature, whilst the Said markedly resembled that of prose along the first slope of literature. It becomes readily apparent that these two Blanchotian slopes mutually contaminate and imbricate themselves and that, quite as in the case of the Saying and the Said, their relation is inherently chiasmatic. On the one hand, poetry desires to reveal what is anterior to the prosaic and punctual order of the merely apparent. This however fails, since the poem is itself inexorably a revelation and thereby at once conceals what it attempts to reveal—for things-in-themselves are not of the order of the manifest, the ostensive, the radiant. The poet has the Midas touch, which simulataneously renders things precious and kills them. This second slope of literature likewise entails a desire to fuse with the sheer materiality of things in a total subjectlessness, a pure night. The first slope of literature, on the other hand, wants to subsume all of reality within the radiant and unadulterated daylight of subjective consciousness through the incessant labor of the negative. Blanchot, however, avers that both of these attempts are inevitably thwarted: "By negating the day, literature reconstructs the day and fatality; by affirming the night, it finds the night as the impossibility of the night. This is its discovery."[156] That is to say, the second slope of literature desires the night, a pure night, but only runs up against the impossibility of such a blissful night of peaceful repose and extinction. Instead of a rapturous poetic merging or fusion with the night of pure unconsciousness, consciousness's ability to sleep is held in check by what Blanchot terms the essential or *other* night that is none other than the impossibility of sleep. Rather than granting refuge in soporific narcosis, the night is the neutrality of fatigue.

The first slope of literature, on the other hand, desires a total grasp of the day, as a world of absolute freedom, but discovers the day not as freedom but as fatality. This is the insomniac's experience of the day: the day stupefied by lack of sleep, the day as something to which one is riveted, what Blanchot calls the madness of the day. Literature is therefore the discovery of the world of the insomniac, as the double impossibility and double necessity of the day and the night. Writing resides in or is shared between the two slopes of literature that are simultaneously necessary and impossible. The truth of literature thus consists in the experience of being suspended between day and night, of watching with eyes open in the night, of eyes stupefied by the specter of insomnia in the day. The experience of literature is thus *aporetic*.

Nonetheless, Blanchot goes quite a bit further than this, tracing affinities between the two slopes, on the one hand, and the work, worklessness, and—most importantly—law, on the other. Let us explore this. The first slope of literature—avers Blanchot—is the province of law and of the work, of that which is complete, punctual, ostensible, manifest. The movement of literature, in writing, however, has its source in the infinite movement of dying along the second slope of literature—a movement which is also one of worklessness and expropriation: "To write is to produce absence of the work (worklessness)."[157] The first slope of literature, however, inevitably supervenes upon this movement, replacing the vertiginous energy of exteriority with the externality of law, ensconsing worklessness within the confines of the work, incarcerating writing within the covers of the book.

However, the second slope makes its return, for argues Blanchot, "writing is the absence of the work as it *produces itself* through the work and throughout the work," and "passes through the book, but the book is not that to which it is destined (its destiny)."[158] Blanchot therefore speaks of the book as "a ruse by which writing goes towards the *absence of the book*" and concludes that the work hesitates "between the work as presence and the absence of the work that is constantly escaping and in which time as time is disturbed."[159] There is an interminable hesitation between the two slopes of literature, an *aporia*.

Writing also holds a certain relationship to law. Of the former Blanchot avers: "Writing, (pure) exteriority, alien to every relationship of presence, and to all legality."[160] This is the second slope. However, "In the book, exteriority" authorizes itself "under the space of Law. The exteriority of writing, spreading itself out in layers in the book, becomes exteriority as Law."[161] Law establishes a mediate relation to the pure exteriority of writing. This is the first slope. Yet, this mediation and establishment of legality finds itself perpetually visited upon and disturbed by writing, since "As we write, we always write in the name of the exteriority of writing and against the exteriority of the law."[162] There is hence a perpetual vacillation between the exteriority of writing and Law.

12. The *Akedah*

The uneasy chiasmatic intrication of the two slopes of literature in Blanchot—as equivalently that between the Lévinasian Saying (*le dire*) and Said (*le dit*), and that between Rosenzweigian *sprechende Sprache* and *unausgesprochene Sprache*—reappears in a very similar form in Kierkegaard's discursus on the binding of Isaac.

John D. Caputo views the deleterious—yet ultimately salvific—vicissitudes narrated in the twenty-second chapter of the book of Genesis as "a story of madness, of a mad economics, an aneconomics, a radical and literal case of death-dealing in an economy of sacrifice," a maddening sequence of actions "defying the most elemental command of ethics and the law, which is not to deal in death, above all—God forbid—with one's own son."[163] "Abraham deals in death; he courts death without flinching. When death calls—or God, God or death—Abraham does not blink. God or death: either way, *tout autre*. The *tout autre* under one name or another: God or death."[164] Abraham's averted sacrifice of his own son at the bidding of God entails what Kierkegaard terms the "teleological suspension of the Ethical" and the paradoxical character of Faith inherent to such a vertiginous suspension.

The Kierkegaardian "Ethical" is not the same as what Lévinas terms the "ethical relation to the Other"—and, by implication, the ethical Saying. Rather, as

Caputo maintains, it is "the concrescence of concrete ethical and historical subsistence," and is thus homologous to Kantian *Moralität* as the universalizability of the law—and likewise to Hegelian *Sittlichkeit,* conceived as the concrete ethical community, as the *polis*, together with the eudaimonic teleology herein implied.[165]

Kierkegaard conceives the Ethical as the "universal," in the sense that it applies to all human beings, and does so at every moment. The Ethical is strongly marked by a certain self-sufficiency and self-sojourning, which resides in the fact that it is endemically its own *telos*, whereas even that which purportedly falls outside its aegis notwithstanding finds its *telos* in it by virtue of the Ethical's inherent tendency to incorporate such exteriority into itself. More specifically, it is the particular individual who has his *telos* in the Ethical, in the sense that he is charged with expressing himself in the universal so as to neutralize his own particularity and "become the universal." On the contrary, from the vantage-point of the Ethical, any movement by the individual whereby he would strive to assert his own particularity over and above the universal is immediately deemed recalcitrant and tantamount to "sin." In fact, the very impulse on the part of the individual to rise to such self-assertion after hitherto having securely ensconced himself in the universal leads him into temptation [*Anfechtung*]—a predicament that may be assuaged only through his renewed penitential self-abandonment in the Ethical as the universal.[166]

In that its inherent tendency is to subsume particularity as a mere function of itself, the Kierkegaardian Ethical bears upon the singularity of a human individual in a manner homologous to the subsumption of the absolutely singular and idiosyncratic pre-ontological speech of the Other—whether this be the Lévinasian Saying (*le dire*) or Rosenzweigian *sprenchende Sprache*—by the ontological order of the Said (*le dit*) or of *unausgesprochene Sprache*, through the mediation of an implicit third party. Common to all of these is the elision of singularity by the generalizing proclivities and dictates of thematization and manifestation.

For Kierkegaard, it is the tragic hero—not to be conflated with the tragic hero whom Rosenzweig exalts as the prototypical embodiment of Character—who moves squarely within the ambit of the Ethical. The tragic hero does so, in the sense that any departure on his part from the Ethical as the universal, though at the moment of divergence rightfully deemed unethical, notwithstanding ultimately finds itself resolved when the hero esconces himself in higher levels or expressions of the Ethical. Kierkegaard: "The tragic hero still remains within the ethical. He lets one expression of the ethical find its *telos* in a higher expression of the ethical; the ethical relation between father and son, or daughter and father, he reduces to a sentiment which has its dialectic in its relation to the idea of morality."[167]

Vasts apart from the Ethical, Kierkegaard delineates the utterly refractory notion of *Faith*. What looms here is the Kierkegaardian Religious register. Crucially, the relation between the Ethical and Faith is not one of a contrary movement of two externally opposed terms. Rather, it is a chiasmatic imbrication of mutually intricated terms. Faith, for Kierkegaard, resides in the strange fact that the individual as the particular, far from occupying a lower and subordinate position to the universal, is in fact exalted above it. That is to say, the individual in his singularity is higher than, superior to, and justified over and against the universal. Notwithstanding, this elevation of the singular individual to a higher plane than that of the universal, paradoxically, comes to pass by way of his having hitherto been subordinated to the universal. Through the universal, he is extolled above the universal—to the extent that he now stands in a direct and absolutely vertiginous relation to the Absolute. His standing in an "absolute relation to the absolute" places him wholly outside the Ethical as the universal, and is therefore sustained without recourse to mediation—for mediation falls exclusively within the province of the universal. This eternal paradox—that of Faith—is, moreover, outside all discursive thought.[168]

What we thus have here is that, firstly, Faith consists in the elevation of the individual as the particular above the realm and reach of the universal into a position in which he stands in direct relation to the Absolute. In the process, the individual comes to assume a position intrinsically recalcitrant to the strictures of mediation—in that mediation is wrought by and holds only for the Ethical as the universal. The Lacanian analogue of Faith would be the dereliction of the interpellating big Other and consequent direct and terrifying exposure of the subject to the Freudian Thing, or *das Ding*, in a vertiginous encounter with the Real. In Lévinasian terms, Kierkegardian Faith would be that which thrusts the individual outside the mediation of the Said (*le dit*)—which latter otherwise served to elide the singularity of the pre-ontological Saying (*le dire*) through its subsumption by ontological discourse—and into a direct and untrammeled exposure to the face *(le viage)* of *illeity* as the Neighbor. In Rosenzweigian terms, something homologous to Kierkegaardian Faith is prepared by the forcible and untimely upsurge of Character, as subtracted from the vapid third person relations and prosaic generality of the quotidian social world—and arrived at with the birth of the Soul from out of the meta-ethical Self, by way of the violent solicitations of divine love.

Secondly, however, it must not be forgotten that—in the Kierkegaardian conception of Fatih—this superiority of the individual (standing as he does in direct relation to the Absolute) to the universal is, precisely and at once, wrought *through* the intermediary and mediation of the universal. This is precisely the paradox to which Kierkegaard alludes. That is to say: *through* mediation, the individual as the particular finds himself outside all mediation; *through* the universal, the individual is beyond the universal; *through* the Ethical the individual exceeds the very ambit of the Ethical.

Thus, it may be contended that the individual as the particular is only able to reside at some indeterminable point *between* the Ethical, on the one hand, and the Beyond of the Ethical, on the other; that is to say, in the paradoxical zone of dehiscence twixt the foregoing two—or, conversely, *at once* in the Ethical *and* outside of it.

We may also rephrase the paradox as follows. As Žižek maintains, on the one hand, for the individual to be within the grip of vicissitudes that are truly and unconditionally ethically binding, his ethical engagement must rely on an accepted *doxa* (this latter being in the province of the Kierkegaardian Ethical register). On the other hand, however, these doxic ethical norms (which serve to constitute the Ethical as the universal) ultimately and imminently show themselves to be irrecovably insufficient to ground a proper ethical disposition (and, still worse indeed, may well serve as the very source of an atavistic fall *beneath* the Ethical, to inauthenticity); which implies that the Ethical cannot attain fulfilment from within its own ambit alone.[169] Hence, it is necessary for the Ethical to suspend itself—that is to say, betray itself—so as to consummate itself in the first place. In Lévinasian terms, this ineluctable suspension of the Ethical would assume the form of the untimely recrudescence of the Saying (*le dire*) within the Said (*le dit*), effectuating the temporary dereliction of the alignment of the Saying upon the Said and doing so in the form of a stubborn return of skepticism—outside of which return, the ironclad conjunction of these two heterogenous orders of discourse would degenerate into tyranny.

But there is more. Not only does the Ethical inevitably fail to fulfill itself from within the ambit of its own parochial resources; but, as the normative dimension that it itself is, the Ethical is always already, from its very inception, a crude truncation and betrayal of the singular relation in which the individual stands before the Absolute—a betrayal of "the Otherness from which every ethical call/injunction emanates."[170] Of course, it is not the case that—in the Religious—the Ethical is *as such* definitively consummated in some kind of "higher" dialectical synthesis; rather, the vacillation between the Ethical and its suspension continues indefinitely. Thus, avers Žižek, the Kierkegaardian Religious register—far from being a mediating synthesis of the Ethical and the Aesthetic—is instead "the radical assertion of the parallax gap (the 'paradox,' the lack of common measure, the insurmountable abyss between the Finite and the Infinite)"; the immediate upshot of which is that one is "never safely within the Religious," and therefore that "doubt forever remains," and that "the same act can be seen as religious or as aesthetic, in a parallax split which can never be abolished, since the 'minimal difference' which transubstantiates (what appears to be) an aesthetic act into a religious one can never be specified, located in a determinate property."[171]

The paradox in question (around which this disconcerting vacillation transpires) is also one that exceeds the bounds of understanding; for understanding is a form of mediation between the object bethought and the concept in which the object gleaned from the manifold of perception finds its repose (and wherein it is thus cognized by being re-cognized, thought as that which it "is" in its annexation under the taxonomic categories of the Kantian Understanding (*Verstand*)—a subsumption whereby it constitutes itself as an object in the first place). This paradox thus leads Kierkegaard to concede that "I can understand the tragic hero but cannot understand Abraham, though in a certain crazy sense I admire him more than all other men."[172] This disparity between the hero of Greek tragedy and Abraham recalls a similar comparison earlier seen to be made by Rosenzweig between Character, as prototypically exemplified by the tragic hero, and the mystic. Here, Rosenzweig had maintained that—though the hero as meta-ethical Self, in all the arctic solitude that endemically pertains to Character, was marked by an acutely obdurate solipsistic refractoriness to the world—he was, notwithstanding, still *visible* to society as the spectable that he was. Through this visibility, the tragic hero—in spite of his headlong and stubborn defiance of the order of the world—still remained to a certain degree calculable from the vantage-point of the world. In vast contradistinction, if we assimilate Abraham—at least in part—to the mystic as delineated by Rosenzweig, then this aspect of visibility forthwith gives way to a rather obscure spectrality, to concealment, secrecy, and other such descriptive terms.

All of these our peregrinations of course bring us now to the torturous ordeal of Abraham itself, he who *primus inter pares* finds himself precariously abandoned to the maddening zone of indeterminacy between the Ethical and its abrogation—an interstitial desertion that is self-identically and from the very outset the paradox of Faith itself.

But what precisely is Abraham's vertiginous *aporia*? On the one hand, from the perspective of the Ethical, conceived of here as including inter alia the Decalogue, Abraham is *stricto sensu* in his heart—and thus for all intents and purposes—a murderer. He, after all, violates, at the level of intention, the commandment forbidding killing—sixth according to the Septuagint, Talmud, and the Reformed Christians; and fifth according to the Samaritan Pentateuch, St. Augustine, Roman Catholicism, and Luther. On the other hand, from the vantage point of Faith, however, Abraham's actions regarding Isaac stem from an absolute duty toward a utterly incalculable God who categorically demands the sacrifice in question. That is to say, from the sedentary purview of the Ethical, Abraham's actions are tantamount to murder—whereas, from the far more vertiginous vantage-point of the Religious, we instead speak here of "sacrifice."

But there is indeed more. For, in that the universal—in the same sense as may be maintained of the Lévinasian Said (*le dit*) and the Blanchotian first slope of literature (that of the Sadeo-Hegelian murder of the thing)—is also the arena of the *manifest*, whilst the paradoxical Faith of Abraham skirts the very ambit of the latter; Abraham nowise moves within the domain of the apparent, the revealed, or the ostensive. Rather, his prerogatives are *concealment*, silence, and *secrecy*—which are precisely certain of the salient aspects of the Deleuzoguattarian passional regime, and likewise those of the Lévinasian Saying (*le dire*), Rosenzweigian *sprechende Sprache*, and the crepuscular viscosity of Blanchot's second slope of literature, or axis of interminable dying, in which things are left to abide in their own nocturnal silence and anonymity. In stark contrast, the tragic hero of Aeschylus and Sophocles—ensconcing himself as he does in the universal—must make his Ethical courage manifest. Kierkegaard: "The tragic hero displays his ethical courage precisely by the fact that it is he," that is, Agamemnon, "who himself announces to Iphigenia her fate. If the tragic hero does this then he is the beloved son of ethics in whom it is well pleased."[173]

Abraham, in vast contradistiction, is not even able to properly *enter* the realm of the ostensive. This is so for two distinct, though related, reasons. Firstly, in regard of the binding of Isaac, God himself retains well-nigh sempiternal secrecy about the appalling sacrifice to be brought about. This is so because he abstains from giving any cogent reasons for it, remains perfectly silent and withdrawn, invisible to Abraham. This calls to mind Deleuze and Guattari's averment that in the passional regime God averts his countenance, precipitating the subject into roiling dread. On the other hand (this is the second reason), Abraham himself maintains the strictest secrecy. He discloses his (and likewise God's) secret to nary a soul: not Sarah, not Eliezer, let alone Isaac. Indeed, this secret is one that, avers Derrida, Abraham cannot but keep: above all since "he doesn't know it, he is unaware of its ultimate rhyme or reason. He is sworn to secrecy because he is in secret."[174] Abraham cannot but retain his cosmic and vertiginous secret because this secret is, more primordially, that of his God. Indeed, God is in secret before Abraham insofar as the latter is riven by a mere unsubstantiated injunction, a pure, unadulterated command enjoined upon Abraham in the fullest fulguration and violence of Rosenzweigian divine love.

But let us pursue the further corollaries of the chiasmatic entwinement of the Ethical and the Religious. How does the tragic hero express the universal in his actions? Through speech. For instance, Agamemnon's announcement to Iphigenia of her pending death is an expression of the superiority of State law to the ethical relations governing the family within the latter's own narrower and far more circumscribed ambit. In this connection, Kierkegaard maintains that the tragic hero would unfailingly be condemned by ethics should he resolve upon maintaining silence and residing within the equivocal, for it is precisely through one's own speech that one acknowledges the Ethical as the universal—and must do so.[175] In the universal, everything is manifest—and manifested through speech. This includes all the counter-arguments issued against the tragic hero by other protagonists and antagonists, and even the chorus of Aeschylean drama. The hero of tragedy has given due consideration to all of these, "to every living being, to every voice from the heart of humanity, to every cunning, every alarming, every accusing, every compassionate thought," and thus "can be sure that everything that can be said against him has been said, unsparingly, mercilessly," unambiguously—and sans prevarication.[176] That is to say, the entire striving of the tragic hero and the counter-striving of his foes against him is always expressed in ostensive and unequivocal form. Even when in distress, the tragic hero can still seek the comfort of tears—for a protagonist such as Agamemnon is able lament with his wife Clytemnestra and daughter Iphigenia, and hereupon draw his heretofore bedraggled self together so as to galvinize himself for imminent, forward propelled action.[177]

The predicament of Abraham, on the other hand, is vasts apart. Unlike the strife of Agamemnon, Abraham's trial and tribulation cannot be categorized as a case of courageous masculine struggle against stolid antagonists whose contrary positions are well delineated through the mechanism of speech and other forms of manifestation. Abraham moves at once outside the zone of unequivocal menace and that of stable sources of comfort, whether the latter be tears or the higher Ethical law of the State. Instead, as Caputo maintains, Abraham's inundation by the violent irruption of divine solicitation "cuts him off from all *Sittlichkeit*, from the reassuring community of practices that shapes communal life, from the family, society, nation. The deal Abraham cuts with God cuts Abraham off from his *oikos*, from his son and from Sarah. Sacrifice is strictly men's business, between God and him (not to mention Isaac). Sarah is being cut off, sacrificed no less than Isaac (not to mention the ram); the whole family, the most immediate sphere of *oikos* and *Sittlichkeit*, will be severed and sacrificed by this secret."[178] More precisely, since the paradoxical and immediate relation to the Absolute is one which unavoidably skirts the ambit of the universal, quite unlike the tragic hero, Abraham is terminally and intrinsically unable to express the latter. He is condemned to remain sempiternally silent. He *cannot speak*.

Two things may readily be mentioned about this silence. Firstly, as Kierkegaard maintains, "silence is also the mutual understanding between the Deity and the individual."[179] That is to say, endemic to the vertiginous and direct relation in which the individual (as the particular) stands before the Absolute—is the absolute necessity of silence. Secondly, the fact that Abraham does not speak ("he did not speak to Sarah, nor to Eleazar, nor to Isaac, he passed over three ethical authorities") may readily be adduced from the simple fact that "the ethical had for Abraham no higher expression than the family life."[180] Thus, all of Abraham's dealings further afield than the immediate ambit of his familial constellation would immediately fall strictly outside the Ethical. Such was, therefore, his relation to the Deity—a vertiginous relation bereft of ostentation and speech, conducted in a crepuscular dusk under whose carapace Abraham remained concealed.

In this quite formal sense, Abraham may be seen as forcibly reduced to a condition remarkably akin to *aphasia*—another feature proper to the Deleuzoguattarian passional regime, and a condition ubiquitous among God's semiophagous prophets. This aphasia is, however, most strange. For, in a fairly palpable sense, far from exhibiting the well-nigh catatonic aphasia of the mute, Abraham does indeed say quite a lot. Words do undoubtedly gush forth from his oral aperture. Indeed, Kierkegaard notes that Abraham, in fact, "is able to utter everything," "is able to say the most beautiful things about how he loves Isaac. But it is not this he has at heart to say, it is the profounder thought that he would sacrifice him because it is a trial. This latter thought no one can understand, and hence everyone can only misunderstand the former."[181] Thus, though Abraham is constantly speaking, the fact that he fails to make himself understood—indeed, *cannot* make himself understood—is tantamount, at least at the level of the Ethical, to his *not speaking at all*. More precisely, he does not speak *by* speaking, obfuscates and conceals *in the mode of* a corrupted/disrupted verbal form of pseudo-manifestation and pseudo-ostentation.

This returns us to and confirms our earlier point wherein it was contended that, though Faith consists in the individual standing outside the Ethical as the universal in a direct and unmediated relation to the Absolute, this latter vertiginous relation is wrought *through* the universal. Hence, we have a peculiar intertwinement of the universal and the Absolute, with Abraham suspended vertiginously between the two (if, that is, one can speak of such an "in-between"). This interstitial suspension is most poignantly in evidence in the immediate run up to what would otherwise have been a tragic denounement, that is, at the moment atop Mount Moriah when Isaac asks Abraham about the whereabouts of the lamb to the sacrificed. Here, Kierkegaard maintains that if Abraham had replied unequivocally to Isaac that Isaac was in fact the sacrificial lamb, then he would have suspended the paradox by sinking it back into the manifest light of the universal. To be sure, Abraham does in fact reply to Isaac. He does so, notwithstanding, in the form of irony, that is, by saying something whilst at the same time not saying it.[182] Abraham's reply is: "God will Himself provide the lamb for the burnt offering, my son."[183]

There is a deep ethical exigency in Abraham's aphasia—not in the sense of the Ethical as the universal, but in the sense of the vertiginous relation to Other than fulgurates in the Lévinasian Saying (*le dire*) and in Rosenzweigian *sprechende Sprache*. Derrida's contention in this regard is that Abraham cannot speak (or, at least, is unable to thematize his actions in unequivocal linguistic form)—precisely because, by way of an inevitable and forcible ensconcement within the universal (the Kierkegaardian Ethical register), speaking as discourse *necessarily violates and occludes* the singularity of other persons and their decisions. Thus, so as to maintain fidelity to his vertiginous responsibility, Abraham cannot but act in accordance with that which renders his individuality irrevocably free and responsible: his absolute singularity or *jemeinigkeit.* He does, in fact, respond to the vertiginous call of divine love. His response, however, categorically fails to adequately thematize his take on the unthinkable secret which must forever remain his and God's alone. The peculiarity of his aphasia consists in the paradox of his both responding and not responding, of speaking without speaking. More precisely, Abraham, by speaking "without saying anything either true or false, says nothing determinate that would be equivalent to a statement, a promise or a lie ..."[184] Here, quite paradoxically, Abraham's secret is maintained precisely *through* its being spoken.

Such a conclusion would be flatly nonsensical if it were not for the fact that the Abrahamic secret is not a secret in any conventional sense. The peculiar nature of Abraham's secrecy resides in the fact that it—as Derrida would have it—nowise "hides itself forever in an indecipherable crypt or behind an absolute veil"—but quite contrarily "exceeds the play of veiling/unveiling, dissimulation/revelation, night/day, forgetting/anamnesis, earth/heaven, etc."[185] As with the "face" (*le visage*) of the Lévinasian Other, secrecy conceived in the foregoing sense is neither of the order of phenomenality nor negatively opposed to the latter. Rather, it is *ab initio* fundamentally heterogenous to and infinitely older than these orders and likewise exceeds the ambit of their binary opposition. Thus, the fact that one "can always speak about it is not enough to disrupt it. One can speak of it ad infinitum, tell stories about it, utter all the discourses which it puts to work and the stories which it unleashes or enchains"; "And the secret will remain secret, mute impassive"; "*It is not more in speech than foreign to speech* [my emphasis]."[186] Exceeding as it does the opposition between manifestation and concealment, the Abrahamic secret is such that perorations on it analogously exceed the polarity between speech and non-speech, allowing Abraham thereby to speak without speaking, to utter his secret without uttering it, and for all intents to hide in plain sight—a sort of purloined letter. *This* is the nature of his most bizarre aphasia.

To be quite sure, Abraham's intersticial position between the opposed extremes speech/non-speech, manifestation/concealment, *Moralität-Sittlichkeit*/singularity makes for a maddening equivocation, the inexorable indeterminacy and vacillation of which is truly awful. Indeed, the tragic hero, though he may even struggle against the entire world, does not know the "terrible responsibility of solitude" endured by Kierkegaard's Abraham.[187] In the case of the former, the "tears and cries" to which the tragic hero recurs are "assuaging."[188] The "unutterable sighs" of Abraham are, however, "torture"; the thankless conclusion being that "to strive against the whole world is a comfort, to strive with oneself is dreadful."[189]

What, however, as Caputo maintains, is most crucial in Abraham's "unresponsiveness" is his fidelity to the Absolute; a responsibility, the precise character of which "absolves" him from both the abstract Kantian universal (*Moralität*) and the concrete Hegelian universal (*Sittlichkeit*). In point of fact, within the context of the singular and vertiginous relation in which he stands to the Absolute, a rectilinearly forthright collusion with the strictures of the Ethical would immediately render Abraham irresponsible. Were he to abide by the dictates of ethical responsibility and give a reason (*rationem reddere*) for his actions by forthrightly responding to his wife and to the community around him, he would automatically betray his absolute responsibility to God. For discharging one's duty merely relates one to a univeral ethical principle, whereas the responsibility to God is absolute—and far exceeds duty. This elevates Abraham beyond the Ethical, beyond simple duty—allowing him to transcend both *Moralität* and *Sittlichkeit* through the Religious, which is "the absolute relationship to the absolute" and "the singular relationship to singarity."[190]

Yet, the price of this fidelity to the Absolute is a betrayal of a different kind. By not speaking the secret, Abraham transgresses the Ethical order—the most profound expression of which is in terms of that which binds one to family, community and State. For the Ethical intrinsically demands manifestation and strict disambiguation, and nowise a singular subversion of thematic speech. He likewise betrays his own son, since his endless equivocation in no sense serves to save Isaac. Through his resolution to contravene the sixth commandment, he moreover betrays God—by infracting against divine law. On the other hand, temporarily abrogating his own commandment, God betrays Abraham, man in general—and, indeed, himself. God betrays man, who in turn betrays God and man, and thereby God again. Crucially, this betrayal is not an accidental, supernumerary, or supervening feature. Nor is it an externally visited malignancy attendant upon an otherwise responsible course of action. Rather, it is a betrayal internal to responsibility itself—in a regime, the passional semiotic of subjectivity, in which secrecy, concealment and betrayal are inherently structural and absolutely ineluctable.

Abraham's betrayal is *necessary.* For, he must heed the pre-thematic divine injunction fulgurating in the countenance of the Other—the too-muchness-of-address that inexorably breaks with parochial law (even with *God's* own law as hitherto thematized in the Decalogue). This is doubtless what occasions Derrida to distinguish between two distinct forms of responsibility; that is to say, between responsibility *in general* and *absolute* responsibility. The *aporia* wrought by the sempiternal chiasmatic coexistence of these two is inherent to the very nature of responsibility itself. Derrida: "For responsibility (we would no longer dare speak of 'the universal concept of responsibility') demands on the one hand an accounting, a general answering-for-oneself with respect to the general and before the generality, hence the idea of substitution, and on the other hand, uniqueness, absolute singularity, hence nonsubstitution, nonrepetition, silence, and secrecy"; "This is ethics as 'irresponsibilization,' as an insoluble and paradoxical contradiction between responsibility *in general* [justice as law] and *absolute* responsibility [justice as the aneconomic incalculability of the gift]."[191]

Abraham is not a mere tragic hero such as Agamemnon, for whom it suffices—securely and with ostensive tears—to irrevocably esconce himself within the radiant light of the Ethical. For the necessary secrecy of absolute responsibility "declines the autobiography that is always auto-justification, *egodicity.*"[192] Egodicity is precisely what would return the gift to the ego, to the economy of narcissism, to the gift's perdition. The Abrahamic paradox of the knight of faith, on the other hand, consists in the fact that "interiority remains 'incommensurable with exteriority.' No manifestation can consist in rendering the interior exterior or show what is hidden."[193] Such nondisclosure, as observed, betrays the Ethical.

Yet, this is not all. For the betrayal of the Ethical entailed within the Kierkegaardian Religious likewise implies a necessary *hatred* on the part of the betrayer. This is not, however, a hatred of what is properly-speaking hateful. Indeed, Abraham must come to hate what he *loves*. Derrida: "I must come to hate what I love, in the same moment, at the instant of granting death. I must hate and betray my own, that is to say offer them the gift of death by means of sacrifice, not insofar as I hate them, but insofar as I love them. I must hate insofar as I love them."[194] Furthermore and more pointedly: "Abraham is faithful to God only in his absolute treachery, in the betrayal of his own and of the uniqueness of each one of them, exemplified here in his only beloved son. He would not be able to opt for fidelity to his own, or to his son, unless he were to betray the absolute other: God, if you wish."[195]

Fidelity to the call of the Other fulgurating in the seething violence of divine love requires of Abraham that he betray the law of God and of man. As espied a little earlier, fidelity to the Absolute and that held with regard to the Ethical nowise constitute two externally opposed moments ready for some sort of crude sub-Hegelian sublation. Rather, their relation is that of an inwardly involuted chiasmus of law and justice, of economy and aneconomy—wherein, in the paradoxical instant, along the very same vector that at once unites and likewise separates Abraham and God, Abraham betrays God in heeding his command and heeds God in betraying him.

13. Law, force and justice

This chiasmatic relation between, on the one hand, law—as we have elaborated within the framework of the Kierkegaardian Ethical as the universal, the Lévinasian Said (*le dit*), or Rosenzweigian *unausgesprochene Sprache*—and, on the other, justice—as the absolutely singular Abrahamic relation to the Absolute, the Saying (*le dire*) unfolding in the limitless exposure to the evanescent passing of *illeity* as the Neighbor, or the solicitations of *sprechende Sprache* fulgurating in the violence of divine love—form a large portion of the focus of Derrida's essay *Force of Law: The "Mystical Foundation of Authority."* Here, the preponderant preoccupation is with what he sees as the unfailingly awkward relation between law (in the sense of *droit*) and justice, but, moreover, how these relate to *force*—and, ultimately to violence. In an initial gloss Derrida contends that, though justice is not to be conflated with law, it can notwithstanding nowise be thought outside of its inevitable relation to the latter. What, as we have already observed, appeared in Lévinas as the forcible subsumption of the anterior ethical Saying (*le dire*) under the punctual auspices of the ontological Said (*le dit*), in Derrida figures under the rubric of the "mystical foundation of authority." This subsumption/conflation is taken by Derrida to be absolutely originary. In and through this primordial *violence*, there comes to pass a forceful conflation of pre-distributive justice and justice articulated as law—which, in the very same movement, establishes the "truth of justice." More precisely, it is the moment of establishment of justice as *droit*, that is to say, as law.[196] This performative force at once founds, justifies, and institutes the law. Force and law are here, however, nowise external to one another. For, in this originary *coup de force*, law is in no sense a mere instrument of force, whereas force is at once a force of interpretation—which at once furnishes law with its dominant interpretation. Therefore, force and law maintain a far more internally intertwined and non-trivial relation with one another: a complex imbrication of law, force, power, and indeed violence.[197]

This violent founding act is precisely the "mystical foundation of authority" mentioned at the outset, whereby justice as *droit* is established. It is an originary violence in which force and law are mutually imbricated and in which force is at once the force of interpretation of law and that which violently founds the law. This origin is, moreover, groundless, since—as the foundation of authority and positioning of law—it cannot rest upon any ground other than itself, and thus—much like the infinitely primordial and semptineral Schellingian pure divine willing from whence God has always already arisen as subject—functions only as an *Un-/Ur-/Ab-grund*, or groundless ground of the "truth of justice."[198]

This originary revolutionary founding moment of the law is precisely what Walter Benjamin terms "mystical violence"—an inundation of non-law in accordance with which the very foundation of law finds itself suspended over the abyss by a pure performative act that is itself in no sense answerable to any legal instance, or "before a law not yet determined, before the law as before a law not yet existing, a law yet to come, *encore devant et devant venir*."[199] This groundlessness of the originary violent institution of the order of legality, of justice as *droit*, means that it is itself in excess of the binary distinction legal/illegal—since its originary character by definition precludes any anterior law-founding origin according to which the originary founding act could itself be adjudged as either legal or illegal. Moreover, in that the "truth of justice" only finds itself established through the violent founding of the order of legality, this originary performative is likewise refractory to the opposition between just and unjust.

What must now be pursued is how this relation between justice *as* law (*droit*) and force, bodes for justice *as such*. Firstly, Derrida declares from the outset that justice is certainly *not* law. For law (*droit*)—much as in the case of the ontological and discursive order of the Lévinasian Said (*le dit*), Rosenzweigian *unausgesprochene Sprache*, the Kierkegaardian Ethical register, and Blanchotian's first slope of literature—is defined by calculation, in the sense that it has to do with the stolid application of a sound rule to a singular case that it co-opts as an instance to be correctly subsumed in accordance with a determinate judgment. If this verily comes to pass, then law may well be accounted for—but not justice. For justice is—as may be said in regard of the Kierkegaardian Religious register, Lévinasian Saying (*le dire*), Rosenzweigian *sprechende Sprache*, and the Blanchotian second slope of literature or axis of interminable dying—not *sensu stricto* of the order of calculation. Rather, it requires that one calculate with that which is fundamentally incalculable—something which Derrida attributes to aporetic experiences: that is to say, experiences that—and we may add, much as does the untimely soliciting fulguration of divine love—at once shatter the framework of the possible and impinge upon the subject with the inexorable force of necessity. Such vertiginous experiences countenance one with the irremissible injunction to decide between the just and the unjust without recourse to any rule whatsoever—or, in Lacanian terms, without the cover of any figure of the big Other, and likewise sans the manifest discursive edifications and measures overwise afforded by the Kierkegaardian Ethical.[200]

So, whereas justice as law (*droit*) pertains to determinant judgments related to rules, justice *as such* concerns itself with the absolute singularity of irreplacable individuals and groups alike as these are encountered in situations that are in each case unique. The uneasy conflation of justice and law wrought by the originary violent foundation of the order of legality, thus, foists upon one the ineluctable necessity of applying general rules to the singular. Such an application, of course, necessarily violates singularity, in that this latter—being *sui generis*—is of its very nature refractory to the order of generality. We are thus countenanced with the necessity of doing something which is inherently impossible—and therefore led to a veritable impasse, an *aporia*, a way that is blocked. This impasse—spawned as it is by the inexorable call of justice—would, as Derrida maintains, be obviated if one was able to address the Other in the Other's own idiosyncratic language.[201] Such circumvention is, however, impossible since one is only able to speak the language of the Other by assimilating and appropriating it by recourse to the "law of an implicit third," that is to say, through an "appeal to a third party who suspends the unilaterality or singularity of the idioms."[202]

Evidently, Derrida is here using Lévinasian terminology, in the sense that the anterior pre-ontological and pre-discurve ethical Saying (*le dire*) in the vertiginous exposure of what one may take here take as the Rosenzweigian meta-ethical Self to the fulgurating divine violence of the Other can only appear in the apophatic language of ontology as the Said (*le dit*)—wherein it is at once betrayed, its idiomatic singularity elided by the turgid and prosaic generality of discourse itself. On the other hand, however, the Lévinasian *le tiers* (the third), who is the Other of the Other, is already implicit in the relation to the Other (since in this latter relation all Others other than the Other already obsess me) and thus implies the instant imbrication of justice as *droit* (which entails the generality of laws and hence calculation) within the singular relation to the Other (which is justice "as such"). So, even though the idiosyncratic and absolutely singular relation to the Other, as fulgurates in the Saying (*le dire*) is irrevocably betrayed in its singularity by the generalizing strictures of the Said (*le dit*), in which the ethical relation is thematized as law, this very relation to the Other implicitly calls out for such a thematization. This is the irresolvable *aporia* between Saying and Said discussed earlier. Furthermore, as observed, Lévinas also speaks of the "Saying in the Said"—a form of excessive indwelling of the Saying in the Said as the latter's disruption. Importantly, the Said is at the service of the Saying (in the sense of manifesting it), yet only in the form of a betrayal; whilst at the same time the Saying calls out for the Said in the form of enjoining justice in society as law. Likewise, for Derrida, it "turns out that *droit* claims to exercise itself in the name of justice and that justice is required to establish itself in the name of a law that must be 'enforced'."[203] This precarious relation of imbricated terms (law and justice) in mutual violation of one another is swiftly seen to throw up a series of *aporias*.

14. The *aporias* of justice

Derrida takes it as an axiom that being just or unjust presupposes that one is free and responsible for one's actions and decisions. Notwithstanding, any free decision must, on the other hand, necessarily follow a certain law, prescription, or rule. As observed, such appropriate and correct application of law through calculation and rules suffices only to make a decision legal. It is, however, in no sense a sufficient condition for rendering the decision just. Indeed, the exact opposite could verily be the case. For Derrida avers the narrow conception and practice of justice as the mere subsumption of "cases" under the auspices of predetermined and determinable rules to be "the surest, the most reassuring definition of *responsibility as irresponsibility*, of ethics confused with juridical calculation ..."[204] In order to be at all just, a decision must go far beyond merely following, conforming to, and simply reproducing a rule of law—whether this rule or law be general or specific. The decision of a judge must, in addition, re-approve and confirm the rule or law by actively and interpretively reinstituting and reinventing it, "as if ultimately nothing previously existed of the law, as if the judge himself invented the law in every case"—as it were, *ex nihilo*.[205]

Therefore, in order for a decision to be just, it must accomplish two mutually contradictory movements. On the other hand—much as the tragic hero is obliged to subsume his individuality under the universal aegis of the Kierkegaardian Ethical so as to be a mere function of it—justice must operate in conformity with the law, and do so by following legal rules, statutes, and due process. This aspect entails *conserving* the law. On the other hand, however, since the unproblematic, untrammeled and prosaic application of the law would effectively confuse ethics with mere juridical calculation, each decision must at once suspend and reinvent the law upon which it rests. The "moment" of decision must, therefore, telescope both of these opposed movements into itself. Thus, for a decision to be just it must, as Derrida maintains, "in its proper moment if there is one," at once conserve the law and at the same time suspend it to a sufficiently large degree to have to reinvent and rejustify it in each case, so as to at least freely reaffirm the principle upon which it rests.[206]

More succinctly put, for a decision to be just, it must—in the very same present moment in which it is made—both follow and not follow law. This is, of course, impossible—at least insofar as this "moment" of justice is conceived as an obdurate present that firmly holds onto itself, as in the case of the strictly chronological time of the Deleuzian first temporal synthesis: that of "habit" or *Chronos*.

The *aporia* just mentioned spawns within its ambit an additional *aporia*, which—as Derrida avers—is merely a somewhat different form of the first. This second *aporia* Derrida terms the "the ghost of the undecidable." What the latter does *not* pertain to is any hesitation between two or more determinate rules, each of which are equally imperative and pressing. It is not a state of indecisiveness between two decisions. Rather, it issues from the tension that exists between the law and its suspension, and resides in the paralysis endured between the dual necessity to, on the one hand, calculate in line with laws and rules and, on the other hand, be outside of calculation. That is to say, the "undecidable" is the aporetic experience of that which, though entirely heterogenous and refractory to the order of the calculable and the rule-bound, notwithstanding still remains obliged to take law and rules into consideration when giving itself up to a decision that is impossible yet absolutely ineluctable. A decision that would not pass through the ordeal of the undecidable would be mere calculation, perhaps lawful, but by not being free, never just.[207]

The upshot here is that there is never a present moment in which one can forthrightly say that a decision or judgment is just, or in which justice can be said to have been served—or that man can in any present moment be described as just. This means that justice has no proper moment, no moment that is properly its own. There is no "time of justice"—at least not in the sense of *Chronos*. On the one hand, a decision cannot be just in the very moment of suspense of the undecidable, since at that point no decision has as yet been made. However, it is also not just in any present moment subsequent to a decision having been made, since the decision would have followed a rule—whether inherited or reinvented—and thereby sided with calculation. More summarily, a decision is never presently just because "either it has not yet been made according to a rule, and nothing allows us to call it just, or it has already followed a rule—whether received, confirmed, conserved or reinvented—which in its turn is not absolutely guaranteed by anything; and, moreover, if it were guaranteed, the decision would be reduced to calculation and we couldn't call it just."[208]

15. Justice as *à venir*

That justice has no present moment does not, of course, mean to say that it simply does not exist. Rather, it would be more accurate to aver that justice, of its very nature, simply belongs to a different order of temporality. In that justice is at once justice as the singular ethical relation to the Other and justice as law (*droit*), it is likewise homologously the paradoxical imbrication of the Lévinasian Saying (*le dire*) and Said (*le dit*). Here, we may recall Critchley's averment that the confluence of these two fundamentally heterogenous orders of discourse—the former pre-ontological and the latter ontological—does occupy a certain temporal moment. This moment is, however, nowise a punctual self-collected present, but rather an internally distented instant perpetually falling out of phase with itself. It is thus a diachronic instant or, equivalently, the instant devoid of thickness which incessantly internally fragments any punctual present into an infinitely recessive virtual past that never was and an irremissibly vertiginous future that itself shall never be. The order of temporality of justice is, therefore, that of Deleuze's third synthesis of time—that of *Aion*, in which the present moment is forever elided and indefinitely thrust into an interminably deferred and impossible future.

This is, indeed, the exact same temporality that Derrida accords to justice, except under the name of the "to come" or *à venir*, according to which justice is that which of its very nature is perpetually deferred into an infinitely distant future. This quality of indefinite postponement pertains to an aspect of Benjaminian "divine" violence. Divine violence may be seen alongside and in contrast with "mythical" violence. Mythical violence is, as already observed, the violence that founds the law (*droit*), yet—in addition—conserves it by refounding it. It is, therefore, at one and the same time a law-founding and law-conserving violence. More precisely, it is *die rechtsetzende Gewalt* ("law making violence") which institutes the law and, on the other hand, *die rechtserhaltende Gewalt* ("law preserving violence") which maintains the permanence and enforceability of law and confirms it—where *Recht* ought to be read as "law" and not so much as "right." As Derrida maintains, the inherent structure of mythical violence itself implies that it necessarily repeats itself by founding that which it conserves and bequeaths to a heritage and tradition. For this reason, one cannot speak of a pure law-founding violence in isolation from a purely law-conserving violence. For the very founding of the law is always already its own iterability in the form of self-conserving repetition. Conservation itself in turn refounds, so as to conserve that which it founds.[209]

Benjamin regards mythical violence as typically Greek, aligning it with the Greek *mythos*. Divine violence, in constradistinction, is the violence which straightforwardly annihilates the law, and is Judaic. These two disparate kinds of violence also maintain a different relation to life, death, and blood. Mythical violence exacts itself against life itself, causing it to bleed. It is, in a sense, homicidal. It threatens and demands expiation. On the other hand, divine violence is violence against life—but only insofar as it, along the very same trajectory, exalts life. Whereas mythical violence satiates itself by sacrificing the living, divine violence destroys life only so as to save the living and life itself. Paradoxically, it is divine violence at its most destructive that commands the respect of the living being, and does so in a manner which transcends law and goes beyond judgment. It is precisely in this that divine violence sacralizes life *even beyond man himself*. Crucially, however, this sacralization of life by divine violence is never a sacralization wrought for the sake of mere unadorned life. For the worth of man consists *primus inter pares* in the fact that he contains within himself the impossible possibility for justice. This Derrida terms the "yet-to-come" (*avenir*) of justice, the "yet-to-come" of man's "being-just," and at once his "having-to-be just."[210] Divine violence, or divine justice, thus bloodlessly wipes out the unjust man such that the just man may live as a possibility—albeit an impossible one.

This latter infinitely temporarily removed and impossible emergence of the just man as portended by divine justice, is strangely homologous to something discussed previously in connection with the Deleuzoguattarian primitive regime: the movement of disappearance of pre-historical human species activity in its post-historical product, that is, in the Nietzschean sovereign self-legislating individual (*souveräne Individuum*). This emergence, of course, necessarily entails the disappearance of man or, more precisely, *homo historia.* It requires the passing out of existence of man, and the human-all-to-human world as such, so as to facilitate the untimely upsurge of a fundamentally different subject than man. As shall be seen in our very final chapter, this apocalyptic extrusion will necessarily presuppose the moribund and melancholy movement of man through the absolute nadir of the Nietzschean Last Man and his leaden exhaustion and desuetude, whose disappearance shall be marked by the Man Who Wilt Perish. The latter shall commit suicide at midnight.

In this vein, divine violence may be seen as residing at the very helm of the *aporia* between divine jusice and the order of ontology itself, as evinced from Abraham's complaint to God in Thomas Mann's *Joseph and His Brothers*, who complains stridently to God that if he desires to still have an extant world, then he cannot demand justice in the absolutely unconditional manner in which he does. On the other hand, if God continues to solely concern himself with justice to the exclusion of all else, then the world will inevitably be destroyed. He unbraids God for wanting it both ways at once, for grasping both ends of the chord and desiring that there be both a world and justice in the world—a unilateral stance which can only lead to the world's terminal annihilation.[211]

The "story" of justice—if one may in any sense speak of one—is for this reason homologous to the story of God as Mann relates it. This story, he avers, is "the future of the future," a future that must necessarily be so absolutely exceeding for the Deity that his own present can in no sense match it. Yet, it is precisely this comparative poverty and utterly vanishing quality of God's present moment in relation to his future that lends to his grandeur and power the quality of constant expectation and of unfulfilled promise.[212] This links us directly to our much earlier discussion of the nature of the Deity in the book of Exodus wherein God was nowise averred at base to be a substantive being that *primus inter pares* "is"—but rather a pure *infinitive* whose Being inherently resides in the future.

In this connection, Jack Miles was observed as maintaining that God's disclosure of his name to Moses at the burning bush as *'ehyeh 'ašer 'ehyeh*, or "I am what I am," could also very plausibly be read as "I am what I shall be"—in that *'ehyeh* can mean both "I am" or "I *shall be*." Moreover, Miles had also observed that the original Hebrew text, in fact, only contains the consonants of God's name, that is, *hyh 'šr 'hyh*. With the change of but one letter in the last of these three words, one obtains the archaic Hebrew root *hwh*—from which the word *yahweh* is derived. Thus, changing *hyh 'šr 'hyh* to *'hyh 'šr 'hwh*, and thereupon interposing slightly different vowels, one ultimately arrives at the phrase *'ehyeh 'ašer 'ahweh*—in which the last word *'ahweh* is almost identical to *yahweh*. As likewise observed, Miles wagered the latter word to mean "become" or "cause to become" or "make happen" or simply "act." Hence, the whole phrase *'ehyeh 'ašer 'ahweh* was seen to mean "I am what I do." Thus, it was established that God is defined primarily by what he does and that he is a mystery that at base unfolds through his actions.[213] The God of Exodus was, therefore, concluded to be a pure *infinitive*.

It was, moreover, Ernst Bloch who was seen as being in concord with this ontological characterization of the God of the Mosaic Exodus, avering that God's very Being is defined in terms of the *futurum* and as open possibility. Here, Yahweh appearred before Moses at the burning bush not as an overburdening presence, nor as a dread haunting absence, as a *Deus absconditus*—but as a *becoming*, as intrinsically open within himself to the not-yet,[214] whose very mode of Being is the *futurum*.[215] The fiery vision on Moriah had disclosed the Exodus God as a redeemer defined solely and exclusively in the future tense, a spatial and temporal wandering. This God is an unfixed futurity inherently and *ab initio* ahead of itself, a symbol and flag of liberation from servitude, a source of leadership through time with a distant goal—a utopian God of human *revolt*.[216] He is a God with no fixed abode, a God, as Bloch was seen as contending, "who is himself not yet what he is: who *is* only in the future of his promise-to-be—if he should keep his word—and in no other way."[217]

This quality of incessant deferral of God and of justice—which is also interminably not yet what it is—was likewise, as observed in a much earlier chapter, that residing within the third of Titorelli's three possibilities in the law, as encountered in Kafka's *The Trial*: that of indefinite postponement. In accordance with this latter, justice was never to be represented in the present moment of a hierarchical tribunal but instead found itself interminably deferred to the room next door and so forth ad infinitum—sending K on a his endless peregrinations through the constantly unfolding rhizomatic mazework of a subterranean architectural state burgeoning within the very innards of the despotic State apparatus.

Justice is, thus, not *sensu stricto* of the order of power as *pouvoir*—roughly, bureaucratic power, or the ideological interpellation of the subject wrought by the Lacanian big Other—but that of *puissance*. It is intimated by the prophets' discernment of the *puissances* of an ineluctably vertiginous future, a future that—as we shall soon see—distorts the countenances of Francis Bacon's Pope Innocent X and Carlos Fuentes' El Señor alike. It is to these *puissances* that we presently turn.

16. El Señor

A tale as vertiginous and singularly bizarre as what follows, perhaps only Fuentes is able to tell. *Terra Nostra* is the novel wherein it can be found. Let us embark where Fuentes' narrative evokes the moment at which, waking as he does from a dream in a stench-ridden bed of canine slobber, animal placenta, urine, blood, semen and grease; El Señor (King Philip II of Spain) suddenly espies a flash of light and, turning his head thither, sees his countenance reflected in a small hand mirror near the head of his bed. In the mirror he sees himself with his mouth open like that of a man yelling. But in fact no scream at all escapes from his breathless and choking throat.[218] This his reflection reminds one of Francis Bacon's painting of a screaming Pope Innocent X.

Instantly fleeing from the muffled horror of the filth-riddled bedchamber, he picks up the mirror and hurries forth into the adjoining chapel. There he slowly ascends a stairway whilst continuously beholding his reflection in his hand-held mirror. Having ascended the first step, he observes that his mouth as reflected in the mirror is slightly more open, indicating that breathing had become a little more difficult. Then he ascends to the second step. In the mirror, he sees that the wrinkles about the eyes are more tightly woven, and the eyes themselves slightly more sunken and hollowed out. He climbs to the third step. Here he discovers a further change. The front teeth are missing. The mesh of wrinkles around the eyes and mouth are now impenetrable. He climbs the fourth step. Here his beard and hair are now completely white. The mouth is now agape, anguised with the inability to acquire enough air. The eyes are bloodshot, having seen far too much. Hereupon, he reaches the fifth step. What he now sees in the mirror is his own asphyxiated face, the shocking image of his own resignation, the resignation that precedes death. In this horrid hallucination, he sees that his neck is bandaged, pus exuding from his ears, worms infesting his nostrils. At this point, he feels that he is already dead. But, still worse, he feels himself already dead in life. Now, out of strange and disgustingly morbid curiosity, he spurs himself on to climb to the sixth step. Things were to become worse. The face in the mirror is completely motionless, neck bandages now shrouding his bedraggled jaw. Fleeing in terror from this image and ascending further, he now finds it very difficult to peer through the darkness of the reflection. The bandages are now being destroyed by the slow but incessant eating of worms. The jaw itself is consumed by the tepid humidity and inexorable weight of the earth. The mouth is now finally forever closed. It no longer pleads for air.[219]

Here we see the foregrounding of the Blanchotian diachronic axis of interminable dying (or the second slope of literature), of senescence, as also that of the essential mode of temporalization of the existent. Its recrudescence, though certainly somewhat disquieting, has thus far only just begun to cast El Señor into the dread of the *il y a.* Moreover, though the disconcertingly ageing countenance in the mirror is El Señor's own, does not the disfigurement it suffers render it "other" (through a "self-othering" whereby El Señor is other than, and to, himself) to such an extent that it approaches the condition of being the apparition of the face (*le visage*) of the Other as the Lévinasian Neighbor (*le prochain*)?

Still further, the sheer suffering, wizening and withering of the countenance in the mirror, the choked scream it cries in utter silence when El Señor first notices it, its asphyxiated and desperate fight for air, its ongoing failure with words, extreme discomfort, etiolation, dying, and ultimate defacement—does this all not render this paling countenance precisely the face of the *Muselmann* mentioned earlier? Is not this face that of the *Muselmann* as the zero-level Neighbor, in whose "faceless" face the Other's call is discernible and solicites us with the most forceful ethical injunction—most starkly, most traumatically? If so (and I contend that this is indeed the case), then this face is precisely the dehumanized, or indeed "inhuman," face that Žižek attributes to the Other (as the Neighbor) in his/its most extreme debasement, abnegation, and—indeed—otherness, as "a grotesquely distorted face, a face in the grip of a disgusting tic or grimace, a face which, precisely, confronts us when the neighbour 'loses face.'"[220]

As we have seen from Deleuze's reading of Francis Bacon, this Other—none other than El Señor himself—is a bestialized humanity whose gaze, formerly supported by the face (or faciality) now being dismantled, has etiolated and approaches the condition of being but a head subject to becomings-animal. With the desuetude of its facial megaphone and solicizing gaze alike, the sole intimation of this nonhuman humanity's suffering is its flesh, its meat.

But now, let us proceed to the continuation of El Señor's dreadful journey as narrated by Fuentes. With the dawning realization in the final stages of El Señor's ascent that the subject—whose ageing has been progressing with the gradual movement up the staircase—has now died, the nauseous forebodings of the *il y a* seem to have receded in the wake of a death punctually and conclusively attained. In the fact of there being here an ascent along discrete steps, each one counted, one may read the continued supervention onto this diachronic axis of interminable dying of the first slope of literature of Blanchot wherein the present arrives punctually and reposes firmly in itself, thus rendering death as the death of the subject itself (momentarily and purportedly) possible.

But El Señor shall ascend further. Let us follow him. He moves onto the seventh step. Here he sees a number of mirrors reflected in his own hand mirror, multiplying a now whitish face slowly ceding its flesh to the bareness of bone. On the eighth step it is now only bone that he sees. This skull, however, frightens him less than all previous apparitions, for it need not necessarily have been his own. Indeed, he asks himself how one is verily to distinguish one skull from all other skulls, if death always entails the loss of one's face. Now, he ascends quickly. Throughout the seeming eternity of the ensuing four steps, the skull persists. On the thirteenth step, the darkness circumvallating a central configuration of bones is dissipated, yielding before an simultaneously opaque and transparent sky, like "the metallic domed sky of solar eclipses clouded by the oval of the mirror." Only here does El Señor come to the retrospective realization that the faces had not appeared in a vacuum, but were accompanied by sounds. For there were always birds, footsteps, the rustle of cloth, fragments of music—all too speedy and evanescent to either be heard or judged. There were voices so low and hushed and thunderbolts so tremendously loud that they could only be recovered by memory ex post factum. There was even the sound of grass growing at a range that was too proximate to perceive. There were sounds that were also too distant, such as bleating, neighing, braying, lowing, barking, howling, and buzzing. All of these, once more, only existed in memory. In the moment of nothingness to which he had ascended, even sound ceased.[221]

But soon these vistas give way to a full disgorging of the *il y a* as the exposure to things in their unknown, free, and silent existence, as the radiant passivity of mineral substances, as the density of the void, as an uncanny and maddening swarming of points. Indeed, on the next step, the dome of the sky parts, and the "metallic light" disintegrates. But the gusts of wind and the flashing lights, resolving themselves into a heretofore unseen coruscating phantasmagoria of terror-inducing and shifting semi-forms effectively obstruct El Señor's ability to situate himself within the associated spatial totality. Evanescent and infinite, they existed—albeit in the form of fragmented, though reconstitutable, dust. This would be the very face of madness beholding something without origin or end.[222]

Soon, however, the milieu—though still possessed of a miasmal diffuseness anterior to the obdurate solidity of fully constituted objects—advances closer to the condition of having reconstituted itself as at least a sort of proto-reality. For, ascending the following step he sees in the mirror an image of lush vegetating depths where no sun can penetrate, bereft of even the petrified and archaic foliage of dead flowers. The blossoms attained life, became "aqueous, marine, plastic, undulating, only when he stepped onto the next stair." Now, in the very center of this viscous and flesh-like vegetation, there scintillated a small spot that, though not recognizable, El Señor knew was in actual fact himself. This tiny dot was a small white drop. He knew it to be possessed of life, a life that he rabidly desired and willed to be his own.[223]

Suddenly, then, we find ourselves imbroiled right in the midst of what appears to be desiring-production on the BwO, where the glowing dot is—if not quite the nomadic subject arising from the conjunctive synthesis of consumption-consummation—assimilable to what Massumi in our Prologue termed the larval subject, or at least the fledging subject. El Señor's desperate willing that the dot is (and/or will be) himself recalls the emergence of the nomadic subject as *inter alia* coextensive with an unbridled affirmation of itself or—more precisely—as affirmation *as such*. Seeing that the nomadic subject is a pure evanescence that instantly consumes itself in its own moment (in the moment that is itself), the more lingering nature of the glowing dot espied by El Señor implies that the latter (post-nomadic) larval/fledging subject possesses sufficient obduracy for the movement from affirmation itself (as being the nomadic subject itself) to affirmation as redoubled and therein tied to a proto-subject (whose originary evanescent moment has passed into the temporal persistence of that which comes to hold onto and repose in itself) to have taken place. More importantly, however, what is in the process of transpiring here—much as was the case with the threatening colony of eggs writhing beneath the transparent cornea of his own deceased self that he in dread beheld and here again chanced to behold—is the burgeoning of nascent and pullulating life in the very wake of the purportedly irrevocable and terminal death of himself whom he sees as having died.

Soon, however, El Señor is terrified as he sees reflected in the mirror a newly born wolf cub with "those eyes, one cruel and the other tender, that gaping snout gasping for air, those sharp teeth."[224] But it gets worse. For suddenly he beholds to his alarm the same wolf as a fully grown beast being hunted by his own men on his own lands. This hunted beast is none other than *himself*.[225]

What El Señor has detected, as Deleuze would have it, are "the diabolical powers of the future knocking at the door."[226] It is these forces which produced the silently screaming countenance that he beheld in the mirror before commencing his ascent of the staircase, the face that would forthwith undergo progressive defacement and distortion during the ascent itself. For if "we scream, it is always as victims of invisible and insensible forces," forces that "convulse the body until they emerge at the mouth as a scrubbed zone" and thereby effect the recrudescence of the head from under the face which they progressively contort and dismantle.[227] And it is Bacon in his painting of a screaming Innocent X who, according to Deleuze, "establishes a relationship between the visibility of the scream (the open mouth as a shadowy abyss) and invisible forces, which are nothing other than the forces of the future."[228] Bacon's painting of the silently screaming Pope is that of a *Muselmann*, in that the curtain in front of him shields his face from our view whilst at the same time obscuring his view and thus deadening his gaze. He sees nothing and screams before the invisible.[229] "Innocent X screams, but he screams behind the curtain, not only as someone who can no longer be seen, but as someone who cannot see, who has nothing left to see, whose only remaining function is to render visible these invisible forces that are making him scream, these powers of the future."[230] As in the case of El Señor, what he screams at—is death.[231]

Horrified by these vertiginous *puissances* of the future, El Señor recoils in absolute terror from the dread of the *il y a*, hurling the mirror down the entire stairway as he races down the steps gasping for breath to meet the now-shattered mirror on the granite flooring of the crypt below—whereupon he throws himself violently in crucifix-like prostration below the altar of God.[232] So horrified is he of not merely the swarming, inauspiciously crepuscular, and disquietingly labile, miasmatic and quasi-amorphous roilings of the *il y a* and all of its diverse vertiginous aspects; but yet more so by the portended impossibility of any exit from the suffocating mire of Being. For the prefigured punctual death which seems to supervene and function as a salubrious *telos* to the movement of interminable dying or senescence along the diachronic axis, that for a while had restored a modicum of equanimity, soon gives way to a renewed recrudescence of life as a fetid, seething, antediluvian morass. Fuentes clearly figures this living-on as metempsychosis, though it is the sheer fact of the mere insistence within Being that is sufficient for our present purposes—metempsychosis or otherwise. So acute is El Señor's exposure to the dread of the *il y a*, that his only resort remains to curse the very experience as quite literally *demonic*. Thus, in supplication before the altar, he proceeds to beseech God:

> In my weakness I beseech your aid in my struggle not to be vanquished by the importune and astute temptations devised against me by that most ancient of Serpents ...[233]

He prostrates himself, he weeps, and is choked by feelings of confusion and incipient guilt. It becomes clear that the recrudescence and encroachment of the *il y a* has been and is an all-too regular visitant upon his beleaguered soul. The *il y a*, as Satan and this latter old (yet far from senile) serpent's dread diachrony, indeed insinuates itself with particular acuteness at the time of El Señor's devotions.

He complains bitterly to God that the Devil strives rabidly and furiously to obstruct the holiness of his inward prayers, to this end deploying any and all means and forms of confusion that he can contrive in his indefatigable malice. Even when he is thwarted in these his sordid attempts, he suddenly modifies his tactics and interjects a slew of cunning deciets into the very midst of his prayers. Prostrate before the altar and filled to the brim with horrifying visions, the despairing monarch, thus, ardently beseeches God not to allow Satan to take advantage of the intense fervor of his prayers, to instantly assure him that his love of God is no less pure that it is, and likewise to assure him that his current state of total abjection and forlorness will nowise serve the Devil as an opportunity to deceive his grieving soul by invading every single minute, occasion, place, and the holy practice of prayer as such. Striken by tormenting doubt, he relates to God that he suspects his occasions of penitence may in fact be the Devil's very greatest opportunity for leading him into the sordid mire of sin. He complains that the venomous serpent strikes especially in moments of utter silence, and that the Devil has the most mendacious and malevolent mind of all, completely bereft of noble thoughts. Above all, he complains that the Devil's mind may well unveil to him a picture of his *future*.[234]

This intimation of the future that precipitates the subject (as the Laclauian discursively consistituted and Althusserian ideologically interpellated subject—not as the Lacanian split subject ($), but as one if its chief symbolic-phantasmatic sutures) out of its repose within the punctual present moment that is its proper province as imposed by the synchronic order, is properly speaking *diabolical*. The sudden imputation to El Señor of the uncanny capacity to remember not merely his past but indeed his future is tantamount to the fact of his being inhabited by Satan. He calls out in his distress to God—the master of the synchronic order (or is God unequivocally that?)—fervidly contending that God would not do such a thing, since he has granted unto mortals the beneficence of not being able to know what lies ahead, reserving for himself this clairvoyant wisdom in the absence of which God would not be God. He continues to pray to God, averring that God has merely reserved for mankind knowledge of the certainty of death, but in no sense where, when, how or for what reason anyone's death would come to pass. Nor would God be God, he continues, if he revealed to men upon their birth the actual course and final end of their lives. For men would not be God's loving creatures if he did so, since such foresight could only be the mendacious and most sordid gift of Satan himself.[235]

But are these visions of which El Señor speaks truly from the Devil and from the Devil alone? El Señor expresses his deeply disquieting concerns as to how one is to tell whether a given revelation emanates from God or from Satan. Gazing with intense passion, anxiety and doubt at the Christ in the painting from Orvieto, El Señor continues to implore the Most High. He implores God as to how he may possibly distinguish the true and Godly interior voices, those supernatural and divine ecstasies and raptures through which God lovingly communicates with the soul, from the sinister techniques of the Devil, who, by imitating the works of God, strives to mimic and effectively counterfeit God's works.[236]

Suddenly, El Señor resolves within his mind that the visions are definitively *not* from God, but indeed from the Devil—and unequivocally from the Devil alone. He ostensively refuses his soul to be deceived into imagining that God speaks to him and rendering him privy to visions, thinking instead that in these it is not God who speaks but his own fevered and palsied imagination. He utterly rejects that it is God who speaks to him:

> … I accept that the Devil has feigned these raptures and ecstasies, that he has caused visions to appear to me, that he has taken advantage of the fact that my mind is but weak clay, and that if Your Majesty permits he may transform himself into the Angel of Light, appear even in the form of Jesus Christ himself.[237]

These last thoughts, however, serve only to bring back his lingering doubts, as they proceed to noisomely persist amid a tangle of sophistries once again wrought in the wake of the endlessly equivocating tendencies of the *il y a* where all dialectic and identity alike dissolve and one thing is always already another. He beseeches God to tell him how he is to distinguish the voices of the Creator of the Universe from his own voices and, above all, from the speech of Satan lodged into every human being since the primordial fall of Adam. He pleads to know how the doctrine of the Church can enable him to avoid the moment of communication with God being transmogrified into communication with Satan, how he is to really distinguish God's visions from his own and both alike from those insinuated by the Devil:[238]

> And how do I know whether I should accept, and suffer, and understand the Devil's visions, his demonaic fantasies, since you have permitted them, and for some reason from Your High Omnipotence you permit the Devil to act instead of crushing him forever beneath your Divine Foot? How am I to know?[239]

The dilemma here is a fundamental one, and indeed well-nigh incircumventable: it partakes of the aforementioned indiscernibility in Lévinas between alterity as *illeity* (of God as Wholly Other in his levity) and alterity as the *il y a* (as the mute, viscous, dread realm anterior to ontology). Ideally, *illeity* is God, whilst the *il y a* may be assimilated to the Devil. However, the immediate corollary of the blurring of these two is the inability to extricate from the Lévinasian Saying (*le dire*) a univocally Divine or an unequivocally diabolical dimension. The Saying, apart from being equivocal in relation to the ontological level of discourse (the Said (*le dit*)), equivocates radically between God and the Devil. It is at once the enigmatic speech prior to all ontology of the Most High and also the dread intimations of what Thomas Mann afore had termed God-as-the-ram, as the ancient sacrificial animal older than God-as-the-Elohim, older than Judaism.

Indeed, it is Lacan who directly assimilates the Good to *das Ding*, that is, the absolute Good with unmitigated evil. More precisely, as Žižek contends, "this identification of the highest Good with the evil Thing can be properly understood only as involving a parallax shift: the very thing which, viewed, from a proper distance, looks like the supreme Good changes into repulsive Evil the moment we come too near it."[240] This nauseating paradoxical equivocation of the ethical Saying and of *das Ding* is at once that of the *Muselmann*'s harrowing countenance in the mirror that so terrifies and haunts El Señor; a monstrous inhuman face whose ambiguity is that of "the Real embodied, the extreme/impossible point at which opposites coincide, at which the innocence of the Other's vulnerable nakedness overlaps with pure Evil."[241]

The equivocation of the Lévinasian Saying (*le dire*) between the demonism of the *il y a* and the Divine levity of *illeity* may likewise be adduced from the view shared by certain strains of kabbalistic thought, such as the Gerona kabbalists and the Zohar, that Good and Evil are both sourced in the Godhead itself. As early as Isaac the Blind it was concluded that there was a positive root of evil which coexisted within the Godhead with the root of goodness. The *Sefer ha-Bahir* speaks of the *Sefirah Gevurah* as an attribute of the Godhead defined as the left-hand of the Holy One, a quality whose name is evil. Below the *Sefirot*, the *Sefirah Gevurah* produced within itself the *sitra ahra*, or "the other side," in the manner of an overflowing vessel spilling its superfluous liquid on the ground. The *sitra ahra* is the domain of dark emanations and demonic powers. Though having separated itself from God, it did nonetheless arise from one of the attributes of the Godhead. According to the so-called "Gnostics" of Castile and the Zohar, the *sitra ahra* is the power of uncleanliness that is active in creation and incessantly receives fresh strength not only from the sinful deeds of man but above all from the *Sefirah Gevurah*, which is in the holy order of the *Sefirot*.[242]

Nathan of Gaza, as do certain other followers of Shabbatai Zevi, likewise traces the origin of evil back to *Ein-Sof* itself. He posits the primordial existence of two "lights" within *Ein-Sof*. These are: "the light which contained thought" and "the light which did not contain thought." Whereas the first light always already contained the thought of creation, the second had no such thought and inherently strove toward remaining concealed within the mysterious obscurity of *Ein-Sof*. The first light was entirely active, the second entirely passive and submerged within its own depths. When the thought of creation arose in the first light and it—like the Schellingian Godhead—contracted to make space for the former, the second passively resisted the ensuing structure of emanation. This resistance henceforth turned the second light into the ultimate source of evil in the unfolding of creation. The light without thought, the root of evil, is a principle within *Ein-Sof* which at once holds itself aloof from creation and likewise seeks to undermine the actualization of created forms. Its sole desire is that nothing but *Ein-Sof* exists. Evil as such is, in fact, the inner dialectic between the two lights within *Ein-Sof*. This struggle between the two lights is incessantly renewed at every stage of creation and never comes to an end. What is produced by the light without thought in this struggle is precisely the *sitra ahra* of the Zohar, the dark side of diabolical powers. Wheras the light containing thought actively strives to impress its holy forms upon the light without thought, the latter's passive resistance tends toward the preservation of the amorphousness of the formless, primordial realm vacated by the first light's contraction. This primordial realm, which Nathan of Gaza calls *golem*, is a formless *hyle*, a viscous and abysal swarming of points in which all stable forms suffer dissolution.[243] In this sense, it is much like the Lévinasian *il y a* and the Blanchotian axis of interminable dying (the second slope of literature). Would not then the internal struggle between the light containing thought and the light without thought not be that which lies at the basis of the vacillation between *il y a* and *illeity?*

But let us return to El Señor. His incessant pleading and prostration before the altar and then before the figure of Christ in the painting is ultimately a desperate attempt at definitively halting this vacillation of the Saying (*le dire*) between the Divine and the demonic, a frenzied tirade of supplications bent on eliciting a Said (*le dit*) from amid the confusion of El Señor's fevered mind that would finally succeed in generating in unequivocal ontological language a discursive frame that, having triumphantly expunged the Devil, would render up a set of ethical injunctions univocally settled upon the side of the Divine. Succinctly: if one is not able to expel Satan from the ethical Saying, then at least his ejection from the ontological Said can surely be elicited through repeated appeals to God himself. So it would seem.

And so, in this vein does El Señor continue quite relentlessly and truly without surcease. Dragging himself forward toward the altar, his body still in the form of a crucifix, with bloodless fingers he touches the painting as his fingertips trace the outlines of the figure of Christ. A few moments later, one readily observes the renewal of his desperate supplications to the Godhead, that he (God) purge his soul of all baseness and all intramundane contact with the sources of debasement of the soul, all that exposes his brittle spirit to the dread equivocations of the Saying and its inexorable propensity to contaminate the psyche with the potential incantations of the Devil. He desparately implores God that he may be away from his palace, far away from the demands of his "dead but living father" and "living but dead" mother, far from the power and cruelty that his male progenitor demanded of him, far from the honor and death that his female progenitor demanded of him.[244] Indeed:

> In your mysteries, Jesus, such unwanted duties of political legitimacy are dissolved and forgotten; in you, and not—as she believes—in the satanic black hole of the very virgin Señora, my wife.[245]

Yet, the viscous encroachment of the *il y a* as the properly demonic aspect of the ethical Saying rises to the fore again, for—in lividly paranoid fear—seeing the Devil lurking surreptitiously underneath well-nigh all possible quoditian vicissitudes, El Señor now sees only one way of being able to finally terminate once and for all time the endless and dread equivocation of the Saying and its incessant Abrahamic tribulations: that of instant immolation in the Godhead. To root out the demonic *il y a* from the very gut of the Saying (*le dire*), he seems now to have eschewed even any entertainment of hope in the possiblity of arriving at a univocally-Divine Said (*le dit*)—or at least a palatably salubrious one from the standpoint of safeguarding equanimity. So unsettling are the recrudescences of the *il y a*, so acute and harrowing his subjection to an Abrahamic ordeal that seems to never be capable of ending and the back of which he is convinced he *will never see*, that he wishes to perforce and posthaste jettison all of this maddening equivocation with nothing short of *death*. Death via immediate immolation of the soul in the Divine substance. But death nonetheless.[246]

What had been intimated to El Señor in his ascent of the staircase is precisely the most fundamental sense of immortality; that is, immortality in the form of the impossibility of the accomplishment of death in or by the (Laclauian or Althusserian) subject. And it is precisely this primordial immortality, as Žižek avers, which belongs to and is that of *evil*—that is to say, evil as a spectral dimension which threatens to return forever, that inexplicably and uncannily survives even its physical annihilation and never ceases to haunt the living.[247] This explains why El Señor is now so desperate to die: thanatophilia (inexorable desire for death as annihilation) arising precisely out of his own thanato*phobia* (lingering dread of death as decay, as exactly its own impossibility). Lacan: the sought annihilation of $\$\lozenge a$ due to the stubborn reemergence of \$ as cleaved by *a*. Žižek: "This is why the victory of good over evil is the ability to die, to regain the innocence of nature, to find peace in getting rid of the obscene infinity of evil."[248]

But it is impossible to be rid of this evil. Returning to the small white dot that El Señor saw burgeoning forth in the wake of his own portended future death, in the subsequent fulgurations of desiring-production, the dot that we afore assimilated to the fledgling subject or larval subject arising from the conjunctive synthesis of consumption-consummation; may we not also (or perhaps rather) identify this dot—conspicuous in its incessant survival—with Odradek? That is to say, with Odradek as an object which—invented by Kafka—is transgenerational, immortal, beyond the strictures of sexual difference, outside time, ateleological, devoid of purpose, bereft of utility; as *jouissance* incarnate, an immortal excess, the Amfortas wound that will never heal but will likewise never allow what is dying from it to actually die; a dread partial-object that exhibits a mock-Hegelian nighmarish "bad infinity" bereft of any *Aufhebung* that would otherwise have sutured the wound and resolved the existent's imbroglio of existential miredness in the salubrious beatitude of a death finally attained.[249]

But it is not attained. For the wound never closes or kills the wounded; Odradek forever eludes our grasp and, at the same time, cannot be dislodged—since *jouissance* is both unattainable and inexorable.[250] Odradek is what Lacan "developed as the *lamella*, the libido as an organ, the inhuman-human 'undead' organ without a body, the mythic presubjective 'undead' life-substance, or, rather, the remainder of the Life-Substance which has escaped the symbolic colonization, the horrible palpitation of the 'acephalic' drive which persists beyond ordinary death, outside the scope of paternal authority, nomadic, with no fixed abode."[251] There is nothing that can be done about Odradek.

For it is here that the bedraggled Spanish monarch has entered the very "zero-level," or pure void, of subjectivity: the absolutely unendurable abyss in which fully constituted objects and subjects dissolve and yield before a miasmal swarming of spectral partial objects, of exemplifications of the *lamella*, of the undead, immortal, libido-as-object that is self-identically the Freudo-Lacanian death-drive (*Todestrieb*). This paronomastic modality of the subject is merely one "degree" removed from the pre-subjective noumenal Real. It is a return to, or recrudescence of, the impossible moment of upsurge of subjectivity in the form of a withdrawal, a recession from the light of phenomenality and its differential degrees of appearing in a world in accordance with its transcendental regime, which assumes the form of a contraction through which constituted reality yields before the surreal phantasmagoria of *membra disjecta*, of organs devoid of body functioning as stand-ins for the dread immortality of the object-libido in its semi-pernally stuck fixity and eternal psychotic circulation around the void. In Kantian terms, this crepuscular and vanishing mode of subjectivity is that of the pre-synthetic imagination gone mad, psychotic, and utterly demented—the impossible domain of absolutely purified and kenotically evacuated transcendental freedom and spontaneity. In this domain, the subject finds itself vertiginously suspended between direct unadorned animality, on the one hand, and the subordination of its freedom to the moral Law, on the other. In this gathering murk, far from being able to galvanize itself in the direction of forward-directed action in the world or attunement to the promptings of the Deity, the self is submitted to the horrifying experience of dread organs without body, of itself as the Lacanian fragmented body, of its *membra disjecta* floating around in a nauseating and disgusting miasmal drift, the "night of the world," the "valley of the shadow of death," the hell-on-earth of Kundry in Wagner's *Parsifal* who, having committed the blasphemy of laughing at Christ dying on the Cross, is condemned to an eternity of wandering as an abject, multiply befouled, purely excremental self entirely rejected by the world, abandoned by God, and subject to involuntary, psychotic automatisms, spasmodic jabberings, slavering aphasia, dyslexia, *and the absolute horror of never being unable to die.*

But let us return to the scene unfolding around El Señor. In the crypt there are rows of open sepulchres. And El Señor fervidly desires to be united with Jesus in the same manner as a corpse is immured in its sepulchral stone. With carved effigies and marble bodies of the Señors and Señoras, stone husbands and wives sleeping alongside in their death beds, prostrate nobility awaiting the arrival of these corpses as his witnesses; El Señor breaks out yet again into frenzied prayer, imploring Jesus to grant him his divine presence, divine touch, and his sovereign encircling arms. He can no longer live apart from his Lord, and beseeches God to grant him a short life so as to hasten his nuptial bond with the Divine. To this end, he implores God to put an end to his waiting and subjection to the tyranny of time, by way of an instant ascent to eternal glory:[252]

> I pray you, allow me to quit this unchanging world, more like its initial sin and pain, more like unto itself, the more it changes, and let me join with you in the delicious variability of a promised Heaven. Come, Jesus, come to me, come, come now, now, now ...[253]

We must not take literally the "delicious variability" extolled here: El Señor is dead-set against the diachronic *tout court*. This purported variability would be at best a mere semblance of categorical change as simulated from within the strictures and confines of the synchronic order.

But now, at the highest pitch of his infinite longing for death via instant immolation in the Divine substance, at the zenith of his feverish desire to suppress and violently beat back the foreboding-riddled burgeoning forth of the *il y a*, with its interminable dying, dread viscosity, prebiotic pullulation, its uncanny and amorphous swarming of anonymous points, its forbidding density of the void; the dreaded and all the aforementioned dread suddenly returns in a series of nauseating anamorphoses and demonic equivocations. The naked men in the painting from Orvieto, originally having stood with their backs to the viewer, begin to turn around to reveal their faces; whilst the figure of Christ commences slowly to turn his face away from El Señor. His gut roiling with nausea and unholy terror, El Señor screams and immediately starts to flog his back, hand and face with a penitential whip—the statues of his ancestors staring at him with voids for eyes. Bleeding from his self-inflicted wounds, he proceeds to mutter through clenched teeth that he positively does not want the world to change, and that he does not want his body to perish and be transmogrified and reborn as an animal that would be hunted in his own lands by his very own descendants. He implores God to literally stop the world and to release his resurrected body into the eternity of Paradise so that it can be with his Lord:[254]

> When I die, I do not want—please, have mercy—I do not want to return again to the world. I want the eternal promise: to ascend to the Kingdom of Heaven and there forget the unchanging world and lose for all time the memory of the life I led, forget that there is life on earth.[255]

Suddenly it comes to be revealed that this acute dread occasioned by the maddening equivocation and vacillation of the *il y a* has harrowed El Señor to the very marrow since time immemorial and has been—and remains—at the helm of the vast majority of his harebrained acts, projects, predilections, neuroses, and metaphysical notions. So has the *il y a* irremissibly imbricated with the Divine intimations of *illeity* in the Lévinasian ethical Saying harrowed El Señor, that he has long since resorted to effect—at the level of cognition—a veritably Puritanical evacuation of all traces of the sacred from the intramundane world: hereby reducing this realm not merely to the prosaic and servile realm of practice, but unequivocally consigning the human world to the domain of the incontrovertibly *demonic*. El Señor's paranoid mode of jettisoning the equivocation of the anterior ethical Saying (*le dire*) assumes the form of a preemptive decision in favor of an iron conviction that occludes any possibly that a single trace of *illeity* remain extant within the ambit of *this* world. Thus, for El Señor, *illeity* is heard only in Heaven, whilst earth and Hell are in essence diabolical. Not only *are* they diabolical; they *must* be diabolical—and El Señor's acts have and are aimed to not only maintain them as Evil embodied but also to make them EVEN MORE EVIL. In fact, to resolve in his mind the equivocation between *il y a* and *illeity* inherent to the ethical Saying, the world must be made AS EVIL AS POSSIBLE. El Señor continues his increasingly more demented rant, contending that in order to attain Heaven, and in order for the said Heavenly abode to actually exist, his world and the world as such must not change—and this all for the reason that the world's infinite horror must necessarily be the contrasting shade from which the pure supernal goodness of Heaven may be born. Hereupon, he immediately proceeds to declare to God that he had, as a youth, LITERALLY MURDERED those who dared to offer him Heaven on earth. It was, he continues, also the reason why he ceases not to ACTIVELY ENCOURAGE EVIL upon the earth; that is to say, so that Heaven can still have a meaning. Having said the foregoing, he goes on to beseech God:[256]

> Let there be a Heaven, God, your Heaven; do not condemn us to a Heaven on earth, to a Hell on earth, to a Purgatory on earth, for if the earth contains in itself all the cycles of life and death, my destiny is to be an animal in Hell. Amen.[257]

Let us be quite assured that the evil avowed and purportedly embraced here by the mind of El Señor in the foregone fervid supplications is *not* the same as the evil associated with and fomented by the immortal excess that is *das Ding.* Rather, it consists in the incipient predilection for determinate negation; that is to say, the desire *to vanquish.* But the sheer inexorability of this immortal excess and its "bad infinity" ensures that determinate negation is in crisis—and will *fail.* Thus, El Señor's fervid—and indeed rabid—supplications again come to naught: the averted countenance of the Christ-figure, the naked men in the same painting, and the sepulchral statues, all continue to stare blindly or blankly into the distance. Realizing this, and quite in consonance with his prior demonization of the entire world, he raises his whip and cusses out the entire motley crew in a delirious frenzy of execration:

> ... Devil.... Devil in disguise ... You Devil, assuming at will the figures of other men, of phantoms, of the One God ... Oh, cruel God, bestowing or withdrawing your gifts at will, permitting that Lucifer himself usurp your figure to deceive my poor soul.... Show

> yourself to me, my God, let me know whether it is you who touch me or whether it is the Devil ... Why do you sumbit us Christians to the severe test of never knowing, on the mystic heights, whether we are speaking with you or with the Enemy? Oh, you bastard Jesus, show yourself to me, give us one single proof that you hear us and that you think of us, one single proof! ... Son of shit, God of shit, how shall I know when you speak to me! Let me enjoy mystic ascension with neither doubts nor visions, for only in this epiphany may I resolve the conflict of my poor soul ... only by your side may I leave all that behind—but you do not wish to tell me whether by sacrificing power, honour, and sex I shall know you, or whether I am embracing the Devil![258]

By this point, the incessant equivocations of the ethical Saying (*le dire*), of the demonic *il y a* insinuating itself midst the very interstices of subjectivity itself, have driven El Señor to the point of delirium. Suddenly, to his own surprise, he garners the strength to rise up and immediately proceeds to lash out with his whip at the painted bodies, imagining himself to be causing the canvas of the painting to bleed. He turns with fury upon the averted Christ, but—as he attempts to strike—his arm is immediately paralyzed as his penitential whip begins to writhe of its own accord like a venomous black serpent from Hell, as if imbued with demonic force, as if having itself become a personification of Satan; the figure of Christ forthwith begins slowly to turn back toward El Señor whilst resounding with a sovereign laughter that instantly leaves El Señor "frozen like a statute, looking almost like another of the thirty sepulchres in this crypt, while the figures in the painting rotated, showing an infinite variety of forms."[259]

Is this derisive laughter of the Christ figure in the painting from Orvieto that terrifies El Señor to the bone not precisely that which Žižek describes as the "dark excess of ruthless divine sadism," an excessive cruelty that is "the necessary obverse of Christian Love"; and is not the parallax-shift between this love and its obscene underside homologous to that between *illeity* and *il y a*?[260] This would, indeed, be a somewhat heretical conclusion. It causes one to be fearful.

Notes

1. Cohen 1994, 59–61.
2. Lévinas 1978, 95.
3. Santner 2001, 9.
4. Cohen 1994, 59–61.
5. Ibid. 57, 58.
6. Ibid. 61, 62.
7. Ibid. 62, 63.

8. Santner 2001, 88, 89.
9. Lévinas 1981, 86.
10. Cohen 1994, 158.
11. Lévinas 1981, 12.
12. Lévinas 1979, 78.
13. Ibid. 79.
14. Rosenzweig 2005, 177.
15. Cohen 1994, 183.
16. Lévy 2004, 207.
17. Ibid.
18. Lévinas 1981, 88.
19. Žižek 2006, 112.
20. Ibid. 113.
21. Lévy 2004, 207.
22. Ibid.
23. Ibid.
24. Ibid. 248.
25. Ibid.
26. Ibid.
27. Ibid.
28. Deleuze and Guattari 1988, 170–172.
29. Ibid. 171.
30. Ibid. 190.
31. Ibid. 171.
32. Ibid. 172.
33. Ibid. 175.
34. Ibid. 167.
35. Ibid. 175, 176.
36. Ibid. 181.
37. Ibid. 168.
38. Ibid. 188.
39. Ibid. 181.
40. Ibid. 179.
41. Ibid. 190, 191.
42. Deleuze 2003, 19.
43. Ibid.
44. Ibid.
45. Ibid. 19.
46. Agamben 1999, 47.
47. Žižek 2006, 79, 80.
48. Derrida 1996b, 57.

49. Barbusse 2004, *passim*.
50. Derrida 1996a, 6–8.
51. Ibid.
52. Lévinas 1981, 11.
53. Cohen 1994, 244.
54. Ibid.
55. Ibid. 272, 273.
56. Batnitzky 2009, 160.
57. Ibid. 161, 162.
58. Rosenzweig 2005, 197.
59. Scholem 1978, 155, 156.
60. Cohen 1994, 185.
61. Pollock 2009, 204.
62. Rosenzweig 2005, 180, 181.
63. Ibid. 181.
64. Ibid. 184, 185.
65. Cohen 1994, 152, 153.
66. Ibid.
67. Ibid.
68. Debray 2004, 44.
69. Ibid. 41.
70. Lévy 2004, 211, 212.
71. Ibid. 245, 246.
72. Ibid.
73. Ibid.
74. Ibid. 247.
75. Agamben 1999, 55, 70, 75, 81, 82.
76. Blanchot 1995, 386.
77. Ibid. 384.
78. Ibid. 387, 392.
79. Lévy 2004, 224.
80. Ibid. 224, 225.
81. Ibid. 225.
82. Ibid.
83. Ibid. 232.
84. Ibid.
85. Ibid. 208.
86. Ibid.
87. Ibid.
88. Ibid. 209.
89. Ibid. 228.

90. Ibid.
91. Mann 2005, 148.
92. Bataille 1991, Vols. 2 and 3, 92, 93.
93. Ibid.
94. Ibid. 95.
95. Ibid. 214–217.
96. Mann 2005, 514.
97. Ibid.
98. Ibid. 928, 929.
99. Ibid.
100. Bataille 1991, Vols. 2 and 3, 96, 97.
101. Fuentes 2003, 138.
102. Ibid. 139.
103. Bataille 1991, Vols. 2 and 3, 80–82.
104. Ibid.
105. Rosenzweig 2005, 174.
106. Gibbs 1994, 27, 28.
107. Mosès 1992, 113.
108. Ibid. 118.
109. Gibbs 1994, 30.
110. Badiou 2009, 33.
111. Batnitzky 2009, 50.
112. Ibid. 113.
113. Blanchot 1995, 386.
114. Critchley 2004, 50, 51.
115. Cohen 1994, 156.
116. Ibid. 124.
117. Ibid.
118. Rosenzweig 2005, 200.
119. Mosès 1992, 127.
120. Ibid. 128, 129.
121. Ibid.
122. Lévinas 1981, 13.
123. Mosès 1992, 128, 129.
124. Ibid. 130, 131.
125. Pollock 2009, 216.
126. Rosenzweig 2005, 228.
127. Ibid. 229, 230.
128. Ibid. 216.
129. Ibid. 230.

130. Ibid. 231.
131. Ibid. 216.
132. Ibid. 232.
133. Batnitzky 2009, 118.
134. Rosenzweig 2005, 234.
135. Batnitzky 2009, 118.
136. Rosenzweig 2005, 234.
137. Ibid.
138. Lévinas 1979, 213.
139. Lévinas 1981, 157.
140. Batnitzky 2009, 118.
141. Lévinas 1981, 158.
142. Ibid. 160.
143. Cohen 1994, 190, 191.
144. Ibid. 126.
145. Ibid.
146. Ibid.
147. Rosenzweig 2005, 191.
148. Lévinas 1981, 6.
149. Ibid. 168.
150. Ibid. 168, 169.
151. Ibid. 44.
152. Rosenzweig 2005, 259.
153. Ibid. 279.
154. Lévinas 1981, 169, 170.
155. Critchley 1992, 165.
156. Blanchot 1995, 385.
157. Ibid. 473.
158. Ibid.
159. Ibid.
160. Ibid. 474.
161. Ibid.
162. Ibid. 475.
163. Caputo 1997, 197.
164. Ibid.
165. Ibid. 200.
166. Kierkegaard 1994, 45.
167. Ibid. 50.
168. Ibid. 47.
169. Žižek 2006, 87.

170. Ibid.
171. Ibid. 105.
172. Kierkegaard 1994, 48.
173. Ibid. 76.
174. Derrida 1996b, 59.
175. Kierkegaard 1994, 98.
176. Ibid. 101.
177. Ibid.
178. Caputo 1997, 199.
179. Kierkegaard 1994, 76.
180. Ibid. 100.
181. Ibid. 101.
182. Ibid. 105.
183. Ibid. 106.
184. Derrida 1996b, 74.
185. Derrida 1995, 26.
186. Ibid. 26, 27.
187. Kierkegaard 1994, 106.
188. Ibid.
189. Ibid.
190. Caputo 1997, 200.
191. Derrida 1996b, 61.
192. Ibid. 62.
193. Ibid. 63.
194. Ibid. 64.
195. Ibid. 68, 69.
196. Derrida 1992a, 12.
197. Ibid. 13.
198. Ibid. 14.
199. Ibid. 31.
200. Ibid. 16.
201. Ibid. 17.
202. Ibid.
203. Ibid. 22.
204. Derrida 1992b, 72.
205. Derrida 1992a, 23.
206. Ibid. 23.
207. Derrida 1992a, 24.
208. Ibid.
209. Ibid. 38.
210. Ibid. 52–54.

211. Mann 2005, 347.
212. Ibid. 350.
213. Miles 1995, 98, 99.
214. Bloch 1972, 56.
215. Ibid. 81.
216. Ibid. 92–94.
217. Ibid.
218. Fuentes 2003, 148, 149.
219. Ibid. 149, 150.
220. Žižek 2006, 113.
221. Fuentes 2003, 149, 150.
222. Ibid. 150, 151.
223. Ibid.
224. Ibid.
225. Ibid. 151.
226. Deleuze 2003, 51.
227. Ibid.
228. Ibid.
229. Ibid. 34, 35.
230. Ibid. 51, 52.
231. Ibid. 52.
232. Fuentes 2003, 151, 152.
233. Ibid. 152.
234. Ibid.
235. Ibid.
236. Ibid. 153.
237. Ibid.
238. Ibid.
239. Ibid.
240. Žižek 2006, 187.
241. Ibid. 113.
242. Scholem 1978, 123.
243. Ibid. 126, 127.
244. Fuentes 2003, 153, 154.
245. Ibid. 154.
246. Ibid. 155.
247. Žižek 2006, 65, 66.
248. Ibid.
249. Ibid. 115.
250. Ibid.
251. Žižek 2006, 117, 118.

252. Fuentes 2003, 155.
253. Ibid.
254. Ibid.
255. Ibid.
256. Ibid. 155, 156.
257. Ibid.
258. Ibid. 156.
259. Ibid. 156, 157.
260. Žižek 2006, 187.

Part Four

The Capitalist Regime

17

The Eviscerations and Entrapments of Capital

1. Gargantuan deterritorialization

Whilst the despotic regime erected itself via a truly massive movement of deterritorialization, sweeping away and subsuming the primitive territorialities of the savage machine; the capitalist machine in turn ensconces itself upon the vestiges of the despotic regime through a still more vertiginous movement of deterritorialization.[1] In the pervasive atmosphere of advanced ferment characterizing this epochal shift, as Marx has it, "new forces and new passions spring up in the bosom of society, forces and passions which feel themselves to be fettered by that society. It has to be annihilated; it is annihilated."[2]

The annihilation of the already brittle despotic regime at the behest of hitherto unprecedented and headlong deterritorializations is at once that of "the expropriation of the great mass of the people from the soil, from the means of subsistence and from the instruments of labour," a "terrible and arduously accomplished expropriation" which constitutes nothing less than "the pre-history of capital."[3] The dissolution of the already ailing despotic regime, Marx continues, "was accomplished by means of the most merciless barbarism, and under the stimulus of the most infamous, the most sordid, the most petty and the most odious of passions."[4] This history of the expropriation of the great many is "written in the annals of mankind in letters of blood and fire."[5]

On the other hand, however, in the wake of these momentous changes which came at length to subvert enormous swathes of humanity, a great amount of freedom was injected into the world. For the "immediate producer, the worker"—forthwith or in due time—"ceased to be bound to the soil, and ceased to be the slave or serf of another person."[6] He "escaped from the regime of the guilds, their rules for apprentices and journeymen, and their restrictive labour regulations."[7] And indeed, the "industrial capitalists, these new potentates" that rose to preeminence in the tumultuous wake of the tremendous deterritorializations breaking out on all sides of the body politic; moved with a liberated surge of unbridled energy to swiftly displace not only "the guild masters of handicrafts, but also the feudal lords, who were in possession of the sources of wealth."[8] From this purview, their meteoric rise indeed appears "as the fruit of a victorious struggle both against feudal power and its disgusting prerogatives, and against the guilds, and the fetters by which the latter restricted the free development of production and the free exploitation of man by man."[9]

Examples of these massive capitalist deterritorializations and land expropriations abound. In the sixteenth and seventeenth centuries, in England, the development of commerce and capitalism through the enclosure of land for the production of wool was accompanied by huge movements of deterritorialization which filled the cities with vagrant poor and beggars who were whipped bloody, expelled from the city to workhouses or transported to the colonies.[10]

With the permission granted by the governor of the Massachusetts Bay Colony John Winthrop to take Indian land, the Puritan colonists declared war on the Pequot Indians in 1636 and immediately piloted massive movements of deterritorialization that swept the Pequots out of what is now southern Connecticut and Rhode Island.[11]

Forty years later, in 1676, the English Puritans effected deterritorializations which cavorted the Wampanoag Indians out of the south shore of Massachusetts Bay, killing three-thousand in the process.[12]

During the nineteenth century, massive deterritorializations systematically removed Indians between the Appalachians and the Mississippi for the cultivation of cotton in the South, grain in the North, canals, railroads, new cities, and the building of a huge continental empire clear across to the Pacific Ocean. From the time of his inauguration as president in 1800, Thomas Jefferson committed the federal government to the future removal of the Creek and Cherokee from Georgia. As Howard Zinn maintains, "Indian removal was necessary for the opening of the vast American lands to agriculture, to commerce, to markets, to money, to the development of the modern capitalist economy."[13]

From 1814 to 1824, a series of massive movements of deterritorialization, inter alia through Machiavellian treaties, removed southern Indians off of three-fourths of Alabama and Florida, one-third of Tennessee, one-fifth of Georgia and

Mississippi, and parts of Kentucky and North Carolina, in the process laying the basis for the cotton kingdom and slave plantations.[14]

In 1814, Andrew Jackson's massive military deterritorializations at the Battle of Horseshoe Bend drove thousands of Creeks off their land, freeing it up for cotton cultivation in the single largest Indian cession of southern American land.[15]

The gargantuan deterritorializations of the 1818 Seminole War, during which Jackson campaigned across Florida, burning Seminole villages and seizing Spanish forts, swepted the Seminoles off their land, leading to the Florida Purchase of 1819.[16] With the Treaty of Camp Moultrie foisted upon them in 1823, further movements of deterritorialization swept all Seminoles out of northern Florida and every coastal area into the swamps of central Florida.[17]

As Jackson assumed office in 1829, the discovery of gold in Cherokee territory in Georgia precipitated a series of deterritorializations through which the Indian population was driven off their land after their property was destroyed and their land and stock seized by thousands of marauding whites.[18]

In late 1831, a gargantuan movement of deterritorialization drove out thirteen-thousand Choctaws who, marshalled by guards, hustled by agents, herded like a flock of sheep, and dying by the hundreds of pneumonia and cholera, were ferried westwards across the Mississippi River so as to free up their lands for United States capitalist enterprise in accordance with the sly Treaty of Dancing Rabbit Creek.[19]

In 1832, Jackson moved to speed up Indian removal. Soon, a white invasion of Creek lands—looters, land seekers, defrauders, whiskey sellers, and thugs—drove thousands of Creeks from their homes into the swamps and forests. A huge movement of deterritorialization led a virtually interminable, stumbling procession of more than fifteen-thousand starving, plague-ridden and dying Creeks trailed by wolf packs and circled by flocks of buzzards across the whole of Arkansas westward across the Mississippi.[20]

In 1838, following the signing of the Removal Treaty and having been rounded up and crowded into stockades, a massive movement of deterritorialization ushered seventeen-thousand Cherokees dying by the thousands of sickness, drought, heat and exposure, west across the Mississippi.[21]

Between 1860 and 1910, huge movements of deterritorialization piloted by the United States Army evicted the Indians of the Great Plains, wiping out their villages and paving the way for the railroad and millions of white farmers, who seized the vast tracts of land for capitalist machinations.[22]

Yet, this bloody expropriation of the land and means of subsistence of entire populations belongs not merely, as Marx would have it, to the pre-history of capitalism. It continues to this very day.

2. The immiserations and eviscerations of neoliberalism

The rabid deterritorialization of the particularly aggressive neoliberal variant of capitalism, as Naomi Klein asseverates, demanded that the world itself would have to be erased to make way for the purist invention of Milton Friedman and his Chicago School-style hyper-laissez-faire economics. Rooted in harebrained analogies with biblical floods and apocalyptic visions, and geared to the end of effectuating a clean slate for business and profit, these eviscerating deterritorializations were fraught with violence and cataclysm. They were a great rupture.[23]

Freemarket capitalist economics, in a massive movement of deterritorialization and its proclivity toward large-scale disaster, likewise desires this grand unmaking to entail the depatterning of the individual human brain.[24] Here, in a manner in which psychologist Ewen Cameron was desirous of deterritorializing the human mind to the point at which it became a veritable *tabula rasa*, Milton Friedman was intent upon deterritorializing human societies to the point where they were animated by an utterly purified capitalism—in which all hindrances, such as government regulations, trade barriers and entrenched interests are eviscerated.[25]

One of the examples hereof is the neoliberal revolution wrought upon Bolivia in 1985, in which Chicago School economic shock therapy and its deterritorializations caused entire families to be displaced into rickety tents, thousands of children to be shunted away from their original dwelling places and subsist on one piece of bread and cup of tea per day, hundreds of campesinos to be driven off their lands to beg on urban streets, and hundreds of thousands of full-time jobs with pensions to be eliminated and replaced with precarious ones with no protections whatsoever.[26] These neoliberal deterritorializations likewise effected mass layoffs of workers in the tin mines and the economic erasure of whole sectors of workers, shunting them off to the shantytowns and slums surrounding La Paz.[27]

Within ten years, the neoliberal deterritorializations of South Africa's transition to democracy cut millions of people off from newly connected water and, unhinged the rand from the gold standard, stripped controls and trade barriers, rendering most trading helter-skelter short-term speculation.[28] Exacerbated by Mbeki's June 1996 introduction of neoliberal shock therapy (GEAR), increased labor "flexibility" led to still more anarchic trade and paltry controls on money flow.[29] The results of these capitalist deterritorializations are that since the ANC assumed power in 1994, the number of people living on less than one US dollar a day has doubled from 2 million to 4 million in 2006; that between 1991 and 2002, the unemployment rate for black South Africans has increased from 23 percent to

48 percent; that roughly 2 million people have lost their homes; whereas close to 1 million have been evicted from farms. This was the first decade of democracy.[30]

In the wake of the gargantuan Chicago School deterritorializations accompanying the transition of Russia from communism to capitalism and piloted inter alia by Harvard neoliberal economist Jeffrey Sachs and Soviet economics minister Yegor Gaidar, by December 1991 Yeltsin had effectively dissolved the Soviet Union. Through the lifting of price controls, imposition of free trade, abrupt cuts to subsidies, and rapid-fire privatization of the county's almost quarter-million State companies, these coruscating neoliberal deterritorializations ensured that within a single year millions of Russians had lost their life savings, millions of workers remained unpaid for months, and a third of the population fell below the poverty line. In the brutal imposition of neoliberal capitalism, Yeltsin abolished the constitution, dissolved parliament (the equivalent of the US president unilaterally dissolving Congress) and all city and regional councils, set fire to the Russian White House with thousands of soldiers, dozens of tanks, killing five-hundred people, wounding a thousand. Many were arrested and interned in a sports stadium or taken to police stations where they were beaten to a pulp.[31]

By 2006, these years of Chicago School-style neoliberal deterritorialization had bankrupted more than eighty percent of Russian farms, closed seventy-thousand factories, pushed seventy-four-million Russians into poverty, killed off ten percent of the population, and left three-and-a-half-million homeless children wandering the streets.[32]

During the mid-nineties, gargantuan neoliberal deterritorializations at the bidding of the IMF emptied the treasuries of Asian Tiger economies of the southern Pacific, stripped them of trade and investment protections, imposed deep budget cuts, causing a staggering twenty-four-million people to lose their jobs and many to commit suicide, hurling twenty-million into poverty, and forcing droves of children from rural families in the Philippines and South Korea into the sex trade and into the arms of human traffickers.[33]

To this day, with foreign owners demanding ever-higher levels of profit for their investments, millions of public sector workers, small business owners, subsistence farmers and trade unionists from Indonesia, Malaysia and South Korea are being thrown overboard—many ending up in slums now housing one-billion people, brothels, or cargo ship containers.[34]

In both the most rapid-fire military siege in history and the most ferociously deterritorializing economic blitzkrieg ever seen, entailing as it did the dropping of thirty-thousand bombs and the firing of two-thirds of the total of all precision-guided cruise missiles *ever made*, the American invasion of Iraq cleaned the state for neoliberal economic experiments on a hitherto unprecedented scale.[35]

Starting with the destruction of the major Baghdad telephone exchanges with massive bunker-busters on March 28, 2003, American neoliberal deterritorializations severed Baghdadis from millions of phones across the country and plunged the city of five-million into endless night. Gargantuan deterritorializations—entailing a full frontal lobotomy of an entire multi-thousand year culture through the looting of the National Museum of the very records of the first human society, the destruction of every book and doctoral thesis ever published in Iraq through the reduction of the national library to a blackened ruin, the theft of thousand-year-old Qur'an's from the burned-out shell of the Ministry of Religious Affairs, the stripping of Iraqi universities and its education ministry, and the evisceration of the country's borders, import tariffs, inspections and taxes—transmogrified Iraq into the most wide-open market anywhere.[36]

Within five months of the invasion, the US neoliberal deterritorializations had hurled two-thirds of Iraqis into unemployment and pushed the country onto the very brink of mass starvation.[37] Within a month of the arrival of top US envoy L. Paul Bremer, Chicago School-style deterritorializations effected the privatization and looting of two-hundred of Iraq's largest companies.[38] Within four months of his arrival, Bremer would sack approximately half-a-million government workers—most of them soldiers, but also doctors, nurses, teachers and engineers. In what was a very faithful application of unrestrained Chicago School capitalist ideology, the neoliberal deterritorializations and mass corporatist pillage to which Iraq was subjected would cost the country 665,000 lives.[39]

From 1974 and 1975, massive neoliberal capitalist deterritorializations displacing masses of people from their properties and means of livelihood would start to be precipitated by the International Monetary Fund (IMF), closely followed by the World Bank, when these bodies came to hone in their predatory focus on Third World countries reeling from the deleterious impact of soaring oil prices. In this rather sordid game, the IMF would proceed to ratchet up the intensity and scope of its coercive "structural adjustment" programs on unsuspecting client nations. In the eighties, the IMF and World Bank were co-opted as the primary instruments of the irrepressible neoliberal revolution touted by Reagan, Thatcher and Helmut Kohl. According to the 1985 Baker Plan, the fifteen largest Third World debtors would be prised open and forced to bow before the tyrannous global economic hegemony of the "Washington Consensus." The effects of these measures on the countries hit were more severe and long-lasting than the Great Depression. Indeed, what Mike Davis terms the "brutal tectonics of neoliberal globalization" since 1978 would constitute a repetition on a far grander scale of the catastrophic vicissitudes of Victorian imperialism in the late nineteenth century, which created the Third World in the first place and during which time the forcible incorporation into the world market of the subsistence peasantries of Africa and Asia fomented the deaths of millions and the uprooting of tens of millions.[40] One of the most shocking effects of these post-1978 neoliberal capitalist deterritorializations would be, as Davis fails not to reiterate, the prodigious and indeed quite terrifying explosion in the proliferation of slums around the globe. The subjection of one country after another to the structural adjustment programs of the IMF, with its deregulation of national markets and attendant processes of currency devaluation and State retrenchment, is contributing to the displacement and collective immiseration of persons on a scale that dwarfs any such phenomena within all previous history combined.[41]

To add to this, it is slum dwellers themselves that are being forcibly evicted on a colossal scale from Third World neighborhoods. As part of neoliberal capitalist urban "beautification" projects and under the pressures of foreign investors such as the IMF, repetitions of Baron Haussmann's evictions of *sans-culottes* from the *quartiers* of Paris in the 1860s would ensue at a relentless pace. These nomads would often be in a perpetual state of relocation. In the Philippines under Imelda Marcos, altogether one-hundred-and-sixty-thousand squatters were forcibly dumped on the squalid outskirts of Malina over the course of the 1974 Miss Universe Pageant, Gerald Ford's visit of 1975, and the IMF-World Bank conference of 1976. The still more ruthless deterritorializations of Corazon Aquino evicted over six-hundred-thousand squatters, usually without relocations sites. Under Joseph Estrada, in preparation for the Association of Southeast Asian Nations (ASEAN) Summit of November 1999, demolition squads, with the added brunt of SWAT teams who injured and killed recalcitrants, annihilated the slum of Dabu-Dabu in Pasay, scattering and relocating its residents along the banks of a sewer were children contracted deadly gastrointestinal viruses. Between 1986 and 1992, as part of a mass-scale Haussmannization drive, the Dominican Republic's ruler Juan Balaguer ensured the brutal eviction of one-hundred-and-eighty-thousand residents from forty *barrios* of Santo Domingo in a hellish reign of government terror during which houses were demolished with people still inside, and paramilitary shock troops kidnapped people and physically violated women and children. Much as in the case of the 1936 Berlin Olympics and the Olympics held in Mexico City, Athens, and Barcelona, so as to boost profits, Korean authorities forcibly relocated as many as seven-hundred-and-twenty-thousand squatters and tenants from Seoul and Injon in the run up to the 1988 Seoul games, whereas prior to the 2008 Beijing games three-hundred-and-fifty-thousand were evicted to make room for stadium construction alone. Finally, between 1989 and 1994, in the most Dantesque "urban beautification" in recent Asian history, so as to facilitate lucrative economic projects aimed at luring profits from Western tourists and Japanese businessmen, the heroin-financed Burmese dictatorship deployed State-sponsored arson to drive and ship out a dumbfounding one-and-a-half-million residents to hastily improvised bamboo-and-thatch huts on the periphery of Rangoon and Mandalay, even evicting the dead from the cemeteries.[42]

Over and above this, the draconian neoliberal regime of imposed labor flexibility is, especially in India and the Middle East, fomenting an unprotected and labor rights-bereft informal sector characterized by the antediluvian semi-feudal recrudescence of mafioso practices such as kick-backs, bribery, and sectarian violence verging upon a Hobbesian *bellum omnium contra omnes*.[43]

The other grim atavism fomented and fostered by these neoliberal deterritorializations is the resurgence writ large of the squalid underworld of exploited labor as portrayed by Zola, Dickens and Gorky, and the savagery of child labor in particular. In Varanasi in Uttar Pradesh, children and even entire families are kidnapped by predatory textile contractors, held in captivity, tortured and forced to work twenty hours a day. In the glass industries of Firozabad, also in Uttar Pradesh, enslaved children labor in deadly factories in near-touching distance of molten fires and live electrical wires while traipsing on glass shards with bleeding bare feet. Their fathers form part of the three-million strong Asian army of rickshaw-pullers who live far away from their families in sheds or stables and die of heart attacks or tuberculosis after a few years as draught animals. The periphery of Chennai sprawls with farms engaged in the mass trafficking of kidneys of deserted women unable to support their children through other means.[44]

The most Sadean, or perhaps even *Guyotatian*, immiseration of human labor precipitated at the behest of neoliberal vectors of deterritorialization was that wrought upon the subalterns of Congo-Zaire. The IMF and World Bank, supported by the French Quai d'Orsay and periodically prodded by the US State Department, imposed structural adjustments upon the country in 1977 through which ordinary Congolese were forced to pay off vast preexisting loans made by Mobutu Sese Seko from Swiss banks. The effect was the dereliction of the entire welfare state infrastructure, the complete decimation of the civil service, and the retrenchment of a quarter-million employees. A second IMF structural adjustment in 1987 rendered the informal sector a milch cow, whilst the Club of Paris rolled over Mobutu's foreign debt in exchange for further public sector layoffs, privatization of State companies, and evisceration of exchange controls. This prising open of the market for the export of diamonds spelled the terminal desuetude of home industries and rampant hyperinflation, which completely nullified all relations between money, labor and production, effectively hurling the populace into an economy of pure chimerical fortune and a vicious mentality of outright sorcery.[45]

Here we enter a phantasmagoric realm of the maccabre, apoplectic superstition, and out-and-out witchcraft. In September 1991, marauding hordes of slum dwellers pillage the factories and warehouses of Kinshasa. This is repeated in January 1993 by the army. The banking system collapses. Public administration disappears. All trade is reduced to pre-Statist barter. The IMF and World Bank withdraw in November 1993. All locked up in Swiss bank accounts, the Congolese economy falls away from the world. The overthrow of Mobutu in 1997 leads to endless civil war with three-million dead from starvation and disease by 2004, and a colossal influx of refugees into a gargantuan archipelago of overcrowded slums. The trashy mendacity of cheap American-style Pentecostalist heresy miscegenated with atavistic tribal superstition metastatizes in the form of a ubiquitous necromantic glossolalia of mass-hysterical séances at which the literal belief in Harry Potter unleashes a concerted manhunt for child witches believed to be flying everywhere on broomsticks. The emaciated droves of child witches wandering these soulless and semi-dead wastelands represent a state dehumanization beyond which lies only the concentration camp and the graveyard.[46] All that is seen is spectral ghouls and cadavers in various stages of decomposition, and around them excrement, excrement, excrement and excrement.

3. Conjunction of deterritorialized flows

But let us return to the Marxist view of bloody capitalist deterritorialization in general and specifically as pertains to the transition from pre-capitalist modes of production to capitalism proper. As Chris Harman maintains, the immediate upshot of the forcible and almost invariably violent expropriation and separation of vast masses of individuals from any direct access to none other than their very means of livelihood and subsistence, the severance of all ties between the now-evicted workers and their means of production; was that these masses had to—from this moment onwards and under pain of starvation—seek their livelihood elsewhere. Reduced to a condition well-nigh identical to vagrancy, these newfangled itinerants, these bedraggled mercenaries, are now at once free and forced for the first time to sell themselves as labor-power to those who—through a marked accumulation of wealth—had emerged as the possessors of the means of production and, of course, liquid capital.[47] These latter are of course the newly preeminent knights of industry.

Marx describes this situation as "the confrontation of, and the contact between, two very different kinds of commodity owners; on the one hand, the owners of money, means of production, means of subsistence, who are eager to valorize the sum of values they have appropriated by buying the labour-power of others; on the other hand, free workers, the sellers of their own labour-power, and therefore the sellers of labour. Free workers, in the double sense that they neither form part of the means of production themselves, as would be the case with slaves, serfs, etc., nor do they own the means of production, as would be the case with self-employed peasant proprietors. The free workers are therefore free from, unencumbered by, any means of production of their own."[48]

Deleuze and Guattari describe this historically unprecedented encounter between the free worker and capitalist as at base entailing the conjunction of a decoded and deterritorialized flow of labor with a likewise decoded and deterritorialized flow of money-capital.[49] Their present use of the term "deterritorialized" is strongly homologous to Marx's foregoing usage of the term "free." That is to say, the worker is free in that, being dispossessed of land and means of production alike, he has become free of overly specific forms and objects of labor;[50] whereas, at least for Deleuze and Guattari, liquid capital is deterritorialized insofar as it stems from the unmooring of money from the parochialism of local price structures and its consequent transformation into a general equivalent of exchange. Otherwise put, the capitalist wage-laborer is deterritorialized in the sense that he is the result of a movement which unhinges the precapitalist worker from the ownership and conditions of his own labor; whilst capital itself is deterritorialized inasmuch as it is the outcome of a transformation that decouples the means of labor from its immediate and overly—territorialy or otherwise—restricted usage by direct producers.[51]

Crucially, it is not—in the most fundamental sense—the wage-laborer himself that enters into the conjunction of the two deterritorialized flows upon which we presently dwell. For, in speaking of a deterritorialized flow of labor, one would certainly have in mind not so much the laborer, but indeed pure impersonal labor-power. For, as Marx has it, the deterritorialization herein entailed is at once a dissolution of "the relations in which the *workers themselves*, the *living labour capacities* themselves, still belong *directly among the objective conditions of production,* and are appropriated as such—i.e. are slaves or serfs."[52]

To be sure, the worker in his own objective person was a direct economic factor in pre-capitalist modes of production. However, in the case of the capitalist mode of production, "the worker is not a condition of production, only work is. If it can make machines do it, or even water, air, so much the better."[53] Indeed, in the conjunction upon which it stands, capital "does not appropriate the worker, but his labour"—which is to say that it appropriates the worker only "as object-less, purely subjective labour capacity" and naught besides.[54]

The deterritorialized flow of labor—that is, labor-power—is describable as being such in the sense that "as the use value which confronts money posited as capital, labour is not this or another labour, but *labour pure and simple,* abstract labour; absolutely indifferent to its particular *specificity* [*Bestimmtheit*], but capable of all specificities."[55]

Capital, as it enters into relation/conjunction with labor, is itself deterritorialized in that it "*as such* is indifferent to every particularity of its substance, and exists not only as the totality of the same but also as the abstraction from all its particularities," and for this reason "can come into relation with every *specific* labour," for it "confronts the *totality* of all labours" of which "the particular one it confronts at a given time is an accidental matter."[56]

From this, Marx thus moves to conclude that the capitalist "economic relation—the character which capitalist and worker have as the extremes of a single relation of production—therefore develops more purely and adequately in proportion as labour loses all the characteristics of art; as its particular skill becomes something more and more abstract and irrelevant, and as it becomes more and more a *purely abstract activity,* a purely mechanical activity, hence indifferent to its particular form; a merely *formal* activity, or, what is the same, a merely *material* [*stofflich*] activity, activity pure and simple, regardless of its form."[57]

Obviously, as Marx renders pellucid, it is not simply the fact that the above conditions and pivotal relations of the capitalist mode of production merely stand at its very origin. Rather, capitalist production—in order to sustain itself and to flourish—does not only simply arise from but indeed repeatedly *reproduces* "in the course of its own process the separation between labour-power and the conditions of labour. It thereby reproduces and perpetuates the conditions under which the worker is exploited. It incessantly forces him to sell his labour-power in order to live, and enables the capitalist to purchase a labour-power in order that he may enrich himself. It is no longer a mere accident that capitalist and worker confront each other in the market as buyer and seller. It is the alternating rhythm of the process itself which throws the worker back onto the market again and again as a seller of his labour-power and continually transforms his own product into a means by which another man can purchase him."[58]

4. The capitalist *socius*

The inevitable consequence of these gargantuan deterritorializations and their rapid evisceration of the already moribund despotic regime, is—as seen—the bloody demise of the full body of the despot and the erection in its place of a new and hitherto unprecedented socius. This new socius, or social machine, is the full body of capital—and more specifically, that of industrial capital.[59] Capital is essentially industrial in its nature in the precise sense that it "is the only mode of existence of capital in which not only the appropriation of surplus value or surplus product, but also its creation, is a function of capital."[60] It is only industry that requires production to be fundamentally capitalist in character; that is to say, it is only industrial production that rests upon the fundamental capitalist conjunction between a deterritorialized flow of labor and a deterritorialized flow of capital, or in other words, the encounter and antagonism between the wage-laborer and the capitalist.[61]

This is nowise to say that the revolutionary upsurgence of the capitalist regime is self-identically that of the first historical emergence of capital itself. For certain more primitive forms of capital do indeed antedate the capitalist regime. Archaic forms of capital did exist under the auspices of the despotic machine: that is to say, the tribute and mercantile capital. Now, whereas the extraction of tribute—that is, the infinite debt owed by the toiling subject down below to the incomparably extolled despot far above—was coextensive with the filiation between the despot and his subject peoples, and thus constituted a filiative form of capital; precapitalist mercantile capital was basically a capital of alliance. In addition to these two forms of primitive capital, there was of course usurer's capital—the archaic forebear of interest-bearing or financial capital and likewise an alliance form of capital. Merchant capital was essentially a capital of alliance because, as Eugene Holland—for one—has it, mercantile capitalists plying their trade in, or in the midst of, precapitalist regimes derived their profits from their exploitation of the heterogenous price structures of different localities such that their modus operandi was essentially reducible to purchasing goods cheaply in one place and thereafter selling them dearly in another. This means that the archaic form of mertancile capital was one of alliance insofar as it transpired in a lateral and relatively open-ended fashion, and—above all—without any relation to a general equivalent of exchange that could serve to transcend the monetary parochialisms between which merchants travelled.[62]

The emergence of such a general equivalent comes to pass only with the upsurge of the capitalist mode of production itself—in which production is at base destined for and co-opted by the overriding prerogative of exchange. Precapitalist merchantilism, on the other hand, is a motion which takes "place within peoples, or between peoples for whose production exchange value has by no means yet become the presupposition."[63] This movement "only seizes upon the surplus of their directly useful production, and proceeds only on its margin."[64] Even though it is the case that entire trading peoples exist in antiquity, the position which these mercantile people do assume is one "between peoples whose mode of production is not yet determined by exchange value as the fundamental presupposition."[65]

Now, with the emergence of the capitalist machine, capital ceases in its primary determination to be one of alliance and becomes—in the most fundamental sense—filiative capital. This is so in the sense that, under capitalism, money comes to serve as a general equivalent applied to the entire social field—and is thereby no longer a mere handmaiden of barter within the circuits of a mercantilism operating within strictly localized price structures.

As Žižek describes it, in the precapitalist scenario, money was "a mere substance of wealth," wealth's "universal embodiment" and no more. In contradistinction, under properly capitalist conditions, money—subject as it is to the pressures of decoding and deterritorialization—undergoes a disembodiment whereby it is able to assume a multiplicity of guises. Capital—in the full bore sense of filiative capital—is in fact a *process* possessed of the following fundamental facets. In the first aspect of the capital process, money appears as a bare means for the exchange of commodities wherein, instead of the bartering and haggling of precapitalist mercantilism, the product is simply exchanged for the universal equivalent of all commodities. This latter universal equivalent may then be exchanged for any commodity whatsoever. Put more succinctly: in this first aspect, money appears as the mere means of exchange of use-values; that is to say, as a subordinate moment of the latter. Now, onto the second aspect of the capital process. This latter foregrounds itself once the circulation of capital is set in motion. It is here that a dialectical reversal transpires: from being a mere means of exchange of fungible commodities, money assumes the position of an end in itself. That is to say, the production and exchange of use-values now comes to be "posited as a moment of what is substantially the self-movement of capital itself—from this moment onward, the true aim is no longer the satisfaction of individual needs, but simply more money, the endless repeating of the circulation as such …"[66] With this dialectical reversal having been brought about (and repeatedly so), money becomes truly autochthonous, seemingly generating increasingly more value—from out of itself.

As Marx maintains, the said dialectical reversal is that through which value becomes "the subject of a process in which, while constantly assuming the form of money and commodities," augments itself, "throws off surplus-value from itself considered as original value, and thus valorizes itself independently. For the movement in the course of which it adds surplus-value is its own movement, its valorization is therefore self-valorization [*Selbstverwertung*]. By virtue of being value, it has acquired the occult ability to add value to itself. It brings forth living offspring, or at least lays golden eggs."[67] This self-positing and self-valorizing movement is precisely the process of capital.

In another—somewhat more theological formulation—Marx avers that in the capital process value, "instead of simply representing the relations of commodities," now "enters into a private relationship with itself" by which it "differentiates itself as original value from itself as surplus value, just as God the Father differentiates himself from himself as God the Son, although both are of the same age and form"—and indeed consubstantial.[68] "Value therefore now becomes value in process, money in process, and, as such, capital. It comes out of circulation, enters into it again, preserves and multiplies itself within circulation, emerges from it with an increased size, and starts the same cycle again and again."[69]

Bataille quotes Benjamin Franklin: "Remember, that money is of the prolific, generating nature. Money can beget money, and its offspring can beget more, and so on. Five shillings turned into six, turned again it is seven and threepence, and so on, till it becomes a hundred pounds. The more there is of it, the more it produces every turning, so that the profits rise quicker and quicker."[70]

5. Reconfiguration of filiation and alliance

But there are many more changes that bring themselves to bear in the transition from the despotic regime to the capitalist regime. Amongst these are the alterations wrought at the level of filiation and alliance.

In the transition from precapitalist alliance capital to properly filiative capital, the older filiation under the towering countenance of the despot now finds itself transmogrified into a filiation under the still more formidable figure of capital itself. More specifically, this new filiation is that of industrial capital, whereas the alliances existing under despotism are in turn replaced by the alliance that comes to prevail between industrial capital and commercial capital, on the one hand, and that between industrial capital and financial capital, on the other.[71] Here Deleuze and Guattari argue that while it is "true that capitalism is industrial in its essence or mode of production, it functions only as merchant capitalism" and "through its alliance with commercial and financial capital."[72]

What this amounts to is the observation that, as Harman maintains, though capital is fundamentally productive (industrial) capital, it also necessarily exists in two other forms: as commodity (or merchants') capital and as money (or financial) capital. The foregoing alliance between industrial capital, on the one hand, and mercantile capital and financial capital, on the other—is reducible to the fact that the capital process, under the conditions of mature capitalism, entails continuously repeated changes from one form of capital to another. More specifically, financial capital is used to purchase means of production, raw materials and labor-power; whilst these latter enter into conjunction in the production of commodities; whereas commodities are, in turn, themselves exchanged for money; and money finds itself subsequently used to purchase more means of production, raw materials and labor-power; and so on in an endless spiral.[73] This would be Harman's more Anglophone clarification of a point which Deleuze and Guattari make rather more abstractly.

Harman continues, maintaining that notwithstanding the incessant movement of perpetual metamorphosis of the three forms of capital into one another, there also exists a partial separation of these forms, in which the organization of direct industrial production, the vending of commodities and the supply of finance, respectively, devolve upon distinct groups of capitalists.[74] Firstly, as Marx has it, commercial capital, which is at base "absolutely nothing more than the commodity capital of the producer which has to go through the process of transformation into money," attains a certain independence from industrial capital when its being sold on the market ceases to be "an incidental operation carried out by the producer himself" and now "appears as the exclusive operation of a particular species of capitalist," that is, the merchant.[75] Secondly, the movement whereby financial capital becomes independent of industrial capital may be argued as follows. Evidently, a "part of the industrial capital, and more directly also of the commercial capital, exists throughout not only in the money form, as money capital in general, but as money capital" specially tasked with or geared toward the facilitation of industrial and commercial capital's processes.[76] Much in the same way as commercial capital had gained autonomy from industrial capital, likewise does it come to pass that a "definite part of the total capital now separates off and becomes autonomous in the form of money capital, its capitalist function consisting exclusively in that it performs these operations" of money-lending "for the entire class of industrial and commercial capitalists."[77]

But we must understand in what precise sense capital is indeed filiative. As opposed to the horizontal, or lateral, sort of relation entailed by alliance; filiation is, in contradistinction, a vertical relation that moreover entails a degree of subordination. As regards capital, it is not merely its subjugation of labor that is here the case, but also the subordination to industrial capital of commercial and financial capital. These latter, Marx's contends, are "subordinated to" industrial capital "and correspondingly altered in the mechanism of their function," for they now "move only on its basis, thus live and die, stand and fall together with this basis."[78] Indeed, money capital (financial capital) and commodity capital (commercial capital), "in so far as they appear and function as bearers of their own peculiar branches of business alongside industrial capital, are now only modes of existence of the various functional forms that industrial capital constantly assumes" in its process.[79]

6. The capitalist axiomatic

In consonance with the capitalist machine's entirely codeless axiomatic, money (as money-capital) functions as an abstract quantity indifferent to the qualified nature of the flows of labor and those of capital that the said axiomatic conjoins. This encounter of decoded and deterritorialized flows is therefore a "direct relation between decoded flows whose respective qualities have no existence prior to the differential relation itself."[80] That is to say, "*dx* and *dy* are nothing independent of their relation [*dy/dx*], which determines the one as a pure quality of the flow of labor and the other as a pure quality of the flow of capital."[81]

Therefore, the fundamental distinguishing feature of such a codeless axiomatic resides in the fact that "the axiomatic deals directly with purely functional elements and relations whose nature is not specified, and which are immediately realized in highly varied domains simultaneously; codes, on the other hand, are relative to those domains and express specific relations between qualified elements that cannot be subsumed by a higher formal unity (overcoding) except by a transcendence and in an indirect fashion."[82] This entirely codeless nature of the purely abstract capitalist axiomatic bears the crucial corollary that the surplus value produced by the capitalist machine, in contradistinction to the surplus value of code generated by precapitalist regimes, is a surplus value of *flux*.[83]

7. Purely economic extraction of productive surplus

Moreover, the capitalist regime's apparatus of antiproduction itself—that is, the mechanism by which the surplus value of production is appropriated by the socius—becomes well-nigh entirely immanent to production itself.[84] This means that the capitalist socius is no longer a transcendent political instance like the despotic full body, but becomes directly economic. The capitalist socius is the first such directly economic socius in history and for this precise reason has no need of interposing extraeconomic means such as coding or overcoding between itself and its productive forces.[85] This would seem to imply that the extraction of productive surpluses by the capitalist socius—unlike in the case of the despotic regime—needs not recur to political or coercive means.

In this connection, as Ellen Meiksens Wood argues, the capitalist worker is moved to sell his labor-power by a compulsion that is—at least at some level—purely economic in nature, a compulsion categorically at variance with "the direct political or military powers that enabled lords and states in non-capitalist societies to exact rent, tax or tribute from direct producers."[86] This fundamentally economic compulsion behind the worker's selling of his labor-power, contends Wood, is—by comparison to that which operated in precapitalist regimes—impersonal, imposed as it is not by men but by markets.[87]

To be sure, workers do indeed enter into relationship with capitalists—meaning that the said economic depersonalization has its limits. Nonetheless, in stark contrast with the juridically cemented relationship of strict domination and subordination between feudal lord and serf, or the more physically wrought relation of brutal subjugation between slave-holder and slave, the relationship between capitalist and worker appears as one between individuals that enter into it in a manner that is entirely free.[88] Any compulsion that exists, it seems, arises from market imperatives and is therefore of a purely economic nature. So it would seem.

But there is more. Wood contends that one of the most monumental consequences of capitalism's detachment of the economic stranglehold that it maintains over labor-power from all forms of direct political or military coercion (what may be called the becoming-economic of power as engineered by the capitalist axiomatic) is that the hegemony of capital is able to extend itself far beyond the limits of direct political domination. This is the sense in which the capitalist regime distinguishes itself from—and shows itself to be vastly in advance of—all other social machines: it is able to exercise and extend the dominance of capital by purely economic means. In heavy contradistinction, the economic hegemony—and indeed hegemonic potential *tout court*—of precapitalist regimes extended no further than the ambit of deployment of their extra-economic force; indeed, only as far as the range and reach of their political, military, or juridical capacities. In point of fact, as regards all such precapitalist machines, "no matter how much surplus was actually produced, accumulation by exploiting classes was limited by what their extra-economic power was able to extract from direct producers."[89]

8. The capitalist State apparatus

Though it is true that the capitalist machine's appropriation of surplus labor is at base effected in accordance with the logic of its entirely codeless axiomatic, this latter is nevertheless *unable to function in isolation.* In point of fact, it must work in tandem with a secondary mechanism: the capitalist State apparatus. The capitalist State, aver Deleuze and Guattari, is "the regulator of decoded flows as such, insofar as they are caught up in the axiomatic of capital."[90] More precisely, it is via the inexorable adjunct of the State apparatus that the capitalist machine effects the recoding and reterritorialization of the flows of production, in order to thwart these decoded and deterritorialized flows from escaping on all sides of the social axiomatic. Indeed, it would seem that "the flows of capital would willingly dispatch themselves to the moon if the capitalist State were not there to bring them back to earth."[91]

As regards the origin of the capitalist State apparatus itself, it is born of none other than the resurrection of the old *Urstaat* from amid the detritus of the vanquished despotic regime.[92] This resuscitation of the despotic State, however, nowise preserves the actual archaic form of the *Urstaat*. Indeed, as pertains to the latter, "from being at first the transcendent unity" that it was in the form of the overwhelming and mightily extolled archaic State, as now resurrected under full bore capitalist conditions, the *Urstaat* henceforth "becomes immanent to the field of social forces, enters into their service, and serves as a regulator of the decoded and axiomatized flows."[93] Elsewhere, Deleuze and Guattari aver that capitalist States are "immanent models of realization for an axiomatic of decoded flows" whose role is to "moderate the superior deterritorialization of capital and to provide the latter with compensatory reterritorializations."[94]

But what form does this newfangled immanence of the resurrected *Urstaat* to the social field assume? Wood may be seen as providing substance in this regard when she avers that "any capitalist enterprise has at its disposal an array of disciplinary mechanisms, as well as internal organizational hierarchies, to keep workers in line and at work"—the most extreme of these sanctions being capital's "ability to deny the worker access to means of labour, that is, its ability to deny the worker a job and a wage, to dismiss workers or to close enterprises altogether."[95] These are, of course, bureaucratic mechanisms borrowed from the State bureaucracy of the despotic regime and resuscitated in transmogrified form at the level of private and public capitalist enterprises more or less distinct from the capitalist State apparatus. This is the form in which the despotic *Urstaat* may be seen to have become immanent to the capitalist socius. It is here that surplus labor is, for the most part, appropriated and labor held in check by disciplinary mechanisms and sanctions.

However, this nowise implies that the State apparatus has become *entirely* immanent to the capitalist full body. For there still remains the capitalist State itself which, if anything—and here one is forced to oppose at least the potentially misleading implications of Deleuze and Guattari's asseverations on this point—assumes under capitalism a *still more extreme* transcendence in relation to the domain of social production than it had at any prior point in history. Now, a little earlier we maintained that—under developed capitalist conditions—the role of extra-economic coercion in the appropriation of productive surplus had come to be downgraded virtually to the point of nullity. This is strictly-speaking incorrect. For it is hardly the case that the force of political and military coercion has categorically ceased to have any relation to the capitalist economy. Rather, under capitalism, the appropriation of surplus labor by the exploiting class, on the one hand, and the coercive power that serves to enforce it, on the other, have become *divided* among two distinct spheres: capitalist enterprises and the State, respectively. (In precapitalist regimes, on the other hand, these two moments were more or less united). The upshot of this is that, though it is true—as Deleuze and Guattari aver—that power in the capitalist regime tends to become directly economic, the economic and political spheres themselves undergo an increasing diremption.[96]

And it is precisely this latter increasingly remote political realm from whence emanates the distant purview and long coercive arm of the capitalist State. Therefore, in the transition from the precapitalist (say, despotic) to the capitalist regime, the former coincidence of surplus appropriation and politico-military coercion is transmogrified in such a way that the latter—far from making a terminal disappearance—merely drifts into the background, that is, into a state of clandestine vigilance. Here, whereas the sheer imminence of prospective extra-economic brutality does indeed pale, the spectral aura of the capitalist State's coercive arm nonetheless remains as the ultimate sanction that—ever ready and poised as it is—shies not away from forceful intervention if need be.[97] As Wood makes pellucid, the State's extraeconomic arm "operates not by intervening directly in the relation between capital and labour," but rather "by sustainting the system of economic compulsions, the system of property (and propertylessness) and the operation of markets."[98] Moreover, in the case of imperialism, though such imperialism is at base economic, the inevitable expansion of capital's domain of operation cannot be wrought without the imposition, maintenance and manipulation of capitalist markets. It is here that the extraeconomic vigilance of what is otherwise a purely economic apparatus of surplus-absorption may come to involve more than a smidgen of brutality. For indeed, "just as workers had to be made dependent on capital and kept that way, so subordinate economies must be made and kept vulnerable to economic manipulation by capital and this can be a very violent process."[99]

What we have in fact observed may be termed the *dual resuscitation of the despotic Urstaat by the capitalist machine*. Firstly, in its becoming-immanent to the full body of capital, the *Urstaat* appears as the full bureaucratic and disciplinary paraphernalia of the capitalist enterprise. Here, one also espies the first resurrected guise of the despotic Damocles' sword, that is, as the capitalist enterprise's capacity to fire the worker and thereby deprive the latter of his means of labor and sustenance. This is the more immediate manner in which the resurrected Damocles' sword insinuates itself. In a more schizoanalytic vein, this latter is its insinuation as the *market form of the death instinct*. But—yet this time in a more clandestine manner—the despotic Damocles' sword is also resurrected in a more distinctly draconian guise. In this its second mode of resurrection, the despotic Damocles' sword reappears as the ever-present imminence of death held *in potentia* by the full military and police brunt of the capitalist State's coercive organs. This second mode of appearance of the said sword is, in fact and as is fairly evident, a near-identical repetition of the despotic regime's sword of Damocles—a well-nigh perfect resurrection occasioned by those moments at which the anachronism of a far more archaic and direct sovereignty is seen to recrudesce in the very midst of a State apparatus whose function had hitherto remained largely of a managerial nature.

But what of the relationship between the apparatus wielding the Damocles' sword of forced unemployment and the apparatus brandishing the Damocles' sword of physical death? That is to say, what is the precise relation between the capitalist enterprise and the capitalist State apparatus? Otherwise put, what is the relationship between capitals and States—or, equivalently, between the capitalist class and the capitalist State? As a first gloss, Chris Harman describes this relationship as one between those engaged in the economic exploitation of workers—via the extraction of surplus or unremunerated—and those in charge of bodies of armed men. Here the nation-State and nationally based capitals arise more or less contemporaneously and develop roughly coextensively. Also—and quite unavoidably—they derive sustenance from one another in the sense that individual capitals depend on the State apparatus for the juridical framework outside of which they can scarcely operate, whilst the State depends on the individual capitals under its jurisdiction for its financial resources—particularly as regards revenues from taxation. Thus, whereas under developed capitalist conditions it is no longer the case—as was indeed true with the despotic regime—that well-nigh all economic enterprises were always already and indistinguishably those undertaken by the capitalist State, individual capitals circumscribed by nation States and the States doing the circumscription nonetheless retain a certain ineluctable *structural interdependence*.[100]

9. Social subjection and the recrudescence of machinic enslavement

The particular form assumed by the resuscitated despotic *Urstaat*, in conjunction with the emergence of a socius that operates in accordance with a purely quantitative axiomatic of decoded flows, has a fundamental bearing upon the role undertaken by the subjection of the subject, on the one hand, and its enslavement, on the other—and moreover on the relative preponderance of these two within the capitalist regime.

Firstly, and most obviously, these two are nowise the same concept. To be more terminologically accurate, the difference at stake here is between *machinic enslavement* and *social subjection.* Deleuze and Guattari maintain that one may speak of machinic enslavement when human beings as such become constituent pieces of a machine that they collectively compose together with other things such as tools, under the control, aegis and general direction of a higher unity. On the other hand, there is social subjection when the same higher unity constitues the human being as a subject joined to what is now an exterior object, whether this object be a tool, a machine, or whatever. In this second scenario—that of social subjection—man ceases to be a mere component of the machine but is henceforth a worker, and a user. Crucially, he is now subjected *to* the machine and no longer enslaved *by* it.[101]

The authors aver, moreover, that the first of these two regimes—that of machinic enslavement—pertains to the archaic imperial formation of the *Urstaat* (and thus to the State formations which Marx includes under the Asiatic mode of production)—a regime in which human beings are, in fact, *not* (or at least, not yet) subjects at all but merely pieces of a machine that overcodes the said aggregate.[102] In contradistinction, the second of these two regimes—that of social subjection—applies to the capitalist regime. For indeed, the capitalist worker is not so much enslaved by technical machines but rather subjected to them—whereas, moreover, the modern capitalist State, by way of technological development, has effectively replaced machinic enslavement with an increasingly preponderant social subjection.[103]

Much is at stake here and we must thus tread with care. It has been the habit of this book to include all proto-capitalist State formations appearing within the ambit of known history under the rubric "despotic regime." However, it is crucial not to pass by the fact that Deleuze and Guattari consider ancient slavery and feudal serfdom to already be—to a certain extent—procedures of social subjection. Within the broader compass of despotic regimes, therefore, machinic enslavement—at least in its pure form—is operative only under the auspices of grand empires that antedate not only feudalism but also antiquity. (Here one has in mind the great dynastic molochs of ancient Egypt, Persia, Babylonia, and China.)[104]

Now, as regards social subjection, it is maintained that the case of the naked or "free" worker of capitalism takes such subjection to its most extreme expression, in that the processes of subjectification cease to even enter into partial conjunctions that interrupt the productive flows. The result of this exacerbation of social subjection in the passage from antiquity and feudalism to capitalism, and moreover within the ambit of developed capitalism itself, is that *capital itself* effectively functions as the point of subjectification that consitutes all human beings as subjects—with the notable qualification that certain men, the capitalists, function as subjects of the enunciation that form the private subjectivity of capital, while the others, the proletarians, are subjects of the statement, subjected to the various technical machines in which constant capital finds itself effectuated.[105]

This all being said, it is vital for us to pause temporarily. In discussing social subjection, we have referred to the subject of the enunciation, subject of the statement, points of subjectification, and of course processes of subjectification themselves. As observed much earlier, these paraphernalia strictly-speaking belong to and originate from the Deleuzoguattarian postsignifying regime of subjectivity, or passional regime. Therefore, in that social subjection roughly-speaking begins at least with the more private forms of subjection exercised within the ambit of ancient slavery and feudal serfdom, these later despotic regimes are themselves each already miscegenated with a passional regime of their own. That is to say, they are—in point of fact—despotic-passional regimes. This becomes still more evident when it is realized that Deleuze and Guattari characterize, for example, Christianity as precisely such a despotic-passional regime. More precisely, however, in despotic-passional regimes, social subjection requires two coextensive processes: that of subjectification and that of subjection itself. The former, as observed, is the business of the postsignifying semiotic whilst the latter is wrought by despotic overcoding.[106]

What is of present concern, however, is the manner in which social subjection (as the concomitance of subjectification and subjection), on the one hand, and machinic enslavement, on the other, are wrought by the capitalist regime. For both of these are indeed present in the developed capitalist scenario. Firstly, let us look at social subjection. As is evident, the preponderant point of subjectification within the feudal Christian regime was God himself. As the postsignifying semiotic component of the latter regime gives way to the capitalist axiomatic, it is capital—the principal deity amongst all commodities—that henceforth assumes the function of point of subjectification. That is to say, under capitalism processes of subjectification are entrusted to the codeless capitalist axiomatic itself. Social subjection, on the other hand, as the inexorable correlate of subjectification, emerges much more in the capitalist axiomatic's models of realization than in this axiomatic itself. As espied earlier, these models of realization of the axiomatic are the various capitalist States. Indeed, it is within the framework of the nation-State, or—more precisely—of national subjectivities, that processes of subjectification and their associated subjections are rendered manifest.[107]

Therefore, though it is true that the capitalist axiomatic provides the kernel (capital as point of subjectification) from whence processes of subjectification and subjection arise, it is only able to bring these latter to bear with the aid of its realizations—the nation States (and other Statist structures). More precisely, overcodes (overcoding) borrowed from the resuscitated despotic *Urstaat* and henceforth serving as the *sine qua non* of what would otherwise be a merely vacuous and strictly nonexistent subjection-in-itself are co-opted by the capitalist axiomatic, with the result that the latter's purely formal subjectifications arise from the very outset solely in conjunction with renascent despotic procedures of subjection in the guise of concrete processes of social subjection. This is how social subjection is brought to bear in the capitalist regime.[108]

But, in addition to this, the capitalist axiomatic also resurrects or recreates, in novel and henceforth technical forms, a whole system of machinic enslavement. This reemergence of machinic enslavement nowise entails a self-identical recrudescence of the despotic regime in the form of the archaic imperial State machine since, as observed, under developed capitalist conditions one is in the immanence of an pure codeless axiomatic, and not so much under the transcendence of a fomal unity.[109] It is, notwithstanding, the reinvention of a social machine of which human beings are constituent working parts instead of being workers subjected to their machines. This strange reemergence of machinic enslavement disgorging itself at the very heart of a hitherto predonderant social subjection occupies a fairly clearly identifiable point within the history of capitalist technological development. The passage from social subjection to machinic enslavement is seen to be roughly coterminous with the transition from the second age of the technical machine—that of the subjection to the motorized machine—to the third such age (the information age) characterized by the preeminent rise of cybernetic and informational machines. This latter entails a regime in which recurrent and non-reversible humans-machines systems take the place of the old non-recurrent and non-reversible relations of subjection between the said two elements. In this newfangled scenario, the relation between human and machine is now based on internal, mutual communication, and no longer so much on usage or action.[110]

Franco "Bifo" Berardi, for one, inveighs with massive polemical force against what seems to be entailed within the machinic enslavement of the third technological age of capitalism. He refers to it as a new form of slavery, roughly coincident as it happens to be with the burgeoning economic imperialism of the neoliberalism of the 80s, 90s and beyond, in which—unlike in the case of the machinic enslavement of archaic oriental despotism—it is no longer so much the body that is incarcerated, but rather the *soul*.[111]

What would Berardi, as the materialist that he no doubt is, consider the soul to be? His definition would surely be no different to that of the madman Daniel Paul Schreber. In the memoirs of his mental illness, Schreber maintains that the "human soul is containted in the nerves of the body," averring of these nerves that "they are extraordinarily delicate structures—comparable to the finest filaments—and that the total mental life of a human being rests on their excitability by external impressions."[112]

The soul, Berardi contends, was hitherto a far freer, more viscous, and more errant entity. Presently, however, so as to align itself with the system of operative exchanges structuring capitalist production, it has been forced into following certain clearly delineated functional paths. In the process, forced as it now is by the cyber-capitalist machine to place its intelligence, sensibility, creativity and lingistic capacity at the service of the great "immaterial factory" of the information age, the soul has notably hardened and lost its former suppleness and plasticity.[113] This ossification of the soul has come to pass coextensively with a significant shift in the focus of capitalist domination in the information age from the sphere of corporeal, mechanical and political discipline to that which excercises itself through logical, psychological and biogenic automatisms. In this shift, the subject of techno-social domination is increasingly less so the body and ever the more so the soul. This diffuse network of techno-linguistic automatisms is what essentially underpins production in globalist capitalist production, to the extent of increasingly unhinging capitalist valorization from conscious activity and—still worse—from the very possibility of human political action.[114] We shall return to and chart the full consequences of the post-industrial age's machinic enslavement of the soul a little later.

Presently, more must be said about the relation between machinic enslavement and social subjection in the capitalist machine. It is of course true that, in spite of the axiomatic's reinvention of generalized machinic enslavement in the post-industrial age of capital, social subjection still remains operative. The coexistence and relative preponderance of these two processes may, in turn, be viewed from the perspective of the organic composition of capital (the ratio of constant to total capital). Here, whereas variable capital defines a regime of the worker's being subjected (that is, human surplus value), the fundamental framework of which is the business enterprise or the factory; the spread of automation, attended as it is by the progressive rise in the organic composition of capital, ushers in and cements a form of enslavement wherein surplus value henceforth becomes machinic.[115]

Most crucially, the relation between social subjection and machinic enslavement exhibits itself, as Deleuze and Guattari observe, to be paradoxical to a beguiling degree. The entailed paradox resides in the fact that a relatively small amount of subjectification takes one away from machinic enslavement, whereas a comparatively large amount thereof brings one back to it.[116] This large amount of social subjection evinces itself particularly in the continuing advance of the information age under which machinic enslavement is suddenly seen to be recrudescing at an unprecedented rate and over a formerly unthinkable expanse.

Maurizio Lazzarato sheds a good deal more light on the dual functioning of social subjection and machinic enslavement in the capitalist regime—and particularly under neoliberal capitalism. Here, social subjection and machinic enslavement are the two heterogeneous power apparati at the intersection of which the production of wealth, and production period, operate. Economy is the assemblage of this dual investment of subjectivity.[117] Firstly, social subjection equips us with a subjectivity, assigning one an identity, a sex, a body, a profession, a nationality, and so forth. It manufactures individuated subjects, their consciousness, representations, and behavior.[118] In so doing, it distributes places and roles within and for the social division of labor. Through language it creates a signifying and representational web from which no one escapes. As Marx already intimated, social subjection in capitalism is a process whereby capital relations become personified. For example, the capitalist and the worker act as personified functions derived from capital flows.[119]

Secondly, machinic enslavement couples itself to social subjection though, unlike social subjection, operates through *desubjectivation*. It dismantles the individual subject, consciousness and representations, functioning as it does on both the pre-individual and supra-individual levels.[120] It does so by mobilizing non-representational and asignifying, rather than linguistic and representational, semiotics. In machinic enslavement, the individual is no longer instituted as an individuated subject, economic subject, or citizen. He is instead a cog, a gear, a component part in the business and financial system assemblages, in the media assemblage, or the welfare State assemblage with its collective institutions such as schools, hospitals, television, internet, and so forth. Machinic enslavement is the management or government of the components of a system. It is the mode of control and regulation of a technical or social machine such as a factory, business, or communications system.[121]

Whereas social subjection produces and subjects individuals, with machinic enslavement individuals in fact become what Deleuze and Guattari term "dividuals," whereas masses become samples, data, markets, or "banks." Social subjection manufactures a subject in relation to an external object—such as a machine, a communications apparatus, money, public services, and so forth—which the subject utilizes and with which he acts. In social subjection, the individual works or communicates with another individuated subject by way of an object-machine, which functions as a means or mediation of his actions or use. Machinic enslavement, in contradistinction, eschews subject/object, words/things, or nature/culture dichotomies. The dividual does not stand opposite machines or make use of an external object. It is rather contiguous with machines, together constituting a humans-machines apparatus in which humans and machines are but recurrent and interchangeable parts of a production, communication, consumption, and so on, process that well exceeds them.[122]

As Lazzarato argues, with machinic enslavement there is no individual subject which acts or makes use of something. Instead, we constitute mere inputs and ouputs, points of conjunction and disjunction in economic, social, or communicational processes. The subject/object, human/machine, or agent/instrument relationship gives way to a total configuration in which there is a convergence/assemblage of forces that do not split into living and dead, subjective or objective. Rather, all forces—such as physical and sub-physical forces of matter, human and subhuman forces of body and mind, machine forces, signs, and so forth—are variously animated. In machinic enslavement, relations among agents cease to be intersubjective, and those among signs are no longer representational. Human agents, like non-human agents, function as points connecting, conjoining and disjoining flows and as networks constituting corporate collective assemblages and communications systems, and so forth.[123]

Not only is the dividual of a piece with the machinic assemblage. He is likewise himself internally fragmented by it, since the component parts of subjectivity—such as intelligence, affects, sensations, cognition, memory, and physical force—are no longer unified in a single "I." They no longer refer to an individuated subject. Intelligence, affects, sensations, cognition, memory, and physical force are now components whose synthesis no longer resides in the person but in the assemblage or process: corporations, media, public services, eductation, and so on.[124]

At this point, we may venture a striking analogy. The individuated subject constructed by social subjection and the dividual engineered by machinic enslavement may, respectively, be aligned with what Charles Taylor, on the one hand, terms the "buffered" self and, on the other, the "porous" self.

Taylor associates the "porous" self roughly with the Medieval world prior to the emergence of disciplinary society in the sixteenth or seventeenth century. The "porous" self exists within an "enchanted" world in which the line separating personal agency and impersonal force was not clearly drawn.[125] The boundary between mind and world was far hazier in this world. Once the "porous" self would enter the zone of an exogenous force or object, it would think of this force or object as including it, or even penetrating it. The force or object can, therefore, no longer be placed simply within the self, nor can it be located exclusively outside it. Rather, the force or object is in a kind of interspace which straddles what would otherwise have been a clear boundary. Or it may be said that the boundary is porous. The inside is no longer just inside, as it is also the outside. That is, emotions and thoughts which are in the very depths of the self reside in a space which at once takes the self beyond itself, porous as it is to some outside force or object.[126]

So, the "porous" self is less than truly individuated. It is not a subject bearing a relation to an external object or force. One may merely speak of a self-object or a self-force amalgam. Are these not features shared by the dividual engineered by machinic enslavement? After all, through de-subjectivation, the dividual stands outside of the subject/object and human/machine dichotomies. Moreover, in that the "porous" self's thoughts and emotions are both inside and outside itself, as in the case of the dividual, its sensations and cognition do not refer to an individuated subject. Therefore, the "porous" self and the dividual, not being individuated, both function at the pre-individual and supra-individual levels. Hence, they are strongly homologous in the said essential ways. The somewhat strange, and even arresting, conclusion that we may draw from this is that the dividual engineered by machinic enslavement appears strangely *atavistic*.

What Charles Taylor terms the "buffered" self, on the other hand, is simply everything that the "porous" self is not. The "buffered" self is characteristic of disciplinary societies subsequent to the enchanted Middle Ages. It is typically modern. We shall dwell upon its precise genealogy at a later point. For now, suffice it to say that the "buffered" self has a clearly demarcated boundary between itself and objects and forces which are now categorically outside of it.[127] It is, thus, exactly the individuated subject wrought by social subjection. But there is more. For not only is there a firm boundary between the interior and exterior of the self which abides in a disenchanted world. Further barriers are erected against strong physical desires and the fascination with the body. The barriers are raised by and in the name of the central identity as agent of disengaged discipline, keeping its distance from the zone of abandon. Outside the narrow circles of intimacy which remain, the "buffered" self is trained to relate to others as dignified subjects of rational control. This is the route and form which a formerly purely Protestant piety eventually embarks upon.[128]

10. The recrudescence of sovereignty at the heart of governmentality

We have seen that in the capitalist regime extraeconomic coercion broadly-speaking assumed two distinct—though not unrelated—forms. Firstly, we have the disciplinary mechanisms and State-supplied juridical framework under whose constant conjunction the worker is held within bounds. Secondly, the political, police and military organs responsible for the enforcement and maintenance of the ostensibly free market imperatives outside of which capitalist conditions of production would cease to exist, are likewise present. Evidently, the *raison d'être* of these two forms of extraeconomic coercion is the more or less clandestine maintenance of pure economic compulsions.

It is, however, upon the second form of extraeconomic coercion that we shall now dwell at somewhat greater length. As observed, though military coercion for the most part remains a poised clandestine vigilance, its ultimate sanction—otherwise held *in potentia*—nonetheless remains the actual deployment of direct military force. Of particular interest here is the scenario within which the capitalist State's direct deployment of military force operates at once covertly and *extra-judicially*. For it is such an eventuality that evinces the advent of a far more archaic form of sovereignty extruding itself within the midst of a State apparatus whose prevalent mode of coercion, as observed earlier, is and remains largely managerial in nature. This newfangled State sovereignty and its attendant recourse to direct politico-military coercion is itself occasioned by the effective suspension of the law in the wake of what executive State organs may, for instance, deem a threat to security. With this suspension of the law—in both its national and international forms—the capitalist State begins to exercise a form of sovereignty that, operating outside the law and through a hastily improvised panoply of administrative bureaucracies, devolves to newfangled officials who suddenly assume and forthwith wield a prerogative power enabling them to decide precisely which particular individuals concretely embody the said purported threat. These petty officials are henceforth authorized to decide which recalcitrants are to be detained—forcibly or otherwise. Indeed, these recalcitrants may be detained indefinitely.[129] We are here following the thoughts of Judith Butler.

Up until this moment of self-induced exception, the conjunction of State and non-State bureaucracies with the clandestine vigilance of the State's politico-military arm amounted to a mode of political power which Foucault terms "governmentality." This characteristically modern mode of power, in constradistinction to premodern forms of sovereignty based on the subjugation of masses swearing fealty to a higher instance of Law emanating from and concretely sanctioned by the imminence of pulverising coercion under the aegis of an eminent despot, proceeds by way of the management of populations and the circulation of goods insofar as this latter maintains and restricts bodies and persons. Governmenality operates through bureaucratic institutions, forms of State power, and likewise through law—with the exception that law is here deployed, at base, as a "set of tactics."[130] It functions through both state and non-state institutions and discourses. These latter are, however, in no sense legitimated either by direct elections or through established forms of authority. There is, thus, no single source or unified sovereign subject from which governmentality acquires its meaning and purpose. Rather, it functions through the deployment of a diffuse array of strategies and tactics, so as at once to dispose and order populations, and likewise to produce and reproduce subjects, as well as their beliefs and practices—all in relation to its specific policy aims.[131] Thus understood, governmentality is—amongst other things—a form of social subjection; or, more precisely, the particular form social subjection assumes in modernity.

This being said, Lazzarato in fact opts to work both social subjection and machinic enslavement into the Foucauldian concept of governmentality. For in addition to governmentality being a "pastoral" power exercised on individuals, he maintains that it is likewise a different type of power and control acting on dividuals—exercised in particular by private enterprises. Moreover, since the early twentieth century, governmentality has increasingly meant precisely the government of dividuals. Google and Facebook, for instance, gather, select and sell information which in fact concerns dividuals, whose profiles, composed of the convergence of data, are mere relays of inputs and outputs in production-consumption machines. Dividuals have a statistical existence controlled by apparatuses whose operations differ from the individualization carried out by social subjection proper, which is exercised on real individuals. The governmentality of dividuals, managed by flows, networks, and machines, not only plays a part in the individual's representations and conscious behavior but in the desires, beliefs, and sub-representational reality of subjectivity. As a whole, governmentality is practiced at the junction of the individual and the dividual, the individual as the dividual's subjectivation.[132]

Now, what of the relation between governmentality and law? As already intimated, for Foucault, laws are regarded by governmentality as mere tactics, in the sense that their deployment and operation is in no sense justified by recourse to any set of pre-existing principles and legitimating functions, but solely through their aim. Though legitimating functions may well be in place, the field of governmentality is in no ultimate sense animated by them. All of this serves to render the operations of governmentality for the most part extra-legal—although not necessarily illegal. In its co-optation as a mere tactic of governmentality, law no longer serves as a ground of legitimation. For this reason, governmentality is the precise sphere in which power is in no sense reducible to law.[133]

Now, it is within the ambit of governmentality that the resurgence of the mentioned newfangled sovereignty takes place. Here, governmentality's innate tendency to reduce the law to a mere tactic deployed within an extra-legal, though not illegal, domain may be exacerbated to the point where the law is itself suspended. It is this suspension of the law—as the ever present possibility intrinsic to governmentality's relegation of law to the arena of the strictly extra-legal—which occasions the abrupt recrudescence of sovereignty at the very heart of governmentality itself. And it does so—as Butler fails not to observe—"with the vengeance of an anachronism that refuses to die."[134] As observed, this resurgent sovereignty—and we are speaking here in the context of the present and ongoing "war on terror"—manifests itself primarily as the exercise of prerogative power: that of the executive branch of government and of the managerial officials to whom it delegates. Indeed, this assumption of judiciary power by the executive branch of government whereby the President himself is invested with final and unilateral power to decide the place, time and eventuality of a military trial, almost resembles an atavistic return to political pre-modernity—that is, to a time in history prior to the separation of powers, to an epoch in which sovereignty was still indivisible.[135] Here, governmentality's suspension of the law at the behest of executive State organs serves to strengthen the discretionary powers of officials charged with relying on their own judgment in deciding upon the most fundamental matters of justice, life, and death.[136] It is a suspension, therefore, wrought to the ends of a renascent despotism.

But we must not err. Born as it is of nascent tendencies inherent to governmentality, the newfangled sovereignty spawned as both extension and bastard progeny of governmentality murders not and cannot for that matter kill its progenitor. Rather, atavistic progeny and bastardizing progenitor enter into immediate symbiosis. That is to say, governmentality is both the condition of emergence of neo-sovereignty and that outside of which sovereignty could never maintain itself in existence. This process is as follows. Firstly, governmentality is sovereignty's condition of possibility. Here, the state of exception/emergency inaugurated by a hitherto purer governmentality and heralding the renascence of sovereignty is that by which power ceases to operate in terms of a set of juridical laws and henceforth does so in accordance with a set of governmental rules—rules which serve to forthwith reinstate sovereign power insofar as they have ceased to be binding in terms of extant law or modes of legitimation. These rules are now quasi-arbitrary and deployed at the pure discretion of officials who not only interpret them unilaterally, but are empowered even to decide upon the conditions and form of their invocation.[137]

Governmentality is indeed the condition of the foregoing resurgent sovereignty because it is with its ambit that law is first reduced to a mere tactic, that is to say, to something whose value is now purely instrumental, and in no sense "binding" as law in the full-bore juridical sense. Notwithstanding, this newfangled sovereignty, wrought as it is by the self-annulment of law within the context of a state of emergency, revitalizes the anachronistic "sovereign" not in his/its original form but as the newly galvanized subject of a power which remains *managerial*. That is to say, the petty officials to which a heightened sovereign power devolves, before acquiring and assuming the mantle of the petty sovereignty that is henceforth theirs, are more fundamentally mere instruments deployed within an exercise of power that remains at base that of governmentality. Indeed, as Butler makes pellucid, these petty officials possessed of unusual discretionary prerogatives are—after all—not true sovereigns, since they are nowise in full control of the broader aims animating their actions. Rather, power is antecedent to them and, only through being delegated, does it effectively constitute them as the "sovereign's" that they have become.[138]

However, within the domain of their jurisdiction—and somewhat beyond—these petty sovereigns, though wrought as such within the milieu of governmentality, still remain the ones with whom the decision rests as to who will undergo detainment and who will or will not be released. Hence, though the power they wield is a delegated one, it nonetheless remains immensely consequential—and all the more so in that their actions are in no sense subject to review by any higher judicial entity.[139]

This is precisely the paradox of the resurgence of sovereignty at the heart of governmentality which Judith Butler renders clear. Though the petty sovereigns proliferated within the firmament of late capitalist neosovereignty are instrumentalized and deployed by tactics of power outside their proper competency, this latter heteronomy nowise precludes them from projecting power in a manner that serves to reanimate a form of sovereignty that heretofore appeared to have been foreclosed by the governmentalized constellation of power. Though these miniscule despots in a certain sense know not what they do, they do nonetheless perform their actions unilaterally and with enormous ramifications. Be their actions self-evidently conditioned by the immanence of managerial power; in that they are at once final and subject to neither review or appeal, they nonetheless remain in the fullest sense *unconditional* judgments. To be sure, these unilateral and absolutely implacable actions—whether they be those of functionaries at the US government's considerable archipelago of black sites, those in the State Department, or even those of the Commander in Chief—insofar as they ultimately devolve from a diffuse set of practices and policy aims resident within a much broader field of governmentality, are not *sensu stricto* sovereign in the same sense as those of oriental despots and precapitalist monarchs in general. Nonetheless, when viewed in the isolation of their own tightly circumscribed ambit, they do assume the semblance of sovereign actions emanating from what in each case appears a self-grounding and unconditional basis for unilateral decision.[140]

In the age of the war on terror, this conjunction of governmentality and neo-sovereignty is brought to bear most graphically in the new war prison: Abu Ghraib, Bagram, Guantanámo, and the dark sites in Solamia and Yemen—to name but a few. Firstly, in that it literally manages populations, the new war prison functions as an operation of governmentality. At the same time, however, it exploits the extra-legal dimension of governmentality to assert a lawless sovereign power over life and death. In other words, the new war prison constitutes a form of self-justificatory governmentality that projects its own form of sovereignty by animating and deploying its extra-legal dimension. For, in the end, it is not representatives of courts bound by law, but rather "officials" who—in this strange hybrid of governmentality and sovereignty—deem suspected terrorists or combatants "dangerous" and review the cases of those indefinitely detained. Similarly, the courts themselves are conceived explicitly as "an instrument" used in the service of national security, the protection of principality, and the continuing and augmented exercise of State sovereignty.[141]

The new war prison may be seen as presenting the managerial tactics of governmentality in an extreme mode. Here, instead of being inexorably tied to law, trial, legal punishment, and to inmates' rights, the new war prison evinces "an effort to produce a secondary judicial system and sphere of non-legal detention that effectively produces the prison itself as an extra-legal sphere."[142] And this is the sense in which governmentality and sovereignty are seen to converge in the war prison: sovereignty appears as an extreme mode of governmentality and operates via the management of populations outside the law.

Later on, we shall, however, see that the extra-legal sphere produced and placed under the aegis of capitalist neo-sovereignty shall extend far beyond merely the war prison or detention centre. For it has, in a sense, come to potentially encompass the entire globe.

11. The return of *homo sacer*

For now, however, we shall dwell upon a crucial concomitant of the extra-legal management of populations: the preemptive and ongoing constitution of such populations as—in some sense—*less than human* or, more precisely, as bereft of the stature of a legal subject. What this extra-legal management produces is, as Butler maintains, a subject which is no longer a subject. Such an *unter*-subject is, from the purview of the law, one which is neither fully constituted as a subject nor, moreover, completely "deconstituted" in death. This means to say that the subject here "wrought" is not only less than a subject, but also neither alive or dead. Crucially for what will follow, the extra-legal management of populations, in fact, entails two concomitant processes: that of the production of the subject through regulatory power, and that of the de-subjectivation of subjects.[143] It therefore becomes clear that the resurgence of anachronistic sovereignty—though occasioned and sustained by governmentality—notwithstanding, as we have already seen apropos the despotic regime proper, likewise resurrects sovereignty's attendant term: what Giorgio Agamben terms *homo sacer*.

Within the strange arena that is the war on terror, this de-subjectivization and dehumanization of man is wrought at the behest of the petty sovereign officials' discretionary power to deem particular subjects "dangerous." The individual thus deemed henceforth undergoes the familiar processes of degradation, abjection, bestialization, and so forth. The latter is what allows renowned theorists of the subhuman such as Donald Rumsfeld to credibly maintain those deemed dangerous to be quite unlike other human beings and therefore not *stricto sensu* punishable by law—but indeed in dire need of immediate and sustained incarceration. Or worse. Rumsfeld's rhetoric vis-à-vis these beastly *untermenschen*, these men on the cusp of the human and the animal; envinces, moreover, a further pointed characteristic of the resurgent sovereignty transpiring within neoliberal times: *paranoia*. Indeed, when asked why the detainees at the various detention centers were being forcibly restrained and held without trial, Rumsfeld expained that if they were not restrained they would certainly kill again—implying that it was the restraining measures alone which prevented them from killing. Clearly then, we are dealing with beings whose very propensity is to kill, who kill as a matter of course. Without doubt, these creatures are pure unadulterated killing machines. Moreover, as is evidently the case, if these beings are pure killing machines then they are not strictly speaking humans replete without cognitive function, conscience, and sufficient rational faculties that would otherwise render them adept at knowing and understanding the charges (or lack thereof) made against them. And precisely for this reason, they are less than human or at the very least represent an equivocation of the human.[144] They are man-dogs in the manner of Diogenes of Sinope. Thus, it becomes clear that the initial paranoiac operation whereby the recalcitrant is preemptively deemed dangerous bears the attendant effect of the recalcitrant's being reduced to Agamben's *homo sacer*.

The striking homology between the prerogative power wielded by the petty sovereigns who enact this operation of deeming and the omniscient purview of the despotic regime's forewarning eye immediately underscores itself. As observed in the chapter on the signifying (despotic) regime, *homo sacer*—he who may be killed with impunity but not be sacrificed—was found to be equivalent to the second figure of the scapegoat delineated by Deleuze and Guattari: the one driven out into the desert or wilderness along a centrifugal line of flight. This second scapegoat represented an increase in entropy in the signifying regime in excess of the regime's operations of reterritorialization. In that it was irrecuperable by the despotic regime's reterritorializing operations, the centrifugal line of the scapegoat was necessarily marked with a negative sign—forthwith coming to represent all that was malign and decrepit in the regime, all that foreboded *in potentia* its terminal desuetude.

The effect of real becomings-animal in excess of the coefficient of deterritorialization of the signs constitutive of the regime, the second scapegoat henceforth bore the insignia of the properly beastial and stood as the index of a protended flight of working organs from the purview of the despot. That is to say, the second scapegoat played verso to the recto of despotic paranoia. In much the same way, its equivalent—*homo sacer*, the counterbody of the despotic full body—serves as the index of the paranoia of governmentality's newfangled petty sovereign (and less-than-petty sovereign, such as Bush, Rumsfeld or Obama) and moreover of the capitalist State apparatus at such. The paranoia of the despotic regime, owing to the pervasive resonance of all concentric signifying chains, was such that but one overly steady look, but one highly localized perturbation was instantly felt to reverberate throughout the entire regime, that is, across the entire surface of the despot's numinous body. Under so ubiquitous a threat, equivalent as it was to an incipient cancer of the body politic soon to metastatize, despotic justice had no choice but to assume the form of preemptive vengeance. The imminence of blastoma portended by the slightest local lesion of the despotic full body meant that, in properly Stalinist fashion, suspicion alone was sufficient ground for the establishment of guilt. (We shall see this soon enough in Jeremy Scahill's discussion of the Obama administration's concerted intensification of Bush and Cheney's war on terror.)

Likewise, with the resurrection of sovereignty by governmentality and the well-nigh permanent state of exception (Agamben) herein instated, justice becomes a matter whose urgency forestalls law and its forebearing order of reasons. For, as Butler maintains, when the petty sovereigns of late capitalist neosovereignty simply "deem" a recalcitrant dangerous, this is nowise done on the basis of deciding whether criminal acts had actually occurred. Rather, it is a wholly unsubstantiated and preemptive judgment, specifically designed to obviate—or at least forestall—the need for evidence. It is likewise a license to brand, categorize and detain individuals on the basis of suspicion alone, and to do so on a potentially enormous scale.[145] It is also paranoid.

This properly paranoiac operation by governmentality's petty despots is precisely that through which *homo sacer* precipitates itself into existence. Here, since the State deems individuals dangerous without reference to conclusively established nefarious acts, it effectively evicts these individuals from the jurisdiction of the law—thereby depriving them of all the legal protections of national and international law alike.[146] In this sense, these populations detained upon the paranoiac whim of a petty despot's gesture of "deeming" cease to be regarded as subjects. In that they are no longer thought of as underwritten by legal entitlements and law, they cease—moreover—to be regarded as truly human.[147] As observed, this de-subjectivization is part of a process of degradation that fairly swiftly reaches the point of bestialization.

But, as already maintained, this precipitation into existence of *homo sacer*—that is to say, the operation whereby politics is brought to bear on bare life—is by no means restricted to the new war prison and its analogues. For *homo sacer* is produced on a far vaster scale—indeed, on a scale the vastness of which rapidly proceeds to the point of virtually extending its domain to the entire human species. Otherwise put, neosovereignty implies that we are—in point of fact—all potentially *homines sacri.* Moreover, the temporal expanse of this political colonization of bare life exhibits itself to be immeasurably more comprehensive than that indicated by recent recrudescences of sovereign anachronism.

Indeed, as Agamben contends, the inherent possibility of the recrudescence of a politics of bare life is already presaged by a document generally placed at the very foundation of modern democracy, that is, the 1679 writ of *habeas corpus.* Effectively deployed from as early as the eighteenth century, the latter's intent was to ensure the physical presence of a person in a court of law. The significance of the document is that it is the inaugural salvo of a form of politics at whose center lay neither the older subject of now-dissolving feudal relations and liberties nor what the future would purport to be the *citoyen*, but the pure and unadorned body. The upshot of this latter, as Agamben renders plain, it that it is indeed "not free man and his statutes and prerogatives, nor even simply *homo*, but rather *corpus* that is the new subject of politics. And democracy is born precisely as the assertion and presentation of this 'body': *habeas corpus ad subjiciendum,* 'you will have to have a body to show.'"[148] The case of European democracy in its nascent upsurge from the ruins of moribund absolutism was the same. At the center of its polemic against the latter it placed "not *bios,* the qualified life of the citizen, but *zoe*—the bare, anonymous life that is as such taken into the sovereign ban."[149] Here we mean the sovereign as *sovrano.*

The relation between the operation brought to bear by modern democracy and that wrought by archaic despotism, thus, nowise entails a difference in kind. For democracy by no means forgoes the precipitation into existence and the centrality accorded to the bare life of *homo sacer.* Rather, the difference between the two is that democracy, whilst maintaining sacred life, "shatters it and disseminates it into every individual body, making it into what is at stake in political conflict."[150] To be sure, the particular idiosyncracy of modern democracy is the unprecedented advent of the national citizen as the newfangled sovereign subject and bearer of legal rights that he is. However, this novel democratic subject only emerges as a supernumerary adjunct who "can only be constituted as such through the *repetition* [my emphasis] of the sovereign exception and the isolation of *corpus*, bare life, in himself."[151] That is to say, the much extolled and liberally endowed subject of modern democracy is only himself on condition that he is first and foremost wrought as *homo sacer.* And this political production of *homo sacer* is brought to bear only by a repetition—that is to say, resurrection—of a sovereignty whose primordial form was that of archaic despotism. In any case, the bare life of *homo sacer* that serves here as the basis of the modern democratic subject emerges as a geminate being: in the sense that it is at once the bearer of subjection to sovereign power and possessor of individual liberties. This peculiar duality is most perspicuously espied by Hobbes when he avers that it is the body's capacity to be killed that—in one and the same stroke—establishes both the purported natural equality of men and the endoconsistency of the body politic or Leviathan wrought out of these imminently disposable bits of *zoe*. As Agamben puts it: "The absolute capacity of the subjects' bodies to be killed forms the new political body of the West."[152]

It is mindful to pause momentarily and note that the Declaration of the Rights of Man of 1789 defines the locus of individual sovereignty in the pure fact of birth. At the same time, the third article of the same document locates the principle of all political sovereignty in the "nation." Alluding to the obvious etymological link between "birth" and "nation" (where the latter is seen to derive from *nascere*—"to be born"), Agamben avers that—according to the founding document—the biopolitical circle opened by the former is closed and consolidated by the latter. What is implicit in the document is the preconceived fictional convergence between *birth* and *nation*; that is to say, the convergence between bare life and political sovereignty.[153]

Indeed, the progressive advancement of capitalism and the modern State finds itself attended by the increasing politicization of bare life, to the extent that the biopolitical reach of nation-States has long since moved inside the life of every citizen: "Bare life is no longer confined to a particular place or a definite category. It now dwells in the biological body of every living being."[154]

As already observed in the chapter on the despotic regime, the degradation of man to *homo sacer*—that is, the inclusion of bare life under the sovereign ban—is effected by the sovereign's (*sovrano*) precipitation into being of a state of exception wherein the law is suspended and under which the said bare life is politicized. It is this state of exception inhabited by bare life that—since the time of the primordial archaic despotic regime—served as the point of contact between life itself and politics. With the progressive advancement of biopolitics and the ever greater pervasion of life by politics, this point of contact maintained by the state of exception swells to the point where life and politics become well-nigh indistinguishable—a convergence in which "all life becomes sacred and all politics becomes the exception."[155]

This incessant process of conflation reaches a degree of saturation so extreme that it ushers in a newfangled juridico-political paradigm in which the norm henceforth ceases to be distinguishable from the exception. This new paradigm is none other than that of the *concentration camp*. In this precise sense, Agamben defines the camp as "the structure in which the state of exception—the possibility of deciding which founds sovereign power—is realized *normally*."[156] In this zenith of biopolitics, sovereignty exacerbated undergoes an apotheosis so extreme that the sovereign (*sovrano*) henceforth "de facto produces the situation as a consequence of his decision on the exception."[157] This is, moreover, "why in the camp the *quaesto iuris* is, if we look carefully, no longer strictly distinguishable from the *quaestio facti,* and in this sense every question concerning the legality or illegality of what happened there simply makes no sense. *The camp is a hybrid of law and fact in which the two terms have become indistinguishable*."[158] Agamben is therefore right to conclude that the camp—as preeminent paradigm of modern biopolitical power—is the "threshold in which law constantly passes over into fact and fact into law, and in which the two planes becomes indistinguishable."[159] Still more pointedly, the camp "is the [biopolitical] space of this absolute impossibility of deciding between fact and law, rule and application, exception and rule, which nevertheless incessantly decides between them."[160]

The capitalist production of man as *homo sacer* far antedates the war on terror. It is to be found at least as far back as the early colonial regime in America. All throughout the eighteenth century, for instance, the Virginia slave code decreed that runaway slaves who did not return immediately could be killed with impunity by anyone whomsoever and in any way whatsoever and that the county court was fully empowered to execute the said slave by dismemberment or any other method. The situation in Maryland was essentially the same starting in the 1660s when slavery had been written into law.[161]

In the summer of 1919, the suspension of the positive legal order and concomitant homosaceralization of aliens was in evidence when a bomb exploded in front of the home of US Attorney General Alexander Mitchell Palmer. Six months later on December 21st, in the first of a series of mass raids by the American government, 249 aliens (including Emma Goldman and Alexander Berkman) were picked up and deported to Soviet Russia, even though the Constitution gave no right to Congress to do so. In January 1920, four-thousand persons were rounded up all over the country, held in seclusion for long periods of time, brought into secret hearings, and ordered deported. In the spring of 1920, a typesetter and anarchist named Andrea Salsedo was arrested in New York by FBI agents and held for eight weeks in the FBI offices, without being allowed any recourse to lawyers. His crushed body was found on the pavement below the building.[162]

In the wake of the Pearl Harbour attack, President Franklin D. Roosevelt signed Executive Order 9066 in February 1942, authorizing the army to arrest—without warrant, indictment or hearing—every one of the 110,000 Japanese-Americans on the West Coast, take them from their homes, transport them to camps far into the interior and keep them there under prison conditions. In 1944, the Supreme Court upheld the forced evacuation on the grounds of military necessity. The Japanese—both Nisei (those born of Japanese parents in America and hence US citizens) and Issei (those born in Japan of Japanese parents and hence barred from US citizenship)—were held without charge in the camps for over three years.[163] Once more, the political production of *homines sacri.*

On March 22, 1947, as part of a broader official witchhunt aimed at suspected Communists, President Harry Truman issued Executive Order 9835, initiating a program to search out any "infiltration of disloyal persons" in the US government. Over the next five years, over six-and-a-half-million persons were investigated. All of this avalanche of interrogations of these *homines sacri* were conducted with secret evidence, secret and often paid informers, and—most crucially—with neither judge nor jury; that is to say, extra-judicially.[164]

In 1950, when Republicans sponsored an Internal Security Act for the registration of organizations deemed to be "Communist-action" or "Communist-front," liberal Senators—in particular Hubert Humphrey and Herbert Lehman—proposed the setting up of detention centers (in reality concentration camps) for suspected subversives who, when the President declared an "internal security emergency," would be held without trial. The detention-camp bill was appended onto the Internal Security Act and the camps were fairly swiftly set up, ready to house those deemed *homines sacri.* The Act was only repealed in 1968.[165]

When President Ngo Dinh Diem was losing his grip on most of the South Vietnamese countryside to local villagers organized by the NLF, his homosaceralization was rapidly pending. American Ambassador Henry Cabot Lodge met secretly with Lucien Conein, a CIA operative closely in collusion with Vietnamese generals plotting a coup to overthrow Diem, and enthusiastically reported to Kennedy's assistant McGeorge Bundy on October 25, 1963. Bundy approved the coup and on November 1, 1963, the generals attacked the presidential palace, apprehended a fleeing Diem and his brother, took them out in a truck, and executed them.[166]

When in the spring of 1968, when Dr. Martin Luther King began speaking out against the war in Vietnam, maintaining that the war effort's drain on US resources was aggravating poverty at home, he became a chief target of the FBI, which tapped his private phone conversations, sent him fake letters, threatened him, blackmailed him, and even suggested once in an anonymous letter that he commit suicide. King had become *homo sacer*. FBI internal memos discussed somehow removing him and finding a more amenable black leader to replace him. A Senate report of 1976 maintained that the FBI tried to destroy Martin Luther King. Standing on a balcony outside his hotel room in Memphis

Tennessee, where he was supporting a strike of garbage workers in the city, he was shot to death by an unseen marksman.[167]

12. Neoliberalism and extra-judicial murder

The rabid and hyper-deterritorializing Friedmanesque capitalism of the Chicago School of Economics under the aegis of the US State apparatus, operating as it often has through certain brutal ruling juntas around the globe via which it has imposed neoliberal economic policy and practice on many a weaker country than itself, has ushered in a vast production of subject populations as *homines sacri*.

In 1976, Orlando Letelier, in a searing article for *The Nation*, wrote that "The violation of human rights, the system of institutionalized brutality, the drastic control and suppression of every form of meaningful dissent" in Chile was associated with the "unrestrained 'free market' policies that have been enforced by the military junta."[168] Letelier averred that Milton Friedman, ideological father of neoliberal rapacity, and "the intellectual architect and unofficial adviser for the team of economists now running the Chilean economy" shared responsibility for Pinochet's crimes.[169] "The economic plan has had to be enforced, and in the Chilean context that could be done only by the killing of thousands, the establishment of concentration camps all over the country, the jailing of more than 100,000 persons in three years."[170] There was, he wrote, "an inner harmony" between the "free market" and unlimited terror.[171] Klein herself maintains that what happened in Chile under Pinochet was genocide.[172]

There are almost graphic connections between the political killings effected by the South American juntas and the free-market revolution imposed upon the same countries. Here, the propinquity between neoliberal enterprises and dictatorial death squad activity is almost bizarre. An abandoned torture center was unearthed underneath the Galerías Pacífico mall in Buenos Aires. During the dictatorship, the First Army Corps had hidden some of its disappeared in the bowels of the mall. This mall is today a jewel of globalized capital.[173]

In Brazil, several multinational corporations following neoliberal doctrine and practice coalesced to finance their very own privatized torture squads to be deployed in South American countries colonized by Chicago School economic ideology. In the same country, in mid-1969, just as the CIA-backed junta entered its most brutal phase, an extra-legal State police force launched Operation Bandeirantes. They were funded by various multinationals, including Ford and General Motors. Since these companies were outside official military and political structures, they enjoyed wonderful flexibility as regards the deployment of interrogation techniques on their *homines sacri*. Their sadism was unparalleled. The collusion between neoliberal capitalist rapacity and the draconian machinations of the various South American dictatorships is inter alia evinced by the fact that Ford had supplied the Brazilian and Argentinian juntas with cars with which they could conduct their abductions. As a consequence, the Ford factory in suburban Buenos Aires was turned into an armed camp. In the factory, the military hooded union members and dragged them to a detention facility inside the factory gates where they were beaten, kicked and electroshocked. Thereafter, these men were taken to outside prisons to be tortured for months.[174]

The mass production of opponents to neoliberal despotism as *homines sacri*, would indeed spread to many countries within the Southern Cone subjected to the gargantuan Chicago School laboratory. If the neoliberal revolution would at all succeed, maintains Klein, it would have to destroy the seed that was heretofore sown during Latin America's prior leftward surge. The Allendes of the continent would have to be annihilated—extrajudicially. In this connection, the Chicago School despotic client regimes in Brazil, Chile, Uruguay and Argentina expressed this intent with terms such as cleaning, scrubbing and uprooting. In Brazil, for instance, the junta's roundups of pernicious leftists were code-named *Operação Limpeza*, or Operation Cleanup. Within the genocidal intent of these regimes, the constitution of the anti-Statist recalcitrants as *homo sacer* is palpable. On the day of the coup, Pinochet referred to Allende and his cabinet as "that filth that was going to ruin the country." One month later, he pledged to "extirpate the root of evil from Chile," to effect a "moral cleansing" of the nation, "purified of vices." This calls to mind the asseveration of Hitler's henchman Alfred Rosenberg: "a merciless cleansing with an iron broom."[175]

These cleanup operations of latter day *homines sacri* would even target economists recalcitrant to neoliberal hegemony. Deemed "pink," they had to die. These were the men the Chicago School could not get rid of prior to the various CIA-backed coups in Chile, Argentina and Brazil, and so forth. Six students at the University of Chile were shot dead on sight at the main entrance to the School of Economics. Yes, economists do kill each other. One type of capitalist kills another variety of capitalist. When the junta siezed power in Argentina, soldiers imprisoned seventeen academics at the University of the South in Bahía Blanca on the basis of "subversive instruction." Most were from the economics department. "It is necessary to destroy the sources which feed, form and indoctrinate the subversive delinquent," said one of the generals at a press conference. Operation Clarity purged eight-thousand "ideologically suspect" leftist educators.[176]

Under the Argentinian Chicago School-backed junta, leaders of the Argentine Agrarian Leagues spreading ideas about the right of peasants to own land were hunted down and tortured, often in the fields in which they worked. They were tortured in situ with electric cattle prods. In the slums, philanthropic community workers were tortured with electric shocks—and almost certainly killed. Six high school students were tortured and murdered in September 1976 for pleading lower bus fares.[177]

In the wake of the neoliberal CIA-backed coup in Chile on September 11, 1973, roughly 13,500 civilians were arrested by Pinochet's minions, loaded onto trucks and imprisoned without charge. Thousands were taken to the two main football stadiums in Santiago, where death replaced football as a public spectacle. Purported subversives were hauled off to locker rooms and skyboxes transformed into impromptu torture chambers. Hundreds were executed. Lifeless bodies lay strewn on the side of major highways. Corpses floated in murky canals. In a four day trail of blood led by Pinochet's most ruthless general Sergio Arellano Stark, *homines sacri* held in prisons in the northern provinces were singled out as many as twenty-six at a time and executed.[178]

Rather than formally arresting their prey, the soldiers of Pinochet's neoliberal US-backed regime would snatch them, take them to clandestine camps, torture, kill them and throw their bodies into mass graves, or cut their stomaches open and throw them into the ocean from aircraft.[179]

All militaries and State police of the dictatorships of the Chicago School's neoliberal laboratory in the Southern Cone, such as Chile, Brazil and Uruguay, were trained by the CIA.[180] By the mid-70s disappearances would become the primary enforcement mechanism of Chicago School juntas in the Southern Cone. None embraced this technique more zealously than the Argentinian junta, throwing their victims from planes into the muddy waters of the Rio de la Plata. By the end of its reign more than thirty-thousand had been disappeared.[181]

This strictly fascist constellation of neoliberal geopolitical colonization, and its nefarious depredations, calls to mind one of the forms of evil whereby a Badiousian truth-event is distorted, occluded, obfuscated, or mendaciously wrought. It falls under the rubric of "simulacrum and terror."[182]

In the case of the simulacrum of an event, Badiou argues that fidelity does not take the universality of the void of a situation as its point of reference but rather the particularity of a closed ensemble, such as a particular race, land, culture, or ideology. Therefore, even though fidelity to the simulacrum often alludes strongly to, for instance, "the National Socialist Revolution," or in our case "the Free Market Revolution," it illegitimately substantializes the void in accordance with a certain overwrought plenitude which it painstakingly strains to augment. In fact, no singular and evanescent event has occurred at all. Chicago School economics has no evental status, as it is merely the purely instrumental passing of goods through a vastly more expedited mechanism of circulation. It merely fabricates a pseudo-event, or simulacrum. The conjunction of extreme financial capitalism with brutal militarism by mercenary proxies produces cadavers, not truths.[183]

In the case of a real event, fidelity assumes the form of the patient labor of construction of a generic subset which emerges as the being of the truth issuing from the event concerned. Fidelity to a simulacrum of an event, on the other hand, engages in a labor of construction of the "event-substance" which fascism extols to the wholesale exclusion of the void. The void is, in the case of fascism, "avoided" and, so that fidelity to the simulacrum can construct the fallacious ideological substance it promotes, returns as the viciously necrophilous universal voiding of all that exists around the ideological plenum under construction.[184]

The pervasive operation of voiding wrought by fidelity to a simulacrum, in that fidelity of its very nature is necessarily universally addressed, therefore immediately assumes the form of a universal dealing of death across a significant expanse. This death, as we have seen in the case of neoliberal market fundamentalism conjoined with the inexorable adjunct of foul militarist juntas and other client regimes of the great hegemon, comes in the guise of war, assassination and massacre. These bloodcurdling vicissitudes, far from being the mere means of this form of fascist fidelity, are its very real.[185] Much as under Nazism when the void recrudesced under the name of the "Jew," so too under the draconian global regime of neoliberal free market fundamentalism the void returns as the collection of all those deemed recalcitrants to or saboteurs of its financial and military imperialism. To void all such categories of sedition, contemporary capitalist imperialism must resort to increasingly sophisticated forms of extra-judicial liquidation. That it is to say, it must resort to assassination. Here, we may again cite Robespierre's dementedly vitriolic invective, with the caveat that its application has been reversed as the prerogative of neoliberal Statist power: "Slander, arson, poison, atheism, corruption, starvation, and murder—they have been prodigal in every sort of crime: but there still remains assassination, assassination and again assassination."[186]

By the mid-70s, the spate of assassinations perpetrated by CIA-backed death squads in various parts of the world was veering into unmanageability. In fact, this situation had become so chaotic that it moved President Gerald Ford to issue an executive order explicitly banning the United States from carrying out "political assassinations."[187] One of the reasons for this was that a scandal had rocked the Nixon White House in the 1970s with the secret bombings of Laos and Cambodia, and revelations of a domestic "enemies" list.[188] Ford's ban was later made more comprehensive by President Jimmy Carter, who extended the assassination ban to US proxies and contractors.[189]

Notwithstanding these curbs, Reagan, George W. Bush and Bill Clinton did not find it particularly difficult to find ways of circumnavigating the ban. Reagan, for instance, would authorize a strike on the home of Libyan dictator Muammar Gaddafi in 1986.[190] In the early 1980s, the Reagan administration covertly supported right-wing Contra death squads against the leftist Sandinistas in Nicaragua.[191] George H.W. Bush would authorize a strike on Saddam Hussein's palaces during the 1991 Gulf War, as would Clinton during Operation Desert Fox in 1998. Already under the Clinton administration, plans were drafted for killing or capturing Osama bin Laden and other al-Qaeda leaders. The caveat introduced by Clinton was that the assassination ban was nowise applicable to foreign terrorists engaged in plotting attacks against the United States. Indeed, in the wake of the bombings of US Embassies in Kenya and Tanzania in late 1998, Clinton authorized cruise missile strikes against alleged al-Qaeda camps in Afghanistan.[192]

Moreover, the CIA's by now renouned rendition program did have its beginnings in the mid-90s under the Clinton Administration. The suspension of the law endemic to the production of *homines sacri* was likewise here in evidence. For the program stemmed from a presidential directive authorizing the CIA and US Special Operations Forces, in conjunction with the FBI, to apprehend terror suspects from across the globe. This was, of course, done without regard of bilateral extradition agreements or international conventions. Clinton's directive also allowed terror suspects to be spirited off to locations at a distant remove from the reach of US law and due legal process. They were, thus, sent to inter alia Egypt, where they had no legal rights and were interrogated by *mukhabarat* agents unhindered by US prohibitions against torture.[193] Notwithstanding, under Clinton, although structures were created and "lethal findings" authorizing the killing of terrorists across the globe signed by the President, this was still all restricted to very surgical and rare cases.[194]

All of this was, however, to change rather drastically. In the wake of the September 11 terrorist attacks, the scope of the United States executive to suspend the law and thereby effect the homosaceralization of individuals was vastly expanded. The bureacracy of oversight and legal review, painstakingly erected over a succession of past administrations, was soon to be progressively dismantled at the behest of Cheney, Rumsfeld and their neoconservative deputies. This would fling open the door for the deployment of nefarious tactics on an unprecedented scale. These would include covert action, black operations, the use of secret prisons, snatch operations and so-termed "High Value Targeting"—that is to say, straight-forward assassinations.[195]

Cheney and Rumsfeld resolved to remove what they saw as the noisome legal restraints and suffocating congressional oversight that for too long had served to hobble covert action. Forthwith, it was devised to reduce lawyers' role to the mere rubber-stamping of secret policies and to consult only a few selected congressmen in such matters. The White House would now internally censor and redact briefings to Congress, implying that US lawmakers would receive vastly diluted versions of these. The express aim in the months following September 11 was to break all bureaucratic obstacles barring the way of the most thanatic homosaceralizing deployment of military force. It was desired that the interagency reviews of proposed lethal actions that were routine under the Clinton Presidency be gutted. Soon after the September 11 attack, the White House tasked a group of senior administrative lawyers with legally concocting legal justifications of torture (one of the recrudescences under the capitalist regime of the despotic regime's trial by ordeal), kidnapping and assassination.[196]

On September 14, 2001, the House of Representatives and the Senate passed the Authorization for Use of Military Force (AUMF). This gave President George W. Bush unprecedented latitude in the prosecution of what would be an increasingly global war. It authorized the deployment of force against individuals whom the president determined as having planned, authorized, committed, or aided the September 11 attacks, in addition to those deemed to have harbored organizations and persons linked to the said atrocity. The administration took the authorization as an endorsement for homosaceralization via assassination. On September 18, the AUMF was signed publically by the president. Something of significance had, however, transpired the day before. Bush had signed a secret presidential directive authorizing the CIA to capture and detain suspected militants across the globe. This would soon enable the rapid proliferation of a network of "black sites" for the detention and interrogation of present-day *homines sacri*. The directive also eviscerated much of the mechanisms of oversight and interagency review that had formerly vetted the process of authorizing targeted killings.[197] All of these foregoing measures led to a sudden and massive exacerbation in the homosaceralization of swathes of humanity.

The extra-legal, or even illegal, nature of the foregoing acts of resurgent sovereignty undertaken by the White House may be see in light of the National Security Act of 1947. According to the latter, not only must the president issue a finding before commencing covert action, but the action arising from the finding must itself in no sense contravene US Law or the Constitution. The presidential finding signed by George W. Bush on September 17, 2001, in vast contradistinction, declared all covert homosaceralizing action to be *pre*-authorized. This declaration, moreover, led to the creation of a highly classified and secret program code-named Greystone (GST) which, in the view of not a few, violated the National Security Act. GST was, in effect, a global assassination and kidnap operation. Still further, its operations no longer even required direct presidential approval for each separate action undertaken. Cofer Black, head of the Counterterrorism Center and whom Judith Butler would likely term a petty sovereign, could henceforth directly order deadly hits.[198]

What had begun to transpire in this burgeoning resurgence of neosovereignty unfolding with the "global war on terror" was a recalibration of the very meaning of combat as prosecuted by the State. Statist violence burst through its traditional limits, in the process becoming indefinite in both space and time. In this connection, Grégoire Chamayou contends that the world had become not so much a battlefield, but rather a "hunting ground." This is so because warfare is defined as combat. Hunting, on the other hand, is essentially pursuit. These two are associated with two distinct types of geography. Combat transpires wherever opposing forces clash, whereas hunting occurs wherever the quarry pursued by the State chances to go. Thus, from the purview of the hunter State, armed violence ceases to be defined within a demarcated zone, but is instead constituted by the simple presence of the homosaceralized prey who carries his mobile zone of hostility with him.[199]

This dismemberment of the notion of a "zone of armed conflict" as geographically circumscribed in favor of one demarcated by the bodies of targeted *homines sacri* had by now likewise increasingly come to be touted by a whole contingent of US lawyers. According to their logic, the zone of conflict, now strictly one of homosaceralization, would have to move with the quarry, irrespective of geography. This amounted to a justification of the deployment of extra-legal lethal force in violation of the geocentric interpretation of the laws of warfare.[200]

On February 7, 2002, President George W. Bush issued an executive order declaring that the Geneva Convention did not apply to al-Qaeda or Taliban operatives. Swiftly hereupon, circumvention of both US and international law would be complemented by circumvention of traditional US military command structures. Such intentions would be communicated by Donald Rumsfeld on July 22, 2002, in a secret directive to SOCOM commander General Charles Holland. The directive envisioned a far more decentralized manhunt prosecuted in a manner resembling the machinations, and depredations, of a private hit team. Rumsfeld's objective was not so much the legal arrest of terrorists, but their extra-legal capture or impunious liquidation.[201] Accordingly, US Special Operations Forces would have to be restructured so as to render lethal global operations faster and leaner, and to free the latter from needless bureaucratic meddling. The Special Mission Units (SMUs) of JSOC would no longer need to coordinate with conventional command authorities. Instead, they were permitted to conduct their discrete global operations more-or-less autonomously. Congress would henceforth have little or no oversight of lethal covert operations conducted by Special Operations.[202] Colonel Lang referred to the post-September 11 homosacralization formula: "You know, the famous 'Cheney one percent' thing – if there's any doubt, you kill 'em. That's basically it, either capture them, or kill them."[203]

The decentralization and autonomization of these lethal operations would be reflected at the level of deployment of the so-termed "kill box." A kill box is a temporary autonomous zone of slaughter of *homines sacri* projected onto a featureless geopolitical space and specifically designed to allow air assets, such as drones, to destroy surface targets without further coordination with command centers. Once a kill box is opened and activated, the operator may fire at will, after which the box is closed.[204] In a 2005 memo to secretary of defense Donald Rumsfeld, the president of the RAND corporation described the efficacy of kill boxes as follows: "Kill boxes can be seized for open terrain or urban warfare and open or closed quickly in response to a dynamic military situation."[205] Chamayou adduces from this susceptibility of the kill box to intermittence and modulation as regards scale that it can easily be extended far beyond zones of declared conflict, that is, into extra-legal areas. For such lethal microcubes can be opened anywhere in the world where targets deemed *homines sacri* are seen to lurk.[206]

One of the theaters for the deployment of the US executive's newfangled kill/capture machine would be Iraq. It would be run by Stanley McChrystal and be centered on JSOC. Most crucially, it would be accountable to no one but a small group of White House and Pentagon insiders, that is, to only the upper echelons of the executive. Its lethal machinations would be a clear instance of the recrudescence of archaic sovereignty out of the ambit of governmentality. Within months, the targeted kill/capture program would begin to resemble the CIA's Phoenix Program from the Vietnam War, wherein the CIA, supported by US Special Operations Forces and colluding with indigenous militias, waged a vicious campaign to assassinate Vietcong members. The Phoenix Program was, in essence, a well-organized death squad.[207]

Already here war had degenerated into the simple act of putting individuals to death in what was approaching a unilateral relationship of necrophilous death-dealing.[208] For Chamayou, this dialectical shift entails at once the decisive coextensive transition from a military ethic of combat in the direction of what he terms a "necroethic," wherein traditional principles of *jus in bello* find themselves co-opted to the ends of justifying decriminalized murder. This resurgent archaism is, of course, nothing but an ethic for executioners or butchers.[209]

The supporters of such outright necrophilous elimination would contrive "legal" ballast for this homosaceralization through an unorthodox hybridization of police and penal praxis, a forceful distortion of the law of warfare whereby unilateral murder came to be condoned through the monopolistic subsumption of *jus in bello* under the condition of *jus ad bellum*.[210] Hence, the emergence of the "just" warrior armed with the right to kill freely without the commission of a crime.[211]

In May 2003, this homosaceralizing necro-ethic would gain a surge in momentum when a new regime of death was put into operation in Iraq by Paul Bremer's chief security adviser James Steele. Termed "the Salvador Option," this regime was, in fact, not particularly new at all.[212] For, in scenes reminiscent of earlier happenings in El Salvador, death squads operated by marauding Iraqi commandos would roam around murdering purported opponents of the Occupation regime. Mutilated and lifeless bodies littered the roadsides. These corpses had bullet holes in the head, hands still cuffed behind their backs, or holes in the skull made with drills. Countless bodies poured daily into the Baghdad morgue, including scores of corpses with wrists bound by police cuffs.[213] The extra-judicial nature of all these depredations conducted at the behest of the executive by its minions is self-evident, and would continue. The suspects detained at Camp NAMA—a small compound rounded by barbed wire near Baghdad International Airport—were accorded no rights whatsoever as prisoners of war (POWs). They were instead classified as unlawful combatants, debarred from receiving legal aid, and prevented from being visited by the Red Cross. Indeed, they were not even charged with any crimes.[214]

In that they were nowise subject to legal oversight, US special operations forces in Iraq could operate with absolute impunity. They were simply angels of death directly answerable only to the highest levels of the US executive, sans legal or bureaucratic intermediaries. JSOC entirely skirted all intermediate levels, such as local command, Baghdad command, CENTCOM and the Pentagon; instead receiving their orders directly from Undersecretary of Defense Stephen Cambone and Secretary of Defense Rumsfeld. As an inexorable adjunct to this homosaceralization process, the law was suspended in myriad ways. A number of the normal rules concerning detention and interrogation ceased to be applicable to JSOC.

For instance, the Special Access Program under which the JSOC Task Force at NAMA operated condoned practices that not only deviated from normal military practice, but also violated military law and policy.[215] The Special Mission Unit at NAMA engaged in unlawful conduct and systematic violations of the Geneva Conventions. Here, the usual homosaceralization argument of post-industrial neosovereignty was touted: being unlawful combatants, the detainees were not entitled to Geneva Convention protection to begin with.[216] Meanwhile, in Abu Ghraib when on November 4th 2003 Manadel al-Jamadi was beaten to death by members of SEAL Team 7, nobody at all was charged with homicide.[217]

In early 2004, this zone of homosacralization would be extended from the ambit of merely the Iraqi and Afghan populace to a far wider sphere. The opening salvo would be Rumsfeld's signing of the al-Qaeda Network Execute Order, or AQN ExOrd. This secret order would forthwith authorize JSOC to pursue their al-Qaeda quarry "anywhere in the world" where it was suspected of operating or receiving sanctuary. This now vastly expanded zone of depredation would include Pakistan, Syria, Somalia, Yemen, Saudi Arabia, and other Gulf nations. The order would also allow quarry to be pursued across national boundaries. The extra-legal nature of these depredations is on display. For according to new guidelines issued by Cambone, Special Ops Forces were endowed with the "right" to conduct clandestine HUMINT operations even before bringing this to the attention of Congress. Consequently, the latter body was now held at a still greater remove, law being even more effectively suspended. Adding more fuel to the fire, National Security Presidential Directive-38 (NSPD-38) gave JSOC hitherto unimaginable lattitude, going so far as effectively pre-authorizing lethal operations outside of any stated battlefield. Through the instrumentality of JSOC, the White House could now conduct assassinations of "irreconcilables" across the entire globe.[218]

The implied broadening of the terrain of lethal depredation to the entire surface of the planet may be adduced from Derek Gregory's observation that the "legal logic through which the battlespace is extended beyond the declared zone of combat in Afghanistan is itself infinitely extendable."[219]

In this connection, Chamayou, referring again to the aforementioned redefinition of the terrain of armed conflict as the mobile zone attached to homosaceralized individuals singled out for eradication; contends that this ultimately amounts to the conferral and justification, under the cover of laws of armed conflict, of the right to execute quarry anywhere in the world.[220] And, of course, illegally and without recourse to further administrative procedures.[221] Self-evidently, as Human Rights Watch would put to President Obama in 2010, this "notion that the entire world is automatically by extension a battleground in which the laws of war are applicable is contrary to international law."[222] This statement appears as a truism under the condition that the classical notion of armed conflict as legally circumscribed in terms of space holds.[223]

The legal implications for especially the drone-driven eradication of those deemed *homines sacri* are here vast. For international law in no sense recognizes the right to kill outside actual armed conflict. This is so because, as a legal category, it deems warfare, in Chamayou's words, a "geographically defined object." Therefore, in that the laws of war pertain only to the physical place were fighting occurs, outside such spatial loci States have no right to belligerence. In accordance with these strictures, the "global war on terror" is not an armed conflict at all. It is, therefore, completely illegal.[224]

The preeminence of executive power, extra-legality and homosaceralization would in no sense be attenuated by the transition from the George W. Bush to the Obama administrations. In fact, quite the reverse would transpire. During his first year in office, Obama would aggressively embrace assassination, firmly maintaining and consolidating its central position within US national security policy. The emphasis would, however, shift somewhat. Whereas hitherto high-value-targetting had been a capture or kill operation, now it was decided simply to kill recalcitrants.[225] Obama would, in fact, up the ante quite prodigiously on assassinations, and in so doing fulfil Rumsfeld's vision of the world as a battlefield more faithfully than ever before. Soon after being elected, he would give carte blanche to JSOC and the CIA to wage a global manhunt. JSOC would not only be tasked with killing al-Qaeda's top leadership, but with decimating its support infrastructure, killing its way down the chain.[226]

Within weeks of assuming office in early 2009, Obama made clear his intention to keep the most aggressive counterterrorism policies of the Bush Administration very much intact. Targeted killings would be retained, as would secret prisons. The evisceration of *habeas corpus* rights for detainees would continue, as would indefinite detention, and extraordinary rendition. As we shall see, drone strikes would be stepped up prodigiously.[227]

This escalation under Obama was already presaged by certain pivotal decisions made in the closing months of the Bush Administration. The CIA had expanded the basis for targeting purported terrorists. Concrete intelligence was no longer the sine que non of pursuing quarry; malign intent extrapollated from more vague and disparate patterns of life would now suffice. These would be the so-termed "signature strikes," made against individuals whose precise identity remains unknown and who would be targeted solely on the basis of behaviors merely suggestive of association with terrorist networks. The designated *homines sacri* would then be found, fixed, and finished.[228]

Obama escalated Bush's already concerted deployment of drones, beginning to strike in Pakistan almost weekly.[229] On September 30, 2009, Obama convened a meeting in the Situation Room of the White House, at which General Petraeus signed the Joint Unconventioal Warfare Task Force (JUWTF) Execute Order. This obviated the need for US Special Operations teams, conducting clandestine capture/kill operations outside the hitherto stated battlefields of Iraq and Afghanistan, to provide regular reports to Congress. Through this directive, capture/kill operations prosecuted at a far remove from regular battle zones would become more systematic and long term. The small JSOC teams tasked under this order were ravenous.[230]

The scope for homosaceralization encompassed by JSOC's High Value Target list would be expanded. This increasingly comprehensive global reach of lethal depredations unleashed by the US State apparatus was only bolstered by the improving precision of the death-wielding drone technology deployed. Chamayou refers here to what he sees as a paradoxical double-movement serving to completely dislocate the "spatiolegal" conception of a zone of armed belligerence. On the one hand, in what he terms the principle of "precision or specification," the zone of conflict is fragmented into miniaturizable kill boxes virtually reducible to as small a space as the individual body of pursued quarry. On the other hand, in accordance with the principle of "globalization or homogenization," the said lethal microcubes must be deployable in any geographical place whatsoever, to the extent that the entire planet becomes a theater of necrophilous Statist machinations.[231] The extra-legality here is extreme. For, the conjunction of these two principles implies that *homines sacri* can now be eradicated outside *any* declared war zone.[232]

Obama was turning the screws so tightly that, by October 2009, he had already authorized as many drone strikes in ten months as Bush had in his entire eight years in office. At this time, he would go still further. The area in Pakistan in which the CIA could pursue *homines sacri* was expanded, and it was authorized to acquire more drones with which to kill. Homosacerization had moved into complete overdrive.[233] One of Scahill's interviewees maintained that, under Obama, things were happening that were unthinkable even under Bush. JSOC was allowed to run wild with its assassination campaign. Obama not only freely approved of lethal operations; he spurred his charges on to be even more aggressive, innovative and proactive in taking down their quarry.[234] The expansion and intensification of these operations would spill into Yemen, where JSOC would hunt down and murder al-Qaeda linked militants.[235]

Awesome as this roam and reach of homosaceralization may be, the prognostications of some US military strategists are simply frightening.[236] In twenty-five years time, it is averred, autonomous robotic nano-drones, capable of marauding in a swarm and miniaturized to navigate through the tiniest of spaces, will extra-legally destroy the individual bodies of future generations of *homines sacri*.[237] Your own bedroom could be a war zone. The microcube of death would be your body. The dead person, by implication, would be yourself.[238]

By the end of 2010, lists of present-day *homines sacri* to be assassinated were maintained by at least three entities linked to the US government: the National Security Council, the CIA, and the US military itself. On almost every Tuesday of the week, more than one-hundred members drawn from these security agencies would convene to mull over the biographies of terrorist suspects so as to recommend to Obama which targets on the list were to be eliminated. This "kill list" was thereupon sent to the White House for the oral approval of the president. The condemned individuals would then be annihilated by drones.[239]

The preeminence accorded to the executive characteristic of the recrudescence of archaic sovereignty within the ambit of governmentality is clearly echoed in CIA director Michael Hayden's averment that the administration's homosaceralization program rested upon the pure personal legitimacy of the president and was nowise legally sustainable. In general, it was the very structure of the assassination system created by Obama that precluded knowledge of evidence or how the situation could be redressed. This is what CIA case officer Philip Giraldi maintained. According to one military official, the Obama administration had authorized certain actions which the Bush administration never would have approved.[240] Part of the rationale behind this extra-legality, as Chamayou avers, resides in the fact that the administration had by now long since departed from standard counter-insurgency policy and moved to anti-terrorism. Whereas in the former paradigm insurgents were viewed as representatives of certain deeper claims within the body politic, in the latter they are regarded as aberrant individuals, incarnations of pure evil, or quite simply insane. They are no longer adversaries to be opposed, but criminals to be annihilated. Anti-terrorism aims not to defeat insurgents' strategy, but to destroy as many individual suspects as possible.[241]

The Agambenian couplet comprised of the despotic sovereign and bare human life placed at opposing ends is here in clear evidence. The United States government, as well-nigh divine sovereign, functioning at once as investigator, policeman, judge, jury and executioner, would liquidate *homines sacri* with the mere extention of an index finger. Those homosaceralized could well be US citizens. All noisome obstacles, such as the right to trial and democratic safeguards, passed into desuetude.[242] This all enabled the administration to attain its aim of neutralizing new *homines sacri* as they emerged on the kill list. In a pattern of regular elimination, the US executive was in the process of periodic reaping approaching infinite eradication.[243]

What had come into existence under the Obama administration was, in fact, an alternative legal structure. Evidently, this structure was of questionable legality, if not strictly extra-legal. With the functions of prosecution, judge, jury and executioner all telescoped into the United States executive; Obama, as high sovereign, would preside over the homosaceralization and annihilation of unqualified bare human life.[244] What this alternative legal structure amounts to is strongly homologous to what law professor Kenneth Anderson terms "naked self-defense": a new regime of ad hoc law for lethal operations that dispenses with the ordinary legal strictures confining national self-defense.[245]

As antecedent, Anderson cites State Department Abraham Soafer's doctrine of the late eighties that: "Targeted killings in self-defense have been authoritatively determined by the federal government to fall outside the assassination prohibition."[246]

Chamayou maintains this reasoning to imply that simple government fiat can now effectively serve to "decriminalize" targeted assassinations. What the latter likewise brings into existence is a very peculiar legal hybrid at once amalgamating the liberties encountered in both conventional warfare and simple policing, whilst at the same time being unhindered by the constraints of either. Here, militarized man-hunting assumes purportedly plausible legal form in the guise of the global right of lethal policing.[247]

This juridical casuistry allows Anderson to conclude that "self-defense gives the discretionary ability to attack anywhere in the world where a target is located, without having to make claims about a state of armed conflict everywhere and always across the world."[248] Chamayou is, however, right in averring that the foregoing solution is in no sense distinguishable from an unbridled license to kill.[249]

It is also what would be concretely enacted. For already during Obama's first term in office, the administration had transformed formerly disparate and quasi-improvisatory elements into a comprehensive institutionalized counterterrorism assassination infrastructure capable of sustaining a seemingly permanent war. By the time Obama had been inaugurated for his second term on January 21, 2013, and John Brennan had replaced David Petraeus as CIA director, assassinations of present-day of *homines sacri* had become so routine that the framework and processes sustaining it had to be streamlined. Pertinent in this connection is a US Department of Justice "white paper" which surfaced in early 2013. The document delineates the "Lawfulness of Lethal Operations Directed Against a US citizen." It regards determination from a "well-informed high-level administration official" that a given individual represents a "imminent threat" to the US as sufficient ground for ordering the assassination of even an American citizen.[250] What was here meant by "imminent" was, however, quite significantly expanded by the Justice Department. They wrote: "The condition that an operational leader present an 'imminent' threat of violent attack against the United States does not require the Uniter States to have clear evidence that a specific attack on US persons will take place in the near future."[251] Jameel Jaffer of the ACLU called the white paper a "chilling document," saying that "it argues that the government has the right to carry out the extrajudicial killing of an American citizen."[252] Jaffer added: "This power is going to be available to the next administration and the one after that, and it's going to be available in every future conflict, not just the conflict against al-Qaeda. And according to the administration, the power is available all over the world, not just on geographically cabined battle-fields."[253]

Michael Boyle of LaSalle University: "The result of the 'guilt by association' approach has been a gradual loosening of the standards by which the US selects targets for drone strikes,"[254] he contends. In a similar vein and in connection with what we may here term the *future of neosovereignty's homosaceralization of generic humanity*, Scahill avers that this neatly streamlined and vastly expanded framework and process for the assassination of real and perceived enemies of the United States shall stand as the inheritance of future American presidents—whether Republican or Democratic. Rationalized in the name of national security, future administrations will inherit an executive branch festooned with enormous means for destroying life.[255] "Obama has routinized and normalized extrajudicial killing from the Oval Office," charges Boyle. "Without the scrutiny of the legislature and the courts, and outside the public eye, Obama is authorizing murder on a weekly basis, with a discussion of the guilt or innocence of candidates for the 'kill list' being resolved in secret."[256] Under this regime, we are all potentially *homines sacri.*

13. The return of trial by ordeal

This return of *homo sacer* under neoliberal capitalism is at once accompanied by the concomitant resurgence of another feature of the Deleuzoguattarian despotic regime, that is, trial by ordeal. We recall that, as Foucault avers of judicial torture in the European Classical Age, the pain applied to the victim's body was under those particular despotic regimes necessary to produce the "truth" of his transgression of State law. Judicial torture would be used to elicit the accused's confession of his transgression. In the prosecution under the Bush administration of the "war on terror," the US, so as to produce the criminal "truth" of high-value detainees extra-legally snatched through its rendition program, would draw on its long-standing Survival, Evasion, Resistance, Escape (SERE) program. This effective repository of much of humanity's knowledge on the subject of torture, formerly deployed to train US soldiers to survive torture at the hands of enemy armies or their special forces, was reverse-engineered after the September 11 strikes and came to be extensively deployed on captured terror suspects.[257]

The CIA had been obsessed with Soviet techniques of extracting false confessions from purported enemies of State and, to this end, meticulously studied torture techniques and modes of interrogation employed in preparation for the show trials of Stalin's Great Terror. In an attempt at eliciting true, rather than false, confessions from modern day *homines sacri*, the CIA would co-opt two specific facets of Stalinist interrogations: "self-inflicted pain" and "sensory deprivation." The first aspect entailed forcing the detainee to assume various stationary stress positions, such as forced standing, that had to be maintained for prodigious periods of times and so as to stealthily induce unbearable pain and discomfort. The second aspect, in that it likewise necessarily included immense periods of isolation from all other human beings, turned the detainee into a highly malleable psychological entity. For, out of an unearthly desperation for human contact, he would almost invariably develop a hopeless codependency on his interrogator, to the extent of tenderly and lovingly looking upon the latter as a veritable father figure who was herein enabled to easily induce him into spilling his guts.[258]

The entire gamut of knowledge and experience the CIA had acquired about coercive interrogation was compiled into what may be regarded as the holy scriptures of this dark genre: the 1963 *Kubark Manual*, and its companion volume, the 1983 *Human Resource Exploitation Training Manual*.[259]

The preponderant mechanism through which the purported truth of the prisoner's anti-Statist infractions was produced, was the manufacture of as high a level of anxiety as possible in the body and mind of the detainee, without of course raising this stress level to such a degree as to render the culprit useless or paralytic. To this end, experiments in which SERE-affiliated scientists charted the fluctuations in the cortisol levels of test guinea pigs by analyzing their blood and saliva, were sedulously conducted.[260] The most valued method of priming the body and mind of the prisoner as a site for the production of purported criminal truth at the behest of anxiety, was the creation of an environment of radical uncertainty for captives. To induce such a state of mind, Scientists and psychologists working for the SERE program would employ such means as disruption of sleep patterns, prolonged starvation, and forced nakedness.[261] Sleep deprivation and the constant distraction induced by intense pain would raise anxiety levels to a significant enough level to severely undermine the captive's capacity to exert control over his own behavior, thereby loosening his tongue. SERE-affiliated psychologists steadfastly maintained that to extract truth out of terror suspects one would necessarily have to seriously hurt them both psychically and psychologically.[262]

Indeed, as averred by former Israeli Prime Minister, Menachem Begin, who was tortured by the KGB as a young adult, sleep deprivation is absolutely the most unbearable form of torment and at the same time the most effective in eliciting confessions from the prisoner. Frazzled by a haze-addled mind, his legs unsteady, and his spirit wearied to death, the desire and utterly incorrigible longing of the prisoner to sleep is so strong that it exceeds even the ravages of hunger and thirst—to the extent that, even in the event of the interrogator positively refusing the captive liberty yet at the same time granting him the opportunity to sleep for even a short time, the prisoner will inevitably confess as he has completely lost the ability to think rationally.[263]

The beguiling aspect of this post-September 11 resurgence of sleep deprivation as a extra-judicial technique of producing the, albeit highly distorted, truth of *homo sacer*'s crime against the State apparatus is that it, in point of fact, marks a bizarre antediluvian recrudescence of a practice espoused in medieval times solely by witch hunters to elicit evidence of collusion, and perhaps even fornication, with the Devil.[264]

Related techniques used by CIA-linked interrogators on rendered detainees arose from the pioneering experiments conducted at the University of Pennsylvania under Professor Martin Seligman and geared to the end of breeding humans as confessing machines through their induction into a state of "learned helplessness."[265] The beastialization required to render up human subjects as *homines sacri* is here in graphic evidence, in that Seligman's experiments of the sixties involved administering powerful electric shocks to dogs so as to break their resolve, and even ability, to escape from a cage.[266]

"Learned helplessness" was adopted as the overarching paradigm for the production of captives as freely talking beasts by James Mitchell, the retired army psychiatrist massively influential in shaping the CIA's interrogation techniques. Much as in the case of the KGB model adopted and perfected by the CIA, the aim was to eliminate the detainee's capacity to forecast even his own short-term future—so as to inculcate dread, disorientation and dependency on interrogators, and thereby breed a confessing animal.[267]

From October 11, 2002, the list of "truth"-eliciting techniques inspired by the CIA's Cold War-era *Kubark Manual* would include prolonged periods of isolation, reduction to nakedness, hooding, shackling in stress positions, religious and sexual humiliation, exploitation of phobias, and death threats.[268]

It is this blueprint that would underwrite the grisly fate of the CIA's first "high-value" detainee, Abu Zubayda. In a mock burial, he was locked like a cur for hours on end in a suffocating black coffin so small that he had to double up his limbs in an agonizing fetal position; thrust headfirst against a concrete wall with a dog collar around his neck; bombarded by bright lights, eardrum-shattering noises, and the sounds of crying babies for twenty-four hours a day; ultimately to confess merely to the most absurdly fictional gibberish.[269]

Another captive, Mohammed al-Qahtani, was subjected to intensive isolation for three months in a cell completely flooded with light and heated to a temperature approaching one-hundred degrees Fahrenheit;[270] bolted to the floor; allowed to sleep for only four hours a day for a continuous period of forty-eight days; stripped naked, forced to wear a leash and perform dog tricks; prevented from defecating; dehydrated to the point that doctors had to implant a shunt in his hand for intravenous force-feeding; forcibly given an enema; profaned with pornographic pictures hung around his neck to the point of not being able to pray; told that his mother and sister were whores and that he was a homosexual; surrounded by marauding rats; and forced to watch the September 11 attacks and grisly images of its victims repeatedly.[271] In the end, after having endlessly soiled himself in his own excrement, he begged his interrogators to permit him to commit suicide.[272]

But let us return now to an earlier point. That is to say, how is this ever-burgeoning homo sacralization of life related to the relative preponderance of machinic enslavement and social subjection?

14. Decoding-deterritorialization/ recoding-reterritorialization

The massive deterritorialization effected by the capitalist machine consists likewise in the incessant development of the social productivity of labor by way of its continuous revolutionizing of the means of production—a deterritorialization of labor and acceleration of productivity whose innate tendency is toward a productivity for productivity's sake. If left unhindered, this deterritorialization and decoding of flows will lead to a ubiquitous liberation of the flows of desire to the point of their running amock in a total destruction of the capitalist socius, hereby bringing desire back to the BwO disgorging itself from beneath the very innards of the socius.[273] This is so because capital, as Marx has it, "posits the *production of wealth* itself and hence the universal development of the productive forces, the constant overthrow of its prevailing presuppositions, as the presupposition of its reproduction. Value excludes no use value; i.e. includes no particular kind of consumption etc., of intercourse etc. as absolute condition; and likewise every degree of the development of the social forces of production, of intercourse, of knowledge etc. appears to it only as a barrier which it strives to overpower."[274]

This well-nigh inexorable deterritorializing tendency is endemic to capital as self-valorizing value. For capital is naught outside of its incessant autochthonian expansion impetus. Marx expresses this tendency of capital as its perpetual transgression of all boundaries with which it is faced. For indeed, "as representative of the general form of wealth—money—capital is the endless and limitless drive to go beyond its limiting barrier. Every boundary [*Grenze*] is and has to be a barrier [*Schranke*] for it. Else it would cease to be capital—money as self-reproductive. If ever it perceived a certain boundary not as a barrier, but became comfortable within it as a boundary, it would itself have declined from exchange value to use value, from the general form of wealth to a specific, substantial mode of the same."[275] And for this precise reason, capital "is the constant movement to create more of the same. The quantitative boundary of the surplus value appears to it as a mere natural barrier, as a neccessity which it constantly tries to violate and beyond which it constantly seeks to go."[276]

The capitalist socius, however, continually—as Deleuze and Guattari maintain—averts such self-destruction by repeatedly frustrating the unlimited development of the productivity of labor toward which its deterritorializing tendency strives. It does so by way of the incessant reterritorialization of labor onto capital. More precisely, it may be said that capitalism does indeed endlessly develop its productive forces—though it does so only to the ends of capital iself. These latter ends are, of course, those of the endless precipitation and absorption of surplus-value. And since capital is the movement of its own valorization, this limit to the potentially boundless revolutionization of production that it unleashes is none other than capital itself. As Marx maintains, the "*true barrier* to capitalist production is *capital itself.* It is that capital and its self-valorization appear as the starting and finishing point, as the motive and purpose of production; production is production only for *capital,* and not the reverse, i.e. the means of production are not simply means for a steadily expanding pattern of life for the *society* of the producers."[277] That is to say, capital's deterritorializing tendency wherein "the methods of production that capital must apply to its purpose and which set its course towards an unlimited expansion of production, to production as an end in itself, to an unrestricted development of the social productive powers of labour"; constantly finds itself countervailed and held in abeyance by reterritorializations whereby capital's expansion unceasingly runs up against the aforesaid "barrier within which the maintence and valorization of the capital-value has necessarily to move …"[278] More succinctly: the centripetal force of limitation that capitalist reterritorialization brings to bear upon its own centrifugal deterritorializing movement may be understood as arising from the inherent contradiction whereby "the means—the unrestricted development of the forces of social production—comes into persistent conflict with the restricted end, the valorization of the existing capital."[279]

The danger inherent to this contradiction is that capitalism incorrigibly produces more than can be consumed and converted back into liquid capital. As Philip Goodchild argues, to circumvent this, capitalism resorts to the maintenance of anachronistic sectors of reterritorialization so as to absorb this excessive production. To effect the latter, capitalism will often deploy mechanisms such as increased taxation and the prosecution of spurious and unjust wars in order to turn this its intrinsic weakness into a strength. Another ubiquitously resorted to technique is the artificial creation of wants and needs within the consciousness of the consuming public. Counter-intuitively, recessions in fact serve to bolster capitalist valorization and do so in the sense that instability of employment can increase profits through the lowering of wages. Another method of counteracting the deleterious effects of overproduction is through the invention of new technologies and pseudo-knowledges for the creation of new needs for consumption, with the caveat that these knowledges are very frequently a form of stupidity and out-and-out triteness. In all of this, the capitalist regime repeats the despotic regime's strategy of tethering the individual subject to debt through the ostensibly liberal and generous circulation of credit.[280]

Matters are, however, still more dire. For indebtedness becomes one of the primary means of man's economic submission. Žižek, following Maurizio Lazzarato, avers in this connection that the lives of capitalist subjects, instead of being Heideggerian *Da-Sein* (being-there), are now ontologically defined as "being-towards-debt."[281] Here, neoliberal ideology, extending the logic of market competition to all aspects of social life (such as health, education and voting), renders these facets investments by the individual in his own personal capital. Governance ceases to be that of disciplinary or biopolitical regimes, and is now instead that of the "entrepreneur-of-the-self." Risks are outsourced to private individuals, whereas provisions and protections are made contingent upon positive personal credit ratings and employability. Individuals become poorer and their financial situations more precarious. Instead of ensuring wages, pensions, housing and education through inalienable rights and securities, neoliberalism provides individual access to consumer credit, personal share portfolios, housing/mortgage credit, and student loans. In this way, workers become indebted to their employers, consumers become indebted, and citizens become indebted by having to take responsibility for their share in their State's foreign debt. One may recall how, in the primitive regime, debt was imminently assuaged and capable of being discharged through the cruel inculcation of a lexical memory of the future. In the despotic regime, however, debt became effectively infinite, unpayable and internalized as irrevocable guilt before God. This debt of the very existence of the subject was a formidable governmental tool of social subjection. With the transition to the capitalist regime being made, this debt is still infinite even though it is largely secularized. Here, the subject performs wage labor, but likewise works upon the self so as to produce himself as one who can assume guilt for indebtedness. As indebted, he is constantly subject to moral evaluation through which he is compelled to assume individual guilt for any prospective or actual defaulting on debt. Despite his ostensibly greater autonomy in comparison with the subject interpellated under disciplinary and welfare regimes, the preponderance of creditor over debtor is here still more extreme than at any prior moment in human history. For, deprived of the capacity to govern his own time or evaluate his own conduct, the purported autonomy of the "entrepreneur-of-the-self" is radically curtailed. Debt effectively becomes an openly ridiculous excess. One enters into the realm of obscenity: the debtor is from the outset never expected to repay the credit which he is so lavishly accorded. At the same time, debt becomes a crushing means of control and outright domination. This is also true at the level of indebted groups or countries, which are accused by creditors for feeling insufficient guilt for their national debt. The unreasonable and dictatorial pressures to impose austerity measures on countries such as Greece is a case of a hypertrophied renascence and unprecedented virulence of the Freudian superego, which bombards the self with impossible demands whilst at the same time beholding the debtor's failure to comply with them with an obscene enjoyment that verges on the pornographic. Not even expected to shed the shackles of his debt, the subject is effectively kept in permanent subordination.[282]

With the transition from Keynesian to neoliberal economic practices, the relation between decoding-deterritorialization, on the one hand, and recoding-reterritorialization, on the other, undergoes an extreme contortion—with potentially sinister portents. As Berardi argues, the said transition was roughly coextensive with two other closely imbricated transformations: the passage from the Fordist to the post-Fordist wage regime, and the movement from the industrial to the information age. This transition effected an increasing evisceration of all stable relations between socially necessary labor time and the determination of value, as the value of immaterial production and cognitive labor is difficult to properly gauge. Thus, the increasing preponderance and hegemony of semiocapitalism not only plunges the relation between existing units of measure and productive forces into a regime of indeterminacy, but likewise injects randomness into the sphere of law, moral judgment, social relations and ethics.[283]

The opening salvo and first dramatic move kick-starting this vertiginous and relentlessly accelerating process of decoding and deterritorialization was Nixon's 1971 decision to delink the dollar from the gold standard. This served to untether American capitalism from the constitutional framework agreed upon at Bretton Woods in 1944. International currencies would henceforth cease to have any reference to reality, as all correspondence between their value, conventional standards, and economic referents fell into sempiternal desuetude. With all relation to external economic materiality eviscerated, neoliberalism's intensive financialization of the economy would usher in a world in which money would create reality through a bizarre and endlessly proliferating series of acts of parthenogenesis.[284]

From the year 1971, we move on to the year 1977—the year that Berardi views as the turning point at which promiscuously pervasive neoliberal dynamics of decoding and deterritorialization effected an epochal shift from the age of human socio-political evolution to our prevailing times of devolution and atavistic civilizational erosion. At the head of this depressing process of decline stand financialization's nefarious vectors of de-realization, through which modernity's centuries-long legacy of public education, health care, and social welfare systems wrought by the conflictive alliance between bourgeoisie and industrial proletariat has suffered well-nigh terminal desuetude through the eviscerations of market fundamentalism. By the time of the second decade of the twenty-first century, the product of two-hundred years of industriousness and collective intelligence underwent immolation at the behest of a rabid deterritorializing process, whereby concrete social civilization was degraded through the ferocious force of mathematical abstraction: physical forms were dissimulated into evanescent images; visual art degraded to mere proliferation of spam; and poetic language debased into the depressing sub-sentient poverty of fake advertising. The very substance of real life had been devoured by the sheer voracity of financial accumulation.[285] With history emaciated and replaced by an incessant procession of infinitely permuting and recombining images, all that remains of collective solidarity, empathy and autonomy, is the merely evanescent upsurge of frantic and precarious pseudo-activity utterly devoid of political awareness and strategy. Berardi seriously doubts whether there is any hope or future beyond this hideous black hole.[286]

As Lazzarato maintains, capitalist decoding and deterritorialization is associated with machinic enslavement. For the latter works on the deterritorialization and decodification of "subjects" and "objects," that is to say, with the molecular and non-individuated, intensive, subhuman potentialities of subjectivity, matter and machines alike. Money and finance deterritorialize and decode social "matter," as exemplified by the last thirty or so years of neoliberalism. They undermine and circumvent laws, such as those dealing with labor, as well as the codes of social, economic, and political subjects established under Fordism. Employees and their institutions, employers and their factories, and the capitalist State and its welfare apparatus have been radically transformed by pervasive processes of deterritorialization. Machinic enslavement works with decoded flows, that is to say, abstract work flows, monetary flows, sign flows, which are not centered on the individual and human subjectivity but on enormous social machinisms such as corporations, the collective infrastructures of the welfare State, communications systems, and so on.[287]

However, this neoliberal emancipation of capitalist economy from what had hitherto been an overly constraining set of economic laws, at the same time potentially voids all means of stabilizing the monetary sign and of imposing monetary rule. Indeed, left in their isolation, these vicissitudes of capitalist decoding-deterritorialization would have precipitated utter catastrophy. Capital must therefore somehow strive to engineer compensatory recodings and reterritorializations to effectively damn up the flows bent upon escaping its purview. Here the particularly rabid nature of neoliberal decoding-deterritorialization is such that the resurrected *Urstaat*'s recoding-reterritorialization must increasingly recur to more and more blatantly direct military coercion. Citing the despotic exercise of American hegemony as the preeminent exemplar, Berardi avers that we have now entered into an epoch in which solely force vouchsafes the meaning of the monetary sign. In an aleatory regime bereft of any standard whereby the value of a currency such as the dollar could be established, and with the advanced financialization of capital proceeding at an inexorable pace, it is now solely violence that can secure this value. The brutal and hegemonic militarism in which financialization and violence conjoin is, furthermore, in no sense merely contingent. It is, rather, unswervingly structural. In fact, the maintenance within the ambit of existence of runaway financial capitalism and its rampant imperialism absolutely requires that the history of mankind become one of complete, total and utter violence, in which deregulation, financial globalization, total competition and infinite war coexist in mutual and reciprocal symbiosis or, better, infinitely proliferating parasitism. Neoliberal economics' submission of the species to the unitary and monomaniacal goal and existential basis of competition is tantamount to the inauguration of a Hobbesian *bellum omnium contra omnes*, that is, a condition describable as quite simply *fascism*. The precise definition of fascism, as given by Deleuze an Guattari, is the state of affairs in which a war machine is to be found in each and every niche of daily socio-political and economic life. In this quite formal sense, *neoliberalism is the perfect fascism*.[288] Within neoliberalism's aleatory regime of fluctuating values, value is henceforth established through an amalgam of lies, swindling and violence wherein the overriding source of law is pulverizing force. Herein entailed is not only capital's increasingly violent imposition and maintenance of markets, but also the more obscenely graphic extra-legal violence associated with neosovereignty's production of *homo sacer*.

The foregoing extra-legal coercion, however, assumes not merely the sort of violence associated with the new war prison and its analogues. For the biopolitical resonances of neoliberal decoding-deterritorialization cum recoding-reterritorialization entail, moreover, the exacerbation of an earlier-mentioned form of slavery. For the inexorable obverse and corollary of neoliberal deregulation's progressive evisceration of legal ties within the domain of production and its untethering of juridical persons from regulations, is the linguistic, physchological and *technological* capture of more and more living social time.[289]

In this connection, it was observed that the demise of the law of value and the fall of labor into an aleatory time was *inter alia* wrought and accompanied by the increasing pervasion of immaterial or microelectronic technologies. This latter was an adjunct of capitalist decoding-deterritorialization. Correlative to the latter is the attendant machinic enslavement of the human by the said technical machines of the information age. What Deleuze and Guattari had earlier discussed as the increasing preponderance of machinic enslavement over social subjection in the post-industrial age and its "shift from mechanical to digital, from reproducible to simulative"—implies moreover a transition from "the limited to the viral dimension of power."[290]

In the industrial age, capital's velocity of decoding-deterritorialization was the more or less on par with the mechanical speed of the assembly line, that of transportation networks, and the printing press. The move to preeminence of microelectronic technologies, however, serves to "equip capital with absolute velocity in the real time of simulation"—the upshot being that "acceleration becomes the domain of hyper-exploitation."[291] Here it is through the pervasion of micro-technologies that capital's velocity of decoding-deterritorialization—which is at once that of the increasingly unbounded and decentralized distribution of production and that of globalization—is forthwith able to outstrip that of workers' organizations, which remain comparatively slow and localized. Alluding to Baudrillard's notion of a well-nigh absolute velocity that renders asunder every form of social communication on the part of nascent resistance to the biopolitical colonization and machinic enslavement of life itself, Berardi inveighs heavily against the purported indomitability of the "multitude" in the face of the dictatorship of capital. In this regard, Berardi eschews Hardt and Negri, Virno, and their epigones, in favor of Baudrillard—who is seen as having shown that any such notion is merely the facile obverse of "the constitutive passivity of the masses."[292] But this now brings us somewhat further afield to a terrain that we shall plough a little later.

15. The strange concomitance of cynicism and piety

The dual tendency effected by capitalist axiomatization—that of recoding-reterritorialization cum decoding-deterritorialization—manifests itself *inter alia* in the form of a peculiar and perennial oscillation between the ostensible opposites of *cynicism*, on the one hand, and *piety*, on the other.[293]

(i) Capitalism's "cynical" tendency

Cynicism results from the tendency of capitalism to decode the flows of desire—a concomitant of capitalist deterritorialization—whose endless expansion of the means and forces of production tends toward a state well-nigh indistinguishable from production for production's sake. Decoding accompanies this deterritorialization, moving as it does in the direction of the wholesale dissolution of all social codes—codes that would otherwise have served to confer meaning upon the quotidian existence of the capitalist subject.

As Lazzarato has it, capitalist sign machines do not function at the level of social representations or in order to produce meaning. Instead, they entail modes of far more abstract and deterritorialized semiotization in the economic, scientific, and technical sphere than do signifying semiotics. In this sense, sign machines operate "prior" and "next" to signification, producing as they do sense without meaning, that is, a merely "operational sense." Their operations are "diagrammatic" in that the subject, consciousness, and representation are left in the background.[294]

In the same vein, Žižek maintains that the breakneck decoding-deterritorialization of the capitalist machine implies that it "is the first socioeconomic order which *de-totalizes* meaning"—by which he means to say that the capitalist machine is the social machine which cannot but fail to be "global at the level of meaning" and therefore that there cannot *stricto sensu* be such a thing as a global capitalist *weltanschauung*, or "capitalist civilization" in the strict sense; and that, for this reason, capitalism's "global dimension can be formulated only at the level of truth-without-meaning, as the 'real' of the global market mechanism."[295]

In the same vein, Alain Badiou has gone as far as to describe advanced capitalist "civilization" as *devoid of world*, that is to say, that the capitalist machine—its axiomatic being the entirely neutral economico-symbolic machine that it is—"sustains a *stricto sensu* 'worldless' ideological constellation, depriving the great majority of people of any meaningful 'cognitive mapping.'"[296] Badiou's position on this point is, notwithstanding, still stronger than this would seem to imply. For capitalism's bereftness of world is not solely maintained here as a mere point of fact; it is contended to be so at the most fundamental *ontological* level.[297] Badiou terms such a worldless world *atonic*, maintaining that in such a world "there's no truth, nothing but objects, nothing but bodies and languages." In this cynical atonic world "nothing will happen any more, so it is impossible to decide anything." All that remains is "happiness through asthenia," "euthanasia," and a "'pointless' existence." Atonic "worlds are simply worlds which are" "so quiescent and homogenous" that "no figure of decision, is capable of evaluating them."[298] Clearly, such an apparently extreme conclusion loses all its seemingly capricious extravagance when viewed from the perspective of a social machine that drives deterritorialization and decoding to the very limit.

According to Baudrillard, one of the effects of capitalist decoding is the annihilation of the referential dimension in all signification. Signs are here eviscerated of any real content, whilst referentiality gives way to an autonomous structural play of signs among themselves. Instead of representation there is now a total relativity, general commutation and combination of liberated signs. The sign no longer designates anything and becomes entirely free, indifferent and completely indeterminate.[299] With this ever more virulent destruction of reference, semblance and designation, what emerges is the "digital" and "programmatic" sign, whose value is purely "tactical." Here, the rational destinations of signs, all finality and the entire order of ends, cede their places to the molecular play of infinitesimal signifiers shorn of their signifieds.[300] Signifieds come "unthreaded" [*se défiler*], whereas the "parades of the signifier" [*les défilés du signifiant*] cease to lead anywhere.[301] Advertising, for instance, fails to constitute a symbolic network, being instead a wall of functional signs whose effects are exhausted in their decoding.[302]

This indeterminacy likewise occurs at the level of labor power and the production process, in which there transpires an annihilation of any goal as regards the contents of production. Similarly, the monetary sign slips into infinite speculation, beyond any reference to the real of production or—as we have seen earlier with Berardi—to any gold standard. These phenomena Baudrillard terms the flotation of money and signs, of the needs and ends of production, and of labor itself. It, of course, all leads to disaffection and general disenchantment. There has transpired an extermination of the real of both production and signification, and indeed the death of the real as such.[303] The real, he maintains, becomes "the ecstasy of denegation and its own ritual extermination: the hyperreal."[304]

This destruction of the hitherto extant regimes of designation, representation and production plunges capitalism into pure simulation. Simulation spells the end of the dialectic of signifier and signified at the helm of the accumulation of knowledge and meaning. This same proliferation and procession of simulacra ushers in a generalized *cynicism*, wherein the distinction between the beautiful and the ugly in aesthetics, the left and the right in politics, the true and the false at the level of the media, and the useful and the useless at the level of objects collapses. All humanist criteria of value, the entire civilization of moral, aesthetic and practical judgment are dissolved.[305] The political sphere is emptied and all social substance disappears.[306] Historical solidarity, whether that of the factory worker or class solidarity, vanishes.[307] There is a dismantling of reason [*démantèlement de la raison*].[308]

Advanced hyper-financialized neoliberal capitalism displays a definite tendency toward a frivolous and cynical immorality in relation to all criteria. Berardi, following the thread of Peter Sloterdijk's thesis in *Critique of Cynical Reason*, argues that *mass cynicism* has become the prevailing state of mind of the past few generations and is linked specifically to the catastrophic failure of the recodings and reterritorializations attempted by the multifarious utopian regimes and ideologies of the twentieth century. Perceiving violent competition, the precarization of labor, and military belligerence—all wrought by blistering movements of neoliberal decoding, deregulation, and deterritorialization—as inevitable, workers have largely eschewed attempts at social solidarity in favor of the pragmatic cynicism of pure survival. Cynicism entails a dissociation of language and behavior from consciousness, as well as pivoting upon a false relation with interlocutors, and is the consequence of a disenchantment with what was hitherto believed true. In view of this, cynicism is the movement of disaffection from truth in the direction of a universal pandemic of strategic lying.[309]

As we have seen with Lazzarato's discussion of machinic enslavement, capitalist decoding effects a flotation of all the categories of consciousness itself, entailing a loss of the subject in which whatever authorities under whose aegis the subject could attain a stable identity in accordance with definite rules have fallen into desuetude. Baudrillard maintains that today there is a general desubjectivation and dereliction of the individual in which the subject, robbed as he is of fixed relations, merely drifts in relation to other subjects. This he terms "the flotation of identities."[310]

The consequences for the individual subject of the rabid decoding and deterritorialization effected by neoliberal capitalism, as Charles Taylor would put it, is that its actions, goals, and achievements have a lack of weight, gravitas, rotundness and substance. The subject lacks a deeper effervescence, which it feels should be there. He suffers from what Taylor terms "the malaises of immanence." This entails a sense of fragility, uncertainty, the sensation that one's chosen path is no longer compelling, one's inability to justify it to onself or to others, a sense of a lack of meaning and significance, the sense that the quotidian is devoid of deeper suffusion, is dry, flat, that the things which surround us are dead, ugly, empty, that our manner of organizing them, shaping and arranging them, in order to live have no meaning, beauty, depth or sense.[311]

These sentiments about the modern world are in fact already to be seen in Matthew Arnold, who opined that the modern world lacked depth, and the modern self wholeness. The modern subject tends to live on the surface, and is therefore separated from the greater currents of meaning: "you must plunge yourself down to the depths of the sea of intuition; all other men are trying as far as in them lies to keep you at the barren surface."[312] This sense of being cut off from some greater source is likewise felt as a division from self: "The misery of the present age is not in the intensity of men's suffering—but in their incapacity to suffer, enjoy, feel at all, wholly and profoundly … in their having one moment the commencement of a feeling, at the next moment commencement of an imagination, & the eternal tumult of the world mingling, breaking in upon, hurrying away all … The disease of the world is divorce from oneself."[313] Arnold's contention was that modern man lacks a deep identity, suffers from disorientation and ennui, shifting and unsatisfying feelings, shallowness of being, dissatisfaction with his own endeavors, and debilities caused by the lack of any compelling spiritual authority. Arnold already thought that the capitalist society of his day was mechanical, materialist, encouraged narrow specialization for advantage, and individual action without a sense of the whole. This civilization was both philistine and atomistic—as indeed is our own. It is a fragmented society in which the self is itself fragmented. For Matthew Arnold, this fragmentation and loss of profundity is the product of what he sees as the ending of the Christian era.[314]

> While we believed, on earth he went,
> And open stood his grave,
> Men called from chamber, church, and tent;
> And Christ was by to save.
> Now he is dead! Far hence he lies
> In the lorn Syrian town;
> And on his grave, with shining eyes,
> The Syrian stars look down.
>
> In vain men still, with hoping new,
> Regard his death-place dumb,
> And say the stone is not yet to,
> And wait for words to come. (II. 169–180)
> …
> Its frame yet stood without a breach
> When blood and warmth were fled;
> And still it spake its wonted speech –
> But every word was dead. (II. 193–196)[315]

In a very similar vein, extreme Romanian pessimist Emil Cioran avers that, whereas prior to the advent of modern Western society one could bury one's lugubrious sighs in God, now we are virtually inconsolable since we have no one to confide in except for mere mortals. Cioran maintains that herein entailed is effectively a splitting in two of history: into a former time in which people lived in the divine vibrancy of God, and the present time in which the world is utterly voided of the divine.[316] As we shall soon see, however, this neat division of history cannot possibly be maintained. For the modern capitalist machine does indeed pilot a movement which countervails its own desacralization of the world.

(ii) Capitalism's "pietistic" tendency

This brings us to the other side of the capitalist axiomatic. That is, in its aforementioned, clandestine tendency to rescusitate the transcendent *Urstaat*, capitalism borrows despotic codes in order to artificially confer strange forms of belief upon the social body. This is the pietistic moment of capitalism: a movement fundamentally inextricable from processes of recoding which accompany capitalism's reterritorialization of production and labor onto capital.[317]

Holland cogently argues that, whereas capital's movement of decoding-deterritorialization properly-speaking stems from its economic component (production for production's sake and the freeing of productivity from fixed codes), capitalist recoding stems from capitalism's power component (that is, the resuscitation of the despotic *Urstaat*) and serves to "extract and realize privately appropriable surplus-value," in which "the evolving apparatuses of production and consumption alike are tied down to what is now obsolete capital-stock, solely in order to valorize it and realize profit on previous investment."[318]

At the level of processes of subjectivation, Lazzarato maintains that capitalist recoding and reterritorialization functions on the plane of social subjection. Here, signifying semiologies such as language and discourses are used and exploited as techniques of control and management to countervail the cynical deterritorializations and decodings undermining established communities, social relations, politics, and their former modes of subjectivation. They serve to model, format and reconfigure the subjectivation process in accordance with the individual subject. Language is one of the preponderant apparatuses for the recoding and reterritorialization of the decoded flows of indiviuals, persons, and individuated subjects. It operates according to a "rudimentary psychology," leading one to *believe* in the "I," the "I" as full bore being, as substance, and moreover projects this belief in the "I" onto all things. It implies, as Nietzsche would have it, a metaphysics of the subject and the object.[319]

As seen, in general, social subjection does function with the aid of a signifying semiotic and therefore at the level of meaning, and not that of asignifying signs. Its aim shall thus be to reconstitute meaning, the real, truth and finality formerly eviscerated by the decoding pole of the capitalist axiomatic. Inter alia, it functions to salvage the hitherto defunct binary distinctions between true and false, beautiful and ugly, and so forth. Whereas capitalist decoding threatened to usher in a fundamentally immoral and cynical world beneath good and evil, capitalist recoding strives to reinstate the good, the moral and the unequivocal. Like pre-capitalist modes of social subjection, the recoding effected by the capitalist regime assumes inter alia the form of the imposition of utopian projects on society and on the individual so that both the former and the latter are remade.

Now, even when these utopian projects are ostensibly secular, they are more precisely secularized versions of earlier strictly religious utopian projects. In this sense, one may readily maintain that they are always inherently pietistic. As John Gray contends, it has particularly been the early Christian faith in an Eschaton initiated by God that has been transmogrified into the secular—or at least secularized—belief that human action can create a utopian society. The pietistic project of universal democracy, for instance, is derived from the early Christian belief in an impending Apocalypse.[320]

In general, utopian projects are all products of a *Weltanschauung* once encountered only in religious cults. The secular utopian faith in a condition of societal stability is a Christian inheritance, as is the pietistic idea of progress. Though ostensibly at variance with the belief in the world as irredemably evil and close to coming to an end, an idea of progress has in fact been latent to Christianity from the very start. In the book of Revelation, for example, one finds the faith that good is incrementally destroying evil, that is, the first expression of the fervid notion of history as progress. Later, Puritans served as a vehicle for the idea—termed post-millennialism—that human effort could hasten the arrival of a perfect new world. They believed that Christ would rule over the world after the arrival of a Millennium ushered in by human endeavor. Even secular theories of progress which ostensibly reject any belief in a final state of perfection ultimately turn out to rest upon religions that do.[321]

This, of course, includes secular beliefs functioning under the aegis of the capitalist axiomatic. Enlightenment thought and much modern politics in general, for instance, latently maintain the millennarian view of history as a battle between light and dark, in which the light of knowledge is by its very nature destined to overcome the darkness of ignorance.[322] Moreover, modern projects of universal emancipation are merely versions of the eschatological Christian belief in universal salvation.[323] They are therefore, as already observed, forms of secular piety.

An example of this was the belief in Manifest Destiny formulated in the mid-nineteenth century, in which the early Christian faith in a messianic savior found itself translated into the belief in the United States as the Redeemer Nation. This is evidently the substance of Alexis de Tocqueville's view that, though overtly secular, American exceptionalism is to be sourced in religious roots. Woodrow Wilson's belief that national self-determination should be extended throughout the world, in turn, arose from his belief in America as the designated Redeemer Nation. Likewise, the George W. Bush administration's disastrously pious attempts at the extension of liberal democracy to the Middle East were a repetition of the utopian programme that Wilson promoted in central and eastern Europe after the First World War.[324]

Toward the end of the last century, the pursuit of utopia entered the late political mainstream in a particularly assidious form. The overwhelming belief came to be that only American-style democratic capitalism would be legitimate in the future. Armed with the eschatological faith that this regime was inevitably the final form of human government, western governments committed themselves to the massive recoding and reterritorializing project of installing democracy throughout the world. This came to be particularly the case in the United States after the September 11 attacks.[325] In this its militant faith in progress, the American Right was merely resurrecting in secular form the absolutely fundamental Christian belief that sinful human nature could be remade.[326]

The belief, a tenet of neoconservative thought in general, that history will inevitably end with a global capitalist system is of course derived in part from Kojève's reading of Hegel. Fukuyama, the principle purveyor of this view and who believes (or rather, believed) the United States to be the first post-historical society, contributed to the pious neoconservative belief that the only legitimate form of government is democracy and naught besides. Again, this proposition concerning western liberal democracy is a confession of eschatological faith by other means.[327] To export this form of government to all parts of the globe is therefore nothing short of a moral duty.

We may thus see that liberal democracy, far from being a skeptical creed and abiding on the decoding and deterriorializing pole of the capitalist axiomatic, with all its missionary zeal is one of the instruments of capitalist recoding. This is hardly surprising. For, being a lineal descendant from Christianity, it shares the militancy of its parent faith. This is what underlies the twenty-first century neo-conservative Enlightenment crusade for universal democracy.[328]

Untrammeled faith in human progress, the unlimited exportability of democracy, and the impossible dream of universal emancipation are, however, not the only forms of secular piety existing within and fostered by the capitalist regime.

Another form of capitalist piety is a more ostensively economic one: the belief that the central condition of individual liberty is, quite simply, the free market. This pious belief was held by Milton Friedman and Friedrich Hayek and continues to be held by today's neoliberal economists. Much as is the Enlighenment faith in human progress, this neoliberal belief has theistic roots. For neoliberalism is merely a late twentieth-century parody of classical political economy going back to Adam Smith himself. Smith viewed the emergence of commercial society as the work of divine providence. As John Gray observes, Smith's conception of the "invisible hand"—the system of hidden adjustments whereby the various exchanges of the market promote the common good—was expressed in overtly theistic terms. This invisible hand was God functioning through the medium of human sentiments, whereas the market itself Smith viewed as a by-product of instincts God had implanted in human beings.[329]

Moreover, the predominant argument for free trade in the early nineteenth century was that tariffs thwart the divine design. Here, it is argued that God scattered resources throughout the world so that widely separated peoples could come into proximate relations through trade and in so doing recognize one another as brothers. That is to say, free trade was a means to brotherhood under the law of God. The neoliberal exaltation of the free market is, therefore, a form of capitalist piety in that it has its ultimate basis in religion.[330]

Up until now we have been dwelling exclusively on various forms of secular piety existing under the aegis of the capitalist regime. However, one may observe in the case of the neo-conservatives in the United States the burgeoning recrudescence, from under its secular variant, of Christian piety in its original form. This is the Christian fundamentalist faith that evil can be defeated. In fact, neo-conservatives have been able to mobilize both the religious belief in imminent catastrophe and the secular faith in continuing progress. Overall, however, in the past generation the American Right has espoused recodings embodying a less secular pietism. In the early 1990s, neoconservatives paired up with Christian fundamentalists, whereas in the aftermath of the September 11 attacks American politics assumed the unmistakably apocalyptic tone characteristic of millenarian movements of yore. Here, many of the theo-conservatives in the Bush administration believed that the Eschaton would arrive by divine intervention, that the desuetude of the dichotomy between good and evil effected by rabid capitalist decoding would end with a final Rapture in which recoding would effect the vanquishment of evil and the triumph of good. In their pious belief that the blemishes of humanity can be wiped away by a benign catastrophe, George W. Bush and the Armaggedonite wing of the theocons were thoroughly Manichean and Augustinian.[331]

More specifically, many of the leaders in the Bush Administration espoused the views of Christian Reconstruction, or Dominion Theology. This post-millenial fundamentalist movement is bent upon countervailing the massive decoding effected by the capitalist axiomatic through the establishment of a Christian form of government in which every aspect of life will be subject to divine law and man's subjugation of the planet.[332] Another form of religious pietism abiding under neoliberal capitalism and adhered to by many fundamentalists on the Christian Right is Dispensationalism. Dispensationalists advocate the hastening of the Rapture, which shall transpire after the victory of Christ over the host of the Antichrist on the plain of Armageddon in modern Israel.[333]

We recall from Lazzarato that the absolute fulcrum of piety is the, as Nietzsche put it, belief in the self, in the obdurate, sanctimonious "I." This cultivation of the ego is effected through the individuation wrought by social subjection. Though, as we have seen in Gray's discussion, modern capitalist piety may assume either a secular or a religious guise, the actual originating point of even secular piety is religious. Indeed, we have traced the genesis of secular forms of piety at the political level as arising from millennialism.

Modern piety, at the level of the individual subject, however, originates roughly from the Protestant Reformation and the disciplinary social formations and practices the latter gave rise to. As Charles Taylor argues, this disciplining and individuation of the subject under Protestantism arose from the eschewing of the special renunciatory vocations hitherto prevalent under Catholicism, in tandem with rendering all vocations equal and, moreover, more exigent in pitch. What is at stake here with the Protestant revolution which overcame large swathes of the world, is the movement from discipline for an elect few to the generalization of disciplinary techniques to vast sectors of the hominid population. Protestant piety principally assumed the form of the eschewing of disordered conduct, doing so by rendering certain pious practices absolutely general. Of course, this enterprise could not be accomplished without recourse to God, to whom humans would have to turn. What must be understood about this Protestant piety, and of Protestant belief, is that it emerges upon the ruins of an earlier Medieval enchantment. Here, the "porous" self of the enchanted world gives way to the "buffered" self.[334]

And yet, one would add, it is only within the disenchanted world of the "buffered" self that belief as such emerges. Within the enchanted world, religion seems not a matter of belief but simply direct and factual, although admittedly magical-realist, experience. In the enchanted world, belief cannot be the case, as within that context there is no possibility that religion is not the case. Only with disenchantment is belief possible or even necessary. Piety as we know it—that is, as something pervading large swathes of society—only begins with both the Catholic and Protestant reforms.

Taylor maintains that the move which was made from this Protestant piety toward secularism, or what he terms "exclusive humanism," merely required the reference to God to be lopped off. This sloughing off entailed a redefinition of the social and individual goal, previously conceived as born of glorifying God, as one henceforth pertaining solely to purely human flourishing; whereas, the capacity to pursue it ceased to be seen as arising exogenously from God, but rather as generated endogenously from within man himself.[335]

Early Protestant piety, necessarily entailing as it did the integral disciplining of society, assumed the form of a humorless determination to castigate sin and disorder, the abolition of carnivalesque practices, the mixing of Christian and pagan elements, and so forth. Madness is no longer seen as the site of vision, but instead as the fruit of sin.[336]

As part and parcel of this generalized disciplining of society at large is the badgering, bullying, pushing, preaching, drilling and organization of the lower and more scabrous echelons of the body politic. Yet, this led to more disciplinary measures, such as a heightened concern with economic productivity and industry. The self was individuated as sober, ordered and industrious. In the reforms of Jean Calvin in Geneva and of Charles Borromeo in Milan, issues of religion and those of good civic order were indistinguishable. They were merely parts of a single programme of reform.[337]

This confluence of piety and work ethic is most in evidence in the Puritan societies of late sixteenth and seventeenth century England and America. The Puritan notion of the "saint" includes precisely a lack of indolence, a seemly trade, and the refusal to be mortified by idleness. This means not just any activity, but one which the disciplined subject has given himself as a lifetime's vocation. An idle man's brain quickly becomes the shop of the devil, swiftly coming to be seen as the principle cause of social disorder, sedition, and other fissiparous tendencies corrosive of the body politic.[338]

Similar programmes were implemented in the developing absolutist or dirigiste State structures of France and Central Europe. The goal was the formation of industrious and production-oriented subjects. Society was to be disciplined, but with the aim of inducing self-discipline. This piety via discipline also included the imposition of an ideal of civility.[339]

In seventeenth century Netherlands, Justus Lipsius set the tone for instilling in citizens a personal ethic of austerity and self-discipline. Subjects were made to internalize the values of self-control and industriousness. Something like Max Weber's "protestant work ethic" and "inner-worldly asceticism" (*innerweltliche Askese*) was inculcated. An amalgam of neo-Stoicism with Calvinism and Pietism engineered the dedicated self-denying entrepreneur. The latter would combine within himself godliness and virtue with professional and productive labor.[340]

As Charles Taylor reminds us, however, this entire project of supposedly purely secular, productive discipline, was at first compatible with a continuing belief in God. Indeed, it was accompanied by a more conscious and zealous dedication to God. This was a conflation of piety and industriousness, in which piety was industriousness and industriousness was piety. The movement here toward the erection of a pious disciplinary society was, however, a *secularizing* tendency. One would think this counterintuitive. Yet, the Reformation paradoxically inaugurated an intensification of religious faith in tandem with a concerted movement toward the secularization of society, in which the concomitant profanation of the sacred and the sacralization of the profane allowed the Christian to worship God through the prosaic vicissitudes of industrious quotidian life and the personalized practice of religion—paving the way for an escape from religious faith into the immanence of a pure secularism founded upon a disciplinary society of isolated individuals.[341] Hence, the "I" was subjected to a congealing in which its egoic obduracy came to undergo increasing reinforcement. And, as we have observed a little earlier vis-à-vis Lazzarato's recourse to Nietzsche, piety ultimately rests upon a preeminence accorded to the ego. The more consolidated, individuated and buffered the self becomes (the "buffered" self), the more pious it comes to be.

Ironically, however, discipline—as the unfailing mark of Godliness—somehow came to infiltrate the very essence of Godliness, gradually becoming indistinguishable from it. Through this conflation, recourse to divine grace for the self's attainment of moral fullness became less essential as grace came to be immanentized and transmogrified into something endogenous to the individual himself.[342] That is to say, one may henceforth attain moral plenitude within the increasingly reinforced, obdurate, individuated, solipsistically self-sufficient, and therefore pious "I," wrought by the reterritorializing and recoding processes endemic to social subjection.

Hereupon, God was replaced by universal human justice, the Enlightenment's rights of man, and rudimentary notions of the welfare state—with the caveat that these paradigms were still pious, in that they implicitly drew on the resources of Christian *agapē* in the form of a community-transcending altruism that even modern exclusive humanism believes to be innate to each person.[343]

In a subsequent move, modern humanism, increasingly reinforcing the endemic obduracy of the "buffered" self, sourced the capacity for universal beneficence in the powers of disengaged instrumental reason of the rational subject which, through its dispassionate impersonality, was believed to transcend parochialism so as to encompass the societal whole for the greater happiness of all individuals.[344] This is what has often been termed "asocial sociability"—particularly that of the Anglo-Saxon. Some would term it the end of love.

Closely related to the immanentizing of moral power in disengaged reason is the Kantian notion of sourcing the power of universal beneficence in the capacity of the individual human will to proceed in accordance universal moral law, which latter invites infinite respect.[345] Yet, this is still the asocial, egoic sociability, wrought by social subjection's individuation of the self.

Be this all as it is, one may argue that disengaged instrumental reason is that through which, through a curious reversal, a counter-movement away from the pious belief in universal beneficence can be forged—as it is the fulcrum upon which the dissociation of language and behavior from consciousness may be effected, a dissociation which defines the very essence of cynicism. Accordingly, it would seem that the late capitalist cynical subject is the very opposite of being pious or ideologically interpellated, and that he moves within a strictly post-ideological constellation. As Žižek avers, the cynical subject is entirely aware of and consciously maintains the hiatus between the ideological mask that he wears and the maligned social reality it masks. Contemporary cynical reason is a kind of "enlightened false consciousness," completely privy to the mendacity, vested egotistical interests, violence, and brutal claims to power lingering behind the ideology within which it operates, yet still refuses to renounce it. Still more bizarrely, contemporary mass cynicism, though immoral, is not so in any direct sense. Rather, it places morality itself at the service of immorality and, in a mind-numbingly crooked perversion, execrates integrity as a supreme form of dishonesty and morality as the ultimate form of profligacy. Adding to its twisted perversity, it co-opts legality itself as its main and most effective means of morally reprehensible enrichment through exploitative plunder.[346]

Contemporary cynicism, therefore, cannot be treated as a case of individuals of whom it may be maintained: "Forgive them, for they know not what they do." They know fully well what they do, yet still do it. For this reason, the traditional critique of ideology, which unearths by way of a "symptomatic" interpretation that which ideology must repress in order to constitute and preserve its consistency, fails—as cynical reason has already taken this hiatus between knowledge and praxis into account in advance.[347]

This is all so because late capitalist ideological capture does not operate on the level of knowledge and symbolic misrecognition masking the real state of affairs. Rather, its pious recoding functions at the level of the *unconscious ideological fantasy* which structures social reality itself: not at the level of obfuscating the real state of affairs, but at the level of a state of affairs that is itself distorted. In this sense, the late capitalist cynic still believes: not at the internal level of disembodied rational knowledge, but at the external level of praxis. In fact, under the universal reign of commodity fetishism, it is now commodities themselves that believe *for* him. Though he has ostensibly expunged himself of all superstitions at the level of rational knowledge, in that social relations between things now believe for him, he notwithstanding *remains pious*—and depressingly so. So, not only is neoliberalism the perfect fascism; neoliberal cynicism itself is the ultimate form of piety.[348]

This implies, as Berardi maintains, that, even though the late capitalist subject may well be entirely skeptical of and consistently maintain an "ironic" distance with regard to the pious mendacity of neoliberal ideology, he still firmly desires to be on the side of power and unfailingly submits to it. Bereft of even a scintilla of imagination, he is the ultimate and most lamentable conformist. Through gargantuan recodings and reterritorializations, power has inexorably claimed him.[349]

Whereas authentic irony is a strategy of suspending the semantic value of the despotic and dictatorial master signifier on the basis of a common sympathetic autonomy from it; cynicism does the same, except against the backdrop of mutual antipathy and the dissociation of ethics from utopian possibility. The ironist is a romantic unflaggingly devoted to a emancipatory Cause and interminably sustained by the immortality of the Freudo-Lacanian death-drive (*Todestrieb*) and its compulsive object-libido. The cynic, on the other hand, in his immoral pragmatic capriciousness, remains a slave to power.[350] Summing up, we may conclude that, in the neoliberal cynic, the brazen caprice of immoralist indifference and piously ideological praxis coexist within a single subject whose ethical integrity, on this account, veers depressingly toward pure unadulterated nullity.

16. The coincidence of decoding and recoding within a single subject: Daniel Paul Schreber

The alternation or co-implication between decoding and deterritorialization, on the one hand, and recoding and reterritorialization, on the other, may be glimpsed in Judge Daniel Paul Schreber's memoirs of his own nervous illness. The difference here is that we glimpse the coexistence of these two opposing tendencies within a single human subject.

The aspect of decoding and deterritorialization is evinced in the fact that Schreber sees all the inmates and personel at the asylum at which he is incarcerated as "fleeting-improvised men," that is to say, not real human beings but merely Baudrillardian simulacra produced by "divine miracles." These fleeting-improvised men lead "a dream life."[351] Even his wife, who would occasionally visit him, fell under this interdict.[352] When she came to see him at Sonnenstein, he believed she was only "fleetingly-improvised" for the occasion and that she would "dissolve" directly after taking her leave from the asylum.[353] According to the superintendent of the asylum, Dr. Weber, this decoding and its attendant proliferation of mere simulacra led Schreber to view everything he saw around him to be a sham, and himself and the persons in his vicinity as mere lifeless shadows.[354]

This process of decoding effecting the procession of unreal or hyperreal phantom beings was likewise accompanied by one responsible for gradually discombobulating discursive language and in its wake leaving only an inchoate hallucinated murmur of voices in his head. These voices were:

> … an empty babel of ever-recurring monotonous phrases in tiresome repetition; on top of this they were rendered grammatically incomplete by the omission of words and even syllables.[355]

The content of the voices which emanated from the lower God (Ariman), in that they entail a pure asignifying semiotic, "amounts to a kind of neutral nonsense," to "predetermined concoctions of thoughts spoken into my head by senseless voices in tiresome, monotonous repetition."[356] Yet another tirade of simulacra experienced by Schreber were "miraculously produced birds" which reeled off phrases "without knowing the sense of the words."[357]

Deterritorialization also assumed the form of Schreber's being "unmanned," transformed into a woman, after having come into contact with "divine nerves" or "rays."[358] This becoming-woman is a "blessedness" accompanied by an increased feeling of voluptuousness.[359] In a strange adoption of Zoroastrian theology, Schreber maintains that it is specifically the rays of the lower God which have the power of producing the miracle of unmanning. He confesses to having twice experienced the miracle of unmanning on his own body.[360]

He lay in bed for weeks with his clothes removed in order to make himself more amenable to voluptuous sensations, which could be stimulated in him by "the female nerves which had already started to enter my body."[361] These "female nerves of voluptuousness" had "given my body a more or less feminine stamp; they have in particular given my skin a softness peculiar to the female sex."[362] These "female nerves" began to produce within him "a sensation of timidity and to 'represent' me as a human being trembling with feminine anxiety."[363] He maintains that God himself had decided to hand Schreber's body over "in the manner of a female harlot."[364] His becoming-woman he would experience inter alia in the guise of various alterations in his member such as its retraction, detumescence approaching the point of complete dissolution, the removal of hair from his beard and mustache, and the diminution in his stature to the point that it would approximate the size of a female body.[365] Indeed,

> … the signs of a transformation into a woman became so marked on my body, that I could no longer ignore the imminent goal at which the whole development was aiming …
>
> … I could see beyond doubt that the Order of the World imperiously demanded my unmanning, whether I personally liked it or not, and that therefore it was *common sense* that nothing was left to me but reconcile myself to the thought of being transformed into a woman.[366]

The becoming-woman, effected by "soul voluptuousness," became so pronounced that he felt that he had received the impression of feminie characteristics on various parts of his body.[367] The divine rays impressed female breasts and genitalia as he lay in bed, and female buttocks whenever he bent down.[368] As Schreber declares:

> For myself I am *subjectively certain* that my body—as I have repeatedly stated in consequence of miracles—shows such organs to an extent as only occurs in the female body.[369]

Suffused with divine ray-induced becomings-woman for several years, he steadfastly concludes:

> I consider it my right and in a certain sense my duty to cultivate feminine feelings which I am enabled to do by the presence of nerves of voluptuousness.[370]

According to Dr. Weber, Schreber would take pleasure in feminine toilet articles, in minor feminine occupations, in decorating himself with gay ribbons and bows and so forth in a womanly manner.[371]

All of these decodings and deterritorializations, of course, receive in the consciousness and world of Schreber countervailing recodings and reterritorializations. Hereby, the cynical tendency unleashed by the capitalist axiomatic is counteracted by its pietistic moment. In the case of Schreber, the said pietistic tendency takes upon itself a manifestly cosmic and theological tincture. It is in no sense a secular form of piety, such as the capitalist axiomatic likewise does generate. As in the case of the cynical, decoding and deterritorializing movement, the pietistic recoding and reterritorializing movement in Schreber is likewise occasioned by divine rays. As Schreber avers:

> Other rays which conducted themselves in the manner described above as if they were God's omnipotence itself, carried names such as "the Lord of Hosts," "the Good Shepherd," "the Almighty," etc., etc.[372]

Within the general gamut of these pietistic swirlings, he undergoes inter alia strange apotheoses such as the following:

> At another time I felt as if I myself were raised up to Blessedness, it was as though from the heights of heaven the whole earth were resting under a blue vault below me, a picture of sublime splendor and beauty; I heard some expression like "God-be-together-view" to describe this picture.[373]

In this pietistic recoding and reterritorialization, Schreber in fact almost becomes Christ himself:

> My head was frequently surrounded by a shimmer of light owing to the massive concentration of rays, like the halo of Christ is pictured, but incomparably richer and brighter: the so-called "crown of rays."[374]

In his delusional psycho-theological system, Schreber refers in pietistic fashion to so-termed "holy times," that is to say, times when nerve-contact or communication with the rays or voices transpired. From a certain point in time, there would only be "holy times."[375] Schreber's pietistic paean to the splendor of the divine continues forthwith:

> I saw God's omnipotence in its complete purity. During the night—and as far as I can remember in one *single* night—the lower God (Ariman) appeared …
>
> … I could not but express my feelings repeatedly in words such as "Oh how pure!"—towards the majesty of the divine rays.…
>
> On the following day and perhaps on one or two more days … I saw the upper God (Ormuzd), this time not with my mind's eye but with my bodily eye.[376]

One may see that Schreber's "soul-voluptuousness" associated with decoding-deterritorialization and becoming-woman is, within his delusional system, ultimately a movement to be subsumed within the greater arc of his piety. For he regards his unmanning as a duty to be fulfilled to God.[377] The becoming-woman of Schreber—which entails an initial loss of the real in the form of decoding and deterritorialization and the universe of simulacra these movements proliferate, he ultimately feels as an ineluctable propaedeutic for recodings and reterritorializations through which the real would be reclaimed in the form of a regeneration of mankind, which species he thought had passed into nonexistence:

> I lived in the certain expectation that one day my unmanning (transformation into a woman) would be completed; this solution seemed to me absolutely essential as preparation for the renewal of mankind, particularly while I thought the rest of mankind had perished.[378]

The result of his unmanning he maintains will be his fertilization by divine rays and the spawning of a new race of real men, as opposed to "fleeting-improvised" men. This very special relationship he feels that he has with God will reclaim the real by way of the continuation of creation on earth.[379] As part of this pietistic project of recoding and reterritorialization, he harbors:

> ... the hope that I will be the middleman, through whose personal fate the knowledge I have gained will spread fruitfully far and wide, and that in this way it will be granted me to help mankind to correct views about the relationship between God and the world, and the revelation of religious truths and salvation, even long after my death.[380]

17. The subsumption of productive labor by capital and the collective disinvestment of working organs

There are a number of further corollaries that emerge from the inherent nature of the capitalist axiomatic. The reterritorialization of labor onto capital, in the appropriation of all productive forces by the full body of capital, implies that labor becomes commodified as exchange value, as a pure abstract quantity. There are two consequences generated here. Firstly, since it is the very nature of the socius to attract all forces of production to itself such that they are miraculated by the said socius as their quasi-cause, labor shall come to emanate and be derived from the abstract capitalist axiomatic as value. In a fairly nice formulation, Marx describes the subsumption of labor by capital as the "incorporation" of wage-labor into capital, an incorporation whereby these laborers collectively comprise a single "working organism" and hereby "merely form a particular mode of existence of capital."[381] This incorporation is so comprehensive that "the productive power developed by the worker socially is the production power of capital."[382] Here, in that the socially productive power of labor costs "capital nothing, while on the other hand it is not developed by the worker until his labour itself belongs to capital, it appears as a power which capital possesses by its nature—a productive power inherent in capital."[383]

Secondly, in that it is abstract labor and the forces of production that are attracted and miraculated by the capitalist socius, unlike in the primitive and despotic regimes, there is no longer a collective investment of productive human agents and human organs by the said socius: the capitalist axiomatic, as Deleuze and Guattari aver, needs not write in bare flesh in the marking of bodies and organs, nor does it require to create a memory for man. Rather, the parts of the social machine are now technical machines as constant capital attached to the full body of the socius, and no longer simply men. For the latter have become merely adjacent to the technical machine. Thus, the inscription no longer bears directly, or at least it theoretically speaking has no need of bearing directly, upon the bodies of men.[384] Hence, it is no longer the producers or nonproducers that are inscribed or marked, but rather the forces and means of production.[385] Marx's formulation of this becoming-peripheral of human working organs in relation to capital is as follows. Firstly, there is an increasing autonomization of the machine itself in relation to the worker: "in machinery the motion and the activity of the instrument of labour asserts its independence *vis-à-vis* the worker."[386] Then it is the machine that is averred to attach itself to the full body of capital whilst at the same time man is progressively sidelined: "Because it is capital, the automatic mechanism is endowed, in the person of the capitalist, with consciousness and a will. As capital, therefore, it is animated by the drive to reduce to a minimum the resistance offered by man, that obstinate yet elastic natural barrier."[387] Elsewhere, this move to the forefront of the technical machine and the simultaneous shunting of human working organs to the periphery is explained by Marx's asseveration that the "automaton itself is the subject, and the workers are merely conscious organs, co-ordinated with the unconscious organs of the automaton, and together with the latter subordinated to the central moving force."[388] This latter force is capital.

The first organ to fall away from the socius, aver Deleuze and Guattari, is the *anus*.[389] It is on the model of the latter that the other organs suffer collective disinvestment. We have already alluded to Berardi's observation that, in the post-industrial age, it is no longer the body that suffers machinic enslavement, but indeed the soul. In a similar vein, Berardi avers that within the sphere of informational labor, it is no longer people as such that are recruited by capital. The worker is almost completely depersonalized and, in the wake his near total disappearance, capital instead co-opts abstract fragments of time. Digitization of the labor process internally fragments the personal continuity of work. Labor time is at once fractalized and cellularized. The packets of time recruited by capital are entirely separated and interchangeable with their contingent bearers. This is inevitably accompanied by the precarization of the worker, as businesses purchase as many cells of time as desired without the obligation to provide any sort of social protection to the worker. Instead, depersonalized time becomes the real agent of the valorization process, with the real physical intellectual laborer divested of legal recognition. The problem here is that depersonalized time cannot possibly be possessed of any rights and, by implication, any capacity for the formation of unionist or proto-unionist recalcitrance or subversive political consciousness. In the process, this purported cognitariat, unable to negotiate or protect their labor time, are effectively expropriated of the time of which they are formally the proprietors, as this time is entirely separated from the real social existence of the persons who make it available in the first place. Capital, on the other hand, by reducing time to minimal recombinant units, is always able to extract surplus value from the cognitive laborer. Though such fractalized labor can effect evanescent impulses of quasi-sedition; given as it is to well-nigh infinite fissiparity, this labor-power is hopelessly inept at mounting any unified front of resistance. This means that the temporal, spatial and physical proximity of co-workers absolutely crucial for any revolutionary upsurge is completely foreclosed by the hyper-exploitative sphere of info-labor.[390]

In this same movement one has the constitution of private persons as individual centers of organs and, as functions derived from the abstract quantities forged by the axiomatic.[391] In this connection, we have *inter alia* the capitalist serving as a function derived of capital, and the worker as a function derived from labor capacity, and so forth.

18. Machinic enslavement in the information age

We have observed earlier how, in the information age, machinic enslavement becomes predominantly that of the soul. The consequences of this shall prove to be absolutely decisive. In this connection, Berardi mentions two mass psychopathologies whose increasing prevalence has definitively been linked to the ever greater enmiring of the soul by cybernetic and other informational machines in the last decades of neoliberal hyper-capitalism: panic disorder and depression.[392]

Panic results when consciousness finds itself at a loss regarding the reception of the vast panoply of stimuli thrown at it by what seems a well-nigh infinite universe. Etymologically, the term is traceable to the Greek word παν, meaning "everything existing," whilst the god by the same name was he who afflicted mortals with a sublime psychic condition roughly equivalent to delirium. As regards depression, we shall turn to this second condition in due course. The present context within which panic disorder appears is precisely the machinic enslavement of the soul already attributed to cyber-capitalism.[393]

Yet it is not merely the bare fact that the soul is thus enslaved that is of pertinence. It is also that the co-optation of the soul's energies and capacities is rapidly and continuously exacerbating itself in tune with the relentless expansion of cyberspace—and of the infosphere *tout court*—in the teeth of the subject's comparatively limited capacity for the elaboration of such myriad sensory inputs.

As Robert Colvile maitains apropos of Brigid Schulte, the soul enslaved by the neoliberal capitalist infosphere is submitted to "The Overwhelm"—a constant scattering and fragmentation of the mind at the behest of informational over-saturation. The ineluctable problem is that man's stress system, tasked as it is with dealing with the incessant informational bombardment of cyberspace, remains essentially a Paleolithic organism and thereby utterly unsuited to this twenty-first century ordeal. Countenanced with an incessant deluge of attention-grabbing and often fear-inducing signs to which evolution has made us acutely vigilant for the sake of our very survival, our amygdala immediately causes hormones to coruscate through the soma: adrenalin accelerates the heart, dopamine floods the brain, cortisol launches sugary glucose into the muscles. The problem is that, whereas this fight-or-flight mechanism of the human organism is designed to be triggered on fairly rare occasions, the cybernetic enslavement of the soul triggers it virtually from minute to minute. In that this stress mechanism takes quite a long time to return to quiescence, information overload inevitably snares the perception-consciousness system in a perpetual feedback loop which inexorably maintains the soul in a continued state of agitation as cortisol and other hormones incessantly frazzle the organism. Apart form the repeated shutting down of the faculties of higher reason, this unbearable stress placed on the perceptive and cognitive faculties foments a perilous acceleration of the cardio-vascular functions that may very easily lead to outright collapse.[394]

As observed, machinic enslavement—whether that of the soul or otherwise—pilots de-subjectifying deterritorializations rendering up the Deleuzo-guattarian dividual. As we likewise saw, the dividual was seen to be strongly homologous to Charles Taylor's "porous" self inhabiting the enchanted Middle Ages. This means that the greater the preponderance and amount of machinic enslavement that there is, the more closely shall the working soul approach the condition of the "porous" self. Now, Charles Taylor maintains that the porousness of the "porous" self is most clearly in evidence in the fear of possession by an evil spirit. By virtue of its porosity, the "porous" self is vulnerable to a malevolence which appears more than human, which resides in the cosmos or even beyond it. It is vulnerable to spirits, demons and cosmic forces. In tandem with this go particular fears which can grip the self under certain circumstances. Possession by an evil spirit or god—like Pan—would entail a full taking over of the person, various kinds of domination and fusion, and thus the total eclipse of the higher faculties and powers. The result would be delirium—or a racing, uncontrollable panic.[395]

One may also aver that the panic disorder induced by the machinic enslavement of the subject to a rapidly expanding and ever more accelerated infosphere exhibits marked similarity to the anxiety attacks experienced by patients suffering nervous disorders and dementia. These patients exhibit the porousness of the self observed above.

Though Schreber lived in the latter part of the nineteenth and the early part of the twentieth centuries, well before the arrival of cyber-capitalism, his symptoms remain largely the same as the ones of those suffering from cybernetically induced sociopathology. He complains in his memoirs how the rush of blood to his heart perpetually caused attacks of anxiety (panic disorder), except of course that these he feels to have been the effect of his nerves having been stimulated by "divine rays." The latter assailed him in the form of a relentless informational stammering of voices, in the form of "compulsive thinking"—forcing him to think incessantly and to constantly have to know what he was thinking about.[396]

These voices harangue him, constantly firing queries into his nervous system that unendingling:

> … compel my mind to work by setting the corresponding nerves in vibration, in a way that they cannot possibly escape the impulse to think.… what I directly feel is that the talking voices (lately in particular the voices of the talking birds) as *inner voices* move like long threads into my head and there cause a painful feeling of tension through the poison of corpses which they deposit.[397]

As he continues to aver concerning the almost cybernetic information overload to which he was subjected:

> … for almost seven years—except during sleep—*I have never had a single moment in which I did not hear voices* … They accompany me to every place and at all times; they continue to sound even when I am in conversation with other people, they persist undeterred even when I concentrate on other things … I can distinguish individual words hardly at all or only with the greatest difficulty …
>
> …
>
> The threads which are pulled into my head—they are also the carriers of the voices—perform a circular movement in it, best compared to my head being hollowed out from inside with a drill.[398]

All in all, the incessant babble of the voices caused "mental unrest, accelerated breathing, palpitation, etc …"[399] Schreber likewise terms the divine rays overtaxing his nervous system "divine miracles." He maintains the effects of this informational overstimulation and panic disorder to even be palpably physical, and that hardly a single limb or organ of his escaped being damaged. In fact, during his first year at Sonnenstein Asylum, the miracles were of such a threatening nature that he feared for his health, his reason, and even his life. Miracles were directed against the organs of the thoracic and abdominal cavities. His lungs underwent painful inflammation, or pulmonary phthisis. He felt that his diaphragm was raised up to just under his larynx, to the point that he could barely breathe. He felt that his chest was compressed and his ribs smashed. His gullet and intestines were torn repeatedly. His pharynx and seminal cord likewise suffered painful miracles. His abdomen underwent putrefaction. He felt that his spinal cord would be pumped out. He felt as though the nerves were being pulled out of his head, causing an unpleasant tension in the head. He felt as though his skull was repeatedly sawn asunder in various directions with its body material partly pulverized by the destructive rays. "Souls" or their single "nerves" assumed tiny human shapes, causing mischief on all parts of his body. "Little devils" walked around on his head continuously and seemed to compress his head as with a vice. His foot and coccyx bones underwent decalcification.[400]

All through this, the "senseless twaddle of voices in my head causes an absolutely unbearable mental martyrdom …"[401] In addition, overstimulation of the muscles of the respiratory system by the lower God (Ariman)—as in Zoroastrian theology—caused him to emit bellowing noises so frequently that it was impossible to remain in bed at night. God's "nerves" would incessantly cry for help, causing untold anxiety. These phenomena assailed him hundreds of times every day and hundreds of thousands of times over the course of ten years.[402]

Returning to Berardi, far from positing it as a merely individual psychopathology, he regards this cybernetically and informationally induced panic disorder that attaches to the increasing preponderance of cybernetic machinic enslavement in present times as a widely spread and quasi-generalized sociopathological syndrome. It is "a social psychopathology induced by the Infosphere's excess, by the hyper-stimulation and the endless cognitive stress affecting the social organism and caused by permanent electrocution."[403]

Vastly exacerbating this situation, as Lazzarato argues, is the fact that neoliberal capitalism pushes not only machinic enslavement to the absolute extreme, but also does so in the case of social subjection. The latter is brought to bear by neoliberalism's forging of the individual subject as human capital in the form of, as seen a little earlier, the "entrepreneur of the self." Conferring unprecedented "freedom" and "autonomy" upon the individuated laboring subject, neoliberalism now solicits the latter to continuously arbitrate between both external situations and himself in accordance with the logic of gain and loss, supply and demand, as well as profitability. It that the individual is well-nigh entirely responsible for his own fate, he is forced to incessantly negotiate between his economic ego and superego. Here, as in Fordism, the norm is still that of the external socioeconomic system, with the exception that it appears for all intents to originate solely from the individual himself. In this sense, neoliberal social subjection subjects the individual to a virtually infinite evaluation, in which the subject acts as his own primary judge, is forced to obey his own injunctions, and engages in a permanent and intensive negotiation with himself. The deleterious aspect of this extreme mode of individuation is that it inevitably leads the subject into a deadlock in which self-realization, freedom, and autonomy rub up against a stubborn reality which systematically nullifies these latter. This quandary, however, does not become a failure of capitalism itself, for the latter annuls it by foisting the same failure upon the subject through the interiorization of the conflict within the closed confines of the individual ego. Thus, instead of being directed against capitalist power relations, the laboring soul turns against itself in the form of resentment, guilt, bad conscience and loneliness, in which the absolute sovereignty of the individual coincides with his complete and utter self-alienation. The result is that the soul falls into the brackish mire of depression. Moreover, in that the subject's failure is not tolerated by a system which views itself as pristine, it comes to transpire solely within the context of psychopathology.[404] Within capitalism, socioeconomic failure appears only as a pathology and naught besides.

Meanwhile, further aggravating this already extreme situation, the contemporary neoliberal capitalist machine likewise drives cybernetic machinic enslavement to the very limit of the exploited soul's elaborative capacities. The phenomenon appertaining here is what Berardi diagnoses as semiotic overproduction: a proliferation of signs increasingly in excess of the collective attention span of the cognitive labor army. This disparity between an ontologically limited field of perception and the virtually infinite stimuli with which it is incessantly bombarded has led, and still does lead, to an assortment of means employed by the cognitariat whereby it strains to extend its attention, chief among these being stimulants such as cocaine and amphetamines. This was in evidence during the course of the dot-com era of the 1990s, at which time the cognitariat's collective elaborative capacity for the relentless influx of economically relevant signs was stretched to its very limit.[405]

Evidently, with the incessant uping of the ante, a breaking-point must loom forth and eventually be reached. Indeed, with the soul already frazzled by panic disorder and the organism overtaxed to a degree that soon exhibits itself to be literally unbearable, the inevitable result will be either complete nervous breakdown or—far more commonly—the organism's detachment from the flow of information and associated competitive tensions. Here the soul falls into *depression*, that is to say, it suffers an abrupt loss of motivation attendant upon its sudden disinvestment of the energy it had previously employed to the ends of its hitherto extant narcissistic stimulation. This unilateral withdrawal of the organism results in what henceforth becomes "a sort of zero degree of the exchange relation between the conscious organism and its world."[406]

This unilateral withdrawal of the cybernetically enslaved laboring soul is most dramatically evinced in the Japanese phenomenon of the *hikikomori*. In Japan in 2010, it was officially found that seven-hundred-thousand individuals had severed all relations with the outside word, so as to live a completely and utterly isolated existence locked in their rooms. These depressants spend most of the day of almost every day voluntarily confined to their homes and stubbornly avoid all social situations, yet exhibit no other mental disorders that could account for their complete and utter abdication from society. Estimates released by the Japanese Ministry of Health indicate that a further one-and-a-half-million individuals are on the cusp of becoming hikikomori.[407]

We recall that the dividual engineered by machinic enslavement was observed to be homologous to Charles Taylor's "porous" self. Now, the "porous" self is not merely, due to its porousness, susceptible to delirium and panic through experiencing itself as possessed by demons. It is likewise highly susceptible to depression. In this connection, Taylor maintains that the "porous" self is vulnerable not merely to animate spirits. It is also vulnerable to inaminate substances, things which have no will. Such as black bile. Here, in contradistinction to the far more obdurate Taylorian "buffered" self, or individuated subject, constructed by social subjection—who would consider depression as something merely *caused* by black bile; the "porous" self, lacking disengagement, necessarily maintains that depression *is* black bile. This is a colossal difference.[408]

As William Styron maintains in his memoir of his own depression, it is a disorder of mood that is painful in a rather uncanny and elusive manner, to the extent of being difficult to describe. For those who have not experienced it, it is quite simply incomprehensible. The gloom, or "the blues" which people occasionally associate with the hassle of everyday existence gives merely a hint of it.[409] What we may, however, maintain of depression is that it is not simply sadness. Styron describes how in his own depressed states he would be assailed by a sense of dread, alienation, stifling anxiety and excruciating near-paralysis.[410] This would likewise be accompanied by confusion, failure of mental focus and lapse of memory in the "gathering murk."[411] As Kay Redfield Jamison contends:

> In its severe forms, depression paralyzes all of the otherwise vital forces that make us human, leaving instead a bleak, despairing, desperate, and deadened state. It is a barren, fatiguing, and agitated condition; one without hope or capacity; a world that is, as A. Alvarez has put it "airless and without exits." Life is bloodless, pulseless, and yet present enough to allow a suffocating horror and pain. All bearings are lost; all things are dark and drained of feeling. The slippage into futility is at first gradual, then utter. Thought, which is as pervasively affected as mood, is morbid, confused, and stuporous. It is also vacillating, ruminative, indesicive, and self-castigating. The body is bone-weary; there is no will; nothing is not an effort, and nothing at all seems worth it. Sleep is fragmented, elusive, or all-consuming. Like an unstable gas, an irritable exhaustion seeps into every crevice of thought and action.[412]

Elsewhere, Jamison maintains that:

> Profound melancholia is a day-in, day-out, night-in, night-out, almost arterial level of agony. It is a pitiless, unrelenting pain that affords no window of hope, no alternative to a grim and brackish existence, and no respite from the cold undercurrents of thought and feeling that dominate the horribly restless nights of despair.[413]

The close association of depression with panic disorder mentioned by Berardi is corroborated by Jamison when she relates in her own memoir of her manic depression how she was assaulted by a bewildering and grotesque cognitive overload such as that fomented by the information overload of the cyber-capitalist infosphere:

> My mind was beginning to have to scramble a bit to keep up with itself, as ideas were coming so fast that they intersected one another at every conceivable angle. There was a neuronal pileup on the highways of my brain, and the more I tried to slow down my thinking the more I became aware that I couldn't.[414]

Moreover:

> I became confused, scared, and disoriented … my behavior was frenetic, and my mind more so. Slowly the darkness began to weave its way into my mind, and before long I was hopelessly out of control. I could not follow the path of my own thoughts. Sentences flew around in my head and fragmented first into phrases and then words; finally, only sounds remained.[415]

This insane panic disorder inducing hyper-saturation by information would continue:

> My thoughts were so fast that I couldn't remember the beginning of a sentence halfway through. Fragments of ideas, images, sentences, raced around and around in my mind like the tigers in a children's story. Finally, like those tigers, they became meaningless melted pools. Nothing once familiar to me was familiar. I wanted desperately to slow down but could not.[416]

Elsewhere, Jamison writes:

> On occasion, these periods of total despair would be made even worse by terrible agitation. My mind would race from subject to subject, but instead of being filled with the exuberant and cosmic thoughts that had been associated with earlier periods of rapid thinking, it would be drenched in awful sounds and images of decay and dying: dead bodies on the beach, charred remains of animals, toe-tagged corpses in morgues.[417]

These moribund images, as we shall see later, presage the potential passage from the depression-panic cycle to the death of the depressant himself. Styron describes the imbrication of depression and panic disorder in his own soul as follows:

> … my brain had begun to endure its familiar siege: panic and dislocation, and a sense that my thought processes were being engulfed by a toxic and unnameable tide that obliterated any enjoyable response to the living world.[418]

Styron describes the panic side of the panic-depression hybrid as "a sensation close to, but indescribably different from, actual pain."[419] For him, the pain was closely associated with the feeling of drowning or suffocation. He quotes William James, who also battled depression for many a year: "It is a positive and active anguish, a sort of psychical neuralgia wholly unknown to normal life."[420] In view of the systematic stress aptly described by Berardi, which effects a severe depletion of the neurotransmitters norepinephrine and serotonin, and an often prodigious increase in cortisol levels, Styron avers that "it is no wonder that the mind begins to feel aggrieved, stricken, and the muddied thought processes register the distress of an organ in convulsion."[421] This injection of panic into the psyche's systematic infospherical information overload and its chemical corollaries moves Styron to aver that "Soon evident are the slowed-down responses, near paralysis, psychic energy throttled back close to zero. Ultimately, the body is affected and feels sapped, drained."[422] This, evidently, is the moment when panic disorder veers into its familiar cousin: depression. The foregoing entails, as Berardi would also contend, a disconnection from the processes of the active life. Styron:

> I began to conceive that my mind itself was like one of those outmoded small-town telephone exchanges, being gradually inundated by floodwaters: one by one, the normal circuits began to drown, causing some of the functions of the body and nearly all of those of instinct and intellect to slowly disconnect.[423]

Of these disconnects, Styron lists the near disappearance of his voice which became faint, wheezy, and that of a ninety-year-old. He also observes the disappearance of his libido and appetite.[424]

Styron likewise describes the oscillation between the depressive lull side and the outright panic side of his mental illness:

> … the shadows of nightfall seemed more somber, my mornings were less buoyant, walks in the woods became less zestful, and there was a moment during working hours in the late afternoon when a kind of panic and anxiety overtook me, just for a few minutes, accompanied by a visceral queasiness—such a seizure was at least slightly alarming after all.[425]

The latter is likewise borne out in the following passage:

> I felt a kind of numbness, an enervation, but more particularly an odd fragility—as if my body had actually become frail, hypersensitive and somehow disjoined and clumsy, lacking normal coordination. And soon I was in the throes of a pervasive hypochondria. Nothing felt quite right with my corporeal self; there were twiches and pains, sometimes intermittent, often seemingly constant, that seemed to presage all sorts of dire infirmities.[426]

There would be "the rhythmic daily erosion of my mood—anxiety, agitation, unfocused dread," "dreadful, pouncing seizures of anxiety."[427] He was "riveted with fear," "helpless, shivering."[428] As to the depressive side of the panic-depression hybrid, Styron would all too frequently be "nearly immobilized and in a trance of supreme discomfort," "a condition of helpless stupor."[429] "The weather of depression is unmodulated, its light a brownout."[430] Further phenomena were also:

> … failure of even forced laughter and, at last, virtually total failure of speech. At this point the ferocious *inwardness* of the pain produced an immense distraction that prevented my articulating words beyond a hoarse murmur; I sense myself turning wall-eyed, mono-syllabic …[431]

As his depression became progressively worse, Styron would experience bouts of "nausea, a desperate and unpleasant wooziness, a sinking sensation and ultimately a distinct revulsion."[432]

To be sure, whether the descent into full-blown depression has or has not taken place, the panic-riddled ambit of neoliberal hyperexploitation—from whose purview the latter appears as an imminent eventuality—nonetheless remains a grand immaterial factory suffused by senility and exhaustion. As Robert Colvile argues, the relentless and unremitting cybernetically-induced flooding and saturation of the human organism with cortisol etiolates the immune system, rendering it susceptible to a plethora of severe ailments, such as cardiovascular disease, high blood pressure, type 2 diabetes, arthritis, osteoporosis, obesity, cancer, chronic fatigue syndrome, and Alzheimer's disease.[433] Here, the fraying of human DNA and atrophying of the telomeres of human cells, with the attendant wrinkling of the skin, graying of hair, impairment of sight, and lowering of life expectancy, gives rise to the prevalence of premature senility amid society as an increasingly normal condition. Over and above this, cybernetically-induced stress and panic leads to a shrinkage of the pre-frontal cortex and a consequent evisceration of the soul's attention span. The adaptation of the mind to a "diet of empty calories" bulks up the amount of bulbous "white matter" in executive centers of the brain whilst vastly atrophying its neuron-rich "gray matter" responsible for speech and memory—in the process not only vastly undermining post-industrial man's cognitive and emotional effectiveness, but likewise effecting a pandemic of conditions such as attention-deficit disorder, dyslexia, cultural cretinism, inane narcissism, and autism. What is being brought into existence is a generation of insipid and trite aphasiacs.[434]

We may also approach the problem of cybernetic machinic enslavement and its exhaustion of the organism from the perspective of Baudrillardian simulation. Simulation, avers Berardi, "is a projection of signs that neither reproduce nor record any facts, but the effect of the projection of ghosts never preceded by a body."[435] Evidently, such a scenario is brought into existence under the pressure of the capitalist axiomatic's massively accelerated processes of decoding and deterritorialization. For it is the latter that serve *inter alia* to sever signifiers from their signifieds and referents, and from concrete objects and extant states of affairs in general. Moreover, being the effect of a thoroughgoing dematerialization, the precipitation into being of new simulacra does not require any investment of energy. And for this reason, the endless replication of synthetic images (simulacra) may be said to have a boundless and indeed viral character. As observed, this seemingly endless pervasion of simulacra invades the lived experience of the soul. The invasion by the virtually limitless proliferation of simulacra is at once a metastatizing cancer that penetrates the very depths of the collective libido—the origin of cyber-capitalism's pathology of desire in which libidinal energy finds itself mercilessly assailed by malignant semiotic parasites.[436]

As seen, the first most tangible effect of this virus, or libidinous cancer, is exhaustion. Simulation—as the infinitely replicating virus that voraciously absorbs libidinal energy—effects a sort of "semiotic inflation" in which the proliferation of signs is simply too great, too fast, and much too chaotic and, for this reason, bursts the very circuits of collective sensibility. This acceleration to which the sensible body is subjected renders it impossible for it to consciously decodify and sensibly perceive the myriad inputs with which it is bombarded.[437] Human attention and imagination finds itself saturated by this well-nigh infinite proliferation of signs to such an exceeding degree that the libidinous energies of society are completely absorbed. This serves to reduce the organism's sensibility to the pulsations of quotidian reality virtually to the point of nullity. The sheer velocity of the infosphere's semiotic proliferation, unleashed by digital simulation, is ratched up to such a degree that it effectively saturates all circuits of collective sensibility to the hilt.[438] As observed, the cumulative effect of this now-well advanced assault is panic disorder, and then of course depression.

The perilous semiotic over-saturation attendant upon simulation may also be viewed from the vantage point of capitalist overproduction. As is well known, overproduction is endemic to the capitalist mode of production, in that exchange value preponderates over use value whilst the entire process as a whole is increasingly that of the self-valorization of capital itself—wherein all other moments of the capital process are subordinated to the self-begeting of money. In this scenario, the adequation of supply and demand is relegated into the distance, whereas the rapid drive toward capital valorization disproportionately privileges the supply side of the equation. Hence, the tendency to overproduction. Now, with the transition to cyber-capitalism being made, overproduction increasingly moves into the domain of the production of signs circulating within the infosphere. That is to say, overproduction assumes a semiotic character.[439]

But there is more. The move toward semiotic overproduction, with its extreme oversaturation of the infosphere and massive drain on the sensibility and libidinal reserves of the soul, is at once attended by a fundamental transmogrification in the dominant social psychopathology afflicting the great many. When the capitalist regime was still in its Fordist stage and the societal paradigm was broadly-speaking disciplinary in character, power was brought to bear on the subject predominantly by way of repression. Thus, the social psychopathology corresponding to this form of power was correctly diagnosed by Freud to be neurosis. Now, in the case of the Post-Fordist paradigm of the information age, power no longer functions via straight-forward repression. Rather, what is now foregrounded is the infosphere's hyper-stimulation of the subject's attention. Thus, the dominant social psychopathology has shifted from repression-induced neurosis and moved far closer to being an ever more pervasive psychosis driven by the panic disorder attendant upon cyber-capitalism's energetic and informational excess.[440] In Berardi's formulation, this present-day teetering of the soul on the very brink of psychosis means that current psycho-libidinous ailments "are not pathologies of repression, but of the *just do it.*"[441]

And it is precisely in direct relation to the foregoing that one may allude to Žižek's contention that the primary mode of subjective interpellation of today's permissive late capitalist liberal democracies, far from proceeding by way of prohibition, functions through an excessive and overbearing superegoic injunction to enjoyment. In this context, the enjoyment herein enjoined—exactly as accords with the hyper-stimulation of the cyber-capitalist infosphere—is beyond the pleasure principle. It is, in fact, a merciless command on the part of the superego which forces the soul to limitlessly expose itself to a *jouissance* far beyond its sensible and libidinal resources. And, paradoxically as it may indeed seem, it is precisely the foregoing that consistutes the essence asceticism assumes when the soul is enslaved to the cybernetic machines of the post-industrial age.

Late capitalist asceticism—and for all that, asceticism in general—is therefore the matter to which we must now turn.

19. The production of the ascetic capitalist subject

Needless to say: whether one speaks of post-industrial or earlier epochs of the capitalist mode of production, the capacity to sustain the increasingly unforgiving demands on productivity presupposes a certain fastidiously imbued asceticism on the part of the capitalist worker, and thus the attendant need that these ascetic subjectivities themselves be produced and cultivated.

That the work of Max Weber, Richard Henry Tawney and others immediately comes to mind at this point is hardly surprising. Certainly, the almost natural affinity between capitalist industriousness and its endless pursuit of profit, on the one hand, and what Bataille terms the "prosaic severity" of the Protestant religion, on the other, is fairly well established. After all, Calvinism's zone of geographical influence corresponds closely enough with those parts of the world characterized by a higher degree of capitalist development.[442] Indeed, one finds in Calvinism and kindred forms of the reformed religion the simultaneous condemnation of idleness and superflous luxury, coupled with the firm affirmation of and exhortation to enterprise. This stance comes to preponderate under Calvinism, in that for it all forms of ostensive and immediate use and consumption of infinite wealth are reserved solely for God himself. With the domain of non-productive consumption thus being rendered the exclusive preserve of the divine, man for his own part was henceforth to be dedicated to strictly productive labor and therefore to the development of productive forces.[443]

The effect of the reformers' acceptance of the most extreme consequences of their own demand for religious purity was the increasingly vast and comprehensive destruction of the Bataillean "sacred" world; that is to say, the world of non-productive consumption.[444] The world now ceases to be so much the domain where personal salvation is sought, and instead emerges as the ambit in which the divine is glorified not so much through prayer but rather via the active sanctification of the world by labor. In this sense, the Calvinist repudiation of good works and personal merit is to be read to mean that these latter cannot be the ancedent causes from whence salvation shall be obtained.

This nowise means that works are hereby jettisoned. Rather, they become the indelible signs that salvation has *already* been attained. To be sure, works are in a certain sense maintained. But they are so maintained only insofar as works involving nonproductive expenditure give way to strictly productive ones. In the passage from Catholicism to Calvinism and its epigones, practices entailing a needless expenditure of wealth are negated. Value is withdrawn from contemplative idleness, ostentatious luxury and forms of charity serving merely to prolong the existence of non-productive poverty; and given over to virtues based in utility. With Calvinism, the reformed Christian's *modus vivendi* had to therefore be suffused by his humility, nonprodigality, and industriousness.[445] In the most general formulation, however, the spread of the reformed religion entailed a greater and greater limitation of man to useful works whose nobility moreover came to increasingly converge and coincide with the descralization of the world; that is to say, with the increasing reduction of the intramundane world to that of *things*.[446]

A crucial upshot hereof, contends Bataille, is that Protestant man henceforth ceases to seek encountering himself in his intrinsic excessiveness in his own actions. That is, he does not look for himself directly when he acts and ever the more resolutely detaches sentiment from action. His excessive truth is referred to the supramundane, divine world. Calvinism in particular, therefore, tends towards an increasingly stark divergence between the realms of economy and things in general, on the one hand, and religious and affective life, on the other. At the same time, there is an analogous tendency toward the complete diremption between man's productive action and his self-reflexive movement of return to the profundity of his own intimate being.[447] Thus, Calvinism reaches the point where, in a self-denial that is at once the affirmation of God, an increasingly meagre spirituality (at once the purest and the poorest) serves to counterbalance the ever greater subjection of the entire body—and activity as such—to the domain of things.

Now, with the burgeoning development of the capitalist mode of production, this conjunction of a prosaic and minimal spirituality with a generalized reification of action at length crosses a threshold at which the utter nullity of Protestant spirituality gives way to—no spirituality whatsoever. (This, as we have seen, is exactly Charles Taylor's thesis.) That is to say, the principle of ascetic servitude nurtured under the conjunction of reformed religion and a then-still fledgling capitalism, with the increasing erosion of the former and the galloping hypertrophy of the latter; gives way to a well-nigh purely capitalist asceticism that henceforth maintains itself quite soundly even in the absence of God.[448] The renunciation of life continues, though now solely at the behest of the imperatives of capitalist production for production's sake. There is, however, no end to the piety of capitalist man. For, rather than simply vanishing into the void, hitherto moribund Protestant piety merely loses its overtly religious integument. Becoming secular, man nowise ceases to be pious. This, of course, has merely brought us back to an earlier point.

So, let us return to and dwell more concertedly upon asceticism itself. Žižek maintains the inseparability of asceticism and the idea of justice as equality, which latter in turn is seen to source itself in envy—the envy the subject feels for the Other for what the Other possesses and enjoys and the subject himself does not. Justice as arising out of envy thus consists in the demand that this excessive enjoyment of the Other be undercut in such a way that every subject's access to *jouissance* is henceforth made equal. It is precisely here that asceticism is seen to arise. But such a direct imposition of equal *jouissance* is quite simply impossible. What is, therefore, imposed in its stead can only be an equally shared *prohibition*. In present advanced capitalist times, however, avers Žižek, this ascetic prohibition paradoxically assumes the form of its opposite: a generalized superego injunction for the subject to expose itself without reserve to an entirely excessive *jouissance*. Yet, far from the latter entailing a liberation from the purportedly draconian prohibitionism of a bygone epoch, the superegoic injunction's overbearing and hypertrophied character ultimately renders it an even greater hindrance to the subject's enjoyment than straightforward prohibition itself.[449]

This dialectical reversal endemic to present-day "nonrepressive" hedonism, wherein a purported lack of constraint ultimately reveals itself to be a still more severe form of coercion, resides in the transmogrification of merely permitted *jouissance* into *obligatory jouissance*. Here, although the postmodern superegoic injunction in the more immediate and explicit sense accords with the preponderance of the pleasure principle and its maintenance of homeostasis, the injunction's actual functioning in fact explodes this ostensive framework and moves rapidly toward an excessive enjoyment that henceforth crushes the subject with the full force of inexorable necessity.[450] For this reason, arising as it does out of an overwrought and ubiquitous postmodern permissiveness that is at once experienced as overbearing, the resultant subjective condition may be termed *hedonistic asceticism.* Žižek conjectures that the prototypical sufferers of this malady are today's yuppies—perhaps the consumate embodiments of the lowest and most decrepit stage of nihilism: the passive nihilism of Nietzsche's last men.[451]

But apart from these particular *untermenschen*, there are others. For capitalist asceticism bears another variant: that of the miser. We have thus far dealt with the envious subject, he who envies the Other's possession and *jouissance* of the object. The miser, in contradistinction, is he who in fact possesses the object but is nowise able to consume or enjoy it. He nonetheless rests content with merely possessing the object and moreover extolling it as a well-nigh sacred, untouchable and thereby prohibited entity which may under no circumstances be consumed—not even by himself. The miser, therefore, raises the object he possesses to the status of an object of desire. And it is precisely this latter elevation and sacralization of the possessed object that thrusts it outside the ambit of possible consumption.[452]

The miser himself, possessed as he is of the so-called "anal character," is seen by psychoanalysis to ultimately arise from his own much earlier infantile recalcitrance to defecate on demand. Obstinately bent upon ceding his excrement only on his own terms so as not to deprive himself of the surplus-enjoyment of solitary defecation, an infant of such stubbornness almost inevitably grows up to become a miserly adolescent and adult. That is to say, he becomes a lifelong compulsive hoarder who, notwithstanding, in the process suffers the consequence of falling under the permanent sway of a renunciation infinitely stronger than infantile anal retentiveness. For he henceforth allows himself no nonproductive consumption or pleasurable indulgence whatsoever—at least none such as fails to serve the higher ends of accumulation. Here Žižek avers that the paradox of this miserly asceticism consists in the fact that "when the small child refuses 'castration' (ceding of the privileged detachable object), he takes the path that will end in his total self-castration in the Real; that is to say, his refusal to cede the surplus-object will condemn him to the prohibition on enjoying any other object. In other words, his rejection of the demand of the real parental Other (to behave properly on the toilet) will result in the rule of an infinitely more cruel interiorized superego Other that will totally dominate his consumption. And this brings us to Chesterton's principle of Conditional Joy: by refusing the founding exception (the ceding of the excessive object), the miser is deprived of *all* objects."[453]

But let us step back a little. For capitalist asceticism and its vicissitudes—and indeed asceticism *tout court*—may also be approached from the angle of what Nietzsche terms the "ideal" by the same name; that is to say, from the perspective of the *ascetic ideal.* As observed, Žižek's pivotal contention was that the transition from regimes of the more traditionally repressive variety to (ostensively) permissive regimes of the present cyber-capitalist sort entailed at once the movement from an asceticism induced by prohibition to one precipitated into being by the imperative of excessive *jouissance*. However, in Alenka Zupančič's Lacanian reading of Nietzsche, it emerges that asceticism *in all its historical forms* fundamentally derives from an underlying superegoic injunction to surplus enjoyment. This is the case for two reasons.[454]

Firstly, the ascetic ideal itself evidently long antedates all forms of capitalism. Secondly, the asceticism entailed in the ascetic ideal—from the very outset—*never* at base involves the straightforward renunciation of enjoyment. For it can indeed—in point of fact—be said to herald and even coincide with the very first inundation of enjoyment itself; that is to say, an enjoyment that of its very nature differs from pure and simple pleasure and thereby lies beyond the Freudian pleasure principle.[455]

This is of course not to say that Žižek is oblivious to the foregoing observation. True: even regimes of a far more ostensively repressive character are at base animated by the imperative to excessive *jouissance*. However, in the case of present-day "liberal democratic" regimes, the difference resides in the simple fact that this always-present superegoic injunction to excessive enjoyment now foregrounds itself to the extent of standing at the very brink of eschewing more traditional forms of repression that had hitherto been its inexorable adjunct.

But what, more precisely, is this surplus-enjoyment enjoined by the ascetic ideal? As already seen, it does not in the first instance consist in any sort of direct renunciation of satisfaction/pleasure. This latter may—and indeed does—come to be the case only as a corollary or after-effect of the imperative to excessive *jouissance*. Rather, it is that which incessantly compels the subject to strive after the "immediate Real" that lies beyond/beneath all "apparent" and always partial satisfactions. It is only when this excessive "it-itself" of satisfaction comes to be experienced as impassably fettered by the partial satisfactions that inevitably attend it, that the subject begins to renounce both the excessive enjoyment and its accoutrements. Zupančič illustrates this point in her discussion of anorexia as a poignant example of the intimate link that exists between the ascetic ideal and the imperative to enjoy. According to Lacan, it is not so much—in the primordial sense—that the anorexic refuses to eat. Mundanely-speaking, of course, the anorexic subject does indeed refuse food. But he does so only in a derivative sense. More fundamentally, the anorexic *eats nothingness itself*. But what is this "nothingness"? It is the uncanny surplus owing to which any given form of food is able to become an object of desire to begin with. With the superego injunction to *jouissance* as excessive and overbearing as it is, the aneroxic subject is impelled to strive directly for this excess. This, however, entails an attempt on the part of the subject to isolate and distill the said surplus from the food itself. When brought into effect, the resultant bifurcation between excessive and partial satisfaction—being at once the diremption of desire and need—destroys the very structure of desire itself. Unable to countenance the partial satisfactions sullying the path toward the pristine *jouissance* on account of which food is desirable, the anorexic has no choice but to eschew this surplus satisfaction as well.[456]

And thus—much as the exhausted, panic-disordered soul enslaved by the cybernetic machines of the infosphere ultimately detaches itself from the overtaxing stimulus of galloping simulation—so does the anorexic withdraw from food and embark upon the path of self-induced starvation. In the same vein, it may be averred that the subject submitted to the machinic enslavement of cyber-capitalism runs the imminent risk of contracting anorexia of the soul.

20. Nihilism

As seen, the anorexic soul strives to directly attain the immediate Real behind the partial satisfaction of mere transitory need. Of course, in that this libidinal short-circuit necessarily obviates and eschews all partial satisfactions and inasmuch as the attainment of the Real is impossible, the anorexic soul in effect desires nothingness itself. Here we are at liberty to exchange the psychoanalytical concept of desire for the Nietzschean notion of will—or will to power. And it is to Nietzsche that we are thus brought. It may thus be said that the will of the anorexic soul hereby exhibits itself to be a *will to nothingness*. In Nietzsche, the will to nothingness is none other than the animating force or motor of the ascetic ideal. On this reading, the merciless Lacanian superegoic injunction that impels the subject to expose itself without reserve to a limitless *jouissance* entirely coincident with the pure excess of the impossible Real shows itself to be homologous to the Nietzschean will to nothingness.

However, as observed, the soul rendered anorexic by cyber-capitalist hyper-exploitation—when overwrought by informational overstimulation to the point of panic disorder—falls into the mire of depression. It is precisely here—with desire itself having fallen into dereliction—that the soul, no longer able to abide merciless superegoic demands, rather than willing nothingness, forthwith *ceases to will*. Now, if the will to nothingness falls under the rubric of what Nietzsche understands by *active nihilism*, then the anorexic soul's descent into depression accords with what the latter terms *passive nihilism* (nothingness of the will). Therefore, within the context of post-industrial capitalist hyper-exploitation, active nihilism would stem from the excessive pressures of an overly draconian superego born of cybernetic hyperstimulation and its tendency to induce panic in the soul; whilst passive nihilism would be the condition ushered in by the soul's withdrawal from the said hyperstimulation into a depressive lull wherein it submits itself to its own self-induced spiritual starvation—or anorexia. Realistically, however, this passive nihilism of the now rapidly atrophying depressive soul does not quite reach a terminus whereat the will is as such cast into nonexistence. Rather, it etiolates to the point of being employed almost exhaustively for the ends of self-restraint. This self-restraint aims at maintaining the soul in its present state of withdrawal from the treacherous ambit of superegoic tyranny and informational overstimulation.

However, in the vast majority of cases—faced as it is with no plausible alternative—the cybernetically enslaved soul will inevitably return to the panic-riddled arena from whence it had afore fled. Ultimately, a cycle ensues—in which the soul is seen to alternately dissent from and accede to informational hyper-stimulation and superegoic despotism. This is the panic-depressive cycle of which Berardi speaks. From the vantage point of nihilism and the ascetic ideal, would the foregoing not entail an analogous oscillation between active and passive nihilism? To a certain extent, yes. But the matter turns out to be somewhat more convoluted than this would initially seem to imply. Here Zupančič observes that, in Nietzsche, the relation between active and passive nihilism is in fact twofold. In the first sense, Nietzsche appears to suggest that active nihilism ultimately leads to passive nihilism when it exhausts itself. In the second sense, passive nihilism is more a response to active nihilism than it is a consequence of active nihilism's inner dynamism. In this latter case, passive nihilism functions in the manner of a sedative and defense against the overly stimulating character of active nihilism—a defensive formation in which the will to nothingness appears as a nothingness of the will.[457]

Returning to Berardi, this two-tiered relation that exists between Nietzschean active nihilism and passive nihilism may in fact also be seen to hold between panic and depression when it is realized that, in addition to depression being the inevitable aftermath of a panic that the soul can no longer bear, panic-disorder and depression are always virtually present within each other from the very outset. Notwithstanding, the upshot gleaned by Zupančič in her observations on nihilism is that it may nowise be considered a general category which subsequently appears in active and passive variants. Rather, it consists in the very *tension* that prevails between its active and passive forms and is thus naught outside of this antagonism. It is therefore the mutual interdependence of active and passive nihilism that consistutes nihilism from the very outset. Within this tensile ambit, active nihilism accords with the tyrannical superegoic injunction to an excessive *jouissance* coincident with the Real. This overbearing superego imperative—as is the case with the hyperstimulation of the soul by the info-sphere—is what holds the subject in a mortifying grip, a state of paralysis or panic which, though it may assume the form of rabidly frenetic activity, nonetheless remains what it is. In tension with the foregoing, passive nihilism serves as a defense against the mortifying panic induced by active nihilism's excessive agitation. It is thus a form of prophylaxis that dampens the surfeit of *jouissance* by way of a narcotization of the will.[458]

Passive nihilism's tranquilization of active nihilism's hypertrophied stimulation, may be viewed as analogous to the manner in which central-nervous system depressants dampen cybernetically induced panic. What remains, however, of crucial import is that, far from forging a way out of the quagmire of nihilism, passive nihilism's etiolation of excessive *jouissance* and likewise will—in fact serves only to preserve active nihilism's will to nothingness in a virtual/deactivated mode. The implicit danger here is that active nihilism may very well find an alternative way of recrudescing from beneath the soporific vapors of its counterpart, that is, passive nihilism. And it may come to do so in a singularly violent and indeed catastrophic way.[459] This occurs when the tyranical superegoic imperative of the Real—rather than precipitating the renunciation of nothingness as the constitutive element of the will and thereby voiding will and desire alike—links itself to a destructive *passage à l'acte*. This dramatic and hitherto unforeseeable event, this violently atavistic upsurgence, may show itself to be not merely destructive. It may indeed prove to be *self*-destructive.[460]

We seem to be heading for suicide.

Notes

1. Deleuze and Guattari 2004, 35.
2. Marx 1990, Vol. 1, 928.
3. Ibid.
4. Ibid.
5. Ibid. 875.
6. Ibid.
7. Ibid.
8. Ibid.
9. Ibid.

10. Zinn 2005, 42, 43.
11. Ibid. 13, 14.
12. Ibid. 15, 16.
13. Ibid. 126.
14. Ibid. 128, 129.
15. Ibid. 127, 128.
16. Ibid. 129.
17. Ibid. 144.
18. Ibid. 134.
19. Ibid. 138, 139.
20. Ibid. 142, 143.
21. Ibid. 147.
22. Ibid. 283.
23. Klein 2007, 19–21.
24. Ibid. 29.
25. Ibid. 50.
26. Ibid. 149, 150.
27. Ibid. 153.
28. Ibid. 206, 207.
29. Ibid. 209.
30. Ibid. 215.
31. Ibid. 221–229.
32. Ibid. 237, 238.
33. Ibid. 269–273.
34. Ibid. 277.
35. Ibid. 331, 332.
36. Ibid. 334–339.
37. Ibid. 342.
38. Ibid. 345.
39. Ibid. 350, 351.
40. Davis 2007, 152, 153, 155, 174.
41. Ibid. 15–18.
42. Ibid. 98, 104–107.
43. Ibid. 185.
44. Ibid. 186–190.
45. Ibid. 192–194.
46. Ibid. 194–198.
47. Harman 2009, 39, 40.
48. Marx 1990, Vol. 1, 874.
49. Deleuze and Guattari 2004, 36.
50. Ibid. 154.

51. Ibid. 244, 245.
52. Marx 1993, 498.
53. Ibid.
54. Ibid.
55. Ibid. 296, 297.
56. Ibid.
57. Ibid.
58. Marx 1990, Vol. 1, 723, 724.
59. Deleuze and Guattari 2004, 246.
60. Marx 1992, Vol. 2, 135.
61. Ibid.
62. Holland 1999, 84.
63. Marx 1993, 253.
64. Ibid.
65. Ibid.
66. Žižek 2006, 59, 60.
67. Marx 1990, Vol. 1, 255.
68. Ibid. 256, 257.
69. Ibid.
70. Cited in Bataille 1988, Vol. 1, 126.
71. Deleuze and Guattari, 248.
72. Ibid. 250.
73. Harman 2009, 107.
74. Ibid.
75. Marx 1991, Vol. 3, 382, 383.
76. Ibid. 431.
77. Ibid.
78. Marx 1992, Vol. 2, 135, 136.
79. Ibid.
80. Deleuze and Guattari 2004, 270.
81. Ibid.
82. Deleuze and Guattari 1988, 454, 455.
83. Deleuze and Guattari 2004, 270.
84. Ibid. 256.
85. Ibid. 270, 271.
86. Wood 2003, 3.
87. Ibid.
88. Ibid.
89. Ibid. 12.
90. Ibid. 273.
91. Ibid. 280.

92. Ibid. 283.
93. Ibid. 273.
94. Deleuze and Guattari 1988, 455.
95. Wood 2003, 16.
96. Ibid.
97. Ibid.
98. Ibid. 4, 5.
99. Ibid. 20–22.
100. Harman 2009, 109, 110.
101. Deleuze and Guattari 1988, 456, 457.
102. Ibid.
103. Ibid. 457, 458.
104. Ibid.
105. Ibid.
106. Ibid.
107. Ibid.
108. Ibid.
109. Ibid.
110. Ibid.
111. Berardi 2009, 192.
112. Schreber 2000, 19.
113. Berardi 2009, 192.
114. Ibid. 200.
115. Deleuze and Guattari 1988, 457, 458.
116. Ibid.
117. Lazzarato 2014, 23, 24.
118. Ibid. 12.
119. Ibid. 23–25.
120. Ibid. 12.
121. Ibid. 25, 26.
122. Ibid. 26.
123. Ibid. 26, 27.
124. Ibid. 27.
125. Taylor 2007, 32.
126. Ibid. 36.
127. Ibid. 37, 38.
128. Ibid. 142.
129. Butler 2006, 51.
130. Ibid. 51, 52.
131. Ibid.
132. Lazzarato 2014, 37, 38.

133. Butler 2006, 94.
134. Ibid. 54.
135. Ibid.
136. Ibid. 54, 55.
137. Ibid. 62.
138. Ibid.
139. Ibid.
140. Ibid. 65, 66.
141. Ibid. 94, 95.
142. Ibid. 97, 98.
143. Ibid.
144. Ibid. 73, 74.
145. Ibid. 76.
146. Ibid. 77.
147. Ibid.
148. Agamben 1998, 123–125.
149. Ibid.
150. Ibid.
151. Ibid.
152. Ibid.
153. Ibid. 127, 128.
154. Ibid. 139, 140.
155. Ibid. 148.
156. Ibid. 170–173.
157. Ibid.
158. Ibid.
159. Ibid.
160. Ibid.
161. Zinn 2005, 34.
162. Ibid. 375, 376.
163. Ibid. 416.
164. Ibid. 429.
165. Ibid. 432.
166. Ibid. 474, 475.
167. Ibid. 461, 462.
168. Cited in Klein 2007, 99.
169. Cited in Ibid.
170. Cited in Ibid.
171. Cited in Ibid.
172. Klein 2007, 101.
173. Ibid. 114, 115.

174. Ibid. 106–109.
175. Ibid. 102–104.
176. Ibid. 104–107.
177. Ibid. 109.
178. Ibid. 76, 77.
179. Ibid. 90.
180. Ibid. 92.
181. Ibid. 90.
182. Badiou 2002, 72.
183. Ibid. 73.
184. Ibid. 74.
185. Ibid.
186. Ibid. 75.
187. Mazzetti 2013, 45, 46.
188. Mayer 2009, 19.
189. Mazzetti 2013, 43, 44.
190. Ibid. 57.
191. Ibid. 51.
192. Ibid. 88, 89.
193. Scahill 2013, 26, 27.
194. Ibid. 5, 6.
195. Mazzetti 2013, 4, 5.
196. Scahill 2013, 23, 24.
197. Ibid. 19, 20.
198. Ibid. 24, 25.
199. Chamayou 2015, 52.
200. Ibid. 57.
201. Scahill 2013, 93.
202. Ibid. 98, 99.
203. Cited in Ibid. 101.
204. Chamayou 2015, 54, 55.
205. Cited in Ibid. 55.
206. Chamayou 2015, 56.
207. Scahill 2013, 114.
208. Chamayou 2015, 162.
209. Ibid. 163.
210. Ibid. 164.
211. Ibid.
212. Scahill 2013, 365, 367.
213. Klein 2007, 371.
214. Scahill 2013, 147, 148.

215. Ibid. 149–151.
216. Ibid. 154.
217. Ibid. 157.
218. Mazzetti 2013, 128, 129.
219. Cited in Chamayou 2015, 57.
220. Chamayou 2015, 57, 58.
221. Ibid. 58.
222. Cited in Ibid.
223. Ibid.
224. Ibid. 58, 59.
225. Scahill 2016, 42–44.
226. Ibid. 43, 44.
227. Scahill 2013, 245, 246.
228. Chamayou 2015, 47.
229. Scahill 2013, 250, 251.
230. Mazzetti 2013, 206, 207.
231. Chamayou 2015, 56, 57.
232. Ibid. 57.
233. Scahill 2013, 251.
234. Ibid. 277.
235. Ibid. 268, 269.
236. Chamayou 2015, 56.
237. Ibid.
238. Ibid.
239. Shane 2016, 210, 211.
240. Scahill 2013, 354, 355.
241. Chamayou 2015, 68.
242. Mazzetti 2013, 5, 6.
243. Chamayou 2015, 71.
244. Scahill 2013, 454.
245. Chamayou 2015, 172.
246. Ibid.
247. Ibid.
248. Cited in Chamayou 2015, 172.
249. Chamayou 2015, 173.
250. Scahill 2013, 513, 514.
251. Cited in Ibid.
252. Cited in Ibid.
253. Cited in Ibid.
254. Cited in Ibid. 516, 517.
255. Scahill 2013, 516, 517.

256. Cited in Ibid.
257. Mayer 2009, 157, 158.
258. Ibid. 159.
259. Ibid. 160.
260. Ibid. 161.
261. Ibid.
262. Ibid.
263. Ibid. 170.
264. Ibid.
265. Ibid. 163.
266. Ibid. 164.
267. Ibid.
268. Ibid. 200.
269. Ibid. 165–169.
270. Ibid. 207.
271. Ibid. 206–208.
272. Ibid. 208.
273. Deleuze and Guattari 2004, 36.
274. Marx 1993, 541, 542.
275. Ibid. 334, 335.
276. Ibid.
277. Marx 1991, Vol. 3, 358, 359.
278. Ibid.
279. Ibid.
280. Goodchild 1996, 99.
281. Žižek 2015, 41.
282. Ibid. 42–46.
283. Berardi 2015b, 76, 77.
284. Berardi 2012, 30.
285. Berardi 2015b, 6.
286. Ibid. 7.
287. Lazzarato 2014, 27, 28.
288. Berardi 2012, 88, 94, 95.
289. Berardi 2009, 188.
290. Ibid. 153–155.
291. Ibid.
292. Ibid.
293. Deleuze and Guattari 2004, 245.
294. Lazzarato 2014, 41.
295. Žižek 2006, 182.
296. Ibid. 318.

297. Ibid.
298. Badiou 2009, 420.
299. Baudrillard 1993, 6, 7.
300. Ibid. 57–59.
301. Ibid. 87.
302. Ibid. 79.
303. Ibid. 7.
304. Ibid. 72.
305. Ibid. 8, 9.
306. Ibid. 66.
307. Ibid. 78.
308. Ibid. 87.
309. Berardi 2012, 161–167.
310. Baudrillard 1993, 23.
311. Taylor 2007, 307, 308.
312. Cited in Ibid. 380.
313. Cited in Ibid.
314. Taylor 2007, 381.
315. Cited in Ibid. 382.
316. Cioran 1998, 5, 6.
317. Deleuze and Guattari 2004, 245.
318. Holland 1999, 82.
319. Lazzarato 2014, 42.
320. Gray 2008, 3, 4.
321. Ibid. 28–31.
322. Ibid. 35.
323. Ibid. 39.
324. Ibid. 156–158.
325. Ibid. 40.
326. Ibid. 45.
327. Ibid. 173–175.
328. Ibid. 270, 271.
329. Ibid. 120–122.
330. Ibid. 122, 123.
331. Ibid. 46–48.
332. Ibid. 161, 162.
333. Ibid. 166.
334. Taylor 2007, 82, 83.
335. Ibid. 84.
336. Ibid. 86, 87.
337. Ibid. 102–104.

338. Ibid. 106.
339. Ibid. 110, 111.
340. Ibid. 118, 119.
341. Ibid. 143–146.
342. Ibid. 244, 245.
343. Ibid. 246, 247.
344. Ibid. 250, 251.
345. Ibid. 251.
346. Žižek 2008c, 25, 26.
347. Ibid. 26, 27.
348. Ibid. 30, 31.
349. Berardi 2012, 161, 162, 166.
350. Ibid. 168, 169.
351. Schreber 2000, 133.
352. Ibid. 53, 77, 117.
353. Ibid. 119.
354. Ibid. 330.
355. Ibid. 152.
356. Ibid. 167.
357. Ibid. 191.
358. Ibid. 53.
359. Ibid. 59.
360. Ibid. 61.
361. Ibid. 63.
362. Ibid. 90.
363. Ibid. 125.
364. Ibid. 66.
365. Ibid. 142.
366. Ibid. 163, 164.
367. Ibid. 164.
368. Ibid. 211.
369. Ibid. 245.
370. Ibid. 248.
371. Ibid. 335.
372. Ibid. 75.
373. Ibid. 79.
374. Ibid. 80.
375. Ibid. 81, 82.
376. Ibid. 131, 132.
377. Ibid. 250.
378. Ibid. 254, 255.

379. Ibid. 258.
380. Ibid. 292.
381. Marx 1990, Vol. 1, 451.
382. Ibid.
383. Ibid.
384. Deleuze and Guattari 2004, 251.
385. Ibid. 285.
386. Marx 1990, Vol. 1, 526, 527.
387. Ibid.
388. Ibid. 544, 545.
389. Deleuze and Guattari 2004, 157.
390. Berardi 2015b, 138–141.
391. Deleuze and Guattari 2004, 157.
392. Berardi 2009, 101.
393. Ibid.
394. Colvile 2016, 53, 54.
395. Taylor 2007, 35–39.
396. Schreber 2000, 55.
397. Ibid. 200, 201.
398. Ibid. 274.
399. Ibid. 302.
400. Ibid. 141–151.
401. Ibid. 183.
402. Ibid. 188.
403. Berardi, 2009, 101–103.
404. Lazzarato 2015, 185–187.
405. Berardi 2012, 98, 99.
406. Berardi 2009, 101–103.
407. Berardi 2015a, 103, 104.
408. Taylor 2007, 37.
409. Styron 2004, 5.
410. Styron 2004, 9, 10.
411. Ibid. 12.
412. Jamison 2000, 104.
413. Ibid. 114.
414. Jamison 1997, 72.
415. Ibid. 79.
416. Ibid. 82.
417. Ibid. 45.
418. Styron 2004, 14.
419. Ibid.

420. Cited in Styron 2004, 15.
421. Styron 2004, 46.
422. Ibid. 46, 47.
423. Ibid. 47.
424. Ibid.
425. Ibid. 41.
426. Ibid. 42, 43.
427. Ibid. 44, 45.
428. Ibid.
429. Ibid. 15.
430. Ibid.
431. Ibid. 17.
432. Ibid. 40.
433. Colvile 2016, 55.
434. Ibid. 47, 48.
435. Berardi 2009, 157.
436. Ibid.
437. Ibid. 158, 159.
438. Ibid. 162, 163.
439. Ibid. 178.
440. Ibid. 179, 183.
441. Ibid. 180.
442. Bataille 1988, Vol. 1, 115.
443. Ibid. 124.
444. Ibid. 127.
445. Ibid. 123.
446. Ibid. 124.
447. Ibid. 134, 135.
448. Ibid. 136–138.
449. Žižek 2008d, 89, 90.
450. Žižek 2006, 310, 311.
451. Žižek 2008d, 90.
452. Ibid. 90, 91.
453. Žižek 2003, 40.
454. Zupančič 2003, 47.
455. Ibid.
456. Ibid. 128, 129.
457. Ibid. 64, 65.
458. Ibid. 66, 67.
459. Ibid. 68.
460. Ibid. 66.

18

From the Self-Crucifixion of the Capitalist Subject to the Empyrean

In an October 1854 letter to Franz Liszt, Richard Wagner polemicizes with fury, averring the world to be completely and utterly evil, and that it—as a consequence—must be treated with nothing but contempt and, indeed, be discarded. The preponderant reason for this, Wagner contends, is that it belongs to Alberich, and to Alberich alone.[1] Alberich is, of course, the purely pragmatic, utterly callous and conscience-bereft über-rogue in Wagner's opera *Der Ring des Nibelungen*, who mercilessly drives his slavish Nibelungs to the very limit of their laboring capacity. Yet, is not Alberich a mere mite in comparison to our own present-day Alberichs—especially when viewed beside the mega-Alberich that is today's roiling cyber-capitalist machine?

Toward the very end of the preceding chapter, it appeared as though we were heading for suicide. This is so, because we are.

1. Suicide

On September 11, 2001, the black hole of stochastic death, long presaged by the maddening resonance of the collective jihadi subject of nineteen young men with that of their reclusive commander, found its rabid consummation in the lividly infernal white heat of consciousness cavorting to molten auto-immolation in the rapidly vanishing epicenter of lashing fires in excess of one-thousand-three-hundred degrees, in which the far-flung hopes, dreams, and memories of two-thousand-seven-hundred-and-ninety-seven suddenly but completely vanquished lives disappeared from the earth. In the same terminal morass, the voluntary moribunds were annihilated in the singular void of their eternal longing, dying in a scorchingly evanescent instant of physical torment, panic attacks and rioting atoms exploding in every nerve and brain cell, as the boiling blood of their own lives, which they had so hated, absented itself in a final palsied echo.

This monumental act, the single most astonishing act of terrorism in all human history, inaugurated the twenty-first century. It was its sickening opening salvo. Though a hideous and diabolical act of mass murder, this foul crime was even more so an absolutely sovereign act of self-annihilation. Though certainly indoctrinated to the very hilt, the overriding prerogative and unquenchable desire of these autothanatophilous haters of their own lives was to die. Long before and right up until the time of their gargantuan apoplectic fit of self-immolation, these men had tormented and gnawed at their own souls, transformed themselves into a squalid workshop of masochistic necromancy and slavering self-laceration, an ever more voracious abyss and suppurating ulcer of searing, self-impregnating pain, a veritable torture chamber. Ever since this stratospheric act of demented evil, suicide has become the leading actor on the stage of third millennium history. This is so since suicide is now, as Berardi asseverates, perceived by increasingly burgeoning droves of the oppressed as the sole and ineluctable recourse for dispelling tremulous anxiety, dank and brackish depression, and the all-pervading sense of complete and utter impotence that marks the present. Suicide is now becoming a quasi-normalized condition in an age veering away from romantic self-sacrifice ennobled by patriotic ideals in the direction of the parched aridity of *anomie*.[2]

Berardi views this syndrome as underpinned by a vertiginous chasm between the cybernetically enslaved soul's rational elaboration and its act. In a strictly fascist environment replete with obscene superegoic bombardments and promiscuous incitements to frenetic action, the depressed soul's inability to mobilize its energies with competitive bravura mires the subject in resentment, frustration, guilt, and other such progeny of the swamp. In this predicament, the sole mode of transcending depressive paralysis may well be by way of a violent act entirely divorced from reflection, conscious elaboration, and any consideration of the future. The subject reduced to depression and catatonia suddenly rushes toward spectacularly murderous self-immolation. In this psychotic and demented acting out, the line of demarcation between self and other, murder and suicide, is blurred to the point of extinction. It is specifically in the atavistic delirium of runaway panic that the borderlines between self and environment, past, present and future, all fall into terminal dereliction. Today's hyper-stimulated cybernetic milieu is psychotic. The self explodes.[3]

At the helm of this demented psychotic realm is a newfangled form of nihilism: that fomented by advanced neoliberal financial capitalism. Berardi maintains that a shift has come to pass from a fairly recently surpassed "hermeneutic nihilism" toward what he terms "annihilating nihilism"—which now actively produces nothingness as its effect. It destroys previously accumulated communal, moral and economic values, so as to affirm and accumulate abstract value as its overriding goal. It dismantles factories, destroys jobs, brings about the deaths of people, causes cities to crumble, and ushers in a generalized denigration of the world. It is the most perfect form of criminality and establishes suicide at the epicenter of the systematic enslavement that it fosters.[4] This is, of course, the mass machinic enslavement of the soul by an ever-burgeoning infosphere that we have been describing all along. Here, the panic-depressive cycle in which the machinically enslaved soul finds itself immured is forcibly broken by a violent atavistic recrudescence of active nihilism in which—rather than desire falling into desuetude at the behest of the subject's renunciation of both the nothingness of excessive *jouissance* that attaches to the Real and all partial satisfactions—the will, on the contrary, wills nothingness itself (the will to nothingness) to the unilateral exclusion of all else. Stated in terms of the nomenclature of subjectification, the articulation of the subject of the enunciation (now the unlimited semiosis of the infosphere) and the subject of the statement (the soul enslaved by cybernetic machines) around capital (as point of subjectification)—is such that the associated subjective double (semiosis-soul) reaches a pitch of resonance that inexorably drags the jaded subject into the black hole of stochastic death wherein the latter *destroys itself.*

We have seen with our conjunction of Zupančič's Lacanianized Nietzsche and Berardi's observations that passive nihilism accords with depression proper and the active nihilism that extrudes itself from the latter's dour ambit is the movement toward suicide. This movement from depression to suicide is clearly to be seen in observations made as early as in Robert Burton's *The Anatomy of Melancholy*, here quoted by Kay Redfield Jamison: "There is … in this [melancholic] humour, the very seeds of fire … In the day-time they are affrighted still by some terrible object, and torn to pieces with suspicion, fear, sorrow, discontents, cares, shame, anguish, etc., as so many wild horses, that they cannot be quiet an hour, a minute of time, but even against their wills they are intent, and still thinking of it, they cannot forget it, it grinds their souls day and night, they are perpetually tormented.… In the midst of this squalid, ugly, and such irksome days, they seek at last, finding no comfort, no remedy in this wretched life, to be eased of all by death … to be their own butchers, and execute themselves."[5] Jamison likewise maintains that depression is at the heart of most suicides, averring that it causes thinking to become more constricted and rigid, and perceived options to narrow to the point where death is seen as the only alternative and, moreover, as a highly seductive one. Suicide is seen as a tranquil surcease of pain, a lulling alternative to the pain of existence and the agonizing, suffusing hopelessness pervading all reaches of the mind.[6]

In William Styron's memoir, similar moribund feelings are expressed. Things simply become too painful:

> I had now reached that phase of the disorder where all sense of hope had vanished, along with the idea of futurity; my brain, in thrall to its outlaw hormones, had become less an organ of thought than an instrument registering, minute by minute, varying degrees of its own suffering.... I'd feel the horror, like some poisonous fogbank, roll in upon my mind ...[7]

And then:

> Death, as I have said, was now a daily presence, blowing over me in cold gusts.... In short, I was keeping the possibility of suicide at bay. But plainly the possibility was around the corner, and I would soon meet it face to face.
>
> What I had begun to discover is that, mysteriously and in ways that are totally remote from normal experience, the gray drizzle of horror induced by depression takes on the quality of physical pain. But it is not an immediately identifiable pain, like that of a broken limb. It may be more accurate to say that despair, owing to some evil trick played upon the sick brain by the inhabiting psyche, comes to resemble the diabolical discomfort of being imprisoned in a fiercely overheated room. And because no breeze stirs this cauldron, because there is no escape from this smothering confinement, it is entirely natural that the victim begins to think ceaselessly of *oblivion* [my emphasis].[8]

Styron was here moving rapidly to actual suicide. He felt he had had enough. He hated life and wanted to die:

> ... many of the artifacts of my house had become potential devices for my own destruction: the attic rafters (and an outside maple or two) a means to hang myself, the garage a place to inhale carbon monoxide, the bathtub a vessel to receive the flow from my opened arteries. The kitchen knives in their drawers had but one purpose for me ...[9]

And later:

> ... there was an annihilation of self appropriate, as always, to melancholia's fecund self-humiliation. I felt my heart pounding wildly, like that of a man facing a firing squad, and knew I had made an irreversible decision.[10]

Finally:

> … I coudn't shake off a sense of melodrama—a melodrama in which I, the victim-to-be of self-murder, was both the solitary actor and lone member of the audience. I had not as yet chosen the mode of my departure, but I knew that that step would come next, and soon, as inescapable as nightfall.[11]

Kay Redfield Jamison had likewise reached this point. She had also had it. She was moving inexorably from depression (passive nihilism) to suicide (active nihilism). In her memoir of her long-standing manic-depression, she intimates her experience. Note the manner in which her depression was strongly accompanied by anhedonia, the deseutude of cogent thought, *anomie*, and exhaustion:

> … my mind had turned on me … it no longer found anything interesting or enjoyable or worthwhile. It was incapable of concentrated thought and turned time and again to the subject of death: I was going to die, what difference did anything make? Life's run was only a meaningless one, why live? I was totally exhausted and could scarely pull myself out of bed in the mornings.[12]

She would then sit:

> … virtually inert, with a dead heart and a brain as cold as clay.[13]

Furthermore:

> Each day I awoke deeply tired … Then a gray, bleak preoccupation with death, dying, decaying, that everything was born but to die now and save the pain while waiting. I dragged myself around a local cemetery, ruminating about how long each of its inhabitants had lived before the final moment. I sat on the graves writing long, dreary, morbid poems, convinced that my brain and body were rotting, that everyone knew and no one would say.[14]

"For several weeks … I thought obsessively about killing myself."[15] The movement toward suicide would forthwith accelerate:

> I understood very little of what was going on, and I felt as though only dying would release me from the overwhelming sense of inadequacy and blackness that surrounded me.[16]

It was both the panic-riddled manic side of Jamison's depression and the darkly melancholy lull of its purely depressive side which occasioned the desire to commit suicide. Firstly, the autothanatophilia induced by panic-suffused mania:

> At one point I was determined that if my mind—by which I made my living and whose stability I had assumed for so many years—did not stop racing and begin working

> normally again, I would kill myself by jumping from a nearby twelve-story building. I gave it twenty-four hours. But, of course, I had no notion of time, and a million other thoughts—magnificent and morbid—wove in and raced by.[17]

Then, however, the entire issue complicates itself. There is an increasing commingling of the panic-manic pole and the flat and morbid depressive pole of her composite condition. This oscillation itself moves her to think of killing herself:

> My manias were occurring more frequently and, increasingly, were becoming more "mixed" in nature (that is, my predominantly euphoric episodes, those I thought of as my "white manias," were becoming more and more overlaid with agitated depressions); my depressions were getting worse and far more suicidal.[18]

Yet, it is ultimately from the descent to the pure dread lull of the depressive side, the more strictly passive nihilist side, of her dual condition that the movement toward suicide, as the nascent recrudescence of self-obliterating active nihilism, would spawn itself into the light of day. The miring of her soul in the murky desuetude and terminal dereliction of Nietzschean passive nihilism is likewise evident:

> A floridly psychotic mania was followed, inevitably, by a long and lacerating, black, suicidal depression; it lasted more than a year and a half. From the time I woke up in the morning until the time I went to bed at night, I was unbearably miserable and seemingly incapable of any kind of joy or enthusiasm. Everything—every thought, word, movement—was an effort. Everything that once was sparkling was now flat. I seemed to myself to be dull, boring, inadequate, thick brained, unlit, unresponsive, chill skinned, bloodless, and sparrow drab. I doubted, completely, my ability to do anything well. It seemed as though my mind had slowed down and burned out to the point of being virtually useless. The wretched, convoluted, and pathetically confused mass of gray worked only well enough to torment me with a dreary litany of my inadequacies and shortcomings in character, and to taunt me with the total, the desperate, hopelessness of it all. What is the point in going on like this?[19]

Hereupon, the thought and movement toward suicide ensues:

> Over and over and over I would say to myself, if I can't feel, if I can't move, if I can't think, and I can't care, then what conceivable point is there in living?
>
> The morbidity of my mind was astonishing: Death and its kin were constant companions. I saw Death everywhere, and I saw winding sheets and toe tags and body bags in my mind's eye. Everything was a reminder that everything ended at the charnel house.[20]

Jamison eventually attempted suicide by way of a lithium overdose.[21]

In Jean Améry's book on suicide, he avers that the movement toward suicide is presaged and precipitated by what he terms the échec, or deeply imbued sensation of failure in life. The échec is, notwithstanding, deeply imbricated with a roiling and pervasive feeling of being disgusted with existence itself and with life—which latter Améry takes to be a malignant tumor of Being itself. This is Sartre's *la nausée*. In order for the forward momentum toward killing oneself to be made, échec and existential digust must be extant.[22]

In the overwhelmingly disgusting, swamping and suffocating movement which life becomes in depression and its terminal dread, suicide becomes an answer to:

> ... the oppressive provocations of existence, especially the passage of time, in the stream of which we are swimming along and watching ourselves drown; bit by bit our ego starts to be washed away, when memories grow pale and when the reality of our person finally gets caught in a whirlpool that sweeps it into bottomless depths. What is suicide as natural death? A resounding no to the crushing, shattering échec of existence.[23]

Moreover, the fact of deep depression means that "a suicidal constitution—disgust with the world" is one in which the ego "has denied the world and, along with that, its own self: it has to get rid of itself and partly feels itself as a has-been, already decomposing." At this point, "it is already losing itself, giving itself up," "turning itself into a pale shadow-existence that no longer is." The final denouement of the foregoing is, of course, "death itself, as the most extreme form of irreversibility ..."[24] Améry continues, averring that "to suffer a new échec and yet another, again and again" would imply that "the highly private entropy will be completed and accelerated into madness." Madness it is, for "when the process of temporal compression becomes from second to second unbearable, like a burden, a mental burden, a bodily burden," "it is not that death pursues the suicide, but that the suicide snatches it to his breast, closing tight all the doors through which help could enter."[25]

Améry, who eventually came to put his money where his mouth was by actually killing himself, goes on to speak that the "necrophiliac tenderness for the dying body" and the "disgust and inclination to death" means that "someone who is suicidal is determined to be a suicide" and one whose prerogative is "never again to trust ridiculously everyday life and its alienation ..." Such a man "wishes to give in to the inclination toward death, when he is no longer willing to offer resistance to his disgust with being," where he is desirous to "dispose of the matter neatly," "to disappear ..."[26] For him, "liberation is destruction" and "finds its most extreme possible confirmation in voluntary death."[27] Voluntary death, he contends a little bit later, "exists and removes us, delivers us from a state of being that has," doubtless due to the depression-panic cycle, become an exit "from a state of being that has become a burden ..."[28]

Améry alludes to, what in Sartre, is an extrusion from échec (as passive nihilism, or sempiternal depressive desuetude) of the active nihilism of the hunt for the pure freedom of *ex-sistere* (what we have been calling pure Lacanian *jouissance*, or the desire for the Real). The latter, which wants to "wrest itself from being," or passive nihilism, "is again and again overtaken by it, finds its end in death ..."[29] Améry:

> The day about to begin is an oppressive weight. Our own body is a weight, a body that certainly carries us but one that we also have to carry ... Work is a burden, leisure is burdensome.... Four walls are also constantly closing in on us. They will crush us to pieces and will be a burden. How does one say it? My heart is heavy, *j'ai le coeur lourd.*
>
> ...
>
> Disgust with life arises, as well as disgust with what remains, even when it is a tender memory. And one flings burdens from oneself, both in freedom and to liberate oneself.
>
> ...
>
> The burden? It only has to be carried for a few more yards; throwing it away is anticipated in an ecstasy that stands high above every other kind of intoxication.[30]

This sudden recourse to and eventuality of violent suicide from échec may also be approached from the vantage point of simulation. In line with Berardi's contentions, what cyber-capital's runaway processes of decoding and their attendant semiotic oversaturation long served to occlude—was the very possibillity of the event itself. In lieu of the event, what transpired was an incessant precipitation of pseudo-events—a well-nigh interminable proliferation identical to the repeated non-occurence of the event in the authentic sense. However, when simulation's hyper-stimulation and hyper-exploitation of the machinically enslaved soul reached the point of breakdown, the cumulative weight of non-events precipitated the sudden and utterly unexpected return of the event—except of course in the most hideously *catastrophic* form. The result, as is evident, is aggressive *suicide*. In this connection, Berardi maintains that the "immense concentration of decision-making power put into play by semiocapitalism already lends itself to catastrophic events. That intolerable suicidal action unveiled the vanity of power's infinite strength, confronting it with a [sic] form of escape that reduces it to zero, to its ashes." The conclusion here is that death, "suicide more precisely, is the unforeseeable event that restores the chain of events."[31]

2. Micro-manifesto: the refusal of work

But instead of committing suicide, why don't we just refuse to work? Lazzarato avers that, rather than accelerating neoliberal financial capital's already blistering movement of decoding and deterritorialization, we should instead slow down. And markedly so. He suggests that it is precisely the demented flux of hypertrophied communication, consumption, and production that must be deserted. It is from this latter macerating wheel of Ixion that we must simply abdicate and do so, as one is strongly tempted to add, following the august injunctions of a sort of Fernando Pessoa-style "aesthetics of abdication," with the caveat that the seductions of solipsistic nihilism must likewise be guarded against. Otherwise put, we must seek to arrest neoliberalism's cannibalistic process of valorization, firstly by refusing to be part of it. This would be the inescapable first move to be made to recover a basis for what would only at a later stage emerge as authentic revolutionary political organization. Such organization necessitates forging, by way of a rupture, a time that suspends the functioning of the vast machinery of exploitation and domination. In that the present vampirical regime is one that at base devours discrete depersonalized packets of time, what must established—following the maxims laid out by inter alia Paul Lafargue and Kazimir Malevich—is an *idle* time. We must not so much seek to claim, but rather directly exercise the right to idleness. The dogma of constantly having to be plugged into the voracious and unconscionable cybernetic machine and its autistic panoply of trite attention-deficit-disorder-inducing gadgetry—not to mention the draconian mechanism of infinite debt and infinite guilt, resentment, frustration, limitless alienation and existential desiccation, in which the incorrigibly deluded "entrepreneur of the self" blindly mires himself—must be eschewed. The blithe, sheepish, cynical, and unthinking daily surrender to the roles, functions, and significations of the prevailing division of labor must be sternly resisted. We must refuse this conformist miring in the enchanted circle of production, productivity and producers. So as to open the space for political praxis and action worthy of the name, the first step must be refusal and withdrawal. The first move is to depart, to go away, to leave, and to leave now. Following Bartleby the Scrivener, we must say that we would genuinely prefer not to be part of the prevailing obscene economic and socio-political fascism. We must, at least for now, leave this madness be and disassociate from it, for those who support and drive it have no love for us—and very little for themselves. This is not to say that our necessary gesture of withdrawal is in any sense one of pure passivity and non-action. Quite contrarily, it takes effective aim at the most fundamental power relation of the capitalist regime as such: wage labor. Lazarrato maintains, moreover, that the refusal to work does not solely entail the pure abdication from the mere physical and social particularities of work itself. Rather, it is above all a refusal to be assigned a role, function and identity predetermined and defined solely at the behest of the social division of labor. That is to say, we must abdicate from the market relations of exploitation and domination that impoverish us as subjects through the foul mechanisms of expropriation, normalization and standardization[32]—mechanisms that, one is ethically bound to add, make us autistic.

It is the very notion and actuality of friendship that is under threat. Far from being some sort of natural or cultural given, friendship was invented by inter alia Greek and Jewish eccentrics. Therefore, in that it had to be concretely produced within conditions favorable to its upsurge, friendship can quite easily fall into dereliction, terminal desuetude, and ultimately disappear in emergent conditions unfavorable to its continuation, or ones in which the imperialism of pure expediency eviscerates the exigency for it. Eccentricity, or the ex-centric, entails the willingness to stand outside one's own narcissistic circle of anal-sadistic egoic closure. Eccentricity's, or ex-centricity's, inclination to render its boundaries porous, whilst of course still maintaining at least a semblance of enclosure, is precisely the politics of friendship. Neoliberal capitalism's relentless evisceration of empathy and authentic communication at the behest of cybernetic machinic enslavement's narcissistic stimulation of the soul, in conjunction with the gargantuan proliferation of simulated or ersatz pseudo-friendship, by increasingly undermining the ex-centric conditions of its continued existence and rendering mendacious copies of love and authentic friendship, can only lead to the annihilation of the original. That this process will be driven to its hated end is very likely.

3. Toward a global suicidal State

The suicidal propensity of the soul enslaved by the cybernetic machines of the late capitalist infosphere (machinic enslavement), in fact, finds itself replicated at the level of the State. In fact, as Terry Eagleton maintains, "For the first time in history, our prevailing form of life has the power not simply to breed racism and spread cultural cretinism, drive us into war or herd us into labour camps, but to wipe us from the planet. Capitalism will behave antisocially if it is profitable for it to do so, and that can now mean human devastation on an unimaginable scale. What used to be apocalyptic fantasy is today no more than sober realism."[33] From at least as early as the "war of terror," military action by the US or by its allies—as

Ellen Meiksens Wood puts it—came to require no specific aim at all. This implied a new principle of war *without end*, either in purpose or in time. Even prior to this, Henry Kissinger and Madeleine Albright advocated an unpredictable deployment of military force, that is, the use of force for political purposes far more diffuse and inchoate than the simple achievement of some specific military goal. Similarly, the Bush (George W. Bush) administration's "Operation Infinite War" called for an open-ended war with no limits in time or geography. Moreover, eschewing long-standing military doctrines of containment and deterrence, since the Bush years the US now abides by the policy of "defensive intervention," according to which it claims the right to mount massive preemptive strikes, whenever and wherever it feels so inclined, and not for any definable reason, and certainly not only in the face of an existing threat, but merely in anticipation of some possible future danger—or even no danger at all. The result, as Wood argues is: total and infinite war, that is, war indefinite in its duration, objectives, means and spatial reach.[34] This Bush doctrine was of course unfettered by any international agreements.[35] It would only be a matter of time before the whole world would come to be regarded as a potential adversary.[36]

In this connection, Hardt and Negri maintain that war appears to have flooded the entire social field—to the extent that the Agambenian state of exception, through which the international rule of law concerning belligerency is suspended, has become permanent and general. The exception has become the rule.[37] The degree of saturation of the social field by war has reached the point of becoming the primary organizing principle of modern capitalist society, with politics merely as one of its means or guises. War has become a regime of biopower, that is to say, a form of rule aimed not only at controlling the population but of producing and reproducting all aspects of social existence.[38] War in the present era has passed form deploying lethal force as a last resort of power to being the first and primary element, the foundation of politics itself. Present-day State sovereignty creates order not by stifling a Hobbesian "war of each against all," but through a regime of disciplinary administration and political control based directly on continuous war action.[39]

War in the present era has, moreover, become absolute due to the availability of weapons of mass and even global destruction. The capacity to inflict genocide and nuclear destruction impinges directly on the very structure of life, corrupting and perverting it. The sovereign power that controls this means of destruction is a form of biopower that rules directly over death—the death not merely of an individual or group but of humanity itself and perhaps of all Being. In fact, when genocide and atomic weapons place life itself at the center, war becomes *ontological.* More precisely, war in late capitalism is moving simultaneously in two opposite directions: on the one hand, it is reduced to police action whereas, on the other, it is raised to an absolute, total, ontological level by the technologies of global destruction. These two oppositional movements, however, ultimately prove not to be contradictory: the reduction of war to police action does not remove but actually confirms its ontological dimension, in that the war police maintain the ontological stigmata of absolute annihilation, genocide and nuclear destruction as their ultimate foundation.[40]

That war has become absolute and ontological means that our current geopolitical constellation is stricto sensu *fascist.* As Deleuze and Guattari maintain, fascism arises when a war machine takes over the State. Fascism erects itself on an intense line of flight or absolute deteritorialization, which it transforms into a line of pure destruction. Not only is absolute annihilation the *raison d'etre* of fascism. But fascism, moreover, renders the State properly speaking *suicidal.* It tasks the war machine with channeling the State into a flow of absolute war where the sole result is the self-destruction of the State itself. Fascism is the will to wager everything and to stake one's own death against the death of others. It presents suicide as "the crowning glory of the death of others." It always maintains "the 'stupid and repugnant' cry, *Long live death!*" Deleuze and Guattari argue that this even becomes the case at the economic level, where the expansion of arms replaces growth in consumption whilst investment shifts from the means of production toward the means of pure destruction.[41]

In a similar vein to Hardt and Negri, Paul Virilio and Sylvère Lotringer maintain that under the present capitalist regime there is no longer an absolute monarch, but instead an absolute weapon, that the center is no longer so much occupied by political power but by the capacity for absolute destruction. They contend that we are approaching the end of history and are living under a state of emergency, which "*sanctions the absolute power of the instant.*" "History as the extensiveness of time—of time that lasts, is portioned out, organized, developed—is disappearing in favour of the instant, as if the end of history were the end of duration in favour of instantaneousness, and of course, of ubiquity." And this instantaneousness is nuclear power.[42]

We are heading toward a universal State which is the result of pure war, that is, of the intensity of the means of absolute destruction. Today, the true enemy is not so much external as internal. That is to say, it is our own weaponry and technology that threatens the end of society. Strategies of deterrence are being superceded by the possibility of instant destruction. This could lead civilization to apocalypse, that is, absolute annihilation. And this apocalypse would not be the revelation of the immortality of the soul, but rather the extermination of all bodies and all species. Drawing on Clausewitz, Virilio and Lotringer maintain that, whereas previously it was politics that prevented war from entailing a complete release of destructive force, now—in that politics is rapidly losing its role as arbiter of armed conflicts—this end of politics, far from entailing a reduction of ideological antagonism, portends the total destruction and the end of our civilization.[43] The possibility of suicide, thus, has been opened not only to individuals, but to a society, a civilization, and an entire species.[44]

Crucially, Virilio and Lotringer contend that "*Absolute destruction will correspond to absolute speed. At that point, there will be no room left for man.*"[45] In this connection, we recall that fascism bases itself on a movement of absolute deterritorialization which it transforms into a line of pure annihilation that turns back against the State it has seized. We have also seen that the machinic enslavement of the soul is based on deterritorialization and decoding. Now, it is the intensification and increasing preponderance of machinic enslavement which accelerates the panic-depression cycles of the soul, thus opening the passage to the latter for suicide. Clearly, if the deterritorializations effecting the machinic enslavement of all souls become absolute, any countervailing reterritorialization and recoding as effected by social subjection would be completely left in the lurch. This would imply the end of individuation, and hence what Virilio and Lotringer intimate above as the effective disappearance of man.

4. Cybernetic machinic enslavement, cynicism, depression, passive nihilism, and the Last Man

But let us return to the languid depression and consequent suicide of the individual subject machinically enslaved by cyber-capitalist hyper-exploitation of the information age.

This entire existential ambit touches heavily upon what nihilism precisely entails in Nietzsche. Nietzsche avers that nihilism is the fact of life assuming a value of nil, insofar as it is denied and depreciated. Nihilism inaugurates and sustains a movement in opposition to life itself.[46] In the more recent configuration, the man of passive nihilism, the depressant fomented by the soul machinically enslaved by cybernetic hyper-exploitation, necessarily depreciates life itself. However, and as already mentioned, the nothingness of the will of the cybernetically depressed late capitalist man of passive nihilism is merely a symptom of the deeper element that is the will to nothingness itself. We shall return to this thought quite repeatedly subsequently. For now, we may follow Nietzsche in contending that this nihilism shows that the quality of the will to power presently being taken on is that of negation. This negative will to power is a will to annihilation—including the possibility of self-destruction, of out-and-out suicide. Though, in the more superficial and outwardly phenomenal sense, nihilism in its passive and final modality is party to an evacuation from life itself of all value, the deeper instance animating it shall go elsewhere—as we shall see. This deeper instance of the nothingness of the will—that is, the will to nothingness—unequivocally desires the complete annihilation of all Being. Yet, this its tendency is still encumbered and thwarted by the more outward and superficial modalities of nihilism to which it is tethered throughout its transhistorical odyssey. These modalities are, of course, its own.

The genealogical antecedent to passive nihilism is the prior move from what in Nietzsche figures as the denial of values higher than life—ones whose ultimate source is in a supersensory world beyond the immanent ambit of terrestrial existence itself. In this move, the so-termed *negative* nihilist who depreciated life from the vantage point of the Beyond yields to the *reactive* nihilist who denies God and with this denial of God simultaneously eschews the Good and the True.[47] This transition may be seen as one from Karl Barth who, as we saw much earlier in the part of this book devoted to the despotic regime, expunged all possible immanence from the Deity and contended that God utters a resounding "No!" into the world from the Beyond, thus causing the essence of religion to consist primarily in "man's disgust with his own self"; to Mao Zedong who, contemptuous of any notions of transcendence, at once eschewed any inkling of goodness and truth in favor of the pure brutal expediency of the Chinese Cultural Revolution. Hereupon, in turn, reactive nihilism flips over into the ubiquitous *taedium vitae* of the man of *passive* nihilism. Willing has ceased:

> Here is snow; here life has grown silent; the last crows whose cries are audible here are called "wherefore?", "in vain!", "nada!"—here nothing will prosper or grow any longer.[48]

According to Nietzsche, it is thus:

> And I saw a great sadness descend upon mankind. The best grew weary of their works. A doctrine appeared, accompanied by a faith: "All is empty, all is the same, all has been!" And from all the hills it echoed: "All is empty, all is the same, all has been!" Indeed we have harvested: but why did all our fruit turn rotten and brown? What fell down from the evil moon last night? In vain was all our work; our wine has turned to poison; an evil eye has seared our fields and hearts. We have all become dry; and if fire should descend

> on us, we should turn to ashes; indeed, we have wearied the fire itself. All our wells have dried up; even the sea has withdrawn. All the soil would crack, but the depth refuses to devour. "Alas, where is there still a sea in which one might drown?" thus are we wailing across shallow swamps. Verily, we have become too weary even to die. We are still waking and living on—in tombs.[49]

Nietzsche maintains that henceforth life lumbers forth in a world bereft of values, stripped as it is of meaning and purpose, cavorting ever more quickly towards its own nothingness.[50] Everything, as in Baudrillard's interminable proliferation of simulacra wrought—as observed—by neoliberal capitalism's rabid deterritorialization and decoding of all stable social, cultural and ethical forms, is now a life immersed in the naked and phantasmagoric dance of depthless simulacra, with all value eviscerated. In this decoded cynical nullity of the post-industrial cyber-capitalist world, it is felt that it is better to have a stagnant, destinerrant herd than a shepherd intent upon leading us too far. It is better to fade away passively.[51]

It is well enough known that depression is not merely sadness. Sadness and malaise is blue. Depression proper is *black*. And it includes dread, metaphysical fear—and absolute *disgust*. Nietzsche likewise cottons onto the intuition that passive nihilism is depression and vice versa. He indeed terms passive nihilism the moment of the "great disgust." The form of desiring taken by the unadulterated nullity that the passive nihilist is is, however, still—as is par for the course—entirely passive.[52] If one is disgusted with society and the world at large, one will be a man of heavy inertia and existential fatigue. Nietzsche allegorizes these vicissitudes via evocation of "the prophet of great weariness," who announces a life bereft of even the will to disappear in any active and resolute sense—far from Yukio Mishima's active desire to die and his actively deployed death via *seppuku*. This nullity of an existence merely longs for a purely passive extinction:

> Everything is empty, everything is past! … All our wells have dried up, even the sea has receded. The earth wants to break open, but the depths will not devour us! Alas, where is there still a sea in which one could drown … Truly we have grown too weary even to die.[53]

The said passive nihilism is likewise that of Job during a certain stage of his travails. As observed much earlier, when they hear of Job's misfortune, his three friends, Eliphaz, Bildad, and Zophar, come to mourn with him, weeping, rending their mantles and sprinkling dust on their heads. They sit down with him for seven days and nights in utter silence (Job 3.11–13).

After these days of silent mourning, Job suddenly opens his mouth and begins a lugubrious lament, cursing the very day of his birth:

> Let the day perish wherein I was born,
> and the night in which it was said,
> "There is a man-child conceived." (Job 3.3)
> ...
> Why died I not from the womb?
> why did I not give up the ghost when I came out of the belly? (Job 3.11)
> ...
> For now should I have lain still and been quiet,
> I should have slept:
> then had I been at rest ... (Job 3.13)
> ...
> as infants which never saw light.
> There the wicked cease from troubling:
> and there the weary be at rest. (Job 3.16, 17)

As already observed above apropos of Job, passive nihilism entails at once the descent into deep depression. The latter is, of course, one of the primary causes of suicide; that is to say *active* suicide—active nihilism brought to fruition. Notwithstanding, when Job's wife—reflecting upon his and her utter destitution, admonishes him to curse God and die, she fails to realize that Job is still at the stage of passive—and not yet at that of active nihilism. His will is not a naked and unadorned will to nothingness whose goal is prompt and resolute annihilation of inter alia the self. His is, rather, a nothingness of the will—which, though it does ultimately desire the ultimate extinction of the melancholic, longs for the latter to be accomplished in a non-violent and languidly passive fading away.[54]

And it is Job's heavy inertia and existential fatigue which actively prevents him from pursuing suicide.

Now, in connection with Job's initial phase of mourning, lamentation, and deep melancholy, two things stand out.

Firstly, there is his weariness and associated desire for eternal repose beneath the earth. This is already at once an etiolation of the will, that is, the nothingness of the will characteristic of passive nihilism, and the associated longing for a passive fading out of existence.

Secondly, Job curses the fact that he was even born at all. This latter we may, in point of fact, view as an even more extreme form of the passive nihilist's desire not to exist. For, far from being the will to nothing's actively propulsive movement to annihilation—and in particular the self-annihilation of violent self-inflicted suicide, Job goes even beyond the suffocating predicament of longing to passively fade out of an overly stagnant, exhausted and futile existence.

For instead of the desire for a will-bereft disappearance from what he experiences as a vain and meaningless life already lived, he laments that he had even entered any form of existence to begin with. For him—at least at this stage of his miredness, active self-destruction, and even passive disappearance will not do. What he desires can only be described as the desire for a retroactive annulment of his own birth, and even of his having been conceived in the womb.

These sentiments are echoed by Emil Cioran in his aptly named *The Trouble with Being Born*, when he complains that, being incessantly uncomfortable in the present, he pines for the infinite stretch of duration prior to his existence during which time he was as yet unborn. He, in fact, proceeds to up the ante, violently asseverating that the present impossibility of finding a single nation or tribe among whom a birth immediately provokes tear-drenched sobs of mourning and lamentation is the most devastatingly incontrovertible proof of the utter regression of the entire human race. Cioran actually goes to the bitter end, to the absolute terminus with this his polemic. For he instantly moves to defy billions of years of heredity—all the way back to the very first cell. It would appear that Cioran is dead set against any form of existence *at all*.[55]

Let us, however, return to Job. Though not quite of the caliber of Cioran's, Job's extreme passive nihilism and nothingness of the will, being a mere modality of the more fundamental Nietzschean will to nothingness of active nihilism, shall soon enough yield to the pure and unadorned recrudescence of the latter—replete with its active destructive proclivities, in particular, the rabid desire of the active nihilist to destroy himself. As we shall see soon enough, Job shall come to eventually—though momentarily—toy with this death wish.

Now, returning to nihilism in general, whether it is expressed in values contemptuous of life, as in the Barthian assassination of terrestrial existence from the Beyond; or in the reactive values of the Communard or of the Soixant-huitaird which assume their place; or of the world without values of neoliberal decoding and the valueless universe of proliferating Baudrillardian simulacra, at which time the Nietzschean Last Man has already arrived—this kingdom of nihilism is always animated by a negative will to power, that is, by the will to nothingness. As already observed, the nothingness of the will of the Last Man—the increasingly de-subjectivized soul enslaved by cybernetic machines and their relentless vectors of decoding and deterritorialization—is itself merely the final avatar of the will to nothingness, the final stage of the becoming-reactive of force (the negative as quality of the will to power).[56]

Now, however, the movement beyond passive nihilism may be made. This transition is, as Deleuze's Nietzsche maintains, in fact already presaged by an genealogically prior event. Evidently, as seen in the chapter on the primitive regime, reactive forces owe their perverse triumph over active forces through their alliance with the will to nothingness—which is their inner motor. However, once this grotesque victory is attained, reactive forces will break off their alliance with the will to nothingness. The latter schism is the case since they come to desire to assert and propogate their base values independently.[57] This is exactly the moment of transition from negative nihilism to reactive nihilism mentioned a little earlier—from Karl Barth to Mao Zedong. The ensuing result of this divorce, as we have likewise seen above, is the movement from reactive to passive nihilism and the secretion into existence of the deprecit anthropological piece of merchandise already known as the Last Man.

5. Cybernetic machinic enslavement and the Man Who Wilt Perish

Yet, this is not all. For, following the above mentioned divorce, the will to nothingness continues its enterprise beyond reactive nihilism and the seething ambit of reactive forces—beyond Mao. The will to nothingness now inspires in man a hitherto unprecedented inclination: to *actively* destroy himself (active nihilism). This is nowise to be conflated and confused with the desire to fade out of existence of the man of passive nihilism—that of the Last Man. The Man Who Wilt Perish *actively* wills his own annihilation, his passing—through resolute and pointed will—out of what has now become the tepid quagmire of a maudlin subsistence within the dank swamp of a pointless world. Nietzsche gives the latter poetic figuration in the form of the young man in *Thus Spake Zarathustra* wandering in the environs outside the town of Pied Cow, who cries: "Yes, Zarathustra, you speak true. I desired my destruction when I wanted to ascend into the heights, and you are the lightning for which I have been waiting!" In his discourse on voluntary death, Zarathustra asseverates: "I commend to you my sort of death, voluntary death that comes to me because *I* wish it." "For many a man, life is a failure: a poison-worm eats at his heart. So let him see to it that his death is all the more a success."[58]

Job again comes to mind. Eliphaz's first extended invective against the latter and Job's lingering suffering forthwith jolts him into suddenly cavorting from his existentially limpid and passive nothingness of the will to an actively self-destructive will to nothingness. He vociferates about his purported desire to actively, though indirectly, commit suicide by imploring God to strike him down so as to abruptly terminate his vapid, destitute, angst-riddled and utterly pointless sojourn on the planet:

> O that I might have my request!
> and that God would grant me the thing that I long for!
> Even that it would please God to destroy me,
> that he would let loose his hand, and cut me off!
> Then should I yet have comfort,
> yea, I would harden myself in sorrow: let him spare,
> for I have not concealed the words of the Holy One.
> What is my strength, that I should hope?
> and what is my end, that I should prolong my life? (Job 6.8–11)

Yet, Nietzsche's Zarathustra praises the man bent on his own active destruction—he who is equivalently the result of cybernetic machinic enslavement's atavistic and quasi-Medieval suicidal delirium, the soul whose utter de-subjectification by the lacerating vectors of cyber-capitalist deterritorialization—as the one who nobly desires to be overcome, to go beyond the merely and mundanely human, to cross the bridge along the path toward the Overman. He, the Man Who Wilt Perish, is indeed the father and ancestor of the Overman. He wants to be done with this overly persistent and stagnant idiocy of merely being human.[59] Zarathustra:

> I love him who lives for knowledge and who wishes to know that one day the Overman may live. *And thus he lives his own downfall.*[60]

Notwithstanding, as regards Job—though he pines for self-destruction at the hand of God for a period of time—his autothanatophilia does not linger and reach consummation. He is, therefore, not a bridge to the Overman. The Nietzschean Man Who Wilt Perish proper, however, is.

6. Midnight: transmutation in the quality of the will to power

This proclivity toward and actuality of the Man Who Wilt Perish is the moment of transmutation in the quality of the will to power. Heretofore having assumed passive nihilism's nothingness of the will, the will to nothingness becomes a will to *active* destruction. With the alliance between reactive forces and the will to nothingness broken, the will to nothingness is transmogrified by way of crossing over onto the side of *affirmation*. Now, as a *power of affirming*, it suddenly comes to rabidly destroy the reactive forces with which it had formerly been married—a veritable uxoricide. In this tirade of destruction, the negative comes to be transmuted into a power of affirmation and, at once, one of joyful annihilation. This is midnight, the point at which the negative is converted into its opposite. Nihilism completes itself in the Last Man and in going beyond the latter to the Man Who Wilt Perish, from the supine and terminally depressed soul fomented by its enslavement to cybernetic machines (machinic enslavement) to the wilfully violent suicide. With the Man Who Wilt Perish, nihilism defeats itself, becoming an affirmative power which moreover already presages and prepares the advent of the Overman.[61]

Coextensively with the above transmutation in the quality of the will to power from a negative will to power to an affirmative will to power, values and their value cease to derive from the negative but from affirmation as such. However, it is not simply the case that noble values now come to merely substitute for base values. It is likewise, and above all, the very *place* and *principle* from whence values are secreted that itself changes. As Nietzsche maintains, this change in values absolutely necessitates a change in the *creators* of values themselves. These new creators must not only be different human beings to the creators of existing values; they must be radically heterogeneous to man himself, an entirely different subject to man.[62]

In what sense, however, does the negative become a power of affirming? It does so by ceasing to become an independent power, and comes instead to be affirmation's handmaiden and adjunct. Henceforth, only affirmation remains as an autonomous power.[63]

Moreover, the negative—as co-opted by the power of affirming or the affirmative will to power—now totally and utterly destroys *all known values*, that is to say, *base* ones. Nietzsche portrays this destruction of all existing values as the second of three metamorphoses required for the reestablishment of the rights of activity and affirmation: that of the loin. A massive, unwieldy and hideous dragon bares the decrepit and ignoble values of a thousand years of human history on its scales, each scale obscenely glittering with the crass inscription "Thou shalt." The lion, counter-posing these foul words with his own "I will!", proceeds to kill the disgusting beast. Heretofore existing values are reversed and become *noble*. At the level of forces, the change in the will to power from negative to affirmative denies reactive forces. All forces become *active*.[64]

The negative, as quality of the will to power, has up until the possible subsequent moment of transmutation constituted man in his sensibility and ethical evaluation. It is in actual fact constitutive of the very essence of man. And together with man the entire world sinks and sickens, all of life is utterly depreciated, everything known slips towards its own nothingness. It is this foul sempiternal desuetude which countenances Zarathustra during the moment of the great disgust. He springs up it terror from his sleep, crying "Disgust, disgust, disgust—woe is me!" In scenes not altogether unlike the sweat and blood-drenched agonies of Christ in the garden of Gethsemane during which he is brutally forced to see all the vile and pestilential sins of the entire human race across all history condensed into one pulverizing sensation, he falls down like a dead man, pale and shivering, remaining in this bedraggled state for seven full days without food or drink. It was man as such that had nauseated him to the very bowels. "The great disgust at man—*it* choked me and had crept into my throat: and what the prophet prophesied: 'It is all one, nothing is worth while, knowledge chokes.'" In a premonition of the eternal return, he is mercilessly inundated with the more than pornographically obscene idea that even the smallest and most palsied man will return to existence for all eternity. "Ah, disgust! Disgust! Disgust!"[65]

Affirmation as quality of the will to power, on the other hand, in point of fact, only manifests itself outside of and *above* and *beyond* man. It is only manifested in the Overman.[66] The Overman is, indeed, unknown and—in fact—*unknowable* to man. In this vein, Zarathustra is "the eternal affirmation of *all things*."[67] "I, however am, am one who blesses and affirms if only you are around me, you pure, luminous sky! You abyss of light!—then into all abysses do I carry my consecrating affirmation."[68] "Supreme constellation of being, that no wish reaches, that no negation can soil, eternal affirmation of being, eternally I am your affirmation."[69]

The issue, however, complexifies itself. For Nietzsche—in accordance with Deleuze's analysis—presents affirmation as inextricable from a *preliminary* negation and likewise from a negating *consequence*. It is straddled twixt these two.[70] In terms of the antecedent negation flowing into affirmation, Nietzsche paints the picture of the lion's "holy no" as necessarily preceding the child's "holy yes."[71] What we have here is active destruction—that of the Man Who Wilt Perish and who desires to be *overcome* as man—as the antecedent condition of affirmation.[72] The antecedent lion and the consequent child, with their varying negations, are absolutely necessary for affirmation to occur.[73] It is averred that "*two negations form part of the powers of affirming which are modes of being of affirmation as such.*"[74]

Man, in fact, can be the lion—yet, at the very most, the lion. For man is *unable* to attain affirmation in its *full* power. That is to say, he is *unable to affirm affirmation itself.* This means that, though he is able to destroy old decrepit values, he is unable to create *new* values.[75] "To create new values—even the lion is incapable of that: but to create itself freedom for new creation—that the lion can do."[76]

7. Dionysus and Ariadne: full affirmation

We have seen that man cannot raise himself to the fullest sense of affirmation. But what is this fullest sense of affirmation? In what does it consist?

Apropos of Nietzsche, Deleuze states that affirmation is, simply, *Being*. Then, of course, qualifications and elucidations ensue. Firstly, it is noted that the foregoing is not to say that Being is the object of affirmation. Affirmation is not the power of Being. Rather, affirmation is *Being itself*, Being in all of its power.[77]

But in what sense, then, is affirmation—in the affirmative will to power—Being? Affirmation is Being inasmuch as it is its own object unto itself. In fact, affirmation is—more precisely—*becoming* (*Werden*). Yet, why have we said that affirmation is Being and becoming at the same time? For it is Being insofar as it is the object of a *second* affirmation which extracts the Being of becoming. Affirmation, at full tilt, is therefore *double* in its nature. This means to say that affirmation, in its fullest sense, is the affirmation of affirmation. There must be another affirmation which redoubles the first affirmation in order for affirmation to attain its highest, most exigent pitch. This affirmation is, for Nietzsche, the business of the third and last of the three metamorphoses: that of the child: "The child is innocence and forgetfulness, a new beginning, a sport, a self-propelling wheel, a first motion, a sacred Yes. Yes, a sacred Yes is needed, my brothers, for the sport of creation: the spirit now wills *its own* will, the spirit sundered from the world now wins *its own* world."[78]

Nietzsche portrays this double affirmation through a motley of bizarre literary figures. Firstly, there are Zarathustra's two animals: the eagle and the serpent. Here, Nietzsche delineates double affirmation from, inter alia, the eternal return. Viewed from the perspective of the eternal return, the eagle is the great cycle of the eternal return, whilst the serpent is the individual destiny inserted into the said grand cycle. The eagle flies in wide circles, whilst the serpent is curled around his neck—likewise in a circular fashion.[79] The serpent is curled around the neck of the eagle "not like a prey but like a friend."[80] We shall see the full implications and meaning of this image quite shortly.

Another intimation of double affirmation which Nietzsche paints is that of the intimate liaison of Dionysus and Ariadne. Here, Dionysus is affirmation as becoming itself, whereas Ariadne is the necessary second affirmation that redoubles the first one. This second affirmation takes the first affirmation as its object.

For Dionysian affirmation and becoming is at the same time Being, yet only insofar as it is itself affirmed by the adjoining affirmation of Ariadne.[81] "Eternal affirmation of being, eternally I am your affirmation."[82] This means to say that Dionysus is becoming itself, whereas Ariadne is the Being of becoming—as becoming's eternal return. There is only Being as the *return* of becoming.

In the aforementioned becoming/Being-of-becoming-as-return couplet, the eternal return is portrayed as a wedding ring: "Oh how should I not lust for eternity and for the wedding ring of rings—the Ring of Recurrence! Never yet did I find the woman by whom I wanted children, unless it be this woman, whom I love: for I love you, O Eternity! *For I love you, O Eternity!*"[83] The wedding ring is the eternal Dionysian cycle. It is also a wedding *mirror* in which—through Ariadne—it admires itself.[84] This all means that, in the eternal return, Dionysus is becoming, whereas Ariadne is the Being of becoming. She is the fiancée, the second affirmation and the loving feminine power.[85]

Another figuration of full affirmation—affirmative will to power driven to its most stratospheric pitch—and the eternal return given by Nietzsche is the labyrinth. The labyrinth designates the eternal return itself. It is becoming and the affirmation of becoming. Being, for Nietzsche, only exists as affirmed of becoming. This Being affirmed of becoming is Ariadne's thread. Dionysus, in fact, teaches Ariadne that he himself is the labyrinth (becoming). Dionysus is becoming; Ariadne is the Being of becoming. Dionysus, as becoming, is the first affirmation; Ariadne, as vouchsafing the Being of becoming, is the second affirmation. Both are necessary such that affirmation, in its full sense, transpires.[86]

For Nietzsche, the eternal return is the Being of becoming (*Werden*). These two are, crucially and essentially, inextricable. This is so because becoming and Being are, ultimately and fundamentally, one single affirmation. True, there is a first affirmation which is Dionysus: the affirmation of becoming; whereas there is a second affirmation which is Ariadne: the affirmation of the Being of becoming as the eternal return of becoming. These two affirmations are, notwithstanding, the recto and verso of one and the same affirmation. This one and single affirmation is simply the affirmation of affirmation.[87]

Yet, before one is able to proceed to this impossible moment, or even to the genealogical juncture at which the maudlin and dessicated man of passive nihilism passes over into the Man Who Wilt Perish, Nietzsche proposes that a certain tendency already inherent in what he had seen in the increasingly systematic homogenization of man must itself undergo concerted exacerbation. In what would seem to us a preposterous and unswervingly wicked succession of thoughts, Nietzsche, averring that the constantly intensified dwarfing of the human species is quite paradoxically the very driving force that can ultimately bring into existence a more noble variant of man or rather something nobler than man, contends that this process must in no sense be countervailed. In fact, he maintains that the process of etiolation of the human race at the behest of an increasingly crass cultural barbarism must, if anything, be hastened so as to create an increasingly vast and vertiginous gulf between a humanity that descends to lows of baseness, decrepitude and supine dereliction so extreme as to ultimately, in a startling dialectical reversal, serve as the desiccated subzoological fragments from which an unprecedentedly excessive, ennobled, and fundamentally affirmative collective subject may emerge after the moment of the greatest and most poisoned self-disgust and generalized perversion has been traversed as the most terrifying of nightmares that it is.[88]

In what would appear a transmogrified version of Ezekiel's vision of fallen Israel as a heap of dessicated bones in a parched and arid landscape generalized to all of humanity, Zarathustra relates in the most profound ill-humor to his disciples: "Truly, my friends, I walk among men as among fragments and limbs of men! The terrible thing to my eye is to find men shattered in pieces and scattered as if over a battle-field of slaughter. And when my eye flees from the present to the past, it always discovers the same thing: fragments and limbs and dreadful chances—but no men!" Yet, in words not entirely distant in import from those by which Ezekiel conjured the hallucinatory reconstitution of the shattered bones of Israel as a resurgent collective Jewish subject, Zarathustra sees himself as a cripple walking along the bridge toward the future, which future is again self-identically none other than himself. Here, he walks among men as "fragments of the future: of that future that I scan,"[89] a future of the unremitting annihilation and concomitant reconstruction of the shattered limbs of Dionysus.

8. Eternal return, Schellingian Godhead, BwO, Plotinus' One, and the self-crucified ego's immolation in Dante's Empyrean

The eternal return is exactly the double affirmation of Dionysus—as becoming (*Werden*) itself—and at once Ariadne as the Being of becoming, that is, as the return of becoming. Yet, we may be still more precise in our delineation of the eternal return. Crucially, the eternal return is—contrary to what many may maintain—in no sense an exact repetition of the same. In this connection, as Deleuze contends, the "eternal return does not cause the same and the similar to return, is itself derived from a world of pure difference."[90]

Originary, primordial, and internal difference is however merely the *will to power* itself. Here we glimpse the internal relation between the eternal return and will to power. For Deleuze, the said relation is, however, more precisely, as follows: the eternal return entails "the assignation of difference as the origin, which then relates different to different in order to make it (or them) return as such"; and is related to the will to power in the sense that "the eternal return is indeed the consequence of a difference which is originary, pure, synthetic and in-itself (which Nietzsche called the will to power)"—an infinitely taut knot in which the internal difference constitutive of the will to power is the "in-itself" and "the repetition in the eternal return is the for-itself of difference."[91]

Here, we have—in fact—returned to the Prologue of our entire book. If will to power is the in-itself of difference, difference in its most pure and originary sense, the pure difference herein being alluded to can only be the purely intensive differen*t*iating difference of the pre-individual individuating factors. These latter, as we observed in the Prologue, are pure intensive quantities. The reprisal of the said free intensive quantities (these latter comprising the vertiginous subterranean world of the will to power) inherent to the eternal return, implies that the latter entails not merely becoming (*devenir*)—but an *intensive* becoming or, more precisely, a becoming-intensive. The ambit of this becoming-intensive is that of the nomadic subject at the level of Deleuzoguattarian desiring-production. This abyssal world of the nomadic subject is, at once, that of eternal return and, equivalently, that of the will to power, "the flashing world of metamorphoses, of communicating intensities, differences of differences, of *breaths*, insinuations and exhalations: a world of intensive intentionatities, a world of simulacra or 'mystries'."[92]

The nomadic subject passes through all the degrees of intensity in an endlessly migratory movement along the surface of the BwO. In this intensive journey, Deleuze and Guattari go on to aver, the nomadic subject "identifies these states" through which it passes "with the names of history: '*every name in history is I* …'"[93] And, as observed, this foregoing vertiginous passage is precisely the Nietzschean eternal return.[94]

In this journey of the vertiginous world of the eternal return, "everything conmingles in these intense becomings [*devenir*], passages, and migrations—all this drift that ascends and descends the flows of time: countries, races, families, parental appelations, divine appelations, geographical and historical designations, and even miscellaneous news items."[95] And all of these things which conmingle on the BwO "designate regions on this body—that is, zones of intensities, fields of potentials. Phenomena of individuation and sexualization are produced within these fields. We pass from one field to another by crossing threshold: we never stop migrating, we become other individuals as well as other sexes, and departing becomes as easy as being born or dying."[96]

Something very homologous is afoot in Dante's *Divina Comedia.* As regards the latter masterpiece, Christian Moevs maintains that, through the movement toward self-knowledge of the human self, the latter assimilates to itself the entire gamut of human experience in history. The *Comedia,* therefore, entails one man's assimilation of other men's experience. Moreover, much as in the case of the Deleuzoguattarian nomadic subject born of and identifying with every intensive state produced within the BwO, the analogous identification by the soul of Dante's Pilgrim with the totality of all possible experience, entails and is only rendered possible by the dissolution of the its fully-fledged individual ego—and its concomitant reduction to a mere extensionless metaphysical point traversing the threshold of the Dantesque Empyrean. The latter—like the Deleuzoguattarian BwO, or Spinozan Absolute Substance, or the Schellingian Godhead—is that which produces and at the same time *is* all things.[97]

Returning to Nietzsche and Deleuze, this world of the eternal return cum will to power, of desiring-production whose immanent principle is the BwO is unremitting in its evanescence, incessant in its sempiternal auto-incineration: for repetition "in the eternal return never means continuation, perpetuation or prolongation, nor even the discontinuous return of something which would at least be able to be prolonged in a partial cycle (an identity, an I, a self) …"[98] This is so because "eternal return, affirmed in all its power, allows no installation of a foundation-ground. On the contrary, it swallows up or destroys every ground," hereby making "us party to a universal *ungrounding.*"[99]

What is, however, the mode of generation of the nomadic subject? We recall from our Prologue that there exist two different and diametrically opposed modes of interaction between the BwO and the organs-partial objects. In the first of the two modes of interaction, when the BwO experiences the organs-partial objects as a form of persecution wherein the partial objects seem to be attempting to break into the BwO, the latter repels the organs and in so doing spawns the so-termed paranoiac machine. Here, the BwO can no longer tolerate the organ-machines. "In order to resist the organ-machines, the body without organs presents its smooth, slippery, opaque, taut surface as a barrier. In order to resist linked, connected, and interrupted flows, it sets up a counterflow of amorphous, undifferentiated fluid."[100] Clearly, the BwO's presenting of itself as a barrier to the organs-partial objects is its becoming taciturn in a movement of self-enclosure.

Now, on the other hand, in a different mode of interaction with the organs-partial objects, the BwO attracts these latter and appropriates them to itself, functioning as quasi-cause of the said organs—in which they appear to emanate from the BwO, seem to be miraculated by it.[101] Since the machine herein spawned—that is, the miraculating machine—is so by way of exactly the opposite process to that by which the paranoiac machine arises, its miraculating is hereby a force of expansion.

But, after this brief propaedeutic, let us return to the mode of generation of the nomadic subject itself. Here, the moments of generation of the paranoiac and miraculating machines give rise to an alternating movement whose oscillatory tension spawns the so-called celibate machine. This pulsating motion constantly being effected by the forces of attraction and repulsion between the BwO and organs-partial objects gives rise to an open series of pure "intensive *nervous states* that fills up the body without organs to varying degrees …"[102] It is these intensive states, taken together, that incessantly and repeatedly precipitate the Deleuzoguattarian nomadic subject.

As espied in the Prologue, something very similar is afoot in Schelling's *Weltalter*. At the level of eternal divine nature, the tension between the first divine potency (which is a contracting-negating force) and the second divine potency (which is an expansive-affirming force) spawns again and again the third divine potency.[103]

We have observed how the nomadic subject passes through all of the intensive nervous states generated within the BwO by the unremitting oscillation between the paranoiac and miraculating machines. In Schelling, something highly homologous transpires. The incessant rotary motion of eternal divine nature serves to constantly generate "images" in an endless profusion.[104] These "images" are "the possibilities or spirits of things" that spirit (the third potency) runs through in its constantly being born and reborn.[105]

In this journey, much as in the case of the pure intensive states traversed by the Deleuzoguattarian nomadic subject along the surface of the BwO, what the Schellingian third potency traverses is an eternal divine nature which "has no actuality" and thus passes all the time "again into becoming" with "nothing abiding, nothing stable," everything being thus "in incessant formation."[106]

What is vital to note is that, though Schelling describes the images generated by eternal divine nature as "archetypes," he deploys the latter term in a strictly non-Platonic sense: for these pseudo-archetypes are neither physical substances, nor vacuous genera, nor static or complete forms; but rather "eternally becoming and in incessant movement and generation."[107] In this sense, the Schellingian third divine potency, as in the case with the Deleuzoguattarian nomadic subject, passes through and is likewise incessantly born and reborn of pure intensive states.

For becoming (*devenir*) is pure internal difference-in-itself, whereas internal difference is self-identically intensity.

Now, as also afore asseverated in the Prologue, the constant generation of the third divine potency out of the reciprocal and incessantly re-engineered opposition between the first potency and the second potency, is at once and precisely the manner in which the Schellingian second eternal will—the will *in* eternity—strives and continues to strive after the first divine will (which is eternity itself). The foregoing is, at once, and self-identically the manner in which the first divine will strives after itself—in order that it may extinguish the originary dissonance at its heart.[108]

But this originary dissonance—a difference-in-itself—cannot be vitiated. If it would be extinguished, the entire universe would simply stop.

The first divine will, of which the second will is merely the mode of its own attempted self-sojourning, is the anteriority of all anteriorities, the eternal and primodial ground of the existence of God. The ground hereto referred—an ultimate and groundless ground (and as such an *Ab/Ungrund*), is the Godhead—an eternal freedom. This will's purity consists in its not being a will to something that it could apprehend as outside of itself, nor is it even a will that wills its own revelation. It neither covets what would otherwise be external objects destined for its apprehension, nor does it desire its own actuality. Rather, it is a will devoid of obsession or craving—at least not in any prosaic, mundane, or quotidian sense. Slavoj Žižek describes this groundless ground as "an abyss [*Ungrund*]—that is to say, 'absolute indifference' *qua* the abyss of pure Freedom which is not yet the predicate-property of some Subject but, rather, designates a pure impersonal Willing [*Wollen*], which wills nothing."[109] The absolutely primordial *Ab/Ur/Ungrund* of God's existence—that is, the Godhead—is thus a willing devoid of *both* subject *and* object. For Alberto Toscano, this purely vertiginous abyss is "the absolute, itself conceived as the unconditioned ground of indifference, the coincidence of producer and product, affirmation and affirmed, intuition and intuited, *natura naturans* and *natura naturata*."[110]

That the Godhead as pure will does not will in any prosaic sense does not in the least identify it with the Nietzschean will to nothingness—let alone with the nothingness of the will that the latter formally assumes in the case of the passive nihilism characterizing existential etiolation, leaden exhaustion and well-nigh sempiternal desuetude.[111] That the "emptiness" and seeming nullity of the Schellingian pure willing nowise arrays the latter with the yonder side of affirmation is expressed by Andrew Bowie as follows: "'the will which wills/wants nothing' can remain negative, in that it is absolutely within itself, because it does not entail any sense of possibility or development, but as such also does not entail any lack in itself, which is the highest affirmation."[112]

The Godhead is "the devouring ferocity of purity" in whose midst all that which bears Being as an attribute is instantly immolated.[113] It is, much as Deleuze avers of aspects of German expressionism, an intensity "raised to such a power that it dazzles or annihilates our organic being, strikes terror into it," an intensity that "culminates in a fire, which burns us and which burns all of Nature," that "unleashes in our soul a *non-psychological life of the spirit*, which no longer belongs either to nature or to our organic individuality, which is the divine part in us, the spiritual relationship in which we are alone with God as light."[114]

These various features of the Schellingian Godhead, or infinite willing, are *mutatis mutandis* shared by Plotinus' One—which Plotinus also interchangeably terms the Good, the Supreme, the First, Principle, Unity, and God.

Firstly, the One—like the Schellingian Godhead—is an unconditional and unconditioned will, in that it autochthonously and vertiginously wills solely itself. As Plotinus maintains: "If, then, we are to allow Activities in the Supreme and make them depend upon will—and certainly Act cannot There be will-less—and these Activities are to be the very essence, then will and essence in the Supreme must be identical. This admitted, as He willed to be so He is; it is no more true to say that He wills and acts as his nature determines than that He wills and acts as his nature determines than that his essence is as He wills and acts. Thus He is wholly master of Himself and holds his very being at his will."[115]

Thus, in that the One, now termed Principle by Plotinus, "must be a consistent self willing its being and being what it wills; its will and itself must be one thing."[116] Moreover, "the will and the essential being must be taken as one identity; the act of will must be self-determined and the being self-caused; thus reason shows the Supreme to be its own Author. For if the act of will springs from God Himself and is as it were his operation and the same will is identical with his essence, He must be self-established. He is not, therefore, 'what He has happened to be' but what He has willed to be."[117]

Furthermore, like the Schellingian Godhead, the Plotinian One is bereft of lack. For the One is "utterly a self-existence, with no concomitant whatever. This self-sufficing is the essence of its unity. Something there must be supremely adequate, autonomous, all-transcending, most utterly without need."[118] Otherwise put, "The sovereign self-sufficing principle will be Unity-Absolute, for only in this unity is there a nature above all need whether within itself or in regard to the rest of things."[119] For this reason, "Unity seeks nothing towards its being or its well-being or its safehold upon existence"; "Nor has it station; it needs no standing-ground as if inadequate to its own sustaining" of itself.[120]

The crucial upshot here is that, since the One requires no accoutrements to supplement it, like the Schellingian Godhead, it desires not objects outside of itself and even requires not and categorically wills not its own revelation. Indeed, the One "need seek nothing beyond himself, he who is over all; to himself and to all suffices by simply being what he is."[121]

Now, though Plotinus has been speaking of the One as "willing his own being and being what he wills," this association of the word "being" with the One is probably a misleading shortcoming of translation or of language in general. For elsewhere it is averred that the One, much as is the Schellingian Godhead, is bereft of and beyond Being. In this connection, Plotinus contends that "The First must be without form, and, if without form, then it is no Being"—adding that "the phrase 'transcending Being' assigns no character, makes no assertion, allots no name, carries only the denial of particular being" as such.[122]

Later on, Plotinus explains the refractory nature of the One to Being in somewhat more detail, averring that the "First is no being but precedent to all Being: it cannot be a being, for a being has what we call the shape of its reality but the Unity is without shape, even shape Intellectual."[123]

Moreover, "The Unity is none of all; neither thing nor quantity nor quality nor intellect nor soul; not in motion, not as rest, not in place, not in time: it is the self-defined, unique in form or, better, formless, existing before Form was, or Movement or Rest, all of which are attachments of Being and make Being the manifold it is."[124]

Thus, like the Schellingian Godhead, the Plotinian One has none of the attributes that would otherwise constitute it as a specific being.[125] More precisely, and again like the Schellingian Godhead, the Plotinian One, or Unity, does not possess Being as an attribute.

One of the corollaries of the One being bereft of any appurtenances of Being, and specifically its being completely outside of both time and place, is that it does not belong to and cannot be characterized as residing in space as *extensio*. Plotinus confirms this elsewhere, reaffirming that the One is spaceless and has no environment or extension.[126] As seen in the Prologue of our book, this is likewise the case with the Schellingian Godhead anterior to the *Ent-Scheidung* through which, coterminously with the latter's upsurge as a thoroughly constituted subject, the non-chronologically antecedant purely intensive *spatium* suddenly explicates itself as extented, volumetric Euclidean space.

The Schellingian Godhead and Plotinus' One are, however, not homologous in every single respect. For instance, in Schelling, the Godhead is always already cleaved or inhabited by a primordial dissonance. This dissonance, as observed, assumes the form of an entirely independent and primordial autogeneration of a second divine will or willing that hopelessly seeks collusion with the first will or willing. Here, as observed in our Prologue, the Godhead strives to eradicate the dissonance that lies at its heart. This latter striving assumes the form of, and is equivalent to, the search by the second will of the first will. This quest by the second will, however, fails since the first will is entirely unbeknowst to it.

Notwithstanding, even though Plotinus nowise posits any sort of inner cleaveage or non-self-coincidence in relation to the One—much as in the case of the Schellingian Godhead—he nonetheless maintains that it has no self-knowing or self-intellection. However, his reason for this assertion is, in fact, entirely the opposite to Schelling's averment of the same: that is to say, the One is unbeknowest to itself not because it fails to coincide with itself, but because it is entirely and seamlessly present to itself.[127] Nevertheless, the case remains that both the Schellingian Godhead and the Plotinian One are, respectively, entirely oblivious to their own selves.

A little earlier (and, of course, much much earlier in our Prologue), we compared the Schellingian Godhead to what Deleuze, apropos of certain aspects of German expressionism, averred to be an intensity which unleashes in the soul a non-psychological life of the spirit which, not belonging to nature or man's organic individuality, is the human being's divine part through which he comes to be alone with God as light.

Something well-nigh identical is afoot in Plotinus. In the wake of authentic mystical contemplation—and *pace* Hegel in his debunking of what he sees as the inevitable impossibility of attempted union with the divine Other entailed within the first phase of the unhappy consciousness (*unglückliche Bewusstsein*)—the soul sees the presence of the One "suddenly manifested within her, for there is nothing between: here is no longer any duality but a two in one; for, so long as the presence holds, all distinction fades: it is as lover and beloved here, in a copy of that union, longing to blend; the soul has now no further awareness of being in its body and will give herself no foreign name, not man, not living being, not being," and thus certainly not—as Deleuze contends above—any organic individuality.[128]

And it is here that Plotinus begins to explicate the meaning of what Deleuze describes as the communing of the divine part of man with the light of God. He starts by dwelling upon seeing, maintaining that—in the state of deeply absorbed mystical contemplation, the soul's "vision is continuous so that seeing and seen are one thing; object and act of vision have become identical."[129]

More precisely, "one becomes to self and to all else, at once seen and seer; identical with Being" and as "the entire living all, we no longer see the Supreme as an external; we are near now" to the One.[130] Finally, the sense in which Deleuze can say that the soul can be alone with God as light, Plotinus maintains to reside in the fact that "the vision floods the eyes with light, but it is not a light showing some other object, the light itself is the vision."[131]

In a very similar sense, though now adapted to a specifically Christian context, Moevs maintans of Dante's *Paradiso* that the blessed themselves shine brilliantly by gazing upon God—the uncreated source of all light. Moreover, they gaze upon God as the source of light sans intermediary. Indeed, the Empyrean—*mutatis mutandis* the Dantesque equivalent of the Schellingian Godhead—is averred to be full of uncreated light, if not light itself.[132]

Bonaventure, for one, delineates the Empyrean as pure light, as infinite luminosity. In this sense, it has no fundamental essence apart from light. This identification of God with pure light, shared by inter alia the Christian patristic and Augustinian-Franciscan theologians, is not a surprising postulation on their part—as it is in fact sourced in Neoplatonic thought. As observed, Neoplatonism *stricto sensu* views God as pure light.[133]

Dante, however, extends Plotinus' characterization of the One. As one should well know, he saddles it within a Christian framework. Here, he fuses the Empyrean with the *coelum Trinitas* (heavenly Trinity) in the uncreated tenth heaven of the *Paradiso*, which he identifies with *la menta divina* (divine mind). Paraphrasing *Paradiso* 13.52–63, God is *lucente* (effulgent light), whereas the Word is *viva luce* (living light). Both are one with *Amor* (love).[134] As regards Christ, being at once divine and human, he is the bridge or union between the spatiotemporal reality of finite Being and the sheer unqualified conscious self-subsistence of the Empyrean.[135]

Now, as Moevs avers apropos of the Dante's metaphysics, it is not only the Godhead that is pure light (*lux*), but similarly the human intellect or soul is itself at base a radiant and spiritual light-energy potentially receptive to being directly infused by divine illumination. Furthermore, since light is the essence of the Godhead, the soul's potential participation in it—that is, in the Empyrean—is a participation in divine Being.[136] More specifically—and again contrary to Hegel's aforementioned averment of its impossibility—this potential of the human soul, a potential espied a little earlier in Plotinus, is one which is variously called *alienato mentis* or *ek-stasis* (contemplative rapture), *deificatio* (deification), or *unio mystica* (divine union).[137]

But the question remains: how is immersion in the Empyrean—the soul's true home and true Being—to be attained? How is the soul—the divine part of man—to lose itself in, and indeed become the Plotinian One, the Schellingian Godhead? The first essential step in this ascent to the Empyrean is, as Moevs maintains, for the human mind to come to know itself. This initial move necessarily entails the soul's reorientation away from the senses and a concomitant turning into the depths of itself. This is attained through a love so intense that the soul becomes desirous—through absolute auto-purification—to renounce everything and, in so doing, itself become love. To attain immersion in the One and thus to become all, the soul must surrender all self-definition and retain nothing of itself.[138]

The above set of Plotinian injunctions for merging with the One are almost the same as those preached by Jesus, who said that only those who give up everything for the kingdom of God would attain it. The Gospels revealed that the path toward God is purification, which entails self-sacrifice and renunciation.[139]

And self-sacrifice and renunciation are, above all, epitomized in the *Cross*. The Cross consists in dying to this life, in order to find life. The latter attainment of life by dying to it is that of the soul living in Christ and Christ living in the soul, and no longer that of the human subject's mere living in itself. The union of human and divine lost in intramundane human experience is here restored in the God-man Christ.[140] And, as we have already seen vis-à-vis Jürgen Moltmann's observation in the Christ-vignette of our chapters on the Deleuzoguattarian passional regime, Christ is at once no more divine and no more human than in his sacrifice on the Cross.

Therefore, in order to find life through Christ, the human subject must voluntarily undergo a similar sacrifice. This, of course, would not necessarily assume the form of a gruesome and harrowing annihilation of the body upon a blood-stained gibbet. Rather, man's sacrifice must be that of the self-crucifixion of his *ego*. For Pseudo-Dionysius the Areopagite, the latter entails an absolute abandonment of oneself, a *complete transcendence of the self* through a radical self-emptying. In the same vein, St. Bernard maintains that the soul must come to love itself without self-interest, turn itself into an empty or broken vessel. For Richard of St. Victor, the mind must turn inward, in the process forgetting and shedding itself. Here, man must *rise above himself* through himself, and climb the mountain of self-knowledge so as to see Christ transfigured. In reaching the summit of this mountain of self-knowledge, man's soul is enlarged, becoming *more than human* and passing *beyond the human* into the divine through a dawning of divine light. In like fashion, Saint Bonaventure contends that man must focus himself on the *apex mentis seu synderesis scintilla* (the highest part of his soul), that is, on the eternal Light that formed his mind. In so doing, man passes *beyond himself* into pure Being. The latter is possible only by man's identification with the Cross of Christ.[141] Finally, Meister Eckhart contends that one must dive into the bottomless divine abyss or light within oneself, doing so through the renunciation, abandonment, and *annihilation* of the self—such that, as Moevs maintains, the mind is no longer petrified by its obsessive *self-identification*.[142]

So, we have seen—by recourse to Pseudo-Dionysius, Bernard of Clairvaux, Richard of St. Victor, St. Bonaventure, and Meister Eckhart—that, in order to move toward the attainment of the divine, man must transcend, rise above, and pass beyond himself through an active and self-willed *annihilation* of his ego.

Is there not something strikingly Nietzschean about the above process? There is. We are, of course, thinking here of a Nietzsche conjoined with Zupančič's Lacanian reading of Nietzsche, and with our earlier characterization of the capitalist machine by Deleuze, Guattari, Berardi, and Lazzarato.

Firstly, the intial step on the path toward the eventual attainment of the Empyrean, is the voluntary auto-dissolution of the ego. Now, as observed by recourse to Lazarrato, it is the social subjection effected by the capitalist machine that engineers the latter by way of manufacturing the self as a self-enclosed, individuated subject. Social subjection produces the solipsistic ego through processes of recoding which accompany its reterritorialization of human labor onto capital. This is the pietistic tendency of the capitalism axiomatic.

Lazarrato maintains that it is signifying semiologies that configure the subjectivation process in accordance with the individual subject. Language itself is one of the preponderant apparati deployed by this recoding and reterritorialization. It leads the self to piously believe in its own "I," its own "I" as a self-sufficient substance. Indeed, the absolute fulcrum of piety, as Lazzarato avers apropos of Nietzsche, is the subject's selfish belief in itself as an impervious, sanctimonious "I." This cultivation of the obdurate ego is effected through the individuation wrought by social subjection.

On the other hand, and as likewise already observed earlier, in tandem with social subjection, the capitalist axiomatic also deploys a mechanism whose tendency is to countervail the latter: machinic enslavement. Unlike social subjection, machinic enslavement operates through de-subjectivation, piloting a movement precipitating the dismantling of the self-collected and individuated subject engineered by social subjection. It does so through processes of capitalist decoding and deterritorialization, rendering up the so-termed dividual.

Now, Deleuze and Guattari contend that—though the capitalist axiomatic's reterritorializing tendency more or less stays in lock-step with its deterritorializing tendency—capitalist decoding tends to outrun capitalist recoding. This latter outstripping of recoding by decoding is still more prominent in the case of our present post-industrial, cybernetic age of the bourgeois mode of production. The upshot here is that, in present-day semio-capitalism, there is an ever-burgeoning preponderance of machinic enslavement over social subjection. As likewise seen, this machinic enslavement is ever more less that of the body and much more that of the soul.

The cybernetic machinic enslavement of the soul under cyber-capitalism, as Berardi has taught us, leads to two main psychopathologies: panic disorder and depression. Panic results when consciousness finds itself at a loss regarding the reception of the vast panoply of stimuli thrown at it by what seems a well-nigh infinite universe. In its extreme form, it approaches delirium. Panic disorder results from the tardiness of the human elaborative capacity compared to the massiveness of a rapidly burgeoning infosphere. Here, informational overstimulation precipitaes a perceptive, cognitive and psychic stress culminating in the dangerous acceleration of all vital functions. Evidently, with the incessant increase in the velocity of capitalist decoding and the implied exacerbation of machinic enslavement, a breaking-point must eventually be reached, to the extent that the soul has to detach itself from the flow of information—thereby falling into *depression*.

As observed, the dividual engineered by machinic enslavement was observed to be homologous to Charles Taylor's "porous" self. Unlike the individuated "buffered" self—who views depression as merely *caused* by black bile—the "porous" self, lacking disengagement and a clearly delineated boundary between itself and the exterior milieu, experiences depression *as* black bile. This depression, caused by the panic-inducing ambit of neoliberal hyper-exploitation, leads to premature senility, exhaustion, and spiritual anorexia.

Earlier, we assimilated cyber-capitalism's informational overstimulation with the merciless superegoic cajoling of the machinically enslaved soul into submission to an excessive *jouissance* coincident with the Lacanian Real. As we observed through Zupančič, this tyrannical superegoic injuction was deemed equivalent to the Nietzchean will to nothingness. However, falling from panic disorder into depression, the sheer exhaustion of the anorexic soul causes its will to nothingness to assume the form of a nothingness of the will—a near-cessation of all willing.

Also as earlier espied, the will to nothingness falls under the rubric of Nietzschean active nihilism; whereas the soul's descent into depression accords with passive nihilism. Therefore, with present-day capitalist hyper-exploitation, active nihilism stems from a draconian superego born of panic-inducing cybernetic over-stimulation; whilst the depressive spiritual anorexia of passive nihilism is inaugurated by the soul's withdrawal from the said overbearing psychosomatic stimulation.

However, the cybernetically enslaved soul will inevitably repeatedly reattach itself to the slavery of informational over-stimulation and its superegoic tyranny. Hence, a cycle ensues, in which the soul alternately dissents from and accedes to these latter. This is Berardi's panic-depressive cycle: a tension between active nihilism (the will to nothingness) and passive nihilism (the nothingness of the will), in which—as Zupančič avers—the latter serves as a defense against the over-stimulations of the former.

However, passive nihilism's etiolation of excessive *jouissance* and will—in fact merely preserves active nihilism's will to nothingness in a virtual/deactivated mode. Therefore, active nihilism may very well find an alternative way of recrudescing from beneath its passive counterpart. And it may do so in a singularly violent way. This occurs when the tyranical superegoic imperative of the Real links itself to a destructive *passage à l'acte*. This dramatic upsurgence can be highly destructive, and even *self*-destructive. It may induce the heretofore depressed passive nihilist to commit suicide.

In this connection, we saw Améry's notion of the échec maintaining that the deeply depressed passive nihilist has denied his own self, feels himself a pale shadow that barely exists. In the midst of all this, he is already possessed of a suicidal constitution. The final denouement of the foregoing is self-destruction.

Of course, according to Deleuze's Nietzsche, the nothingness of the will of the cybernetically depressed late capitalist man of passive nihilism is merely a symptom of the deeper element that is the will to nothingness itself. The latter at once means that this decrepit man is suffused by a will to power whose quality is that of negation—a negative will to power. As observed, he is Nietzsche's Last Man. He is the well-nigh completely de-subjectivized soul enslaved by cybernetic machines (Deleuzoguattarian machinic enslavement) and their relentless vectors of decoding and deterritorialization, the final avatar of the will to nothingness, the ultimate stage of the becoming-reactive of force. His negativity of will to power is a will to fizzling passively out of existence.

As seen, Zupančič avers that the will to nothingness may well surge up from out of its deactivated modality as the nothingness of will. In this *passage à l'acte*—the passage from passive nihilism to active nihilism, the will to nothingness now inspires in man a new inclination: to *actively* destroy himself. Man becomes the Man Who Wilt Perish.

This all seems hopelessly morbid—if not downright psychotic. Yet, as seen, Nietzsche's Zarathustra praises the man bent on his own active destruction; the man who, through cybernetic machinic enslavement and its attendant vectors of decoding and deterritorialization, has undergone well-nigh complete de-subjectification and ego-dissolution, and now stands on the atavistic threshold of the suicidal delirium of Charles Taylor's neo-Medieval "porous" self. He is extolled by Zarathustra since he nobly desires to be overcome, to go beyond the merely and mundanely human, to cross the bridge beyond which stands the Overman. He is praised by Zarathustra, as he is the father and ancestor of the Overman.

The active self-annihilation of the Man Who Wilt Perish, who commits suicide at midnight, is the moment of transmutation in the quality of the will to power from negative to affirmative—a transmogrification which presages the advent of the Overman.

On this golden road between man and the Overman stands Zarathustra himself. He is already affirmation, "the eternal affirmation of *all things*."[143] Zarathustra: "I, however am, am one who blesses and affirms if only you are around me, you pure, luminous sky! You abyss of light!—then into all abysses do I carry my consecrating affirmation."[144]

Nevertheless, Zarathustra is not affirmation in the fullest sense. His is an affirmation of Being. But consummate affirmation is *Being itself.* So, whereas Zarathustra's affirmation takes Being as its object; full affirmation *is* Being, for it is its own object unto itself. It is, as espied earlier, the affirmation of affirmation; that is, double affirmation: the double affirmation of the divine couple Dionysus and Ariadne.

Now, affirmation in Nietzsche is, more precisely, the will of a will to power whose quality is affirmative—the affirmative will to power. Therefore, to say that affirmation at its most exigent pitch is Being itself, is to say that the fully affirmative will is Being. Thus, to aver that full affirmation takes itself as the object of its own affirmation, means to aver that the fully affirmative will wills itself. Hence, in that the fully affirmative will is Being, its willing and its Being are one and the same.

Moreover, since the fully affirmative will to power of Nietzsche only wills its own self, it is almost exactly the Schellingian Godhead—for the latter, as observed in our Prologue, is an unconditional and unconditioned pure will or eternal willing that likewise solely wills itself.

Also, in that the consummately affirmative Nietzschean will is Being and, because it wills itself alone, its Being must at once be its will. Therefore, its will is its Being and its Being is its will. For this reason it is likewise well-nigh identical to Plotinus' One, which Plotinus delineates as "a consistent self willing its being and being what it wills"—meaning that "its will and itself must be one thing."[145]

As noted a little earlier, the Man Who Wilt Perish commits suicide exactly at midnight. Now, the crossing of this bridge at which the quality of the will to power is transmuted from negative to affirmative leads us toward the dawning of *first light.* Zarathustra appears, joyfully exclaiming: "I, however am, am one who blesses and affirms if only you are around me, you pure, luminous sky! You abyss of light!—then into all abysses do I carry my consecrating affirmation."[146] Forthwith, we enter into the full radiant light of *day.*

Zarathustra, as seen, is another metamorphic bridge to be crossed on the way to the full Dionysian affirmation identical to both the Plotinian One and the Schellingian Godhead—the latter whose ferocious incandescence we assimilated to what Deleuze termed an intensity that "unleashes in our soul a *non-psychological life of the spirit,* which no longer belongs either to nature or to our organic individuality, which is the divine part in us, the spiritual relationship in which we are alone with God as light."[147]

Here, our averment that the sense in which Deleuze is justified in asseverating that the soul can be alone with God as light, Plotinus maintains to reside in the fact that "the vision" of the One "floods the eyes with light," but that "it is not a light showing some other object," for "the light itself is the vision."[148]

In a very similar sense, and likewise as already observed, we noted Moevs maintaning of Dante's *Paradiso* that its blessed inhabitants shine incandescently in gazing upon God—the uncreated source of all light. They begaze God as the unmediated source of light. Indeed, the Empyrean—*mutatis mutandis* the Dantesque equivalent of the Schellingian Godhead—was earlier averred to be be full of uncreated light, if not light itself.[149]

Assimilating the Plotinian One to a Christian framework, Dante further pursues and extends the notion of the Godhead as light. He fuses the Empyrean with the *coelum Trinitas* (heavenly Trinity) in the uncreated tenth heaven of the *Paradisio*, which he identifies with *la menta divina* (divine mind). As seen regarding *Paradiso* 13.52–63, God the Father is *lucente* (effulgent light), whereas the Word is *viva luce* (living light). The Word, or Son, flows from the Father—who is the source of all light—but is nowise other than its source. For both are a power of love, which "en-threes" itself with them. Divine love is the power to be all things. This power is the Father. Love is, at the same time, all things upon which conscious Being bestows Being by knowing them as itself. This is love as the Son. Love is, likewise and at once, the qualification of Being as particular beings. This is love as the Holy Spirit.[150]

Dante's pilgrim in the *Divina Comedia* is on the path toward eventual immersion in and fusion with the Empyrean. Let us join him at the point where he is close to having scaled the mountain of Purgatory. In *Purgatorio* 27, the pilgrim stands on the seventh and last ledge of the mountain of Purgatory. He has completed his purgation on all the lower ledges of the mountain, and is now ready to enter into the garden of Earthly Paradise. However, in order to actually enter Eden, he must pass through a wall of purifying fire.[151]

Now, the pilgrim's passing through the purifying fire is associated with Dante's mulberry tree, which evokes Christ's Passion and is planted in his heart (*bono*). Having received the mulberry, which is equivalent to the Cross, the pilgrim is inflamed by the blood of self-sacrifice; that is, by death to his ego in Christ. By passing through the fire, the pilgrim is surrendering his attachment to his body and wordly things (which would certainly include the shedding of capitalist greed and its insatiable gluttony of incessant material consumption), and voluntarily undergoing martyrdom for the sake of a holy *name*, through which he will *escape death forever*. In doing so, he follows Christ on the Cross, turning the mulberry red with his blood. Whilst passing through the fire, Beatrice's *name* calls the pilgrim Dante and, like the fire itself, extricates him from the consuming grip of finite material things. Here, the *name* "Beatrice" is salvific, in that it not only disintricates the pilgrim from his former materialist integuments, but likewise serves to dissolve his ego (the principle obstacle to his attainment of the divine).[152]

The foregoing paragraph is, in fact, littered to the brim with not only Nietzschean but, oddly enough, also with Badiousian motifs. Firstly, in receiving the mulberry (the Cross)—which inflames his heart with the blood of the voluntary sacrifice of his body and self-willed disintegration of his ego, to the ends of the attainment of eternal life; the pilgrim is much like Nietzsche's Man Who Wilt Perish. For the latter likewise shuns both his body and his ego, so that the Overman may be born and live. Now, as earlier observed, the Overman was deemed equivalent to the double affirmation called Dionysus-Ariadne, which was averred to be a will that wills itself. Full affirmation—that is, the completely affirmative will to power—was moreover maintained to be Being itself. Therefore, the fully affirmative will—in that it, much like the Schellingian Godhead, pure will or eternal willing, solely wills its own willing and, since this willing is Being itself, like the One of Plotinus, is a will whose will is its very Being and whose Being is its will. Hence, in that the Schellingian Godhead and the Plotinian One are both eternal, so is the Nietzschean Overman. Thus, both Dante's pilgrim and the Nietzschean Man Who Wilt Perish, through their respective voluntary self-sacrifices, give birth to the *immortal.*

And this very last word brings us to Badiou. What Badiou describes as being "immortal," is the subject born of a truth-event—or the son-subject we espied a lot earlier. As observed, it is Dante's pilgrim who, by passing through the wall of fire, will escape death forever. He achieves the latter through a holy *name*. Now, as observed far earlier apropos of Badiou, when a truth-event is deemed to have occured, this event must be *named* with a proper noun. Through the event and its necessary nomination, a subject of the truth-event is spawned—a subject which embarks upon an infinite truth-procedure, faithfully following and unfolding the consequences arising in the wake of the said event. The name attributed to the truth-event is an enigma to the prevailing situation and finite knowledges and discourses of the world in which it transpires. It punches a hole in the limited material configuration of this world. It is the upsurge of the infinite. And, by way of the event of voluntary self-crucixifion, through which Dante's pilgrim sheds his attachment to his finite body and dissolves his finite ego, he truly becomes the immortal that he fundamentally is.

This enigmatic name, the proper noun "Beatrice," is that which extricates the pilgrim from the hold of the finite—and, in particular, from the obduracy of his heretofore extant ego (the preponderant obstacle to his attainment of the divine and, hence, of immortaly). As Moevs has it, the event of the pilgrim's self-crucifixion is that by which the ultimate ontological reality in the human—that is, Intellect or Being—becomes aware of itself (*Purgatorio* 25.75). This truth-event is, therefore, that through which—transcending as he does the finite via his rebirth as a immortal Badiousian subject-to-truth (or son-subject)—the pilgrim comes to the self-knowledge of himself as divine. Through the event of his passing through the wall of fire, the pilgrim enters into a life-giving reunion with Beatrice and, through this transfiguring salvific love, undergoes a unique liberating metamorphosis that enables him to eventually immerse himself in the pure light of the Godhead.[153]

This immersion in divine light is made possible, since the fire's purging of the pilgrim's attachment to his body is that of rendering it transparent to light, whereas in *Purgatorio* 26.22, 23 his body was still *parete al sol*—that is, a wall to the sun. The pilgrim's passing through the fire is a self-crucifixion which, through his voluntary sacrifice of his identification with his own corporeal form, is at once the self-revelation of the divine. This is the transmogrification through which the gateway to the *Paradiso* opens itself for the pilgrim.[154]

In *Paradiso* 27.88–96, the divine reveals itself to the pilgrim in the form of Beatrice. In *Paradiso* 27.97–99, the power of Beatrice's gaze impels him into the *Primo Mobile,* that is, the source of space-time itself. Entering into the *Primo Mobile* at *Paradiso* 27.100, he now transcends his own finite identity. He is de-subjectified by vertiginous movements of deterritorialization that effect the dissolution of his individuated ego.[155]

What the pilgrim Dante, in fact, sees when in the *Primo Mobile* is merely an extensionless and overpowering point of pure luminosity (*Paradiso* 28.19–21). This point is unfathomably miniscule—an indication of the sheer extent to which this dimensionless light of pure awareness transcends the finite, individuated, mind of image, sense, and human reason. This *punto* is the nexus between spatio-temporal extension and the self-subsistent Being that is the Godhead. Bonaventure terms this point the *apex mentis* or *apex affectionis,* through which the mind enters itself and beyond which it passes beyond itself into pure Being—where it is completely transmogrified into God himself. Meister Eckhart refers to this point as the "uncreated spark" through which the soul and God are one.[156]

This extensionless point is likewise the yoking together of the mortal and the divine through Christ.[157] As such, the *punto* is likewise the bridge between, on the one hand, the limited human mind petrified as an egocentric, obdurate, pious and "buffered" self—as in Charles Taylor's lexicon—wrought by the reterritorializing and recoding forces of Deleuzoguattarian social subjection and, on the other hand, the deterritorializing vectors of de-subjectification of Christic awakening. It is the cusp separating sunset and sunrise, the boundary dividing night from day, darkness from light, eternal damnation and eternal life.[158]

The crucifixion motif, intially espied in the pilgrim's passage through the purging wall of fire separating the last ledge of Purgatory and Eden, likewise returns. The *Primo Mobile,* the nexus and eternally evanescent instant, is the dimensionless point through which human intelligence sacrifices all finite corporeal form and returns to its source—that is, to the Godhead. This is, indeed, the crucifixion of the self. In fact, Moevs furthermore maintains that Dante's entire journey from the earth to the Empyrean—that is, from space-time to the extensionless and atemporal Plotinian One, is itself a self-crucifixion. Here, the pilgrim enters Hell at sunset and, after ascending the mountain of Purgatory, emerges from the latter at dawn. This renders the Crucifixion a timeless instant of revelation and salvation, in which the ground of Being—that is, the Godhead—awakens to itself within and as man. At the same time, the human intellect recognizes all that it sees as itself. It comes to know itself as both eternity and time.[159]

From the *Primo Mobile* and his being "conquered" by the *punto*, Dante moves into the Empyrean. In the Empyrean, his power of sight is conjoined with the Godhead (*Paradiso* 33.94–96) to the extent that it is consumed by the divine light (*Paradiso* 33.82–84). In an ever-intensifying absorption, the pilgrim's intellect passes over into the ground of all Being, until it is completely withdrawn from or suspended (*sospesa*) from space and time.[160]

Since this "experience" transpires outside of extended space and chronological time alike, it leaves no trace, no form, and hence nothing to remember. Thus, as in the Deleuzian third synthesis of time (that of the Nietzschean eternal return), there comes to pass a universal ungrounding of time and—through a becoming-intense—a collapsion of *extensio* back into the pure implex of *spatium* from which it, through the Schellingian *Ent-Scheidung* whereby the spatio-temporal universe emerged into existence coextensively with God's upsurge as a fully-fledged subject, had initially explicated itself coterminously with the emergence of chronological time. The Crucifixion of the all corporeal form and memory effected by the passage from the *Primo Mobile* into the Empyrean is, thus, a reversal of Creation in which the human soul is at once immolated in and deified as the divine. This is a total collapse of all reality into a single dimensionless burning point of all-encompassing love and Being. Nothing whatsoever can be said about this point. One can only see it. And to see it, is to be it.[161]

However, if we eschew what we have extensively been developing as a sort of strange Christianized Nietzscheanism and a Nietzscheanized Christianity—not so much in its militantly Leninist but rather in its Neoplatonist form—and move instead to a Lacanian framework, we may be inclined to suspect both the infinite purification wrought by the Nietzschean eternal return, and even more so the apotheosis of the subject *as* a immortal nullity traced through Dante, as operations implicitly presupposing the Kantian regulative Idea and fantasy of an immortal soul. With this conceptual move and the parallax shift it brings about having been brought to bear, the trajectory of subjective destitution traced here, in which the individual wrought by capitalist social subjection suffers the crucifixion of its obdurate ipseity at the behest of rabidly deterritorializing vectors of cybertenic machinic enslavement, instead yields up a different and far more dreadful type of immortality: the immortality of the organ devoid of body.

In this scenario, the subject enters what Žižek terms the "zero-level," or pure void, of subjectivity: the absolutely unendurable abyss in which fully constituted objects and subjects dissolve and yield before a miasmal swarming of spectral partial objects, of exemplifications of the Lacanian *lamella*, of the undead, immortal, libido-as-object that is self-identically the Freudo-Lacanian death-drive

(*Todestrieb*). This paronomastic modality of the subject is merely one "degree" removed from the pre-subjective noumenal Real. It is a return to, or recrudescence of, the impossible moment of upsurge of subjectivity in the form of a withdrawal, a recession from the light of phenomenality and its differential degrees of appearing in a world in accordance with its transcendental regime, which assumes the form of a contraction through which constituted reality yields before the surreal phantasmagoria of *membra disjecta*, of organs devoid of body functioning as stand-ins for the dread immortality of the object-libido in its semipernally stuck fixity and eternal psychotic circulation around the void. In Kantian terms, this crepuscular and vanishing mode of subjectivity is that of the pre-synthetic imagination gone mad, psychotic, and utterly demented—the impossible domain of absolutely purified and kenotically evacuated transcendental freedom and spontaneity. In this domain, the subject finds itself vertiginously suspended between direct unadorned animality, on the one hand, and the subordination of its freedom to the moral Law, on the other. In this gathering murk, far from being able to galvanize itself in the direction of forward-directed action in the world or attunement to the promptings of the Deity, the self is submitted to the horrifying experience of dread organs without body, of itself as the Lacanian fragmented body, of its *membra disjecta* floating around in a nauseating and disgusting miasmal drift,[162] the "night of the world," the "valley of the shadow of death," the hell-on-earth of Kundry in Wagner's *Parsifal* who, having committed the blasphemy of laughing at Christ dying on the Cross, is condemned to an eternity of terrestrial wandering as an abject, multiply befouled, purely excremental self entirely rejected by the world, abandoned by God, and subject to involuntary, psychotic automatisms, spasmodic jabberings, slavering aphasia, dyslexia, *and the absolute horror of never being able to die.*

Notes

1. Magee 2001, 40.
2. Berardi 2015b, 145, 146.
3. Ibid. 55, 56.
4. Ibid. 86–89.
5. Cited in Jamison 2000, 25.
6. Jamison 2000, 92, 94.
7. Styron 2004, 58.
8. Ibid. 49, 50.
9. Ibid. 52, 53.

10. Ibid. 64.
11. Ibid.
12. Jamison 1997, 38.
13. Ibid.
14. Ibid.
15. Ibid. 39.
16. Ibid. 44.
17. Ibid. 83.
18. Ibid. 102, 103.
19. Ibid. 110.
20. Ibid. 111.
21. Ibid. 115.
22. Améry 1999, 47.
23. Ibid. 60.
24. Ibid. 70.
25. Ibid. 88, 89.
26. Ibid. 91–93.
27. Ibid. 126.
28. Ibid. 128.
29. Ibid. 129.
30. Ibid. 130–133.
31. Berardi 2009, 164, 165.
32. Lazarrato 2015, 245–248.
33. Eagleton 2013, 8.
34. Wood 2003, 148–151.
35. Ibid. 160.
36. Ibid. 163.
37. Hardt and Negri 2004, 7.
38. Wood 2003, 12, 13.
39. Hardt and Negri 2004, 21.
40. Ibid. 18, 19.
41. Deleuze and Guattari 1988, 230, 231.
42. Virilio & Lotringer 2008, 59, 60.
43. Ibid. 61–64.
44. Ibid. 143.
45. Ibid. 72.
46. Nietzsche 1968, 9, 11, WP 2, 11.
47. Ibid. 9, WP 3.
48. Nietzsche 2000, 157, GM III.26.
49. Nietzsche 1977, 245, Z "The Soothsayer."
50. Nietzsche 1968, 12, 13, WP 12.

51. Deleuze 2006, 149.
52. Ibid. 150.
53. Nietzsche 1977, 155, 156, Z II "The Prophet."
54. Nietzsche 1968, 17, 18, WP 23.
55. Cioran 2013b, 4, 5.
56. Nietzsche 1968, 14, WP 13, 15.
57. Deleuze 2006, 173, 174.
58. Ibid. 174; Nietzsche 2003, 69, 70, Z I "Of the Tree on the Mountainside"; 97, 98, Z I "Of Voluntary Death."
59. Nietzsche 1977, 44, Z Prologue 4.
60. Ibid.
61. Deleuze 2006, 174, 175.
62. Ibid. 175; Nietzsche 2003, 85, Z I "Of the Thousand and One Goals."
63. Deleuze 2006, 176.
64. Ibid; Nietzsche 2003, 54, 55, Z I "Of the Three Metamorphoses."
65. Nietzsche 2003, 232–236, Z III, "The Convalescent" I, II.
66. Deleuze 2006, 177.
67. Nietzsche 1977, 6, EH III "Thus Spoke Zarathustra."
68. Nietzsche 1977, 185, 186, Z III "Before Sunrise."
69. Nietzsche 1977, DD "Glory and Eternity."
70. Deleuze 2006, 177.
71. Nietzsche 1977, Z I "Of the Three Metamorphoses."
72. Deleuze 2006, 177, 178.
73. Ibid. 179, 180.
74. Ibid. 180.
75. Ibid. 185.
76. Nietzsche 1977, 55, Z I "Of the Three Metamorphoses."
77. Deleuze 2006, 186.
78. Ibid.; Nietzsche 2003, 55, Z I "Of the Three Metamorphoses"
79. Nietzsche 2003, 52, 53, Prologue 10.
80. Nietzsche 1977, 53, Z Prologue 10.
81. Deleuze 2006, 187.
82. Nietzsche 1977, DD "Glory and Eternity."
83. Nietzsche 2003, 244, 245, Z III "The Seven Seals" I.
84. Nietzsche 1968, WP II. 51, DD "Ariadne's Complaint."
85. Deleuze 2006, 188.
86. Ibid.
87. Ibid. 189.
88. Nietzsche 1968, 477, 478, WP 898.
89. Nietzsche 2003, 160, 161, Z II "Of Redemption."
90. Deleuze 2001, 125.

91. Ibid.
92. Deleuze 2001, 242, 243.
93. Deleuze and Guattari 2004, 22.
94. Ibid.
95. Ibid. 93.
96. Ibid.
97. Moevs 2005, 10.
98. Deleuze 2001, 201, 202.
99. Ibid. 67.
100. Deleuze and Guattari 2004, 10.
101. Ibid. 12, 13, 18.
102. Ibid. 20.
103. Schelling 1997, 143.
104. Schelling 2000, 59, 60.
105. Schelling 1997, 154; 2000, 66, 67.
106. Schelling 2000, 66, 67.
107. Ibid.
108. Schelling 1997, 143–145.
109. Žižek 1996, 14.
110. Toscano 2004, 116.
111. Deleuze 2006, 148, 149.
112. Bowie 1993, 106.
113. Schelling 2000, 24, 25.
114. Deleuze 1986, 53, 54.
115. Plotinus 1962, 606, VI.8.13.
116. Ibid.
117. Ibid. 607, VI.8.13.
118. Ibid. 619, VI.9.6.
119. Ibid. 620, VI.9.6.
120. Ibid.
121. Ibid. 591, VI.7.38.
122. Ibid. 408, V.5.6.
123. Ibid. 617, VI.9.3.
124. Ibid.
125. Ibid. 604, VI.8.11.
126. Ibid. 605, VI.8.11.
127. Ibid. 620, VI.9.6.
128. Ibid. 588, VI.7.34.
129. Ibid. 589, VI.7.35.
130. Ibid.
131. Ibid.

132. Moevs 2005, 19.
133. Ibid.
134. Ibid. 27.
135. Ibid. 34.
136. Ibid. 20.
137. Ibid. 58.
138. Ibid. 60.
139. Ibid. 60, 61.
140. Ibid. 61.
141. Ibid. 61–65.
142. Ibid. 68.
143. Nietzsche 1977, EH III "Thus Spoke Zarathustra," 6.
144. Nietzsche 1977, Z III "Before Sunrise," 185, 186.
145. Plotinus 1962, 607, VI.8.13.
146. Nietzsche 1977, Z III "Before Sunrise," 185, 186.
147. Deleuze 1986, 53, 54.
148. Plotinus 1962, 590, VI.7.36.
149. Moevs 2005, 19.
150. Ibid. 121.
151. Ibid. 91.
152. Ibid. 96–100.
153. Ibid. 104, 105.
154. Ibid. 115.
155. Ibid. 133, 134.
156. Ibid. 141, 142.
157. Ibid. 145.
158. Ibid. 157.
159. Ibid. 160.
160. Ibid. 166.
161. Ibid. 167.
162. Žižek 1999, 51, 52.

References

Achituv, Josef. 2007. Excerpts from "Theology and the Holocaust." Reprinted in *Wrestling with God: Jewish Theological Responses during and after the Holocaust*, edited by Steven T. Katz, Shlomo Biderman and Gershon Greenberg, 319–325. London: Oxford University Press.

Agamben, Giorgio. 1998. *Homo Sacer: Sovereign Power and Bare Life.* Translated by D. Heller-Roazen. Stanford, CA: Stanford University Press.

———. 1999. *Remnants of Auschwitz: The Witness and the Archive.* Translated by D. Heller-Roazen. New York: Zone Books.

———. 2005 *The Time that Remains: A Commentary on the Letter to the Romans.* Translated by P. Dailey. Stanford, CA: Stanford University Press.

———. 2005. *State of Exception.* Translated by Kevin Attell. Chicago: University of Chicago Press.

Alexander, Jeffrey C. 2006. *The Meanings of Social Life: A Cultural Sociology.* Oxford and New York: Oxford University Press.

Alighieri, Dante. 1971a. *The Comedy of Dante Alighieri the Florentine: Paradise.* Translated by Dorothy L. Sayers and Barbara Reynolds. Middlesex: Penguin Books.

———. 1971b. *The Comedy of Dante Alighieri the Florentine: Purgatory.* Translated by Dorothy L. Sayers. Middlesex: Penguin Books.

———. 2003. *The Divine Comedy: Volume I: Inferno.* Translation and Introduction by M. Musa. Middlesex: Penguin Books.

Al-Muqarram, Abd al-Razzaq. *Maqtal al-Husayn (Martyrdom Epic of Imam al-Husayn).* Translated by Yasin T. Ali-Jibouri. al-islam.org/printpdf/book/export/html/27512: Al-Kharsan Foundation for Publications.

Alter, Robert. 2004. *The Five Books of Moses: a Translation with Commentary.* New York: W.W. Norton & Co.

Altizer, Thomas J.J. 1967. *The Gospel of Christian Atheism.* London: Collins.

———. 2003. *Godhead and the Nothing.* Albany: State University of New York Press.

Alvarez, Alfred. 1972. *The Savage God: A Study of Suicide.* London: Weidenfeld & Nicolson.

Améry, Jean. 1980. *At the Mind's Limits: Contemplations by a Survivor on Auschwitz and its Realities.* Translated by S. Rosenfeld and S.P. Rosenfeld. Bloomington and Indianapolis: Indiana University Press.

———. 1999. *On Suicide: A Discourse on Voluntary Death.* Translated by John D. Barlow. Bloomington and Indianapolis: Indiana University Press.

Amis, Martin. 2003. *Koba the Dread: Laughter and the Twenty Million.* London: Vintage Books.

———. 2009. *The Second Plane: September 11: Terror and Boredom.* London: Vintage Books.

Anderson, R. Lanier. 2011. "On the Nobility of Nietzsche's Priests." In *Nietzsche's On the Genealogy of Morality: A Critical Guide*, edited by Simon May, 24–55. Cambridge and New York: Cambridge University Press.

Angela of Foligno. 2006. Excerpts from *The Memorial.* In *The Essential Writings of Christian Mysticism,* edited by Bernard McGinn, 374–378, New York: Modern Library.

Ansell-Pearson, Keith. 1994. *An Introduction to Nietzsche as Political Thinker: The Perfect Nihilist.* Cambridge, Melbourne, New York: Cambridge University Press.

———. 1999. *Germinal Life: The Difference and Repetition of Deleuze.* London: Routledge.

Arenas, Reinaldo. 2001. *Before Night Falls.* London: Serpent's Tail.

Arendt, Hannah. 2006. *Eichmann in Jerusalem: A Report on the Banality of Evil.* London: Penguin.

Aristotle. 2013. *Poetics.* Translated with an Introduction and Notes by Anthony Kenny. Oxford: Oxford University Press.

Artaud, Antonin. 2003. *Heliogabalus: Or, the Crowned Anarchist.* Translated by A. Lykiard. London: Creation Books.

Avineri, Shlomo. 1996. "Labor, Alienation, and Social Classes in Hegel's *Realphilosophie.*" In *Hegel's Dialectic of Desire and Recognition*, edited by J. O'Neill, 187–208. Albany: State University of New York Press.

Badiou, Alain. 1999. *Manifesto for Philosophy: followed by two essays: "The (re)turn of Philosophy Itself" and "Definition of Philosophy".* Translated, edited, and with an Introduction by Norman Madarasz.

———. 2002. *Ethics: An Essay on the Understanding of Evil.* London: Verso.

———. 2003a. *On Beckett.* Translated by Alberto Toscano. Foreword by Andrew Gibson. London: Clinamen Press Ltd.

———. 2003b. *Saint Paul: The Foundation of Universalism.* Translated by R. Brassier. Stanford, CA: Stanford University Press.

———. 2005. *Being and Event.* Translated by Oliver Feltham. Preface by Alain Badiou. London: Continuum Books.

———. 2008. *Conditions.* Translated by Steven Corcoran. London: Continuum.

———. 2009. *Logics of Worlds: Being and Event, 2.* Translated by Alberto Toscano. New York: Continuum Books.

———. 2011. *Second Manifesto for Philosophy.* Translated by Louise Burchill. Cambridge, UK and Malden, MA: Polity Press.

Barbusse, Henri. 2004. *Hell.* Translated by Robert Baldick. New York: Turtle Point Press.

Bartol, Vladimir. 2007. *Alamut.* Translated from the Slovenian and afterword by Michael Biggins. Berkeley, CA: North Atlantic Books.

Bataille, Georges. 1988. *The Accursed Share: An Essay in General Economy, Vol. 1: Consumption.* Translated by R. Hurley. New York: Zone Books.

———. 1991. *The Accursed Share: Vol. 2 & 3: The History of Eroticism and Sovereignty.* Translated by R. Hurley. New York: Zone Books.

Batnitzky, Leora. 2009. *Idolatry and Representation: The Philosophy of Franz Rosenzweig Reconsidered.* Princeton, NJ: Prince University Press.

Baudrillard, Jean. 1993. *Symbolic Exchange and Death.* Translated by Iain Hamilton Grant. Introduction by Mike Gane. London: Sage.

Bauman, Zygmunt. 1989. *Modernity and the Holocaust.* Ithaca, NY: Cornell University Press.

Beckett, Samuel. 2006. *Poems, Short Fiction, Criticism, Vol. 4.* Edited by Paul Auster. Introduction by J.M. Coetzee. New York: Grove Press.

Benjamin, Jessica. 1996. "Master and Slave: The Bonds of Love." In *Hegel's Dialectic of Desire and Recognition,* edited by J. O'Neill, 209–222. Albany: State University of New York Press.

Berardi, Franco "Bifo". 2009. *The Soul at Work: From Alienation to Autonomy.* Translated by Francesca Cadel and Giuseppina Mecchia. Preface by Jason Smith. Los Angeles, CA: Semiotext(e).

———. 2012. *The Uprising: On Finance and Poetry.* South Pasadena, CA: Semiotext(e).

———. 2015a. *And: Phenomenology of the End.* South Pasadena, CA: Semiotext(e).

———. 2015b. *Heros: Mass Murder and Suicide.* London and New York: Verso.

Bergson, Henri. 1988. *Matter and Memory.* Translated by N.M. Paul and W.S. Palmer. New York: Zone Books.

Berkovits, Eleizer. 2007. Excerpts from *Faith after the Holocaust* (New York, 1973) and *Crisis and Faith* (New York, 1975). Reprinted in *Wrestling with God: Jewish Theological Responses during and after the Holocaust,* edited by Steven T. Katz, Shlomo Biderman and Gershon Greenberg, 462–489. London: Oxford University Press.

Bernard of Clairvaux. 2006. Excerpts from *On Loving God.* In *The Essential Writings of Christian Mysticism,* edited by Bernard McGinn, 434–437. New York: Modern Library.

Bialik, H.N. and Ravnitzky, Y.H. 1992. *The Book of Legends: Sefer ha-Aggadah: Legends from the Talmud and Midrash.* Edited by Hayim Nahman Bialik and Yehoshua Hana Ravnitzky. Translated by William G. Braude. Introduction by David Stern. New York: Schocken Books.

Bierce, Ambrose. 2011. *The Devil's Dictionary, Tales, and Memoirs.* Edited by S.T. Joshi. New York: The Library of America.

Blake, William. 1997. *The Complete Poetry & Prose of William Blake.* Edited by David V. Erdman. Commentary by Harold Bloom. Anchor Books.

Blanchot, Maurice. 1995. *The Station Hill Blanchot Reader: Fiction and Literary Essays.* Translated by Lydia Davis, Paul Auster, and Robert Lamberton. Foreword by Christopher Fynsk. Afterword by George Quasha and Charles Stein. Edited by George Quasha. Barrytown, NY: Station Hill Ltd.

Bloch, Ernst. 1972. *Atheism in Christianity: The Religion of the Exodus and the Kingdom*. Translated by J.T. Swann. New York: Herder and Herder.

Bogue, Ronald. 1989. *Deleuze and Guattari*. London and New York: Routledge.

Bonta, Mark, and Protevi, John. 2004. *Deleuze and Geophilosophy: A Guide and Glossary*. Edinburgh University Press.

Bowie, Andrew. 1993. *Schelling and Modern European Philosophy: An Introduction*. London: Routledge.

Buber, Martin. 1958. *Moses: The Revelation and the Covenant*. New York: Harper Torchbooks.

Buchanan, Ian. 2008. *Deleuze and Guattari's Anti-Oedipus: A Reader's Guide*. London and New York: Continuum.

Butler, Judith. 2006. *Precarious Life: The Powers of Mourning and Violence*. London and New York: Verso Books.

Canetti, Elias. 1978. *Crowds and Power*. Translated by Carol Stewart. New York: Seabury Press.

Caputo, John D. 1997. *The Prayers and Tears of Jacques Derrida*. Bloomington, IN: University Press.

Celan, Paul. 2005. *Paul Celan: Selections*. Edited and with an introduction by Pierre Joris. Berkeley, Los Angeles, London: University of California Press.

Cesarani, David. 2016. *Final Solution: The Fate of the Jews*, 1933–1949. London: Macmillan.

Cha, Victor. 2013. *The Impossible State: North Korea, Past and Future*. New York: Ecco.

Chamayou, Grégoire. 2015. *Drone Theory*. Translated by Janet Lloyd. Middlesex: Penguin.

Chidester, David. 1988. *Salvation and Suicide: An Interpretation of Jim Jones, the Peoples Temple and Jonestown*. Bloomington, IN: Indiana University Press.

Cioran, Emil Michel. 1998. *Tears and Saints*. Translated and with an Introduction by Ilinca Zarifopol-Johnston. Chicago: University of Chicago Press.

———. 2012. *A Short History of Decay*. Translated by Richard Howard. Foreword by Eugene Thacker. New York: Arcade Publishing.

———. 2013a. *The Temptation to Exist*. Translated by Richard Howard. Foreword by Eugene Thacker. New York: Arcade Publishing.

———. 2013b. *The Trouble with Being Born*. Translated by Richard Howard. Foreword by Eugene Thacker. New York: Arcade Publishing.

Clastres, Pierre. 1994. *Society Against the State: Essays in Political Anthropology*. Translated by R. Hurley and A. Stein. New York: Zone Books.

———. 2000. *Chronicle of the Guyaki Indians*. Translated and forward by Paul Auster. New York: Zone Books.

Clements, Jonathan. 2007. *The First Emperor of China*. London: Sutton Publishing Ltd.

Cohen, Arthur A. 2007. Excerpt from *The Tremendum: A Theological Interpretation of the Holocaust* (New York, 1981). Reprinted in *Wrestling with God: Jewish Theological Responses during and after the Holocaust*, edited by Steven T. Katz, Shlomo Biderman and Gershon Greenberg, 566–580. London: Oxford University Press.

Cohen, Richard A. 1994. *Elevations: The Height of the Good in Rosenzweig and Levinas*. Chicago and London: University of Chicago Press.

Colvile, Robert. 2016. *The Great Acceleration: How the World is Getting Faster, Faster*. London and New York: Bloomsbury.

Coughlin, Colin. 2003. *Saddam: The Secret Life*. London: Pan Books.

Craney, Glen. 2008. *The Fire and the Light: A Novel of the Cathars and the Lost Teachings of Christ.* Los Angeles, CA: Brigid's Fire Press.

Critchley, Simon. 1992. *The Ethics of Deconstruction: Derrida and Levinas*. Oxford: Blackwell.

———. 2004. *Very Little … Almost Nothing: Death, Philosophy and Literature*. London: Routledge.

Cutter, Kimberly 2011. *The Maid: A Novel of Joan of Arc*. New York: Houghton Mifflin Harcourt.

Davis, Joyce M. 2003. *Martyrs: Innocence, Vengeance and Despair in the Middle East*. New York: Palgrave Macmillan.

de Beistegui, Miguel. 2004. *Truth and Genesis: Philosophy as Differential Ontology*. Indianapolis: Indiana University Press.

Debray, Régis. 2004. *God: An Itinerary.* Translated by J. Mehlman. London and New York: Verso Books.

Davis, Mike. 2007. *Planet of Slums*. London and New York: Verso.

Delanda, Manuel. 2004. *Intensive Science and Virtual Philosophy*. London: Continuum.

Deleuze, Gilles. 1986. *Cinema 1: The Movement-Image*. Translated by Hugh Tomlinson and B. Habberjam. London: The Athlone Press.

———. 1989. *Cinema 2: The Time-Image.* Translated by Hugh Tomlinson and R. Galeta. London: The Athlone Press.

———. 1990. *The Logic of Sense*. Translated by M. Lester. Edited by C. Stivale and C.V. Boundas. London: Athlone Press.

———. 2001. *Difference and Repetition.* Translated by Paul Patton. London: Continuum Press.

———. 2003. *Francis Bacon: The Logic of Sensation*. Translated and introduction by D.W. Smith. Afterword by T. Conley. Minneapolis: University of Minnesota Press.

———. 2004. *The Fold: Leibniz and the Baroque*. Translated and Foreword by T. Conley. Minneapolis: University of Minnesota Press.

———. 2006. *Nietzsche and Philosophy*. Translated by H. Tomlinson. Foreword by M. Hardt. New York: Columbia University Press.

Deleuze, Gilles, and Guattari, Félix. 1986. *Kafka: Toward a Minor Literature*. Translated by D. Polan. Foreword by R. Bensmaia. Minneapolis: University of Minnesota Press.

———. 1988. *A Thousand Plateaus: Capitalism and Schizophrenia.* Translated and foreword by B. Massumi. The Athlone Press Ltd.

———. 2004. *Anti-Oedipus: Capitalism and Schizophrenia*. Translated by R. Hurley, M. Seem and H.R. Lane. Preface by M. Foucault. London: Continuum Press.

De Masi, Franco. 2011. *The Enigma of the Suicide Bomber*. Translated by Philip Slotkin. London: Karnac Books Ltd.

Derrida, Jacques. 1992a. "Force of Law: The 'Mythical Foundation of Authority'." In *Deconstruction and the Possibility of Justice*, edited by Drucilla Cornell, Michel Rosenfeld, 3–67. London: Routledge.

———. 1992b. *The Other Heading: Reflections on Today's Europe*. Indianapolis: Indiana University Press.

———. 1993. *Aporias.* Translated by Thomas Dutoit. Stanford, CA: Stanford University Press.

———. 1995. *On the Name.* Edited by Thomas Dutoit. Translated by David Wood. Stanford, CA: Stanford University Press.

———. 1996a. *Specters of Marx: the State of the Debt, the Work of Mourning and the New International.* Translated by Peggy Kamuf. London: Routledge.

———. 1996b. *The Gift of Death.* Translated by David Wills. Chicago: University of Chicago Press.

———. 2001. *Writing and Difference.* Translated with an Introduction and additional notes by Alan Bass. London and New York: Routledge.

———. 2009. *The Beast and the Sovereign, Volume 1.* Edited by Michel Lisse, Marie-Louise Mallet, and Ginette Michaud. Translated by Geoffrey Bennington. Chicago and London: University of Chicago Press.

Dershowitz, Alan M. 2000. *The Genesis of Justice: Ten Stories of Biblical Injustice that led to the Ten Commandments and Modern Law.* New York: Warner Books.

Dimont, Max I. 2004. *Jews, God and History.* 2nd edition. Edited and revised by Ethel Dimont. Signet Classics.

Dio, Cassius. 1927. *Dio's Roman History, Vol. 9.* With a translation by Earnest Carey, on the basis of the version of Herbert Baldwin Foster. London: Heinemann.

Donnelly, Mark P., and Diehl, Daniel. 2011. *The Big Book of Pain: Torture and Punishment Throughout History.* Shroud, Gloucestershire: The History Press.

Dostoyevsky, Fyodor. 2004. *The Idiot.* Translated by D. McDuff. Introduction by W.M. Todd III. Middlesex: Penguin Books.

———. 2008. *Demons.* Translated by Robert A. Maguire. Edited by Ronald Meyer. Introduction by Robert L. Belknap. Middlesex: Penguin Books.

Droge, Arthur J., and Tabor, James D. 1992. *A Noble Death. Suicide and Martyrdom Among Christians and Jews in Antiquity.* San Francisco, CA: HarperCollins.

Eagleton, Terry. 2005. *Holy Terror.* London: Oxford University Press.

———. 2013. *Why Marx Was Right.* New Haven and London: Yale University Press.

Elberg, Simhah. 2007. Excerpts from "The *Akedah* of Treblinka." Reprinted in *Wrestling with God: Jewish Theological Responses during and after the Holocaust,* edited by Steven T. Katz, Shlomo Biderman and Gershon Greenberg, 192–198. London: Oxford University Press.

Eliade, Mircea. 1978. *A History of Religious Ideas: Volume 1: From the Stone Age to the Eleusinian Mysteries.* Translated by W.R. Trask. Chicago: University of Chicago Press.

———. 1985. *A History of Religious Ideas: Volume 2: From Gautama Buddha to the Triumph of Christianity.* Translated by W.R. Trask. Chicago: University of Chicago Press.

———. 1988. *A History of Religious Ideas: Volume 3: From Muhammad to the Age of Reforms.* Translated by A. Hiltebeitel, D. Apostolos-Cappadona. Chicago: University of Chicago Press.

Emmerich, Anne Catherine. 2007. *The Dolorous Passion of Our Lord Jesus Christ.* New York: Cosimo Classics.

Fackenheim, Emil. 2007. Excerpts from "The Commanding Voice of Auschwitz." Reprinted in *Wrestling with God: Jewish Theological Responses during and after the Holocaust,* edited by Steven T. Katz, Shlomo Biderman and Gershon Greenberg, 434–439. London: Oxford University Press.

Fisk, Robert. 2006. *The Great War for Civilization.* London: Vintage Books.

Frances de Sales. 2006. Excerpts from *The Treatise on the Love of God.* In *The Essential Writings of Christian Mysticism,* edited by Bernard McGinn, 465–471. New York: Modern Library.

Freud, Sigmund. 1925. "A Note upon the 'Mystic Writing Pad'." http://home.uchicago.edu/~awinter/mystic.pdf

Freud, Sigmund. 1968. *The Standard Edition of the Complete Psychological Words of Sigmund Freud: Volume XIII: Totem and Taboo and Other Works.* Translated and edited by J. Strachey, A. Freud, A. Strachey and A. Tyson. London: The Hogarth Press of the Institute of Psycho-Analysis.

Foucault, Michel. 1991. *Discipline and Punish: The Birth of the Prison.* Translated by A. Sheridan. Middlesex: Penguin Books.

Foxe, John. 1965. *Book of Martyrs.* Edited and abridged by Geoffrey Arthur Williamson. Boston: Little, Brown.

Frankl, Victor. 2008. *Man's Search For Meaning.* Part One translated by Ilse Lasch. Precafe by Harold S. Kushner. London, Sydney, Auckland and Johannesburg: Rider Books.

Fromm, Erich. 1997. *The Anatomy of Human Destructiveness.* London: Pimlico.

Fuentes, Carlos. 2003. *Terra Nostra.* Introduction by Jorge Volpi. Translated by Margaret Sayers Peden. Afterword by Milan Kundera. Chicago: Dalkey Archive Press.

Fukuyama, Francis. 2012. *The Origins of Political Order*: From Prehuman Times to the French Revolution. London: Profile Books.

Gadamer, Hans-George. 1996. "Hegel's Dialectic of Self-Consciousness." In *Hegel's Dialectic of Desire and Recognition*, edited by J. O'Neill, 149–170. Albany: State University of New York Press.

Genet, Jean. 1971. *The Miracle of the Rose.* Translated by B. Frechtman. Middlesex: Penguin Books.

Geuss, Raymond. 2011. "The Future of Evil." In *Nietzsche's On the Genealogy of Morality: A Critical Guide*, edited by Simon May, 12–23. Cambridge and New York: Cambridge University Press.

Gibbs, Robert. 1994. *Correlations in Rosenzweig and Lévinas.* Princeton, NJ and Oxford, UK: Princeton University Press.

Gil, José. 1998. *Metamorphoses of the Body.* Translated by S. Muecke. Minneapolis: University of Minnesota Press.

Gilbert, Martin. 1990. *The Holocaust: The Jewish Tragedy.* London: Collins.

Ginzburg, Eugenia Semyonovna. 1995. *Journey into the Whirlwind.* Translated by Paul Stevenson and Max Hayward. San Diego, CA: Harcourt Brace.

Goodchild, Philip. 1996. *Deleuze and Guattari: An Introduction to the Politics of Desire.* London, Thousand Oaks and New Dehl: SAGE Publications.

Graetz, Heinrich M. 2012. *History of the Jews, Vol. 1–5.* London: Forgotten Books.

Graves, Robert. 1989. *I, Claudius. From the Autobiography of Tiberius Claudius.* New York and Toronto: Vintage International.

Gray, John. 2008. *Black Mass: Apocalyptic Religion and the Death of Utopia.* Middlesex: Penguin Books.

Greenberg, Irving (Yitzchak). 2007. Excerpts from "Cloud of Smoke, Pillar of Fire: Judaism, Christianity, and Modernity after the Holocaust" (New York, 1977). Reprinted in *Wrestling with God: Jewish Theological Responses during and after the Holocaust*, edited by Steven T. Katz, Shlomo Biderman and Gershon Greenberg, 499–523. London: Oxford University Press.

Grossman, Vasily. 1995. *Life and Fate.* London: The Harvill Press.

Groves, Christopher. 1999. "Ecstasy of Reason, Crisis of Reason: Schelling and Absolute Difference." http://plijournal.com/papers/christopher-groves-ecstasy-of-reason-crisis-of-reason-schelling-and-absolute-difference/

Gutiérrez, Gustavo. 1988. *A Theology of Liberation: History, Politics and Salvation.* Translated and edited by Sister Caridad Inda and John Eagleson. London: SCM Press.

Guyotat, Pierre. 2003. *Eden Eden Eden.* London: Creation Books.

Habermas, Jürgen. 1996. "Labour and Interaction: Remarks on Hegel's Jena *Philosophy of Mind.*" In *Hegel's Dialectic of Desire and Recognition*, edited by J. O'Neill, 123–148. Albany: State University of New York Press.

———. 2004. "Dialectical Idealism in Transition to Materialism: Schelling's Idea of a *Contraction* of God and its Consequences for the Philosophy of History." In *The New Schelling*, edited by Judith Norman and Alistair Welchman, 43–89. London and New York: Continuum Press.

Hadewijch of Antwerp. 2006. Excerpts from *Vision VII.* In *The Essential Writings of Christian Mysticism*, edited by Bernard McGinn, 102–104. New York: Modern Library.

Hallward, Peter. 2006. *Out of this World: Deleuze and the Philosophy of Creation.* London and New York: Verso Books.

Hardt, Michael, and Negri, Antonio. 2004. *Multitude: War and Democracy in the Age of Empire.* New York: The Penguin Press.

Harlap, Ya'akov Mosheh. 2007. "Waters Sublime: From the Sources of Salvation." In *Wrestling with God: Jewish Theological Responses during and after the Holocaust*, edited by Steven T. Katz, Shlomo Biderman and Gershon Greenberg, 147–151. London: Oxford University Press.

Harman, Chris. 2009. *Zombie Capitalism: Global Crisis and the Relevance of Marx.* London: Bookmarks Publications.

Hartman, Geoffrey H. 1996. *The Longest Shadow: In the Aftermath of the Holocaust.* Indianapolis: Indiana University Press.

Hatab, Lawrence J. 2008. *Nietzsche's On the Genealogy of Morality: An Introduction.* Cambridge and New York: Cambridge University Press.

Hazelton, Leslie. 2009. *After the Prophet: The Epic Story of the Shia-Sunni Split.* New York: Anchor Books.

Hegel, George Wilhelm Friedrich. 1966. *The Phenomenology of Mind.* Translated by J.B. Baillie. London: Allen and Unwin.

———. 1998. *Lectures on the Philosophy of Religion: Volume III: The Consummate Religion.* Edited by Peter C. Hodgson. Translated by R.F. Brown, P.C. Hodgson, and J.M. Stewart with the assistance of H.S. Harris. Berkeley, Los Angeles, London: University of California Press.

Heidegger, Martin. 2001. *Being and Time.* Translated by J. Macquarrie and E. Robinson. Oxford: Blackwell Publishers Ltd.

Henry, Michel. 2003. *I am the Truth: Toward a Philosophy of Christianity.* Translated by Susan Emanuel. Stanford, CA: Stanford University Press.

Holland, Eugene. 1999. *Deleuze and Guattari's Anti-Oedipus: Introduction to Schizoanalysis.* London: Routledge.

Herodian. 1969. *Herodian: History of the Empire, Volume 1: Books I–IV.* London: Heinemann.

Hitchens, Christopher. 2010. *Hitch-22: A Memoir.* London: Atlantic Books.

———. 2011. *Arguably.* London: Atlantic Books.

Hogg, James. 1992. *Confessions of a Justified Sinner.* Introduction by Roger Lewis. London: Everyman's Library.

Holland, Eugene W. 1999. *Deleuze and Guattari's Anti-Oedipus: Introduction to Schizoanalysis.* London: Routledge.

Holmes, Stephen. 2005. "Al-Qaeda, September 11, 2001." In *Making Sense of Suicide Missions*, edited by Diego Gambetta, 131–172. London: Oxford University Press.

Höss, Rudolph. 1996. *Death Dealer: The Memoirs of the SS Kommandant at Auschwitz.* Translated by Andrew Pollinger. Introduction by Primo Levi. Edited by Steven Paskuly. Translation by Andrew Pollinger. Foreword by Primo Levi. New York: Da Capo Press.

Hugh of St. Victor. 2006. Excerpts from *Contemplation and Its Forms.* In *The Essential Writings of Christian Mysticism*, edited by Bernard McGinn, 336–340. New York: Modern Library.

Hyppolite, Jean. 1996. "Self-Consciousness and Life: The Independence of Self-Consciousness." In *Hegel's Dialectic of Desire and Recognition*, edited by J. O'Neill, 67–86. Albany: State University of New York Press.

Jabès, Edmond. 1991. *The Book of Questions: Volume II.* Translated by Rosmaire Waldrop. Hanover, NH: University Press of New England.

James of Vitry. 2006. Excerpts from *The Life of Mary of Oignies.* In *The Essential Writings of Christian Mysticism*, edited by Bernard McGinn, 60–65. New York: Modern Library.

James, William. 1987. *Writings, 1902–1910.* New York: Literary Classics of the United States.

Jamison, Kay Redfield. 1997. *An Unquiet Mind: A Memoir of Moods and Madness.* London: Vintage Books.

———. 2000. *Night Falls Fast: Understanding Suicide.* London: Vintage Books.

John Cassian. 2006. Excerpts from *Conferences 9 and 10.* In *The Essential Writings of Christian Mysticism*, edited by Bernard McGinn, 89–96. New York: Modern Library.

John of the Cross. 2006a. Excerpts from *The Ascent of Mount Carmel.* In *The Essential Writings of Christian Mysticism*, edited by Bernard McGinn, 72–78. New York: Modern Library.

———. 2006b. Excerpts from *The Living Flame of Love.* In *The Essential Writings of Christian Mysticism*, edited by Bernard McGinn, 213–220. New York: Modern Library.

———. 2006c. Excerpts from *The Spiritual Canticle.* In *The Essential Writings of Christian Mysticism*, edited by Bernard McGinn, 460–464. New York: Modern Library.

John of Ruusbroec. 2006. Excerpts from *The Little Book of Enlightenment.* In *The Essential Writings of Christian Mysticism*, edited by Bernard McGinn, 444–450. New York: Modern Library.

Johnson, Paul. 1988. *A History of the Jews.* New York: Harper Perennial.

———. 2010. *Brief Lives: An Intimate and Very Personal Portrait of the Twentieth Century.* London: Hutchinson.

Josephus, Flavius. 1970. *The Jewish War.* Translated with an introduction by G.A. Williamson. Middlesex: Penguin Books.

———. 2006. *Jewish Antiquities.* Translated by William Whiston. Introduction by Brian McGing. Hertfordshire: Wordsworth Editions Limited.

Kafka, Franz. 2015. *The Trial.* Translated and with an Introduction by Idris Parry. Middlesex: Penguin.

———. 1999. *The Complete Short Stories.* Edited by N.N. Glazer. London: Vintage Books.

Kalkavage, Peter. 2007. *The Logic of Desire: An Introduction to Hegel's Phenomenology of Spirit.* Philadelphia, PA: Paul Dry Books.

Kantorowicz, Ernest. 1957. *The King's Two Bodies: A Study in Medieval Political Theology*. Princeton, NJ: Princeton University Press.

Katz, Reuven. 2007. "Torah Portion *Behukotan*: Israel's Misery and Israel's Redemption." In *Wrestling with God: Jewish Theological Responses during and after the Holocaust*, edited by Steven T. Katz, Shlomo Biderman and Gershon Greenberg, 104–109. London: Oxford University Press.

Kelly, George Armstrong. 1996. "Notes on Hegel's 'Lordship and Bondage'." In *Hegel's Dialectic of Desire and Recognition*, edited by J. O'Neill, 253–272. Albany: State University of New York Press.

Kershaw, Ian. 2009. *Hitler: A Biography.* London: Penguin.

Khosrokhavar, Farhad. 2005. *Suicide Bombers: Allah's New Martyrs*. Translated by David Macey. London and Ann Arbor, MI: Pluto Press.

Kierkegaard, Soren. 1994. *Fear and Trembling and the Book of Adler*. Translated by Walter Lowrie. London: Everyman's Library.

Kiernan, Ben. 2007. *Blood and Soil: A World History of Genocide and Extermination from Sparta to Darfur*. New Haven and London: Yale University Press.

Klein, Naomi. 2007. *The Shock Doctrine: The Rise of Disaster Capitalism.* Middlesex: Penguin Books.

Klossowski, Pierre. 1997. *Nietzsche and the Vicious Circle*. Translated by Daniel W. Smith. Chicago: University of Chicago Press.

Kojève, Alexandre. 1996. "Desire and Work in the Master and Slave." In *Hegel's Dialectic of Desire and Recognition,* edited by J. O'Neill, 49–66. Albany: State University of New York Press.

Largo, Michael. 2008. *Genius and Heroin: the Illustrated Catalogue of Creativity, Obsession, and Reckless Abandon Throughout the Ages.* New York: HarperCollins.

———. 2010. *God's Lunatics: Lost Souls, False Prophets, Martyred Saints, Murderous Cults, Demonic Nuns, and Other Victims of Man's Eternal Search for the Divine*. New York: HarperCollins Publishers.

Lawrence, Thomas Edward. 2008. *Seven Pillars of Wisdom: A Triumph*. London: Vintage Books.

Lazzarato, Maurizio. 2014. *Signs and Machines: Capitalism and the Production of Subjectivity.* Translated by Joshua David Jordan. Los Angeles: Semiotext(e).

———. 2015. *Governing by Debt.* Translated by Joshua David Jordan. South Pasadena: Semiotext(e).

Leibniz, Gottlieb Wilhelm. 1898. *The Monadology and Other Philosophical Writings*. Translated, introduction and notes by R. Latta. London: Oxford University Press, Geoffrey Cumberlege.

Leiter, Brian. 2011. "Who is the 'Sovereign Individual'? Nietzsche on Freedom." In *Nietzsche's On the Genealogy of Morality: A Critical Guide*, edited by Simon May, 101–119. Cambridge and New York: Cambridge University Press.

Lengyel, Olga. 1947. *Five Chimneys: The Story of Auschwitz*. Chicago and New York: Ziff-Davis Publishing Company.

Levi, Primo. 2003. *If This is a Man and The Truce*. Translated by S. Woolf. Introduction by P. Bailey. London: Abacus.

Lévinas, Immanuel. 1978. *Existence and Existents*. Translated by Alphonso Lingis. Pittsburgh, PA: Duquesne University Press.

———. 1979. *Totality and Infinity: An Essay on Exteriority*. Translated by Alphonso Lingis. Pittsburgh, PA: Duquesne University Press.

———. 1981. *Otherwise than Being: or Beyond Essence.* Translated by Alphonso Lingis. Pittsburgh, PA: Duquesne University Press.

Lévy, Bernard-Henri. 2004. *War, Evil, and the End of History.* Translated by Charlotte Mandell. New York: Melville House.

Lingis, Alphonso. 1983. *Excesses: Eros and Culture.* Albany: State University of New York Press.

Llosa, Mario Vargas. 2010. *The Feast of the Goat.* Translated by E. Grossman. London: Faber and Faber.

Loebl, Eugen. 1976. *My Mind on Trial.* New York: Harcourt Brace Janovich.

Lukács, Georg. 1996. "Hegel's Economics during the Jena Period." In *Hegel's Dialectic of Desire and Recognition*, edited by J. O'Neill, 101–122. Albany: State University of New York Press.

Madame Guyon. 2006. Excerpts from *Commentary on the Song of Songs.* In *The Essential Writings of Christian Mysticism*, edited by Bernard McGinn, 41–46. New York: Modern Library.

Magee, Bryan. 2001. *Wagner and Philosophy.* London: Penguin.

Mann, Thomas. 2005. *Joseph and His Brothers: The Stories of Jacob, Young Joseph, Joseph in Egypt, Joseph the Provider.* Translated from the German by John E. Woods. New York, London, Toronto: Alfred A. Knopf.

Marat, Jean-Paul. 2003. "Are We Undone?" from *The Friend of the People.* In *The Communist Manifesto and Other Revolutionary Writings: Marx, Marat, Paine, Mao, Gandhi, and Others*, edited by Bob Blaisdell, 82–84. Mineola, New York: Dover Publications, Inc.

Margolin, Jean-Louis. 1999a. "Cambodia: The Country of Disconcerting Crimes." In *The Black Book of Communism: Crimes, Terror, Repression*, edited by S. Courtois, N. Werth, et al. Translated by J. Murphy and M. Kramer, 577–635. Cambridge, MA and London: Harvard University Press.

———. 1999b. "China: A Long March into Night." In *The Black Book of Communism: Crimes, Terror, Repression*, edited by S. Courtois, N. Werth, et al. Translated by J. Murphy and M. Kramer, 436–546. Cambridge, MA and London: Harvard University Press.

Marguerite Porete. 2006. Excerpts from *The Mirror of Simple Annihilated Souls.* In *The Essential Writings of Christian Mysticism*, edited by Bernard McGinn, 172–179, New York: Modern Library.

Markovski, Venko. 1984. *Goli Otok: The Island of Death: A Diary of Letters.* New York: Columbia University Press.

Martin, Bradley K. 2006. *Under the Loving Care of the Fatherly Leader: North Korea and the Kim Dynasty.* New York: Thomas Dunne Books.

Martinez, Raoul. 2016. *Creating Freedom: Power, Control and the Fight for Our Future.* Edinburgh and London: Canongate Books.

Marx, Karl. 1990. *Capital: A Critique of Political Economy: Volume 1.* Translated by Ben Fowkes. Introduction by Ernest Mandel. Middlesex: Penguin Books.

———. 1991. *Capital: A Critique of Political Economy: Volume 3.* Translated by David Fernbach. Introduction by Ernest Mandel. Middlesex: Penguin Books.

———. 1992. *Capital: A Critique of Political Economy: Volume 2.* Translated by David Fernbach. Introduction by Ernest Mandel. Middlesex: Penguin Books.

———. 1993. *Grundrisse: Foundations of the Critique of Political Economy.* Translated with a foreword by Martin Nicolaus. Middlesex: Penguin Books.

Mason, Herbert. 1991. *The Death of al-Hallaj: A Dramatic Narrative.* Notre-Dame: University of Notre Dame Press.

Massignon, Lious. 1982. *The Passion of al-Hallāj: Mystic and Martyr of Islam.* Translated and with a biographical foreword by Herbert Mason. Princeton, NJ: Princeton University Press.

Massumi, Brian. 1992. *A User's Guide to Capitalism and Schizophrenia: Deviations from Deleuze and Guattari.* Cambridge, MA and London: MIT Press.

Maybaum, Ignaz. 2007. Excerpts from *The Face of God after Auschwitz* (Amsterdam, 1965). Reprinted in *Wrestling with God: Jewish Theological Responses during and after the Holocaust,* edited by Steven T. Katz, Shlomo Biderman and Gershon Greenberg, 402–408. London: Oxford University Press.

Mayer, Jane. 2009. *The Dark Side: The Inside Story of How the War on Terror Turned into a War on American Ideals.* New York: Anchor Books.

Mazzetti, Mark. 2013. *The Way of the Knife: The CIA, a Secret Army, and a War at the Ends of the Earth.* New York: Penguin Books.

McDermott, Terry. 2006. *Perfect Soldiers: The 9/11 Hijackers: Who They Were, Why They Did It.* New York: Harper Perennial.

Miles, Jack. 1995. *God: A Biography.* New York: Alfred A. Knopf.

———. 2002. *Christ: A Crisis in the Life of God.* New York: Vintage Books.

Milton, John. 1998. *The Complete Poems.* Edited with a preface and notes by John Leonard. Middlesex: Penguin Books.

Mishima, Yukio. 1966. *Death in Midsummer and Other Stories.* New York: New Directions Publishing.

Moevs, Christian. 2005. *The Metaphysics of Dante's Comedy.* London: Oxford University Press.

Moltmann, Jürgen. 1974. *The Crucified God: The Cross of Christ as the Foundation and Criticism of Christian Theology.* London: SCM Press.

Mosès, Stéphane. 1992. *System and Revelation: The Philosophy of Franz Rosenzweig.* Translated by Catherine Tihanyi. Foreword by Emmanuel Lévinas. Detroit, MI: Wayne University Press.

Mulhall, Stephen. 2011. "The Promising Animal: the Art of Reading *On the Genealogy of Morality* as Testimony." In *Nietzsche's On the Genealogy of Morality: A Critical Guide,* edited by Simon May, 234–264. Cambridge and New York: Cambridge University Press.

Musurillo, Herbert. 1972. *The Acts of the Christian Martyrs.* Introduction, texts and translations by Herbert Musurillo. Oxford: Clarendon Press.

Nathan, Jonathan. 1975. *Mishima: A Biography.* London: Hamilton.

Negri, Antonio. 2009. *The Labor of Job: The Biblical Text as a Parable of Human Labor.* Foreword by M. Hardt. Translated by M. Mandarini. Commentary by R. Boer. Durham and London: Duke University Press.

Newport, Kenneth G.C. 2006. *The Branch Davidians of Waco: The History and Beliefs of an Apocalyptic Sect.* London: Oxford University Press.

Ngoc, J., and Huynh, Q. 1994. *South Wind Changing.* Minneapolis, MN: Graywolf Press.

Ngor, H.S., and Warner, R. 1998. *Surviving the Killing Fields: The Cambodian Odyssey of Haing S. Ngor.* London: Chatto and Windus.

Nietzsche, Friedrich. 1968. *The Will to Power.* Translated by W. Kaufmann and R.J. Hollingdale. Edited by W. Kaufmann. London: Vintage Books.

———. 1977. *The Portable Nietzsche.* Translated by W. Kaufmann. Middlesex: Penguin Books.

———. 2000. *Basic Writings of Nietzsche*. Introduction by P. Gay. Translated by W. Kaufmann. New York: Modern Library.

———. 2003. *Thus Spake Zarathustra: A Book for Everyone and No One*. Translated and with an Introduction by R.J. Hollingdale. London: Penguin.

———. 2007. *On the Genealogy of Morality*. Edited by Keith Ansell-Pearson. Translated by Carol Diethe. Cambridge and New York: Cambridge University Press.

———. 2014. *Beyond Good and Evil: Prelude to a Philosophy of the Future*. Translated by R.J. Hollingdale. Introduction by Michael Tanner. London: Penguin.

Nomberg-Przytyk, Sara. 1985. *Auschwitz: True Tales from a Grotesque Land*. Translated by R. Hirsch. Edited by E. Pfefferkorn and D.H. Hirsch. Chapel Hill, London: The University of North Carolina Press.

Oliver, Anne Marie and Steinberg, Paul F. 2005. *The Road to Martyr's Square: A Journey into the World of the Suicide Bomber*. London: Oxford University Press.

Parr, Adrian, ed. 2005. *The Deleuze Dictionary*. New York: Columbia University Press.

Payne, N., and Romanoff, R. 1975. *Ivan the Terrible*. New York: Thomas Y. Cromwell Company.

Pegg, M.G. 2008. *A Most Holy War: The Albigensian Crusade and the Battle for Christendom*. New York: Oxford University Press.

Pessoa, Fernando. 2015. *The Book of Disquiet*. Edited and translated by Richard Zenith. Middlesex: Penguin Books.

Peterson, Amy Rachel. 2004. *Perpetua: A Novel*. Relevant Books.

Plato. 1972. *The Trial and Execution of Socrates*. Translated with an introduction by Peter Georges. London: Folio Society.

Plotinus. 1962. *The Enneads*. Translated by Stephen MacKenna. Third edition revised by B.S. Page. Foreword by E.R. Dodds. Introduction by Paul Henry, S.J. London: Faber & Faber Ltd.

Poellner, Peter. 2011. "*Ressentiment* and Morality." In *Nietzsche's On the Genealogy of Morality: A Critical Guide*, edited by Simon May, 120–141. Cambridge and New York: Cambridge University Press.

Pollock, Benjamin. 2009. *Franz Rosenzweig and the Systematic Task of Philosophy*. New York: Cambridge University Press.

Procopius. 1954. *Procopius: With an English Translation, Vol. I: History of the Wars, Books I & II, The Persian War*. Translated by H.B. Dewing. London: Heinemann.

———. 1960. *Procopius: With an English Translation, Vol. VI: The Anecdota, or Secret History*. Translated by H.B. Dewing. London: Heinemann.

Reginster, Bernard. 2011. "The Genealogy of Guilt." In *Nietzsche's On the Genealogy of Morality: A Critical Guide*, edited by Simon May, 56–77. Cambridge and New York: Cambridge University Press.

Rejali, Darius. 2007. *Torture and Democracy*. Princeton, NJ: Princeton University Press.

Reuter, Christoph. 2006. *My Life is a Weapon: A Modern History of Suicide Bombing*. Princeton, NJ: Princeton University Press.

Richard of St. Victor. 2006. Excerpts from *The Four Degrees of Violent Charity*. In *The Essential Writings of Christian Mysticism*, edited by Bernard McGinn, 155–161. New York: Modern Library.

Ricolfi, Luca. 2005. "Palestinians, 1981–2003." In *Making Sense of Suicide Missions*, edited by Diego Gambetta, 77–129. London: Oxford University Press.

Ridley, Aaron. 2011. "Une Promesse de Bonheur? Beauty in the Genealogy." In *Nietzsche's On the Genealogy of Morality: A Critical Guide*, edited by Simon May, 309–325. Cambridge and New York: Cambridge University Press.

Reiterman, Tim and Jacobs, John. 2008. *Raven: The Untold Story of the Rev. Jim Jones and His People.* New York: Jeremy P. Tarcher/Penguin.

Robertson, Geoffrey. 2006. *Crimes against Humanity: The Struggle for Global Justice.* London: Penguin.

Rosenzweig, Franz. 2005. *The Star of Redemption.* Translated by Barbara E. Galli. Madison, WI: University of Wisconsin Press.

Ruo-Wang, B., and Chelminski, R. 1975. *Prisoner of Mao.* London: André Deutsche Limited.

Sachs, Nelly. 2011. *Collected Poems, 1944–1949.* Translated by Michael Hamburger, Ruth and Nathan Mead, and Michael Roloff. Introduction by Hans Magnus Enzensberger. København, Los Angeles: Green Integer Books.

Sade, Marquis de. 1990. *The 120 Days of Sodom and Other Writings.* Compiled and translated by Austryn Wainhouse and Richard Seaver. Introductions by Simone de Beauvoir and Pierre Klossowski. New York: Arrow Books.

Santner, Eric. 2001. *On the Psychotheology of Everyday Life: Reflections on Freud and Rosenzweig.* Chicago: University of Chicago Press.

Saramago, José. 2012. *Cain.* Translated from the Portuguese by Margaret Jull Costa. New York: Mariner Books.

Sartre, Jean-Paul. 1983. *Saint Genet: Actor & Martyr.* Translated by B. Frechtman. New York: Pantheon Books.

Scahill, Jeremy. 2013. *Dirty Wars: The World is a Battlefield.* New York: Nation Books.

Scahill, Jeremy and the Staff of the Intercept. 2016. *The Assassination Complex.* Foreword by Edward Snowden. Afterword by Glenn Greenwald. London: Serpent's Tail.

Schama, Simon. 2014. *The Story of the Jews: Finding the Words: 1000 BCE – 1492 CE.* London: Vintage Books.

Schelling, Friedrich Wilhelm Joseph. 1997. *The Abyss of Freedom/Ages of the World: An Essay by Slavoj Žižek with the Text of Schelling's* Die Weltalter *(second draft, 1813) in English Translation by Judith Norman.* Translated by Judith Norman. Ann Arbor, MI: University of Michigan Press.

———. 2000. *The Ages of the World (Fragment): From the Handwritten Remains, Third Version (c.1815).* Translated and introduction by J.M. Wirth. Albany: State University of New York Press.

Schimmel, Annemarie. 1975. *Mystical Dimensions of Islam.* Chapel Hill: University of Carolina Press.

Schneersohn, Yosef Yitshak. 2007a. "First Proclamation: 26 May 1941." In *Wrestling with God: Jewish Theological Responses during and after the Holocaust,* edited by Steven T. Katz, Shlomo Biderman and Gershon Greenberg, 172–175. London: Oxford University Press.

———. 2007b. "Fourth Proclamation: 11 September 1941." In *Wrestling with God: Jewish Theological Responses during and after the Holocaust,* edited by Steven T. Katz, Shlomo Biderman and Gershon Greenberg, 181–185. London: Oxford University Press.

———. 2007c. "Let Our Master Teach Us: 11 September 1942." In *Wrestling with God: Jewish Theological Responses during and after the Holocaust,* edited by Steven T. Katz, Shlomo Biderman and Gershon Greenberg, 185–187. London: Oxford University Press.

Scholem, Gershom. 1978. *Kabbalah*. New York: Plume.

Schreber, Daniel Paul. 2000. *Memoirs of My Nervous Illness*. Introduction by Rosemary Dinnage. Translated and edited by Ida Macalpine and Richard A. Hunter. New York: New York Review Books.

Schweid, Eleizer. 2007. Excerpts from "Does the Idea of Jewish Election Have Any Meaning after the Holocaust?" Reprinted in *Wrestling with God: Jewish Theological Responses during and after the Holocaust*, edited by Steven T. Katz, Shlomo Biderman and Gershon Greenberg, 229–243. London: Oxford University Press.

Scurr, Judith. 2006. *Fatal Purity: Robespierre and the French Revolution*. New York: Henry Holy and Company.

Semelin, Jacques. 2007. *Purify and Destroy: The Political Uses of Massacre and Genocide*. Introduction by S. Hoffman. Translated by C. Schoch. New York: Columbia University Press.

Service, Robert. 2004. *Stalin: A Biography*. Cambridge, MA and London: Belknap Press of Harvard University Press.

Shakespeare, William. 1994. *Histories: Volume 2*. London: Everyman's Library.

Shane, Scott. 2016. *Objective Troy: A Terrorist, A President, and the Rise of the Drone*. New York: Tim Duggan Books.

Shepkaru, Shmuel. 2006. *Jewish Martyrs in the Pagan and Christian Worlds*. New York: Cambridge University Press.

Shklar, J.N. 1996. "Self-Sufficient Man: Dominion and Bondage." In *Hegel's Dialectic of Desire and Recognition*, edited by J. O'Neill. 289–304. Albany: State University of New York Press.

Sloterdijk, Peter. 2010. *Rage and Time: A Pychopolitical Investigation*. Translated by M. Wenning. New York: Columbia University Press.

Soloveitchik, Joseph B. 2007. Excerpts from *Kol Dodi Dofek*: The Voice of My Beloved Knocks. Reprinted in *Wrestling with God: Jewish Theological Responses during and after the Holocaust*, edited by Steven T. Katz, Shlomo Biderman and Gershon Greenberg, 382–393. London: Oxford University Press.

Solzhenitsyn, Alexandr. 1999. *The First Circle*. Translated by M. Hayward, M. Harari and M. Glenny. London: Harvill Press.

———. 2007. *The Gulag Archipelago: An Experiment in Literary Investigation, Vol. 1–3*. Foreword by A. Applebaum. Translated by T.P. Whitney (vol. 1 & 2) and H. Willets (vol. 3). New York: Harper Perennial Modern Classics.

Spinoza, Baruch. *2002. Complete Works. Translated by S. Shirley. Edited, introduction and notes by M.L. Morgan*. Indianapolis, IN and Cambridge: Hackett Publishing Company Inc.

Stevenson, Robert Louis. 1992. *Dr. Jeckyll and Mr. Hyde and Other Stories*. Introduction by Nicholas Rance. London: Everyman's Library.

Stokes, Henry Scott. 2000. *The Life and Death of Yukio Mishima*. New York: Cooper Square Press.

Styron, William. 2004. *Darkness Visible: A Memoir of Madness*. London: Vintage Books.

Suetonius, Gaius. 1958. *The Twelve Caesars*. Translated by R. Graves. Middlesex: Penguin Books, Harmondsworth.

Sumption, Jonathan. 1978. *The Albigensian Crusade*. London: Faber.

Taylor, Charles. 1975. *Hegel*. New York: Cambridge University Press.

———. 2007. *A Secular Age*. Cambridge, MA and London: Belknap Press of Harvard University Press.

Taylor, Mark C. 2009. *After God*. Chicago: University of Chicago Press.

Teresa of Avila. 2006. Excerpts from *The Interior Castle 7.1–2*. In *The Essential Writings of Christian Mysticism*, edited by Bernard McGinn, 451–459. New York: Modern Library.

Thibodeau, David (with Leon Whiteson). 1999. *A Place called Waco: A Survivor's Story*. New York: PublicAffairs.

Toscano, Alberto. 2004. "Philosophy and the Experience of Construction." In *The New Schelling*, edited by Judith Norman and Alistair Welchman, 106–127. London and New York: Continuum Press.

———. 2006. *The Theatre of Production: Philosophy and Individuation between Kant and Deleuze*. New York: Palgrave Macmillan.

Venezia, Shlomo (with Béatrice Prasquier). 2009. *Inside the Gas Chambers: Eight Months in the Sonderkommando of Auschwitz*. Translated by Andrew Brown. Edited by Jean Mouttapa. Foreword by Simone Veil. Historical notes and additional material by Marcello Pezzetti and Umberto Gentiloni. Cambridge, UK and Malden, MA: Polity.

Virilio, Paul and Lotringer, Sylvère. 2008. *Pure War*. Translated by Mark Polizzotti and Brian O'Keeffe. Los Angeles, CA: Semiotext(e).

wa Thiong'o, Ngũgĩ. 2006. *Wizard of the Crow*. Translated by the author. London: Harvill Secker.

Walker, Paul E. 2009. *Caliph of Cairo: Al-Hakim bi-Amr Allah, 996–1021*. Cairo, New York: The American University of Cairo Press.

Walzer, Michael. 1985. *Exodus and Revolution*. New York: Basic Books.

Wasserman, Elhanan. 2007. Excerpts from *Tractate: The Onset of the Messiah* (New York, 1938). Reprinted in *Wrestling with God: Jewish Theological Responses during and after the Holocaust*, edited by Steven T. Katz, Shlomo Biderman and Gershon Greenberg, 30–38. London: Oxford University Press.

Wiesel, Elie. 1981. *Five Biblical Portraits: Saul-Jonah-Jeremiah-Joshua*. Notre-Dame: University of Notre Dame Press.

———. 2005. *Messengers of God: Biblical Portraits and Legends*. Translated from the French by Marion Wiesel. New York: Simon & Schuster Paperbacks.

———. 2006. *Night*. Translated by Marion Wiesel. London: Penguin.

Wilson, Colin. 1989. *The Misfits: A Study of Sexual Outsiders*. Glasgow: Grafton Books.

Wood, Ellen Mieksins. 2003. *Empire of Capital*. London and New York: Verso Books.

Wyschogrod, Michael. 2007. Excerpts from "Faith and the Holocaust." Reprinted in *Wrestling with God: Jewish Theological Responses during and after the Holocaust*, edited by Steven T. Katz, Shlomo Biderman and Gershon Greenberg, 456–461. London: Oxford University Press.

Zinn, Howard. 2005. *A People's History of the United States*. New York: Harper Perennial Modern Classics.

Žižek, Slavoj. 1996. *The Indivisible Remainder: An Essay on Schelling and Related Matters*. London and New York: Verso Books.

———. 1999. *The Ticklish Subject: The Absent Centre of Political Ontology*. London and New York: Verso Books.

———. 2003. *The Puppet and the Dwarf: The Perverse Core f Christianity.* Cambridge, MA and London: MIT Press.

———. 2006. *The Parallax View.* Cambridge, MA and London: MIT Press.

———. 2008a. *In Defense of Lost Causes.* London and New York: Verso Books.

———. 2008b. *The Fragile Absolute: Or Why is the Christian Legacy is Worth Fighting For?* London and New York: Verso Books.

———. 2008c. *The Sublime Object of Ideology.* London and New York: Verso Books.

———. 2008d. *Violence: Six Sideways Reflections.* Picador.

———. 2010. *Living in the End Times.* London and New York: Verso Books.

———. 2013. *Less than Nothing: Hegel and the Shadow of Dialectical Materialism.* London and New York: Verso Books.

———. 2015. *Trouble in Paradise: From the End of History to the End of Capitalism.* Middlesex: Penguin Books.

Žižek, Slavoj and Milbank, John. 2009. *The Monstrosity of Christ: Paradox or Dialectic?* Edited by Creston Davis. Cambridge, MA and London: MIT Press.

Žižek, Slavoj and Schelling, Friedrich Wilhelm Joseph. 1997. *The Abyss of Freedom/Ages of the World: An Essay by Slavoj Žižek with the Text of Schelling's* Die Weltalter *(second draft, 1813) in English Translation by Judith Norman.* Translated by Judith Norman. Ann Arbor, MI: University of Michigan.

Zupančič, Alenka. 2003. *The Shortest Shadow: Nietzsche's Philosophy of the Two.* Cambridge, MA and London: MIT Press.

———. 2011. *Ethics of the Real: Kant and Lacan.* London and New York: Verso Books.

Index

A

B

C

D

F

G

H

J

K

M

N

O

P

Q

R

S

T

U

V

W

Y

Z

www.ingramcontent.com/pod-product-compliance
Ingram Content Group UK Ltd.
Pitfield, Milton Keynes, MK11 3LW, UK
UKHW021028041026
15470UKWH00013B/348

9 781433 134654